Evaporites Through Space and Time

The Geological Society of London
Books Editorial Committee

Chief Editor
BOB PANKHURST (UK)

Society Books Editors
JOHN GREGORY (UK)
JIM GRIFFITHS (UK)
JOHN HOWE (UK)
PHIL LEAT (UK)
NICK ROBINS (UK)
JONATHAN TURNER (UK)

Society Books Advisors
MIKE BROWN (USA)
ERIC BUFFETAUT (France)
RETO GIERÉ (Germany)
JON GLUYAS (UK)
DOUG STEAD (Canada)
RANDELL STEPHENSON (The Netherlands)

Geological Society books refereeing procedures

The Society makes every effort to ensure that the scientific and production quality of its books matches that of its journals. Since 1997, all book proposals have been refereed by specialist reviewers as well as by the Society's Books Editorial Committee. If the referees identify weaknesses in the proposal, these must be addressed before the proposal is accepted.

Once the book is accepted, the Society Book Editors ensure that the volume editors follow strict guidelines on refereeing and quality control. We insist that individual papers can only be accepted after satisfactory review by two independent referees. The questions on the review forms are similar to those for *Journal of the Geological Society*. The referees' forms and comments must be available to the Society's Book Editors on request.

Although many of the books result from meetings, the editors are expected to commission papers that were not presented at the meeting to ensure that the book provides a balanced coverage of the subject. Being accepted for presentation at the meeting does not guarantee inclusion in the book.

More information about submitting a proposal and producing a book for the Society can be found on its web site: www.geolsoc.org.uk.

It is recommended that reference to all or part of this book should be made in one of the following ways:

SCHREIBER, B. C., LUGLI, S. & BĄBEL, M. (eds) 2007. *Evaporites Through Space and Time*. Geological Society, London, Special Publications, **285**.

TURNER, P. & SHERIF, H. 2007. A giant late Triassic–Early Jurassic Evaporitic Basin on the Saharan Platform, North Africa. *In*: SCHREIBER, B. C., LUGLI, S. & BĄBEL, M. (eds) 2007. *Evaporites Through Space and Time*. Geological Society, London, Special Publications, **285**, 87–106.

GEOLOGICAL SOCIETY SPECIAL PUBLICATION NO. 285

Evaporites Through Space and Time

EDITED BY

B. C. SCHREIBER
University of Washington, USA

S. LUGLI
Università degli Studi di Modena e Reggio Emilia, Italy

and

M. BĄBEL
Warsaw University, Poland

2007
Published by
The Geological Society
London

THE GEOLOGICAL SOCIETY

The Geological Society of London (GSL) was founded in 1807. It is the oldest national geological society in the world and the largest in Europe. It was incorporated under Royal Charter in 1825 and is Registered Charity 210161.

The Society is the UK national learned and professional society for geology with a worldwide Fellowship (FGS) of over 9000. The Society has the power to confer Chartered status on suitably qualified Fellows, and about 2000 of the Fellowship carry the title (CGeol). Chartered Geologists may also obtain the equivalent European title, European Geologist (EurGeol). One fifth of the Society's fellowship resides outside the UK. To find out more about the Society, log on to www.geolsoc.org.uk.

The Geological Society Publishing House (Bath, UK) produces the Society's international journals and books, and acts as European distributor for selected publications of the American Association of Petroleum Geologists (AAPG), the Indonesian Petroleum Association (IPA), the Geological Society of America (GSA), the Society for Sedimentary Geology (SEPM) and the Geologists' Association (GA). Joint marketing agreements ensure that GSL Fellows may purchase these societies' publications at a discount. The Society's online bookshop (accessible from www.geolsoc.org.uk) offers secure book purchasing with your credit or debit card.

To find out about joining the Society and benefiting from substantial discounts on publications of GSL and other societies worldwide, consult www.geolsoc.org.uk, or contact the Fellowship Department at: The Geological Society, Burlington House, Piccadilly, London W1J 0BG: Tel. +44 (0)20 7434 9944; Fax +44 (0)20 7439 8975; E-mail: enquiries@geolsoc.org.uk.

For information about the Society's meetings, consult *Events* on www.geolsoc.org.uk. To find out more about the Society's Corporate Affiliates Scheme, write to enquiries@geolsoc.org.uk.

Published by The Geological Society from:
The Geological Society Publishing House, Unit 7, Brassmill Enterprise Centre, Brassmill Lane, Bath BA1 3JN, UK

(*Orders*: Tel. +44 (0)1225 445046, Fax +44 (0)1225 442836)
Online bookshop: www.geolsoc.org.uk/bookshop

The publishers make no representation, express or implied, with regard to the accuracy of the information contained in this book and cannot accept any legal responsibility for any errors or omissions that may be made.

© The Geological Society of London 2007. All rights reserved. No reproduction, copy or transmission of this publication may be made without written permission. No paragraph of this publication may be reproduced, copied or transmitted save with the provisions of the Copyright Licensing Agency, 90 Tottenham Court Road, London W1P 9HE. Users registered with the Copyright Clearance Center, 27 Congress Street, Salem, MA 01970, USA: the item-fee code for this publication is 0305-8719/07/$15.00.

British Library Cataloguing in Publication Data

A catalogue record for this book is available from the British Library.

ISBN 978-1-86239-232-8

Typeset by Techset Composition Ltd, Salisbury, UK

Printed by The Cromwell Press Ltd, Wiltshire, UK

Distributors

North America
For trade and institutional orders:
The Geological Society, c/o AIDC, 82 Winter Sport Lane, Williston, VT 05495, USA
Orders: Tel. +1 800-972-9892
            Fax +1 802-864-7626
            E-mail: gsl.orders@aidcvt.com

For individual and corporate orders:
AAPG Bookstore, PO Box 979, Tulsa, OK 74101-0979, USA
Orders: Tel. +1 918-584-2555
            Fax +1 918-560-2652
            E-mail: bookstore@aapg.org
            Website http://bookstore.aapg.org

India
Affiliated East-West Press Private Ltd, Marketing Division, G-1/16 Ansari Road, Darya Ganj, New Delhi 110 002, India
Orders: Tel. +91 11 2327-9113/2326-4180
            Fax +91 11 2326-0538
            E-mail: affiliat@vsnl.com

Contents

SCHREIBER, B. C., BĄBEL, M. & LUGLI, S. Introduction and overview — 1

Tectonics, basin evolution and evaporites

KARNER, G. D. & GAMBÔA, L. A. P. Timing and origin of the South Atlantic pre-salt sag basins and their capping evaporites — 15

BERTONI, C. & CARTWRIGHT, J. A. Clastic depositional systems at the base of the late Miocene evaporites of the Levant region, Eastern Mediterranean — 37

AL-JUBOURY, A. I., AL-TARIF, A. M. & AL-EISA, M. Basin analysis of the Burdigalian and Early Langhian successions, Kirkuk Basin, Iraq — 53

RAHIMPOUR-BONAB, H., SHARIATINIA, Z. & SIEMANN, M. G. Role of rifting in evaporite deposition in the Great Kavir Basin, central Iran — 69

TURNER, P. & SHERIF, H. A giant Late Triassic–Early Jurassic evaporitic basin on the Saharan Platform, North Africa — 87

Working depositional models

BĄBEL, M. Depositional environments of a salina-type evaporite basin recorded in the Badenian gypsum facies in the northern Carpathian Foredeep — 107

LOPEZ, P. L. & MANDADO, J. M. Experimental evaporation of superficial brines from continental playa–lake systems located in Central Ebro Basin (northeast Spain) — 143

Post-depositional evolution of sediments

ALBERTO, W., CARRARO, F., GIARDINO, M. & TIRANTI, D. Genesis and evolution of 'pseudocarniole': preliminary observations from the Susa Valley (Western Alps) — 155

LUGLI, S., DOMINICI, R., BARONE, M., COSTA, E. & CAVOZZI, C. Messinian halite and residual facies in the Crotone basin (Calabria, Italy) — 169

Ancient basins

LUGLI, S., BASSETTI, M. A., MANZI, V., BARBIERI, M., LONGINELLI, A. & ROVERI, M. The Messinian 'Vena del Gesso' evaporites revisited: characterization of isotopic composition and organic matter — 179

MATANO, F. The 'Evaporiti di Monte Castello' deposits of the Messinian Southern Apennines foreland basin (Irpinia–Daunia Mountains, Southern Italy): stratigraphic evolution and geological context — 191

BĄBEL, M. & BOGUCKI, A. The Badenian evaporite basin of the northern Carpathian Foredeep as a model of a meromictic selenite basin — 219

BUKOWSKI, K., CZAPOWSKI, G., KAROLI, S. & BĄBEL, M. Sedimentology and geochemistry of the Middle Miocene (Badenian) salt-bearing succession from East Slovakian Basin (Zbudza Formation) — 247

HRYNIV, S., PARAFINIUK, J. & PERYT, T. M. Sulphur isotopic composition of K–Mg sulphates of the Miocene evaporites of the Carpathian Foredeep, Ukraine — 265

VOVNYUK, S. V. & CZAPOWSKI, G. Generation of primary sylvite: the fluid inclusion data from the Upper Permian (Zechstein) evaporites, SW Poland — 275

GANDIN, A. & WRIGHT, D. T. Evidence of vanished evaporites in Neoarchaean carbonates of South Africa — 285

Regional reviews

HRYNIV, S. P., DOLISHNIY, B. V., KHMELEVSKA, O. V., POBEREZHSKYY, A. V. & VOVNYUK, S. V. Evaporites of Ukraine: a review   309

HOVORKA, S. D., HOLT, R. M. & POWERS, D. W. Depth indicators in Permian Basin evaporites   335

Index   365

Introduction and overview

B. C. SCHREIBER[1], M. BĄBEL[2] & S. LUGLI[3]

[1]University of Washington, USA (e-mail: geologo@comcast.net)

[2]Institute of Geology, Warsaw University, Al. Zwirki i Wigury 93, PL-02-089 Warszawa, Poland

[3]Dipartimento di Scienze Della Terra, Università degli Studi di Modena e Reggio Emilia, Largo S. Eufemia 19, 41100 Modena, Italy

This volume grew out of the various oral and poster presentations given during the 'Evaporite' session at the International Geological Union conference (2004) in Florence, Italy. It was clear that only a few of the participants or attendees, coming from many countries and various distant parts of the world, were well informed about evaporites outside their immediate area of study, apart from data from the most commonly available literature. Diversity in the languages of publication and the logistical difficulty of making first-hand comparisons acts as a major barrier to study. As a result, the basic concept of evaporites that most geologists have is that deposits are the product of simple, chemically controlled environments and, if evaporitic compounds are chemically the same, then it follows that their lithology and origin are also the same. Based on many studies over the past 30 years, it is now clear that this long-held impression is manifestly untrue. The disparities were very evident in the photographic presentations and descriptions at the 2004 International Geological Congress: the same evaporitic compounds can have distinct and diverse lithology, depositional sources and geological history. They cannot be considered to arise solely from simple, direct chemical origins as explained by a single universal model! This is the same paradigm of sedimentation that functions with other deposits, such as siliciclastics and carbonates, and should also apply to evaporites.

In putting this volume together, we realized there were many questions about evaporites from parts of the world not discussed in Florence, despite the large number of participants. Thus, in order to broaden and balance the topics already in our Congress program, we added two additional review papers. One paper that we have added is a chapter covering the many evaporites in the former Soviet Union territory: Russia, Ukraine, Belarus and Moldova. This area contains many deposits of different ages but is poorly represented in Western literature due to language barriers. This review paper gives those scientists who do not read languages written in the Cyrillic alphabet an opportunity to become acquainted with those evaporites. The second additional paper is also a review, taken from tectonically stable interior basins of North America (the Permian deposits of Texas and New Mexico). This study is in contrast to many other deposits described in this volume as it contains a well-developed, undisturbed basin-fill. As a group, this set of 18 papers gives a good introduction to many diverse environments and styles of deposition and preservation to be found in varied evaporite basins.

In order to make the volume more approachable, we divided the book into five parts: (1) Tectonics, Basin Evolution and Evaporites, (2) Working depositional Models, (3) Post-depositional Evolution of Sediments, (4) Ancient Basins, and (5) Regional Reviews. The first section (five papers) places the formation of evaporite basins and the mechanical behaviour of some evaporites into an observed structural/stratigraphic framework. Karner & Gamboa present the creation and early infilling of the South Atlantic rift basin, while Bertoni & Cartwright address the destabilization and massive reworking of evaporitic sediments into an adjacent ocean basin. Specific tectonic settings with associated depositional styles and stratigraphy are presented in the other three papers in part one, and links tectonics to sedimentary styles. The early and middle Miocene evaporites of Iraq, by Ismail Al-Juboury, Mehdi Al-Tarif & Al-Eisa, and Iran by Rahimpour-Bonab, Shariatinia & Siemann are strongly tied to their rapid geologic evolution. Turner & Sherif, describe a large Triassic–Jurassic basin that stretches across North Africa from Libya to Morocco and its evolution.

The second part of the book (two papers) presents a fairly rigorous treatment of evaporitic water-body behaviour with associated sedimentation. The first paper, by Bąbel, is based on the observed sedimentary section from the middle Miocene of the Carpathian foredeep, and addresses the problem of how evaporite deposition is controlled by water stratification, circulation and mixing. The second paper, by Lopez & Mandado, deals with an observed example of modern evolution of water in several saline lakes from north-central Spain.

Part three contains only two papers both of which deal with unusual aspects of synsedimentary deformation and then diagenesis within two very different basins. The first is a study by Alberto, Carraro, Giardino & Tiranti that addressess an unusual form of gypsum diagenesis present within some tectonized Triassic evaporites that rsults in a much altered facies (rauhwacke or 'pseudo-carnioles') developed within the original evaporite rocks. These truly reveal the complexity caused by both the tectonics and burial/exhumational history of the deposits. The second paper in this section, by Lugli, Dominici, Barone, Costa & Cavozzi, is a geochemical study of the late Miocene (Messinian) Apennine gypsum deposits that considers their deposition in basins which are initially marine-sourced but change over to a largely nonmarine (but concentrated) evolution higher in the stratigraphic section. This observation appears to be valid throughout the late Miocene (Messinian) in many areas of the Mediterranean and may well hold true in other evaporite sequences.

The fourth and largest section of the book (seven papers), presents a series of observation of diverse evaporites ranging in age from the late Miocene to the Neoarchean. The first, another paper on the late Miocene of Italy by Lugli, Bassetti, Manzi, Barbieri, Longinelli & Roveri, discusses isotopic composition and organic matter in the evaporites of the Northern Apennines. Late Miocene geology and stratigraphy of the Southern Apennines is addressed by Matano. Next presented are the middle Miocene gypsum evaporites of the Carpathian region addressed first by Bąbel & Boguckiand then by Bukowski, Czapowski & Karoli. The isotopic composition of the K–Mg sulphates of the same age in one of the Carpathian basins is addressed Hryniv, Parafiniuk & Peryt. The next topic treats the origins of some of the Upper Permian (Zechstein) salts of SW Poland as discussed by Vovnyuk & Czapowski. The last paper in this section, by Gandin & Wright, examines deposits from the Neoarchean and addresses the elusive products of extreme evaporite diagenesis, where only morphological remnants of evaporites are left.

Finally, in part five of this volume, there are the two regional review papers, describing two amazingly evaporite-rich areas of the world. The first, by Hryniv, Dolishniy, Khmelevska, Poberezhskyy & Vovnyuk, covers a broad overview of the evaporites of Ukraine (Devonian, Permian, Jurassic and Miocence ages). The second, by Hovorka, Holt & Powers, reviews the Permian evaporites of West Texas (Palo Duro, Midland and Delaware basins). Together, these five sections give a fair introduction to the problems faced in evaporite study and some of the necessary answers.

An overview of evaporite puzzles

Elucidating the geological history of evaporites is never straight-forward because evaporite deposits record information in many different ways. Some of the most difficult problems encountered by geologists studying evaporities include realistic assessments of the time intervals of evaporite formation and accumulation, the necessary climatic controls, and the rates of water evaporation/ionic concentration required for evaporite formation. These constraints are presented below, followed by a review of the styles of sedimentation imposed on evaporite formation by differing synsedimentary tectonic histories within the formative basins. It seems that tectonically active and passive basins develop significantly different facies assemblages, as with other types of deposits.

Depositional time intervals in evaporite formation

A feature most evaporites share is that the total depositional time-period for formation and accumulation is very short (even in very thick deposits), so short that the entire deposit may be represented by only one or two biozones in the surrounding area. This means that standard biostratigraphical methods cannot date the evaporite formations precisely and some other high-resolution methods are more useful. For example, Anderson et al. (1972), Richter-Bernburg (1963) and Kirkland (2003), based on isochronous correlation of particular sequences of varves in laminated evaporites, estimated that the late Permian evaporites of both West Texas and the European Zechstein represent deposition over a period of only c. 275,000–300,000 years. Based on the detailed magnetostratigraphic study by Krijgsman et al. (1999), we also know that the late Miocene (Messinian) of the Mediterranean took no more than 600,000 years, and sections of it (0.75–2 km thick) probably took far less time. In many other evaporite basins it is much more difficult to estimate the total time interval precisely due to the lack of significant stratigraphical markers (like ash beds) and, consequently, applicable methodology. It seems that the internal stratigraphy of evaporite formations with well-preserved primary features can be resolved by event stratigraphical techniques, based on key beds that can be correlated over long distances (e.g. Bąbel 2005 a, b). Particularly in shallow-water evaporite settings, event or key beds represent extremely short time intervals, and it seems that the life time of many shallow-water evaporite basins is relatively short (evaporite deposition > accommodation).

Modern depositional rates in shallow water are up to 10 m per 1000 years for halite and

1–2 m per 1000 years for $CaSO_4$ (Schreiber & Hsü 1980). For ancient, major deep-basin evaporites, we can only estimate their somewhat slower accumulation rates from biostratigraphic estimates because there is no modern working analogue to be taken as a model. In contrast to their marine counterparts, evaporite deposits forming within non-marine basins, particularly those lying within orographic rain shadows (Roe 2005), and in areas of rapid and uninterrupted subsidence may persist for much longer intervals. Their total thickness is largely governed by available accommodation as well as ionic source rates. These very short depositional intervals for thick sediment packages can only take place under strictly constrained environmental conditions, as outlined in the next section.

Controls on rates of evaporation and ionic concentration

Kinsman (1976) pointed out that, in order to accumulate thick halite deposits, the regional relative humidity must be low. Kinsman also noted that there is a need for regionally lowered water vapour input, causing a low average relative humidity [less than 76% relative humidity (RH) is needed for halite formation], although this value is only a rough estimate (see Walton 1978). For example, if humidity rises either during the night or seasonally, most of the halite that has been formed will go back into solution and there will be no net accumulation. Less-soluble mineral precipitates, such as gypsum and anhydrite, commonly can persist and accumulate, but at less than optimum rates.

Another problem that requires consideration is the decrease in rate of evaporation as ionic concentration (salinity) rises. The evaporation rate of water approaching near-saturation for halite commonly slows, thus it is necessary to have and maintain elevated water temperatures to promote evaporation (particularly to replace energy lost due to evaporation). For example, commercial salt works optimize evaporation conditions by adding a thin layer of new surface water of lower salinity on saline ponds, which insulates the more saline water and also has the effect of heating the water by refracting the infrared (heat) radiation from incoming sunlight back down into the dense bottom waters (heliothermal effect; e.g. Kirkland *et al.* 1983). By raising the temperature of the water and increasing evaporation, the surface water soon reaches saturation, adding to the original mix. The salt works also keep the surface clear of floating salt that slows down evaporation and they make certain that vigorous phytoplankton blooms of halophylic microorganisms are maintained as their red colour helps to increase water temperature by several degrees (Sammy 1985).

This delicate balance, for both air and water, is not readily maintained for geologically prolonged periods of time (only a few hundreds of thousands of years), except within substantial orographic shadows and deep depressions, where evaporites may form for millions of years. In areas at or near sea level (Brutsaert 1982), as well as above mountainous areas (Nullet & Juvik 1994), evaporation rates have been studied extensively. However, in areas that lie in depressions, well below sea level (possibly like the late Miocene in the Mediterranean), comparatively little is known. Because the lapse-rate for the atmosphere is 6–10 °C per kilometre of depression (Brutsaert 1982), the temperatures in deep depressions can become considerably warmer than at sea level or above. The result of this heating is demonstrated in observations of the Dead Sea by Steinhorn (1997), where evaporation proceeds rapidly, forming highly concentrated brines. If, in the past, isolation and drawdown of some large basins such as the late Miocene in the Mediterranean, were even deeper below mean sea level than is the Dead Sea, the temperature and evaporation rates would be higher, despite increased atmospheric pressure. Theoretical calculations by Hay (1996) suggest that descending air could be warmed to more than 50–60 °C at the 2 km-deep, near-dry bottom of the Mediterranean depression, highly increasing the evaporation potential, and if this is true, the deposition of K–Mg salts in such an area is quite possible.

The pycnocline and its effect

The behaviour of water bodies, particularly those with elevated salinity, is fairly complex. In the paper of Bąbel (2007; also Bąbel 2005a, b), the following concepts are clearly developed: (1) evaporite deposition and its morphology reflect the stratification in the formative brine column; and (2) evaporative crystallization of various salts is commonly associated with mixing periods in the basin. Bąbel (2007) has shown that evaporite facies are linked to the particular stratified water zones in which they form. These zones are separated by a pycnocline (or a boundary between a less saline, diluted and lighter upper water mass, with low ionic concentrations (occasionally undersaturated), and more saline denser bottom brine (permanently oversaturated). In a simplified way, Figure 1 demonstrates the physical conditions potentially established in a water body with elevated salinity. Another major control over deposition is exerted by the pycnocline position and permanence. The known depositional facies within a modern salt works reported by Busson *et al.* (1982) and Ortí Cabo *et al.* (1984) yielded enough

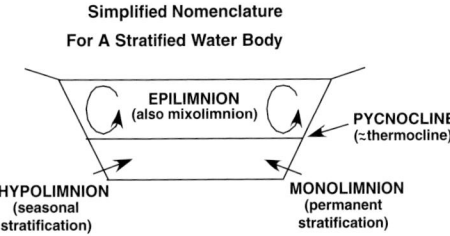

Fig. 1. Simplified view of the key points of stratification in a saline water body.

information to permit Peryt (1996) to realistically interpret some of the Ukranian gypsum of the Badenian (facies shown in Fig. 2a & b), which has now been refined and described more fully by Bąbel (2007).

Even the bedding morphology of thick gypsum layers is controlled by the position of the pycnocline (Fig. 3). Massive selenite beds appear 'truncated' or flattened when they grow to the level of the pycnocline. Truncation is not caused by mechanical erosion or chemical dissolution, but rather the

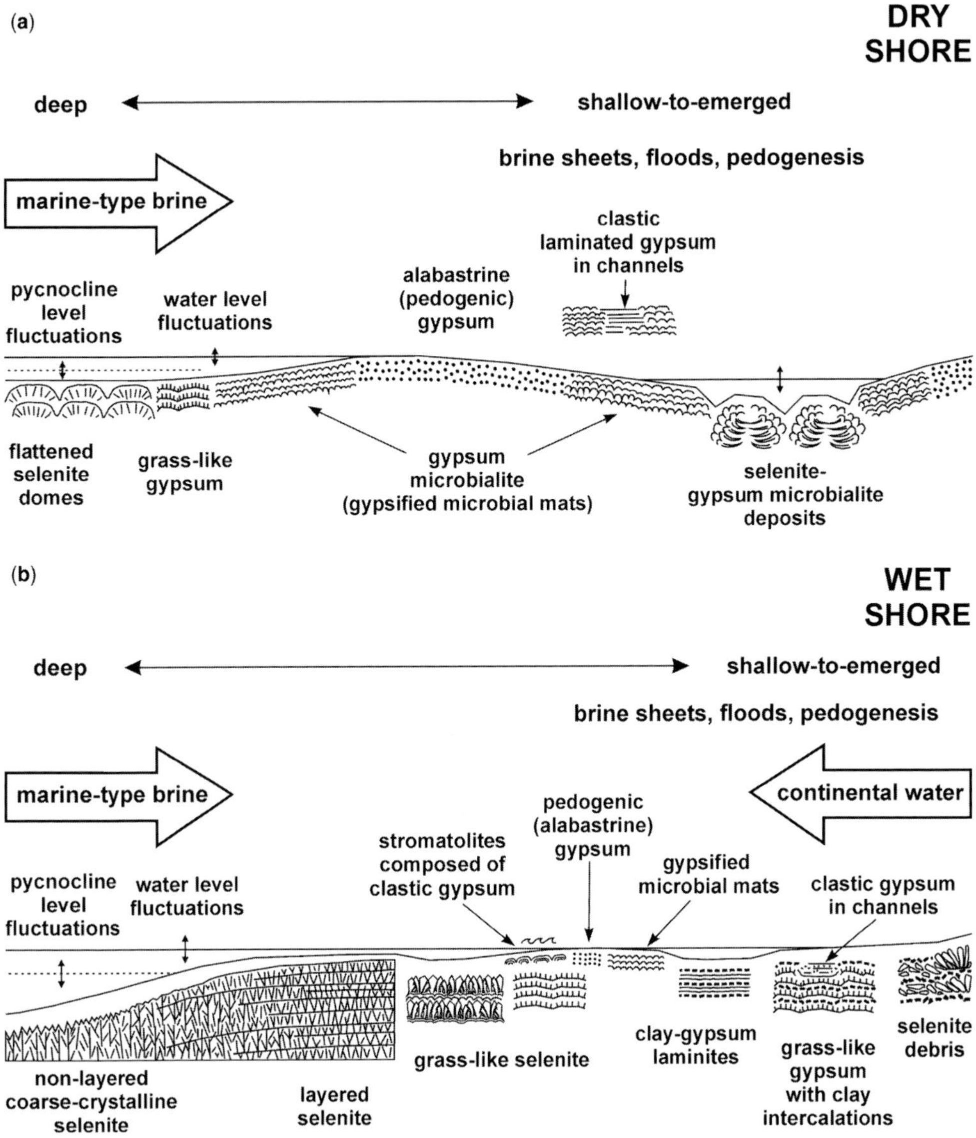

Fig. 2. Scheme showing the typical Badenian selenite and gypsum environments and facies in the Carpathian Foredeep Basin as interpreted by Peryt (1996) and Bąbel (2005b, with references). (**a**) Sketch for the 'dry shore' model, present mostly in Ukraine. (**b**) Sketch for the 'wet shore' model, in the Polish part of the basin.

limiting planation surface (developed subaqueously). This pycnocline boundary is not limited to modern evaporite formation in salinas and saline lakes, but it is readily noted in flattened selenite domes in the primary Messinian gypsum of Sicily (Fig. 3a & b), as well as in the secondary anhydrite (after selenitic gypsum) of the late Silurian in the Michigan Basin (Fig. 3c) and in the Badenian of the Carpathian foreland (Bąbel 2005a, pp. 18–19, and Pl. 5, fig. 1, marker bed h1).

Styles of deposition in evaporite basins

It is possible to subdivide evaporite-bearing basins according to the stability of the underlying crust because this governs the development of sedimentary facies. There are basins which are developed on the vast continental platforms, showing rigid consolidated crust (tectonically passive basins) characterized by very slow subsidence or uplift with a paucity of seismic shocks, and those which are developed on a mobile substrate with unconsolidated rocks or a dense network of moving tectonic blocks (tectonically active basins) which are characterized by high levels of tectonic activity and rapidity of tectonic movements (faulting, folding and overthrusting). It has become evident that evaporite deposits forming in tectonically calm basins (as in continental interior basins, or failed arms of rifts) may receive a significantly different overall assemblage of depositional facies than those formed in tectonically active basins (as in foredeep portions of foreland basins). Deposits within tectonically calm or passive basins may remain comparatively undeformed (unfolded, undomed or otherwise contorted) after formation. However, those in foredeep or active rift basins are normally both mechanically reworked during deposition and deformed afterwards, so that many of their component parts are distorted and even subjected to low-temperature metamorphism relatively early in their histories. Commonly these deposits also act as the décollement for thrust sheets. These factors make accurate depositional interpretation of evaporites very difficult, but possible with care. In both types of basins the halokinetic effects of uneven loads on salt deposits can obliterate and disturb the primary features, but such tectonic effects are apparently more drastic in active basins.

Evaporites deposited in tectonically active regions (for example, the northern margins of the late Miocene Mediterranean; the middle Miocene Carpathian foredeep) vary greatly in facies development both vertically and laterally across the basin, and even along strike. Added to this complexity is the fact that evaporites are readily slumped, and mechanically

Fig. 3. Control of crystal growth by the position of a pycnocline recorded in flattened tops of many gypsum domes. (**a**) Primary selenite domes (flattened) in Sicily (late Miocene, Eraclea Minoa). (**b**) Secondary alabastrine gypsum after primary selenite (flattened domes) in Sicily (late Miocene, Cinciana-Rafffadali road). NB. loss of detail is due to replacement of primary selenite by massive alabastrine gypsum, but much of the original structure is still retained. (**c**) Late Silurian, flattened secondary alabastrine gypsum domes, overlain by bedded, intertidal to subtidal dolomudstone with displacive Ca sulphate nodules: Celotex quarry (Gypsum, IN, USA).

water is less concentrated (undersaturated) above the pycnocline so that the crystals are unable to continue growth above that level. Because a pycnocline is normally horizontal, such flattened surfaces at the tops of selenite beds can be treated as a kind of reworked relatively early in their history. Because many evaporites are lithified as they are deposited, they produce large deposits that are solely composed of mechanically reworked materials (Hardie & Eugster 1971; Parea & Ricci Lucchi 1972; Ricci Lucchi 1973; Schreiber et al. 1976; Manzi et al. 2006; Roveri et al. 1998, 2003, 2006a, b).

In contrast to foredeep areas, large-scale syndepositional or early postdepositional mechanical reworking and deformation is rare within most basins developed on stable continental platforms and within abandoned rifts. These evaporites can remain scarcely altered or deformed even over prolonged periods of time (over hundreds of millions of years), unless deeply buried. Because of this, they can retain most of their primary features and morphologies, and commonly their original fluid inclusions (for example, from the USA, Michigan Basin, late Silurian and the Delaware Basin, middle and late Permian). The facies that developed within these deposits vary little laterally, even over considerable distances. Facies diversity is relatively limited throughout most of the section. Examination of very thick evaporite sections (Anderson et al. 1972) or in some basin margins, a well-developed facies multiplicity may be recognizable (Peryt 1996). These latter variations appear to develop in and fill localized sags and irregularities.

In areas where burial of evaporites has been greater than 2–3 km, evaporites may become considerably altered. Gypsum dehydrates to anhydrite commonly by about 1 km depth (0.4 km to more than 4 km depending on many local conditions; Jowett et al. 1993). Below a depth of 3 km halite also may develop significant secondary porosity and permeability (Lewis & Holness 1996). Such secondary porosity was first recognized by Land et al. (1988), especially in halite, fostering alteration. This does not necessarily occur in areas that are tectonically stable but is well developed in regions of elevated fluid pressure due to tectonics or increasing hydrothermal gradient (Hovland et al. 2006a, b).

Facies diversity

The sedimentation within evaporite environments, as observed in deposits from settings such as marine-marginal (sabkha), salina or a relatively shallow subaqueous marine-marginal (1 to perhaps 10 m water), usually serves as a reasonable representation of most of the evaporite facies assemblage found in the rock record (Busson et al. 1982; Ortí Cabo et al. 1984). However, for most large subaqueous basinal deposits, salinas represent only a partial working model because they are restricted to very shallow water bodies; therefore we are obliged to apply the physical conditions known from deeper lakes to model the conditions for many of the other subaqueous facies (Bąbel, this volume; Bąbel & Bogucki, this volume). Naturally, without a good and tested working model, this projection may provide a major source of error. The only certainty is that, once the physical conditions become suitable for evaporite formation, very rapid sediment accumulation results.

Tectonically passive basins

Many thick evaporites (for example, the Michigan Basin and Delaware Basin; see Hovorka et al. this volume) developed in basins located in continental sags or failed arms of ancient rifts, and, based on their chemistry, were largely marine-fed. They were enclosed within a gradually subsiding portion of the continental plate providing accommodation for rapid sediment accumulation under tectonically stable conditions. While the general pattern of pre-evaporite sedimentation is normal in these basins, marked by fauna indicative of open but gradually deepening marine water, the time interval necessary for the evaporite sedimentation is commonly so short that palaeontologists may not even find a 'break' in the faunal record between the under- and overlying deposits. This observation has led to the idea that reefs could grow freely in a Silurian sea in which thick evaporites also were forming (Droste & Shaver 1977). There are also reliable sedimentological records that demonstrate intervals of evaporite formation between periods of normal reef growth (pers. comm. 2006, Wm B. Harrison), perhaps precursors to the short and dramatic main interval of evaporite deposition. In more recent sediments (for example the late Miocene of the Mediterranean), establishing stratigraphic controls seems easier, using absolute age dating provided by ash layers and magnetostratigraphy of unaltered sediments, and fossil biostratigraphy. It is evident that the interval of extremely restricted climate is very short. Similarly short time intervals appear to have controlled evaporite deposition for major portions of the thick evaporites in the late Permian Zechstein and the Delaware basins.

There are few facies variations within these passive basins. Within the basin centres, sediments are largely laminar, present as singlets (thin carbonate, admixed with or alternating with a very thin layer of organic matter), couplets (composed of carbonate alternating with anhydrite and/or gypsum at the surface) and triplets (made up of carbonate, anhydrite and then a thin layer of halite). An excellent review of the Castile Formation of the late Permian of the Delaware Basin and its facies development may be obtained from Kirkland (2003) and Hovorka (2000), but studies in the late Silurian Michigan Basin (Budros 1974; Nurmi & Friedman 1975; Budros & Briggs 1977; Nurmi 1977) and the late Permian of the Zechstein (Richter-Bernburg 1963) also point out this lithology as well. *Lit-par-lit* correlation in such basins is very distinct and extends across most of the deposit. In all of these basins sporadic thicker beds are present, especially in the upper portions of the section. These thicker beds contain pseudomorphic relics of selenitic gypsum, indicative of shallower water facies (Richter-Bernburg 1985; Kendall & Harwood 1989), shown in Figure 4a & b. Some of the shelfal (marginal) deposits also contain these shallow-water facies. This general assemblage of facies is shown in Figure 5.

In limited areas of these basins, at some margins and also in their upper sections, there is a greater variation in facies. This is probably due to inequities in subsidence and/or filling, as well as sea-level fluctuation. In such non-tectonic areas there are few proximal indicators (such as reworked, downslope breccias), but localized buildups of anhydrite do occur adjacent to some reefs and shallow margins. However, most of the deposits in such stable basins are made up of laminites (forming a 'poker-chip' facies), sometimes intercalated with a few thicker nodular anhydrite beds that are the relics of layers of selenitic gypsum (Richter-Bernburg 1985, figures 12 & 23; Kendall & Harwood 1989).

A totally different type of facies modification, noted in some basinal evaporites, may come about through seismic effects from very distant regional tectonism, with no evidence of local disruption. Earthquakes or even tsunamis otherwise may have little directly to do with the areas of deposition. Seilacher (1969) and then Bachmann & Aref (2005) have demonstrated that seismic shocks have disrupted thin-bedded soft sediments producing 'seismites' in bottom deposits of otherwise calm basins, far from mass-flows, turbidites and other active downslope mechanisms. Comparable deformation and localized brecciation is also noted locally in the Lisan Formation of the Dead Sea (Agnon *et al.* 2006) and possibly in the Castile Formation of West Texas (Permian).

Tectonically active basins

The distinctly broader facies assemblage found within many active basins is readily seen in the exposed portions of the late Miocene (Messinian) Mediterranean, largely because the evaporites are relatively young and have not undergone burial diagenesis. This basin contains many areas composed of clearly developed, shallow-water evaporite facies, largely formed just below the pycnocline of the basin. There are also a number of large areas with definite deeper-water facies. The 200–250 m thick cyclic gypsum section, correlatable between Spain, Sicily and the Apennines (Italy), required less than 300,000 years for deposition. The climatic/water inflow mechanisms for such rapid deposition have been discussed in numerous papers but most recently Meijer (2006) and Meijer & Krijgsman (2005) have presented interesting and useful models for their formation. The biostratigraphy of the under- and/or overlying sediments are from well-dated, open marine facies, hence the time intervals are well constrained.

Associated with these shallow-water evaporitic sediments are areas that are a conglomeration of reworked fragments and blocks of primary, shallow water sediments. A comparison of reworked sections between basins shows that these facies are controlled by source areas and the rate and style of reworking. Many sub-basins are partially filled by huge reworked blocks (kilometres in size) that have slid down into basinal muds. Others were broken, slumped and reduced to sands and silts,

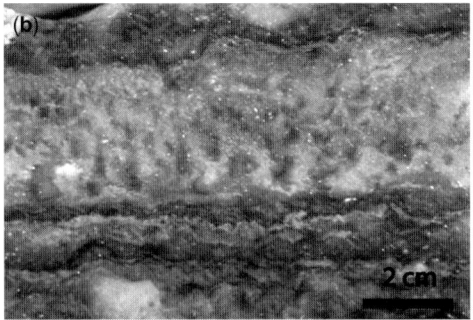

Fig. 4. Development of relic selenite structures within anhydrite. (**a**) Primary selenite layers (late Messinian, Sicily): note 'grasslike' orientation of gypsum crystals in rows separated by thin layers of gypsified, bacterial carbonate. (**b**) Relic structures in anhydrite, reflecting grasslike gypsum structure, Castile Formation (late Permian) ERDC 9 WIPP core, W. Texas, USA.

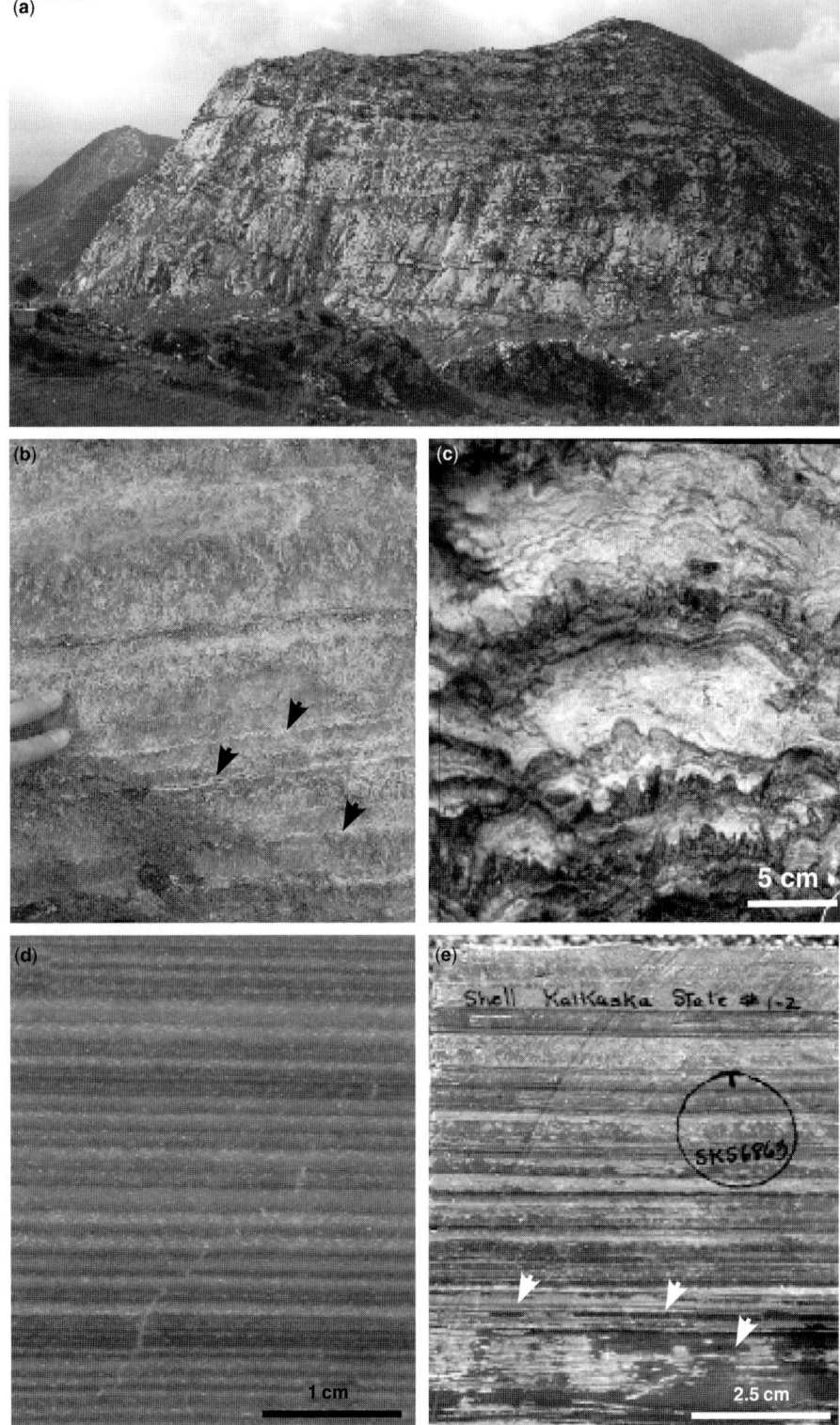

reworked as mass flows and well-developed turbidites (Roveri et al. 1998, 2003; Roveri & Manzi 2006). In some areas, such large volumes of reworked materials were deposited onto the floors of adjacent deep basins that most of the marginal source areas became virtually denuded of their evaporite sections, as in the eastern Apennines and parts of Sicily (Fig. 6a–f). In the Granada Basin and in western Sicily (Gibellina), reworking creates elegantly formed turbidite sections, replete with graded beds having classical scour surfaces and load and prod casts. It has been suggested that many parts of the deep-basin deposits originate as slumps and turbidites fed from the tectonized margins (Roveri et al. 1998, 2003; Roveri & Manzi 2006; Bertoni & Cartwright 2007). Similar reworked facies are known in the middle Miocene of the Carpathian foredeep (e.g. Kolasa & Ślączka 1985). In parts of the late Permian of the Zechstein Basin, Schlager & Bolz (1977), Meier (1977, 1981) and Richter-Bernburg (1985) also describe somewhat similar aspects of reworking, but their general analysis remains somewhat unclear.

In some early stages of rift development basin, margins have a rapidly changing slope due to crustal thinning and rapid subsidence. If there is early salt formation on rapidly subsiding margins, they commonly deform as a result of tilting and increasing slope development, not commonly from any other type of tectonics. These deposits, particularly the chloride facies, readily deform plastically on slopes in response to gravity (see Kehle 1988; Jackson 1995, 1997). The entire overlying non-evaporitic section becomes destabilized, exerting an uneven load, generating diapirs in the salt or allowing the salt simply to flow out from under its irregular burden. Such uneven loading along continental margins commonly controls the position of subsequent deltaic buildups, especially in failed rift arms, exacerbating the regional deformation, and further destabilizing the halite. Having undergone many phases of plastic deformation, the original halite facies that composed the deformed beds become unrecognizable petrologically, or at least very difficult to decipher (Schléder & Urai 2005). In this way, almost nothing remains recorded in the salt to reveal its depositional history. In other parts of a rift, where the evaporitic filling is largely confined to the relatively stable rift basin floor or the marginal satellite basins, the basin remains relatively stable, although subsiding rapidly, and the evaporites may remain undeformed and continuous.

Evaporite deposits: what do the lithologic groupings mean?

With the accumulated observations of evaporites available to sedimentologists, there is now enough data available to sort evaporites into meaningful lithological groupings. These groupings are not just based on compositions, trace elements and isotopes, but also on lithology and stratigraphy. The paper presented by Bąbel & Bogucki (2007, and the references therein) demonstrates the product of the controls exerted by composition, environmental conditions and behaviour of the water that flowed in the Badenian basin(s) more than 11.1 million years ago. These observations are clear enough to draw a realistic model of formation. The general concepts presented step beyond the observations made in modern salinas were reported in Busson et al. (1982) and Ortí Cabo et al. (1984). While Schreiber et al. (1976, 1977) attempted a simplistic environmental reconstruction, the papers in this volume have put primary subaqueous evaporite facies into a stronger and more realistic framework and have pointed the direction of many tectonic and diagenetic changes that are part of the evaporite story. Building on this framework, extending from the clearly understood into the hypothetical, must be done stepwise, incorporating data from other basins. Sedimentation in continental sag-basins, here specifically addressed in

---

Fig. 5. Facies commonly developed in passive basins such as continental 'sags' and/or failed rift-arms. A similar facies assemblage may be present on the stable, marginal portions of tectonically active basins. In both cases lateral facies changes are commonly gradual in this setting and have considerable lateral and vertical persistence. Mechanically reworked evaporites are shallow-water in origin and usually uncommon in this setting. (**a**) Outcrop view of M. Banco (late Miocene, Messinian; near Raffadali, Sicily). Note relative continuity of layers, all composed of relatively shallow-water facies, formed below a somewhat variable regional pycnocline. Total thickness c. 250 m. (**b**) 'Grass-like' selenitic gypsum alternating with very thin microbial carbonate interbeds (late Miocene, Messinian; Passo Funnuto, Sicily). Upon burial this facies becomes regular anhydrite beds alternating with thin $CaSO_4$ cemented carbonate. (**c**) Microbially controlled stromatolite structures composed of carbonate, pervasively cemented and partially replaced by gypsum that is synsedimentary in origin (middle Miocene, Badenian, near Gartwice, Poland). (**d**) Basin-centre anhydrite–carbonate couplets from a cored sample. Small amounts of argillite on bedding planes may cause parting into thin 'poker-chips'. Castile Formation, late Permian, Texas, USA. (**e**) Basin-centre anhydrite–carbonate–halite triplets, with some additional anhydrite cementation of porosity in carbonate layers (grey patches, at arrows). Michigan Basin, A1E; Shell Kalkaska no. 1–2 well (Michigan, USA).

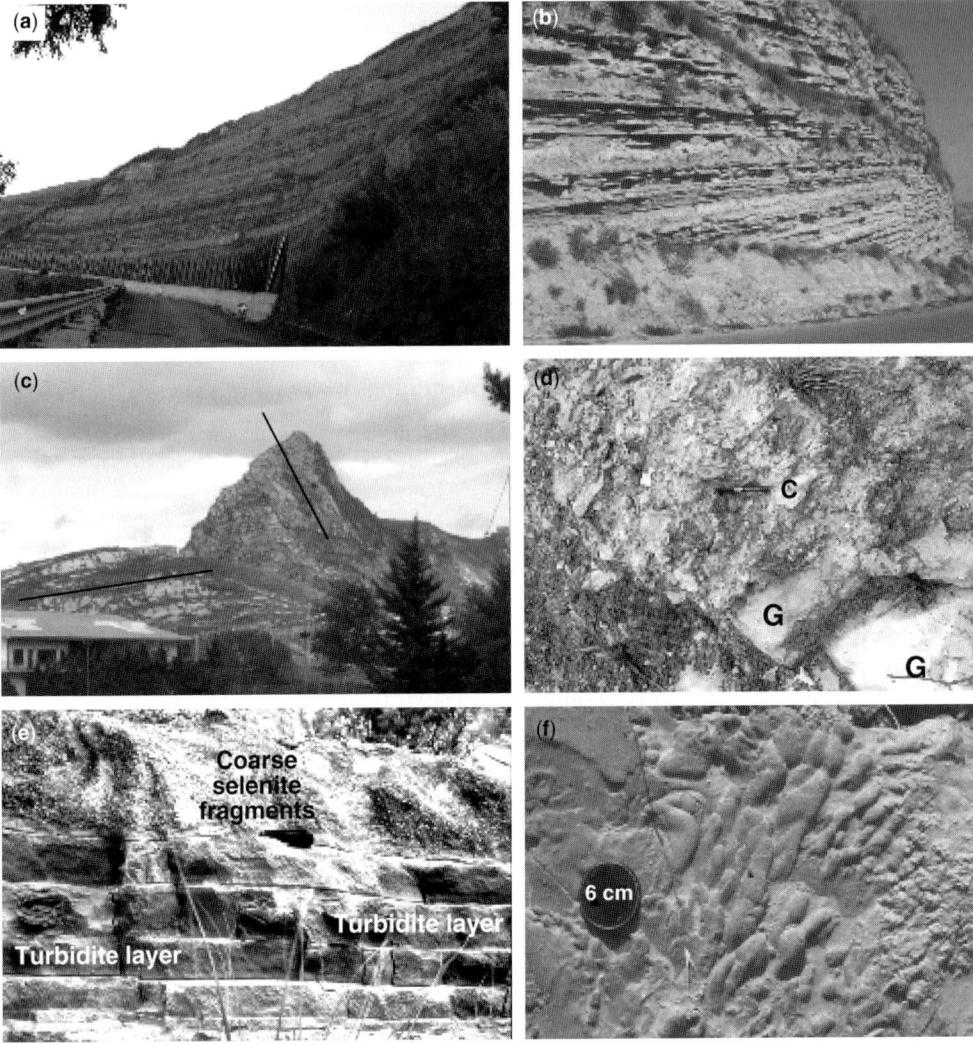

Fig. 6. Facies developed in a tectonically active basin, here in late Miocene evaporite deposits (Tortonian and Messinian), showing mechanically reworked evaporites in an active tectonic setting. (**a**) Turbidite section (Highway 118, above the ruins of old Gibellina, Sicily), about 40 m in height. Sediment is composed of reworked, shallow-water, primary selenitic gypsum clasts (varying from silts to cobble-sized fragments) plus evaporitic carbonates in an organic-rich matrix (up to 7% organic TOC). Many beds are well sorted and monomineralic while others are very mixed in composition. Outcrop covered by a mesh of protective netting. (**b**) Sequence of gypsum turbidite beds, containing a considerable siliciclastic, sand component (late Miocene, near la Malaha, Spain (Granada). (**c**) Irregularly tipped and slid gypsum deposits (shallow-water in origin) in mountain-sized blocks, central Sicily. Bedding is clearly preserved; part of a synsedimentary phase of large-scale downslope reworking, involving the lower beds in the sequence. Messinian, near Raffadali, Sicily (Roveri *et al.* 2006*b*). Lines point up the bedding orientation of adjacent blocks. (**d**) Mass flow at the margin of a long turbidite section near Gibellina, Sicily (Messinan). Composed of evaporative carbonate clasts, large shallow water selenite fragments, and fine sand-size selenite in a gypsiferous argillitic matrix. (**e**) Coarse, selenite sandstone, part of a turbidite sequence. Crystal fragments contain filamentous, bacterial outlines, indicative of original photic zone growth. Near old Gibellina, Sicily. (**f**) Close-up of a bedding-plane marked by load casts, at the base of a turbidite layer. Messinan, near old Gibellina, Sicily.

Hovorka et al. 2007 review of the Permian of west Texas, adds to this depositional framework.

Future investigation

In the past 10 years a number of new concepts have been proposed concerning the evolution of seawater composition that may change our view on some of the sources and modes of deposition of evaporite deposits. The idea that mid-ocean ridges can process and modify circulating seawater as it moves through spreading centre magmas and basalts is key to this. Such circulation can change the composition of the world's oceans over time, but the suggestion is a comparatively new one (Hardie 1996). In addition, a suggestion, presented since the Florence Conference, proposes that large volumes of 'evaporites' may actually be due to and form from rift-sourced and hydrothermal fluids (Hovland et al. 2006a, b). This concept certainly must be addressed in the future. Evaporites still represent an uncertain narrative, with further clarity coming slowly as we gain additional information as we learn more about old and new deposits.

In the future, Martian evaporites will increasingly captivate our thoughts, but we will not be able to address those alien deposits until we have a better hold on our Earthly, comparatively modern deposits. Further, if we can make a stab at the meaning of and controls for those evaporites formed on Earth in the Archean, we will be in a far better position to understand off-Earth deposits. This volume about Earth-bound deposits, is a collection of observations, concepts and ideas that can point us toward a fuller understanding of our own world and the many others waiting to be explored.

References

AGNON, A., MIGOWSKI, C. & MARCO, S. 2006. Intraclast breccias in laminated sequences reviewed; recorders of paleo-earthquakes. In: ENZEL, Y., AGNON, A. & STEIN, M. (eds) The Dead Sea. New Frontiers in Dead Sea Paleoenvironmental Research. Geological Society of America, Special Publication **401**, 195–214.

ANDERSON, R. Y., DEAN, W. E., KIRKLAND, D. W. & SNIDER, H. I. 1972. Permian Castile varved evaporite sequence, West Texas and New Mexico. Geological Society of America Bulletin, **83**, 59–86.

BACHMANN, G. H. & AREF, M. A. M. 2005. A seismite in Triassic gypsum deposits (Grabfeld Formation, Ladinian), southwestern Germany. Sedimentary Geology, **180**, 75–89.

BĄBEL, M. 2005a. Event stratigraphy of the Badenian selenite evaporites (Middle Miocene) of the northern Carpathian Foredeep. Acta Geologica Polonica, **55**, 9–29. On-Line Appendix www.geo.uw.edu.pl/agp/table/appendixes/55-1/

BĄBEL, M. 2005b. Selenite-gypsum microbialite facies and sedimentary evolution of the Badenian evaporite basin of the northern Carpathian Foredeep. Acta Geologica Polonica, **55**, 187–210.

BĄBEL, M. 2007. Depositional environments of a salina-type evaporite basin recorded in the Badenian gypsum facies in northern Carpathian Foredeep. In: SCHREIBER, B. C., LUGLI, S. & BĄBEL, M. (eds) Evaporites Through Space and Time. Geological Society, London, Special Publications, **285**, 107–142.

BĄBEL, M. & BOGUCKI, A. 2007. The Badenian evaporite basin of the northern Carpathian Foredeep as a model of the meromictic selenite basin. In: SCHREIBER, B. C., LUGLI, S. & BĄBEL, M. (eds) Evaporites Through Space and Time. Geological Society, London, Special Publications, **285**, 219–246.

BERTONI, C. & CARTWRIGHT, J. A. 2006. Controls on the basinwide architecture of Messinian evaporites on the Levant margin (Eastern Mediterranean). Sedimentary Geology, **188/189**, 93–114.

BERTONI, C. & CARTWRIGHT, J. A. 2007. Clastic depositional systems at the base of the late Miocene evaporites of the Levant region, Eastern Mediterranean. In: SCHREIBER, B. C., LUGLI, S. & BĄBEL, M. (eds) Evaporites Through Space and Time. Geological Society, London, Special Publications, **285**, 37–52.

BRUTSAERT, W. 1982. Evaporation into the Atmosphere: Theory, History, and Applications. Reidel, Hingham, MA.

BUDROS, R. 1974. The stratigraphy and petrogenesis of the Ruff Formation, Salina Group in northeast Michigasn. Masters thesis, University of Michigan.

BUDROS, R. Y. & BRIGGS, L. I. 1977. Depositional environment of Ruff Formation (Upper Silurian) in Southeastern Michigan. In: FISHER, J. H. (ed.) Reefs and Evaporites – Concepts and Depositional Models. American Association of Petroleum Geologists, Studies in Geology **5**, AAPG, 53–72.

BUSSON, G., CORNÉE, A. ET AL. 1982. Données hydrochimiques, biologiques, isotopiques, sédimentologiques et diagénétiques sur les marais salants de Salin-de-Giraud (Sud de la France). Géologie Méditerranéene, **9**(4).

DROSTE, J. B. & SHAVER, R. H. 1977. Synchronization of deposition: silurian reef-bearing rocks on Wabash Platform with cyclic evaporites of Michigan Basin. In: FISHER, J. H. (ed.) Reefs and Evaporite – Concepts and Depositional Models. American Association of Petroleum Geologists, Studies in Geology, **5**, 93–109.

HARDIE, L. A. 1996. Secular variation in seawater chemistry: An explanation for the secular variation in the mineralogies of marine limestones and potash evaporites over the past 600 m.y. Geology, **24**, 279–283.

HARDIE, L. A. & EUGSTER, H. P. 1971. The depositional environment of marine evaporites: a case for shallow, clastic accumulation. Sedimentology, **16**, 187–220.

HAY, W. W. 1996. Tectonics and climate. Geologischs Rundshau, **85**, 409–437.

HOVLAND, M., RUESLÅTTEN, H., JOHNSEN, H. K., KVAMME, B. & KUTZNETSOVA, T. 2006a. Salt formation associated with supercritical water. *Marine and Petroleum Geology*, **23**, 855–869.

HOVLAND, M., KUZNETSOVA, T., RUESLÅTTEN, H., KVAMMEW, B., JOHNSEN, H. K., FLADMARK, G. E. & HEBACH, A. 2006b. Sub-surface precipitation of salts in supercritical seawater. *Basin Research*, **18**, 221–230.

HOVORKA, S. D. 2000. Deep-water to shallow-water transition in evaporites in the Delaware Basin, Texas. *In*: LINDSAY, R. F., TRENTHAM, R. C., WARD, R. F. & SMITH, A. H. (eds) *Classic Permian Geology of West Texas and Southeastern New Mexico: 75 Years of Permian Basin Oil and Gas Exploration and Development: West Texas*. Geological Society, Geo 2000 field trip guidebook Publication **00–108**, 273–299.

HOVORKA, S. D., HOLT, R. M. & POWERS, D. W. 2007. Depth indicators in Permian Basin evaporites. *In*: SCHREIBER, B. C., LUGLI, S. & BĄBEL, M. *Evaporites Through Space and Time*. Geological Society, London, Special Publications, **285**, 335–364.

JACKSON, M. P. A., 1995. Retrospective salt tectonics. *In*: JACKSON, M. P. A., ROBERTS D. G. & SNELSON, S. (eds) *Salt Tectonics: A Global Perspective*. American Association of Petroleum Geologists, Memoir **65**, 1–28.

JACKSON, M. P. A. 1997, Conceptual breakthroughs in salt tectonics: a historical review, 1856–1993. *The University of Texas at Austin, Bureau of Economic Geology Report of Investigations 246*.

JOWETT, E. C., CATHLES, III, L. M. & DAVIS, B. W. 1993. Predicting depths of gypsum dehydration in evaporitic sedimentary basins. *American Association of Petroleum Geologists Bulletin*, **77**, 402–413.

KEHLE, R. O. 1988. The origin of salt structures. *In*: SCHREIBER, B. C. (ed.) *Evaporites and Hydrocarbons*. New York, Columbia University Press, 345–404.

KENDALL, A. C. & HARWOOD, G. M. 1989. Shallow water gypsum in Castile Formation – significance and implications. *In*: HARRIS, P. M. & GROVER, G. A. (eds) *SEPM Core Workshop: Subsurface and Outcrop Examination of the Capitan Shelf Margin, N. Delaware Basin*, Society for Sedimentary Geologists, Tulsa, Oklahoma. **13**, 451–457.

KIRKLAND, D. W. 2003. An explanation for the varves of the Castile evaporites (Upper Permian), Texas and New Mexico, USA. *Sedimentology*, **50**, 899–920.

KIRKLAND, D. W., BRADBURY, J. P. & DEAN, W. E. 1983. The heliothermic lake – a direct method of collecting and storing solar energy. *Archiv für Hydrobiologie, Supplementband (Monographische Beiträge)*, **65**, 1–60.

KINSMAN, D. J. J. 1976. Evaporites: humidity control of primary mineral facies. *Journal of Sedimentary Petrology*, **46**, 273–279.

KOLASA, K. & ŚLĄCZKA, A. 1985. Sedimentary salt mega-breccias exposed in the Wieliczka mine, Fore-Carpathian Depression. *Acta Geologica Polonica*, **35**, 221–230.

KRIJGSMAN, W., HILGEN, F. J., RAFFI, I., SIERRO, F. J. & WILSON, D. S. 1999. Chronology, causes and progression of the Messinian Salinity Crisis. *Nature*, **400**, 652–655.

LAND, L. S., KUPECZ, J. A. & MACK, L. E. 1988. Louann Salt geochemistry (Gulf of Mexico sedimentary basin, USA): a preliminary synthesis. *Chemical Geology*, **74**, 25–35.

LEWIS, S. & HOLNESS, M. 1996. Equilibrium halite–H_2O dihedral angles: high rock-salt permeability in the shallow crust? *Geology*, **24**, 431–434.

MANZI, V., LUGLI, S., RICCI LUCCHI, F. & ROVERI, M. 2005. Deep-water clastic evaporites deposition in the Messinian Adriatic foredeep (northern Apennines, Italy): did the Mediterranean ever dry out? *Sedimentology*, **52**, 875–902.

MEIER, R. 1977. *Turbidite und Olisthostrome – Sedimentationsphänomene des Werra-Sulfats (Zechstein 1) am Osthang der Eichsfeld-Schwelle im Gebiet des Südharzes*. Veröffentlichungen des Zentralinstituts für Physik der Erde (Akademie der Wissenschaften der DDR, Forschungsbereich Geo- und Kosmoswissenschaften), Podstam, **50**, 1–45.

MEIER, R. 1981. Clastic resedimentation phenomena of the Werra sulphate (Zechstein 1) at the eastern slope of the Eichsfeld swell (Middle European basin) – an introduction. *In*: *International Symposium, Central European Permian, Jabłonna, April 27–29, 1978, Proceedings*. Wydawnictwa Geologiczne, Warsaw, 269–373.

MEIJER, P. TH. 2006, A box model of the blocked-flow scenario for the Messinian Salinity Crisis. *Earth and Planetary Science Letters*, **248**, 471–479.

MEIJER, P. & KRIJGSMAN, W. 2005. Quantitative analyses of sea level, salinity, and strait transports during and preceding the Messinian Salinity Crisis. *Earth and Planetary Science Letters*, **240**, 510–520.

NULLET, D. & JUVIK, J. O. 1994. Generalized mountain evaporation profiles for tropical and subtropical latitudes. *Singapore Journal of Tropical Geography*, **15**, 17–24.

NURMI, R. 1975. *Sedimentology and depositional environments of basin-center evaporites, Lower Salina Group (Upper Silurian), Michigan Basin*. Ph.D. Thesis, Rensselaer Polytechnic Institute, NY.

NURMI, R. & FRIEDMAN, G. M. 1977. Sedimentology and depositional environments of basin-center evaporites, Lower Salina Group (Upper Silurian), Michigan Basin. *In*: FISHER, J. H. (ed.) *Reefs and Evaporites – Concepts and Depositional Models*. American Association of Petroleum Geologists, Studies in Geology **5**, 23–52.

ORTÍ CABO, F., PUEYO MUR, J. J., GEISLER-CUSSEY, D. & DULAU, N. 1984. Evaporitic sedimentation in the coastal salinas of Santa Pola (Alicante, Spain). *In*: ORTÍ CABO, F. & BUSSON, G. (eds) *Introduction to the Sedimentology of the Coastal Salinas of Santa Pola (Alicante, Spain)*. Revista del Instituto de Investigaciones Geológicas, Barcelona, **38/39**, 169–220.

PAREA, G. C. & RICCI LUCCHI, F. 1972. Resedimented evaporites in the periadriatic trough (upper Miocene, Italy). *Israel Journal of Earth-Sciences*, **21**, 125–141.

PERYT, T. M. 1996. Sedimentology of Badenian (middle Miocene) gypsum in eastern Galicia, Podolia and Bukovina (West Ukraine). *Sedimentology*, **43**, 571–588.

RICCI LUCCHI, F. 1973. Resedimented evaporites: indicators of slope instability and deep-basins conditions in Periadriatic Messinian (Apennines foredeep, Italy). In: DROOGER, C. W. (ed.) *Messinian Events in the Mediterranean*. Koninklijke Nederlandse Akademie Van Wetenshappen, Geodynamics Scientific Report no. 7 on the colloquium held in Utrecht, 2–4 March 1973, 142–149.

RICHTER-BERNBURG, G. 1955. Über salinare Sedimentation. *Zeitschrift Deutsche Geologisches Gesellschaft*, **105**, 593–645.

RICHTER-BERNBURG, G. 1963. Solar cycle and other climatic periods in varvitic evaporites. In: NAIRN, A. E. M. (ed.) *Problems in Climatology. Proceedings NATO Conference, Newcastle*, 510–532.

RICHTER-BERNBURG, G. 1985. Zechstein Anhydrite. Fazies und Genese. *Geologisches Jahrbuch, A. E.* Schweizerbartische Verlagsbuchhandlung in Kommission, Stottgart, **85**, 3–82.

ROE, G. H. 2005. Orographic Precipitation. *Annual Review of Earth Planetary Sciences*, **33**, 645–671.

ROVERI, M. & MANZI, V. 2006. The Messinian salinity crisis: looking for a new paradigm? *Palaeogeography, Palaeoclimatology, Palaeoecology*, **238**, 386–398.

ROVERI, M., MANZI, V., BASSETTI, M. A., MERINI, M. & RICCI LUCCHI, F. 1998. Stratigraphy of the Messinian post-evaporitic stage in eastern-Romagna. *Giornale di Geologie*, **60**, 119–142.

ROVERI, M., MANZI, V., RICCI LUCCHI, F. & ROGLEDI, S. 2003. Sedimentary and tectonic evolution of the Vena del Gesso basin (Northern Apennines, Italy): implications for the onset of the Messinian salinity crisis. *Geological Society of America Bulletin*, **115**, 387–405.

ROVERI, M., LUGLI, S., MANZI, V., GENNAI, R., IACCARINO, S. M., GROSSI, F. & TAVIANI, M. 2006a. The record of Messinian events in the Northern Apennines foredeep basins. RCMNS IC Parma 2006 'The Messinian salinity crisis revisited II', pre-congress field-trip guidebook. *Acta Naturalia de 'L'Ateneo Parmense'*, **42**, 65.

ROVERI, M., MANZI, V. ET AL. 2006b. Clastic vs. primary precipitated evaporites in the Messinian Sicilian basins. RCMNS IC Parma 2006 'The Messinian salinity crisis revisited II', post-congress field-trip guidebook. *Acta Naturalia de 'L'Ateneo Parmense'*, **42**, 66.

SAMMY, N. 1985. Biological systems in North-West Australian solar salt fields. In: SCHREIBER, B. C. & HARNER, H. L. (eds) *Sixth International Symposium On Salt*, The Salt Institute, Alexandria, Virginia, **1**, 207–215.

SCHLAGER, W. & BOLZ, H. 1977. Clastic accumulation of sulfate evaporites in deep water. *Journal of Sedimentary Petrology*, **47**, 600–609.

SCHLÉDER, Z. & URAI, J. L. 2005. Microstructural evolution of deformation-modified primary halite from Middle Triassc Röt Formation at Henglo, The Netherlands. *International Journal of Earth Sciences (Geologische Rundschau)*, **94**, 941–955.

SCHREIBER, B. C. & HSÜ, K. J. 1980. Evaporites. In: HOBSON, G. D. (ed.) *Developments in Petroleum Geology*, **2**. Applied Science Ltd., London, 87–138.

SCHREIBER, B. C., FRIEDMAN, G. M., DECIMA, A. & SCHREIBER, E. 1976. The depositional environments of the Upper Miocene (Messinian) evaporite deposits of the Sicilian Basin. *Sedimentology*, **23**, 729–760.

SCHREIBER, B. C., CATALANO, R. & SCHREIBER, E. 1977. Evaporitic facies observed in the Upper Miocene (Messinian) deposits of the Salemi Basin (Sicily) and a modern analog. In: FISHER, J. H. (ed.) *Reefs and Evaporites*. American Association of Petroleum Geologists, Special Publication **5**, 169–180.

SEILACHER, A. 1969. Fault-graded beds interpreted as seismites. *Sedimentology*, **13**, 155–159.

STEINHORN, I. 1997. Evaporation estimate for the Dead Sea: essential consideration for saline lakes. In: NIEMI, T. M., BEN-AVRAHAM, Z. & GAT, J. R. (eds) *The Dead Sea: The Lake and its Setting*. Oxford University Press, New York, 122–132.

WALTON, A. W. 1978. Evaporites; relative humidity control of primary mineral facies; a discussion. *Journal of Sedimentary Petrology*, **48**, 1357–1359.

Timing and origin of the South Atlantic pre-salt sag basins and their capping evaporites

G. D. KARNER[1] & L. A. P. GAMBÔA[2]

[1]*Lamont-Doherty Earth Observatory, Palisades, NY, 10964, USA*
(e-mail: garry@ldeo.columbia.edu)

[2]*Petroleo Brasileiro S.A., Av. Chile, 65, 20035, Rio de Janeiro, RJ, Brazil*

Abstract: Continental extension between West Africa and Brazil was responsible, directly or indirectly, for the development of the pre-salt sag basins and the evaporites of the South Atlantic salt basin. Subsidence mechanisms to explain these basins and their capping evaporites include: (1) deposition on Barremian-aged ocean crust; (2) rift propagation from east to west across the West African margin such that post-rift subsidence commenced in the east while rifting was still occurring to the west; and (3) depth-dependent lithospheric extension. Predicted thermal subsidence of oceanic crust or rifted lithosphere is inadequate to generate sufficient accommodation for the evaporites. Within the Santos Basin, extensional faulting within the pre-salt sag basin occurs up to the base of the evaporites; extension clearly continued to the late Aptian. Time-equivalent onshore and offshore pre-salt sections across the West African margin, and the inability to generate sufficient subsidence if the sections are considered to be post-rift, disqualifies east to west rift propagation as a mechanism for the observed pre-salt basins and evaporites. Barremian–Aptian depth-dependent extension best explains the general rift and post-rift development of the West African and Brazilian margins and the paucity of syn-rift faulting, the strain balance being achieved by the lateral emplacement of lower crust and continental mantle out from under the adjacent continental lithosphere. Regional exposure and truncation of the top pre-salt sag section attests to a climate-induced lake level drawdown during the mid Aptian, and offers a simple mechanism to generate the shallow water environments for evaporite precipitation across the West African–Brazilian rift system. In the subsequent marine transgression the Gabon and Angolan salts and the evaporites within the conjugate Camamu-Almada, Jequitinhonha and Cumuruxatiba basins were deposited. Santos and Campos basin evaporites are younger. The barrier to southern Atlantic marine incursions and the possible delay in Santos and Campos evaporite deposition relates to the magmatic constructions of a proto-Walvis Ridge and the long-lived anomalous topography of the southeastern Brazilian highlands; Campos and Santos basin extension was necessarily superimposed on a broad, high-relief plateau.

The extensive salt province of the South Atlantic is a series of basins separated by basement highs, represented by the Loeme, Ezanga, Ariri, Ibura, Paripueira, Taipus Mirim and Mariricu formations of West Africa and Brazil. Salt deposits are a characteristic of many Brazilian basins (Santos–Campos-Espírito Santo, Cumuruxatiba, Jequitinhonha, Camamu–Almada and Sergipe–Alagoas) and the African basins of northern Angola, Kwanza, Congo, Gabon, Rio Muni and Doula (Figs 1 & 2). Karner *et al.* (2003) suggested that the evaporites, because of the need for restrictive environments and shallow water depths required to deposit the thick (1–2 km) evaporites across the entire West African–Brazilian rift system, need to be of late syn-rift age. The evaporites are the final deposits of a regional mega-regressive package that also comprise the regionally distributed pre-salt sag basin. In contrast, other researchers (e.g. Henry *et al.* 1995; Jackson *et al.* 2000; Marton *et al.* 2000; Davison 2007) argue that the salt basins are post-breakup in age, that their distribution is a consequence of separation by Barremian-aged (magnetic anomaly M3) spreading centres, and that distal salt accumulated on proto-oceanic crust, not thinned continental crust. In this scenario, the evaporites were formed as part of the rapid, early post-rift phase of basin subsidence as the region became inundated by seawater input across the Walvis Ridge, a volcanic edifice generated by the Tristan da Cunha plume.

The syn- or post-rift origin of the South Atlantic evaporites is at the centre of a major controversy concerning the timing and the lateral and vertical distribution of extension across the West African and Brazilian continental lithospheres. Rifting, by definition, deals with brittle faulting of the crust leading to the generation of horst and graben morphologies. In general, rifting is recognized by a number of key observations; normal faulting, divergence and rotation of seismic reflectors indicative of differential subsidence and block rotation, and

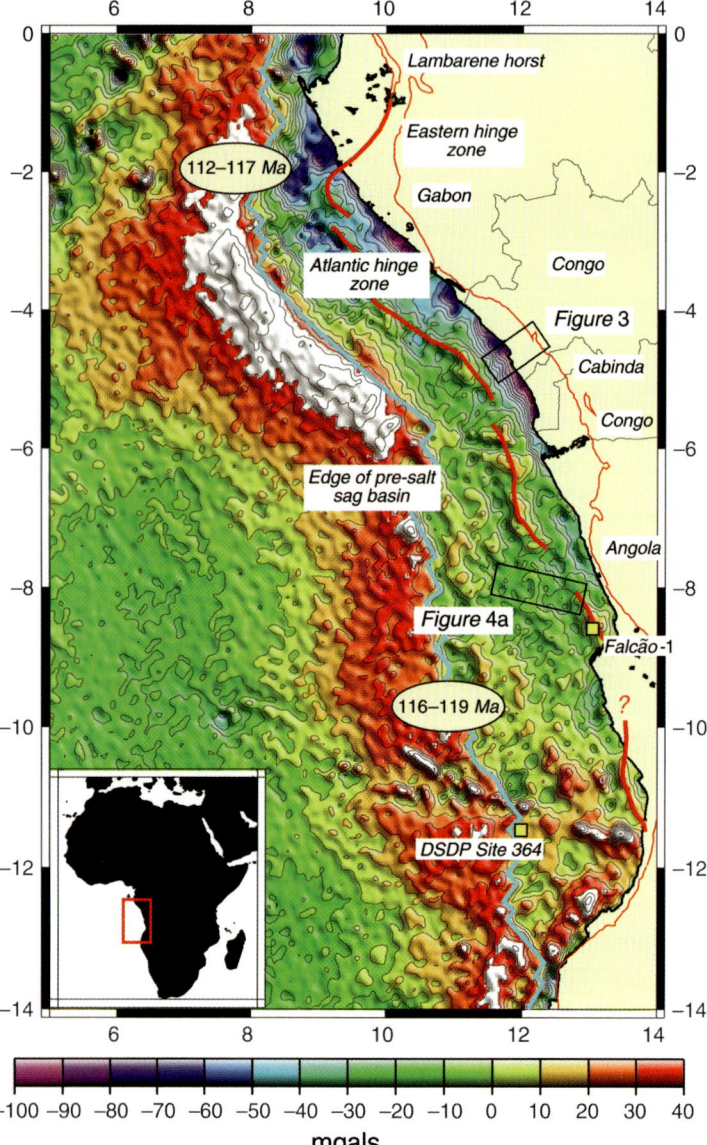

Fig. 1. Location map showing the major tectonic and structural features of the Gabon–Angola continental margin. The base map is a residual Bouguer gravity anomaly. The scale is in mgals with values greater than +40 in white. Seismic reflection profiles are located in the two outlined boxes on the Congo and northern Angolan margins (Kwanza Basin). Two tectonic hinge zones, the Eastern (thin red line) and Atlantic (bold red line), trend subparallel to the margin. The onshore Eastern hinge zone demarcates the eastern limit of Neocomian extension and separates continental margin sediments from Precambrian basement. The Atlantic hinge zone occurs beneath the outer shelf/upper slope transition and consists of a series of en-echelon high-standing blocks. The seaward edge of the pre-salt basin is approximately delineated by the strong positive/negative gradient in the gravity anomaly (bold aqua line). The Falcão-1 well and DSDP Site 364 (solid yellow squares) are located within the pre-salt sag basin and seaward of the Atlantic hinge zone, respectively. The age ranges of the evaporites for the Gabon and Angolan margin are also given (yellow ellipses).

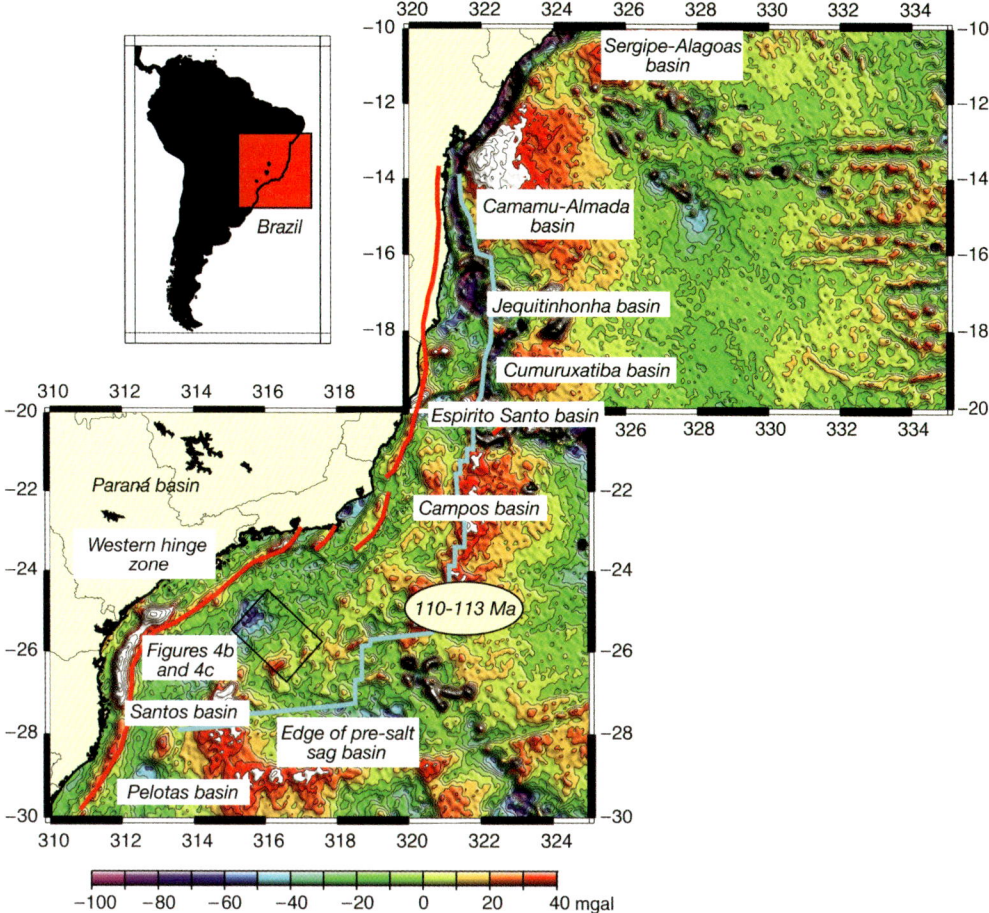

Fig. 2. Location map showing the major tectonic features and basins of the Brazilian continental margin. The base map is a residual Bouguer gravity anomaly. The scale is in mgals with values greater than +40 in white. A major offshore hinge zone, the Western hinge zone (bold red line), trends subparallel to the margin. The edge of the pre-salt basin is approximately delineated by the strong positive/negative gradient in the gravity anomaly (bold aqua line). The age range of the Santos Basin evaporites is also given (yellow ellipse).

sediment wedge geometries. However, studies from the northwest Australian and northeast Australian margins, Grand Banks and Labrador margin, West of Shetland, Vøring Plateau and Møre basin, and the West African and Brazilian margins (Royden & Keen 1980; Roberts et al. 1997; Baxter et al. 1999; Karner & Driscoll 1999; Isern et al. 2002; Karner et al. 2003) have clearly shown that these stratigraphic relationships are rare and the syn-rift subsidence tends to be regionally distributed with only minor normal fault control. While the geological details and sedimentary facies differ between the various margins, the style of deformation and the regional distribution of accommodation are remarkably similar; that is, a notable absence of syn-rift basement faults.

The development of significant post-rift accommodation in the same region characterized earlier by minor syn-rift faulting and shallow depositional environments has been explained by Driscoll & Karner (1998) in terms of depth-dependent extension that is vertically partitioned across a zone of decoupling that results in the development of a relatively non-deforming upper crust (i.e. the upper plate) from a ductile-deforming lower crust and lithospheric mantle (i.e. the lower plate), the boundary between them having a ramp–flat–ramp geometry. These sag packages are clearly syn-rift off northwest Australia because they predate the age of the first sea-floor spreading magnetic anomaly (e.g. Robb et al. 2005). Extension of the lower plate thins the lower crust (generating subsidence) and thins the lithospheric mantle by advecting heat (generating uplift). The competition between the two creates the depositional environments of the syn-rift sag and also explains the rapid plummeting of the margin following break-up. The dominance of 'stacked syn-rift sag packages' is exemplified by the general absence of faults controlling syn-rift accommodation.

A potential problem with depth-dependent extension is that it may create a 'space problem', depending how the extension is distributed with depth (e.g. White & McKenzie 1989). As demonstrated for the Iberian and Newfoundland margins (e.g. Müntener & Hermann 2001; Manatschal et al. 2001; Whitmarsh et al. 2001; Shipboard Scientific Party 2004) and the Exmouth Plateau (Tischer 2006), there are two components of the strain balance: (1) a counterbalancing upper crustal extension leading to late-stage brittle deformation in the vicinity of the ocean-continent transition zone (c. 200 km); and (2) the lateral emplacement and exposure of serpentinized and magmatically modified continental mantle and lower crust out from under the adjacent, extending continental lithosphere (e.g. Lavier & Manatschal 2006). The width of the transition zone balances the width of the region characterized by syn-rift sagging (Tischer 2006).

The preserved syn-rift stratigraphy of the West African margin is the result of a series of discrete rifting events distributed along and across the margin (Karner & Driscoll 1999), the timing of which is controversial. The controversy arises because: (1) seafloor spreading, when it commenced, was during the Mesozoic magnetic quiet zone – the first correlatable seafloor spreading anomaly is Anomaly 34 (Cande et al. 1988); and (2) the syn-rift section is non-marine with ostracod zonations and palynological assemblages being the prime method used for inter-basin correlations and age assignments. Karner et al. (1997) have interpreted the onshore and offshore depositional packages of the Brazilian Sergipe-Alagoas and Camamu-Almada basins and the West African Gabon, Congo and Kwanza basins as a series of regressive sequence packages, each package representing the development of individual deep, anoxic, lacustrine systems. Based on ostracod zonations, the ages of these packages are Berriasian– Valanginian, Hauterivian–early Barremian and early Barremian–late Aptian. Each rift phase resulted in the formation of deep, anoxic, lacustrine systems. Chronostratigraphic charts of the Campos basin (e.g. Feijó 1994) indicate that syn-rift packages are deposited on plume related tholeiitic basalts of the Serral Geral Formation with $^{40}Ar/^{39}Ar$ ages of 129–133 Ma (e.g. Menzies et al. 2002). This observation implies that the syn-rift assemblages of the Campos and Santos basins consist only of the early Barremian–late Aptian package, although it is very possible that the basalts have partially infilled and flowed across sediments associated with earlier Berriasian–Hauterivian events.

An enigmatic sediment package, the pre-salt sag basin, was regionally developed seaward of the Atlantic hinge of the West African margin and is well represented in the residual Bouguer gravity anomaly (Fig. 1). The gravity base maps of Figures 1 & 2 were constructed by determining the Bouguer gravity anomaly from the 1 min × 1 min global free-air gravity grid of the World's oceans (GEOSAT) generated by Sandwell & Smith (1992) and the 1 min × 1 min global GEBCO topographic grid (IOC et al. 2003) using a sediment/water reduction density of $2400-1030 kg/m^3$. From this grid, a least-squares bicubic surface was subtracted to produce the filtered Bouguer gravity maps.

A similar but thinner pre-salt basin exists across the Brazilian margin in the Campos and Santos basins (Fig. 2). The African pre-salt sag basin, with a width of 300 km, strike length of 1000 km and maximum thickness of 7 km (average of c. 3 km, e.g., Henry et al. 2004), consists of a number of individual sag packages. Further, the regional accommodation of the pre-salt sag basin is not a function of normal faulting in the terms of extension across the few observed faults and the collapse of their hanging wall blocks (Karner & Driscoll 1999). As stated earlier, depth-dependent extension characterizes many passive continental margins (e.g. Royden & Keen 1980; Driscoll & Karner 1998; Madon & Watts 1998; Baxter et al. 1999; Davis & Kusznir 2004; Kusznir et al. 2005). Because depth-dependent extension generates regional subsidence with only minor amounts of brittle deformation, this subsidence form is difficult to distinguish seismically from the more familiar thermal subsidence of oceanic and thinned continental crust. Thus, there are a number of viable tectonic mechanisms to explain the form of pre-salt sag basin accommodation: (1) Is the pre-salt sag basin and space for the evaporites yet another example of depth-dependent extension so that basin accommodation and subsidence rates are controlled primarily by the rate of lower crustal extension (Karner et al. 2003)? (2) Do pre-salt sag basin sediments represent deposition onto Barremian-aged oceanic crust augmented by kilometres of water level draw-down and precipitation of c. 2 km of evaporites at the end of the Aptian (Marton et al. 2000; Davison 2007)? In the former scenario, break-up would be late Aptian and in the latter, break-up would be early Barremian. Alternatively (3), is rifting diachronous, propagating from east to west across the West African margin, thereby allowing post-rift subsidence to start in the east while rifting is still occurring significantly to the west (Henry et al. 1995)?

As mentioned earlier, the evaporites are the capping sequence of a regional mega-regressive package that also comprises the regionally distributed pre-salt sag basins. Davison (2007) presents compelling evidence for diachronous depositional ages between the various salt provinces of the South Atlantic. Is this diachroneity due simply to local topographic variations allowing the development of discrete rift basins or is it tectonic in origin, given that the Campos and Santos basins developed in close proximity to the Tristan da Cunha plume? Thus, the purpose of this paper is to investigate the origin of the pre-salt sag basins, and the tectonic significance and ages of the South Atlantic salt province.

Pre-salt sag depositional packages of the West African and Brazilian margins

Large sedimentary basins developed both seaward of the Atlantic hinge and between the Atlantic and Eastern hinge zones across the West African margin (Fig. 1). Similar pre-salt basins underlie the Campos and Santos basins of the Brazilian margin (Fig. 2). Broadly distributed continental rifting, or more precisely, tectonic events represented by rapid and large increases in accommodation space within the Gabon, Congo and Angolan basins and their conjugate Brazilian basins (Camamu–Almada and Sergipe–Alagoas), commenced in the early Cretaceous (early to mid Neocomian) with the rapid creation of deep anoxic, lacustrine systems and the deposition of the Sialivakou, Djeno and lower Bucomazi formations (Braccini et al. 1977; Karner et al. 1997; Grosdidier et al. 1996). The lowermost Sialivakou Formation is characterized by minor thickening into small half graben, the thickening being indicative of minor differential subsidence and block rotation, the only clear evidence for brittle deformation across the West African margin (Fig. 3). The remainder of the accommodation was infilled by coarsing upward (regressive) sedimentary packages: the upper Djeno and Cricaré formations of the West African and Brazilian margins, respectively. The early-mid Barremian uppermost Djeno Formation was subaerially exposed as evidenced by the existence of paleosol horizons with iron-stained sands (Karner et al. 1997). Ostracod dating of the pre-salt sag basin immediately to the west of the Atlantic hinge zone (Falcão-1 well) indicates that the drilled section is early Barremian to late Aptian in age (ostracod biozone AS6 to AS12; Bate et al. 2001), with the basal fluvial sandstones and claystones of the Lukunga Sandstone Formation being a pre-rift unit. The basal Lukunga Sandstone Formation, equivalent to the Lucula sands of southern Gabon and Cabinda, is early Neocomian in age

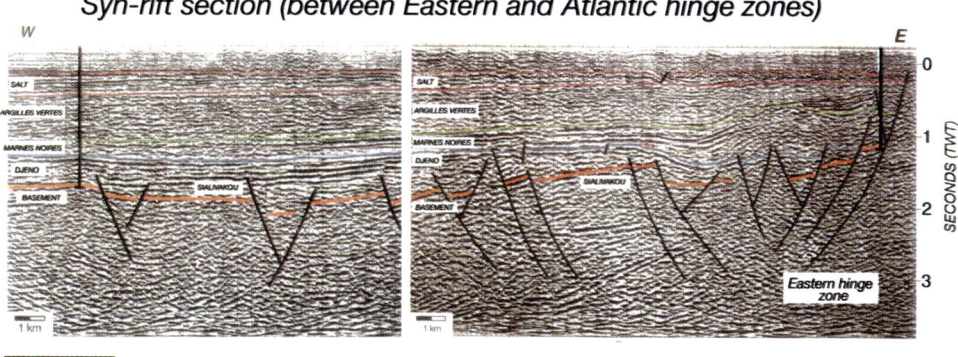

Fig. 3. Seismic reflection profile located between the Eastern and Atlantic hinge zones on the Congo margin showing minor fault-control and the local development of a series of minor half graben structures. Rotated and divergent reflectors indicate differential subsidence and block rotation to allow deposition of the Sialivakou Formation. The brown line delineates basement. Djeno (Top Djeno: blue line) and Marnes Noires (Top Marnes Noires: green line) formations sediments have prograded into the basin, downlapping the underlying Sialivakou Formation. The Argiles Vertes Formation also shows evidence of downlap onto the Marnes Noires. Top and base salt are shown in red. Together, the Djeno, Marnes Noires and the evaporites comprise a syn-rift sag section. Bold vertical lines locate the well control. Clearly, the minor early faulting does not control the generation of the regional accommodation space developed across the region. Seismic data courtesy of HydroCongo.

(ostracod biozone AS5 to AS2; Braccini et al. 1997). Rift initiation varies between ostracod biozone AS42 to 3 boundary in Cabinda (Braccini et al. 1997) and AS2 in northern Angola (Bate et al. 2001). The total age range of extension, mid-Neocomian to late Aptian, is thus approximately contemporaneous with syn-rift deposits developed inboard of the Atlantic hinge zone (Karner et al. 1997). While it is true that these dates are only relative because the ostracod zonations have not been correlated to an international chronostratigraphy, the relative age-range of sediments east and west of the Atlantic hinge are nevertheless similar.

The formation of the Atlantic hinge zone and a series of large offshore sag basins records the development of significant accommodation seaward of the Atlantic hinge zone (Fig. 4a). The hinge zone itself was accentuated by local normal faulting that allowed the sourcing of sands, the Lucula sands, deposited from the west across the Cabinda margin (Karner et al. 1997). Thick sediments, the Erva Formation and its equivalents, were deposited in these rift sub-basins (McHargue 1990), dated as late Neocomian to early Barremian in age (ostracod biozone AS3 to AS6; Braccini et al. 1997). The Erva Formation consists of mudstone and matrix-supported, coarse-grained argillaceous sandstones and conglomerates and poorly sorted, micaceous and feldspathic lithic wacke or arenites (McHargue 1990). This formation is interpreted as debris flows from adjacent high topographic relief regions into deep lacustrine basins. The pre-salt sag basin can be divided into at least two distinct seismic facies units, a lower unit that onlaps onto either Djeno equivalent or basement ostracod biozone AS2 to AS7; and an upper unit located farther to the west ostracod biozone AS7 to AS12; Stratal patterns observed in these units suggest that the sediment source area varied through time (Fig. 4a). The upper boundary of the pre-salt sag package is an erosional unconformity (Fig. 4a), the Chela and pre-Alagoas unconformities of the West African and Brazilian margins, respectively.

Between the Eastern and Atlantic hinge zones, the rapid transition from paleosoil horizons of the upper Djeno Formation to the thick (200–600 m), anoxic, lacustrine black shales of the Marnes Noires Formation in the late Barremian documents the rapid generation of regional accommodation in the absence of any normal faulting (Figs 3 & 4a; Karner et al. 1997). Proximal to the Eastern hinge zone, the Marnes Noires sediments are represented by a series of small deltas and thus tend to be sand prone (Fig. 3). By early Aptian time, the regional accommodation between the hinge zones was basically infilled by prograding deltaic sediments of the Argilles Vertes, Dentale and Aguia formations (Karner et al. 2003). The clinoform amplitude of the Argilles Vertes deltas suggests maximum lake depths of 500–650 m (Karner et al. 1997). An early Aptian lake level fall exposed the topsets of the Argilles Vertes clinoforms; subsequent truncation produced the pre-Chela unconformity. Downdip and during the ensuing transgression the Chela unconformity was produced by onlap of the Tchibota Formation onto the Argilles Vertes Formation. Canyon cutting and incision into the Argilles Vertes and earlier formations is observed on seismic profiles, indicating a relative lake level fall during the development of this regional unconformity (e.g. Teisserenc & Villemin 1990).

The Brazilian basins conjugate to the Congo margin are the Jequitinhonha and Cumuruxatiba basins (Santos et al. 1994). Cross-sections derived from seismic reflection and drilling data indicate that the syn-rift packages of the Cumuruxatiba basin consists initially of high-energy fan-deltas overlain by dark grey to black lacustrine shales of the Porto Seguro Formation deposited during the upper Berriasian (Santos et al. 1994; i.e. coeval with the Gabon, Congo and Angolan basins). Limited exploration drilling in the Jequitinhonha basin has thus far only penetrated early Aptian aged sediments while the first unequivocal syn-rift units of the Sergipe-Alagoas basin are the fluvial sands and lacustrine shales of the Berriasian-aged Serraria Formation (Feijó 1994).

As with the West African margin, the Campos and Santos basins are underlain by relatively thick, pre-salt, syn-rift sedimentary packages. However, unlike the West African margin, the syn-rift sediments tend to be disrupted by high angle normal faults leading to block rotation and the deposition of characteristic syn-rift sediment wedges (Fig. 4b). Above the tilted syn-rift sediments but below the evaporites, a regionally deposited and relatively unfaulted sag sequence can be mapped with thicknesses varying between 200 and 300 m (Fig. 4b & c). The normal faulting gradually decreases upward in the section from the highly faulted early syn-rift sediments near the base of the section to the less faulted sediments just below the base of the evaporites (Fig. 4b & c). The erosional unconformity at the base of the evaporites creates a conspicuous reflection that can be followed throughout the entire Aptian salt basin (Fig. 4b & c).

The capping sequence of the pre-salt sag basin: Loeme and Ezanga evaporites

The transition between the progressive refilling of the freshwater lakes that produced the transgressive lag deposits of the Chela, Gamba, and Cuvo formations and the regional deposition of evaporites

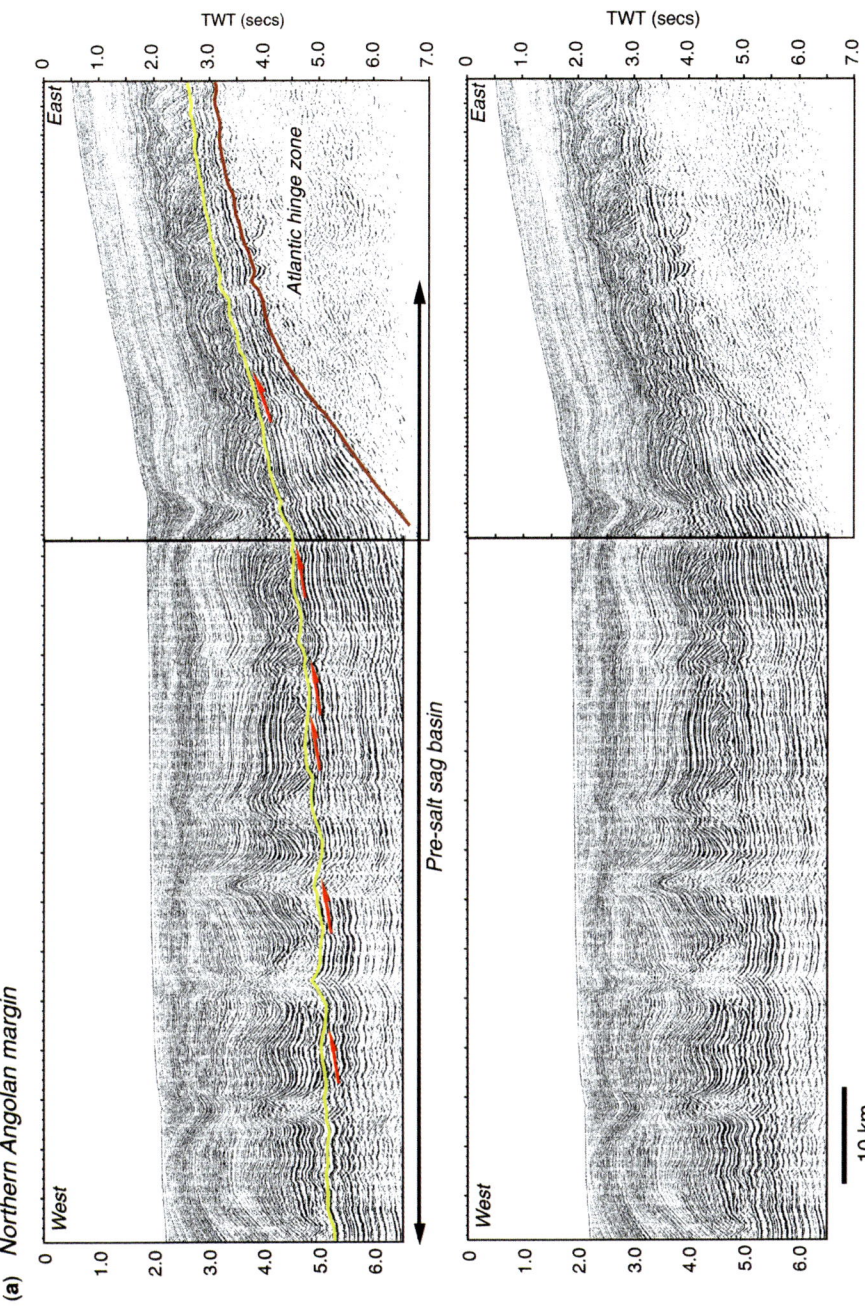

Fig. 4a. Interpreted and uninterpreted multi-channel seismic reflection profiles across the northern Angolan margin (West Africa) and the Santos Basin (Brazil). (**a**) Interpreted and uninterpreted east–west seismic reflection profile located seaward of the Atlantic hinge zone on the northern Angolan margin showing the thick depositional packages of the pre-salt sag basin, regionally developed seaward of the Atlantic hinge of the West African margin and onlapping the hinge zone, and a regional but gentle erosional truncation (red arrows). The base salt/top Cuvo is shown by the yellow line. Salt rafts comprise detached and rotated blocks of Albian shallow marine carbonates and sandstones. The top basement is shown by the brown line. Total width of the pre-salt sag basin width is c. 300 km with a maximum thickness of 7 km and an average of 3 km. The regional accommodation allowing deposition of the pre-salt basin and the overlying evaporites is essentially independent of normal faulting. Modified from Fraser et al. (2005).

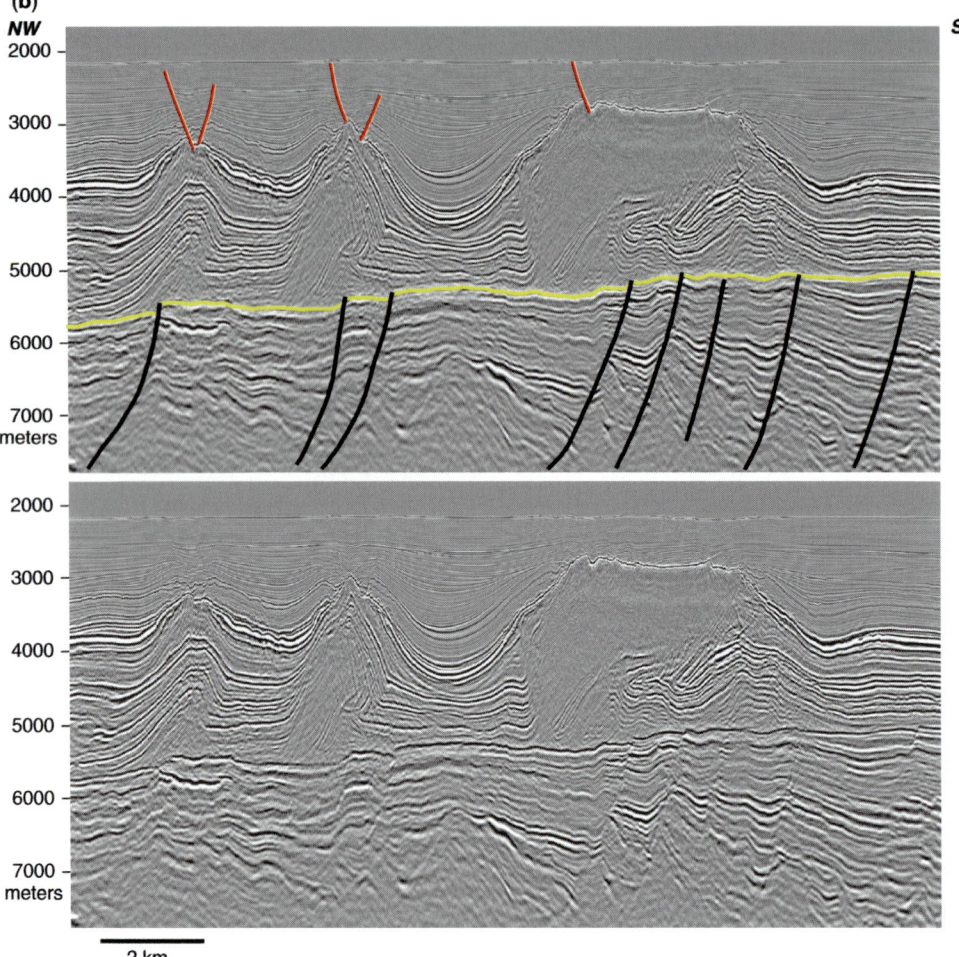

Fig. 4b. Interpreted and uninterpreted seismic reflection profiles across the distal part of the Santos Basin showing the faulted style (bold black lines) within the pre-salt basin and small faults at the crest of salt diapiric structures (red lines). Extensional faulting within the pre-salt sag basin occurs up to the base of the evaporites (yellow line), and in places, offsets the base salt reflector. Evaporites are dated as 110–113 Ma. Sediments directly beneath the unconformity are dated as 116 Ma; extension clearly has continued to at least 116 Ma (late Aptian) and so break-up, seafloor spreading and post-rift subsidence can only have occurred post 116 Ma. Seismic data courtesy of Veritas DGC Inc.

is a crucial period in the final stages of extension between Africa and Brazil, both between the hinge zones (from the craton boundary to the hinge zone) and seaward of the respective West African and Brazilian hinge zones some hundreds of kilometres across the pre-salt sag basins. Continental extension between West Africa and Brazil produced, directly or indirectly, the following sequence of events: (1) the development of the pre-Chela unconformity (and the equivalent Brazilian pre-Alagoas unconformity) as lake level dropped in the mid Aptian, exposing the prograding deltas of the Argilles Vertes and Dentale formations; (2) the regional development of the Chela unconformity and transgressive lag deposits of the Chela, Gamba and Cuvo formations in the mid to late Aptian – these formations comprise fluvial sandstones and conglomerates at their base grading upward, with increasing marine affinity, into lagoonal facies and the evaporites of the Ezanga and Loeme formations (Teisserenc & Villemin 1989; Bate *et al.* 2001); (3) the development of regionally extensive, shallow water, restricted marine conditions across the entire margin (between West Africa, the Gabon, Congo

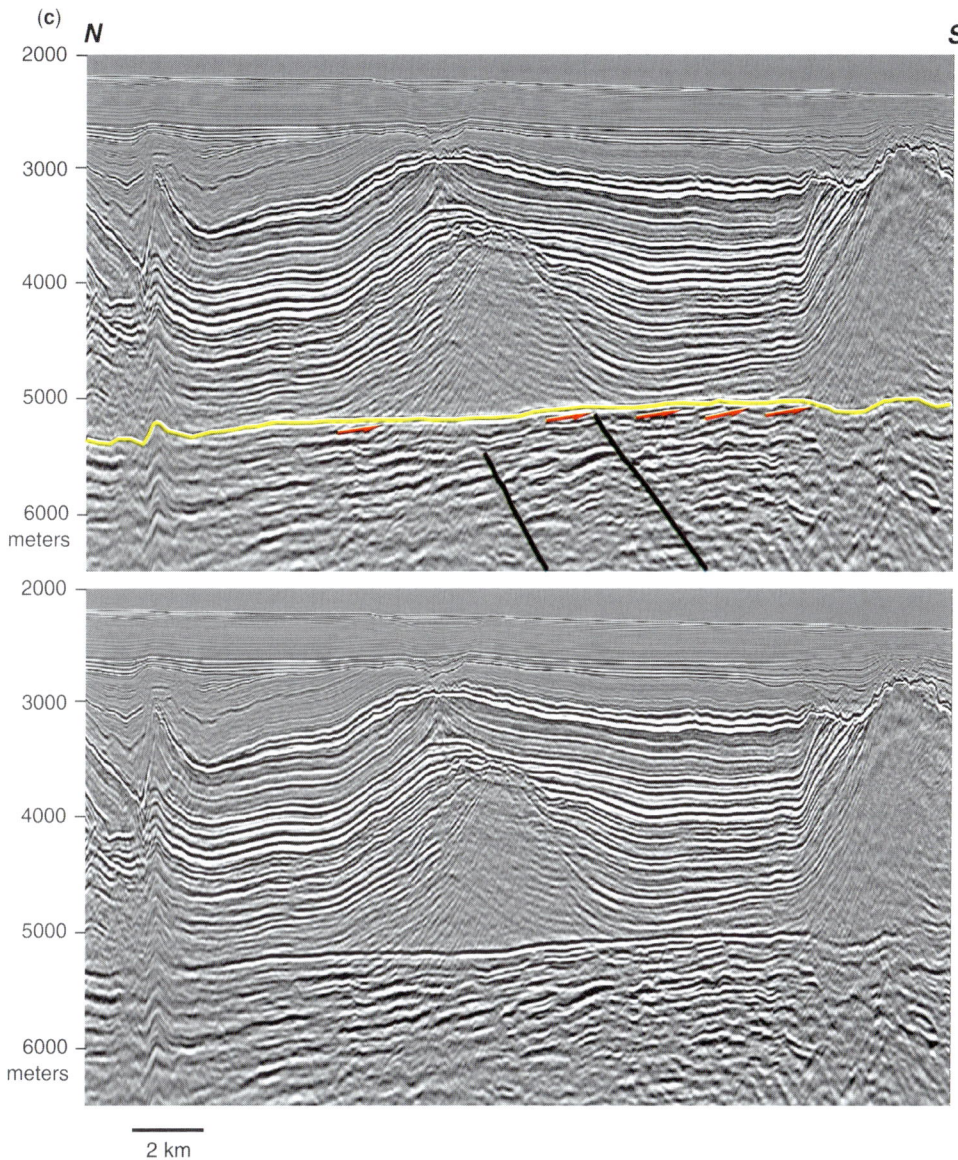

Fig. 4c. Interpreted and uninterpreted north-south seismic reflection profiles across the distal part of the Santos Basin showing the truncational unconformity at the base of the evaporite section (yellow line), faulted sediments (bold black line) of the pre-salt sag basin and reflector terminations at the base of the evaporites (red arrows). This truncation is clear evidence for sub-aerial exposure of the top pre-salt sag section prior to evaporite deposition. Seismic data courtesy of Veritas DGC Inc.

and Kwanza basins, and Brazil, the Cumuruxatiba, Jequitinhonha and Camamu-Almada basins) immediately prior to evaporite precipitation; as such, the Ezanga and Loeme evaporites were deposited during the evaporation of these waters; and (4) the development of significant post-rift accommodation (deposition of the late Cretaceous, Paleogene, and Neogene formations) in the same region previously characterized by minor syn-rift faulting, repeated desiccation cycles (allowing the precipitation of thick evaporites) and minor erosion of earlier syn-rift units.

The evaporites were precipitated across the entire margin, from West Africa to Brazil. For this to occur, the freshwater lakes need to be both shallow and to eventually be saline. The decreasing complexity of ostracod form and reduction in diversity observed within the upper Marnes Noires Formation reflects increasing lake salinity as a function of time (Bate 1999) due presumably to evaporation and diminished fresh water input into the lake because of a drier climate at this time and marine waters gaining entry to the basin. Unambiguous marine fauna do not appear until the deposition of the Chela and Loeme formations (Teisserenc & Villemin 1990). The high potassium and magnesium salts that characterize the Loeme evaporites require a low relative humidity (<35%), suggesting that the paleoenvironment necessarily needed to be exceedingly dry (pers. comm., Schreiber 1999). That is, the river systems appear to no longer exist, consistent with the lack of clastic sediments within the evaporites even in areas that, immediately prior to salt deposition, were major sediment depocentres (e.g. the large deltas represented by the Argilles Vertes and Dentale formations). Late Aptian eustatic variations (e.g. Haq et al. 1988) may have helped to modulate the chemistry of the brines. Open marine conditions were not established across the entire region until the mid-upper Albian (Dale et al. 1992). After deposition of the evaporite sequences, carbonate platforms dominated the stratigraphic development of both the West African and Brazilian margins.

Davison (2007) has suggested that the Brazilian and African salt basins developed diachronously. Summarizing Davison (2007), the oldest salt in the South Atlantic is thought to be the Sergipe–Alagoas Paripuiera salt (P-230 to P-260 palynological zone; Feijó 1994) dated at c. 124.8 Ma (Davison 2007), followed by the Kwanza margin Loeme evaporites at 121–124.5 Ma (using the Gradstein et al. 2005 timescale), and then the Sergipe–Alagoas Ibura Member (upper part of P-270 palynological zone; Feijó, 1994) and Gabon Ezanga Formation at some time prior to 114.5 Ma. That is, there may be a 6.5–10 million years difference between the northern Brazilian and Kwanza margin salts relative to the Gabon, Congo and southern Brazilian margin salts. However, results from other exploration wells indicate that the units immediately overlying the Ezanga evaporites contain various late Aptian palynomorph assemblages belonging to biozone CIX (112–117.2 Ma, adjusted to Gradstein et al. 2005; southern Gabon, undisclosed shelfal wells), which represents the top of the pre-salt sequence with respect to the Loeme evaporites of the Kwanza basin, supposedly one of the oldest salts, the existence of *Globigerinelloides barri* and the first appearance of *Hedbergella gorbachikae* (planktic foraminifers) in sediments immediately above the salts in DSDP 364 well (Caron 1978), indicate an early late Aptian age (c. 119–116 Ma). While the Tethyan planktic foramiferal index species is missing (*G. algerianus*), the faunal association with *G. barri* suggests a salt age that is no older than 116–119 Ma (adjusted to Gradstein et al. 2005; Leckie et al. 2002). These ages are in conflict with Davison (2007), who concludes that the top of the Angolan salt was deposited before 120.8–124.5 Ma based on the age range of the early Aptian planktic foraminifera *Leupoldina cabri* zone found above the salt in a confidential well in one of the shelfal blocks in Angola and in DSDP 364 well. We were not able to confirm the existence of *Leupoldina cabri* in the DSDP 364 well. More work needs to be done to reconcile the fact that *G. barri* and *H. gorbachikae*, foramiferal species also recovered from DSDP 364 well but showing an age range of c. 119–116 Ma and first occurrence at c. 116 Ma, respectively, supposedly co-exist with planktic foraminifera of the *L. cabri* zone (124.5–120.8 Ma). Based on our palaeontological review of well completion reports and DSDP wells, we suggest a revision of the salt ages for the West African and Brazilian margins as summarized in Figures 1 and 2.

The Sergipe–Alagoas evaporites, either the Paripuiera or the Ibura members, may be independent of the main African and the more southerly Brazilian salt basins. These salt members tend to be limited to small onshore depocentres within fault controlled rift basins that locally may be thick but are generally less than 100 m. Seaward dipping reflectors in the deepwater and ultra-deepwater environments characterize the offshore Sergipe–Alagoas basin with a noticeable lack of significant salt deformation structures and therefore, by inference, an absence of thick salts. Although evaporites might exist under the volcanic sequences, it is suggested that the Sergipe–Alagoas salt deposits represent local restrictive environments and sabkhas behind island archipelagos along the western Brazilian hinge zone. Thus, there is no need to link the onshore Sergipe–Alagoas $CaCl_2$-dominated salts with the Gabon, Angolan and southern Brazilian basin salts.

The evaporites in the Santos Basin thicken towards the depocentre, from c. 200 m in the proximal western margin of the basin to c. 2000 m in the depocentre. A regional, erosional unconformity can be clearly mapped at the base of the evaporites (Fig. 4b & c). Above the evaporites, shallow water Albian limestones were deposited in the proximal areas and marls and shales were deposited in the distal basin. The evaporite packages of the Santos and Campos basins are broadly dated based on the sediments above and below the salt. Age information comes predominantly from wells drilled in the proximal part of the basin. Microfauna indicate a maximum age of 116 Ma for the sediments below the evaporites and 110 Ma for the sediments immediately overlying the evaporites (Davison 2007). However, sediments from just below the evaporites were deposited under conditions in which microfossils were poorly preserved. Combined with the fact that the top of the pre-salt sag basin is an erosional unconformity (Fig. 4c), the 116 Ma is thus a maximum age for evaporite precipitation. Farther south in the Santos Basin on the Florianopolis High, anhydrite and carbonates of the Ariri Formation lie unconformably above the Curumim Volcanics in well 1-SCS-3B, which gave an Ar/Ar date of 113.2 ± 0.1 Ma (Dias et al. 1994; Davison 2007). The age of the main salts of the Espírito Santo and Campos basins is not known, but is thought to be coeval with those of the Santos Basin (Davison 2007). To summarize, given the age of all the evaporites, the Gabon and Angolan salts are essentially synchronous and slightly older than the Santos Basin (and Campos and Espírito Santo basins) evaporites (Figs 1 & 2).

It is usually assumed that the Aptian evaporites were sourced by seawater spilled across the Walvis Ridge. However, as noted by Dingle (1996) and Bate (1999) using ostracod data, the Walvis Ridge was not breached until Cenomanian–Turonian times. It was only after this time that South African ostracods were freely able to migrate northwards into the developing South Atlantic. It is thus not possible that the Ezanga and Loeme evaporites are simply the repeated flooding and desiccation of South Atlantic Ocean seawater across a Walvis Ridge sill. These Tethyan faunas are probably introduced by ocean influx through the evolving central Atlantic transtensional shear zone between northeast Brazil and Northwest Africa in the late Aptian or via the Parnaiba–Araripe–Sergipe–Alagoas connection of northeast Brazil (Davison & Bate 2004).

Basement morphology, crustal structure and the Aptian salt basin

If the regional distribution and amplitude of the post-Aptian subsidence is a consequence of continental extension and thinning, then the subsidence is not consistent with the minor amounts of Neocomian–Aptian brittle deformation (i.e. faulting) observed in seismic sections across the West African margin (e.g. Figs 3 & 4). On the other hand, because of the lack of significant faulting, a number of workers have suggested that the pre-salt sag basin is necessarily a post-rift unit deposited on thinned continental or oceanic crust (e.g. Marton et al. 2000; Jackson et al. 2000) or represents thick volcanic flows associated with seaward dipping reflectors (Jackson et al. 2000).

Figure 5 is a summary of the seismic refraction and reflection studies of Contrucci et al. (2004) and Moulin (2003). Clearly identified are the post-rift sediment package, the pre-salt sag basin, thinned crust beneath the pre-salt sag basin and the onshore rift basins, zones of massive salt diapiric structures, and a possible zone of underplating or mantle serpentinization. This anomalous velocity zone (P-wave velocities of 7.2–7.4 km/s) may be an inherited part of the Pan African crust or magmatic underplating generated during extension. If these relatively high velocities are related to extension, then underplating or mantle serpentinization will tend to result in uplift rather than induce basin subsidence. The zone marked as syn-rift sediments have P-wave velocities that range from 4.6 to 5.6 km s^{-1}, indicating that most of this zone represents compacted sediment rather than thick volcanic flows (cf. Jackson et al. 2000). Further, the fact that regional subsidence occurs both landward and seaward of the Atlantic hinge in areas of relatively thick crust disqualifies the regional subsidence in those regions as being associated with oceanic crust (Contrucci et al. 2004).

The continental crust rapidly thins from 30–34 km in the east to less than 10 km over a distance of 50 km (Fig. 5). Seaward of the Atlantic hinge zone, crustal thickness thins to values of 2–5 km over a 150 km-wide zone of parallel reflectors (sediments of the pre-salt sag basin). So, what do these P-wave velocities and crustal thicknesses imply in terms of tectonics? An ultrathin crust is anomalous, albeit oceanic or thinned continental. In a region supposedly dominated by plumes, we should expect relatively thick oceanic crust (10–12 km) rather than the observed small crustal thicknesses. Further, it is interesting to note that in the extreme west of the profile, oceanic crustal thicknesses are normal (6–7 km), indicating the absence of any plume involvement in magma production during the generation of oceanic crust and seriously questioning a sub-aerial environment for seafloor spreading, at least until evaporite precipitation. From the drilling results in the Newfoundland Basin (ODP Leg 210; Tucholke et al. 2004), ultra-thin crust is usually associated with extremely slow oceanic spreading rates or with exposed continental mantle and slow extension rates. So one possibility is that basement to the pre-salt sag basin is Barremian oceanic crust and synchronous with the age of margin break-up to the south of the Walvis Ridge (southern South Atlantic). It is conceivable that the 'true' ocean–continent boundary is juxtaposed directly against the Atlantic Hinge zone.

Predicted water depths for this scenario are 2.5 km for the depth of the spreading centre plus c. 1.0–1.5 km of thermal subsidence from the Barremian to the late Aptian (Fig. 6). Maximum accommodation space developed from rift onset until immediately prior to salt deposition

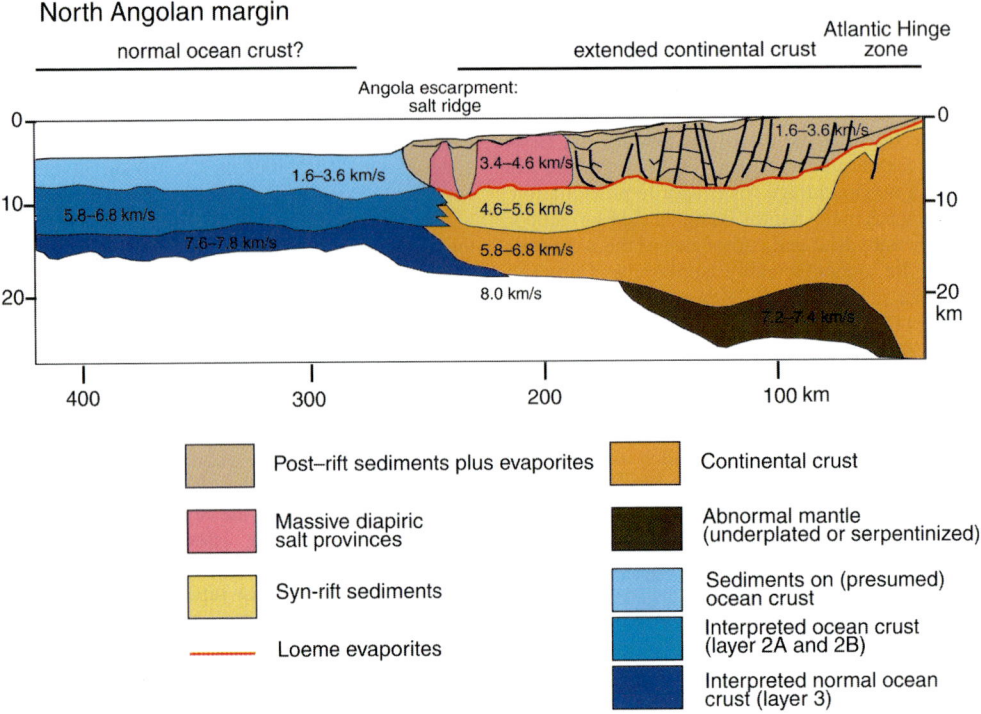

Fig. 5. Crustal-scale cross-section across the northern Angolan margin modified after Séranne and Anka (2005) and based on seismic refraction data from Moulin (2003) and Contrucci et al. (2004). The continental crust thins abruptly over a lateral distance of less than 50 km, from at least 30 km thick below the continental platform to less than 4 km below the pre-salt sag basin. This pre-salt sag basin, with velocities varying from 4.6 to 5.6 km s^{-1}, is located between the Atlantic hinge zone and the oceanic domain. The underlying crust is termed transitional: a crustal upper layer of 3–7 km exists with velocities that increase from 5.8 to 6.8 km s^{-1} at the base. Below this upper crustal layer, an anomalous 7.2–7.4 km s^{-1} velocity layer exists below the eastern side of the basin. The oceanic termination of the pre-salt sag basin is a basement ridge overlain by evaporites. While normal oceanic crust purportedly exists seaward of this basement ridge, crustal thickness and layer 2 and 3 velocities are anomalous (5.8–6.8 and 7.6–7.8 km s^{-1} respectively). Alternatively, this region of interpreted ocean crust may consist of lower crust and serpentinized and magmatically modified continental mantle, the lateral strain balance to the adjacent depth-dependent extension of the continental lithosphere.

was potentially therefore c. 3.5–4.0 km, implying maximum sediment thicknesses of c. 7–8 km within the pre-salt sag basin. Total sediment thickness is calculated from the following flexure equation assuming local isostasy:

$$\frac{\rho_m - \rho_w}{\rho_m - \rho_s} h_{wd}$$

where h_{wd} is the maximum water depth (i.e. 3.5–4.0 km), and ρ_s, ρ_w and ρ_m are the average sediment, water and mantle densities, respectively (2.2, 1.03 and 3.33 g cm^{-3}). However, this total sediment thickness is a maximum (because we assume that the space is completely filled with sediment), and so in adjacent areas where the sediment thickness in the pre-salt sag basin is only 2–4 km, significant paleowater depths are predicted to develop (1.5–3.0 km) immediately prior to salt deposition away from the major pre-salt sag sediment depocentres. So we come to two crucial questions: (1) is it possible to deposit thick salt (2 km) in water depths of 1.5–3.0 km; and (2) how much space can be generated on top of the thick pre-salt sag sections to accommodate the evaporites?

If we assume that the salt was deposited over 2 Ma (although it could have been significantly less), we can expect to generate approximately some 100–150 m of water-filled space for oceanic crust subsidence and 20–100 m for the post-rift subsidence of thinned continental lithosphere (Fig. 6). Thus, if the only space generation mechanism is conductive cooling of the oceanic lithosphere

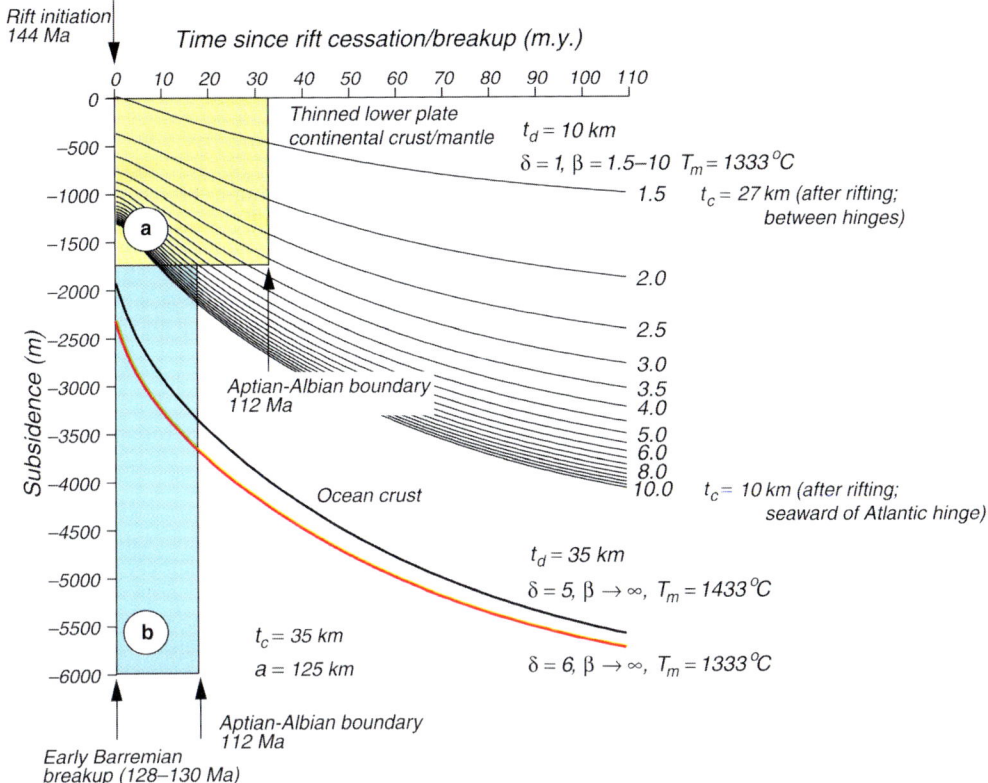

Fig. 6. Predicted water-filled subsidence for the upper/lower plate model of Karner *et al.* (2003) and of oceanic crust for asthenosphere temperatures of 1333 °C (standard model) and 1433 °C (plume model). The upper/lower plate model is for an initial crustal thickness of 35 km and a detachment depth of 15 km. The upper plate is thus 15 km and the lower plate represents a 20 km thick lower crust and a 90 km thick lithospheric mantle. Such a model is required to explain the lack of fault control on syn-rift accommodation and the large post-rift subsidence, either between the Eastern and Atlantic hinges or seaward of the Atlantic hinge. (**a**) Yellow box: Predicted post-rift subsidence of thinned continental lithosphere over 32 Ma (112–144 Ma) for a range of lower plate thinning factors (1.5–10). (**b**) Light blue box: predicted subsidence of oceanic crust over 18 Ma (112–130 Ma).

or extended continental lithosphere, then it becomes extremely difficult to generate sufficient space to precipitate the observed 1–2 km of evaporites, which requires a cumulative accommodation space of 600–1200 m (assuming a salt density of 2.0 g/cm^{-3}); it is simply not possible to generate sufficient space in 2 Ma to accommodate the evaporites.

In an alternative interpretation, Jackson *et al.* (2000) hypothesized that the West African and Brazilian Aptian salt basins formed after continental separation and thus never existed as a single super salt basin, being partitioned by sub-aerial mid-ocean spreading centres installed during the late Barremian–late Aptian. However, as shown for the Santos Basin, extensional faulting within the pre-salt sag basin occurs to the base of the evaporites (Fig. 4b & c), which are dated as 110–113 Ma. Sediments directly beneath the salt are dated as 116 Ma. Extension clearly has continued to at least 116 Ma (late Aptian) within the Santos Basin, questioning the inception of spreading centres and the commencement of post-rift subsidence in the late Barremian–late Aptian. Sub-aerially exposed spreading centres are supposedly consistent with seaward dipping reflectors (SDRs) that occur sporadically along both the Brazilian and West African margins (Jackson *et al.* 2000). As shown by Tischer *et al.* (2003) for the northwest Australian margin, SDRs characterize excessive magma production late in the rifting process and are not necessarily a proxy for oceanic spreading, sub-aerial or otherwise.

Thus, if thermal subsidence of oceanic crust or rifted lithosphere is inadequate to generate sufficient accommodation space and there are no sub-aerial mid-ocean spreading centres, the question remains, is it possible to deposit thick salt in water depth of 600–1200 m? A shallow water environment seems to be a prerequisite for evaporite precipitation, with repeated marine incursions into a subsiding basin, or a basin with adequate accommodation space, to produce the cumulative salt thickness. Drawdown offers a simple mechanism to generate shallow water or sub-aerial environments across the West African–Brazilian rift system (e.g. Karner et al. 2003; Davison 2007). Deepwater lacustrine conditions existed during the late Barremian–early Aptian as evidenced by the prograding systems of the Argilles Vertes and Dentale formations (Karner et al. 1997). Maximum water depths were 500–650 m (Karner et al. 2003). Any drawdown must therefore occur after the deposition of the Argilles Vertes and Dentale deltas. Canyon cutting and incision into the Argilles Vertes and regional truncation of the top pre-salt sag section attests to a relative lake level fall during the mid-late Aptian (Fig. 4a; Teisserenc & Villemin 1990). This same drawdown allowed regional but gentle truncation of the pre-salt sag units in the Santos Basin, even in areas of extreme total subsidence today (Fig. 4c). During the transgression following the drawdown, onlapping Tchibota Formation sediments produced the Chela unconformity. We postulate that the exposure and incision of the Argilles Vertes and Dentale formations and the regional truncation of the top pre-salt sag sediments are registering the drawdown of the lakes, triggered by a climate change, which ultimately results in the generation of at least 500–650 m of accommodation (as this is the estimate of the Argilles Vertes clinoforms, which exist between the hinge zones; water depths should be greater seaward of the Atlantic and Western hinge zones). Repeated marine incursions into this space allowed the Gabon and Angolan salts and the evaporites of the Camamu–Almada, Jequitinhonha and Cumuruxatiba basins to form. Salt loading produces a cumulative thickness of 1000–2000 m.

The Walvis ridge/dynamic topography barrier

We now return to the question of why there is a tendency for the Gabon and Angolan evaporites to be older than the Campos and Santos basin evaporites. Is this tendency related simply to the development of discrete rift basins with variable marine water communication or is it tectonic in origin, given that the Campos and Santos basins developed in the proximity of the Tristan da Cunha plume?

We address possible geomorphic effects of the plume by investigating the anomalous topography

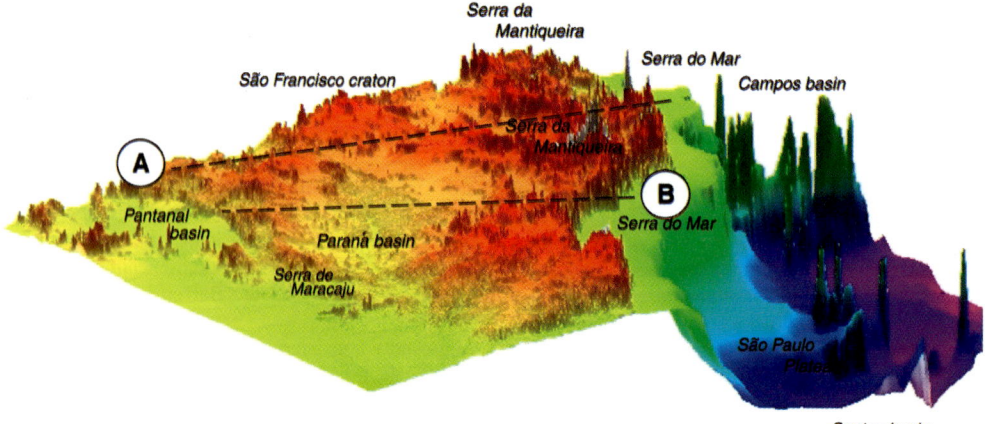

Fig. 7. Southwest perspective view of the topography of southeastern Brazil and the continental shelf and rise of the Campos and Santos basins. Onshore, the dissected topography is characterized by an asymmetric, long-wavelength topography (600 km) from the coastline to the western edge of the Paraná Basin, and a significantly shorter wavelength topography (100–200 km) related to the Serra do Mar and the Serra da Mantiqueira mountains. The Serra do Mar Mountains represent the erosional remnants of the original rift flank topography to the Santos and Campos basins. The Serra da Mantiqueira Mountains, west of the Serra do Mar, formed as part of the slope retreat of valleys initiated by the south-westward headward incision and erosion of the Paraíba river. Profiles A and B are shown in Figure 8.

of the Southeastern Brazilian highlands, which developed across the Paraná Basin and São Francisco craton. A perspective view of the Brazilian topography shows that there are two components (Fig. 7): (1) a relatively short-wavelength (100–150 km), asymmetric component representing the erosional remnants of the rift flank topography, principally the Serra do Mar; and (2) a long-wavelength (600–800 km), symmetric component that encompasses Paraná basin and the São Francisco craton. Actually, the coastal ranges consist of two sub-parallel systems, the Serra do Mar and the Serra da Mantiqueira. However, the latter marginal mountain system is the eastern boundary of an extensive plateau covering most of southeast Brazil and has a width of some 600–800 km (Fig. 7). The relief of this plateau ranges from 1000–1500 m with respect to sea level, and from 500 to 1000 m relative to the surface topography of the Paraná Basin. The Paraná Basin formed in the late Ordovician on continental crust created during the assembly of Gondwana (i.e. the Brasiliano/Pan-African orogeny; Zalán et al. 1990). Reconnaissance apatite fission-track analyses (Gallagher et al. 1994; Harman et al. 1999) indicate that the southeastern Brazilian highlands cannot be an erosional remnant of the late Proterozoic orogeny or Paleozoic basin forming mechanisms.

The Ordovician–early Carboniferous sequences of the Paraná Basin tend to be marine whereas the late Carboniferous–early Permian sequences are primarily the result of Gondwana glaciation. In contrast, the widely distributed but thin overlying Mesozoic sequences are strictly continental, consisting of varied alternations of lacustrine, fluvial and eolian sedimentary facies. Early Cretaceous basic and tholeiitic volcanics cap the entire sedimentary succession. These volcanics have a maximum thickness of 1500–2000 m and cover an area of greater than 1,100,000 km^2. According to Renne et al. (1996a, b), the Paraná volcanic province and (Etendeka volcanics of Namibia) were erupted rapidly at 132–133 Ma. Since these volcanics are pre-rift or early syn-rift, their tholeiitic composition (which implies significant partial melting of the asthenosphere) and timing necessarily imply a plume rather than a rift origin. Campos and Santos basin extension augmented melt generation leading to tholeiitic intrusions into the syn-rift section (Feijó 1994).

The generation of tholeiitic basalts represents large partial melting of the mantle and thus significant thermal modification of the lithospheric mantle across the Paraná-Etendeka volcanic province and, as such, should have thermal and topographic implications. Using a joint inversion of surface wave dispersion and receiver functions across the Paraná Basin, An & Assumpção (2004) concluded that underplating accompanying the tholeiitic basalt extrusion was not widespread, but limited and localized. A recent upper mantle tomography study in southeastern and central Brazil by Assumpção et al. (2004) mapped P-wave velocity anomalies from lithospheric depths down to 1300 km. In this region, higher seismic activity occurs preferentially in areas of low P-wave velocities at 150–250 km depth, suggesting an equilibrium lithosphere thickness of c. 250 km. Assumpção et al, (2004) interpreted these low P-wave velocities as relatively shallow asthenosphere and that the Tristan da Cunha plume generated significant lithosphere/asthenosphere topography that is still returning to thermal equilibrium.

This concept of a lingering thermal anomaly is explored by modeling the thermal thinning of the Paraná, Campos and Santos basin lithospheric mantle prior to significant early Cretaceous extension, representing the thermal modification of the lithosphere by the impingement of the Tristan da Cunha plume at c. 130 Ma (early Barremian). The form of the thermal thinning is assumed to be:

$$\beta(x) = 1.0 + \beta_o[1 + \cos(\pi x/L)]$$

where $\beta(x)$ is the lithospheric mantle thinning factor, x is horizontal distance, and L and β_o are scaling parameters, here assumed to be 1000 km and 1.0, respectively (Fig. 8). Further, the thermal half-life of the lithosphere, τ, which is a useful parameter to indicate how quickly the lithosphere can return to thermal equilibrium is given by $a^2/(\pi\kappa)^2$, where a is the lithosphere thickness and κ is the thermal diffusivity of the lithosphere. For a lithosphere thickness of 125 km, τ is 62.8 Ma. (McKenzie 1978), implying that the lithosphere has effectively returned to thermal equilibrium 125 Ma after the cessation of the thermal modifying event. Likewise, this age also represents the longevity of any transient topography. In marked contrast, if the lithosphere is 200–250 km, then τ is c. 160–250 Ma, implying that the lithosphere will return to thermal equilibrium 320–500 Ma after the causative thermal-modifying event. Using the lithosphere thickness from Assumpção et al. (2004) and the above parameters, it is possible to match the observed and predicted long wavelength topography across the across the Paraná Basin and São Francisco craton (Fig. 8), keeping in mind that the escarpment retreat of the Serra do Mar and the Serra da Mantiqueira represents extensive erosion by the Paraiba River since the late Cretaceous.

Because the modelled topography in Figure 8 represents the remnant transient topography after this region moved off the plume some 130 million

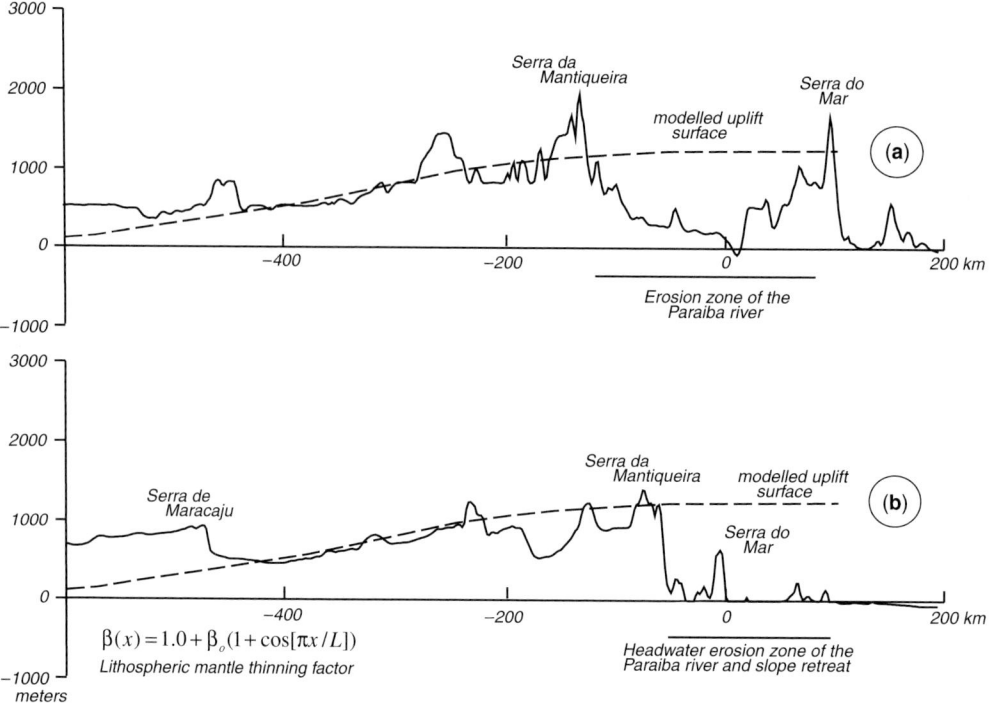

Fig. 8. Modelled topography across the southeastern Brazilian highlands and Paraná Basin; observed topography shown as solid grey line and predicted topography shown as dashed black line. The predicted topography is modelled as a consequence of cooling of the lithosphere after the emplacement of the Serra Geral tholeiitic basalts and the migration of the early Cretaceous Tristan da Cunha plume. Heat input is via a lithospheric thinning/extensional model in which only the lithospheric mantle is involved. The topography is shown 110 Ma after the start of lithospheric cooling. Asthenosphere temperature is assumed to be 1433 °C. Profile A samples the highest topography of both the Serra do Mar and the Serra da Mantiqueira mountains. Profile B, across the city of São Paulo, samples primarily the regional distributed, thermally controlled topography.

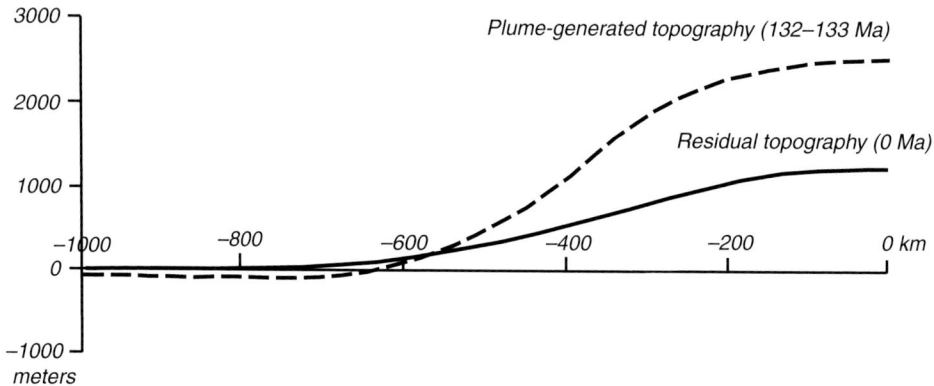

Fig. 9. Comparison of thermal topography at the time of plume emplacement 132–133 Ma; dashed grey line and the residual topography today (solid black line). The long-lived residual, transient topography is the consequence of the impingement of the Tristan da Cunha plume at 132–133 Ma and the thermal erosion of relatively thick (200–250 km) Paraná Basin and São Francisco craton lithospheres.

years ago, then it is also possible to reconstruct the topography during plume impingement and volcanic flow emplacement. Figure 8 shows that the topography, more or less centred over what will become the Santos and Campos basins, has a maximum relief of c. 2800 m and has slowly subsided to the present highland relief of 800–1000 m. From this modelling, it is concluded that the southeastern Brazilian highlands represent transient topography associated with residual heat of the Tristan da Cunha plume within the relatively thick Paraná Basin and São Francisco craton lithospheres. So while there were indeed rift flank topographies adjacent to the Campos and Santos basins, they were superimposed on a long wavelength transient topography. The longevity of the topography is proportional to the long thermal half-life of the Paraná Basin and São Francisco craton lithospheres, but also the inherent difficulty in eroding a broad plateau. Most importantly, Campos and Santos basin extensions necessarily were superimposed on this broad, high-relief plateau, which should form an effective barrier between the southern South Atlantic basins from the newly forming Brazilian and West African basins to the north (Gabon, Congo, Angola, Camamu-Almada, Jequitinhonha and Cumuruxatiba) until late-stage rift structuring or post-rift subsidence reduces the region below sea level. These northerly basins should therefore experience marine conditions before the Santos and Campos basins and may explain the age difference between the Santos and Campos evaporites and the evaporites of the Gabon and Angolan margins. While the northern rift basins were sub-aerial prior to extension, as evidenced by the fluvial sedimentary facies (Feijó 1994), any subsidence analysis of the Santos and Campos basins needs to take into account the fact that the starting elevation may be 2000–2500 m higher compared with basins to the north.

Conclusions: testing the hypotheses

We will now address the original questions posed earlier concerning the origin of the pre-salt sag basin of the West African margin, the pre-salt basin of the Santos Basin, and the evaporites of the West African and Brazilian margins.

First, do pre-salt sag basin sediments represent deposition onto early Barremian-aged ocean crust augmented by kilometres of water level drawdown and precipitation of 1–2 km of evaporites at the end of the Aptian (Marton et al. 2000; Davison 2007)? As stated earlier, the rationale for this hypothesis comes from the recognition of seafloor spreading anomaly M3 south of the Walvis ridge in the Argentinian Basin, and on the western South African margin (e.g. Cande et al. 1988). If break-up was early Barremian, then the West African pre-salt sag basin sediments would range in age from early Barremian to late Aptian, and were deposited on oceanic crust, with the evaporites being part of the post-rift sequence. The 4–5 km crustal thickness interpreted from the seismic refraction data is consistent with the crust being oceanic. Inception of the spreading ridge and 18 million years of post-rift subsidence (early Barremian to late Aptian) implies that the evaporites would have to either form in water depths of 3.25–3.75 km (e.g. Fig. 6), presumably by salts forming in the proximal coastal areas and spilling into the deepwater regions, or by the setup of northern and southern sills allowing drawdown to expose the spreading centres, or by filling completely the early Barremian–late Aptian accommodation so that shallow water conditions exist prior to evaporite deposition by the latest Aptian. However, the idea that a regional 3–4 km exposed relief can exist around the northern South Atlantic for 18 million years without immense fluvial incision and canyon development of earlier syn-rift and pre-rift units seems unrealistic. Freshwater lacustrine systems are known to have existed across the entire rift system, both between the hinges basinward and adjacent to the craton boundaries of the hinges, during this time range. For example, the anoxic, lacustrine black shales of the Marnes Noires Formation were being deposited during the early Barremian–late Barremian while the Argilles–Vertes deltas were being deposited during the late Barremian–early Aptian (Karner et al. 1997). Clinoform amplitudes associated with these formations imply that relatively deep but freshwater conditions (500–650 m) existed across the evolving West African-Brazilian rift until the deposition of the evaporites. That is, sub-aerial conditions were not established until salt time. Further, if the only space generation mechanism following deposition of the early Barremian–late Aptian sediments within the pre-sag basin was by conductive cooling of oceanic lithosphere (i.e. thermal subsidence), then it becomes extremely difficult to generate the necessary 600–1200 m accommodation to produce the observed 1–2 km of evaporites in ≤ 2 million years (assuming a salt density of 2.0 g cm^{-3}).

The age of break-up north of the Walvis ridge is difficult to establish because the Cretaceous magnetic quiet zone was not conducive to the production of correlatable seafloor spreading anomalies until anomaly 34 (Cenomanian). We therefore need to infer the tectonic development of the region from the preserved stratigraphy on the West African and Brazilian margin. The onshore basins of West Africa and Brazil (e.g. Congo and Sergipe-Alagoas basins, respectively) range in age from Berriasian to Aptian. When normal faults are present, they were active over this same range in time (e.g. Hamsi & Karner 2005). For the Santos Basin, extensional faulting of the pre-salt basin sediments occurred until at least 116 Ma (late Aptian). Using these observations, we conclude that the pre-salt sag basin sediments are syn-rift and do not represent deposition onto Barremian-aged ocean crust.

Second, is rifting diachronous, such that the cessation of rifting propagates from east to west across the West African margin, thereby allowing post-rift subsidence to start in the east while rifting is still occurring significantly to the west (Henry et al. 1995)? The Neocomian–Aptian sedimentary units deposited between the Eastern and Atlantic hinges (e.g. Fig. 3) and the similarly aged sediment package sampled by the Falcão well, spudded seaward of the Atlantic hinge (Karner & Driscoll 1999; Bate et al. 2001), seems to disqualify the idea of an east to west rift propagation.

Given these sediment ages, however, we can also explore the viability of rift propagation by estimating the amount of post-rift accommodation amount of across the West African margin as a function of time (Fig. 6) given the crustal thickness variations mapped by the seismic refraction analysis (Fig. 5). Observed maximum sediment thicknesses are 4 km for the onshore basins and 7 km for the pre-salt sag while the associated crustal extension factors are 1.3 (27 km crust at the eastern edge of Fig. 5) and 4–6 (6–10 km crust between 100 and 125 km of Fig. 5), respectively (assuming an initial crustal thickness of 35 km). If we assume an early Berriasian onset of rifting (i.e. age of the Sialivakou Formation) and that the observed pre-salt sediments between the Eastern and Atlantic hinges are post-rift, basement will only subside by 475 m in 32 million years (Fig. 6; early Berriasian–late Aptian water-filled subsidence, extension factor of 1.3), or a total sediment thickness of 1175 m. That is, it is not possible to deposit the observed 3–4 km of sediment between the hinges if accommodation generation is solely by post-rift thermal subsidence.

A similar problem exists for the syn-rift sag basin seaward of the Atlantic hinge zone. If the pre-salt sag basin is a consequence of rift propagation to the west and thus a totally syn-rift unit and that the rift onset here is say early Barremian then the maximum predicted rift subsidence is 1.15–1.3 km (Fig. 6; early Barremian extension, water-filled accommodation, extension factors of 4–6), or a total sediment thickness of 2.34–2.65 km significantly less than observed. Thus again, the predicted rift and thermal subsidences are inadequate to account for the observed sediment thicknesses basinward of the Atlantic hinge zone and so we can reject the hypothesis that the observed pre-salt sediment thickness distribution is the result of varying rift cessation across the West African margin.

Third, is the pre-salt sag basin and space for the evaporites caused by depth-dependent extension so that basin accommodation and subsidence rates are determined primarily by the rate of lower crustal extension (Karner et al. 2003)? Because the sediment thickness of the pre-salt sag basin and the thickness and timing of the evaporites is inconsistent with accommodation generated by the post-rift subsidence of extended continental crust, then we conclude that the pre-salt sag basin and overlying evaporites are necessarily syn-rift (Karner & Driscoll 1999; Karner et al. 2003). This result is consistent with observations from the Santos Basin that extensional faulting within the Brazilian pre-salt basin occurs at least to the base of the evaporites. Thus, the West African and Brazilian passive continental margins are characterized by the stacking of regional syn-rift sag packages, the amplitude and distribution of which are inconsistent with the minor amounts of brittle deformation interpreted from seismic sections across these margins. As hypothesized by Driscoll & Karner (1998) and Karner & Driscoll (1999), the regional distribution and thickness of syn- and post-rift sediment packages that are essentially independent of normal faulting requires major strain partitioning across a depth-dependent, crustal decoupling zone that results in the development of a relatively non-deforming upper crust (i.e. the upper plate) from a ductile-deforming lower crust and lithospheric mantle (i.e. the lower plate). For the West African and Brazilian margins, we hypothesize that the lithospheric strain balance engendered by depth-dependent extension is achieved by the lateral emplacement of lower crust and serpentinized and magmatically modified continental mantle out from under the adjacent continental lithosphere (cf. Lavier & Manatschal 2006). We suggest further that the significant positive gravity anomalies juxtaposed against the West African and Brazilian pre-salt basins, often thought to delineate the ocean–continent boundary, may instead be sourced by lower crust and serpentinized continental mantle (Figs 1 & 2) and thus defines an ocean-lower crust/continental mantle boundary.

This upper plate/lower plate extension style is postulated to have produced the accommodation across the West African and Brazilian margins. Between the hinges, sediment thickness ranges from 2–4 km (total thickness), consistent with an upper plate thickness of 15 km and delta = 1.1 (i.e. minor extension, minimal faulting) and a lower plate extension of 40% (i.e. beta of 1.4). Thick sediments (4–7 km) were deposited in the offshore sag basins, consistent with an upper plate thickness of 5 km and delta = 1 (i.e. no extension, no faulting) and removing most of the lower crust and lithospheric mantle (i.e. beta approaching infinity). Lake level drawdown exposed the top of the pre-salt sag basins (both between the hinges and seaward of the Atlantic hinge) to produce the pre-Chela unconformity. Continuing minor late Aptian extension and the progressive refilling of the lakes resulted in the Chela unconformity across the West African margin and the pre-Alagoas unconformity across the Brazilian margin (e.g. Karner et al. 2003) and the space for the various evaporite formations. Rift cessation and the onset of seafloor spreading, which split the evaporites into separate West African and Brazilian salt basins, was followed by significant post-rift accommodation (deposition of shallow-water Albian carbonates and late Cretaceous, Paleogene and Neogene formations) in the same region previously characterized by minor late Aptian syn-rift faulting, repeated marine incursions and desiccation cycles (allowing the precipitation of thick evaporites), and minor erosional truncation of earlier syn-rift units.

We gratefully acknowledge Charlotte Schrieber, Ian Davison and Anthony Watts for critically reviewing the manuscript and Charlotte Schrieber, Mark Leckie, Gianreto Manatschal, Henk Brinkguis, Roel Verreussel and Cedric John for discussions and information about evaporite chemistry, planktonic foraminiferal and palynological assemblages, absolute ages and the geology of Alpine extensional systems. We are indebted to Veritas DGC Inc. for permission to publish the excellent seismic lines shown in Figures 4b and c. The Wessel & Smith (1995) GMT software was used in the construction of figures shown in the paper. This work was supported by National Science Foundation grant OCE 04-25411. Lamont-Doherty Earth Observatory publication no. 7064.

References

AN, M. & ASSUMPÇÃO, M. 2004. Multi-objective inversion of surface waves and receiver functions by competent genetic algorithm applied to the crustal structure of the Paraná Basin, SE Brazil. *Geophysical Research Letters*, **31**, L05615, doi:10.1029/2003GL019179.

ASSUMPÇÃO, M., SCHIMMEL, M., ESCALANTE, C., BARBOSA, J. R., ROHA, M. & BARROS, L. V. 2004. Intraplate seismicity in SE Brazil: stress concentration in lithospheric thin spots. *Geophysical Journal: International Royal Astronomical Society*, **159**, 390–399.

BATE, R. H. 1999. Non-marine ostracod assemblages of the Pre-salt rift basins of West Africa and their role in sequence stratigraphy. In: CAMERON, N. R., BATE, R. H. & CLURE, V. S. (eds) *The Oil and Gas Habitats of the South Atlantic*. Geological Society of London, Special Publication **153**, 283–292.

BATE, R. H., CAMERON, N. R. & BRANDÃO, M. G. 2001. The lower Cretaceous (pre-salt) lithostratigraphy of the Kwanza Basin, Angola. *Newsletters on Stratigraphy*, **38**, 117–127.

BAXTER, K., COOPER, G. T., HILL, K. C. & O'BRIAN, G. W. 1999. Late Jurassic subsidence and passive margin evolution in the Vulcan Sub-basin, north-west Australia: constraints from basin modelling. *Basin Research*, **11**, 97–111.

BRACCINI, E., DENISON, C. N., SCHEEVEL, J. R., JERONIMO, P., ORSOLINI, P. & BARLETTA, V. 1997. A revised chronostratigraphic framework for the pre-salt (lower Cretaceous) in Cabinda, Angola. *Bulletin du Centre de Recherches Elf Exploration Production*, **21**, 125–151.

CANDE, S. C., LABRECQUE, J. L. & HAXBY, W. F. 1988. Plate kinematics of the South Atlantic: Chron 34 to present. *Journal Geophysical Research*, **93**, 13479–13492.

CARON, M. 1978. Cretaceous planktonic foraminifers from DSDP Leg 40, southeastern Atlantic Ocean. In: BOLLI, H. M. & RYAN, W. B. F. ET AL. *Initial Results of the Deep Sea Drilling Project, Leg 40*. (US Government Printing Office), Washington, DC, 651–678.

CONTRUCCI, I., MATIAS, L. & MOULIN, M. ET AL. 2004. Deep structure of the West African continental margin (Congo, Zaire, Angola), between 5°S and 8°S, from reflection/refraction seismics and gravity data. *Geophysical Journal International*, **158**, 529–553.

DALE, C. T., LOPES, J. R. & ABOLIO, S. 1992. Takula oil field and the greater Takula area, Cabinda, Angola. In: HALBOUTY, M. T. (ed.) *Giant Oil and Gas Fields of the Decade 1978–1988*. AAPG, Memoir, **54**, 197–215.

DAVIS, M. & KUSZNIR, N. J. 2004. Depth-dependent lithospheric stretching at rifted continental margins. In: KARNER, G. D. (ed.) *Proceedings of NSF Rifted Margins Theoretical Institute*. Columbia University Press, 92–136.

DAVISON, I. 2007. Geology and tectonics of the south Atlantic Brazilian salt basins. In: RIES, A. C., BUTLER, R. W. H. & GRAHAM, R. H. (eds) *Deformation of the Continental Crust: The Legacy of Mike Coward*. Geological Society, London, Special Publications, **272**, 345–359.

DAVISON, I. & BATE, R. 2004. Early Opening of the South Atlantic: Berriasian rifting to Aptian salt deposition. *PESGB/HGS 3rd International Joint Meeting Africa: the Continent of Challenge and Opportunity*, 7–8 September 2004.

DIAS, J. L., SAD, A. R. E., FONTANA, R. L. & FEIJÒ, F. J. 1994. Bacia de Pelotas. *Boletim Geociências da Petrobrás*, **8**, 235–246.

DINGLE, R. V. 1996. Cretaceous ostracoda of the SE Atlantic and SW Indian Ocean: a stratigraphical review and atlas. In: JARDINE, S., DE KLASZ, I. & DEBENAY, J.-P. (eds) *Géologie de l'Atlantique Sud*. Elf Aquitaine Memoire, **16**, 1–11.

DRISCOLL, N. W. & KARNER, G. D. 1998. Lower crustal extension across the northern Carnarvon basin, Australia: evidence for an eastward dipping detachment. *Journal Geophysical Research*, **103**, 4975–4992.

FEIJÒ, F. J. 1994. Bacias de Sergipe e Alagoas. *Boletim Geociências Petrobrás*, **8**, 149–162.

FRASER, A. J., HILKEWICH, D., SYMS, R., PENGE, J., RAPOSO, A. & SIMON, G. 2005. Angola Block 18: a deep-water exploration success story. In: DORÉ,

A. G. & VINING, B. A. (eds) *Petroleum Geology: North-West Europe and Global Perspectives – Proceedings of the 6th Petroleum Geology Conference*. Geological Society, London, 1199–1216.

GALLAGHER, K., HAWKESWORTH, C. J. & MANTOVANI, M. S. M. 1994. The denudation history of the onshore continental margin of SE Brazil inferred from apatite fission track data. *Journal & Geophysical Research*, **99**, 18117–18145.

GRADSTEIN, F. M., OGG, J. G. & SMITH, A. G. 2005. *A Geologic Time Scale*. Cambridge University Press, Cambridge, 500.

GROSDIDIER, E., BRACCINI, E., DUPONT, G. & MORON, J.-M. 1996. Non-marine lower Cretaceous biozonation of the Gabon and Congo basins. *In*: JARDINE, S., DE KLASZ, I. & DEBENAY, J.-P. (eds) *Geologie de l'Afrique Sud*. Elf Aquitaine Memoire, **16**, 67–82.

HAMSI, G. P. & KARNER, G. D. 2005. Revisão de secões crustais da sub-bacia de Sergipe através de modelagens tectonofísicas: Implicações quanto aos padrões de preenchimento e quanto à evolução do fluxo térmico (abstract). *X Simpósio Nacional de Estudos Tectônicos, 19–24 June, 2005*, Curitiba, Brazil.

HAQ, B. U., HARDENBOL, J. & VAIL, P. R. 1988. Mesozoic and Cenozoic chronostratigraphy and eustatic cycles. *In*: WILGUS, C. K., HASTINGS, B. S., KENDALL, C. G. St.C., POSAMENTIER, H. W., ROSS, C. A. & VAN WAGONER, J. C. (eds) *Sea-level Changes: an Integrated Approach*. SEPM, Special Publication, 42, 71–108.

HARMAN, R., GALLAGHER, K. L., BROWN, R., RAZA, A. & BIZZI, L. 1999. Accelerated denudation and tectonic/geomorphic reactivation of the cratons of northeastern Brazil during the Late Cretaceous. *Journal Geophysical Research*, **103**, 27091–27105.

HENRY, S. G., BRUMBAUGH, W. & CAMERON, N. 1995. Pre-salt source rock development on Brazil's conjugate margin: West African examples. *1st Latin American Geophysical Conference, Rio de Janeiro*, extended abstracts, 3.

HENRY, S., DANFORTH, A., VENTRAKAMAN, S. & WILLACY, C. 2004. PSDM- sub-salt imaging reveals new insights into petroleum systems and plays in Angola-Congo-Gabon. *Petroleum Exploration Society Great Britain–Houston Geological Society Joint Africa Symposium*, London, 7–8th September 2004, abstract.

IOC, IHO & BODC. 2003. *Centenary Edition of the GEBCO Digital Atlas*. Published on CD-ROM on behalf of the Intergovernmental Oceanographic Commission and the International Hydrographic Organization as part of the General Bathymetric Chart of the Oceans. British Oceanographic Data Centre, Liverpool.

ISERN, A. R., ANSELMETTI, F. S., BLUM, P. *ET AL*. 2002. *Proc. ODP, Init. Repts,* 194, College Station, TX (Ocean Drilling Program), doi:10.2973/odp.proc.ir. 194.2002.

JACKSON, M. P. A., CRAMEZ, C. & FONCK, J. M. 2000. Role of subaerial volcanic rocks and mantle plumes in creation of South Atlantic margins: implications for salt tectonics and source rocks. *Marine and Petroleum Geology*, **17**, 477–498.

KARNER, G. D. & DRISCOLL, N. W. 1999. Tectonic and stratigraphic development of the West African and eastern Brazilian Margins: insights from quantitative basin modelling. *In*: CAMERON, N. R., BATE, R. H. & CLURE, V. S. (eds) *The Oil & Gas Habitats of the South Atlantic*. Geological Society, London, Special Publications, **153**, 11–40.

KARNER, G. D., DRISCOLL, N. W., MCGINNIS, J. P., BRUMBAUGH, W. D. & CAMERON, N. 1997. Tectonic significance of syn-rift sedimentary packages across the Gabon–Cabinda continental margin. *Marine and Petroleum Geology*, **14**, 973–1000.

KARNER, G. D., DRISCOLL, N. W. & BARKER, D. H. N. 2003. Synrift subsidence across the West African continental margin: The role of lower plate ductile extension *In*: ARTHUR, T. J., MACGREGOR, D. S. & CAMERON, N. R. (eds) *Petroleum Geology of Africa: New Themes and Developing Technologies*. Geological Society, London, Special Publications, **207**, 105–125.

KUSZNIR, N. J., HUNSDALE, R., ROBERTS, A. M. & iSIMM Team 2005. Norwegian margin depth-dependent stretching. *In*: DORÉ, A. G. & VINING, B. A. (eds) *Petroleum Geology: North-West Europe and Global Perspectives – Proceedings of the 6th Petroleum Geology Conference*. Petroleum Geology Conferences Ltd., Geological Society, London, 767–783.

LAVIER, L. L. & MANATSCHAL, G. 2006. A mechanism to thin the continental lithosphere at magma-poor margins. *Nature*, **440**, 324–328.

LECKIE, R. M., BRALOWER, T. J. & CASHMAN, R. 2002. Oceanic anoxic events and planktonic evolution: biotic response to tectonic forcing during the mid-Cretaceous. *Paleoceanography*, **17**, 13.1–13.29, doi: 10.1029/2001PA000623.

MARTON, L. G., TARI, G. C. & LEHMANN, C. T. 2000. Evolution of the Angolan passive margin, West Africa, with emphasis on post-salt structural styles. *In*: MOHRIAK, W. & TALWANI, M. (eds) *Atlantic Rfts and Continental Margins*. Geophysical Monographs, **115**, 129–149.

MCHARGUE, T. R. 1990. Stratigraphic development of proto-South Atlantic rifting in Cabinda, Angola – a petroliferous lake basin. *In*: KATZ, B. J. (ed.) *Lacustrine Basin Exploration. Case Studies and Modern Analogs.* American Association of Petroleum Geology Memoirs, **50**, 307–326.

MCKENZIE, D. P. 1978. Some remarks on the development of sedimentary basins. *Earth Planetary Sciences Letters*, **40**, 25–32.

MADON, M. B. & WATTS, A. B. 1998. Gravity anomalies, subsidence history and the tectonic evolution of the Malay and Penyu basins (offshore Peninsula Malaysia). *Basin Research*, **10**, 375–392.

MANATSCHAL, G., FROITZHEIM, N., RUBENACH, M. & TURRIN, B. D. 2001. The role of detachment faulting in the formation of an ocean-continent transition: Insights from the Iberia abyssal plain. *In*: WILSON, R. C. L., WHITMARSH, R. B., TAYLOR, B. & FROITZHEIM, N. (eds) *Non-volcanic Rifting of Continental Margins: A Comparison of Evidence from Land and Sea*. Geological Society, London, Special Publications, **187**, 405–428.

MENZIES, M. A., KLEMPERER, S. L., EBINGER, C. J. & BAKER, J. 2002. Characteristics of volcanic rifted margins. *In*: MENZIES, M. A., KLEMPERER, S.L., EBINGER, C. J. & BAKER, J. (eds) *Volcanic Rifted*

Margins. Geological Society of America Special Papers, **362**, 1–14.

MOULIN, M. 2003. *Etude géologique et géophysique des marges continentales passives: exemple du Zaïre et de l'Angola*. Doctorai Thesis, Universite Bretagne Occidentale.

MÜNTENER, O. & HERMANN, J. 2001. The role of lower crust and continental upper mantle during formation of non-volcanic passive margins; evidence from the Alps. *In*: WILSON, R. C. L., WHITMARSH, R. B., TAYLOR, B. & FROITZHEIM, N. (eds) *Non-volcanic Rifting of Continental Margins: a Comparison of Evidence from Land and Sea*. Geological Society, London, Special Publications, **187**, 267–288.

RENNE, P., DECKART, K., ERNESTO, M., FERAUD, G. & PICCIRILLO, E. 1996a. Age of the Ponta Grossa dyke swarm (Brazil) and implications to the Paraná flood volcanism. *Earth and Planetary Science Letters*, **144**, 199–211.

RENNE, P., GLEN, J. M., MILNER, S. C. & DUNCAN, A. R. 1996b. Age of Etendeka flood volcanism and associated intrusions in southwestern Africa. *Geology*, **24**, 659–662.

ROBB, M. S., TAYLOR, B. & GOODLIFFE, A. M. 2005. Re-examination of the magnetic lineations of the Gascoyne and Cuvier Abyssal Plains, off NW Australia. *Geophysical Journal International*, **163**, 42–55.

ROBERTS, A. M., LUNDIN, E. R. & KUSZNIR, N. J. 1997. Subsidence of the Vøring Basin and the influence of the Atlantic continental margin. *In*: ROBERTS, A. M. & KUSZNIR, N. J. (eds) *Tectonic, Magmatic and Depositional Processes at Passive Continental Margins*. Geological Society, London, **154**, 551–557.

ROYDEN, L. & KEEN, C. E. 1980. Rifting process and thermal evolution of the continental margin of eastern Canada determined from subsidence curves. *Earth and Planetary Science Letters*, **51**, 343–361.

SANDWELL, D. T. & SMITH, W. H. F. 1992. Global marine gravity from ERS-1, Geosat, and Seasat reveals new tectonic fabric. *EOS Transactions of the American Grophysical Union*, **73**, 133.

SANTOS, C. F., GONTIJO, R. C., ARAÚJO, M. B. & FEIJÓ, F. J. 1994. Bacias de Cumuruxatiba e Jequitinhonha. *Boletim Geociências Petrobrás*, **8**, 185–190.

SÉRANNE, M. & ANKA, Z. 2005. South Atlantic continental margins of Africa: a comparison of the tectonic vs climate interplay on the evolution of equatorial west África and SW África margins. *Journal of African Earth Sciences*, **43**, 283–300.

SHIPBOARD SCIENTIFIC PARTY 2004. Explanatory notes. *In*: TUCHOLKE, B. E., SIBUET, J.-C., KLAUS, A. ET AL. *Proc. ODP, Init. Repts*, 210: College Station, TX (Ocean Drilling Program), 1–69, doi:10.2973/odp.proc.ir.210.102.2004.

TEISSERENC, P. & VILLEMIN, J. 1990. Sedimentary basin of Gabon – geology and oil systems. *In*: EDWARDS, J. D. & SANTOGROSSI, P. A. (eds) *Divergent/Passive Margin Basins*, AAPG Memoirs, **48**, 117–199.

TISCHER, M. J. 2006. *The structure and development of the continent-ocean transition zone of the Exmouth Plateau and Cuvier margin, Northwest Australia: implications for extensional strain partitioning*. Columbia University Ph.D. thesis.

TISCHER, M. & TEN BRINK, U. ET AL. 2003. Insights into along-strike passive continental margin variability from seismic reflection, refraction and gravity data, Northwest Australia. *EOS Transactions of the American Geophysical Union*, **84**, Fall meeting suppl., abstract T51F–0218.

TUCHOLKE, B. E., SIBUET, J.-C., KLAUS, A. ET AL. 2004. *Proc. ODP, Init. Repts*, 210: College Station, TX (Ocean Drilling Program), doi:10.2973/odp.proc.ir.210.204.

WESSEL, P. & SMITH, W. H. F. 1995. New version of the generic mapping tools released. *EOS Transactions of the American Geophysical Union*, **76**, 329.

WHITE, R. S. & MCKENZIE, D. P. 1989. Magmatism at rift zones: the generation of volcanic continental margins and flood basalts. *Journal of Geophysical Research*, **94**, 7,685–7,729.

WHITMARSH, R. B., MANATSCHAL, G. & MINSHULL, T. A. 2001. Evolution of magma-poor continental margins from rifting to seafloor spreading. *Nature*, **413**, 150–154.

ZALÁN, P. V. & WOLFF, S. ET AL. 1990. The Paraná Basin, Brazil. *In*: LEIGHTON, M. W., KOLATA, D. R., OLTZ, D. F. & EIDEL, J. J. (eds) *Interior Cratonic Basins*. AAPG Memoirs, **51**, 681–708.

Clastic depositional systems at the base of the late Miocene evaporites of the Levant region, Eastern Mediterranean

C. BERTONI* & J. A. CARTWRIGHT

3DLab, School of Earth, Ocean and Planetary Sciences, Cardiff University, Main Building, Park Place, Cardiff CF10 3YE, UK

*Present address: Repsol-YPF, Paseo de la Castellana, 278, 28046 Madrid, Spain
(e-mail: cbertoni@repsolypf.com)

Abstract: This study investigates the evidence for the presence of clastic sediments at the base of the distal Messinian evaporites in the Levant region (Eastern Mediterranean). Seismic geomorphological analysis of three-dimensional seismic data clearly reveals the occurrence of a well-imaged clastic body composed of two closely spaced channel-mouth lobe deposits, within the basal part of the Messinian evaporites. Comparable seismic facies observed at the same stratigraphic level point to the occurrence of additional clastic deposits and allows their correlation with the El Arish and Afiq canyon systems. The seismic characteristics of the clastic bodies and the analogy with other coeval deposits in the Mediterranean Basin suggest that they deposited in a submarine (shallow- or deep-water) setting. Knowledge of the occurrence and distribution of these clastic deposits has considerable impact on the interpretation of the depositional environment of this basinwide evaporitic system.

The presence of clastic sediments within the Messinian (late Miocene) evaporites in the Mediterranean Basin has been a debated topic since the discovery of this giant evaporitic system. Clastic evaporites interbedded with *in situ* gypsum, halite and anhydrite series have been recorded in outcrop in parts of the Mediterranean area (Ricci Lucchi 1973; Schreiber *et al.* 1976; Vai & Ricci Lucchi 1977; Roveri *et al.* 2001, 2003). Clastic deposits (mainly continental-derived) have also been observed on two-dimensional (2D) seismic and well data on the margins of the Messinian evaporitic basin (Barber 1981; Savoye & Piper 1991; Lofi *et al.* 2005). However, the occurrence of clastic sediments within the thick distal evaporitic series and their location within a subaerial or submarine depositional setting have not yet been conclusively demonstrated (see e.g. Garfunkel & Almagor 1987; Lofi *et al.* 2005).

High-quality three-dimensional (3D) seismic data from the Levant region (Eastern Mediterranean, Fig. 1) how permit the application of seismic geomorphology techniques (Posamentier 2003) to the solution of this scientific problem. Detailed horizon mapping and areal analysis of seismic attributes (e.g. seismic amplitude) reveal the presence of a high-amplitude body within the lower part of the Messinian evaporites. The morphology and seismic character of this body allows its interpretation as composed of two channel-fed clastic lobes, similar in all aspects to other examples recorded worldwide (e.g. Weimer & Link 1991; Collison 1999). We discuss the correlation of this and other similar clastic bodies with a long-lived system of canyons on the Levant continental margin (i.e. the El Arish and Afiq Canyons, Druckman *et al.* 1995).

The main aims of this study are to report the observation of these clastic sedimentary bodies, to argue their origin as submarine or subaerial systems, and to discuss the implications of this discovery for the interpretation of clastic sediment fairways during the deposition of the Messinian evaporites. Ultimately, this study documents the predictive importance of 3D seismic analysis for future scientific investigations (e.g. ultra-deep drilling of the Messinian evaporites) aiming to understand the depositional environment of this giant evaporitic system.

Geological setting

The Levant region is located in the SE part of the Mediterranean Sea (Fig. 1). Since the Oligocene, this area has been characterized by the deposition of large volumes of siliciclastic sediments on the continental margin and basin. Submarine canyons were incised on the shelf and slope areas, primarily by the Afiq, El Arish and Ashdod Canyons (Fig. 1, Neev 1979; Druckman *et al.* 1995).

Normal deep-marine deposition terminated abruptly at the end of the Miocene, during the Messinian Salinity Crisis (MSC), in response to the restriction of seawater supply from the Atlantic Ocean (Hsü *et al.* 1978). Within a time-span of

From: SCHREIBER, B. C., LUGLI, S. & BĄBEL, M. (eds) *Evaporites Through Space and Time.*
Geological Society, London, Special Publications, **285**, 37–52.
DOI: 10.1144/SP285.3   0305-8719/07/$15.00 © The Geological Society of London 2007.

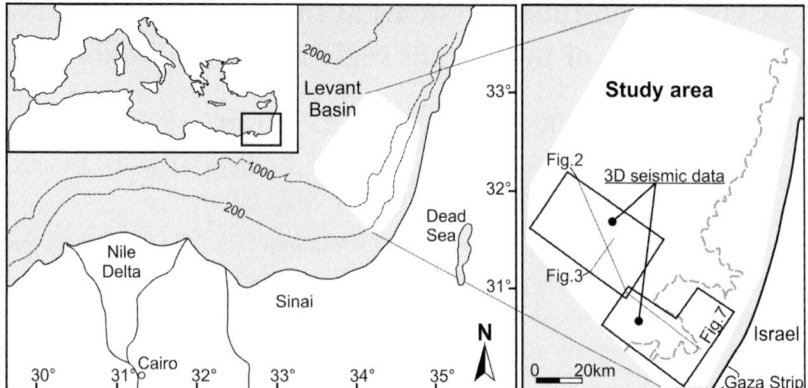

Fig. 1. Location maps for the study area in the Eastern Mediterranean. The areas where 3D seismic data are available are indicated. The grey dashed line marks the locus of pinch-out of the buried Messinian evaporites (from Bertoni & Cartwright 2006).

significantly less than 1 Ma (Clauzon et al. 1996; Krijgsman et al. 1999) a thick wedge of evaporites was deposited on the Levant continental margin and basin, where it reaches a thickness of 2 km (Garfunkel & Almagor 1987). Previous studies document the activity of the submarine canyons before, during and after the MSC in the study area (Buchbinder & Zilbermann 1978; Neev 1979; Druckman et al. 1995). However, the specific role of the canyons as conduits for the delivery of clastic sediments to the distal part of the Messinian evaporitic basin has not yet been addressed.

Since the Pliocene, the restoration of deep-marine conditions led to the rebuilding of the marine siliciclastic wedge on the Levant continental margin (Garfunkel & Almagor 1985; Tibor et al. 1992). During the early Pliocene, a turbiditic basin floor fan (Yafo Sand Member) was deposited within the Afiq and El Arish submarine canyons, which are now infilled and buried under >1 km of overburden (Druckman et al. 1995; Frey-Martinez et al. 2007).

Seismic analysis

The 3D seismic data analysed in this study were acquired in 2000, over an area of approximately 6000 km^2 offshore Israel and the Gaza Strip (Fig. 1). These data were time migrated before stacking to generate a seismic grid with cells of 12.5 by 12.5 m. The general stratigraphic context of the Levant continental margin is displayed with a representative interpreted seismic line in Figure 2, with the focus of this paper being on the Messinian evaporites. The base and the top of this seismic unit are represented by, respectively, Horizons N and M (Fig. 2), i.e. two distinct seismic events that are regionally correlatable across the Mediterranean Basin (Hsü et al. 1973). On seismic data, Horizon M represents a high-amplitude positive seismic reflection generated by the acoustic impedance contrast between the deep-marine siliciclastic sediments of the Plio-Pleistocene unit and the underlying Messinian evaporites (Figs 3 & 4). Horizon N represents a high-amplitude negative seismic reflection at the transition between the Messinian evaporites and the underlying siliciclastic sediments of the Oligo-Miocene unit (Figs 3 & 4). An average velocity of 4000 m/s has been applied for the time–depth conversion of the evaporitic unit. This approximate value is mainly based on previous studies of the distal Messinian evaporites in the area (e.g. Garfunkel & Almagor 1987), on wireline log measurements and on borehole cuttings available for the proximal upper part of the evaporites (Bertoni & Cartwright 2005). The thickness of the Messinian evaporites spans from almost 2 km in the Levant Basin to a few tens of metres towards the continental margin, where Horizon N and M merge into a single seismic horizon (Fig. 4).

The distal Messinian evaporites are seismically composed of alternating transparent facies and continuous seismic reflections (Horizons ME20–ME50 in Fig. 3). In the lower part of this unit, a series of high-amplitude seismic reflections are observed above Horizon N (Fig. 4). These reflections are correlatable laterally for up to 6 km (Figs 5 & 6). The areal extent of these high-amplitude events is clearly imaged by computing the maximum seismic amplitudes over a 120 ms TWT (two-way travel time) window above Horizon N. The

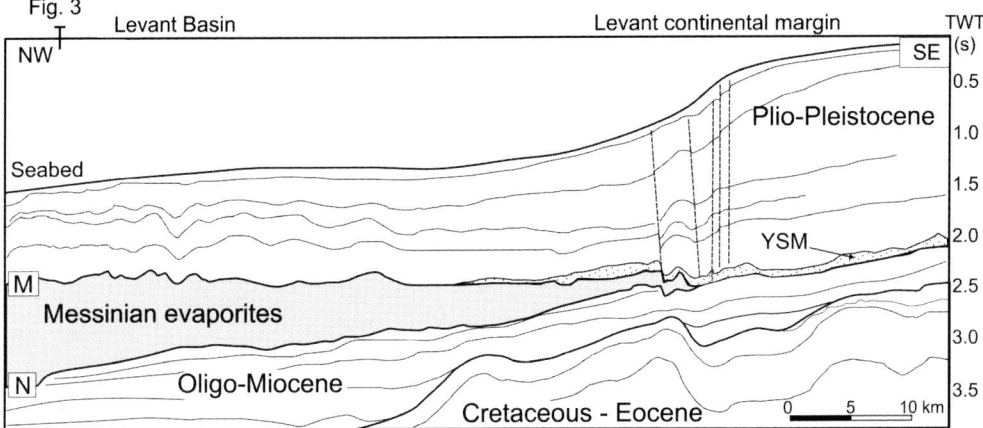

Fig. 2. Schematic geo-seismic section showing the seismic-stratigraphic context of the study area in the Levant Basin and continental margin (see text for detailed explanation). N = Horizon N; M = Horizon M; YSM = Yafo Sand Member (indicated by the dotted fill pattern). The marginal faults in the Plio-Pleistocene unit are marked by subvertical dashed lines. On the vertical scale, TWT is the two-way travel time expressed in seconds.

resulting image (Fig. 7) shows that they correlate with a series of kilometre-scale high-amplitude bodies with a rather irregular and elongated ellipsoidal morphology. These bodies are located basinward of the pinch-out of the Messinian evaporites and fully confined within this unit (Fig. 7).

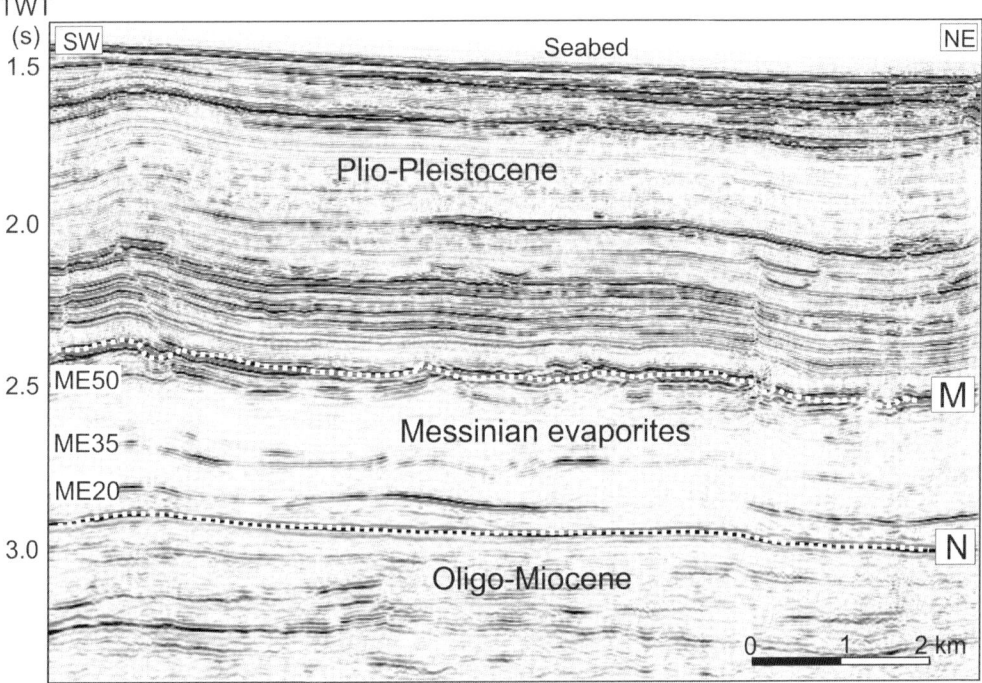

Fig. 3. Seismic section crossing the study area in a NE–SW direction (location in Fig. 1). The stratigraphy of the distal part of the Messinian evaporites is displayed. In this area the Messinian evaporites are seismically composed of transparent facies alternating with medium to low amplitude seismic reflections (Horizon ME20–ME50).

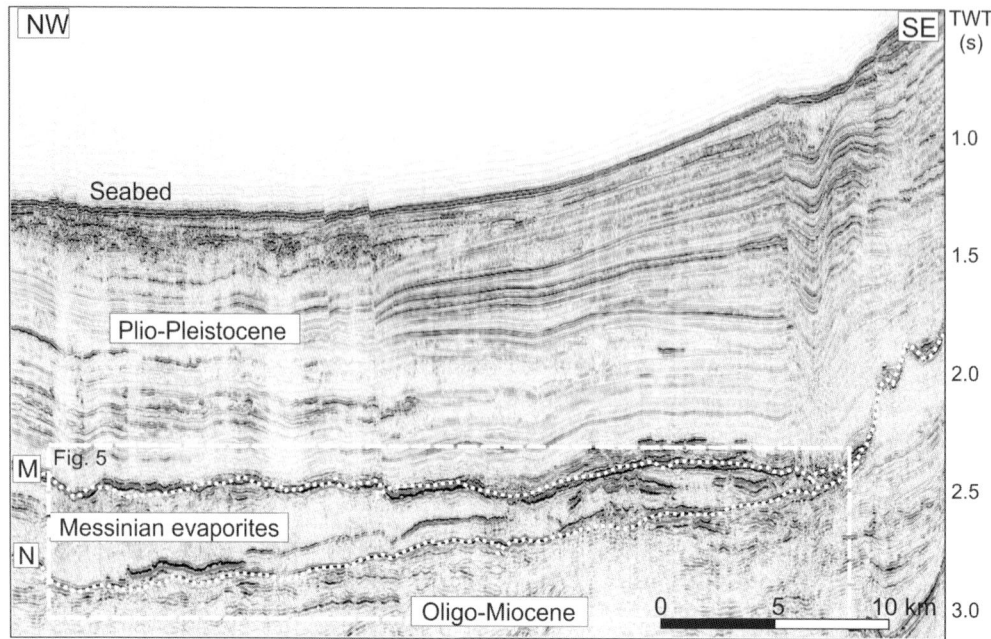

Fig. 4. Seismic section crossing the study area in a NW–SE direction (see location in Fig. 7) displaying the stratigraphy of the Messinian evaporites from their proximal (SE) to the distal (NW) part. Note that the Messinian evaporites wedge out towards the SE, where Horizons N and M merge into a single seismic horizon.

Detailed 3D seismic description of HAB1

The best-imaged high-amplitude body is HAB1. Its geometry in cross-section and in plan view is described in detail with reference to a series of seismic sections, time-structure maps and maximum amplitude window extractions (Figs 8–12). On seismic sections, HAB1 is bounded at the top by Horizon ME2, represented by a high-amplitude negative seismic reflection (Fig. 8). The near zero-phase seismic polarity and negative amplitude character of Horizon ME2 has been assessed by comparison with flat spots observed at similar stratigraphic levels (Brown 1999). This allows the definition of Horizon ME2 as a 'soft' seismic event generated by a decrease in acoustic impedance (Brown 1999). The amplitude of Horizon ME2 is comparable to that of Horizon M, though with negative sign (i.e. the opposite polarity).

The base of HAB1 is represented by a low-amplitude seismic reflection directly overlying Horizon N (Fig. 8). This seismic reflection appears poorly defined probably because it is obscured by the directly overlying high-amplitude seismic reflection, causing its loss in amplitude. The Oligo-Miocene reflections underlying the Horizon are truncated against it (Fig. 8a). This erosional truncation has been interpreted as being related to the incision of the Afiq-El Arish Canyon at the base of the Messinian evaporites (Bertoni & Cartwright 2006). HAB1 is characterized by the maximum thickness at the central parts of the body and by lateral bi-directional pinch-out and downlap at its edges, defining an overall mounded geometry (Fig. 8a & c). On NE–SW directed cross-sections, HAB1 has an asymmetric geometry produced by the pinch-out of the body on a structurally elevated area to the SW and on a structurally depressed area to the NE (Fig. 9a). The locally observed loss of resolution and of clear imaging at the edges of the body could be due to overlapping and interference of the two opposite polarity wavelets representing the top and the base of HAB1. These two reflections will be resolved only as long as their distance is greater than half the wavelength of the incident waveform, and this critical distance is defined as the tuning thickness (Badley 1985). Some internal reflections are observed within HAB1 (Fig. 8a & c), although their internal geometry is not clearly defined because it is at the limits of vertical seismic resolution.

The areal maximum amplitude window extraction displayed in Figure 10a clearly shows that HAB1 has an irregular-elliptic shape elongated in

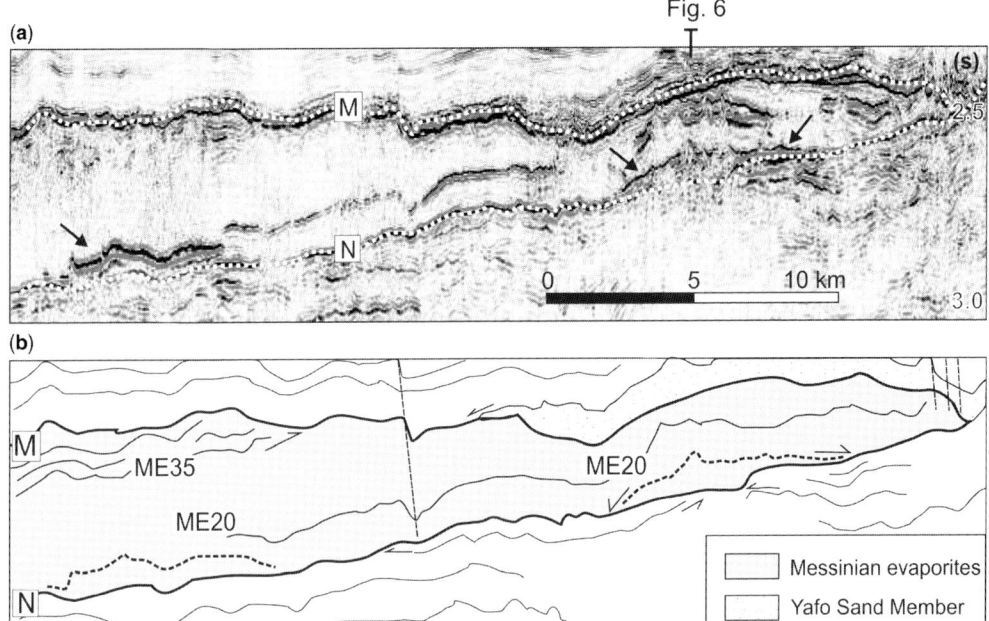

Fig. 5. Seismic section crossing the study area in a NW–SE direction, and interpretation (see location in Fig. 7). (**a**) Close-up of the Messinian evaporites as displayed in the seismic section in Figure 4. In the lower part of this unit, directly above Horizon N, the high-amplitude seismic reflections described in the text are indicated by the black arrows. (**b**) Line-drawing of the seismic section, with interpretation of faults and terminations of seismic reflections (indicated by black arrows).

a NW–SE direction. The maximum radius of the body is approximately 8.5 km and the minimum is 4 km (Fig. 10a), the total area covered being c. 35 km². The thickness of HAB1 is best visualized on the isochron map computed between the base and top of this body (Fig. 10b). This map shows that the thickness gradually increases from the edges of the body towards its central area, where it reaches over 100 ms (Fig. 10b). A synopsis of the nature and distribution of the lateral terminations of HAB1 is also displayed on this figure (Fig. 10b). Horizon ME2 terminates laterally by downlap to the northwest (Figs 8b & 10b). On the northeastern side of HAB1, Horizon ME2 is involved in remobilization caused by tectonic deformation (Fig. 9b), therefore we are unable to judge the nature of its original stratal termination and its possible correlation with the overlying reflections. On the eastern side, HAB1 appears to thin at its edge. However, the decrease in amplitude of Horizon ME2 in this area (e.g. Fig. 8a) prevents a definitive interpretation of this termination as a low-angle downlap of Horizon ME2 on the base of HAB1, or alternatively as an apparent downlap.

Two ribbon-shaped SE–NW oriented high-amplitude bodies are attached to the main part of HAB1, as clearly displayed in Figure 10a. These features are approximately 0.5 km wide and 6 km long (Fig. 10a) and their thickness ranges from 20 to 50 ms (Fig. 10b). In cross-section, these bodies are subtle, convex-upward features (Fig. 11a & b). They appear as alternately confined and filling the underlying lows, or totally unconfined and showing a distinct constructional geometry (Fig. 11a), being laterally shifted as regards the location of the underlying structural depressions (Fig. 11b). Importantly, the isochron map in Figure 10b defines the presence of two loci of increased thickness (up to >100 ms) within the main part of HAB1, located immediately NE of the two ribbon-shaped bodies. These bodies are connected landward to a thicker high-amplitude body (up to 100 ms thick), located to the south of HAB1 (Fig. 10b). In cross-section (Fig. 11c) this body does not present a clear internal geometry and it is highly disrupted by a subsequent deformational phase, thus precluding a more detailed seismic analysis.

In order to define the complex relationship of HAB1 to the underlying basin physiography, we

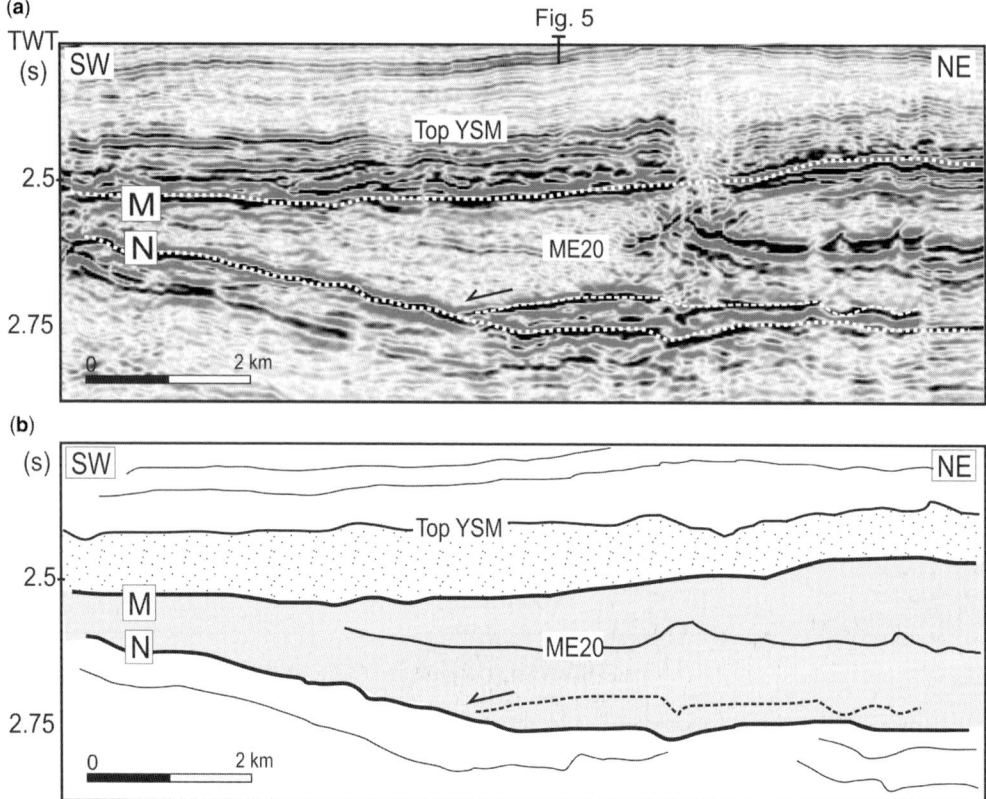

Fig. 6. Seismic section crossing the study area in a NE–SW direction (see Figs 5 & 7 for location). (**a**) Seismic cross-section with indication of the main seismic horizons and reflection terminations. (**b**) Line-drawing of the seismic section: In the lower part of this section, the onlap termination of the high-amplitude seismic reflection against Horizon N (indicated by the black arrow) should be noted.

have overlain the amplitude map of Horizon ME2 to the 3D visualization of the time-structure map of Horizon N (Fig. 12). This visualization shows that the thicker body to the south of the ribbon-shaped features is located on a structurally elevated area (Fig. 12). The main ellipsoidal part of HAB1 is confined to a structurally depressed area to the NW, within the floor of the El Arish-Afiq Canyon (Fig. 12). Significantly for their interpretation, the ribbon-shaped features are positioned on a sloping part of the time-structure map of Horizon N that links the structurally elevated area to the SE to the depressed area to the NW (Fig. 12).

Interpretation

The most important features for the diagnosis of the high-amplitude body HAB1 are its lateral terminations, seismic character and shape. A key observation is that HAB1 presents an overall mounded geometry determined by lateral downlap and bi-directional pinch-out. In evaporitic settings, three main depositional bodies can be recognized as mounded features on seismic data: halite pods (Hodgson et al. 1992; Bishop et al. 1995; Taylor 1998), carbonate build-ups (Taylor 1998) or clastic lobes (e.g. Mitchum 1985). The simplest way to discriminate among these three possible interpretations is to assess the amplitude and polarity of the top of HAB1 in more detail.

The high-amplitude 'soft' seismic event defining the top of HAB1 is a product of the different seismic velocities and densities of the two lithologies as juxtaposed at the boundary. The similarity in amplitude of Horizon ME2 to Horizon M and their opposite polarity (Fig. 8), would be consistent with Horizon ME2 representing a similar magnitude of acoustic impedance contrast, but with reverse relationship (evaporites overlying siliciclastic deposits).

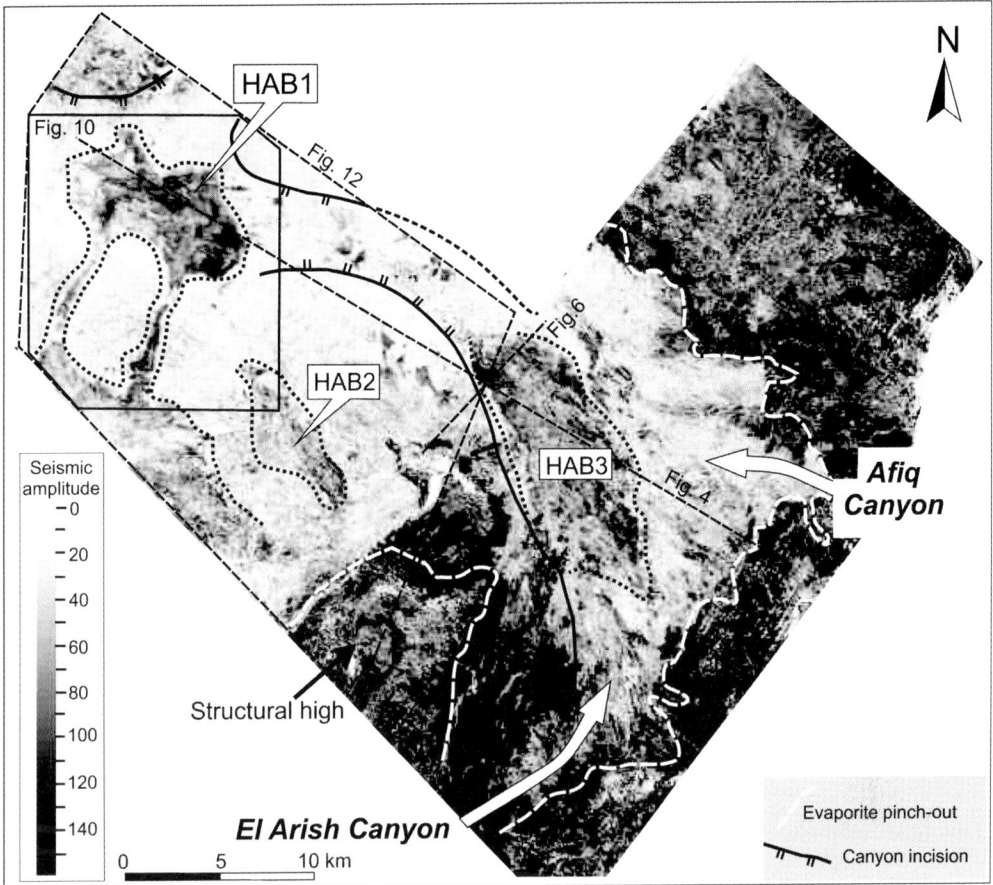

Fig. 7. Maximum seismic amplitudes calculated over a 120 ms TWT (two-way travel time) window above Horizon N. The kilometre-scale high-amplitude bodies correspond to dark-grey coloured areas, marked with a dotted line and are named HAB1, HAB2 and HAB3. These bodies are located basinward of the pinch-out of the Messinian evaporites on seismic data (white dashed line) and are fully confined within this unit.

In addition to the acoustic character, perhaps the most significant diagnostic feature of the body is its morphology in plan view. HAB1 is composed of a main ellipsoidal part and of two attached ribbon-shaped bodies (Fig. 10a). This morphology is remarkably similar to channels feeding a downslope clastic body (Mitchum 1985; Posamentier & Erskine 1991; Weimer & Slatt 2004). Conversely, the feeder channels would not be expected to develop in halite pods or carbonate build-ups. Based on the previous observations, we interpret HAB1 as a constructional clastic depositional body developed downstream of points where laterally confined flows from the feeder channels expand (e.g. Reading 1999). The presence of the two feeder channels suggests a double-point source of the body from the SE (Fig. 10a). This interpretation is consistent with the presence of two depocentres immediately basinward of the two feeder channels (Fig. 10b). The two depocentres define a geometry analogous to clastic lobes generally observed at the termini of the feeder channels (Weimer & Slatt 2004). The main part of the clastic body can thus be described as composed of two closely spaced channel-mouth lobe deposits.

The main part of HAB1 overlies the floor of the El Arish-Afiq Canyon (Fig. 7). However, the location and direction of the feeder channels suggests that the source of sediment supply to the clastic body is lateral to the main canyon path (Fig. 7). The thicker body located to the south of the feeder channels (Fig. 10b) could represent the source, or a locus of accumulation of the clastic deposits in an intervening intra-slope basin. The clastic body appears as an overall constructional feature with only limited erosion observed at the

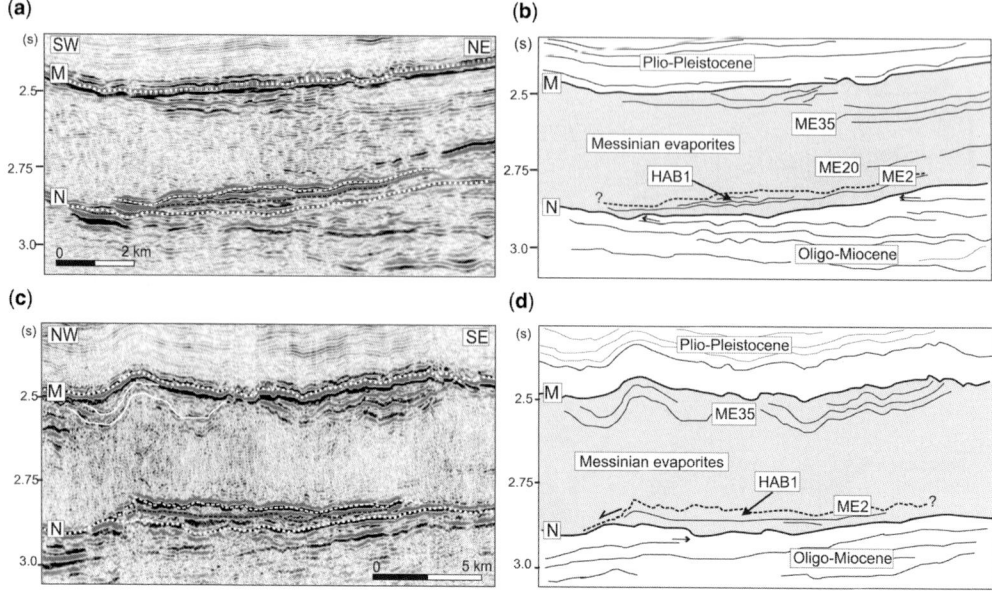

Fig. 8. Seismic sections across the main high-amplitude body HAB1. (**a**) Seismic section oriented in a NE–SW direction (location in Fig. 10). (**b**) Line-drawing of the seismic section displayed in Figure 8a, with interpretation of the main seismic horizons and reflection terminations. The erosional truncation of the Oligo-Miocene reflections against Horizon N is indicated by the black arrows. (**c**) Seismic section oriented in a NW–SE direction (location in Fig. 10). (**d**) Line-drawing of the seismic section displayed in Figure 8c, with interpretation of the main seismic horizons and reflection terminations.

base of the feeder channels and of the main body. The limited thickness of the body hinders the resolution of subtler depositional features such as channels within the lobes, and this precludes a detailed sequential seismic geomorphology analysis (Posamentier 2003). The lack of clear lapout terminations and the uncertain correlation of Horizon ME2 to the north and east suggest that the clastic body could be more extensive than mapped.

The other high-amplitude bodies observed in the study area (HAB2 and HAB3 in Fig. 7) occur at a stratigraphic position and show seismic characteristics comparable to HAB1 (Figs 5 & 6). Although their morphology in plan view is less clearly defined than that of HAB1 (Fig. 7), we interpret them as clastic bodies by analogy with the interpretation of HAB1. HAB3 forms an elongated feature located at the intersection between the El Arish and Afiq Canyons (Fig. 7). Albeit no evidence of feeders is observed, the geographic position of this body suggests that the two canyons acted as the main conduits for the sediment supply. Clastic input from the El Arish–Afiq Canyon is registered in the study area from the Oligocene through the early Pliocene (Druckman et al. 1995; Frey-Martinez et al. 2005). No specific record in the literature exists for the clastic input during the deposition of the Messinian evaporites. It is probable that such clastic supply continued even during the MSC, because the Levant continental margin was actively being eroded at this time (Gvirtzmann & Buchbinder 1978).

The seismic amplitude response of these clastic deposits is complex to interpret as it depends on the rock physics, thickness and fluids in the sediment (Weimer & Slatt 2004). Therefore, further constraints on the lithology of the body (i.e. siliciclastic and/or evaporite or carbonate dominated) are not available simply considering the relative amplitude and polarity of the seismic reflections. Based on the pre-evaporitic dominant siliciclastic setting of the area of interest and on the seismic geomorphological analysis here presented, we consider it more likely that the clastic deposits are dominantly represented by siliciclastic sediments. An evaporite clastic component could be possible if the reworked clastic sediments were derived from marginal evaporitic deposits. This could potentially occur considering the substantial diachronity observed within the Messinian evaporites of different Mediterranean areas, with some marginal evaporitic deposits pre-dating the basinal evaporitic series (e.g. Rouchy 1982; Butler et al. 1995; Clauzon et al. 1996; Riding et al. 1998). We can give an

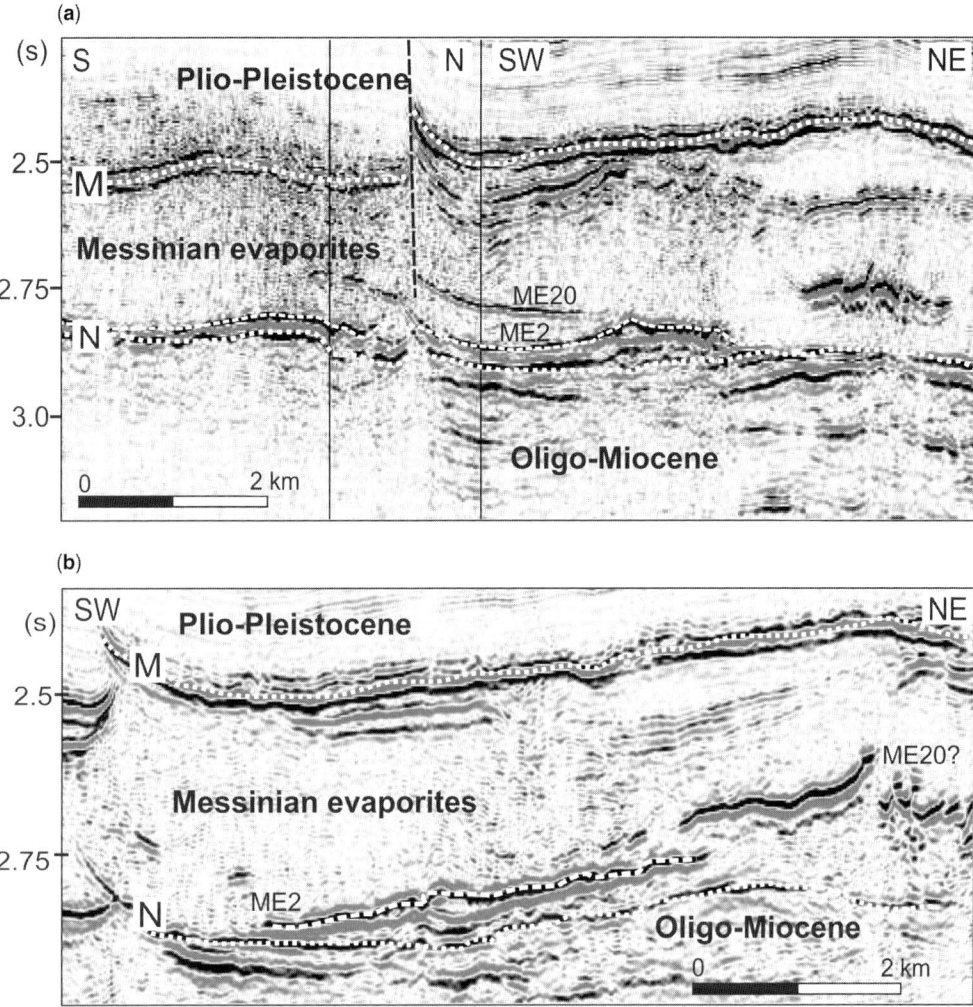

Fig. 9. Seismic sections crossing the main high-amplitude body HAB1 (location in Fig. 10). (a) Seismic section crossing the body HAB1 in a south–north/SW–NE direction. In this section, HAB1 presents an asymmetric geometry that is produced by the pinch-out of the body on a structurally elevated area to the SW and on a structurally depressed area to the NE. The subvertical black dashed line indicates post-evaporitic faulting. (b) Seismic section crossing the body HAB1 in a NE–SW direction. On the northeastern side of HAB1, Horizon ME2 is deformed by remobilization, hindering the interpretation of the nature of its original termination.

estimate of the time–depth converted thickness of the body HAB1, based on a dominantly siliciclastic composition, and on the analogy with submarine siliciclastic fan deposits observed at similar depths in the study area, i.e. the Yafo Sand Member of early Pliocene age (unpublished well reports; Frey-Martinez et al. 2006). The average seismic velocity obtained on wells for these deposits is 2000–2500 m/s, which applied to HAB1 results in a thickness range of 100–125 m. The seismic velocity, and consequently, the thickness of the clastic body HAB1 would be significantly higher if the clastic deposit had instead an evaporitic component.

The available data do not allow us to determine with any certainty whether the evaporites enclosing or overlying the clastic bodies have a primary or clastic origin. Nonetheless, the analysis of the Messinian evaporites directly overlying the high amplitude bodies shows no evidence of any development of similar clastic bodies at higher stratigraphic levels, and the interbedding of clastic units

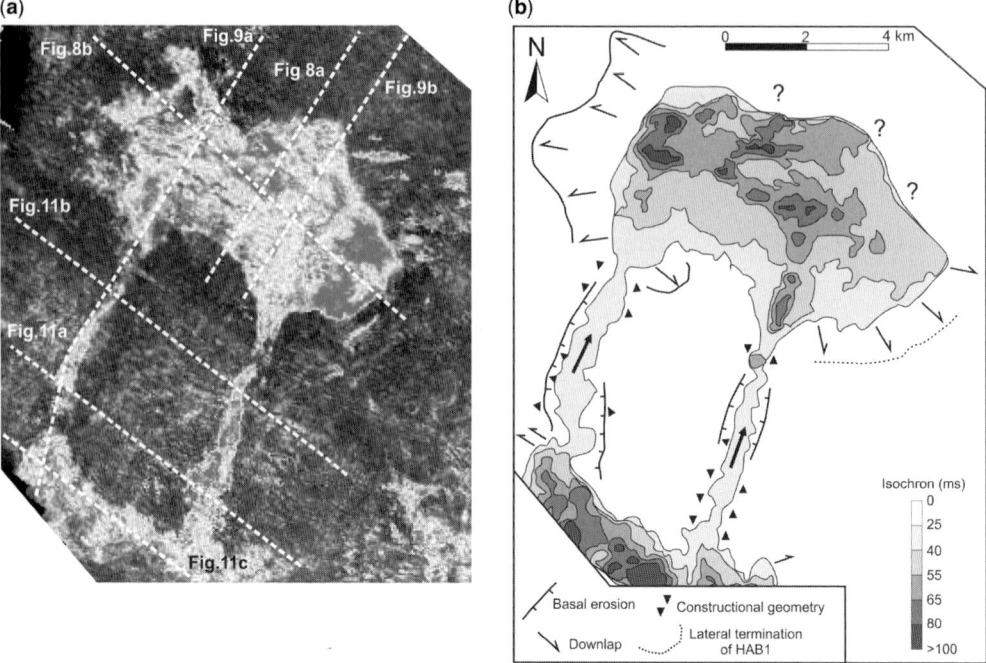

Fig. 10. Morpho-structural maps of HAB1. (**a**) Close-up of Fig. 7, showing details of the morphology of HAB1 with a different colour display. (**b**) Isochron map calculated between the base and the top of HAB1. A synopsis of the nature and distribution of the lateral terminations of HAB1 is displayed in this map.

therefore appears to be confined within the basal part of the evaporitic succession.

Discussion

Distribution of clastic sediments within the Messinian evaporites

Clastic sediments can constitute an important component of basinwide evaporites, as previously stressed in many studies (Schlager & Boltz 1977; Schreiber 1988; Martinez del Olmo 1996; Kendall & Harwood 1999; Peryt 2000; Manzi et al. 2005). In the Western Mediterranean, a subject of current intense debate is whether a conspicuous part of the Messinian evaporitic unit has a clastic rather than evaporitic character (e.g. Lofi et al. 2005).

The clastic lobes described in this study are clearly located within the basal part of the distal Messinian evaporitic unit. Therefore, their occurrence implies that significant volumes of clastic sediments were supplied to the Levant Basin during the early stages of the deposition of the Messinian evaporitic unit. This is clearly shown by the stratigraphic position of the clastic bodies analysed above the erosional surface defining the base of the Messinian evaporites in this basin (i.e. Horizon N). The absence of similar clastic bodies within the overlying evaporites could be due to the fact that they deposited in different areas of the basin, or conversely, it could be linked to the subsequent inactivity of the fairways supplying the clastic sediments. Both hypotheses might be related to a variation in the base-level of erosion and clastic supply or to a change in the efficiency of the distributary system, and would be consistent with the high variability of depositional environments in evidence during the Messinian Salinity Crisis.

Depositional environment – submarine or subaerial?

The morphology and seismic character of clastic bodies are generally similar in either subaerial or submarine settings (Weimer & Link 1991; Collison 1999). Thus, in order to define the depositional environment of the clastic lobes composing HAB1, additional information must be taken into account, regarding any evidence for coeval subaerial exposure in the basin, and the position

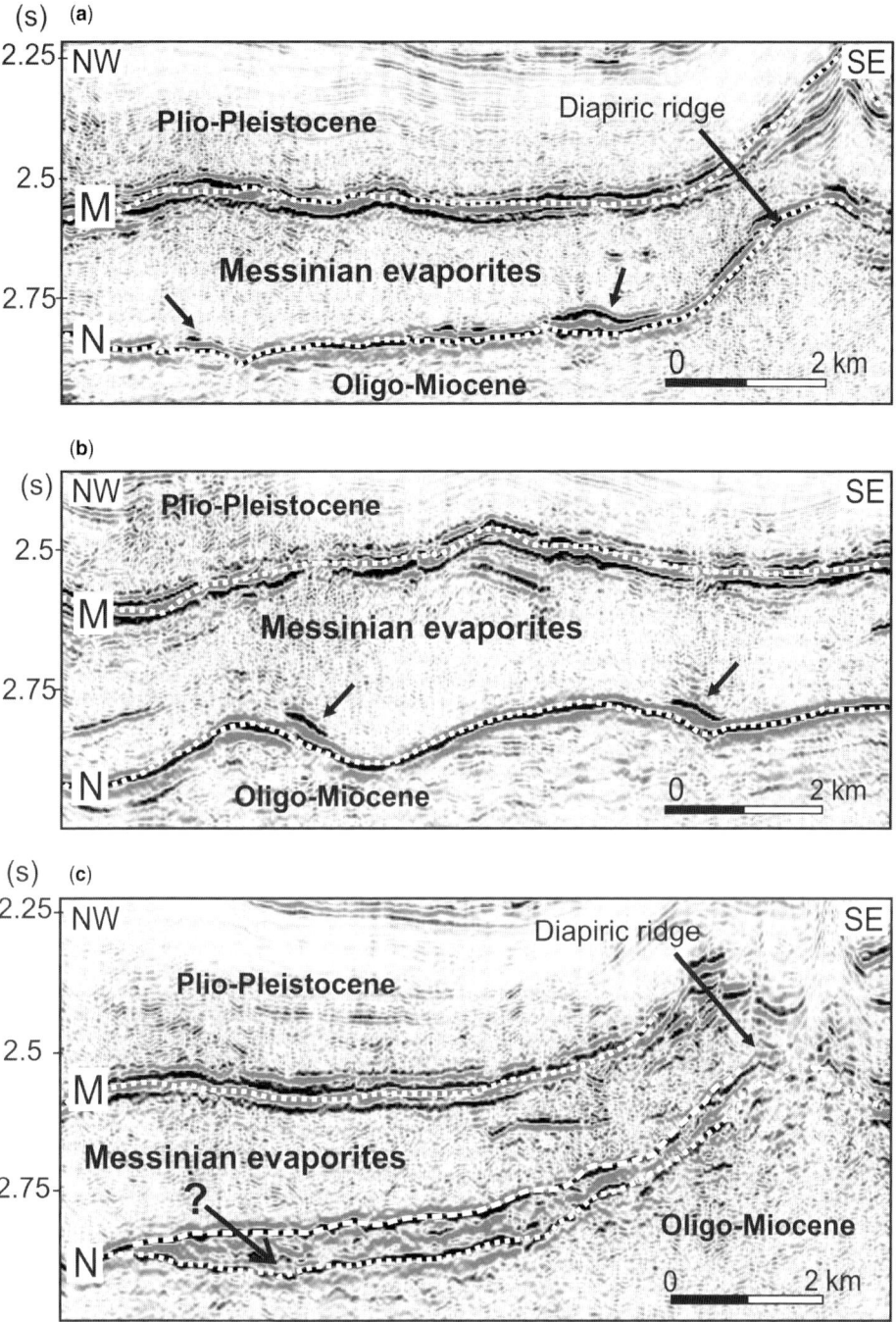

Fig. 11. Seismic sections crossing HAB1 (location in Fig. 10a). (a) Seismic section crossing the two ribbon-shaped high-amplitude bodies attached to the main part of HAB1. The bodies are subtle, convex-upward features and appear as partly confined in underlying lows. (b) Seismic section crossing the two ribbon-shaped bodies, which appear to be laterally shifted with regard to the location of the underlying structural depressions. (c) Seismic section crossing the thicker high amplitude body located landward and to the south of the main part of HAB1 (location in Fig. 10a). In cross-section this body does not present a clear internal geometry because it is highly deformed.

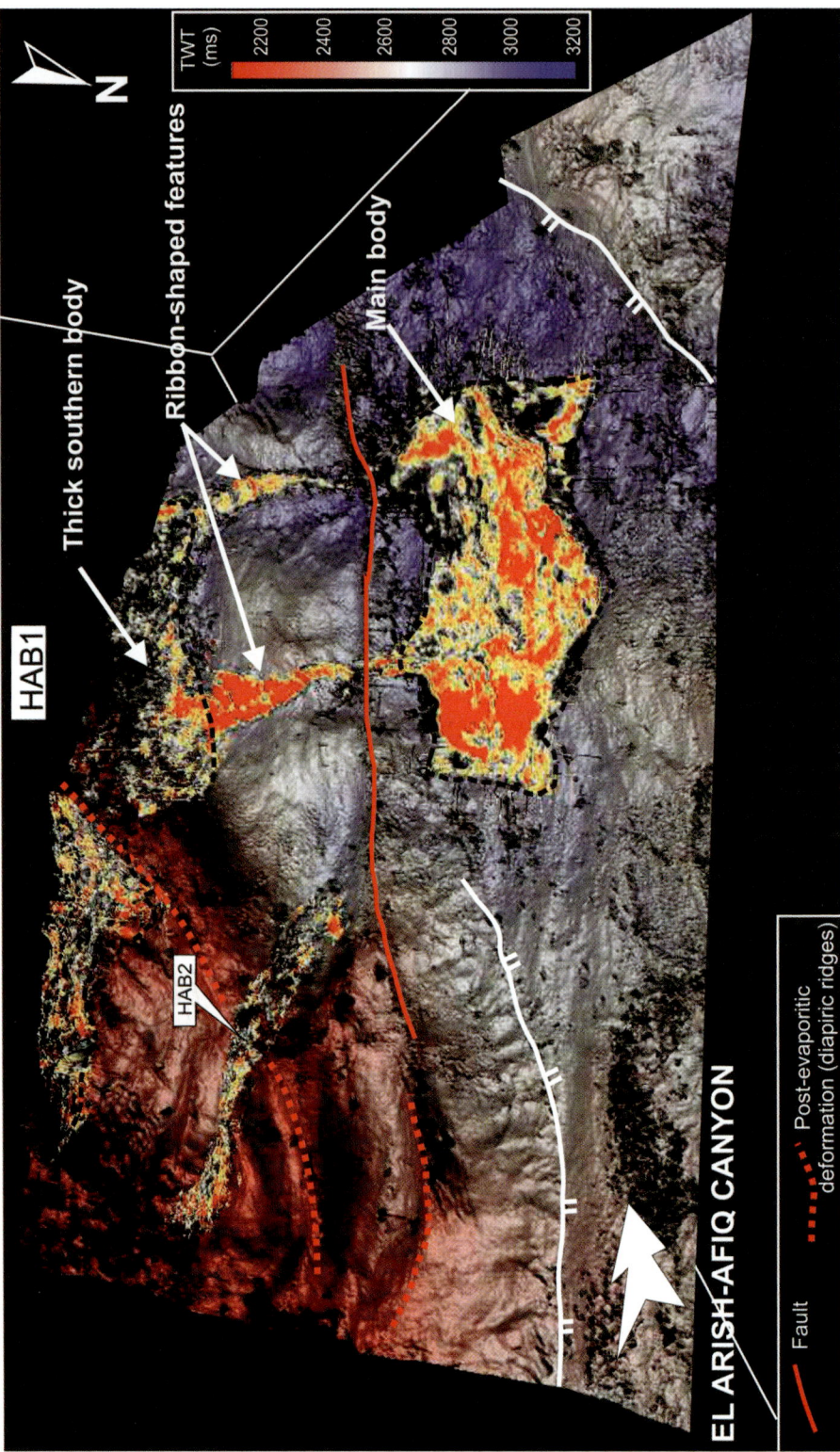

Fig. 12. Three-dimensional perspective of HAB1 (as displayed in Fig. 10) draped over the time-structure map of Horizon N (location in Fig. 7). The maximum amplitude values are shown in red. This visualization illustrates the overall morphology and geometry of HAB1: the thicker body to the south of the ribbon-shaped features is located on a structurally elevated area and the main part of HAB1 is confined to a structurally depressed area to the NW, within the floor of the El Arish–Afiq Canyon. The ribbon-shaped features lay on a sloping part of the time-structure map of Horizon N, between the structurally elevated area to the SE to the depressed area to the NW. Faults and ridges indicated represent post-evaporitic deformational structures.

and geometry of the clastic bodies within the Messinian evaporites.

The physiography of the Levant continental margin during the MSC was characterized by structurally elevated areas landward of the pinch-out of the Messinian evaporites (Bertoni & Cartwright 2006). In these areas evidence of subaerial exposure is recorded as:

- A prominent erosional surface characterised by a dendritic drainage pattern (Gvirtzmann & Buchbinder 1978; Mart & Ben Gai 1982) comparable to the Messinian erosional surfaces described on the Ebro, Gulf of Lions and Nile continental margins (Ryan 1978; Stampfli & Hocker 1989; Guennoc et al. 2000; Frey-Martinez et al. 2004). These erosional surfaces closely resemble a badlands topography and are interpreted as subaerial (Ryan 1978; Stampfli & Hocker 1989; Frey-Martinez et al. 2004).
- A series of marginal scarps (Fig. 13; Bertoni & Cartwright 2006), that are analogous to the wavecut platforms and rejuvenation terraces observed in the nearby Nile delta (Barber 1981). Importantly, the clastic deposits are located in an area basinward of the pinch-out of the Messinian evaporites, where such morphological evidence of subaerial exposure (badlands and/or terraces) is not observed (Figs 7 & 13). Furthermore, the base of the Messinian evaporites directly beneath the clastic bodies shows evidence of erosion by a few confined incisional features related to the El Arish–Afiq Canyons (Figs 7 & 13). This downslope change in the stream pattern from 'badlands' erosion to basinward focused incision would be consistent with a transition to a submarine environment.

Based on data compiled from previous studies, the clastic bodies HAB1, HAB2 and HAB3 appear to be located approximately at the same distance from the pinch-out of the Messinian evaporites (Fig. 13). However, the location of the channels feeding the body HAB1 suggests that the source of sediment supply to this clastic body is lateral to the El Arish–Afiq Canyon (Fig. 13). Therefore, the basinward extension of the El Arish and Afiq Canyons appears to have acted as a bathymetrically depressed area attracting sediments from the SW of the study area, externally to the El Arish and Afiq Canyons. The source of clastic supply could be either represented by a tributary of the El Arish Canyon, or by the Nile Delta area, as suggested by the regional drainage pattern observed at the base of the Messinian evaporites (Fig. 13; Ryan 1978). It is significant to note that, during the Messinian lowstand, the Nile Delta

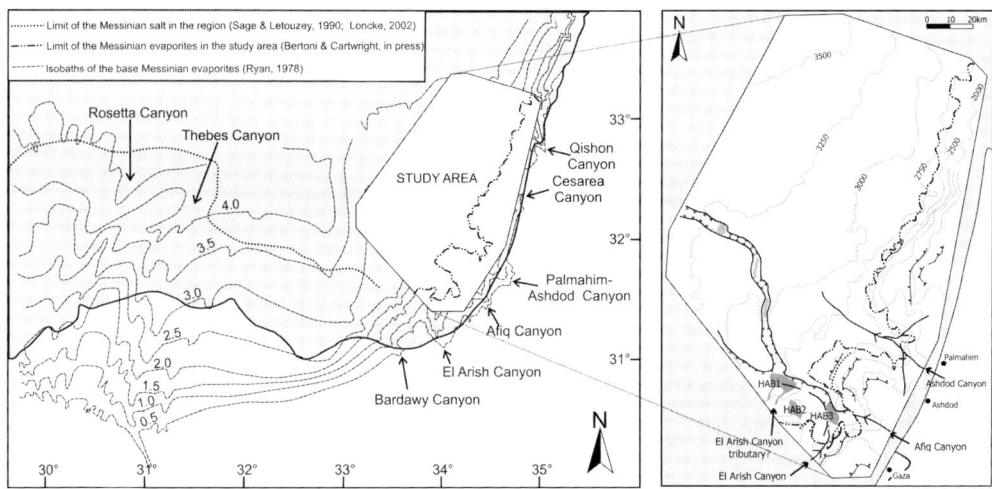

Fig. 13. Map of the sector of the Eastern Mediterranean analysed, showing the regional setting of the basal discordance to the Messinian salt (and associated evaporite and clastic formations) and its marginal continuity with Horizon M (contour lines in seconds TWTT; after Ryan 1978). The landward limit of the Messinian salt in the region is also indicated (after Sage & Letouzey, 1990; Loncke, 2002; Bertoni & Cartwright 2006). In the figure to the right, the distribution of the clastic sediments (in grey) is shown within the paleogeographic context of the base of the Messinian evaporites. Canyon incisions, erosional scarps and contour lines (in seconds TWTT) of the base of the Messinian evaporites are after Bertoni & Cartwright (2006). The grey areas identify the location of the clastic deposits described in this study.

shifted extensively seawards (see e.g. Barber 1981), thus approaching the area of the body HAB1. Consequently, this area would be located in close proximity of the Eonile and Eosahabi Deltas, i.e. the two major sources of clastic material in the Eastern Mediterranean during the Messinian (Griffin 2002). Based on these observations, the pre-Pliocene Nile Delta system could be considered as a likely source of sediment supply for the clastic body HAB1.

A further indication of the depositional setting of the clastic body HAB1 arises from the analysis of its depositional geometry. It is noteworthy that the geometry of the feeders is mostly unconfined and convex upward, i.e. mainly constructional, and this is generally regarded as typical of submarine rather than subaerial settings (Reading 1999). The apparent effect of differential compaction in creating this geometry can be ruled out due to the absence of incision or of downwarped reflections directly underneath the feeders. Based on these observations, we conclude that a submarine (shallow or deep-water) setting is more plausible than a subaerial setting for the clastic bodies analysed.

Sea-level position

Submarine channel-mouth lobes commonly develop at the base of slopes, within turbiditic depositional systems where clastic sediments are transported via gravity-induced processes (Stow 1986). These systems have been extensively studied mainly due to their importance as hydrocarbon reservoirs (Gluyas & Swarbrick 2004). They can deposit from shallow-water (e.g. at the mouth of river deltas) to deep-water settings (beyond the base of the continental slope) (Reading & Richards 1994). Furthermore, the location of basin-floor clastic lobes and feeders can be strongly influenced by active fault lineaments (Rattey & Hayward 1993). Sequence stratigraphy studies showed that the deposition of clastic lobes can occur at any tract of the relative sea-level curve, under the appropriate physiographic and sedimentologic conditions (Posamentier & Erskine 1991). However, most commonly they are deposited due to enhanced erosional processes during intervals of relative sea level fall (Posamentier & Vail 1988). As emphasized by Schreiber (1988), the cut-off in marine inflow necessary to generate the hypersaline Messinian evaporitic basin is likely to have resulted in widespread sea-level fluctuations during the Messinian Salinity Crisis. The consequent subaerial exposure of the continental margins would have driven extensive erosional processes and generated large quantities of reworked material basinward (Schreiber 1988). Therefore, the presence of submarine clastic deposits above the base of the Messinian evaporites is not a conclusive proof but it is nonetheless supportive of a relative sea-level fall before the onset or at the beginning of evaporite deposition in the distal basin.

The location of the clastic bodies basinward of the first onlap of the evaporites (see e.g. Fig. 4) suggests two possible interpretations for the relative sea-level history of the Messinian evaporitic basin: (1) the deposition of the clastic bodies in shallow water, followed by a sea-level rise up to the first onlap of the evaporites – However, this first hypothesis is not supported on seismic data by any evidence of a subaerial exposure recorded between the clastic bodies and the pinch-out of the Messinian evaporites; (2) the deposition of the lobes occurring in a relatively deeper water setting, with the sea level positioned not beyond the first onlap of the evaporites. The validity of either of these hypotheses needs to be tested by additional lithological and biostratigraphic constraints, and studies of basinal architecture of the evaporitic system. Sub-salt boreholes penetrating the clastic system are required in this key area of the Mediterranean Basin to confirm our seismic interpretation and, consequently, to provide a full understanding of the processes acting at the initial stages of the Messinian Salinity Crisis.

Conclusions

The analysis of 3D seismic data from the Levant Basin and continental margin has provided strong evidence of the presence of significant amounts of clastic sediments in the basal part of the distal Messinian evaporites in this area.

- Seismic geomorphology techniques revealed the presence of a remarkably well-imaged clastic body at the base of the Messinian evaporites. This body is composed of two channels feeding two main downslope lobes. Comparable seismic facies observed at the same stratigraphic level suggest the occurrence of additional clastic bodies and allows their correlation with a long-lived system of canyons on the Levant continental margin.
- Based on the seismic geometry and the basin physiography during the deposition of the Messinian evaporites, we conclude that a submarine (shallow or deep-water) setting is more plausible than a subaerial setting for the clastic bodies analysed. Further studies and data are needed in order to define more accurately the water depth at which the clastic bodies deposited and, consequently, the relative sea-level history in this part of the Messinian evaporitic basin.

Thanks are due to BG-Group for kindly providing the seismic and well data upon which this work is based. In particular, we would like to express our gratitude to I. Campbell for his help in obtaining permission to publish this paper. Schlumberger Ltd is acknowledged for the use of Geoquest seismic interpretation software. Referees J. Clark, D. Griffin and M. Roveri are thanked for their insightful reviews of the manuscript, which led to significant improvements. We are also grateful to R. Davies, M. Huuse and J. Frey-Martinez for discussions and to editors B.C. Schreiber and S. Lugli for their comments and the opportunity to present this paper at the 32nd International Geological Congress.

References

BADLEY, M. E. 1985. *Practical Seismic Interpretation*. International Human Resources Development Corporation, Boston, MA.

BARBER, P. M. 1981. Messinian subaerial erosion of the proto-Nile delta. *Marine Geology*, **44**, 253–272.

BERTONI, C. & CARTWRIGHT, J. A. 2005. 3D seismic analysis of circular evaporite dissolution structures, Eastern Mediterranean. *Journal of the Geological Society of London*, **162**, 909–926.

BERTONI, C. & CARTWRIGHT, J. A. 2006. Controls on the basinwide architecture of Messinian evaporites on the Levant margin (Eastern Mediterranean). *Sedimentary Geology*, **188–189**, 93–114.

BISHOP, D. J., BUCHANAN, P. G. & BISHOP, C. J. 1995. Gravity-driven thin-skinned extension above Zechstein Group evaporites in the western central North Sea: an application of computer-aided section restoration techniques. *Marine and Petroleum Geology*, **12**, 115–135.

BROWN, A. 1999. *Interpretation of Three-Dimensional Seismic Data*. American Association of Petroleum Geologists Memoirs, **42**.

BUCHBINDER, B. & ZILBERMAN, E. 1997. Sequence stratigraphy of Miocene–Pliocene carbonate–siliciclastic shelf deposits in the eastern Mediterranean Margin (Israel): effects of eustasy and tectonics. *Sedimentary Geology*, **112**, 7–32.

BUTLER, R. W. H., LICKORISH, W. H., GRASSO, M., PEDLEY, H. M. & RAMBERTI, L. 1995. Tectonics and sequence stratigraphy in Messinian basins, Sicily: constraints on the initiation and termination of the Mediterranean salinity crisis. *Geological Society of America Bulletin*, **107**, 425–439.

CLAUZON, G., SUC, J.-P., GAUTIER, F., BERGER, A. & LOUTRE, M.-F. 1996. Alternate interpretation of the Messinian salinity crisis: controversy resolved? *Geology*, **24**, 363–366.

COLLISON, J. D. 1999. Alluvial sediments. *In*: READING, H. G. (ed.) *Sedimentary Environments: Processes, Facies and Stratigraphy*. Blackwell Science, Oxford, 37–382.

DRUCKMAN, Y., BUCHBINDER, B., MARTINOTTI, G. M., SIMAN TOV, R. & AHARON, P. 1995. The buried Afiq Canyon (Eastern Mediterranean, Israel): a case study of a Tertiary submarine canyon exposed in Late Messinian times. *Marine Geology*, **123**, 167–185.

FREY-MARTINEZ, J., CARTWRIGHT, J., BURGESS, P. M. & VICENTE BRAVO, J. V. 2004. 3D seismic interpretation of the Messinian Unconformity in the Valencia Basin, Spain. *In*: DAVIES, R. J., CARTWRIGHT, J., STEWART, S. A., LAPPIN, M. & UNDERHILL, J. R. (eds) *3D Seismic technology: Application to exploration in sedimentary basins*. Geological Society, London, Memoirs, **29**, 91–100.

FREY-MARTINEZ, J., CARTWRIGHT, J. & HALL, B. 2005. 3D seismic interpretation of slump complexes: examples from the continental margin of Israel. *Basin Research*, **17**, 83–108.

FREY-MARTINEZ, J., CARTWRIGHT, J., HALL, B. & HUUSE, M. 2007. Clastic intrusion at the base of deep-water sands: A trap-forming mechanism in the Eastern Mediterranean. *In*: HURST, A. & CARTWRIGHT, J. (eds) *Sand Injectites: Implications for Hydrocarbon Exploration*. American Association of Petroleum Geologists, Memoirs, **87**, 49–63.

GARFUNKEL, Z. & ALMAGOR, G. 1985. Geology and structure of the continental margin off northern Israel and the adjacent part of the Levantine Basin. *Marine Geology*, **62**, 105–131.

GARFUNKEL, Z. & ALMAGOR, G. 1987. Active salt dome development in the Levant Basin, southeast Mediterranean. *In*: LERCHE, I. & O'BRIEN, J. (eds) *Dynamical Geology of Salt and Related Structures*. Academic Press, London, 263–300.

GLUYAS, J. & SWARBRICK, R. 2004. *Petroleum Geoscience*. Blackwell Science, Oxford.

GRIFFIN, D. L. 2002. Aridity and humidity: two aspects of the late Miocene climate of North Africa and the Mediterranean. *Palaeogeography, Palaeoclimatology, Palaeoecology*, **182**, 65–91.

GUENNOC, P., GORINI, C. & MAUFFRET, A. 2000. Geological history of the Gulf of Lions: mapping the Oligocene–Aquitanian rift and Messinian surface. *Géologie de la France*, **3**, 67–97.

GVIRTZMAN, G. & BUCHBINDER, B. 1978. The late Tertiary of the coastal plain and continental shelf of Israel and its bearing on the history of the eastern Mediterranean. *Initial Reports of the Deep Sea Drilling Project*, **42**. US Government, Printing Office, Washington, DC, 1195–1222.

HODGSON, N. A., FARNSWORTH, J. & FRASER, A. 1992. Salt-related tectonics, sedimentation and hydrocarbon plays in the Central Graben, North Sea, UKCS. *In*: HARDMAN, R. F. P. (ed.) *Exploration Britain: Geological Insights for the next Decade*. Geological Society, London, Special Publications, **67**, 31–63.

HSÜ, K. J., CITA, M. B. & RYAN, W. B. F. 1973. The origin of the Messinian evaporites. *Initial Reports of the Deep Sea Drilling Project*, **13**. US Government Printing Office, Washington, DC, 1203–1231.

HSÜ, K. J., MONTADERT, L. *ET AL*. 1978. *Initial Reports of the Deep sea Drilling Project*, **42**. US Government, Printing Office, Washington, DC.

KENDALL, A. C. & HARWOOD, G. M. 1996. Marine evaporites: arid shorelines and basins. *In*: READING, H. G. (ed.) *Sedimentary Environments: Processes, Facies and Stratigraphy*, 3rd edn, Blackwell Scientific, Oxford, 281–324.

KRIJGSMAN, W., HILGEN, F. J., RAFFI, I. & SIERRO, F. J. 1999. Chronology, causes and progression of the Messinian salinity crisis. *Nature*, **400**, 652–654.

LOFI, J., GORINI, C., BERNE, S., CLAUZON, G., DOS REIS, A. T., RYAN, W. B. F. & STECKLER, M. S. 2005. Erosional processes and paleo-environmental changes in the western Gulf of Lions (SW France) during the Messinian Salinity Crisis. *Marine Geology*, **217**, 1–30.

LONCKE, L. 2002. Le delta profound du Nil: structure et évolution depuis le messinien (Miocène terminal): Thèse de l'université Paris, 6, 184pp.

MANZI, V., LUGLI, S., RICCI LUCCHI, F. & ROVERI, M. 2005. Deep-water clastic evaporites deposition in the Messinian Adriatic foredeep (northern Appennines, Italy): did the Mediterranean ever dry out? *Sedimentology*, 52, 875–902.

MART, Y. & BEN GAI, MART, Y. 1982. Some Depositional Patterns at Continental Margin of Southeastern Mediterranean Sea. American Association of Petroleum Geologists, Bulletins, 66, 460–470.

MARTINEZ DEL OLMO, W. 1996. Depositional sequences in the Gulf of Valencia Tertiary basin. *In*: FRIEND, P. F. & DABRIO, C. J. (eds) *Tertiary Basins of Spain: The Stratigraphic Record of Crustal Kinematics*. World Regional Geology, 6, 55–67.

MITCHUM, R. M. 1985. Seismic stratigraphic recognition of submarine fans. *In*: BERG, O. R. & WOOLVERTON, D. G. (eds) *Seismic Stratigraphy II*. American Association of Petroleum Geologists, Memoirs, 39, 117–136.

NEEV, D. 1979. Deep-water gypsum deposits as indicated by the Neogene geological history of the central coastal plain of Israel. *Sedimentary Geology*, 23, 127–136.

PERYT, T. M. 2000. Resedimentation of basin centre sulphate deposits: Middle Miocene Badenian of Carpatian Foredeep, Southern Poland. *Sedimentary Geology*, 134, 331–342.

POSAMENTIER, H. W. 2003 Seismic geomorphology; new tricks for an old dog. *Reservoir*, 30, 18–19, 28, 30.

POSAMENTIER, H. W. & ERSKINE, R. D. 1991. Seismic expression and recognition of ancient submarine fans. *In*: WEIMER, P. & LINK, M. H. (eds) *Seismic Facies and Sedimentary Processes of Submarine Fans and Turbidite Systems*. Frontiers in Sedimentary Geology. Springer, New York, 197–222.

POSAMENTIER, H. W. & VAIL, P. R. 1988. Eustatic controls on clastic deposition II – Sequence and systems tract models. *In*: WILGUS, C. K., HASTINGS, B. K., POSAMENTIER, H., VAN WAGONER, J., ROSS, C. A. & KENDALL, C. G. ST. C. (eds) *Sea Level Change – an Integrated Approach*. Society of Economic Paleontologists and Mineralogists, Special Publications, 42, 125–154.

RATTEY, R. P. & HAYWARD, A. B. 1993. Sequence stratigraphy of a failed rift system: the Middle Jurassic to Early Cretaceous basin evolution of the Central and Northern North Sea. *In*: PARKER, J. R. (ed.) *Petroleum Geology of Northwest Europe: Proceedings of the 4th Conference*. Geological Society, London, 215–250.

READING, H. G. 1999. *Sedimentary Environments: Processes, Facies and Stratigraphy*. Blackwell Science, Oxford.

READING, H. G. & RICHARDS, M. T. 1994. The classification of deep-water siliciclastic depositional systems by grain size and feeder systems. *American Association of Petroleum Geologists Bulletin*, 78, 798–822.

RICCHI LUCCHI, F. 1973. Resedimented evaporites: Indicators of slope instability and deep basin conditions in Periadriatic Messinian. *In*: DROOGER, C. W. (ed.) *Messinian Events in the Mediterranean*. Amsterdam, North Holland, 136–144.

RIDING, R., BRAGA, J. C., MARTIN, J. M. & SANCHEZ ALMAZO, I. M. 1998. Mediterranean Salinity Crisis: constraints from a coeval marginal basin, Sorbas, southeastern Spain. *Marine Geology*, 146, 1–20.

ROUCHY, J. M. 1982. La genèse des évaporites messiniennes de Méditerranée. *Mémoires du Muséum National d'Histoire Naturelle, Paris*, 50.

ROVERI, M., BASSETTI, M. A. & RICCI LUCCHI, F. 2001. The Mediterranean Messinian Salinity Crisis: an Apennine foredeep perspective. *Sedimentary Geology*, 140, 201–214.

ROVERI, M., MANZI, V., RICCI LUCCHI, F. & ROGLEDI, S. 2003. Sedimentary and tectonic evolution of the Vena del Gesso basin (Northern Apennines, Italy): implications for the onset of the Messinian salinity crisis. *Geological Society of America Bulletin*, 115, 387–405.

RYAN, W. B. F. 1978. Messinian badlands on the southeastern margin of the Mediterranean Sea. *Marine Geology*, 27, 349–363.

SAGE, L. & LETOUZEY, J. 1990. Convergence of the African and Eurasian plate in the eastern Mediterranean. *In*: LETOUZEY, J. (ed.) *Petroleum and Tectonics in Mobile Belts*, Éditions Technips, Paris, 49–68.

SAVOYE, B. & PIPER, D. J. W. 1991. The Messinian event on the margin of the Mediterranean Sea in the Nice area, southern France. *Marine Geology*, 97, 279–304.

SCHLAGER, W. & BOLTZ, H. 1977. Clastic accumulation of sulphate evaporites in deep water. *Journal of Sedimentary Petrology*, 47, 600–609.

SCHREIBER, B. C. 1988. Subaqueous evaporite deposition. *In*: SCHREIBER, B. C. (ed.) *Evaporites and Hydrocarbons*. Columbia University Press, New York, 182–255.

SCHREIBER, B. C., FRIEDMAN, G. M., DECIMA, A. & SCHREIBER, E. 1976. Depositional environments of Upper Miocene (Messinian) evaporite deposits of the Sicilian Basin. *Sedimentology*, 23, 729–760.

STAMPFLI, G. M. & HÖCKER, C. F. W. 1989. Messinian paleorelief from a 3-D seismic survey in the Tarraco concession area (Spanish Mediterranean Sea). *Geologie Mijnbouw*, 68, 201–211.

STOW, D. A. V. 1986. Deep clastic seas. *In*: READING, H. G. (ed.) *Sedimentary Environments: Processes, Facies and Stratigraphy*, 3rd edn. Blackwell Science, Oxford, 399–444.

TAYLOR, J. C. M. 1998. Upper Permian–Zechstein. *In*: GLENNIE, K. W. (ed.) *Petroleum Geology of the North Sea – Basic Concepts and Recent Advances*, 4th edn. Blackwell Science, Oxford, 174–211.

TIBOR, G., BEN-AVRAHAM, Z., STECKLER, M. & FLIGELMAN, H. 1992. Late Tertiary subsidence history of the Southern Levant Margin, Eastern Mediterranean Sea, and its implications to the understanding of the Messinian event. *Journal of Geophysical Research*, 97, 17593–17614.

VAI, G. B. & RICCI LUCCHI, F. 1977. Algal crusts allochtonous and clastic gypsum in a cannibalistic evaporite basin: a case history from the Messinian of the Northern Appennines. *Sedimentology*, 24, 211–244.

WEIMER, P. & LINK, M. H. 1991. *Seismic Facies and Sedimentary Processes of Submarine Fans and Turbidite Systems*. Frontiers in Sedimentary Geology, Springer, New York.

WEIMER, P. & SLATT, R. M. 2004. Deepwater reservoir elements. *In*: *Petroleum Systems of Deepwater Settings*. Distinguished Instructor Short Course, Distinguished Instructor Series (sponsored by SEG and EAGE), 7, 6–1.

Basin analysis of the Burdigalian and Early Langhian successions, Kirkuk Basin, Iraq

A. I. AL-JUBOURY[1], A. M. AL-TARIF[2] & M. AL-EISA[3]

[1]*Research Center for Dams and Water Resources, Mosul University, Mosul, Iraq*
(e-mail: alialjubory@yahoo.com)
[2]*Applied Geology Department, Tikrit University, Tikrit, Iraq*
[3]*Northern Oil Company, Kirkuk, Iraq*

Abstract: Two depositional basins were present in the Kirkuk Basin of northeastern Iraq during the Burdigalian and Early Langhian ages (Late Lower–Early Middle Miocene) and have been studied at several oilfields. Both basins display shallowing upward sequences composed of carbonate and evaporite sediments. Evaporites form part of the sedimentary sequences that begin with normal open marine and shelf deposits and pass upward into thickly bedded and massive- nodular anhydrite that alternates with thinly bedded limestone (dolomitized with anhydrite nodules), and that is unconformably overlaid by massive bedded lagoonal limestone. The first basin of (Burdigalian age) is represented by the Serikagni, Euphrates and Dhiban formations, whereas the second of Langhian age includes the Jeribi and Fat'ha formations. The early part of the Burdigalian age is characterized by a marine transgression that covered a large part of the region, resulting in a broad spectrum of marine environments ranging from the open marine in the uppermost part of the gentle slope environments, evidenced in the Serikagni Formation, to the shallower environments of fore- and back-shoals of the Euphrates Formation. Regression took place at the end of this period and resulted in the deposition of the Dhiban Anhydrite Formation. The basin of deposition of the Dhiban Formation is an inherited basin from the older Serikagni and Euphrates basins. This is reflected in an inhomogeneous distribution, both of type and thickness, of evaporites. The Early Langhian age of the Kirkuk Basin is also characterized by a shallowing-upward sequence begun by sediments rich in planktonic foraminifera for the lower part of the Jeribe Formation and then shallow water and lagoonal carbonates for the upper part of the Jeribe Formation. The environment passed upwards into the evaporitic sequence of the Fat'ha Formation.

This paper is the product of a subsurface study that attempts to characterize the physical setting and environmental conditions in Kirkuk area, northeastern Iraq, during the Late Lower Miocene (Burdigalian) and the early part of the Middle Miocene (Langhian). Two traverses were chosen in the area of study; the first begins from the Miocene shoreline and passes toward the deep environment and is represented by the wells K-317, Ja-30, Ja-32, Pu-5, Qm-1 and Nk-35. The other traverse is parallel to the first traverse and represented by the wells Hr-2, In-5 and Mn-1 (Fig. 1).

The Lower Miocene can be divided into two secondary sedimentary sequences; the first is the Aquitanian sequence that includes the Ibrahim, Azkand, Anah and Hamrin formations, and the second is the Burdigalian sequence that is represented by the Serikagni, Euphrates and Dhiban formations (Al-Eisa 1992*a*). The ensuing Middle Miocene is represented by the Langhian sequence that includes the Jeribe and Fat'ha formations. The Fat'ha Formation was not studied in detail in this paper due to the limitations on information released by the Northern Oil Company, Iraq. The basic stratigraphy of the region during the Miocene is outlined in Figure 2.

This study is primarily focussed on petrographic and facies analyses of the available cores from the studied formations. This focus was made in order to understand the sedimentary environments employed in the basin analysis relative to the data deduced from the associated well logs (Sonic, GR & FDC/CNL) and by comparing the studied lithological sections with the type sections of the studied formations as described by Bellen *et al.* (1959).

The present study aims to address the basin analysis of the Burdigalian and the lower Langhian sequences in one of the economically important basins in Iraq and the Middle East.

Geological setting

The stratigraphy of Iraq is strongly affected by the structural position of the country within the main geostructural units of the Middle East region as well as by the structure within Iraq. Iraq lies in

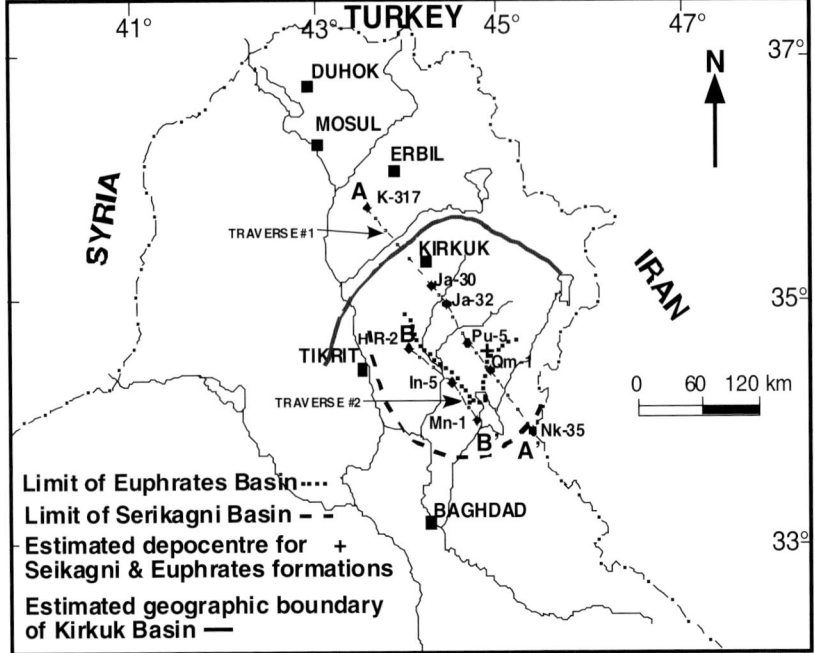

Fig. 1. Map showing the well distribution, and cross sections of the studied traverses.

the border area between the major Phanerozoic units of the Middle East, i.e. between the Arabian part of the African Platform (Nubio-Arabian) and the Asian branches of the Alpine tectonic belt. The platform part of the Iraqi territory is divided into two basic units, i.e. a stable shelf and an unstable shelf. The stable shelf is characterized by a relatively thin sedimentary cover and the lack of significant folding. The unstable shelf has a thick and folded sedimentary cover and the intensity of the folding increases toward the northeast (Buday 1980).

Structurally, the area of study lies in the Foothill and the Mesopotamian zone of the quasiplatform foreland of Iraq, as illustrated in Figure 3 (from Numan 2001). The continental plate collision between the underthrust Arabian Plate and the overriding Turkish and Iranian plates caused the evolution of the structure and deposition in the area of the unstable shelf and the formation of a marginal

Series	Stages	Formations	Sequence		
Miocene	Langhian	Fat'ha Jerlbe			SB
	Burdigalian	Dhiban Euphrates Serikagni	HST TST HST	SB	
		Hamrin	TST	SB	
	Aquitanian	Anah Azkand Ibrahim	HST TST	SB	
Oligocene	Chattian				

Fig. 2. Basic stratigraphy of the Kirkuk Basin during the Miocene, illustrating the main depositional sequences formed during the Aquitanian and Burdigalian. SB marks the sequence boundaries. TST, Transgressive System Tract; HST, Highstand System Tract; and the dashed lines (right-hand column) are the contacts between the TST and the HST systems.

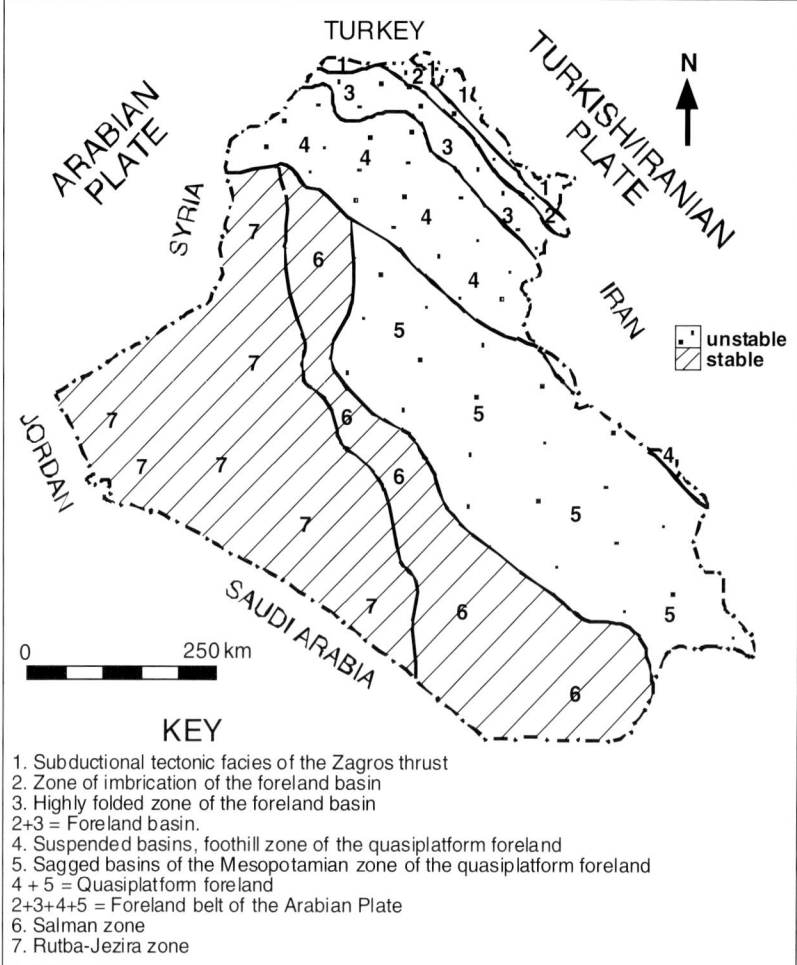

Fig. 3. Tectonic divisions of Iraq, after Numan (2001).

basin. It also caused the creation of accommodation space for rapid sediment accumulation during the Miocene (Numan 1997).

Basin configuration of the Burdigalian period

The upper Lower Miocene sediments (Burdigalian) can be regarded as a portion of a type cycle in the area of study because it represents an ideal shallowing-upward facies succession. The deposit begins with a marine transgression, represented by the deposition of Serikagni Formation in the deeper marine areas of the basin and the Euphrates Formation in the shallower areas. Both are overlain by the Dhiban evaporites in the upper part of the sequence.

Isopach maps of the Serikagni and Euphrates formations

The isopach maps of these formations are studied together due to the interfingering of the two formations (Serikagni and Euphrates), as suggested by Al-Naqib (1960) and Buday (1980). The Serikagni Formation is thickest in the area of the Qm-1, Hr-2, In-5 and Mn-1 wells of the second traverse (e.g. 40 m at Hr-2) and thins towards the Ja-32 and Ja-30 wells of the first traverse (Fig. 4a). The maximum thickness (50 m) of the Euphrates Formation is concentrated in the area of wells Ja-32 and Ja-30 and disappears toward well Pu-5 with no indication of the deposition of the Euphrates in the other studied well sites (Fig. 4b). Therefore, the rocks of these two

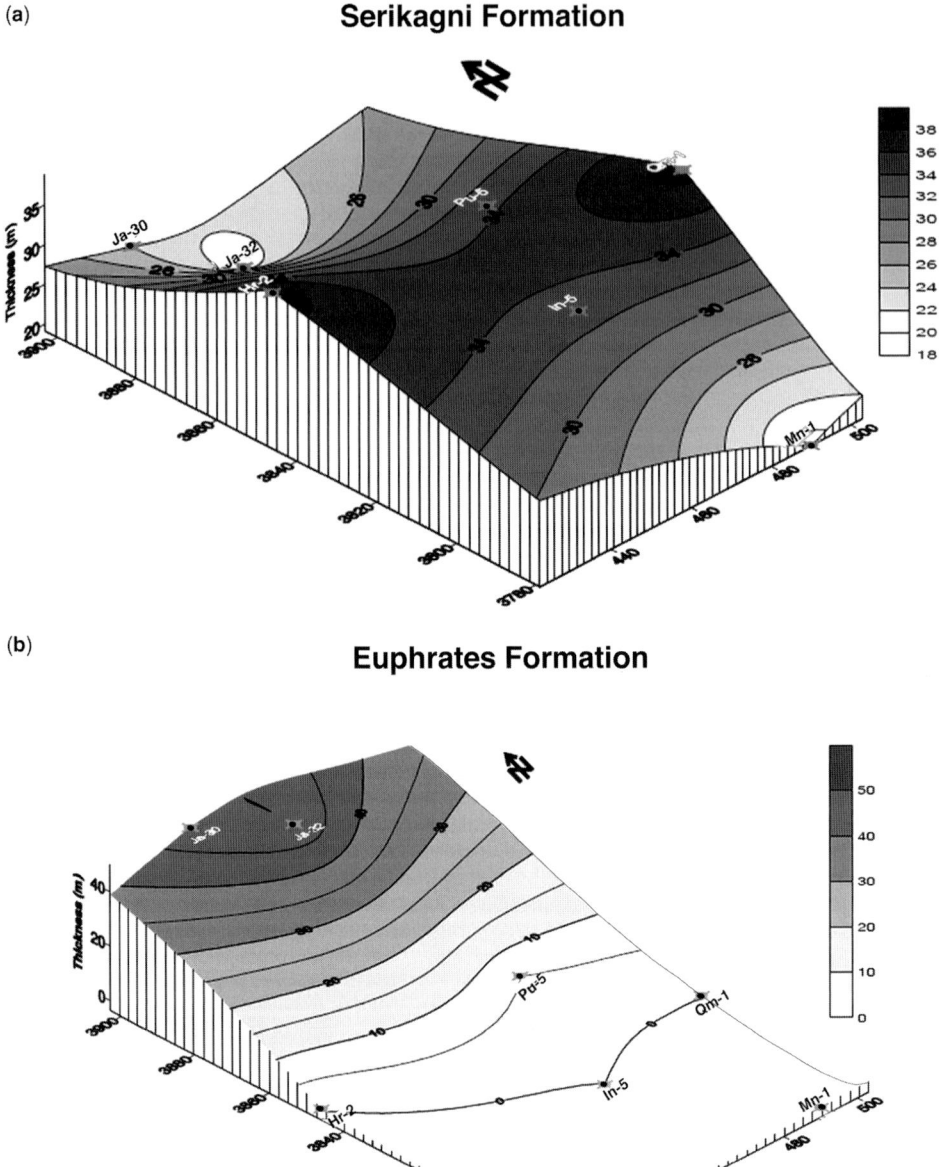

Fig. 4. Isopach map of the (a) Serikagni Formation and (b) Euphrates Formation.

formations were distributed in two distinct areas; the first represents the deep basinal area where there is the greatest thickness of the Serikagni Formation and the second (the Euphrates Formation) is thinner.

Facies analysis of the Serikagni and Euphrates Formations

Biounits used in this basin analysis are based on observation of the biocomponents (both floral and

Table 1. *Faunal/Floral assemblages of the Burdigalian and early Langhian Formations Kirkuk basin, Iraq*

Formation	Lithology	Fauna/Flora
Jeribe	Calcic dolomite	Echinoderms, Ostracods, Bryozoans, Gastropods, Pelecypods, Red algae
	Marly limestone	Planktonic foraminifera (*Orbulina*) Benthonic foraminifera (*Dentritina rengi*, *Peneroplis*, *Borelis melo curdica*, Rotaliid, *Favrina*)
Dhiban	Dolomitic limestone (intercalations)	Rotalids, Miliolids, *Borelis melo curdica* Debris of pelecypods, gastropods, echino-derm spines, ostracods and green algae
Euphrates	Dolomitic limestone	Benthonic foraminifera (*Operculina*, *Borelis melo*, Miliolidae, Peneroplidae, Rotalidae, *Ammonia*) Mollusca, Bryozoan, Echinoderms, Algae, Ostracods
Serikagni upper part	Limestone and marl	Benthonic foraminifera (*Operculina*) Mollusca (pelecypods and gastropods), Bryozoans, Red algae
lower part		Planktonic foraminifera (*Globigerinoides diminutus* Bolli, *Globigerinoides parawoodi* Keller, *Globigerinoides quadrilobatus primordius* Blow & Banner, *Globigerinoides trilobus sacculifera* Brady, *Globigerinoides trilobus bullatus* Chang, *Globigerinoides immaturus* Le Roy Kadar, *Praeorbulina transitoria* Blow, *Catapsydrax stainforthi* Bolli, Loeblich & Tappan)

faunal). Based on these assemblages we can distinguish the deeper regions from the shallower areas of the basin, which in turn reflect the sedimentary environments of the studied formations (Table 1).

The Euphrates Formation comprises mainly of one unit, named the Euphrates fossiliferous unit, that is composed mainly of a fossiliferous limey packstone rich in benthonic foraminifera together with shells of molluscs (mainly gastropods and cephalopods), algae and non-skeletal components such as oolites and pellets (Table 1 and Fig. 5). The distribution of this unit is concentrated in the wells of the traverse, A–A', and disappears towards B–B'. The presence of benthonic foraminifera with fewer red algae and echinoderms may reflect the beach depositional system, whereas the presence of higher amounts of these fossils may

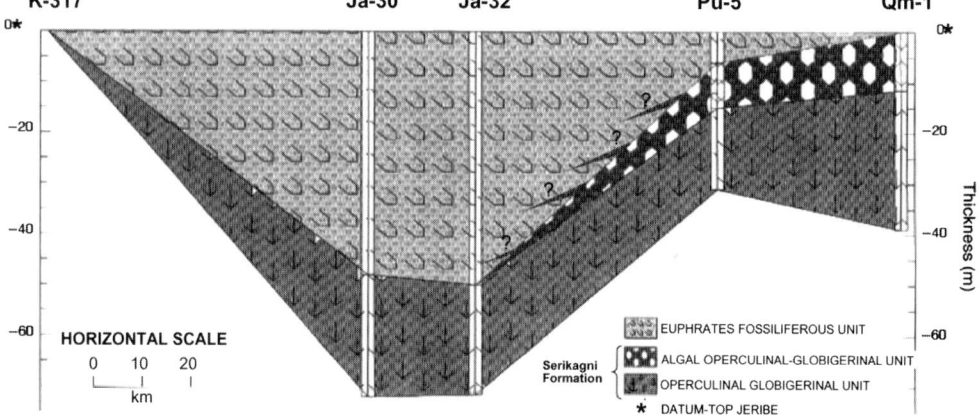

Fig. 5. Biounit distribution of the Serikagni and the Euphrates formations in the first traverse (NW–SE).

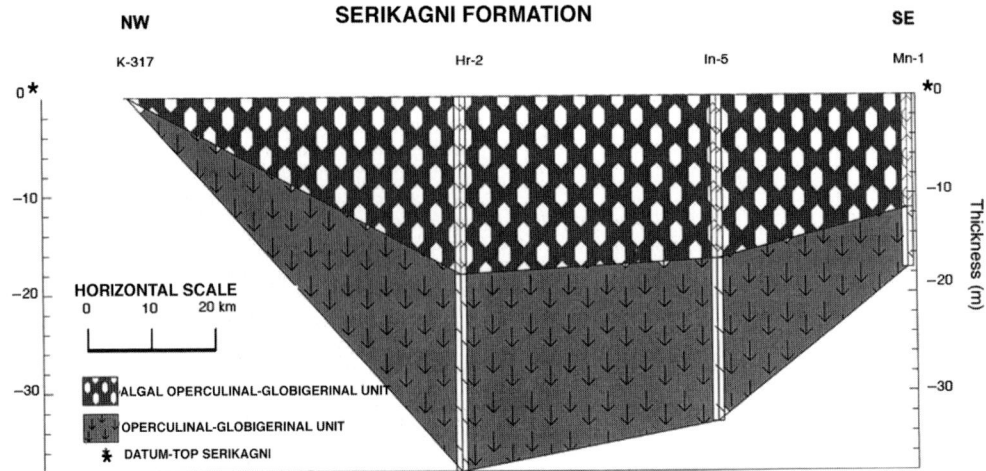

Fig. 6. Biounit distribution of the Serikagni Formation in the second traverse (NW–SE).

indicate the shoal depositional system (Abawi 1989). The different environmental conditions affecting the organisms in the Euphrates Formation indicate that this formation was deposited in a spectrum of sedimentary marine environments ranging from the outer- to inner-shelf as shoals, open and semi-restricted, back lagoon shoals.

Two biounits are observed in the Serikagni Formation; the first is composed of both large benthonic and planktonic foraminifera and red algae (an algal–operculinal–globigerinal unit) in the upper part of the formation, which is not present in the wells Ja-32 and Ja-30 that contain most of the Euphrates biounits (Fig. 6). The other biounit is composed of large benthonic and planktonic foraminifera (operculinal–globigerinal unit)

that is distributed throughout the entire formation and is even better developed in the wells along traverse B–B'. The vertical and horizontal distribution of the biounits in both formations is shown in Figure 7 and indicates that the Euphrates Formation thins and then disappears toward south and southeastern portion of the basin. The areas of non-deposition of the Euphrates Formation received the deep basinal deposits that make up the Serikagni Formation. Uplift of the area and/or isolation from the open sea records the end for both the Euphrates and Serikagni depositions and the beginning of the deposition of the Dhiban evaporates as a result of evaporation of waters within these restricted and semirestricted areas. The presence of the large benthonic foraminifera (*Operculina*) together with

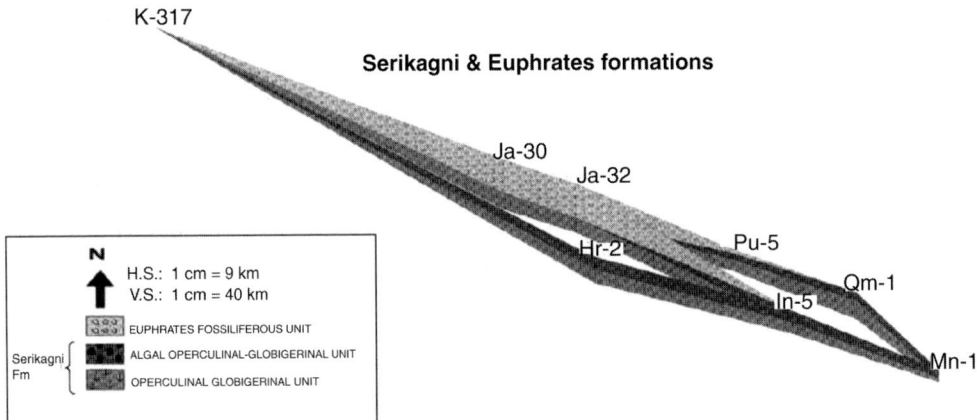

Fig. 7. Panel diagram showing the biounit distribution of the Serikagni and the Euphrates formations in the first and second traverses.

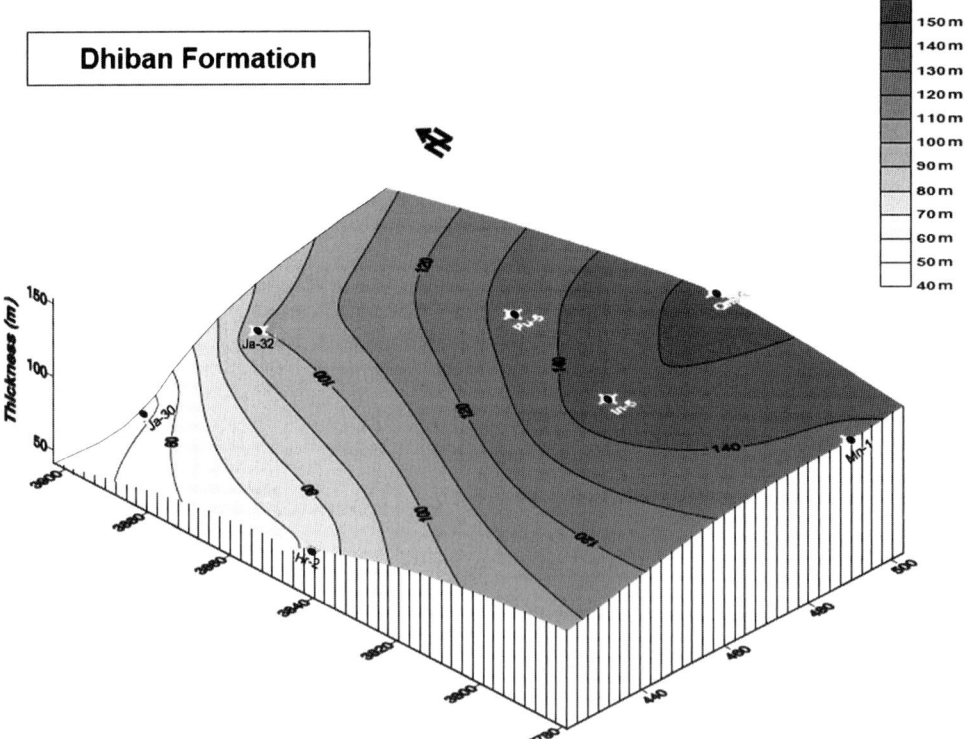

Fig. 8. Isopach map of Dhiban Formation.

planktonic foraminifera and algae may reflect the deeper environment of the Serikagni Formation that lies in the uppermost parts of the continental slope (Boersma 1978).

Isopach map of the Dhiban Formation

The greatest thickness of Dhiban Formation is recorded in wells Qm-1, Inj-5, Mn-1 and Pu-5 (see Fig. 8) and it thins towards wells Ja-32, Ja-30 and Hr-2 (Fig. 8). This distribution indicates that the Dhiban Formation has an inherited nature following the same general trend as the lower part of the underlying Burdigalian sequence. A greater thickness of the Serikagni Formation is present in these wells, with a deeper environment of deposition where there is greater accommodation space, leading to the accumulation and preservation of a great thickness of evaporites and composed mainly of thick gypsum, anhydrite and rich in halite when the conditions were suitable for the precipitation of such evaporites. The reduced thickness of the Dhiban Formation in the areas of the earlier deposition of the Euphrates Formation reflects the shallow nature of this deposition with less accommodation space, and halite is thin and evaporites are mainly gypsum and anhydrite.

The evaporites of the Dhiban Formation alternate with greyish brown dolomitic limestones containing displacive anhydrite nodules. Anhydrite forms most of the evaporite rocks with some residual gypsum (most of the sulphate was altered to anhydrite) and halite (Fig. 9a & b). The salt (halite) is identified by gamma and FDC/CNL logs and is 60 m thick, as recorded in the Dhiban Formation from the well Qm-1 (Fig. 9a).

The presence of marine organisms in the limestone beds that alternate with evaporites in the Dhiban Formation indicates the re-establishment and continuation of restricted marine environmental conditions between evaporative phases. Generally the intercalated limestone beds within the evaporites are dolomitic and are highly affected by anhydrization and dissolution; therefore, only ghosts of the fossils were preserved. The observed fauna include benthic foraminifera *Borelis melo curdica* and relics of pelecypods and gastropods as well as echinoderms and ostracods (Table 1). The limestones are thinly bedded in the lower unit of the Dhiban Formation including oriented parallel anhydrite nodules (Fig. 10), while others are randomly

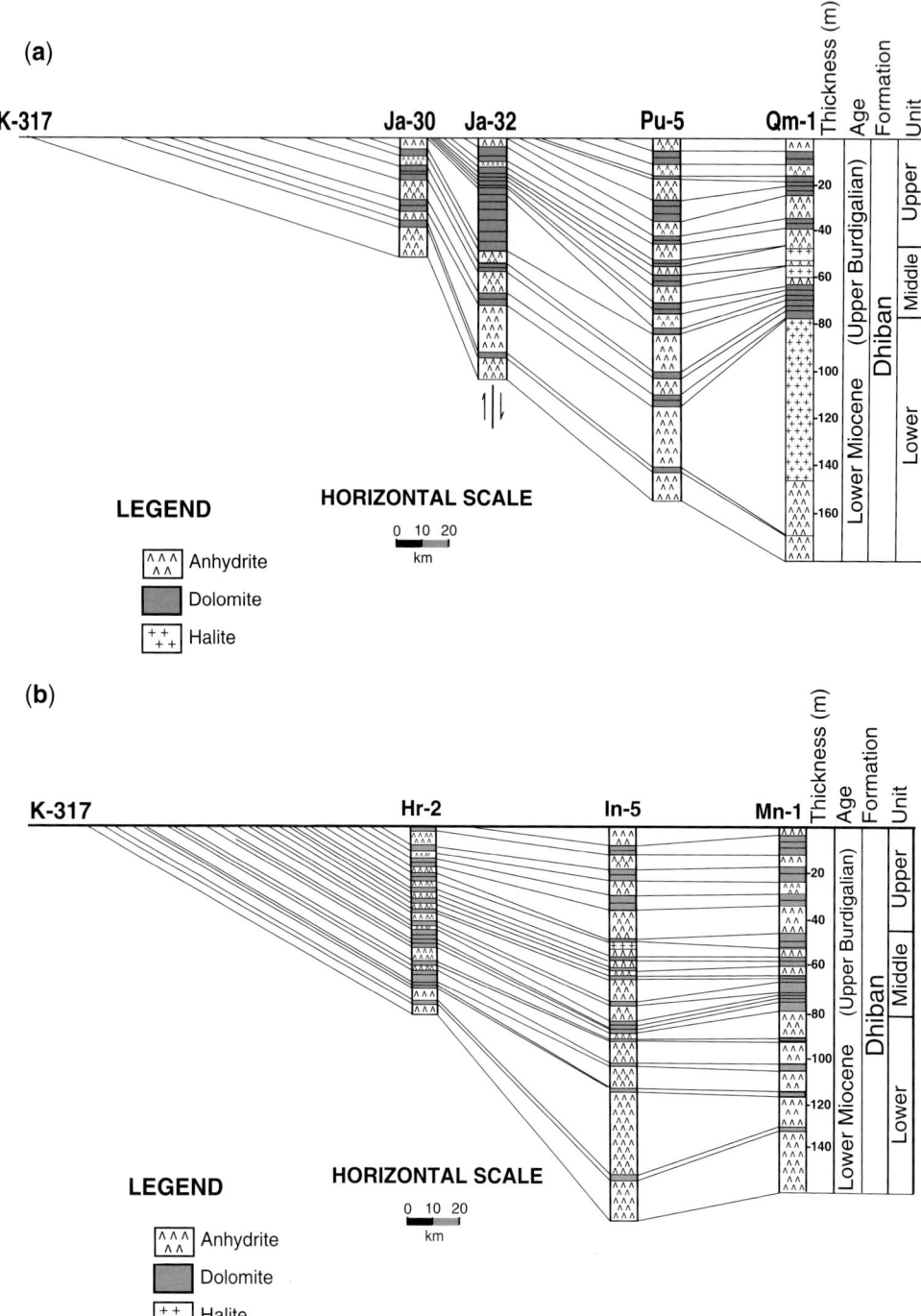

Fig. 9. Lateral correlation of the Dhiban Formation in the first and second traverses (**a** & **b**) respectively. Note the symbol below well Ja-32 refers to the reverse faults affecting the area.

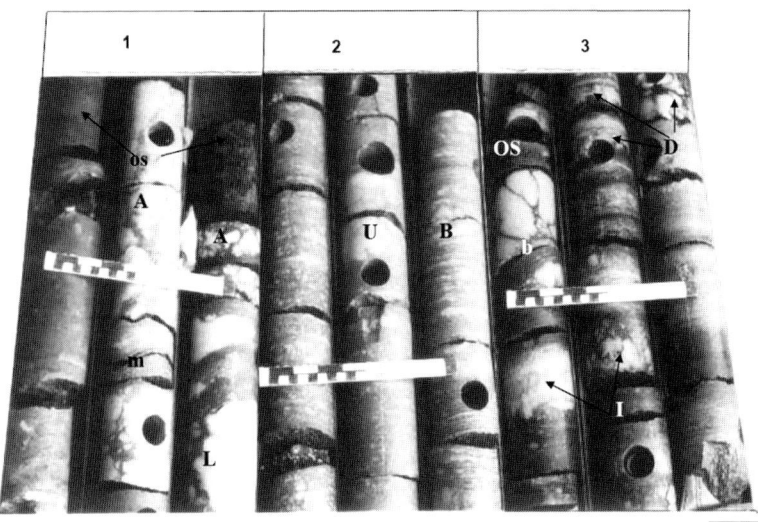

Fig. 10. Different structures of the Dhiban Formation, well Hr-2. (**1**) Anhydrite containing gathered mosaic nodules (A), single large nodules (L), algal mats (m) and oil shows (os). (**2**) Alternation of grey limestone with anhydrite containing mosaic nodules arranged in lines (B) and single mosaic nodules (U). (**3**) Alternation of bedded limestone (b) containing oil shows (OS) with anhydrite showing enterolithic structure (I) and accumulation of bedded mosaic nodules (D). Scale ruler = 20 cm.

oriented with mosaic shape. Local restriction of secondary basins as a result of the tectonic uplift and higher evaporation rates led to the deposition of these evaporites. The longer-term restriction of these basins with higher evaporation rates also may have influenced the precipitation of salts (McCaffry et al. 1987; in Kendall 1992).

Evaporites with no halite accumulation may also precipitate in semi-arid aqueous areas such as lagoons (Schmalz 1969). As a result we find both complete and incomplete depositional series in the Dhiban Formation. The non-deposition of halite in an incomplete series may be related to the invasion of the open sea waters into the restricted area, affecting the concentration of various ions in these restricted areas (Abu-Khadra et al. 1992). It seems, from the nature of evaporite deposition, that successions of thick evaporite beds in the lower unit of the Dhiban Formation (Fig. 9a & b) are characteristic of sub-aqueous deposition with longer term tectonic stability which are represented here by gypsum, anhydrite and halite, whereas thin evaporites of gypsum and anhydrite in the middle and upper units of the formation may be related to sub-aerial, sabkha deposition and/or erosion (Wali & Tabakh 1992; Schreiber & El Tabakh 2000).

Relative humidity may form another factor controlling the presence or absence of halite. Kinsman (1976) pointed out that halite may precipitate but is not preserved in regions of high humidity, but we did not find any relics of dissolved halite in the middle and upper parts of the Dhiban Formation.

Beydoun (1988) and Warren (1989) have commented that all the Tethyan basins of the Middle East were affected by two separate drying phases in the middle Miocene as a result of the tectonic uplift along the Levant Fault System that led to the isolation of the eastern arm of the Tethyan Basin system. In the present study we can consider the Dhiban Anhydrite Formation as the second phase of drying followed by yet a third phase of the evaporite accumulation in the Fat'ha Formation of the Middle Miocene.

The effect of the regional compression and tectonic uplift in the Miocene led to the formation of restricted and semi restricted basins coupled with increased evaporation due to aridity that caused the evaporite deposition in an inherited basin from the older Euphrates and Serikagni Formations (early Burdigalian period). The continuation of tectonic activity throughout the area led to nearly complete isolation of the basin from the open seas (Beydoun 1991) and the Dhiban evaporites were deposited as a result of high evaporation during the upper Burdigalian period. High rates of evaporation probably were the main reason for the water deficit and for chemical sedimentation of Dhiban evaporites from marine water in interconnected or temporarily disconnected sub basins, similar to those discussed for the Badenian evaporite basin of the northern Carpathian Foredeep (Bąbel 2004).

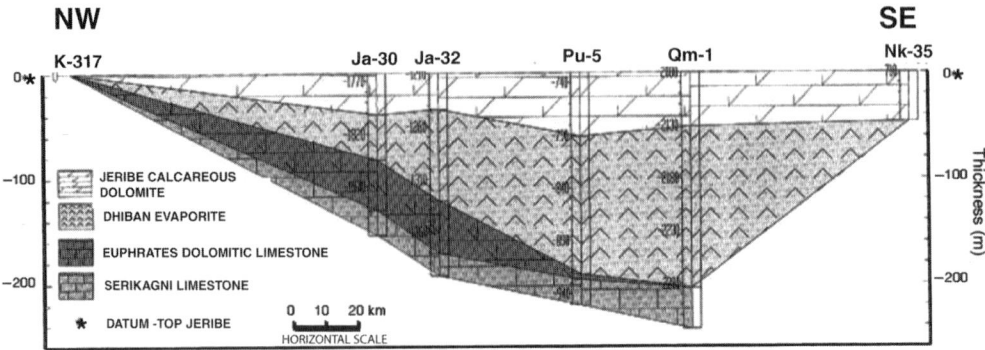

Fig. 11. Stratigraphic correlation of the studied formations in the first traverse (NW–SE).

The inherited nature of the basin resulted in deposits of thin evaporites in the area of wells Hr-2, Ja-30 and Ja-32. These evaporitic areas are sporadically penetrated by marine water in the shallow areas of the Euphrates Formation where the thin evaporites, composed of sulphates and halite, are intercalated with dolomitic limestones and dolomites in the middle and upper units of the Dhiban Formation. This phase of deposition generally had less isolation as compared with the deeper areas of the older Serikagni Formation, represented in the present study by wells Nk-35, Qm-1, Pu-5, In-5 and Mn-1. There the basin was deeper, with greater accommodation space, and thick evaporites (including halite) were deposited alternating with thin limestones in the lower unit of the Dhiban Formation.

The thickness of the Dhiban Formation increases towards the southeast, especially in wells Pu-5, Qm-1, In-5 and Mn-1. The Dhiban has a greater thickness (173 m) as recorded in well Qm-1 but is thinner (43 m) in well Ja-30 (Figs 11 & 12). Thick beds of anhydrite, up to 30 m, are present in the lower unit of the formation (Fig. 9a), which is composed mainly of nodular anhydrite. The nodules size range from 1 to 10 cm with oriented distribution in parallel beds that alternate with thin laminae of dolomitic limestones or randomly distributed beds displaying nodular mosaic texture.

The thickness of the evaporite beds in this unit is up to 30 m for anhydrite and 60 m for halite, whereas in the middle unit the beds are thinner (15 m for the dolomitic limestones and 6–10 m for the anhydrite and gypsum) (well Qm-1, e.g. Fig. 9). The upper unit of the formation is constituted by thick anhydrite mainly having a nodular mosaic texture (Fig. 10).

The difference in thickness and nature of evaporites between the wells Ja-30, Ja-32 and Hr-2 from those lying to the east (i.e. Pu-5 and Qm-1 in traverse A–A′ and In-5 and Mn-1 in traverse B–B′) may relate to the continuation of the tectonic effect that is represented by two reverse faults in the area and led to uplift of the area and formed the horst structure visible in Figure 13. These faults were first noted in unpublished reports in the Northern Oil Company (2002) and are parallel growth faults contemporaneous with sedimentation. These structures exhibited sporadic movement from the Triassic through the Cenozoic.

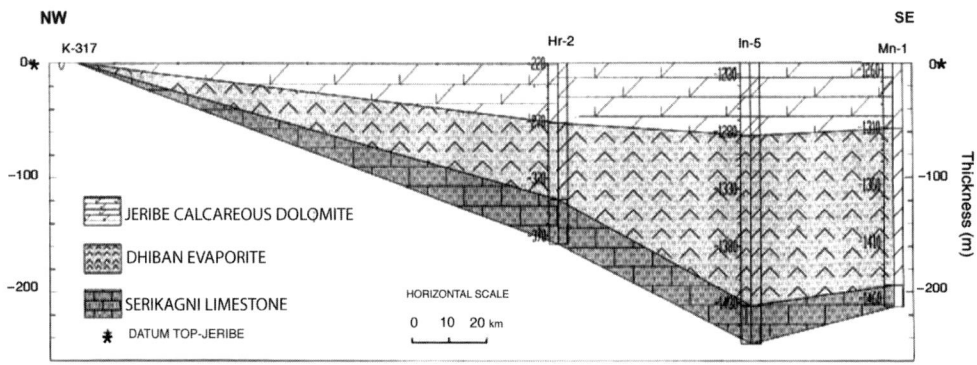

Fig. 12. Stratigraphic correlation of the studied formations in the second traverse (NE–SE).

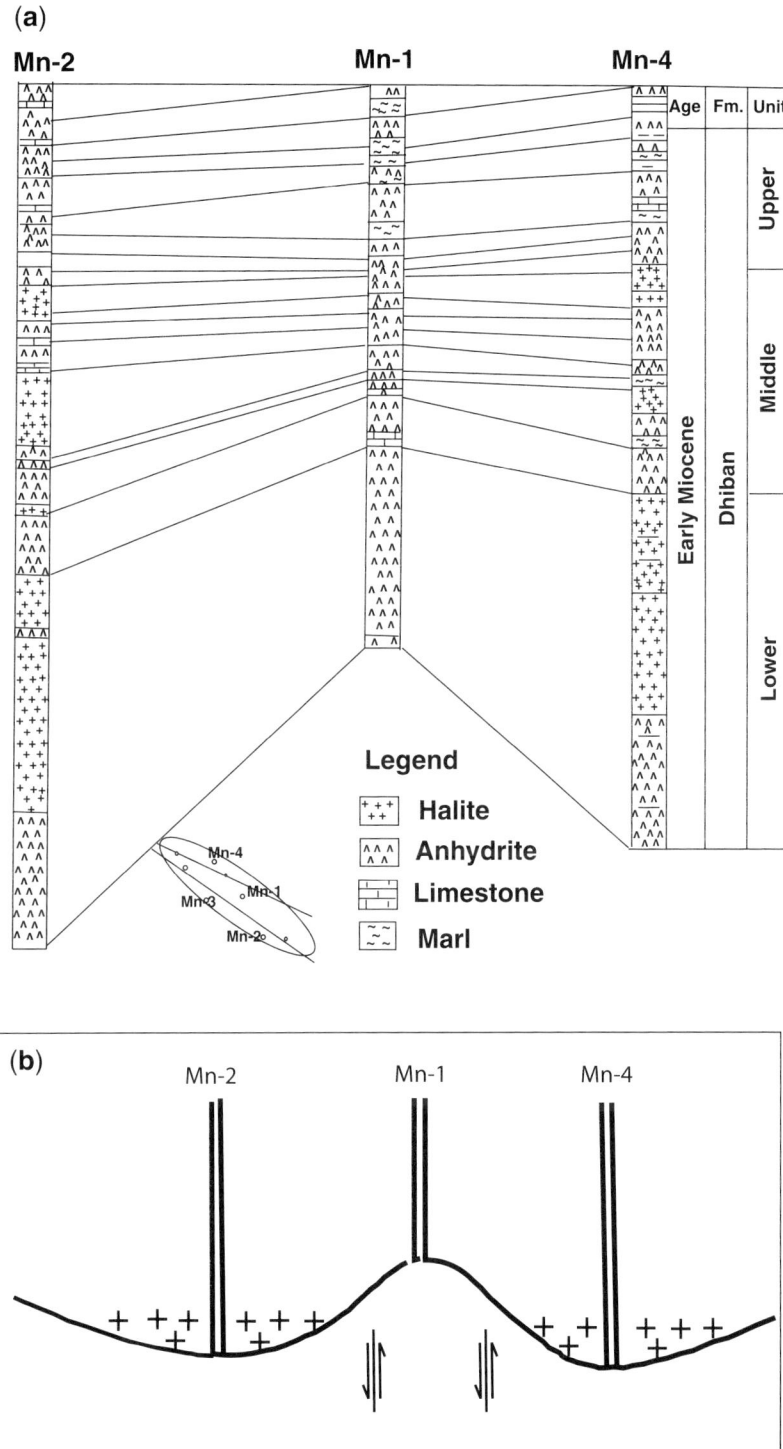

Fig. 13. Nature and configuration of the basin of the Dhiban deposition in the Mansuriya oilfield. (a) Effect of the parallel faults in uplift of the depositional surface of the Dhiban Formation. (b) The horst structure resulting from these faults, unpublished study (Northern Oil Company 2002).

In the lower unit of the Dhiban Formation thick anhydrite beds alternate with thinly bedded limestones. The difference in the depth of the basin is illustrated in Figure 14, and is the result of the variation in accommodation space available for the accumulation of evaporites within the basin of deposition of the Dhiban Formation. In the lower unit this initial depth difference is clearly observed, with limited positive areas, whereas in the middle unit the depths are fairly similar throughout the area and finally attain a fairly uniform basin depth in the deposition of the upper unit, which prograde across the earlier positive areas in the region. These faults controlled the position and formation of the oolitic shoal in the Euphrates Formation (Fig. 14) and reactivation of the faults influenced the evolving differences in basinal depth d noted in the Euphrates Formation rocks, and affect the build-up in the positive area noted in the Jampur oilfield wells Ja-30 and Ja-32.

Generally the lower contact of the Dhiban Formation is unconformable (marked by sharp and clear contacts and represented by the presence of anhydrite beds over the marine carbonates) with the Serikagni Formation but is conformable with the Euphrates Formation, whereas the upper contact of the Dhiban is always lithologically unconformable with the overlying Jeribe Formation.

Basin configuration of the Early Langhian stage

The nature and configuration of the sedimentary basin during the lower part of Middle Miocene in the area of study reflects its inherited nature from the earlier Burdigalian basin and was affected by the same structural and environmental conditions that marked the older sequence. This basin was formed after the transgressive phase that followed the second desiccation period (a regression that was responsible for the deposition of Dhiban Anhydrite Formation at the end of the earlier sequence). This transgression resulted in the formation of a spectrum of marine environments ranging from nearly open marine conditions for the Jeribe Formation that is composed of marly limestone (Table 1), rich in planktonic foraminifera, in the lower part of the formation passing upward into shallow marine limestones with benthonic foraminifera. This wide environmental range may be

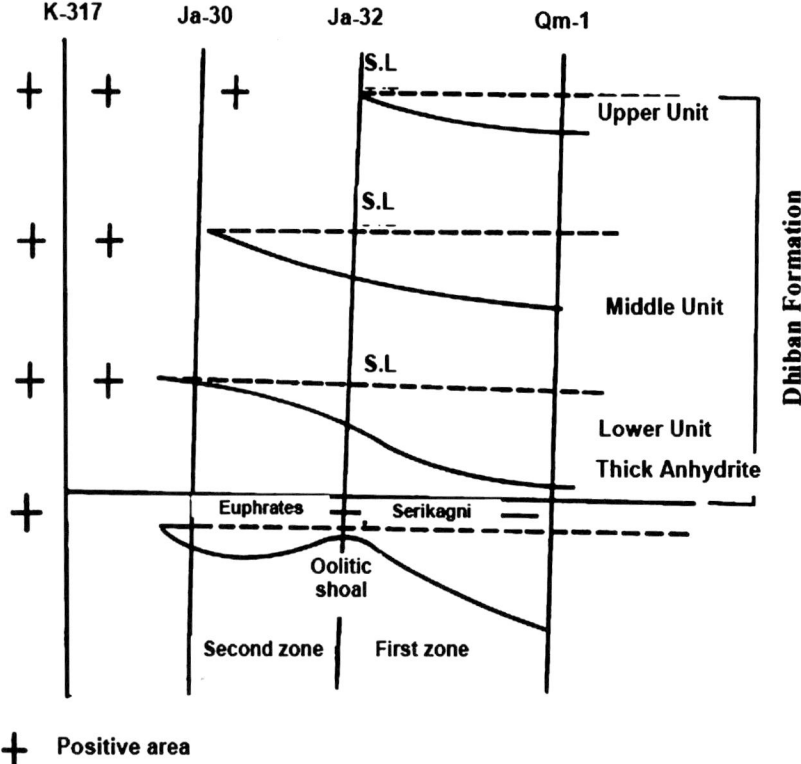

Fig. 14. Evolution of the sedimentary basin of Dhiban Formation.

divided into several sub-environments (Table 1) similar to those observed in the Euphrates Formation and is represented by shoal, lagoonal and tidal flat environments.

Isopach Map of the Jeribe Formation

The maximum thickness (63 m) for the Jeribe Formation is recorded in the well Inj-5 of the B–B' traverse and thins out towards the wells of A–A' traverse (34 m of the well Ja-32) (Fig. 15).

Three biounits are recorded in the Jeribe Formation of traverse A–A'. These biounits are: (1) an echinodermal–gastropodal unit; (2) a miliolidal unit; and (3) an echinodermal unit (Fig. 16). However, the echinodermal and miliolidal units were recorded in the Jeribe Formation of traverse B–B' (Fig. 17).

The lateral variations of these biounits in the sedimentary basin of Jeribe Formation refer to the shallowing-upward trends in the Jeribe environment of deposition toward the Kirkuk oilfield (in well K-317, Fig. 18). Distribution of a miliolid-rich biounit throughout most of the formation may relate to an open lagoonal environment covering the area of deposition. Then with decreasing water depth, the biounits rich in echinoderms and gastropods develop in the upper part of the formation.

The geometry and dimension of the biounits as well as their lateral and vertical variation and the thickness variation within the Jeribe rocks indicate that the sedimentary basin of the Lower Langhian is topographically similar to that of the earlier Burdigalian cycle.

The lower part of the Langhian sequence started with renewed marine transgression responsible for the deposition of the lower part of the Jeribe Formation, composed of marl rich in planktonic foraminifera (*Orbulina*) then the deposition of limestones and marls that contain benthonic foraminifera and shells of mollusca, echinoderms, red algae and ostracods (Table 1). The various sedimentary facies of the Jeribe Formation are related to a variety of sedimentary environments. These environments range from enclosed shallow lagoons represented by packstones and wackstones rich in milliolids and more open, perhaps deeper lagoons of muddy limestones rich in red algae and shells of echinoderms and molluscs similar to those of the aforementioned facies of the Euphrates Formation.

Discussion

The effect of the continental collision between the Eurasian and Gondwanan Plates led to the formation of the Neotethys, where normal faults formed at the Arabian margin through tensional forces during Triassic-Jurassic. During the Cretaceous, compressional and transpressional forces led to the formation of reverse faults as a result of subduction in the area

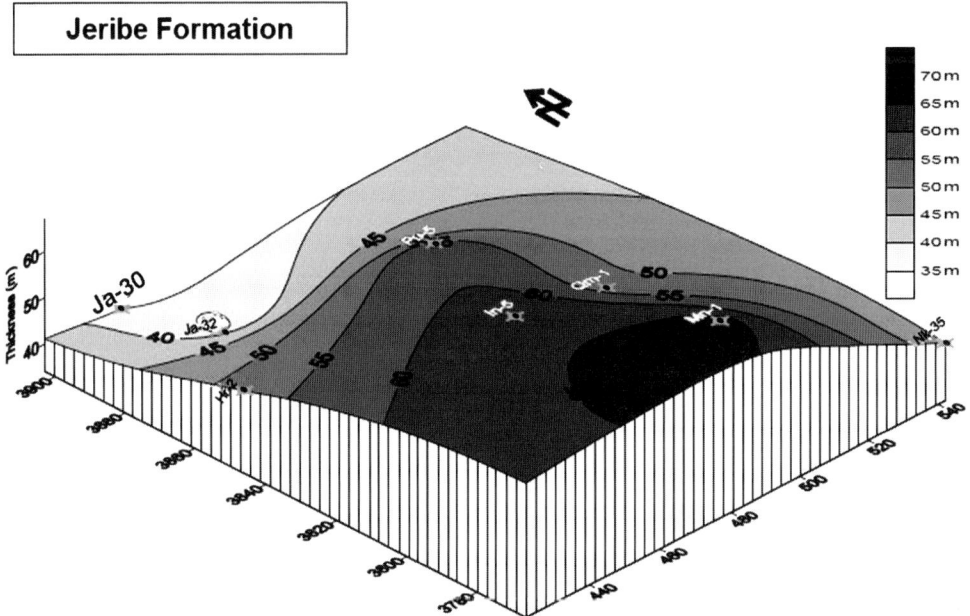

Fig. 15. Isopach map of the Jeribe Formation.

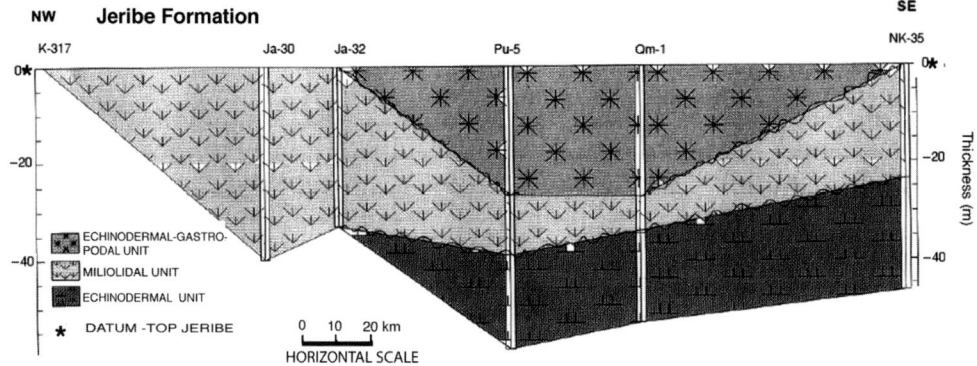

Fig. 16. Biounit distribution of the Jeribe Formation in the first traverse (NW–SE).

near the collision zone while normal faults still persisted in areas still far from the collision, a state that continued until the Eocene (Haynes & Mcquillan 1974; Numan 1997). The reverse faults resulted in the uplift of *some* areas and the formation of marine ridges that facilitated the growth of coral reefs during Oligocene–Early Miocene (Al-Eisa 1992*b*). These reverse faults played an important role in local uplift and formation of local basins in this study area in the Early Miocene (Fig. 19).

Continuation of the collision during Early Miocene (Burdigalian) resulted in volcanic activity in the northwestern portion of Saudi Arabia with basaltic flows in several areas as a result of the opening of the Red Sea. This may have contributed to a marine transgression and the development of deeper environments for the Serikagni Formation and shallower environments for the Euphrates Formation.

The tectonic stability in the area by the end of Burdigalian caused the short-term development of localized, restricted to closed basins, that are responsible for the formation of Dhiban evaporitic successions due to drawdown and/or desiccation.

Repeated periods of local tectonic movement, as well as worldwide sea level variations, during the early Late Miocene (Langhian), resulted in stratigraphic successions similar to those of the Burdigalian. These may have contributed to a marine transgression and deeper environments rich in planktonic foraminifera (*Orbulina*) in the deep part of the basin that then changed to an open lagoonal environment rich in milliolids throughout all of the Jeribe Formation. Perhaps due to the

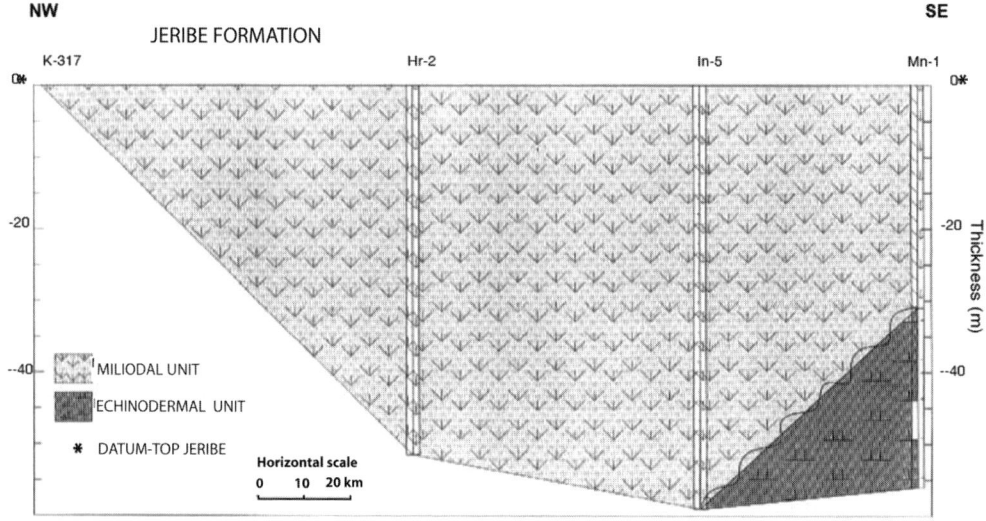

Fig. 17. Biounit distribution of the Jeribe Formation in the second traverse (NW–SE).

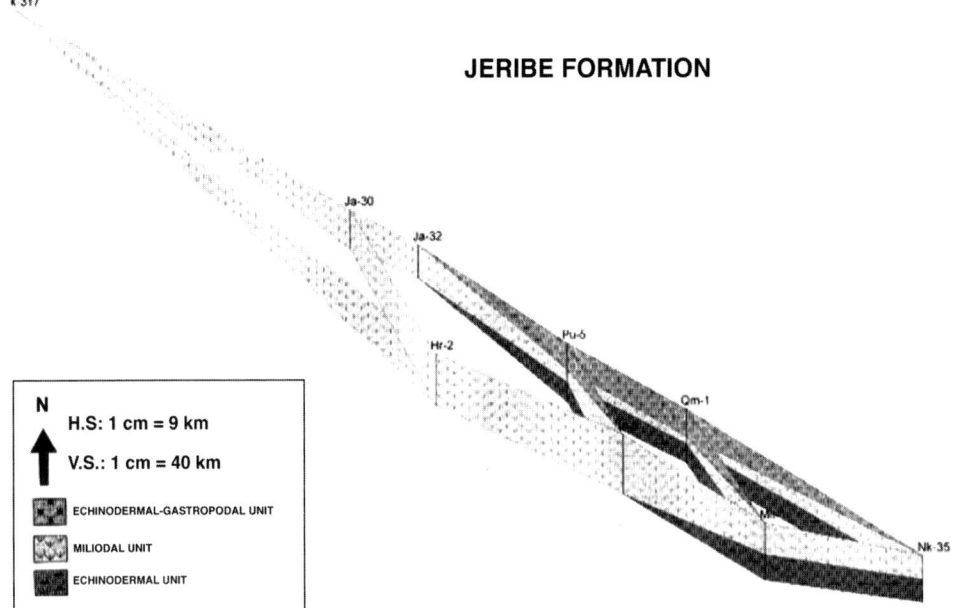

Fig. 18. Fence diagram of the biounit distribution of the Jeribe Formation in the first and second traverses.

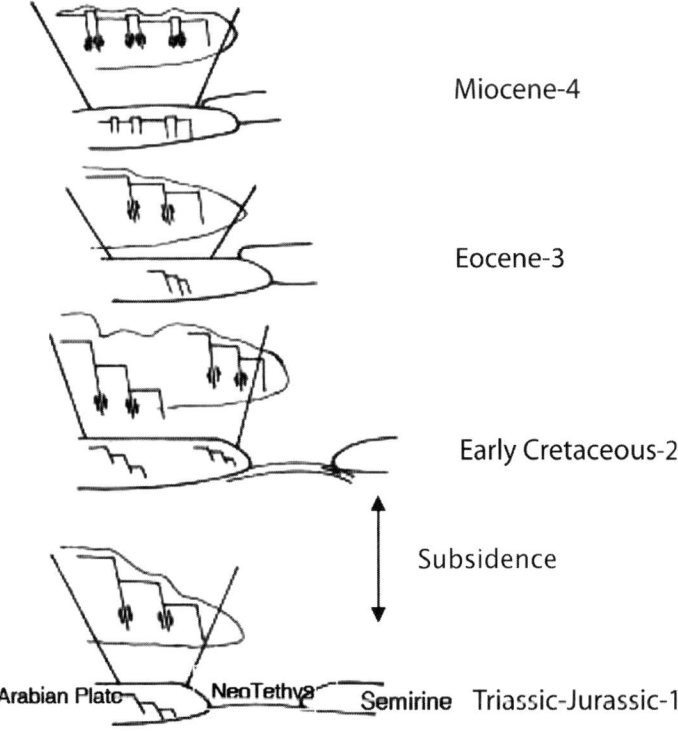

Fig. 19. Diagram illustrating the series of the structural changes from Triassic through Miocene, in the area of study.

continuation of regional uplift, a third phase of drying appears to have taken place and led to the deposition of Fat'ha successions of evaporites and limestones depending on the nature and time period of tectonic stability.

Marine conditions continued throughout the early Miocene, and the Serikagni, Euphrates, Dhiban Anhydrite and Jeribe Formations were deposited in marginal basins. Marine conditions became progressively more restricted, with many small seas and lagoons sporadically replenished with normal marine water. These gave rise to the cyclic deposition of evaporites in the Middle Miocene Fat'ha Formation in (Numan 1997).

The Dhiban successions are variously composed of both complete and incomplete sedimentary series. The complete series are considered as (A) series and are located in the deeper parts of the Dhiban depositional basin. These are composed of thick evaporites and halite beds intercalated with thin limestone beds within the lower unit of the Dhiban Formation. This reflects periods of tectonic stability that provide enough time for evaporate deposition and to concentrate the water to precipitate halite, whereas thin evaporites (without halite) and thick limestone beds characterize the incomplete series in the middle and upper parts of the Dhiban Formation.

The structural nature of the area gives the sedimentary rocks, especially those near the collisional zone in the area of study, an important feature of the petroleum system. Each cycle provides source rocks developed in deeper environments and reservoir rocks in areas of shallower environments and evaporitic successions that provide the cap rocks (the seal). The evolution of the structural changes helped to form the oil reservoirs in each cycle as in the Euphrates and Jeribe Formations that continued throughout the entire formation, as in the Dhiban or the lower transitional unit of the Fat'ha Formation.

The authors wish to thank the Northern Oil Company, Iraq for providing the data and supporting this research activity. An early version of this manuscript benefited from comments and suggestions by editor B. Charlotte Schreiber and reviewers Steven Hageman and Anthony Lomando.

References

ABAWI, T. S. 1989. Foraminifera, stratigraphy and sedimentary environment of the Euphrates Formation, Lower Miocene, Sinjar area, northeastern Iraq. *Newsletter of Stratigraphy*, **21**, 15–24.

ABU-KHADRA, A., WALI, A. & ATTIA, O. 1992. Geologic setting and sediments structures of Abu Khashir Miocene evaporites, Ras Sudr, Sinai, Egypt. *Proceedings of the 1st Conference on the Geology of the Arab World, Cairo*, 303–331.

AL-EISA, M. I. 1992a. The secondary sedimentary cycles for Early Miocene in the oilfields surrounding Kirkuk oilfield, North Iraq. *Iraqi Geological Journal*, **25**, 41–58.

AL-EISA, M. I. 1992b. Coral reefs of the Late Oligocene and Early Miocene in Kirkuk and adjacent areas. *Iraqi Geological Journal*, **25**, 17–23.

AL-NAQIB, K. M. 1960. *Geology of the southern area of Kirkuk Liwa, Iraq*. IPC, Oxford. Technical Publications, **124**.

BABEL, M. 2004. Badenian evaporite basin of the northern Carpathian Foredeep as a drawdown salina basin. *Acta Geologica Polonica*, **54**, 313–337.

BELLEN, R. C. VAN., DUNNINGTON, H. V. & WETZEL, R. 1959. *Lexique Stratigraphic International, Asia*. Fascicula. loa, Iraq–Paris Central National de La Recharde Science.

BEYDOUN, Z. R. 1988. *The Middle East Regional Geology and Petroleum Resources*. Scientific Press, UK.

BEYDOUN, Z. R. 1991. Arabian plate hydrocarbon geology and potential plate tectonic approach. *AAPG Studies in Geology*, 19–29.

BOERSMA, A. 1978. Foraminifera. *In*: HAQ, B. V. & BOERSMA, A. (eds) *Introduction to Marine Micopaleontology*. Elsevier, Amsterdam, 19–79.

BUDAY, T. 1980. *The Regional Geology of Iraq. Stratigraphy and Paleogcography*. Dar Al-Kutib Publishing House, Mosul University, Iraq.

HAYNES, S. J. & MCQUILLAN, H. 1974. Evolution of the Zagros suture zone, Southem Iran. *Geological Society of America Bulletin*, **85**, 739–744.

KENDALL, A. G. 1992. Evaporites. *In*: WALKER, R. G. & JAMES, N. P. (eds) *Facies Model, Response to Sea Level Change*. Geotext, **1**. GeoText Canada, 375–395.

KINSMAN, D. J. J. 1976. Evaporites: relative humidity control of primary mineral facies. *Journal of Sedimentary Petrology*, **46**, 273–279.

NORTHERN OIL COMPANY, 2002. *Geological study of the Mansuriya oilfield*. Unpublished report, Northern Oil Company, Kirkuk.

NUMAN, N. M. S. 1997. A plate tectonic scenario for the Phanerozoic succession in Iraq. *Iraqi Geological Journal*, **30**, 85–110.

NUMAN, N. M. S. 2001. Cretaceous and Tertiary Alpine subductional history in Northern Iraq. *Iraqi Journal of Earth Science*, **1**, 59–74.

SCHMALZ, R. I. 1969. Deep-water evaporate deposition: a genetic model. *American Association of Petroleum Geologists Bulletin*, **53**, 798–823.

SCHREIBER, B. C. & EL TABAKH, M. 2000. Deposition and early alteration of evaporites. *Sedimentology*, **47** (Millennium Issue), 215–238.

WALI, A. O. & EL TABAKH, M. 1992. On the mineralogy and geochemistry of Ras Azbaryan evaporites Gulf of Suez, Egypt. *Proceedings of the 1st Conference on the Geology of the Arab World, Cairo*, 317–329.

WARREN, J. K. 1989. *Evaporite Sedimentology: Importance, Hydrocarbon Accumulation*. Prentice Hall, Englewood Cliffs, NJ.

Role of rifting in evaporite deposition in the Great Kavir Basin, central Iran

H. RAHIMPOUR-BONAB[1], Z. SHARIATINIA[1] & M. G. SIEMANN[2]

[1]*School of Geology, College of Science, University of Tehran, PO Box 14155-6455, Tehran, Iran (e-mail: rahimpor@khayam.ut.ac.ir)*

[2]*Technical University Clausthal, Department of Mineralogy-Mineralogy-Geochemistry-Salt Deposits, Adolph-Roemer-Str. 2A-38678, Clausthal-Zellerfeld, Germany (e-mail: michael.siemann@tu-Clausthal.de)*

Abstract: The thick middle Miocene evaporites of the Great Kavir Basin, Central Iran, are classified as $MgSO_4$-poor potash-bearing deposits. In the northern domain of the Great Kavir Basin these sediments occur as 50 very large salt diapirs, as well as outcrops in several areas, including the Melheh salt pit to the south of Semnan, north Central Iran. Based on the excellent preservation of layering in these sequences, along with well-preserved bottom nucleated primary textures within the halite and sylvite, they are interpreted as largely primary deposits. They also contain some secondary minerals such as langbeinite, d'ansite, polyhalite and anhydrite. The Br content in the halite and sylvinite layers falls within the range of marine evaporites. These $MgSO_4$-poor potash-bearing evaporites deposits, along with other marine and continental sediments, appear to have been precipitated in the marginal marine basins around continental rifts. Seemingly, these evaporites formed in marginal marine rift-related basins which also received input from deep, circulating hydrothermal $CaCl_2$-rich brines. These hydrothermal brines were also rich in Fe and Zn and apparently were driven upward along faults and fracture zones by the thermal gradient along the axis of the extensional trough.

In Iran the evaporite deposits of Cenozoic age are mainly situated in the central Iran tectonic zone and its eastern sector, the northern Great Kavir Basin (Fig. 1). In the latter area these deposits, along with red beds, limestone and other marine and continental sediments, are to be found in tens of very large diapirs as well as extensive outcrops. These deposits may be divided into two stratigraphic units, and are Eocene, Oligocene and Miocene in age. They have been subject of many research projects since 1888 (e.g. Vaughan 1893; Kalhor 1961; Stocklin 1968; Neal 1969; Krinsley 1970, 1974, 1976, 1977; Russo 1976; Rowlands *et al.* 1984; Jackson *et al.* 1990; Sadedin 1990; Dari *et al.*, 1993; Rahimpour-Bonab & Alijani 2003; Rahimpour- & Kalantar-Zadeh 2007). When the Iranian Oil Company was established, one of its initial goals was to explore central Iran and the Great Kavir Basin. In the last decade preliminary potash exploration has been initiated by the Geological Survey of Iran, which has added invaluable insight concerning distribution and lateral facies changes of these deposits.

In earlier studies (Rahimpour-Bonab & Alijani 2003; Rahimpour-Bonab & Kalantar-Zadeh 2007) the origin, depositional model and diagenesis of similar Eocene and Miocene evaporites in the NW and north of the central Iran tectonic zone have been tackled. The main objectives of this study are to reconstruct the sedimentary environment of these evaporites, to examine their relationship with their equivalents in the other areas of central Iran, to decipher the possible origin of brine and potash minerals and to propose an appropriate depositional model for such thick evaporite deposits.

Geology and stratigraphy

The study area is situated in the northeastern sector of the central Iran tectonic zone, in the northern part of the Great Kavir Basin (Figs 1 & 2). The Cenozoic sediments of the central Iran tectonic zone and its eastern sector, the Great Kavir Basin, are more or less similar throughout the area, but with some regional differences. The Cenozoic of the Great Kavir Basin is an intracontinental rift basin filled with several kilometres of Eocene to Recent marine and continental sediments that are mainly evaporites, carbonates and red beds. To the south of the city of Semnan (Fig. 3) these sediments occur as about 50 very large salt diapirs that contain Eocene to Miocene evaporites, limestone and continental red beds.

This basin began to form during an important episode of rifting in the Iranian plateau during the

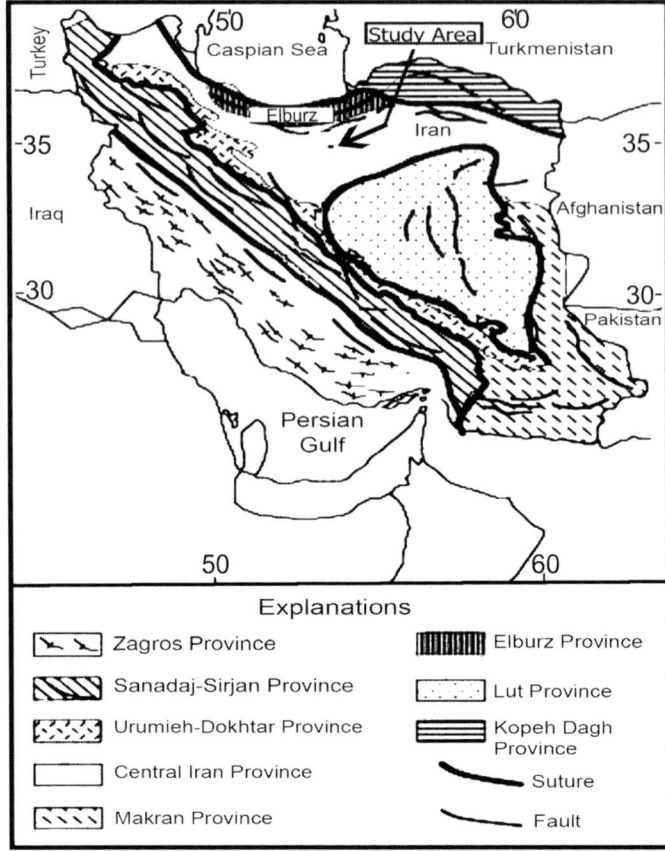

Fig. 1. Tectono-sedimentary provinces of Iran and location of the studied area (map from Geological Survey of Iran; National Iranian Oil Company 1977).

Early Cretaceous and this area became a region of intense and pervasive block faulting during rifting. Evidence for this style of tectonism is present as horst and graben structures that underlay the Kavir Basin. In these rift grabens, best developed in the northeastern border region of the Kavir, the sequence comprises sediments that range from Upper Cretaceous to Recent (Stocklin 1968).

Most authors favour the idea of a wide (Neo-)Tethys ocean passing through Iran from Mesozoic until the Eocene (Belov et al. 1986; Dercourt et al. 1986; Scotese 1987). Accordingly, the ophiolitic mélange zones exposed in the north of the Great Kavir fault are remnants of an old ocean floor that had completely encircled and isolated a Central Iran microcontinent (Fig. 3). Consequently, distribution of the mélange zones and their lateral discontinuity in the Great Kavir Basin suggest a pattern of strongly subsident rift grabens and narrow troughs of ocean crust separated by relatively stable horsts (Jackson et al. 1990). Since that time, at least up to the Eocene, this oceanic crust was largely subducted leaving remnants of oceanic crust and its associated sediments in different parts of Central Iran, including the Great Kavir Basin and the Zagros suture zone. Although evaporites of the Great Kavir Basin first appeared in small, scattered sub-basins in the Early Middle Eocene, significant evaporite deposition began during regional regression toward the end of the Eocene. This regression was related to an important phase of faulting, tilting and regional uplifting that affected all of central Iran. These changes in the relative plate movements, during Eocene times (Dewey et al. 1989), were associated with subduction-related volcanism that affected central Iran. Owing to these tectonic events, the setting changed from the marine sediments of Lower to Middle Eocene and passed up into thick continental red bed facies of the Lower Red Formation (LRF) during the Oligocene (Fig. 4). The LRF, which is composed of red continental clastics and evaporites

RIFTING AND EVAPORITE DEPOSITION IN CENTRAL IRAN 71

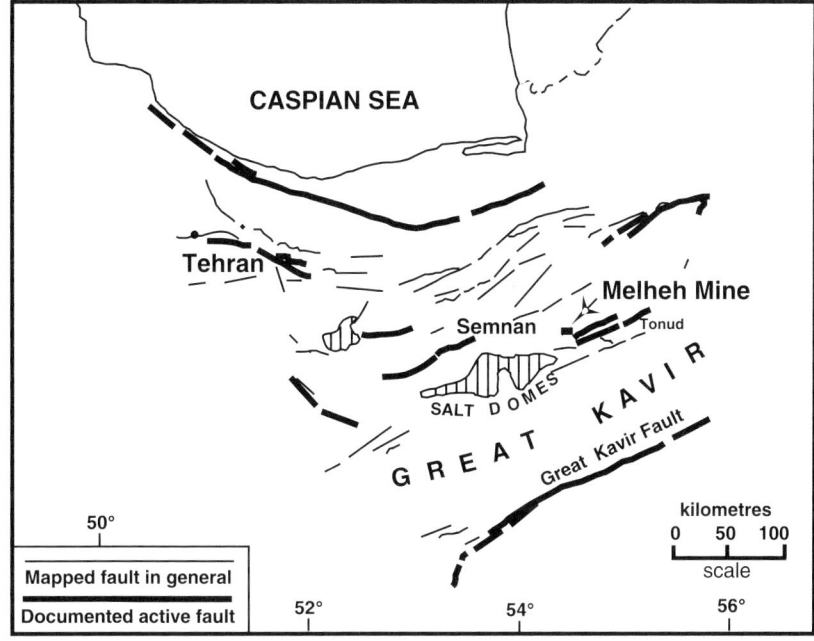

Fig. 2. Main fault system and distribution of diapirs in the Great Kavir Basin (Jackson *et al.* 1990, with permission from GSA).

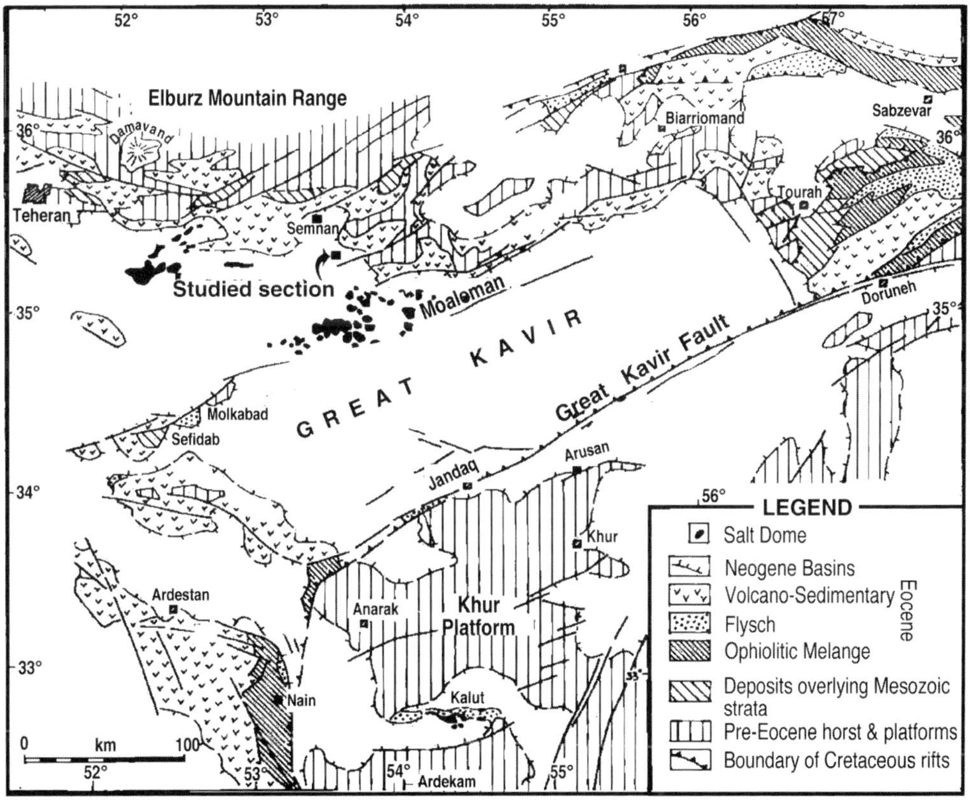

Fig. 3. Diapirs, salt flat and ophiolitic mélange distribution in the study area (Jackson *et al.* 1990, with permission from GSA).

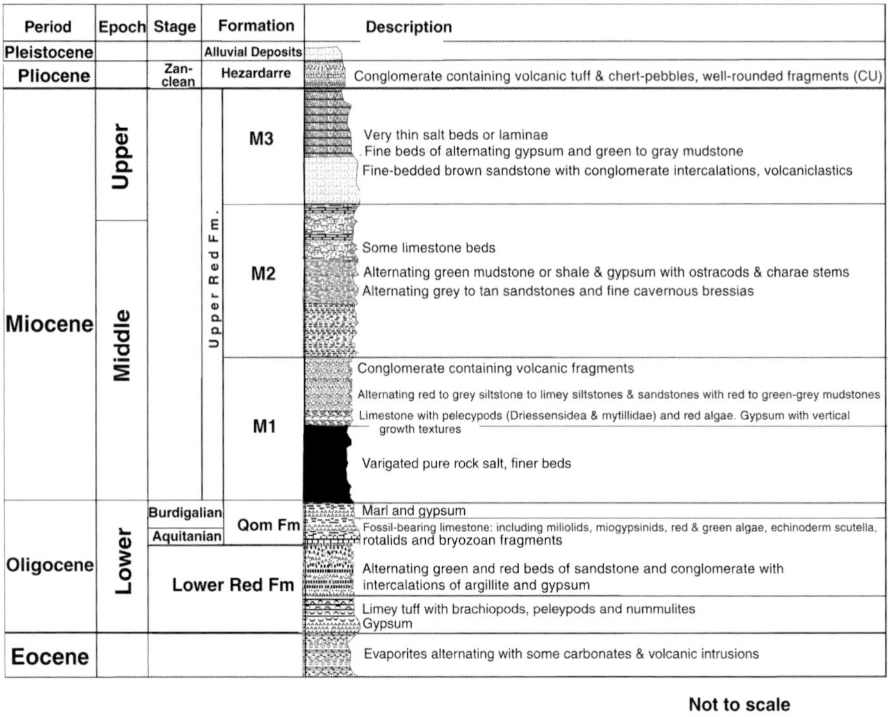

Period	Epoch	Stage	Formation	Description
Pleistocene			Alluvial Deposits	
Pliocene		Zanclean	Hezardarre	Conglomerate containing volcanic tuff & chert-pebbles, well-rounded fragments (CU)
Miocene	Upper		M3 (Upper Red Fm.)	Very thin salt beds or laminae. Fine beds of alternating gypsum and green to gray mudstone. Fine-bedded brown sandstone with conglomerate intercalations, volcaniclastics
Miocene	Middle		M2 (Upper Red Fm.)	Some limestone beds. Alternating green mudstone or shale & gypsum with ostracods & charae stems. Alternating grey to tan sandstones and fine cavernous bressias
Miocene	Middle		M1	Conglomerate containing volcanic fragments. Alternating red to grey siltstone to limey siltstones & sandstones with red to green-grey mudstones. Limestone with pelecypods (Driessensidea & mytillidae) and red algae. Gypsum with vertical growth textures. Varigated pure rock salt, finer beds
Oligocene	Lower	Burdigalian / Aquitanian	Qom Fm	Marl and gypsum. Fossil-bearing limestone: including miliolids, miogypsinids, red & green algae, echinoderm scutella, rotalids and bryozoan fragments
Oligocene	Lower		Lower Red Fm	Alternating green and red beds of sandstone and conglomerate with intercalations of argillite and gypsum. Limey tuff with brachiopods, peleypods and nummulites. Gypsum
Eocene				Evaporites alternating with some carbonates & volcanic intrusions

Not to scale

Fig. 4. Stratigraphic column of the studied area in the Northern Kavir Basin.

of Oligocene age, marks the onset of a new sedimentary cycle. The evaporite units of the LRF, which crop out in the northern border of the central Iran tectonic zone, in the Great Kavir Basin (south of the city of Semnan) and are present in its very large diapirs, are marine in origin (Rahimpour–Bonab & Alijani, 2003). These evaporites are mainly halite with some potash layers that have simple mineralogy, with high bromine content (up to 450 ppm) and are associated with some marine limestone.

The retreat of the sea in the Early Miocene (Burdigalian) led to the establishment of continental wadi conditions, and the marine limestone of the Qom Formation was gradually replaced by thick evaporite and continental red bed facies of the Upper Red Formation (URF). The lithologies of the Upper Red Formation are very similar to the LRF. Thus, in the Middle Miocene development of restricted marine conditions produced a facies change from the shelf carbonates of the Qom Formation to evaporite sequences of the M1 member, Upper Red Formation (URF) (Rahimpour-Bonab & Kalantar-Zadeh in press). This basal evaporite member (M1), is about 200 m thick and is mainly composed of gypsum, halite and in some places potash minerals, in the type section, overlying the uppermost marine marl of the Qom Formation.

The area of the Melheh salt pit is located in the M1 unit and there its thickness is more than 800 m (Figs 5 & 6). Seemingly, this evaporite member is a lateral equivalent of massive reef limestone, which in some areas of the Central Iran is classed as the uppermost member of the Qom Formation. The M1 member demonstrates a partial lateral transition from the Qom limestone to evaporites. Above this basal evaporite are about 2000 m of red-bed continental sediments with distinct repetitive series composed of shale, siltstone, mudstone and sandstone. Each series is from 10 to a few tens of metres thick and comprises from bottom to top: impure salt, gypsum, green gypsiferous mudstone and red claystone. Apparently, the sequences represent a reversal of a typical (marine) evaporite sequence by a periodic influx of relatively fresh water into the highly concentrated brine of a landlocked basin dominated by arid climate.

Member M2 is 2000–3000 m thick and is composed of an alternation of gypsiferous mudstone and siltstone with distinct colour banding. In the middle part of this member numerous intercalations of pale green marl (up to 500 m in thickness) that contain abundant ostracodes and Chara oogonia are also

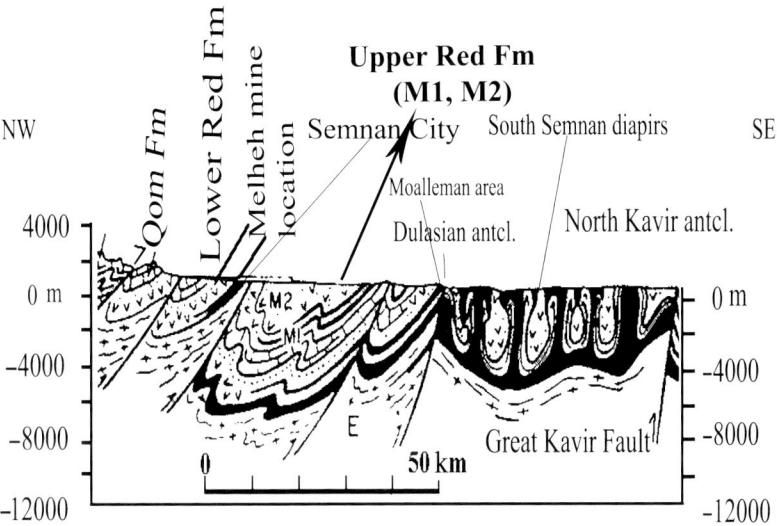

Fig. 5. Schematic cross section across the Great Kavir Basin. Salt units are in black (modified from Jackson *et al.* 1990, with permission from GSA).

present. This part is overlain by fine-grained grey sandstone that contains some conglomeratic (pebble) intercalations.

Member M3, having about 1000 m thickness, consists of sandstone with intercalations of conglomerates overlain by cyclic saline mudstone with grey to green colour.

Overall, the URF sequences are made up of a facies mosaic of lagoonal and salina evaporites (mainly halite beds) admixed with continental siliciclastics that conformably, and in some places gradually, overlies the Qom Formation (Rahimpour-Bonab and Kalantar-Zadeh 2007). Most of the extensive red beds exposed in the Kavir Basin and central Iran belong to the URF. In the northern border of the Kavir, the URF is well exposed in the large Moalleman anticline, southeast of the Melheh salt pit (Fig. 5). This unit overlies the palaeontologically dated topmost Burdigalian beds of the Qom Formation and lies below an unconformable overburden of continental deposits, which in NW Iran contain a freshwater fauna of Zanclean (Early Pliocene) age. Thus the URF in central Iran and Kavir Basin is between 17 and 5 Ma old (Jackson *et al.* 1990).

The structural trends of the area follow the major structural elements of the region, i.e. the Alborz Main Fault and Zagros Trust Fault (Jackson *et al.* 1990; Fig. 7). By mid Miocene time significant orogenic movements ceased in the Middle East (Late Alpine orogeny). This resulted in the development of several shallow basins that then were filled by evaporite and continental sediments (Jackson *et al.* 1990; Darvishzadeh 1991). In the northern province of the central Iran tectonic zone and in the Zagros Basin shallow discontinuous

Fig. 6. Outcrop photo showing well-preserved layering in the halite of the Melheh salt pit.

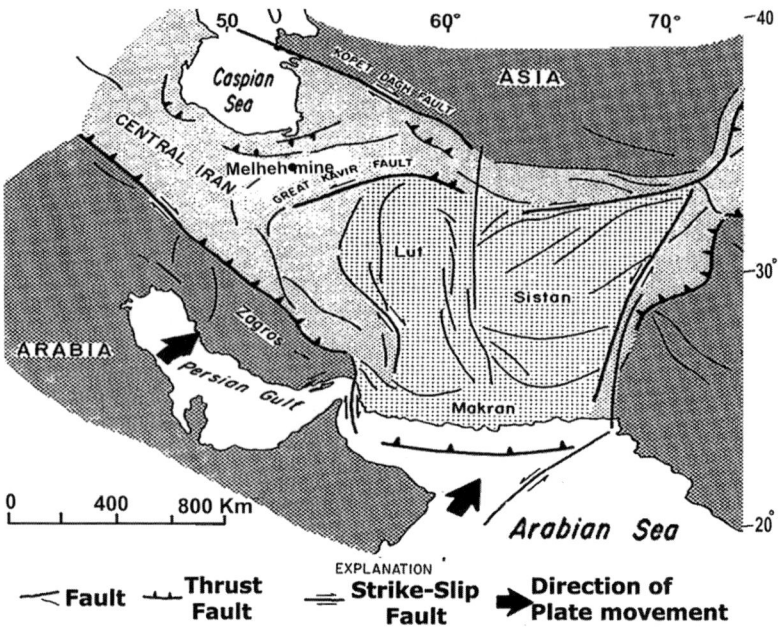

Fig. 7. Map of plate tectonic setting of central Iran and adjacent micro-plates between converging Arabian, Indian and Asian plates. Complied from Gansser (1974); Stocklin (1984); and Dercourt *et al.* (1986) (with permission from GSA).

epicontinental settings were dominated by an arid climate, and became the sites of widespread evaporite deposition and continental wadi sediments. This phase of evaporite deposition was interrupted several times and such deposits are diachronous all over central Iran.

Methods

This paper is based on the field study and petrographic examinations (thin sections and XRD) and geochemical analysis (Br, K, Ca, Mg, Zn, Fe) of the core material and salt pit samples. Accordingly, about 200 samples were systematically collected at regular intervals (1 m) from outcrops and boreholes of the Melheh salt pit in order to decipher the depositional environment and diagenesis of evaporite deposits. Samples that showed minimal post-depositional alteration, with well-preserved syndepositional textures, have been used for geochemical analysis.

The chemical composition of minerals was analysed by ion chromatography. Major anions and cations in the solutions were measured simultaneously after dilution by a factor of 5000–10,000 by use of a Metrohm ion chromatograph with chemical suppression for the anions. All samples were acidified to a pH of 2.6. Accuracy and reproducibility were investigated by analysing the North Atlantic Standard Seawater (NASS4) international reference sample ($n - 50$). The accuracy was better than 1% (relative) for all elements except for K (<4% relative). The relative standard deviation for the analysis of the reference sample was 1% except for K (<3%) (Siemann & Schramm 2000).

Br in the crystals was analysed in neutral solution after they were diluted 1:1000. Accuracy as well as reproducibility was better than 1% measured against an in-house halite reference sample. The calibration for the Br quantification was strongly adapted to the Cl concentration of the crystal. Before dissolving the crystals, they were analysed by laser Raman spectroscopy (Siemann and Ellendorff 2001) and X-ray powder diffraction (Siemann 1993).

Petrography and diagenesis

In this study area the sedimentary sequence, from base to the top, is composed of following units.

Evaporite sequences

These can be subdivided into three units.

A chloride unit This unit, which is very well-bedded (Fig. 6), having more than 800 m thickness, is strongly variegated and shows alternations of various colours in different scales (Fig. 8). In the

Fig. 8. Well-preserved lamination with variegation in halite sample from Melheh salt pit.

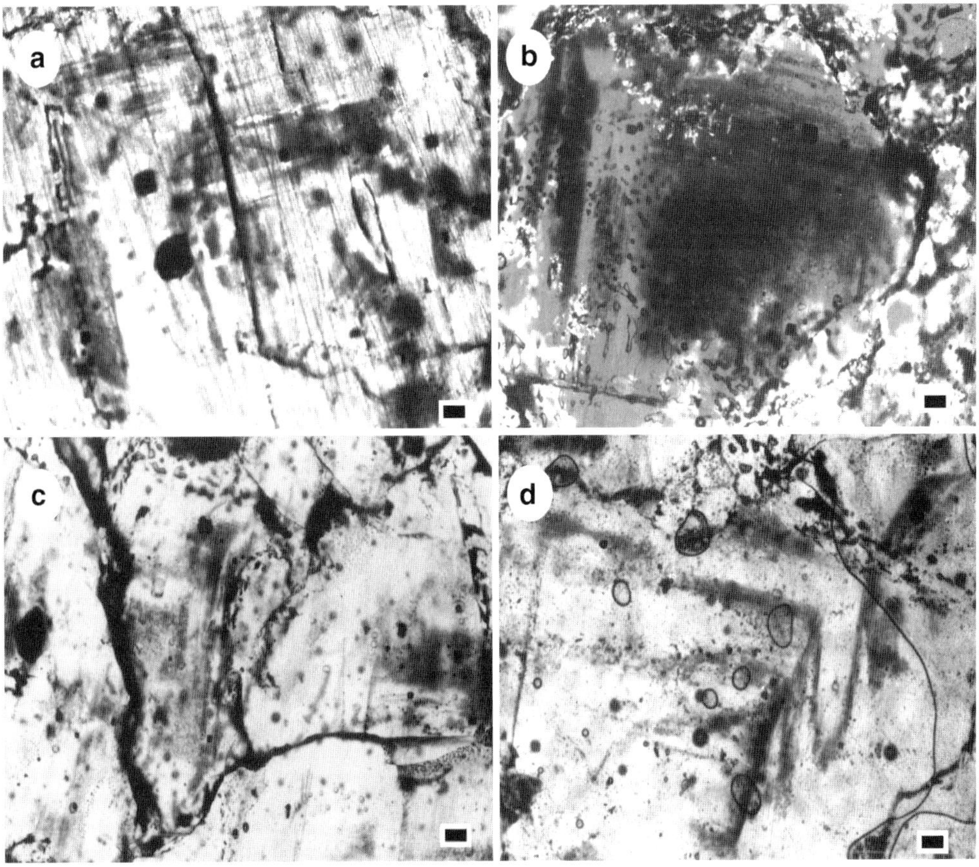

Fig. 9. Salt growth structures (scale bar for all figures 250 μm). (**a**) Vertically elongated chevron textures (plane polarized light). (**b**) Halite cornets (polarized light and gypsum filter) in halite. (**c**) Vertically elongated halitechevron textures and cornets (plane polarized light). (**d**) Halite in sylvite (plane polarized light).

Melheh mine, the blue to violet colours in halite are of great interest, because they are used as key beds for prospecting for potash-bearing layers (sylvinite beds). They are restricted to the contact zone between halite and sylvinite. Discoloration is attributed to water percolation along bedding planes that induces recrystallization (Sonnenfeld 1995). Accordingly, bromine is leached by circulating brine (from NaBr), but metallic Na remains intact and that produces blue colours (Sonnenfeld 1995).

The chloride unit is composed of the following paragenetic assemblage: primary minerals that include halite, gypsum, sylvite (KCl) and carnallite ($MgCl_2 \cdot KCl_2 \cdot H_2O$). However, based on the petrographic examinations, the secondary minerals are polyhalite ($2K_2SO_4 \cdot CaSO_4 \cdot MgSO_4 \cdot 2H_2O$), langbeinite ($2MgSO_4 \cdot K_2SO_4$), d'ansite ($9Na_2SO_4 \cdot MgSO_4 \cdot 3NaCl$), anhydrite ($CaSO_4$), plus minor amounts of dolomite and ankerite. In addition, some tiny particles of reworked volcaniclastic minerals are present in the studied thin sections.

Based on the chemical groupings proposed by Hardie (1984) the studied evaporites can be classified as $MgSO_4$-poor potash-bearing deposits, made up of cycles composed of halite, sylvite and subordinate carnallite, which show delicately, well-preserved, primary depositional bottom-growth textures. In thin sections of halite and sylvinite beds, many halite and sylvite crystals show remarkable preservation of depositional textures, fabrics and primary fluid inclusion banding. Both minerals commonly exhibit vertically elongated chevron textures with cloudy fluid inclusion bands outlining primary crystal growth faces (Fig. 9). Some halite and sylvite layers show well sorted millimetre-sized or larger cumulate crystals (Fig. 10) and centimetre-sized, randomly oriented hoppers, with fluid inclusion banding, now filled with clear halite cement (Fig. 10). These primary textures are evidence of growth in a shallow brine pool (e.g. Philips 1956; Llewellyn 1968).

Thus, sylvinite layers are largely interpreted as primary deposits, because the sylvite is intergrown with halite and both contain unquestionable primary subaqueous textures (Fig. 10) and they show no evidence of an origin by replacement. In many cases they do show polygonal-mosaic textures, with individual halite and sylvite crystals with various sizes. The other common texture observed in sylvinite beds is a framework of euhedral and subhedral halite cubes, poikilitically enclosed by anhedral crystals of sylvite. Carnallite

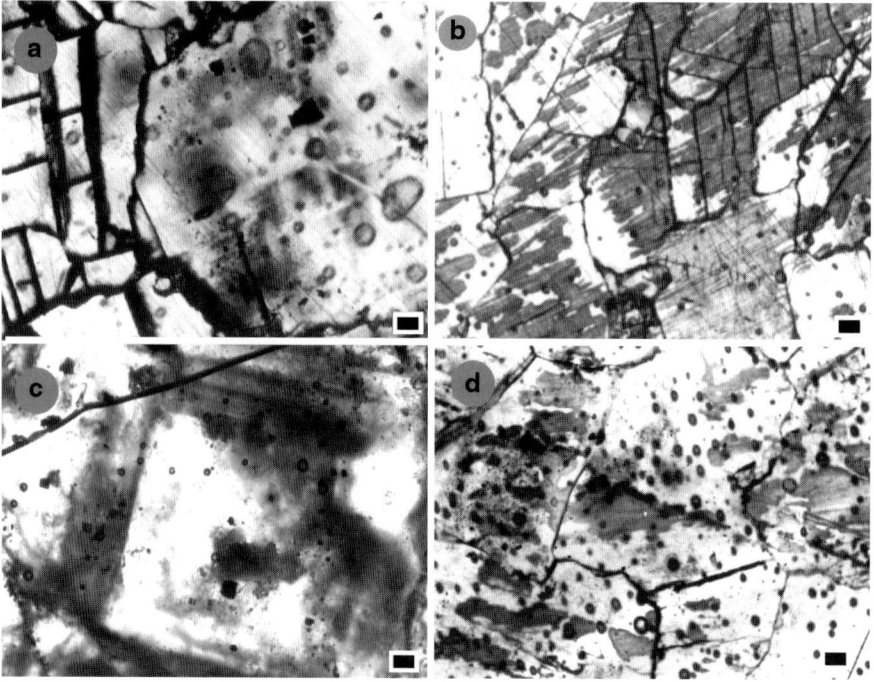

Fig. 10. Salt growth structures (all figures in plane polarized light, scale bar 250 μm). (**a**) Well-sorted millimetre-sized cumulates crystals of halite. (**b**) Well-sorted millimetre-sized cumulates crystals of sylvite. (**c**) Centimetre-sized, halite hoppers with fluid inclusion banding, which are filled with clear halite cement. (**d**) Intergrowth of primary sylvite and halites crystals.

occurs as tiny anhedral crystals between framework crystals of halite and sylvite within the sylvinite layers. Textural evidence suggests a syndepositional cement origin for such carnallite.

Absence of alteration rims around sylvite crystals rules out the possibility of their secondary origin. Surface nucleated halite crystals (such as pyramidal hoppers) are present and commonly associated with primary crystals of sylvite (Fig. 11). This texture not only indicates primary deposition but also denotes either settling in a stratified brine body or very rapid evaporation in an unstratified system (Hardie et al. 1985). Periodic refreshment and renewed stratification may have inhibited saturation of the brine column (Kovalevich 1978). As a result, in the stratified brine pool, co-precipitation of sylvite occurs as bottom nucleated crystals. In addition, the sylvinite layers follow the depositional boundaries of well-preserved halite layers.

In the K-salt-bearing horizons, potash is present both as primary and also secondary minerals that comprise up to one-third of the sequence. Secondary potash minerals such as polyhalite, d'ansite and langbeinite are confined to the crystals boundaries of primary halite crystals (Fig. 11).

Some sylvite crystals form millimetre-sized anhedral mosaics which lack fluid inclusion banding (Fig. 11), but in most instances sylvite is composed of millimetre-sized chevrons. Mosaic textures of sylvite crystals were probably produced by post-depositional modification (diagenetic neomorphism) of sylvite crystal boundaries. One possible modification mechanism for primary textures of the sylvite crystals, during early diagenesis, is via a 'brine curtain' mechanism (Warren 1999, pp. 80–81). Lack of the extensive cementation by primary potash minerals in the halite layers indicates that the primary potash minerals were formed rapidly, before the brine could reach the desiccation stage.

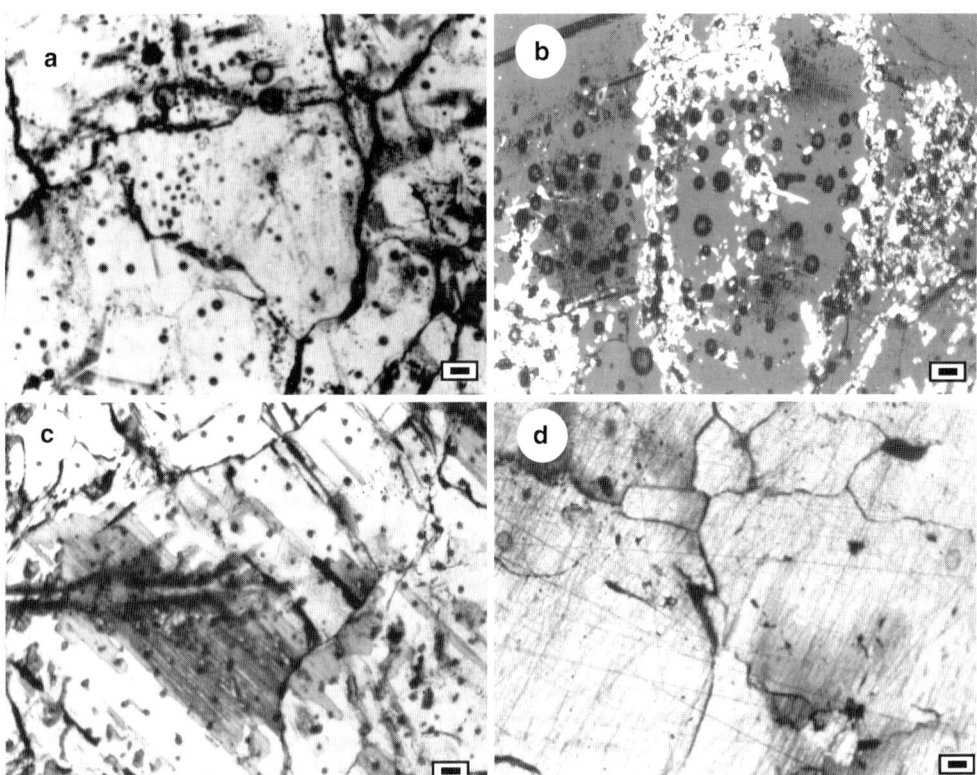

Fig. 11. Photomicrographs of various salts (scale bar for all figures 250 μm). (**a**) Hopper halite crystals associated with primary crystals of sylvite (plane polarized light). (**b**) Secondary potash minerals (such as polyhalite, d'ansite and langbeinite) are confined to the crystals boundaries of primary halite crystals (polarized light and gypsum filter). (**c**) Sylvite crystals form anhedral mosaics which lack fluid inclusion bandings associated with primary sylvite with chevron texture (plane polarized light). (**d**) Halite with drusy mosaic texture showing curved crystal boundaries meeting at triple junctions, (plane polarized light).

Moreover the absence of mud partings supports the latter conclusion.

In some cases halite chevrons are truncated by irregular patches of clear halite that enclose the cloudy traces of the primary textures. Seemingly, this feature is an indication of syndepositional dissolution–precipitation periods and represents partial recycling of the concentrated pore fluids in the micropores of newly precipitated halite crystals (Kirkland *et al.* 2000). Thus, the clear halite crystals are early diagenetic cements (Shearman 1978; Spencer 1987).

In a few samples halite crystals lack any syndepositional features and rather display diagenetic textures such as inclusion-free equigranular mosaic crystals with curved boundaries that meet at triple junctions (Fig. 11). These observations along with the preservation of the primary textures in most samples suggest an early stage of shallow burial diagenesis.

In the chloride unit anhydrite and gypsum and secondary potash minerals, such as polyhalite, d'ansite and langbeinite), occur as alteration haloes confined to the boundaries of the halite and sylvite crystals (Fig. 11). In this unit anhydrite and gypsum are also present as nodules and microcrystalline aggregates. Presumably, secondary potash minerals, such as polyhalite, formed by partial alteration of sulphate nodules. This may suggest that the secondary potash minerals have been formed due to chemical reactions between sulphate nodules and Mg and Ca-rich fluids. Presumably these concentrated pore fluids were released from primary inclusions in the capillary fringe (Hardie *et al.* 1985; Peryt *et al.* 1998). This type of anhydrite and gypsum transformation to polyhalite is an indication of interactions between sulphate-bearing diagenetic fluids with the K-bearing host mineral such as sylvite. Association of diagenetic ankerite and dolomite crystals with polyhalite is another indication of high Ca and Mg content of these diagenetic fluids.

Sulphate beds Toward the uppermost section of the chloride unit the pure halite and sylvinite beds gradually change to the alternations of halite and gypsum that signal the end of halite deposition. In this part of the section about 50 m of gypsum, composed of vertically aligned primary vertical crystals (fibrous), overlies the chloride unit with sharp contact. These sulphate beds are overlain by a marine, fossil-bearing, carbonate unit having approximately 30 m thickness.

Carbonate unit This unit sharply overlies the latter beds and contains copious marine fossils including molluscs such as mytillidea, cardidea and dreisinsidea, microfossils such as some species of milliolidea, and other marine biota such as red algae, green algae and corals. The lower parts of the carbonate unit, which is thick bedded limestone, contains plentiful quantities of marine micro- and macrofossils of shallow-water organisms. However, toward the top, due to facies change, the abundance and diversity of fauna and flora diminishes and thick-bedded marine limestone passes upward into red, sandy, fine-bedded limestone. This suggests that, after deposition of the sulphate beds, open marine conditions returned to the area; however, a gradual facies change from marine fossiliferous limestone to marginal, sandy, non-fossiliferous carbonate signals the onset of another marine regression.

Siliciclastic sequences

The carbonate unit is overlain by very thick sequences (up to 1.5 km) of siliciclastic continental sediments (red beds) that contain red to grey sandstone, mudstone, siltstone, marl and conglomerates, mainly composed of volcaniclastic fragments. These belong to the upper parts of M1 and M2 members. Presumably, they demonstrate permanent marine retreat from the Great Kavir Basin, and considerable subsidence to create the capacity for the accumulation.

Geochemistry

Evidently, potash deposits that contain primary sylvite and carnallite and lack $MgSO_4$ salts have been precipitated from Na–Ca–Mg–K–Cl brines (Hardie 1984). Normally, upon evaporation seawater may reach 70–100 times its original concentration. Usually, at this stage the brine is depleted in SO_4, and Ca (due to formation of either gypsum or anhydrite), and will be relatively enriched in Mg, K and Br. The high Br content of such concentrated brines is due to the high concentration of 0.84 mol Kg^{-1} H_2O in seawater (Braitsch 1971; Bruland 1983) and its long residence time (1.0×10^3 years) in the ocean (Chester 2000). Accordingly, the Br content of the halite could provide invaluable insight about depositional conditions and brine evolution of halite- and sylvite-bearing evaporites. Additionally the Br partition coefficient between halite and brine is about 0.12–0.14 (Schreiber & El Tabakh 2000; Siemann & Schramm 2000, 2002). Thus, at the beginning of the potash precipitation stage the highly concentrated brines are significantly enriched in Br. This partition coefficient is affected by other ionic species in the brine, such as K^+, HCO_3^- and Ca^{+2} (e.g. Braitsch & Herrmann 1963; Siemann & Schramm 2002). Accordingly, the Br content of

Table 1. *Bromine concentration in the halite and sylvite samples from Melheh salt pit*

Sample no.	Br(ppm)	Sample no.	Br(ppm)
1-1	140	26	203
1-2	105	27	256
2	100	28	79
3	112	29	170
4	136	30	195
5	137	31	114
6	124	32	225
7-1	159	34	216
7-2	99	35	202
8	157	36	85
11	197	37	207
12	179	38	184
13	175	39	196
13-1	155	40	215
13-2	162	41	203
14	143	42	305
15	186	43	211
16	232	44	219
17-1	162	45	224
17-2	204	46	49
18	194	47	241
19	181	48	207
20	167	49	196
21	197	54	226
22	190	64	168
23	220	70	169
24	205	72	172
25	199	75	155

the halite deposits could provide invaluable insight about depositional conditions and brine evolution of halite- and sylvite-bearing evaporites. Finally there is one other factor in this consideration, the dissolution of older salts into the brine, as in the case of the Dead Sea (Zak 1997).

The Br content in the primary halite layers, those that show well-preserved depositional textures, is about 60–140 ppm but rises to 200–300 ppm in sylvinite beds (Table 1). The Br content of the secondary diagenetic samples (clear halite), those halites which show triple junction boundaries, is about 49–75 ppm. This secondary texture could have been formed by dissolution during diagenesis (Holser 1979) or by partial recycling of diagenetic fluids released during the gypsum–anhydrite reactions (Kirkland *et al.* 2000). Mean Br values of the sylvinite layers are about 330 ppm, implying that the brine was already sufficiently concentrated to crystallize carnallite. These values demonstrate that the parent brine had evolved to the potash level deposition.

Based on Br values, the bromine profile of the chloride unit may be divided into three parts. Halite samples with less than 60 ppm Br were influenced by late diagenesis while halite crystals with primary textures show Br values of about 65–190 ppm. Higher Br values (up to 305 ppm) belong to sylvinite layers (Fig. 12). Vertical Mg and K profiles approximately correlate with the latter values, which suggests a primary signature for Br (Fig. 13). In most samples Ca values correlate negatively with the other elements, except where secondary Ca-sulphate minerals are present. In halite samples Zn values vary between 200 and 1250 ppm and Fe from 100 to 943 ppm (Table 2). These values, which are higher than the average for marine evaporites, suggest a contribution of hydrothermal brines (Hardie 1990).

Fluid inclusions

Petrographic examination of halite and sylvite reveal well-preserved fluid inclusions along growth bands of halite and sylvite crystals, with their typical rectangular shape (Fig. 10). There are also some secondary anhedral inclusions that show irregular shape and formed during diagenesis. Based on Bodnar's (1994) classification, there are three types of inclusions recognized in the studied samples that include primary monophase fluid inclusions, primary polyphase inclusions and secondary polyphase inclusions, which are located at the grain boundaries. The primary inclusions range in size from 10 to 100 μm, but the secondary inclusions are much larger.

Some sylvite crystals, with well-preserved fluid inclusion bandings (such as chevrons), contain

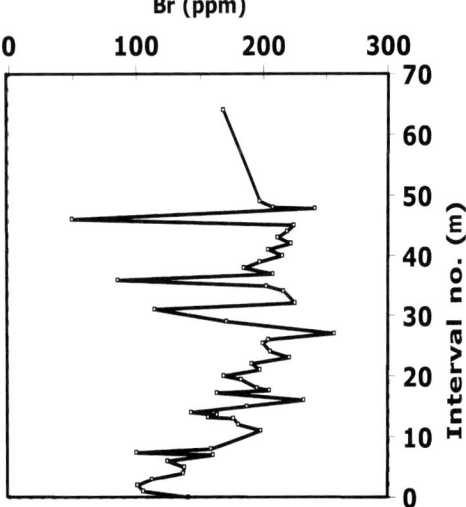

Fig. 12. Vertical bromine profile of the Melheh salt pit (depth from top of the section).

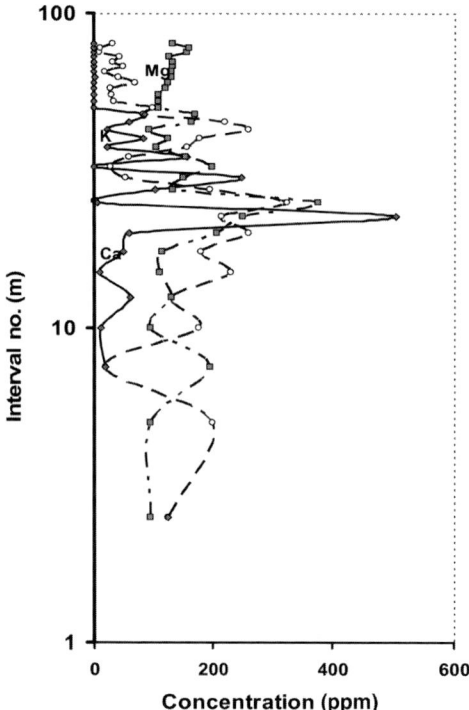

Fig. 13. Vertical Mg, K and Ca profiles of the Melheh salt pit (depth from top of the section).

euhedral carnallite and sylvite daughter minerals in their inclusions with their characteristic crystallographic appearance (Fig. 14). The primary polyphase inclusions, present in the sylvite crystals, represent brine evolution to the stage of potash saturation. Moreover, these observations along with the presence of mono-phase fluid inclusions in the primary minerals suggest an early stage of shallow burial diagenesis. Both the mono- and polyphase secondary inclusions are situated along the crystal boundaries of primary halite and have different compositions such as ferrous oxides, sulphates and diagenetic salt minerals.

Depositional model

Secular variations in the mineralogy of potash-bearing evaporite during Phanerozoic are discussed by many authors (e.g. Hardie 1990, 1991, 1996; Hryniv & Peryt 2003). Accordingly, in some periods $MgSO_4$ salts (such as polyhalite and kieserite) were the main potash minerals while $MgSO_4$-poor potash salts were dominant in other times (absence of $MgSO_4$ salts). The latter deposits, which contain halite, sylvite and carnallite as primary minerals and are usually entirely free from the primary Mg-sulphate salts, make up more than 60% of ancient potash deposits (Hardie 1990, 1991). However, as Hardie (1984) pointed out, these salts cannot be derived by normal evaporation of modern seawater or any similar sulphate-enriched brine. Hence, several hypotheses have been presented for the origin of these $MgSO_4$-poor salts, trying to explain these salts either as unusual marine evaporites or as altered marine evaporites. Among them, burial alteration of normal marine evaporites by groundwaters (Borchert 1977), bacterial sulphate reduction (e.g. Sonnenfeld 1984), seawater mixing with bicarbonate-rich river water (e.g. Valyashko 1972), and finally contemporaneous dolomitization of co-existing marine carbonates by seawater brines in a marine evaporite environment (Holland 1978), are most important. However, Hardie (1978) proposed a quite different explanation for this type of evaporite, particularly those deposited in rift and strike–slip basins. He suggested that upwelling $CaCl_2$ hydrothermal brines were the main source for $MgSO_4$-poor salts, as has been discovered in some modern analogous settings (e.g. Dead Sea trough, African rift basins). A similar concept has been put forward very recently by Hovland et al. (2006) with suggestive experimental data supporting this idea. The tectonic setting of Hardie's area had, presumably, the crucial conditions for evaporite deposition, such as arid climate and hydrologic closure of the depositional basin. In this system uplifted mountains encircled a series of elongated valleys, which were hydrologically closed sedimentary basins.

In this model, proposed by Hardie (1978), the basins were similar to the abundance of modern evaporite basins in central Iran, in which the uplifted mountains act as rainfall traps and barriers and the valley floors are local rain-shadow deserts. Deposition of thick evaporite deposits in these basins required rapid, active subsidence during the rapid evaporite accumulation and substantial inflow of brine (marine and/or non-marine) to the basin (Schreiber & Hsü 1980) with deposition and preservation that outpaces sand and mud influx into the basin (Hardie 1990). Seemingly, the evaporating basins were coastal depressions with their floors below sea level and with a constricted, very shallow inlet or a sort of permeable seal. As could be seen in some modern tectonically active closed basins, such as Dead Sea trough, Salton trough, Danakil Depression and Khur region in the southern border of the study area, deep circulating hot $CaCl_2$ brines are driven upward along faults and fracture zones by thermal anomalies along the axis of the extensional trough or by gravity from uplifted mountains that fed evaporite basins. The Khur region (Fig. 3), an old system of a closed continental rifting zone, which still shows some substantial

Table 2. Br and SO_4^{2-} content in the halite layers (in bulk samples)

Sample no.	Br(ppm)	SO₄%	Sample No.	Br(ppm)	SO₄%
1A	140	0.77	24	205	0.74
1	105	—	25	199	0.75
2	100	—	26	204	—
3	112	—	27	257	0.80
4	137	—	29	171	0.04
5	138	—	30	195	0.60
6	124	—	31	114	—
7	160	—	32	226	—
7a	100	—	34	216	0.21
8	158	—	35	203	—
11	198	—	36	86	0.61
12	179	—	37	208	—
13	176	3.14	38	184	—
13-1	156	—	39	197	2.06
13-2	162	—	40	—	215
14	143	0.89	41	—	204
15	186	—	42	222	6.10
16	232	2.94	43	211	—
17-1	163	—	44	219	—
17-2	205	—	45	224	—
18	195	—	46	50	0.67
19-1	182	—	47	242	1.19
20	168	—	48	208	—
21	197	—	49	197	—
22	190	—	64	168	—
23	220	3.08			

tectonic activity, is filled with about 7 km of Cenozoic marine evaporites, continental sediments and some volcanic extrusives.

In the study area, development of halite evaporites with substantial thickness (hundreds to thousands of metres), which suggests initial inflow of seawater along the narrow continental rift axis of the incipient ocean basin, initiated in Early Eocene time and continued up to the Middle Miocene in the isolated failed rift arms, oblique to the axis of divergence. Intermittent deposition of normal marine sediments (e.g. Qom Formation) on these continental sediments and marginal marine evaporites (LRF) demonstrates the full influx of the sea and the flooding of the subsiding margins of the divergent continents (Fig. 15). However, competition between marine and non-marine environments, at the edge of the encroaching sea, produced several cycles of abrupt and gradual transition from continental wadi sediments to marginal marine evaporite in the study area. This finally led to accumulation of marine evaporites of M1 of URF that, in turn, is gradually replaced by red beds of continental sediments of M2 and M3.

Brine origin and evolution

The brine origin and evolution in the modern and ancient continental rift basin has been addressed by several researchers, among them Hardie (1984, 1990) and Zak (1997). In particular the origin of CaCl₂ brine in the Dead Sea has been discussed by Zak (1997) and ascribed to mixing and chemical evolution of three basic brines. These include diagenetic brine formed by seawater evaporation, then modified by Mg + Ca exchange into Ca–Mg–Na–Cl–Br

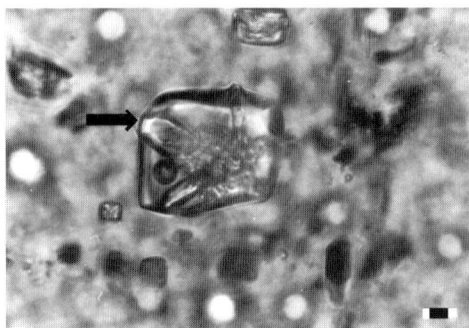

Fig. 14. Primary sylvite crystals that contain carnallite euhedral daughter mineral in its inclusions. Carnallite inclusion (arrow) shows its crystallographic characteristics (plane polarized light, scale bar 600 μm).

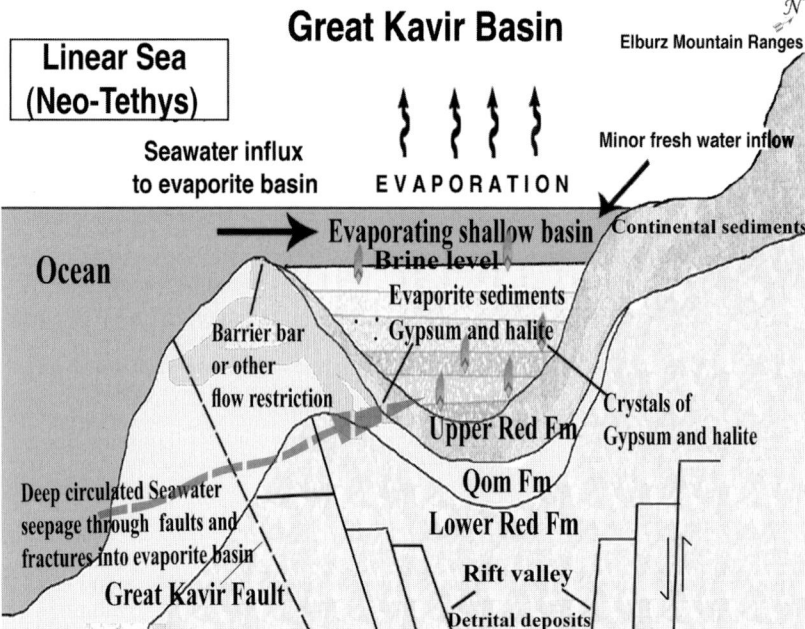

Fig. 15. Depositional model for evaporite deposition and associated marine and non-marine sediments in the Great Kavir Basin. Seawater, intermittently, recharged evaporite subbasins through land barriers and faults systems. In addition, modified seawater ($CaCl_2$ brines) was driven upward into the basin along faults and fracture zones by deep circulation. Arid climate and hydrologic closure enhanced evaporative concentration, and resulted in deposition of halite and $MgSO_4$-poor potash salts.

water. In this marine derived brine, Mg has been exchanged with the Ca from carbonate sediments and then most of the dissolved sulphate has been consumed by calcium sulphate precipitation. The other type of brine is altered meteoric water containing airborne sea salts. The third type is metamorphic brine that resulted from incongruent alteration and dissolution of hydrous evaporite minerals. The mixtures of these solutions experienced several progressive–regressive evaporation cycles with periodic precipitation of halite and sporadically of carnallite (Zak 1997).

In the Great Kavir Basin, which is an old continental rift setting, seawater moved into the subsurface through faults and fractures in the bedrock and after some alteration fluxed into isolated evaporite basins along hydraulic pathways. These marginal basins formed by extensive block faulting during rift opening and closure. These restricted basins, which temporarily were connected with the ocean and filled by normal seawater, were under the domination of the arid conditions and resulting evaporative concentration. On the other hand, in transit the deeper circulating seawater became heated and presumably reacted with the hot igneous and metamorphic bedrocks (Hardie 1990) and/or carbonate rocks (Zak 1997) before surfacing as $CaCl_2$ brine appropriate for precipitation of $MgSO_4$-poor salts. Thus, evaporating surface brines in the closed arid basins originated as mixtures of normal seawater plus the aforementioned hydrothermal $CaCl_2$ brine. As Hardie (1990) pointed out, 'only relatively small additions of $CaCl_2$ brines upwelling into the basin along faults and other permeable pathways are needed to convert a restricted body of seawater into $MgSO_4$-poor brine'. These $MgSO_4$-poor potash-bearing evaporites were deposited in marginal marine basins around the continental rift, which probably received an inflow of hydrothermal $CaCl_2$ brines. These basins were occasionally flooded with seawater, and seawater-derived hydrothermal brines leaked, more or less continually, through bedrock or another type of barrier by thermal or gravitational convection (Fig. 15). Some gypsum precipitated from such mixtures, and then, in the late stages of evaporation concentrated brines formed halite, sylvite and carnallite, that were precipitated with an absence of $MgSO_4$ salts. It seems that in the Great Kavir Basin, a long history of repeated tectonic movements and magmatism (from Early Cretaceous to Middle Miocene) was responsible for a very thick sequence of $MgSO_4$-poor potash evaporite deposits and its associated mixture of both marine and continental sediments. Another indication of the contribution of hydrothermal waters to the brines forming $MgSO_4$-poor potash-bearing deposits in the study area is the higher than average concentration of Zn and Fe in the halite, as, for example the 943 ppm Fe, up to 1250 ppm Zn.

Conclusions

The Miocene evaporites studied in the Great Kavir Basin are $MgSO_4$-poor potash-bearing deposits, made up of cycles composed of halite, sylvite and carnallite with delicately and well-preserved primary depositional bottom growth textures. This conclusively demonstrates that primary, undersaturated inflow waters were $MgSO_4$-poor and also rules out any post depositional alteration mechanism as the cause of the lack of sulphate. However, the very soluble $CaCl_2$ mineral tachyhydrite ($CaCl_2 \cdot 2MgCl_2 \cdot 12H_2O$), which is typical of some of the $MgSO_4$-poor potash deposits, is absent in the study area. This could be ascribed to the moderate meteoric diagenesis that affected these deposits. These deposits are clearly primary precipitates and little altered since their accumulation. This conclusion is fully supported by textual evidence.

The primary potash of the study area formed syndepositionally, during or just after deposition by processes controlled by the existing depositional environment. According to field and petrographic examinations, the syndepositional origin of the potash salts in the Melheh salt pit is substantiated and the potash minerals formed primarily as bedded subaqueous deposits.

The stratigraphic relationships, petrographic textures, fluid inclusions and geochemistry of these potash-bearing evaporites demonstrate their primary origin. These observations lead us to conclude that the sylvite and carnallite are syndepositional in origin and formed as *in situ* subaqueous deposits, and the high Br content of those halite samples with well-preserved primary textures is an indication of a marine signature for parent brines. Moreover, Br values vary concordantly with the K and Mg that present preservation of a primary geochemical signature.

Seemingly, these evaporites were deposited in rift-related sub-basins formed by extensive block faulting during rift opening and closure. The latter were under the influence of upwelling $CaCl_2$ hydrothermal brine. These Zn- and Fe-rich hydrothermal brines were the main source for $MgSO_4$-poor salt deposition. Arid climate and hydrologic closure of this depositional basin enhanced evaporative concentration of the brines and salt deposition.

This paper is produced by a grant to the senior author from Tehran University, which is appreciated. B. C. Schreiber and A. Lomando provided invaluable comments and suggestions on the draft, for which we are sincerely grateful.

References

BELOV, A. A., GATINSKY, YU. G. & MOSSAKOVSKY, A. A. 1986. A précis on pre-Alpine tectonic history of Thetyan paleooceans. *Tectonophysics*, **127**, 197–211.

BODNAR, R. J. 1994. Philosophy of fluid inclusion analysis. *In*: DE VIVO, B. & FREZZOTTI, M. L. (eds) *Fluid Inclusions in Minerals, Methods and Applications*. Virginia Tech, Blacksburg, VA, 1–6.

BORCHERT, T. H. 1977. On the formation of lower Cretaceous potassium salts and tachyhydrite in the Sergipe Basin (Brazil) with some remarks on similar occurrences in West Africa (Gabon, Angola, etc.). *In*: KELMM, D. D. & SCHNEIDER, H. J. (eds) *Time and Strata Bound Ore Deposits*. Springer, Berlin, 94–111.

BRAITSCH, O. 1971. *Salt Deposits: Their Origin and Composition*. Springer, Berlin.

BRAITSCH, O. & HERRMANN, A. G. 1963. Zür geochemie des broms in salinaren sedimenten. Teil II, die Bildungstem-peraturen primärer sylvin und carnallit-gestein. *Geochemica et Cosmochemica Acta*, **28**, 1081–1109.

BRULAND, K. W. 1983. Trace elements in sea-water. *In*: RILEY, J. P. & CHESTER, R. (eds) *Chemical Oceanography*. Academic Press, New York, 157–221.

CHESTER, R. 2000. *Marine Geochemistry*. Blackwell Science, Oxford.

DARI, M., BADAKHSHAN MOMTAZ, Q. & SAYAREH, A. 1993. *Potash exploration and equipping project in Iran for the Semnan and Garmsar Provinces*. Ministry of Mines and Industries, Report, **19**.

DARVISHZADEH, A. 1991. *Geology of Iran*. Neda University Press, Tehran.

DERCOURT, J., ZONENSHAIN, L. P. *ET AL.* 1986. Geological evolution of the Tethys belt from the Atlantic to the Pamirs since the Lias. *Tectonophysics*, **123**, B241–315.

DEWEY, J. F., HELMAN, M. L., TURCO, E., HUTTON, D. H. W. & KNOTT, S. D. 1989. Kinematics of The Western Mediterranean. *In*: COWARD, M. P., DIETRICH, D. & PARK, R. G. (eds) *Alpine Tectonics*. Geological Society Special Publication, **45**, 265–283.

GANSSER, A. 1974. The ophiolite mélange, a world-wide problem on Tethyan examples. Eclogae Geologiae Helvetiae, **67**, 429–507.

HARDIE, L. A. 1978. Evaporites, rifting and the role of $CaCl_2$ hydrothermal brines. *Geological Society of America*, Abstracts with Programs, **10**(7), 416.

HARDIE, L. A. 1984. Evaporites: marine or non-marine. *American Journal of Science*, **52**, 171–200.

HARDIE, L. A. 1990. The roles of rifting and hydrothermal $CaCl_2$ brines in the origin of potash evaporites: a hypothesis, *American Journal of Science*, **290**, 43–106.

HARDIE, L. A. 1991. On the significance of evaporites. *Annual Review of Earth and Planetary Sciences*, **19**, 131–168.

HARDIE, L. A. 1996. Secular variation in seawater chemistry: An explanation for the coupled variation in the mineralogies of marine limestones and potash evaporites over the past 600 my. *Geology*, **24**, 279–283.

HARDIE, L. A., LOWENSTEIN, T. K. & SPENCER, R. J. 1985. The problem of distinguishing between primary and secondary features in evaporites. *In*: SCHREIBER, B. C. & HARNER, L. (eds) *Sixth Symposium on Salt*. Salt Institute, Alexandria, VA, **1**, 11–39.

HOLLAND, H. D. 1978. *The Chemistry of the Atmosphere and Oceans*. Wiley-Interscience, New York.

HOLSER, W. T. 1979. Trace element and isotopes in evaporites. *In*: BURNS, R. G. (ed.) *Marine Minerals. Mineralogical Society of America, Reviews in Mineralogy*, **6**, 295–346.

HOVLAND, M., KUZNETSOVAW, T., RUESLA, H., KVAMME, B., JOHNSEN, H. K., FLADMARK, G. E. & HEBACH, A. 2006. Sub-surface precipitation of salts in supercritical seawater. *Basin Research*, **10**, 1365–2117.

HRYNIV, S. P. & PERYT, T. M. 2003. Sulfate cavity filling in the lower Werra anhydrite (Zechstein Permian), Zrada area, northern Poland: evidence for early diagenetic evaporate paleokarst formed under sedimentary cover. *Journal of Sedimentary Research*, **73**, 451–461.

JACKSON, M. P. A., CORNELIUS, R. R., CRAIG, C. H., GANSER, A., STOCKLIN, J. & TALBOT, C. J. 1990. *Salt Diapirs of the Great Kavir, central Iran*. Geological Society of America, Memoir, **177**.

KALHOR, R. 1961. Geology of Neogene formation in Varamin-Garmsar area and evaluation of Abardej Nose. *NIOC, Geological Report*, **233**.

KIRKLAND, D. W., DENISON, R. E. & DEAN, W. E. 2000. Parent brine of the Castile evaporites (Upper Permian), Texas and New Mexico. *Journal of Sedimentary Research*, **70**, 794–761.

KOVALEVICH, V. 1978. Fiziko-Khimicheskie uslovia formirovania soley Stebnikskogo kaliynogo mestorozhdenia: Kiev. *Naukova Dumka*.

KRINSLEY, D. B. 1970. *A Geomorphological and Palaeoclimatological study of the Playas of Iran*. Final reports for Air Force Cambridge Research Laboratories, part I, 329 and Part II.

KRINSLEY, D. B. 1974. Preliminary road alinement through the Great Kavir in Iran by repetitive ERTS-1 coverage [abstract]. *In*: *3rd Earth Resource Technology Satellite-1 Symposium, section A, Technical presentations*. US National Aeronautical and Space Administration, Special Publications, **351**(1), 823.

KRINSLEY, D. B. 1976. Selection of a road alinement through the Great Kavir in Iran. *In*: *ERTS-1, a new Window on our Planet*. US Geological Survey, Professional Paper, **929**, 296–299.

KRINSLEY, D. B. 1977. *Use of ERTS-1 (Landsat-1) Images for Engineering Geologic Applications in North-central Iran*. US Geological Survey, Professional Paper, **1015**, 113–121.

LLEWELLYN, P. G. 1968. Dendritic halite pseudomorphs from the Keuper Marl of Leicestershire, England. *Sedimentology*, **11**, 293–297.

NATIONAL IRANIAN OIL COMPANY 1977. *Geological map of Iran, sheet 2 north-central Iran, with Explanatory text* (compiled by H. Huber). NIOC, Tehran, scale 1:100,000.

NEAL, J. T. 1969. Playa variation. *In*: MCGINNIS, W. G. & GORMAN, B. J. (eds) *Arid Lands in Perspective*. Arizona University Press, Tucson, **A2**, 14–44.

PERYT, M., PIERRE, C. & GRYNIV, S. P. 1998. Origin of polyhalite in the Zechstein (Upper Permian) Zrada platform (northern Poland). *Sedimentology*, **45**, 565–578.

PHILLIPS, F. G. 1956. *An Introduction to Crystallography*. Longmans, London.

RAHIMPOUR-BONAB, H. & ALIJANI, N. 2003. Petrography, diagenesis and depositional model for potash deposits of the north Central Iran, and use of bromine geochemistry as a prospecting tool. *Carbonates and Evaporites*, **18**, 19–28.

RAHIMPOUR-BONAB, H. & KALANTAR-ZADEH, Z. 2005. Origin of the secondary potash deposits; a case from Miocene evaporites of NW central Iran. *Journal of Asian Earth Sciences*, **25**, 157–166.

ROWLANDS, G. O., DELPAK, R. & GHAEM-MAGHAMI, E. 1984. The engineering properties of the saline soils of the Great Kavir in central Iran. *In*: BOYCE, J. R., MCKECHNIE, W. R. & SCHWARTZ, K. (eds) *Proceedings, 8th Regional Conference for Africa on Soil Mechanics and Foundation Engineering*, **8**, 337–341.

RUSSO, T. N. 1976. *Water Resources in the KAVIR National Park, Iran, 1975*. Iranian Department of Environment, Division of Nature Conservation.

SADEDIN, N. 1990. *Potash Exploration and Equipping Project in Iran for the Semnan Province*. Ministry of Mines and Industries, Reports.

SCHREIBER, B. C. & HSÜ, K. J. 1980. Evaporites. *In*: HOBSON, G. D. (ed.) *Developments in Petroleum Geology*, Applied Science, Barking, **2**, 87–138.

SCHREIBER, B. C. & EL TABAKH, M. 2000. Deposition and early alteration of evaporites. *Sedimentology*, **47**, Millennium Issue, 215–238.

SCOTESE, C. R. 1987. Phanerozoic reconstructions: a new look at the assembly of Asia. *Paleooceanographic Map Project Progress Report* 19–1286.

SHEARMAN, D. J. 1978. Evaporites of coastal sabkhas. *In*: DEAN, W. E. & SCHREIBER, B. C. (eds), *Marine Evaporites*. SEPM Short Course, **4**, 6–42.

SIEMANN, M. G., 1993. A practice-orientated way to produce diffraction reference cards using evaporate minerals. *Material Sciences Forum*, **136**, 27–32.

SIEMANN, M. G. & SCHRAMM, M. 2000. Thermodynamic modeling of the Br partition between aqueous solutions and halite. *Geochemica et Cosmochemica Acta*, **64**, 1681–1693.

SIEMANN, M. G. & ELLENDORFF, B. 2001. The composition of gases in fluid inclusions of Late Permian (Zechstein) marine evaporates in Northern Germany. *Chemical Geology*, **173**, 31–44.

SIEMANN, M. G. & SCHRAMM, M. 2002. Henry's and non-Henry's law behavior of Br in simple marine systems. *Geochemica et Cosmochemica Acta*, **66**, 1387–1399.

SOFFEL, H. C. & FORSTER, H. G. 1984. Polar wander of the Central-east Iran microplate including new results. *Neues Jahrbuch Geologische und Palaontologische Abhandlungen*, **168**, 165–172.

SONNENFELD, P. 1984. *Brines and Evaporites*. Academic Press, New York.

SONNENFELD, P. 1995. The color of rock salt – a review. *Sedimentary Geology*, **94**, 267–276.

SPENCER, R. J. 1987. Origin of $CaCl_2$ brines in Devonian formations, western Canada sedimentary basin. *Applied Geochemistry*, **2**, 373–384.

STOCKLIN, J. 1968. Salt deposits of the Middle East. *In*: MATTOX, R. B. (ed.) *Saline Deposits*. Geological Society of America, Special Paper, **88**, 158–181.

STOCKLIN, J. 1974. Possible ancient continental margins in Iran. *In*: BURK, C. A. & DRAKE, C. L. (eds) *The Geology of Continental Margins*. Springer, Berlin, 873–887.

VALYASHKO, M. G. 1972. Playa lakes – a necessary stage in the development of salt-bearing basin, *In*: RICHTER-BERNBURG, G. (ed.), *Geology of Saline Deposit*. UNESCO, Paris. Earth Sciences Series, **7**, 41–51.

VAUGHAN, H. B. 1893. A journey through Persia. *Royal Geographical Society Journal*, Supplementary Papers, **3**, 89–115.

WARREN, J. K. 1999. *Evaporites, their Evolution and Economics*, Blackwell Science, Oxford.

ZAK, I. 1997. Evolution of the Dead Sea brines. *In*: NIEMI, T. M., BEN-ABRAHAM, Z. & GAT, J. R. (eds) *The Dead Sea: The Lake and Its Setting*. Oxford University Press, Oxford, 133–144.

A giant Late Triassic–Early Jurassic evaporitic basin on the Saharan Platform, North Africa

P. TURNER[1,2] & H. SHERIF[1]

[1]*School of Geography, Earth and Environmental Sciences, The University of Birmingham, Birmingham B15 2TT, UK (e-mail: turnerpetergeos@btinternet.com)*

[2]*Present address: Turner Geoconsultants Ltd, 7 Carlton Croft, Streetly, West Midlands B74 3J7, UK*

Abstract: This paper describes the Late Triassic and Early Jurassic evaporites of the Berkine/Ghadames Basin in eastern Algeria and western Libya, which is important because it forms the regional seal to a number of hydrocarbon reservoirs in the Triassic Argilo-Gréseux Inférieur. This evaporitic succession spans the time interval Norian to Late Liassic and is divided into three contrasting units (S1, S2 and S3). S1 consists of five evaporitic cycles dominated by mudstone and halite, which represent the main salt deposits of the Berkine Basin. At this time the Berkine/Ghadames basin was a restricted evaporitic basin with a barrier to the north (Medenine High) separating the basin from the developing peri-Tethys and North Atlantic. The succession thickens to the west, thins around the Hass Messouad Ridge and continues across the Maghreb into the western High Atlas of Morocco, a distance of some 1500 km. S2 is represented by the basin-wide development of a carbonate platform in Pliensbachian times. This is the product of a relative sealevel rise concomitant with the developing extension of the region and may reflect the opening of an Atlantic gateway. S3 comprises five cycles that are predominantly mudstones, fine-grained carbonates and anhydrites. Mounded structures, evident on seismic sections, may represent the development of patch reefs and the whole succession is interpreted as a shallow-marine carbonate–evaporite ramp. Similar thickness variations to those observed in Sequence 1 continue through Sequence 3 and reflect the on-going influence of differential subsidence along basement lineaments. A number of factors controlled the development of the evaporite basin. These include Late Triassic sealevel rise and flooding of a sub-sealevel basin, combined with globally warm climatic conditions and on-going extensional tectonics in the peri-Tethyan and Atlantic areas. Restricted circulation was caused by an east–west oriented sill to the north (Medenine High in central Tunisia) and the basin was episodically refilled as global sealevel rose during the Early Mesozoic and extensional tectonics continued.

During the Middle to Late Triassic to Lower Jurassic, a number of basins developed along the northern margin of the Saharan Craton. These peri-cratonic basins, which included the Oued Mya and Berkine Basins of Algeria and the Ghadames Basin of Libya, were initially filled with continental clastic (alluvial) deposits. These clastics include the late Triassic Argilo-Gréseux Inférieur (AG-I), which extends from Algeria to southern Tunisia, and its equivalent, the Ra's Hamia Formation of western Libya, deposited in a fluvio-lacustrine system, which flowed mainly from SW–NE towards peri-Tethys (Turner *et al.* 2001).

As the eastward opening of Tethys progressed and relative sealevel rose, the basins were inundated and developed a system of evaporitic basins which stretched from Hassi R'Mel in the west to Jebel Nafusah in the east, straddling northern Algeria, Tunisia and western Libya, a distance of some 750 km (Fig. 1). The northern margin of this basin is not seen because of the overthrust Rif–Tell system of the Atlas Mountains, although contemporaneous Atlasic half graben was formed in Morocco and this also contains significant evaporite deposits. The southern margin is formed by the Palaeozoic rocks of the Tinhert–Qargaf uplift. The Amguid–Biod Ridge forms a major north–south spur, which protrudes into the basin and reflects the basement control of Pan-African lineaments. Thickness variations in the evaporitic successions appear to be controlled by movements on these basement lineaments (Busson & Corneé, 1989; Ait Salem *et al.* 1998; Eschard *et al.* 1999; Guirard *et al.* 2005; Turner *et al.* 2001).

A summary of the regional setting of the Triassic successions in the Berkine Basin is presented by Busson (1971*a*, *b*) and the Triassic section, and the sequence stratigraphy of the Maghreb area (stretching from Morocco in the west to eastern Algeria) has been studied by Courel *et al.* (2003). The oldest Triassic deposits of pre-Ladinian age occur in Tunisia. These are interbedded with marine carbonates and are linked to the Tethyan domain (Ben Ismail 1991; Guiraud 1998; Soussi *et al.* 1998; Soussi 2000). Upper Ladinian–Carnian clastics overlie these deposits and the

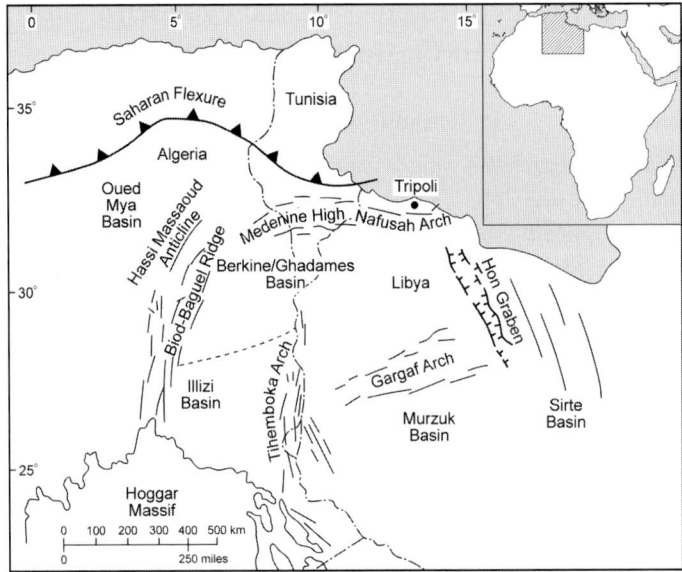

Fig. 1. Location map of the Ghadames/Berkine Basin (Tunisia and Algeria) showing main structural features.

pre-Triassic basement. From near the Algerian/Tunisian border towards Hass R' Mel the clastic/evaporite successions become progressively younger (Mami & Bourmouche 1994; Ait Salem et al. 1998) and by the late Triassic and Liassic, evaporite deposition predominates. Courel et al. (2003) divided the succession into seven Megasequences (I–VII), which thicken eastwards towards the Tunisian border.

Deposition began in the Middle Triassic with the deposition of fluvial and lacustrine clastics of the Triassic Argilo-Greseux Inferieur (TAG-I) (Turner et al. 2001). In the study area, the TAG-I rests unconformably on a subcrop of Carboniferous and Devonian sediments and is overlain by the Triassic Argilo-Carbonaté (or Triassic Carbonates). Farther westwards younger Triassic clastics are known as the TAG-M (Moyen) and TAG-S (Superieur). The Triassic carbonates pass upwards through a series of shales with thin dolomites into a thick carbonate/evaporite succession.

In this paper we describe the Triassic and Liassic evaporites of the Berkine/Ghadames Basin for the first time as an integral unit and extend the basin model and stratigraphic database eastwards from Algeria into Libya. Our results show that this basin is a broad depositional sag, broadly east–west in orientation and thickening away from Palaeozoic basement highs which formed the margins of the basin. The basin was silled, probably by the Medenine High in central Tunisia, and replenished by peri-Tethyan waters which periodically breached the sill. Contemporaneous evaporite basins farther west in Morocco indicate that in Late Triassic/Early Jurassic times the whole of North Africa consisted of a huge evaporitic basin complex (LeRoy & Pique 2001). The Moroccan sections show remarkable similarity with those of the Berkine basin. There are two evaporite successions spanning the Rhaetian–Sinemurian time interval which, like the Algerian sections, show increasing marine influence with time (Ouijdi 1998; Tourani et al. 2000). Furthermore geochemical studies of these evaporites show clear evidence of the mixing of marine and non-marine brines as the Atlantic transgressed southward along the rifted continental margin (Clement & Holser 1988; Horita et al. 2002).

Database and methods

The database for this study comprises two-dimensional seismic data and wireline logs from wells in the Berkine Basin in Algeria and the Ghadames Basin in Libya (Fig. 2). There are no available cores apart from a thin-cored succession at the base of the evaporitic section in the Triassic Carbonaté in the Berkine Basin. However, these do give an insight into the early stages of development of the basin in that they record the initial marine flooding (Turner et al. 2001). A total of 18 wells has been used in our study, including 13 from Algeria (SFNE-1, SFNE-2, BSFN-1, BSFN-2, BSF-1, BSF-2, ROD-1, ROD-2, ROD-4, FAT-1, BTR-1, BMA-1 and RER-2) and five from Libya (L1-NC100, A1-NC118, I1-90, D1-70 and Q2-NC7).

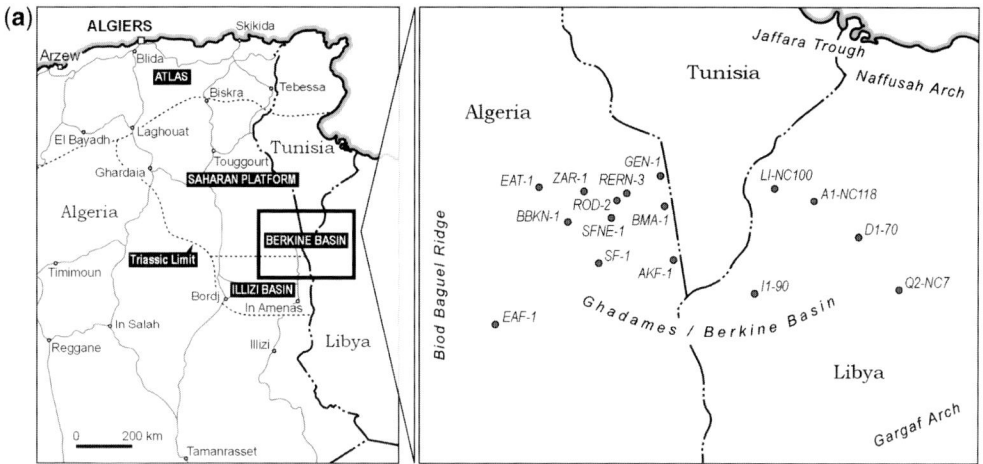

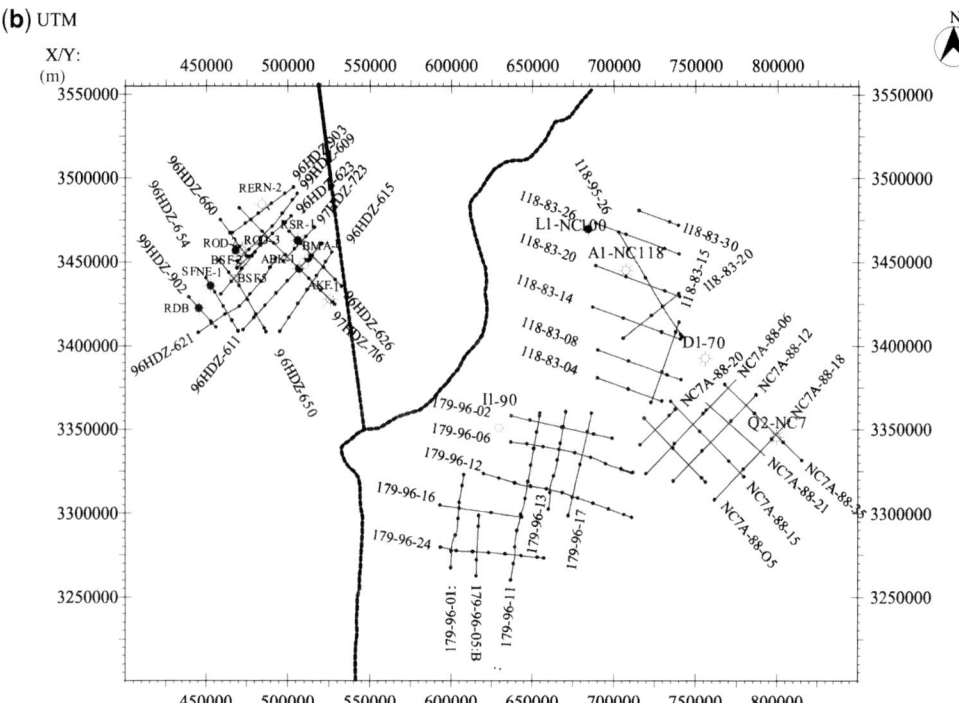

Fig. 2. Regional location of seismic lines and wells used in the study. (**a**) Well locations. (**b**) Distribution of seismic lines.

The seismic data used in the study covered most of area in the west and the northeast of the basin. It comprises 40 two-dimensional migrated lines of various vintages. The quality of the data is generally good. The high resolution displayed by the seismic data in some sections has permitted detailed interpretation of sequence stratigraphic relationships. The evaporite successions in the Ghadames/Berkine basin are displayed between 0.5 and 2.0 s two way travel time and, in general, show that the evaporite horizons are thinner and shallower in the east (Libya) and become thicker and deeper towards the west.

The seismic data were combined with information from the 18 wells (formation tops, interval velocities, synthetic seismograms) together with published data to interpret the evaporitic successions. Our study is limited by poor biostratigraphical control and the fact that no seismic data were available in the central part of the study area (south-central Tunisia), causing a gap in the seismic tie-ins between the Algerian and Libyan areas. This area corresponding to the southernmost part of Tunisia was readily interpolated, by matching the seismic sections and well data for each area and comparing the results with available published information (Soussi et al. 1998, 2000; Soussi & Ben Ismail 2000).

The wireline log data were used as a basis for identification of the main lithologies (shale, carbonate, anhydrite and halite) using gamma, density and sonic logs. From these data cycles were constructed for all the wells and correlated on a well-to-well basis. Isopach maps were constructed from thicknesses derived from the wireline logs and the seismic data for individual cycles and also for identified successions and the basin as a whole. Wherever possible our sequence analysis has been compared with conventional lithostratigraphies for the Ghadames and Berkine Basins.

Lithostratigraphy and seismic stratigraphy

Lithostratigraphy

A generalized Triassic and Early Jurassic stratigraphy for the Berkine/Ghadames Basin is shown in Figure 3. The oldest deposits are clastic units which fill an irregular topography developed on the Hercynian unconformity by erosion during Permian and earliest Triassic times. In Libya this corresponds to the Ra's Hamia Formation, the Kirchaou Sandstone in Tunisia and the TAG-I in Algeria. The relative age relationship of these units is somewhat uncertain because of the lack of biostratigraphical control and the marked lateral facies variations. These are, at least in part, due to the irregular topography of the Hercynian unconformity (Benrabah et al. 1991), but also to the local sediment sources on the adjacent palaeo-highs of the Medenine, Nafusah, El-Biod and Dahar Highs. The oldest of these successions is probably the Kirchaou Sandstone of Tunisia, which is pre-Ladinian in age (Ben Ismail 1991). The Ras Hamia Formation is broadly the same age and has a similar maximum thickness (c. 230m) to the TAG-I (Ben Ismail 1991; Soussi et al. 1998).

In the central part of the basin the Triassic clastic unit is followed by the Triassic Carbonaté, a succession of shales, thin sandstones, dolomites and thin anhydrites. Westwards this succession becomes sandier towards the Amguid–El Biod high and the predominantly shaly succession is replaced by marginal clastics of the TAG-M and TAG-S. In the central part of the Berkine basin and in Tunisia the equivalent-aged strata are dominated by evaporites, including anhydrite and halite. In central Tunisia these deposits correspond to the Rheoius Formation and, further eastwards, towards the Jebel Nafusah, these can be traced into the dolomites of the Azizia Formation. The Azizia Formation is followed by the Carnian–Norian aged Abu Shaybah Formation, a fluvial complex separated from the underlying Aziziah Limestones by a largescale incised valley fill (Rubino et al. 2003). The Abu Shaybah grades upwards into lagoonal carbonates and grades northward to sabkha and shallow-marine carbonates of the Tethyan margin and gradually onlaps to the south–southeast towards the basin margin (Bishop 1975; Hallet 2000).

In the latest Triassic salt deposits are present in the centre of the Berkine Basin (Lower Saliferous Shale and the S4 halite) and also in the Tunisian Atlas (Gaetani et al. 2000). Here these salts were deformed during Liassic rifting (Tanfous Amri et al. 2005). In Algeria the Triassic–Jurassic boundary is often taken at the D2 dolomite (Courel et al. 2003), but this is difficult to determine because of the lack of biostratigraphical control. The base of the Jurassic in Tunisia must be close to the base of the Lower Nara Formation, a predominantly dolomitic formation. This boundary corresponds to the base of the Bir Ghanam Group of Libya Abu Ghaylan of western Libya. The lowermost Jurassic deposits in the Berkine Basin are predominantly evaporitic and include thick halite deposits. Farther eastwards these pass into anhydrite and gypsiferous units (Bhir Formation), which correlate with the Bir Al Ghanam Formation of Libya (Hallet 2000).

This clastic–evaporite succession is terminated by Pliensbachian times by the development of a major carbonate platform complex which corresponds to the well-known 'B marker' from seismic sections in the Berkine Basin. In surface outcrops this corresponds to the Zmilet Haber Formation in southern Tunisia

Above the 'B marker' in Algeria the section is dominated by the Liassic salt and anhydrite deposits which correlate with the Abreghs Formation in the Ghadames area and the Bir El Ghanam Group around Jebel Nafusah. The lithostratigraphy in these areas is complex and difficult to correlate, partly because of the nature of the lithologies, rapid vertical and lateral changes and the lack of biostratigraphical constraints. We believe our sequence analysis goes some way towards clarifying the picture.

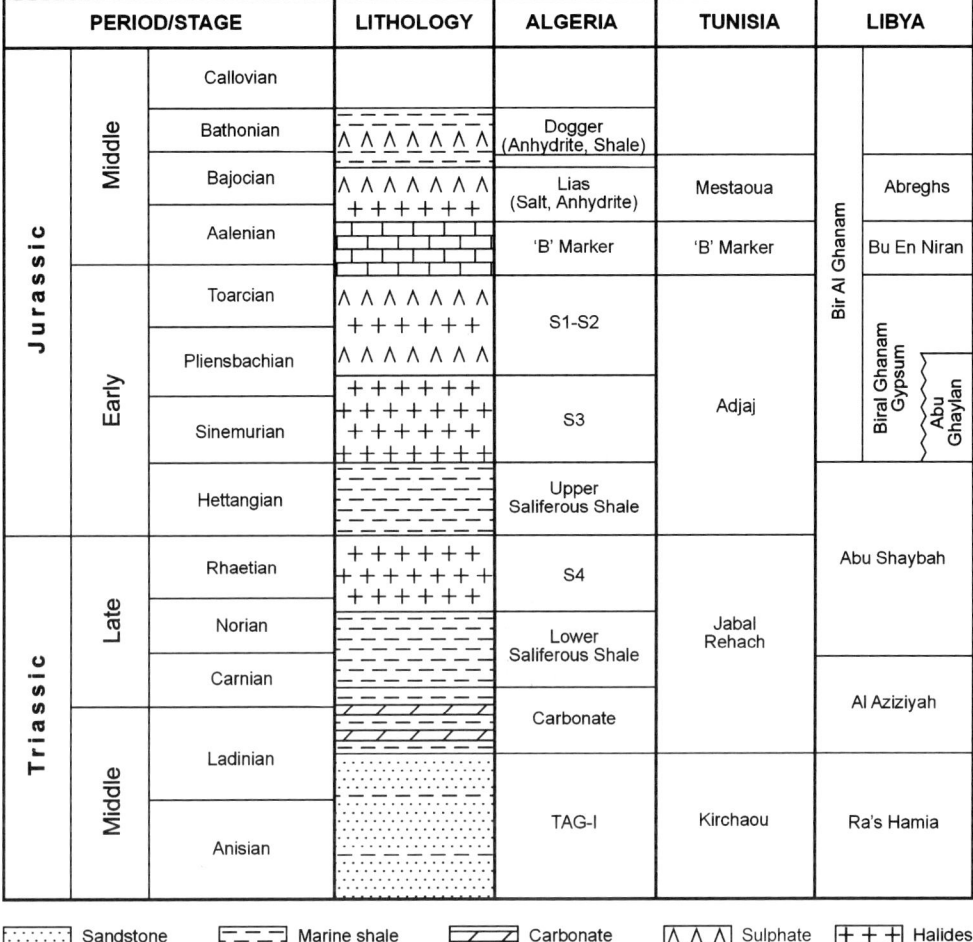

Fig. 3. Comparative lithostratigraphy of the Triassic/Jurassic in the Berkine/Ghadames basins.

Seismic stratigraphy

The evaporite succession in the giant basin has a maximum thickness of about 1.2 km and has been studied using seismic data, boreholes and outcrops from around the basin margin. Four key horizons were picked from the seismic data:

(1) the top of the evaporite succession;
(2) the B-marker horizon;
(3) the base of the evaporite section; and
(4) Hercynian unconformity.

A regional seismic section with geoseismic interpretation showing these features is presented in Figure 4.

The seismic interpretation is consistent with the known lithostratigraphical succession as observed in the well sections. In broad terms, and based on the data presented in Figures 3 and 4 we define the units as follows: the Triassic clastics (TAG-I), Sequence 1 comprising the interval between the base of the Triassic Carbonate, Sequence 2 corresponding to the 'B marker' and Sequence 3 corresponding to the section above the 'B marker' up to the Dogger Shale. In Libya and eastern Algeria these successions dip gently to the west and thicken towards the Hassi Messouad- El Biod Ridge. The thinning and onlap of the succession towards the Nafusah Ridge accounts for the more complex basin–margin stratigraphy and facies seen in this area. This pattern contrasts with that observed by Courel *et al.* (2003), who showed that in the Triassic, at least, the succession thins westward onto the Hassi R' Mel high. This is considerably farther west than the present study area but does confirm that, in general, the Triassic and Lower Jurassic clastic/evaporites form a giant

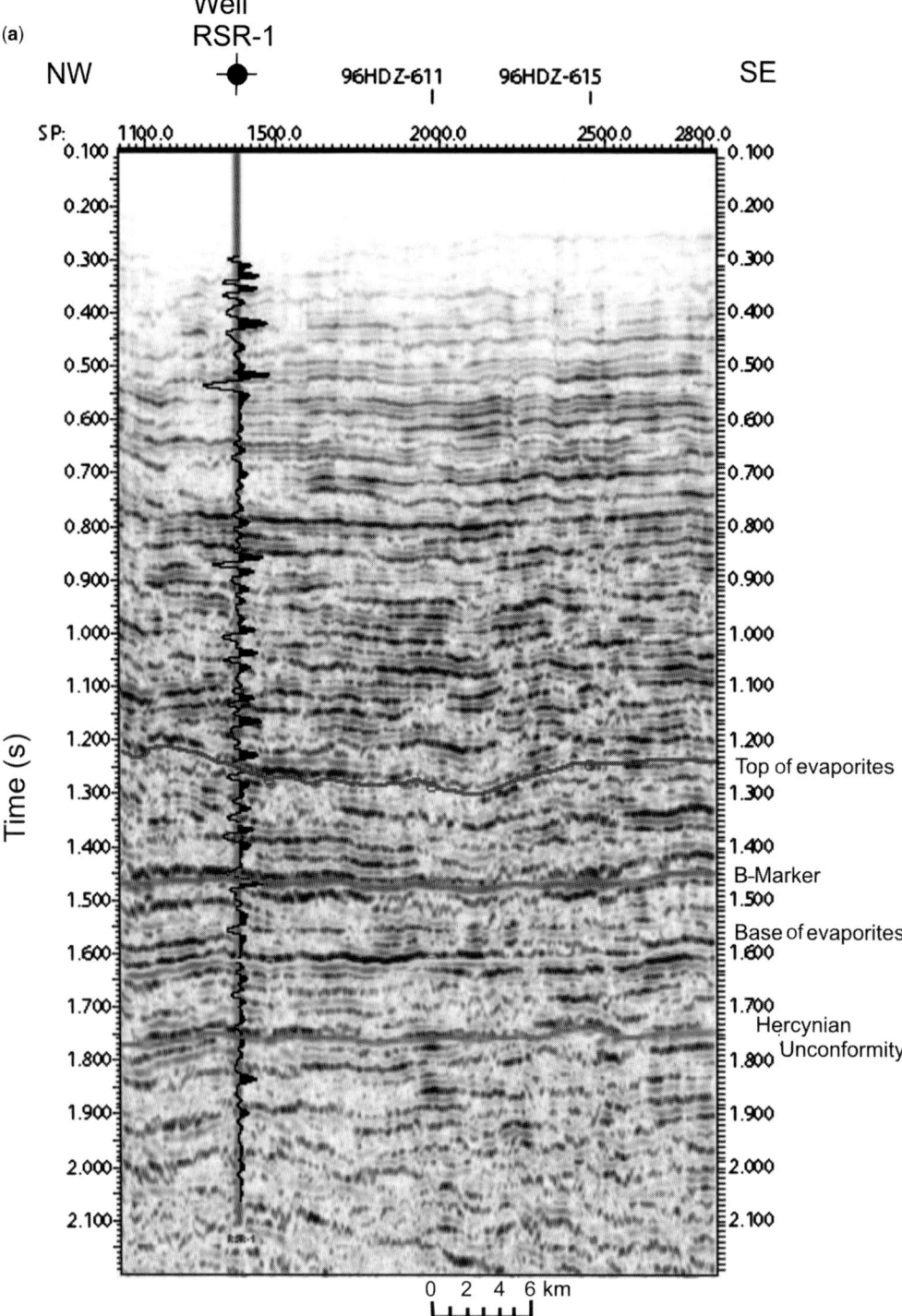

Fig. 4. (a) Regional seismic section showing thickness variations of the main units. (b) Geoseismic interpretation showing overall thickness variations from Jebel Nafusah to the central Berkine basin.

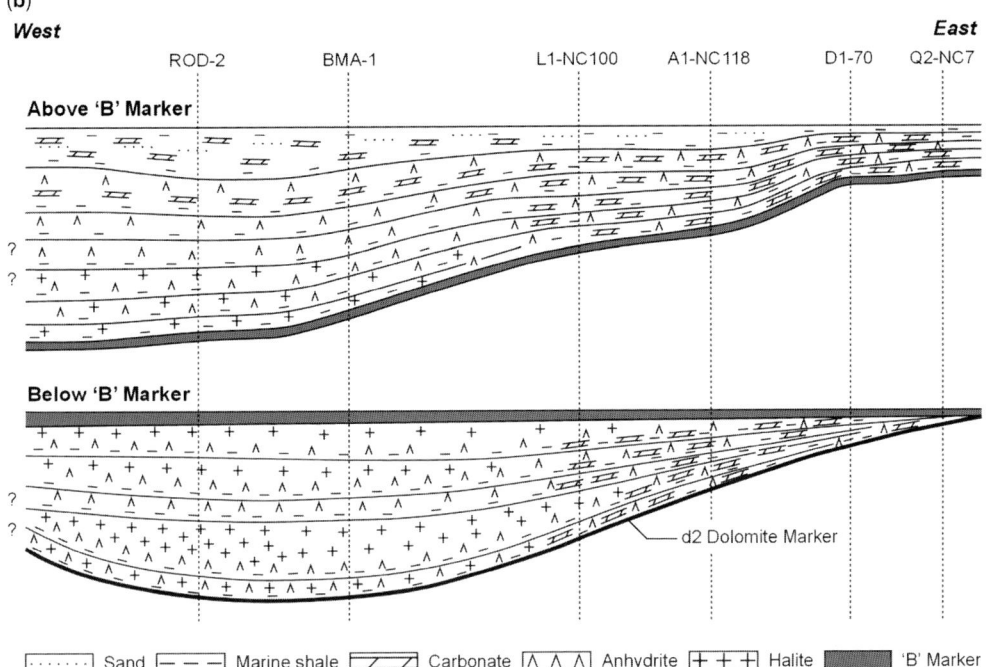

Fig. 4. *Continued.*

basin stretching over a distance of at least some 750 km.

Evaporite cycles

In eastern Algeria four main halite units are recognized but these give little indication of the sedimentological context or detailed stratigraphy of the deposits. The combination of seismic interpretation and wireline log studies shows that the Triassic/Jurassic evaporitic succession can be divided into three distinct major sequences, 1, 2 and 3, which are best developed in the central part of the basin in the Berkine Basin, near study well ROD-2 (Fig. 5)

Sequence 1

Description of Sequence 1 Sequence 1 comprises up to five cycles which include all the main salt units S1–S4 of the conventional lithostratigraphical scheme (Fig. 5). The cycles in Sequence 1 range from 50 to 170 m in thickness and are dominated by mudstone and halite with only very minor anhydrite development. The insets in Figure 5 show typical wireline log responses in Sequence 1 and how these have been interpreted using a combination of gamma and sonic logs. These cycles typically start with mudstone and pass upwards through relatively thin anhydrite into thicker halite deposits.

The first cycle (C1) is relatively thin, but this is immediately followed by the thickest cycle (C2), which shows the maximum halite development of about 120 m (Fig. 5); the upper cycles C3, C4 and C5 are of comparable thickness and show a relatively higher proportion of anhydrite than that seen in C1 and C2.

Interpretation of Sequence 1 cycles These cycles are interpreted as 'deep water evaporites' (Sarg 2002). The mudstones at the base of each of these cycles are interpreted to represent the initial flooding of the basin and consequent evaporation with the precipitation of deep-water anhydrite and halite. At this time marine circulation in the basin must have been restricted with periodic replenishment. Our evidence suggests that the barrier to the basin lay to the north in the vicinity of the Medenine High (Bishop 1975; Soussi & Ben Ismail 2000) and the initial Gahadames/Berkine basin probably lay below sealevel. The evaporite cycles provide a detailed record of relative sealevel changes, which relate to the global sealevel rise in the late Triassic and early Jurassic (Fig. 6).

Sequence 2: the 'B Marker'

Description Sequence 1 is terminated relatively abruptly by the 'B marker'. This unit is a distinctive

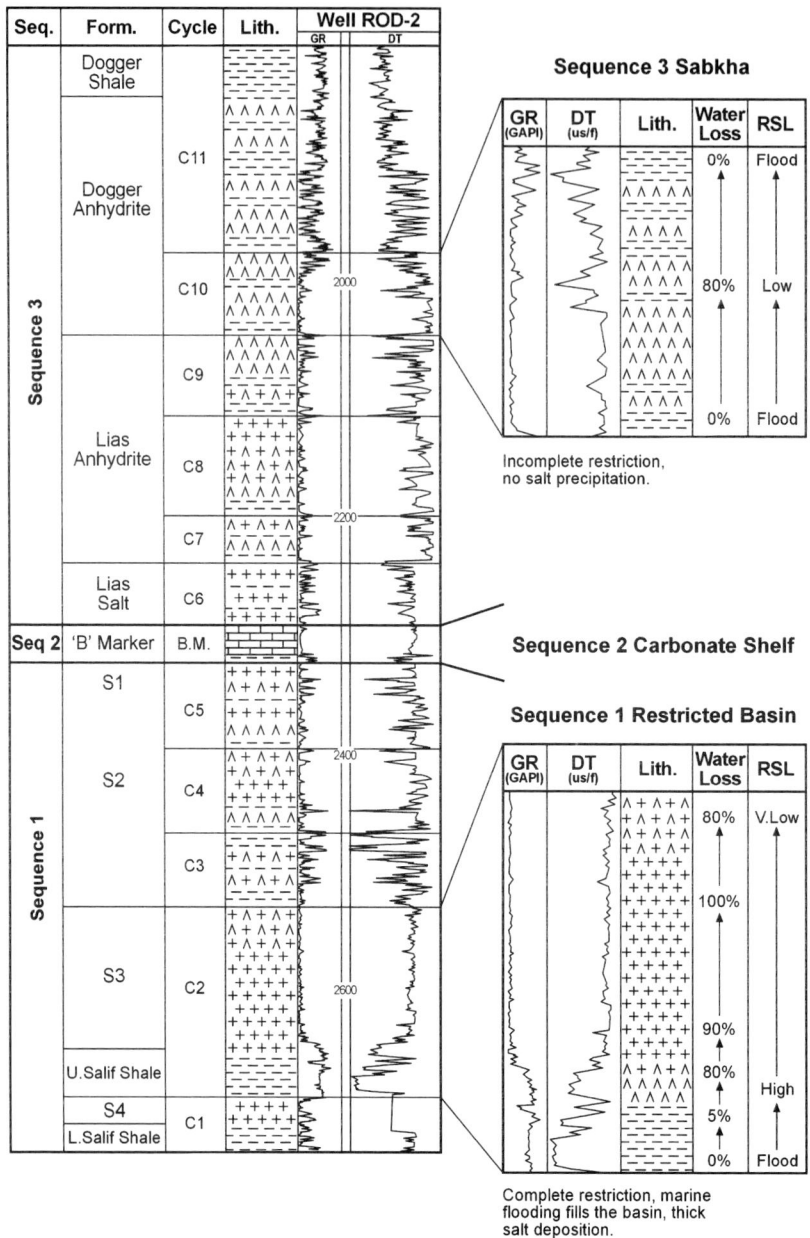

Fig. 5. The ROD-2 well in Algeria showing evaporite stratigraphy and sequence analysis. The details of two individual cycles are shown to demonstrate the methodology used in wireline log interpretation.

marker horizon on seismic sections and has been studied at outcrop on Jebel Nafusah in Libya and in southern Tunisia. It is an intensively dolomitized sequence of carbonate grainstones including bioclastic and oolitic facies. In southern Tunisia the 'B marker' forms part of a thick Liassic platform carbonate succession (Soussi *et al.* 1998, 2000) and the evaporites, as seen farther south in the Berkine/Ghadames Basin, are absent.

Interpretation The precise age of the 'B marker' is not well known. In Libya and central Tunisia it is probably Pliensbachian whereas in eastern Algeria it may be somewhat younger, suggesting

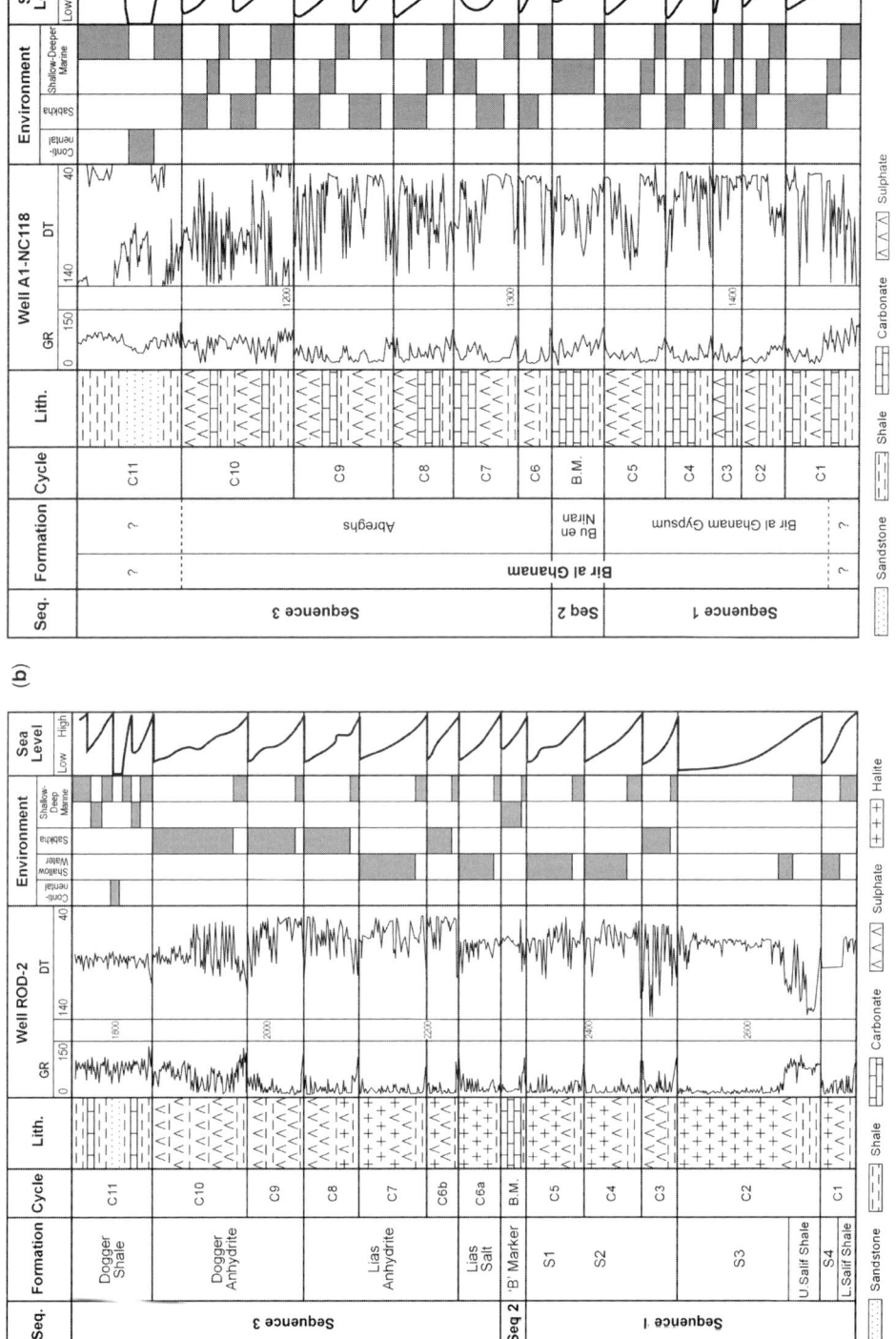

Fig. 6. Stratigraphy and evaporite cycles in Well ROD-2 from the Berkine Basin compared with Libyan Well A1-NC-118.

that the development of this carbonate platform was diachronous, although this is difficult to prove because of poor biostratigraphical control.

Sequence 3

Description Sequence 3 shows significant lithological contrast with Sequence 1 and comprises six cycles of mainly carbonate/anhydrite ranging from 30 to 100 m in thickness (Fig. 5). In the central part of the basin the first of these cycles (C6) contains a minor amount of halite and corresponds to the 'Lias Salt' of previous classifications (Busson 1971*a*; Boudjema 1987). This is the last significant halite development in the basin and the higher cycles (C7–C11) are dominated by carbonate and anhydrite. The topmost part of the succession becomes progressively mud-rich. Detailed inspection of the seismic data shows that higher cycles (cycles 8 and 9) developed mounded topographies along the northern margin of the basin, which indicates the development of low-relief carbonate buildups. These may have a relief of up to 80 m and a lateral extent of over 4 km. The Dogger Shale is considered to represent the termination of the evaporite conditions on the Saharan Platform (Figs 3 & 5)

Interpretation Sequence 3 is dominated by carbonate–anhydrite cycles which are best interpreted as sabkha-type evaporites (Purser & Evans 1973; Sarg 1988; Tucker 1991; Tucker *et al.* 1993). However, the first of the cycles in Sequence 3 (C6) contains substantial halite, indicating that restricted circulation again developed immediately after the collapse of the 'B marker' carbonate platform (Soussi & Ben Ismail 2000). The reason for this is unclear, but the succeeding cycles (C7–11) show no further significant halite and the succession is dominated by dolomites and anhydrites. We interpret this to indicate major changes in the area in terms of tectonics and global sealevel change. After the deposition of C6, marine circulation generally improved and this may be related to the ongoing crustal extension, including the opening of the Atlantic gateway and rising sealevel during the Lower Jurassic (Thierry *et al.* 2000*a*, *b*). By the middle part of Sequence 3 our seismic data show the development of mounded structures which may be similar to the reefs described from the Sinemurian of Morocco by Chafiki *et al.* (2004).

Correlation of evaporite cycles

The two wells shown in Figure 6 show major variations in both lithology and thickness when the succession is traced eastwards into Libya. The key difference is that the Libyan section is much thinner (c. 350m) as opposed to c. 950m in the Berkine Basin (e.g. ROD-2) and there is a complete absence of halite. Sequence 1 comprises five cycles with mudstone, carbonate and anhydrite. The 'B marker' (Bu en Niram) is well developed with a basal shale unit overlain by dolomite and C6–C11 contain relatively thick anhydrites interbedded with thin mudstones and dolomites. In C11 there is a sand accumulation which is not seen in the basin centre succession. Overall the Libyan succession is clearly more marginal than that seen in the Berkine Basin area and this is interpreted as a result of its proximity to the Nafusah High.

The cycles can be readily correlated across the basin. The correlation of the cycles below the 'B marker' is illustrated in Figures 7 and 8, above the 'B marker'. The pattern is essentially the same for both successions and consistent with that observed in the seismic sections. Below the 'B marker' the cycles thin rapidly towards the east and southeast such that they cannot be recognized in the Libyan D1-70 well. Above the 'B marker' the same pattern is observed but the cycles are more readily recognizable farther southeastwards. These correlations clearly indicate the progressive thinning and onlap of the evaporitic sequence onto the Medenine and Nafusah Highs.

Sequence stratigraphy

Courel *et al.* (2003) described the sequence stratigraphy of the late Triassic clastic/evaporite succession in the Maghreb, extending from Morocco to near the eastern border of Algeria (Fig. 9). The Triassic succession is about 500 m thick and thickens eastwards towards the Tunisian border. These authors divided the Triassic succession into seven megasequences based on the identification of maximum flooding surfaces and major lithological changes. We have extended the study of Courel *et al.* (2003) by incorporating well/seismic data from the central part of the Berkine Basin and farther eastwards into Libya. Furthermore, we have incorporated data from the Lower Jurassic in order to investigate the overall geometry and evolution of the Saharan evaporitic basin. Additionally our successions are based on the identification of individual evaporite cycles in the manner described above.

Sequence 1 is a predominantly mudstone–halite succession with individual cycles comprising shale, thin anhydrite and mostly halite. The succession thickens rapidly to the west and individual cycles thin and onlap the Jebel Nafusah high to the east. On the western flank thickness variations show two, step-like increases, which may have been controlled by extensional activity on basement

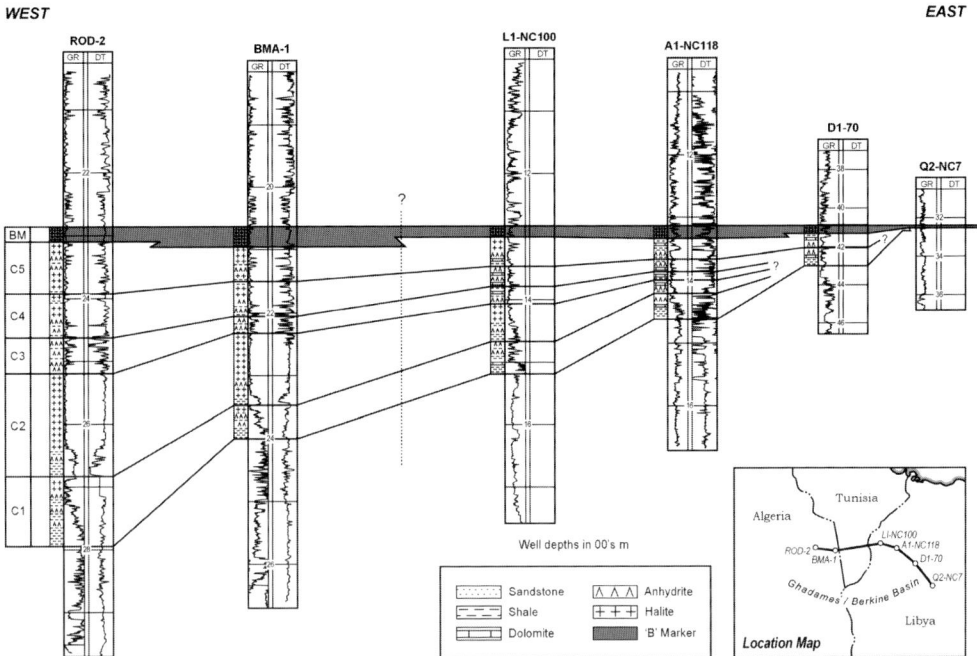

Fig. 7. Correlation of cycles below B marker.

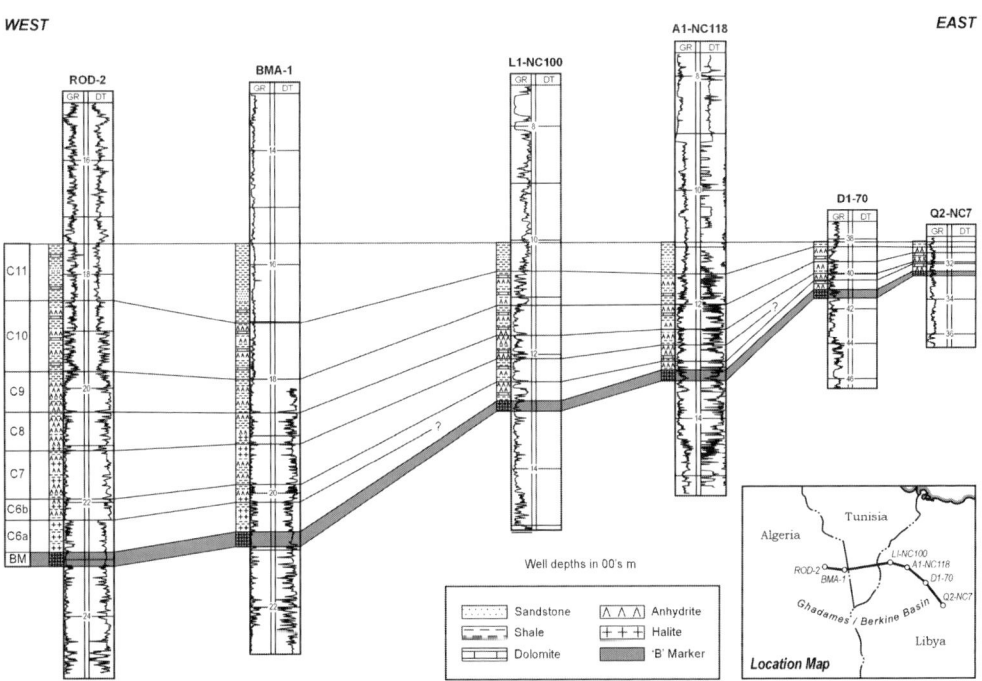

Fig. 8. Correlation of cycles above the B marker.

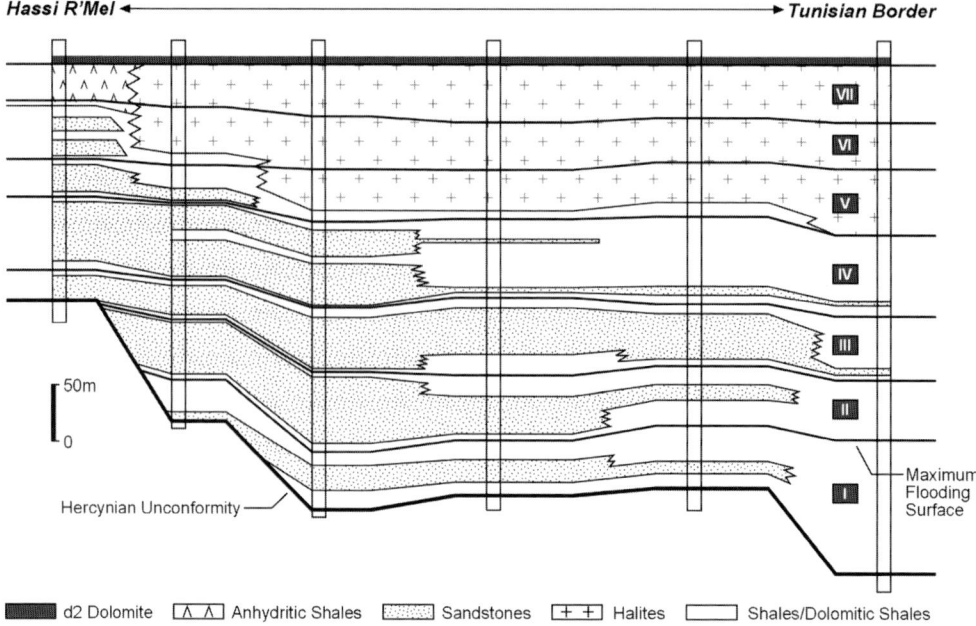

Fig. 9. Sequence stratigraphy of the late Triassic (modified from Courel *et al.* 2003).

lineaments. This Sequence 1 is terminated by the development of the 'B marker' (Sequence 2). This is a basin-wide carbonate shelf unit of relatively constant (about 25 m) thickness, comprising dolomitized oolitic and bioclastic grainstones, which appear to have been deposited during a tectonically quiescent period between the two evaporitic episodes (Ben Ismail 1991).

Sequence 3 shows a return to predominantly evaporitic conditions although in this upper succession anhydrite deposits predominate reflecting less restricted conditions. Thickness variations are similar to those in Sequence 1 with rapid thinning towards the Nafusah High. The thickness variations indicate the renewal of extensional activity. Continued opening of peri-Tethys and the North Atlantic resulted in the development of normal-marine conditions in later Jurassic times.

Palaeogeography and basin evolution

The geometry of the basin has been studied by constructing isopach maps of Sequences 1 and 2. These isopach maps, which compare very closely with those of Bishop (1975) (Fig. 10), are useful because they help delineate the basin margins and provide evidence of the likely position of basin margin sills which were instrumental in the restriction of basin circulation. They also provide direct evidence of potential connections with the peri-Tethys, which was the main marine water supply.

The thickness variations contrast with those seen in the Triassic and further to the west by Courel *et al.* (2003). This defines a giant basin contained by the Precambrian basement lineaments of the Jebel Nafusah, Medenine High, the Hassi Messouad-El Biod ridge and the Gargaf Arch to the south.

Palaeogeographic maps have been constructed for each of the cycles in Sequences 1 and 3. In Sequence 1 (cycles 1–5; Fig. 11), the cyclostratigraphy shows that the basin was restricted throughout with halites developed in most of the cycles; the exception is cycle 3, which is thinner and less extensive and does not (at least in the Berkine Basin area) show any signs of significant halite development. Overall, however, Sequence 1 is terminated by the 'B marker', a carbonate shelf unit, which was established across the Saharan craton. Clearly any restrictions in circulation were breached at this time (Pliensbachian) and relative sealevel rise was probably the main cause. The 'B marker' shows all the attributes of shallow marine carbonate platform deposits including oolites and bioclastic grainstones, but the rocks are intensively dolomitized and good biostratigraphical information is sparse. Interestingly the development of the 'B marker' does not mark a permanent change in the basin; cycle 6a, which occurs immediately above the 'B

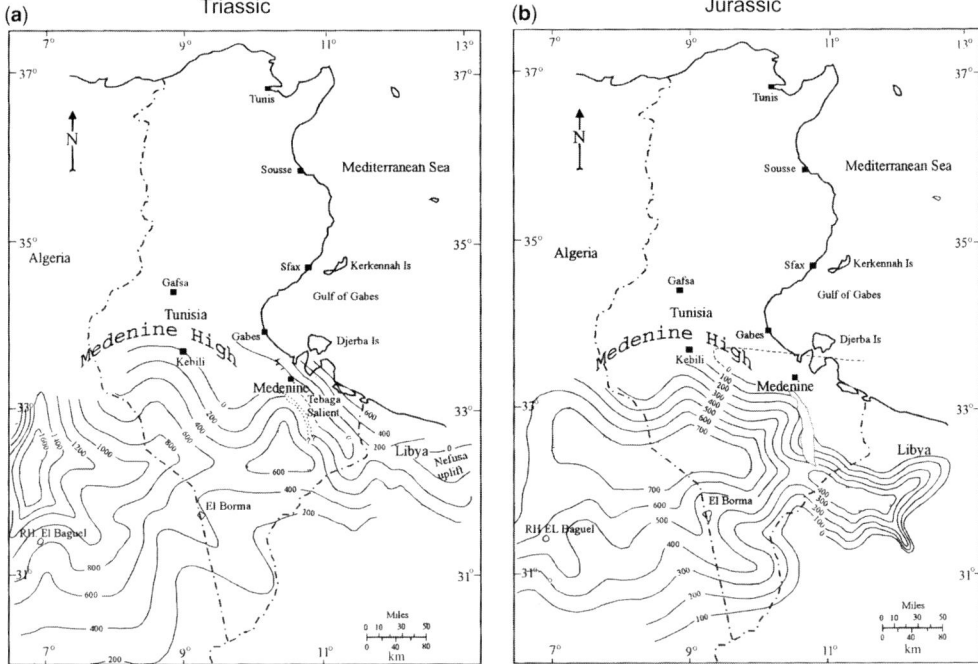

Fig. 10. Isopach map showing overall thickness variations in the Triassic and Jurassic (based on Bishop 1975).

marker' and forms the base of Sequence 3, shows a return to evaporitic conditions and the deposition of the thick 'Liassic Salt'. The reason for this is not clear but indicates renewed restriction of circulation, possible as a result of tectonic activity (uplift along the Medenine High?).

However, in the main part of Sequence 3, deep basinal evaporitic conditions are no longer seen and the cycles are dominated by dolomites and anhydrites. We interpret this to indicate that the influence of the Medenine High is reduced, circulation is only partially restricted, and the cycles represent sabkha depositional environments in an upper ramp setting. The palaeogeography of our study area (Fig. 12) shows more widespread carbonate deposition with time as the basin becomes more fully marine. The opening of the proto-Atlantic gateway in late Sinemurian (Thierry et al. 2000a, b) may have been responsible for the ultimate demise of restricted circulation in the Berkine/Ghadames basins.

Development of the Saharan evaporite basin

In the Berkine basin the total thickness of the evaporite succession is up to 1.25 km and similar deposits can be traced across the whole of North Africa (Pique et al. 2002; Courel et al. 2003). A generalized outline of the basin and schematic cross-section is shown in Figure 13. The Late Triassic–Early Jurassic deposits of the Saharan Platform constitute a major evaporitic basin stretching over a distance of some 750 km. The total thickness of the evaporite succession is up to 1.5 km. A generalized outline of the basin is shown in Figure 14. The factors which controlled the development of this basin include tectonics, global sealevel changes and changes in the prevailing climate. Evaporitic basins require that evaporation exceeds the net inflow of water into the basin and, to achieve this, basin circulation must be restricted, usually by some sill or barrier to the basin (Sarg 2002). The variation in thickness of individual cycles and the repetition of cycles thus has major implications for the evolution of the Berkine/Ghadames basins. Our studies show that a number of factors combined and we envisage three main stages in the evolution of the basin.

Phase 1: the marine transgressive phase

Associated with the break-up of Pangaea Middle Triassic times, in the northern Berkine Basin, the Trias Carbonaté overlies the Trias Argılo-Gréseux Inférieur (TAG-I) (Fig. 1). Recent regional stratigraphic reviews and more extensive drilling have

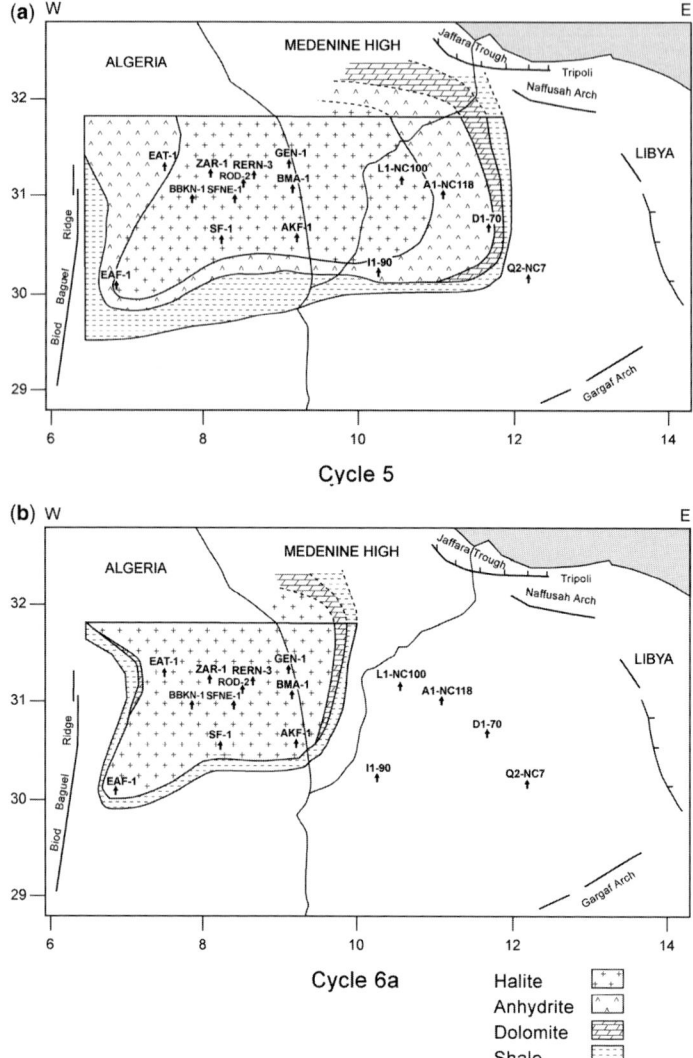

Fig. 11. Palaeogeography of cycle 5 (below the B marker) and cycle 6a (above the B marker). Note that the halite-rich cycle 6a has much less areal extent than cycle 5.

led to the recognition of a regionally persistent event comprising intraclasts, plant debris and reworked bone fragments (Sabaou et al. 2005). This event marks the onset of transgressive marine to marginal marine sedimentation from the northeast (i.e. the Lower Carbonaté) and marks the end of continental fluvio-lacustrine sedimentation sourced from the southwest. This transgressive surface of erosion (TSE) defined by Turner et al. (2001) is characterized by a transgressive lag of phosphatic fish bone and shell debris and is overlain by locally fossiliferous shales interbedded with dolomite, siltstones and sandstones. The sequence becomes increasingly dolomitic both upwards and distally. A shale marker bed represents an associated marine flooding surface. The Triassic Carbonaté has been divided into a Lower and Upper Carbonaté (e.g. Sabaou 2003).

The Lower Trias Carbonaté is a heterolithic unit made up of siltstones and shales, with thin sandstones and locally dolomitic-cemented bands and dolomite beds (Fig. 2). The sandstones are observed in the central part of the basin, their thickness ranging from a few centimetres to 1 m, although thick equivalent sandstones are reported in the southwest of the basin around wells SDL-1 and

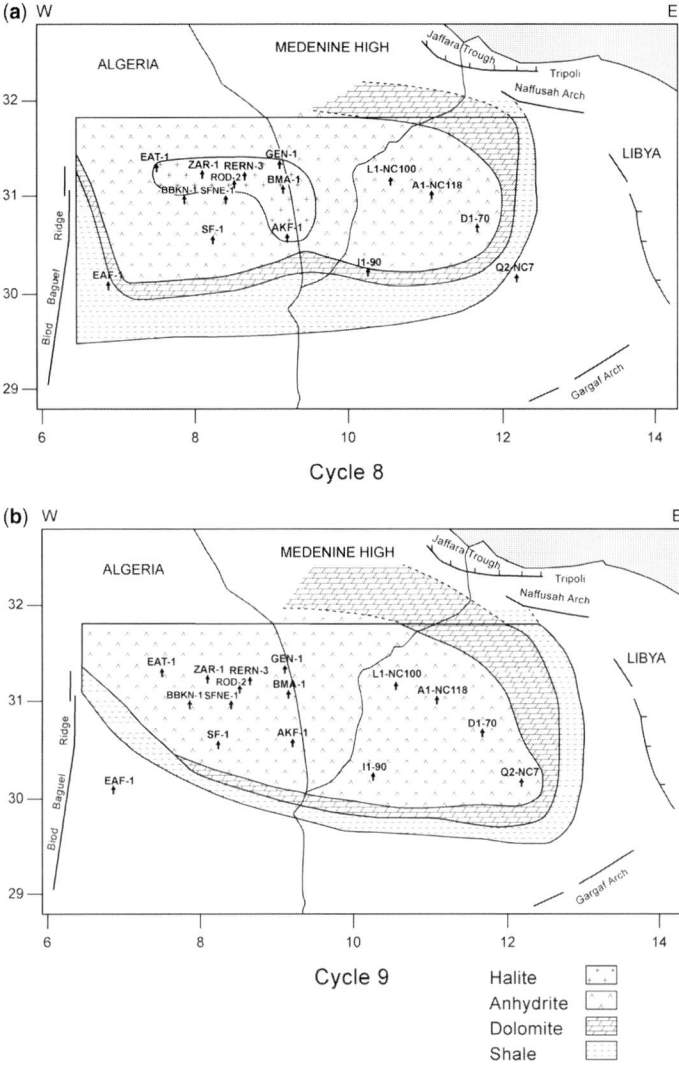

Fig. 12. Palaeogeographic map of cycles 8 and 9 above B marker. Note the relatively large expanse of carbonate sediments compared with the earlier cycles.

RMD-1 (Turner et al. 2001). The sandstones are very fine- to fine-grained, fining-upwards into grey/green siltstones and mudstones. The mudstones are green, rarely grey and dolomitic in most cases, and may contain plant debris. Vertisol-like palaeosoils or hydromorphic palaeosoils are abundant in the Lower Carbonaté. Thin coals, up to 40 cm thick, are also seen (e.g. ROD-2).

Phase 2: the rifting phase

In late Triassic and early Jurassic times active rifting took place as the North Atlantic and peri-Tethyan regions began to separate (see Soussi et al. 2000). Our seismic and well sections show that tectonic subsidence was most active along the Hassi Messoud-El-Biod margin. At this time the Berkine/Ghadames basins became partially isolated from the peri-Tethyan area. The barrier or sill which resulted in restricted circulation within the basin was probably caused by footwall uplift along the Medenine High (Fig. 13), which is composed of a thick succession of Permian carbonates (Bishop 1975). The peak halite precipitation occurs in cycle 2 and this may represent the period of maximum tectonic subsidence with Jebel Nafusah

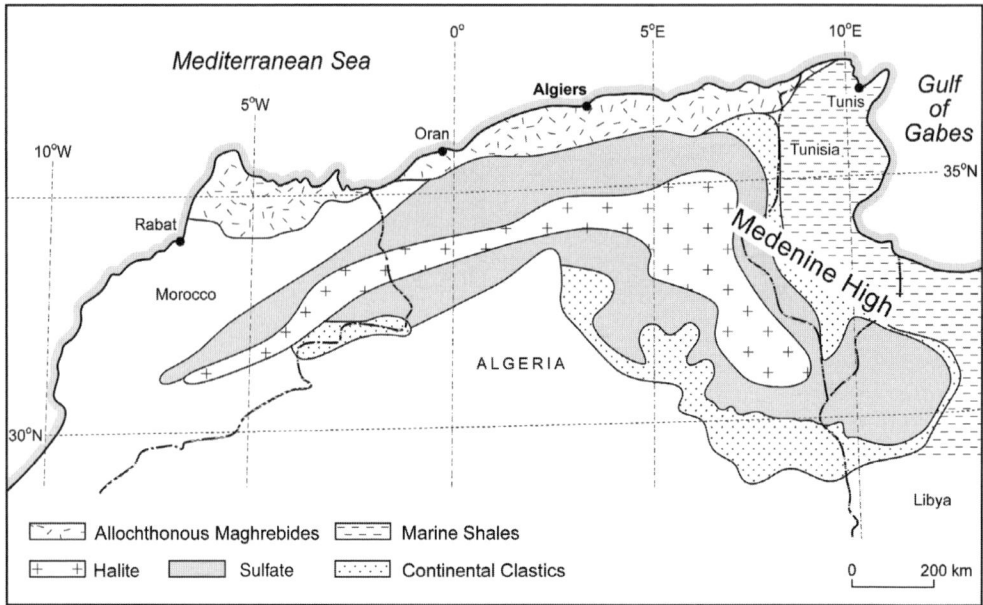

Fig. 13. A general map of the Saharan evaporitic basin. Although the evaporitic areas might not have been completely contiguous, they were intimately connected across the whole of northern Saharan craton.

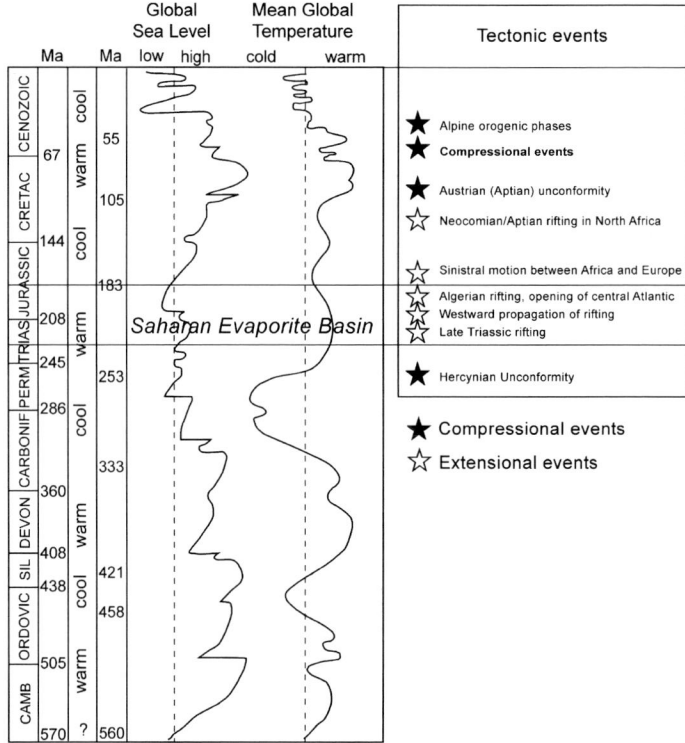

Fig. 14. Stratigraphic diagram showing the external controls on the formation of Saharan basin evaporites during the Late Triassic and Early Jurassic of North Africa.

acting as a stable hinge zone. Development of restriction occurred in other areas further west along the northern margin of the Saharan craton.

Phase 3: connection of Atlantic and peri-Tethys

The opening of the proto-Atlantic gateway in Late Sinemurian (Thierry *et al.* 2000*a, b*) may have been responsible for the ultimate demise of restricted circulation in the Berkine/Ghadames basins that gave rise to the evaporites. At this time a wide carbonate platform was developed and normal circulation conditions resumed. Warm climatic conditions coupled with rising sealevel led to the growth of a wide carbonate shelf in the late Triassic and early Jurassic across the northern part of the Saharan platform when the region moved into the northern tropical zone (Philip 2003). This lead to the widespread formation of neritic carbonates.

Summary

The conditions leading to the formation of the Saharan evaporitic basin are summarized in Figure 14. The Late Triassic and Early Jurassic sediments of the Berkine/Ghadames basin, spanning western Libya, southern Tunisia and eastern Algeria, form part of a major evaporitic basin which stretches over 1000 km to Morocco in the west. The total thickness of the evaporitic sequence reaches 1250 m and shows significant thickness variations which are related to differential subsidence along basement lineaments.

The understanding of this basin has been hampered by the fact that it is almost entirely in the subsurface and the rocks have poor biostratigraphical control. Furthermore complexities in stratigraphic nomenclature have meant that correlations have been our understanding of the origin and evolution of these evaporites deposits is determined almost entirely from two-dimensional seismic data, wireline logs and well description logs.

There are two main evaporitic successions (S1 and S3) which are separated by a thin carbonate platform succession known regionally as the 'B marker'. S1 comprises five cycles (C1–C5) of marine shale, anhydrite and halite. The thickest of these (C2) corresponds to the main phase of salt deposition in the basin. The thickness variation in C2 shows that differential subsidence along basement lineaments was a major control on the accommodation space needed to facilitate thick salt accumulation. The regional pattern of sedimentation changes dramatically changes in Sequences 2 and 3. Following the development of the carbonate platform in Sequence 2, the evaporitic deposits of Sequence 3 are dominated by carbonate–anhydrite rich sabkhas which formed on the upper parts of carbonate ramps which formed part of the peri-Tethyan ocean. The changes from Sequences 1 to 3 reflect the continuing developing of Tethys and its link to the Atlantic Ocean. The overall pattern of sedimentation is thus remarkably similar to that seen in other parts of North Africa, especially Morocco. Early Mesozoic sea floor spreading and global sealevel rise effectively brought about he end of the early Mesozoic giant evaporite basin of the Saharan platform.

The authors wish to express their thanks to Sonatrach, BHPBilliton, ENI and NOC for providing data and giving permission to present these results. The authors wish to thank David Lawton and Mike Jones, Graham Westbrook and Ali D Gayara for their support and assistance in this project. PT would particularly like to thank Charlotte Schreiber for editorial assistance, which greatly improved the paper.

References

AIT SALEM, H., BOURQUIN, S., COUREL, L., FEKIRINE, B., HELLAL, C., MAMI, L. & TEFIANI, M. 1998. Triassic series on the Saharan Platform in Algeria; Peri-Tethyan Onlaps and related structuration. *In*: CRASQUIN-SOLEAU, S. & BARRIER, E. (eds) *Peri-Tethys Memoir 3: Stratigraphy and Evolution of Peri-Tethyan Platforms*. Museum National d'Histoire Naturelle, 177–191.

BEN ISMAIL, M. H. 1991. *Les bassins mesozoiques (Trias-Aptien) du Sud de la Tunisie: stratigraphie integree, caracteristiques geophysiques et evolution geodynamique*. These de Doctoral Sciences, University of Tunis, Tunis.

BENRABAH, B., KERDJIJI, K. & BENCHIEDH, S. 1991. *A propos des charactéristiques roches mères de certains niveaux de Trias*. IIième Séminaire de géologie pétrolière. Sonatrach 21–23 October, Bourmedes.

BISHOP, W. F. 1975. Geology of Tunisia and adjacent parts of Algeria. *American Association of Petroleum Geologists Bulletin*, **59**, 413–450.

BOUDJEMA, A. 1987. *Evolution structurale du bassin petrolier 'triasique' du Sahara nord-oriental (Algerie)*. These, Université Paris XI, Orsay.

BUSSON, G. 1971*a*. *Le Mesozoique sarahien, deuxieme partie: Essai de syntheses des donnees de sondages Algero-tuniiens, Centre de recherche sur le zones arides*. Serie geologie, **11**, Editions de Centre National de la recherche scientific.

BUSSON, G. 1971*b*. *Principes, methodes et resultats d'une etude stratigraphique du Mesozoique saharien*. These, Université Pierre et Marie Curie, Paris.

BUSSON, G. & CORNEÉ, A. 1989. Some data on climatic previous history of Sahara-signification of detrital red beds and evaporites from Trias to Lias-Dogger. *Bulletin de la Societe Geologique de France*, **5**, 3–11.

CHAFIKI, D., CANEROT, J., SOUHEL, A., EL HARIRI, K. & EDINE, K. T. 2004. The Sinemurian carbonate mud-mounds from central High Atlas (Morocco): stratigraphy, geometry, sedimentology and geodynamic patterns. *Journal of African Earth Sciences*, **39**, 337–346.

CLEMENT, G. P. & HOLSER, W. T. 1988. Geochemistry of Moroccan evaporites in the setting of the North Atlantic Rift. *Journal of African Earth Sciences*, **2**, 375–383.

COUREL, L., AIT SALEM, H., BENAOUISS, N., ET-TOUHAMI, M., FEKIRINE, B., OUJIDI, M., SOUSSI, M. & TOURANI, A. 2003. Mid-Triassic to Early Liassic clastic/evaporitic deposits over the Maghreb Platform. *Palaeogeography, Palaeoclimatology, Palaeoecology*, **196**, 157–176.

ESCHARD, R., DESAUBLIAUX, G., BEKKOUCHE, D. & HAMEL, A. 1999. Stratigraphic architecture of the Triassic series in the Saharan Province, Algeria. *AAPG International Conference, Abstracts*, Birmingham, 176.

FRAKES, L. A., FRANCIS, J. E. & SYKTUS, J. I. 1992. *Climate Modes of the Phanerozoic; The History of the Earth's Climate Over the Past 600 Million Years*. Cambridge University Press, Cambridge.

GAETANI, M., GUIRAUD, R. *ET AL*. 2000. Late Norian. *In*: DERCOURT, J., GAETANI, M. *ET AL*. (eds) *Atlas Peri-Tethys, Palaeogeographical Maps*. CCGM/CGMW, Paris, map 7.

GUIRAUD, R. 1998. Mesozoic rifting and basin inversion along the northern African Tethyan margin: an overview. *In*: MCGREGOR, D. S., MOODY, R. T. J. & CLARK-LOWES, D. D. (eds) *Petroleum Geology of North Africa*. Geological Society of London, Special Publications, **132**, 217–229.

GUIRAUD, R., BOSWORTH, W. THIERRY, J. & DELPLANQUE, A. 2005. Phanerozoic geological evolution of Northern and Central Africa: an overview. *Journal of African Earth Sciences*, **43**, 83–143.

HALLET, D. 2000. *Petroleum Geology of Libya*. Elsevier Science, Amsterdam.

HORITA, J., ZIMMERMANN, H. & HOLLAND, H. D. 2002. Chemical evolution of seawater during the Phanerozoic: implications from the record of marine evaporites. *Geochimica Cosmochimica Acta*, **66**, 3733–3756.

LEROY, P. & PIQUE, A. 2001. Triassic–Liassic Western Morocco synrift basins in relation to the Central Atlantic opening. *Marine Geology*, **172**, 359–381.

MAMI, L. & BOURMOUCHE, R. 1994. Résultats palynostratgraphiques des Formations Eriasiqnes d'El Biad. Rapport 44, SONATRACH-CRD, Bourmerdes.

OUIJDI, M. 2000. Triassic marine onlap in the southwestern peritethyan domain, example of Oujda Mountains (Eastern Morocco). *In*: BACHMANN, G. H. & LERCHE, I. (eds) *Epicontinental Triassic*, Volume 2. International Symposium Halle, 21–23 September 1998, Zentralblatt fur Geologie und Palaontologie, 1243–1268.

PHILIP, J. 2003. Peri-Tethyan neritic carbonate areas: distribution through time and driving factors. *Palaeogeography, Palaeoclimatology, Palaeoecology*, **196**, 19–37.

PIQUE, A., TRICART, P., GUIRAUD, R., LAVILLE, E., BOUAZIZ, S., AMRHAR, M. & AIT OUALI, R. 2002. The Mesozoic–Cenozoic Atlas belt (North Africa): an overview. *Geodinamica Acta*, **15**, 185–208.

PURSER, B. H. & EVANS, G. 1973. Regional sedimentation along the Trucial Coast, S. E. Persian Gulf. *In*: PURSER, B. (ed.) *The Persian Gulf, Holocene Carbonate Sedimentation in a Shallow Epicontinental Sea*. Springer, New York, pp. 211–232.

RUBINO, J.-L., GALEAZZI, S. & SBETA, A. 2003. *Late Triassic Abu Shaybah Formation of Northern Libya; Stratigraphic Analysis of a Continental Series Including Braided Stream Rivers and Meandering Complexes*. AAPG abstract.

SABAOU, N. 2003. *Sedimentology, geometry and depositional environments of the Triassic reservoirs of the Saharan Platform, Algeria*. Unpublished Ph.D. Thesis, The University of Birmingham.

SABAOU, N., LAWTON, D. E., TURNER, P. & PILLING, D. 2005. Floodplain deposits and soil classification: the prediction of channel sand distribution within the Triassic Argilo-Greseux Inferieur, Berkine Basin, Algeria. *Journal of Petroleum Geology*, **28**, 3–20.

SARG, J. F. 1988. Carbonate sequence stratigraphy. Sea-level changes: an integrated approach. *In*: WILGUS, C. K., HASTINGS, B. S., KENDALL, C. G. ST. C., POSAMENTIER, H. W., ROSS, C. A. & VAN WAGONER, J. C. (eds) *Sea-Level Changes – an Integrated Approach*. Society of Economic Paleontologists and Mineralogists, Special Publications, **42**, 155–181.

SARG, J. F. 2002. The sequence stratigraphy, sedimentology, and economic importance of evaporite-carbonate transitions: a review. *Sedimentary Geology*, **14**, 9–42.

SOUSSI, M. 2000. *Le Jurassique de la Tunisie atlasique: Stratigraphie, Dynamique Sedimentaire, Paleogeographie et Interet Petrolier*. These de Doctoral Sciences, University of Tunis II, Tunis.

SOUSSI, M. & BEN ISMAIL, M. H. 2000. Platform collapse and pelagic seamount facies: Jurassic development of central Tunisia. *Sedimentary Geology*, **133**, 93–113.

SOUSSI, M., ABBES, CH. & BELAYOUNI, H. 1998. Sedimentary record of sea-level changes and associated organic-rich facies on the southern Tethyan margin: the Upper Triassic of central Tunisia. *African Geoscience Review*, **5**, 275–285.

SOUSSI, M., ENAY, R., MANGOLD, C. & TURKI, M. 2000. The Jurassic events and their sedimentary and stratigraphic records on the Southern Tethyan margin in Central Tunisia. *In*: CRASQUIN SOLEAU, S. & BARRIER, E. (eds) *Peri-Tethys Memoir 5: New Data on peri-Tethyan Sedimentary Basins*. Memoires de la Musee Nationale d'Histoire Naturelle, **182**, 57–92.

TANFOUS AMRI, D., BEDIR, M., SOUSSI, M., AZAIEZ, H., ZITOUNI, L., HEDI INOUBLI, M. & BEN BOUBAKER, K. 2005. Halocinèse precoce associée au rifting jurassique dans l'Atlas central de Tunisie (région de Majoura-El Way). *Compte Rendu Geoscience Geodynamique*, **337**, 703–711.

THIERRY, J., GUIRAUD, R. *ET AL*. 2000*a*. Late Sinemurian. *In*: DERCOURT, J., GAETANI, M. *ET AL*. (eds) *Atlas Peri-Tethys, Palaeogeographical Maps*. CCGM/CGMW, Paris, map 7.

THIERRY, J., GUIRAUD, R. *ET AL*. 2000*b*. Middle Toarcian. *In*: DERCOURT, J., GAETANI, M. *ET AL*. (eds)

Atlas Peri-Tethys, Palaeogeographical Maps. CCGM/CGMW, Paris, map 8.

TOURANI, A., LUND, J. J., BENAOUISS, N. & GAUPP, R. 2000. Stratigraphy of Triassic syn-rift-deposition in Western Morocco. *In*: BACHMANN, G. H. & LERCHE, I. (eds) *Epicontinental Triassic*, volume 2. International Symposium Halle, 21–23 September 1998, Zentralblatt fur Geologie und Palaontologie, 1193–1216.

TUCKER, M. E. 1991. Sequence stratigraphy of carbonate-evaporite basins: models and application to the Upper Permian (Zechstein) of northeast England and adjoining North Sea. *Journal of the Geological Society*, **148**, 1019–1036.

TUCKER, M. E., CALVET, F. & HUNT, D. 1993. Sequence stratigraphy of carbonate ramps: systems tracts, models and application to the Muschelkalk carbonate platforms of eastern Spain. Sequence Stratigraphy and Facies Association. *In*: POSAMENTIER, H. W., SUMMERHAYES, C. P., HAQ, B. U. & ALLEN, G. P. (eds) *Sequence Stratigraphy and Facies Association.* International Association of Sedimentologists, Special Publications, **18**, 397–415.

TURNER, P., PILLING, D. J., WALKER, D., EXTON, J., BINNIE, J. & SABAOU, N. 2001. Sequence Stratigraphy and Sedimentology of the Late Triassic TAG-I (Blocks 401/402, Berkine Basin, Algeria). *Marine and Petroleum Geology*, **18**, 959–981.

Depositional environments of a salina-type evaporite basin recorded in the Badenian gypsum facies in the northern Carpathian Foredeep

M. BĄBEL

Institute of Geology, Warsaw University, Al. Żwirki i Wigury 93, PL-02-089 Warszawa, Poland (e-mail: m.babel@uw.edu.pl)

Abstract: An integrated group of conceptual models for evaporite deposition is presented for a shallow salina-type basin, supplied both with marine and non-marine water, and with a water level separated from and drawn down below world sea-level. Knowledge of, and terminology for, modern limnology is used in these models in order to build a new and more comprehensive link between the hydrography and hydrochemistry of brines and the depositional and stratigraphical record anticipated in such basins. In these modelled basins it is assumed that (as in saline lakes): (1) the evaporite deposition reflects the stratification-mixing cycles in the brine column; (2) the evaporative crystallization of salts culminates during the mixing periods; and (3) the evaporite facies are linked to the position of water zones separated by a horizontal pycnocline. The models are prepared especially for interpretation of subaqueous coarsely crystalline gypsum (selenite) deposition in perennial-to-ephemeral saline pans, and also for fine-grained gypsum deposition (clastic, microbialite and pedogenic) on specific, flat, semi-emerged shoals (majanna-type shoals), commonly flooded by wind-driven brine sheets, in between them. Coarse, crystalline selenites appear to form below the pycnocline and the crystallization of particular well-developed grass-like selenite beds is thought to be connected with the pycnocline highstands. Such beds are both marker beds and ideal datum surfaces. The models presented are used for sedimentological interpretation of the various gypsum facies of the Badenian (Middle Miocene) evaporite basin in the northern Carpathian Foredeep. The observed architecture of these facies suggests that the margin of the basin was occupied by a system of variable saline pans (dominated by selenite deposition) and evaporite shoals (dominated by gypsum microbialite deposition); some of these shoals were majanna-type. Many features of the studied facies can be clearly explained by the models presented, particularly by the rapid and variable fluctuations in water and pycnocline levels incomparable in time scale and size with any global sea-level changes. All these features together consistently suggest that the Badenian gypsum basin was a salina-type basin with a water level below global sea-level, although salts in the basin are/were basically marine in origin.

Evaporite deposition in a shallow salina-type basin (a basin with the water level below world sea-level and supplied by seawater by occasional inflows and/or seepage) is specific. It is controlled by very rapid and large-scale water level changes. The hydrography of brine bodies is comparable to saline lakes. The water and brine chemistry commonly are of seawater composition but some features may become drastically different from it, depending on place in the basin and stage of its evolution.

An integrated group of conceptual models for evaporite deposition in a salina-type basin was introduced in Bąbel (2004a) employing modern limnological knowledge and terminology, building a new and comprehensive link between the hydrography and hydrochemistry of brines and the depositional and stratigraphical record of the formative basin. The basic ideas are the following: (1) the evaporite deposition reflects the stratification-mixing cycles in the brine column; (2) the evaporative crystallization of various salts culminates during the mixing periods in the basin; and (3) the evaporite facies are linked to the particular stratified water zones, separated by a pycnocline.

These models have been applied for sedimentological interpretation of the various gypsum and selenite facies of the Badenian (Middle Miocene) evaporite basin in the northern Carpathian Foredeep (Bąbel 2004b, 2005a, b). Many features of these observed facies are clearly explained by these models and particularly by the rapid and variable water and pycnocline level fluctuations and by different water levels between subbasins.

This paper partly reviews earlier ideas, but also refines them, adding new concepts supplementing previous interpretations. The functional concepts are outlined and then they are applied to sedimentological interpretation of the Badenian evaporite deposits. These hypotheses, naturally, are open for future corrections, modification or withdrawal whenever the new data does not fit.

From: SCHREIBER, B. C., LUGLI, S. & BĄBEL, M. (eds) *Evaporites Through Space and Time.*
Geological Society, London, Special Publications, 285, 107–142.
DOI: 10.1144/SP285.7   0305-8719/07/$15.00 © The Geological Society of London 2007.

Geology of the Badenian evaporite basin in the Carpathian Foredeep

The Carpathian Foredeep Basin is the largest of the several evaporite basins formed in the Carpathian area as a result of the separation of the Central Paratethys from the Mediterranean during the Badenian time (in the Middle Miocene) about 6–8 million years before the well-known Messinian salinity crisis (Figs 1 & 2). The Parathethyan Badenian is a rough equivalent of Langhian–Serravallian and the Badenian salinity crisis took place in Serravallian rather than Langhian (e.g. Oszczypko 2006a; Peryt 2006). The Badenian evaporites comprise mainly calcium sulphate and sodium chloride facies and the most widespread evaporite facies in the northern Carpathian Foredeep are primary gypsum deposits such as bottom-grown gypsum crystals (selenite), gypsum microbialites and clastic gypsum. They crop out along the northern margin of the basin but to the south they are in the subsurface and are commonly transformed into anhydrite (Fig. 2). Halite deposits intercalated with laminated anhydrite and clay occupy the central and southern area of the basin. They are univocally interpreted as deeper facies (Kwiatkowski 1972; Garlicki 1979; Petrichenko et al. 1997; Kasprzyk & Ortí 1998; Kasprzyk 2005) in relation to the more shallow marginal sulphate facies (zero to several metres of water). These deposits are largely in the subsurface and the transition between the two facies is not visible. Additionally many details have been obliterated by the processes of calcium sulphate hydration/dehydration (Kasprzyk 2005). Depending of the stage of basin evolution the marginal sulphate area was a platform in relation to the halite basin (Garlicki 1979; Kasprzyk & Ortí 1998), or a shallow subbasin (or chain of subbasins) separated from the deeper halite subbasin by semi-emerged shoals or islands (Bąbel 2005b). These shoals were periodically inundated and finally entirely covered with gypsum sediments in the later phase of evaporite deposition.

Following some earlier suggestions, it seems that the Badenian gypsum evaporite basin in the Carpathian Foredeep was a depression affected by water-level drawdown below sea-level (Peryt 1996, 2001; Bąbel 2004b). This concept fits, more fully, to the observed geochemical data (Cendón et al. 2004) and works well for the event stratigraphic and facies analysis of the evaporite basin (Bąbel 2005a, b). The Badenian gypsum facies are very similar to those recorded in some modern coastal marine salinas (Kasprzyk 1993b). However the Badenian facies show greater differentiation because the basin was larger than any Recent evaporite basin. The understanding and interpretation of these facies over such an extended area require an application of new, comprehensive models.

Concepts for evaporite and selenite deposition in a salina-type basin

The concepts explaining evaporite and gypsum deposition in the basin with the water level drawdown below sea-level are developed below, step-by-step, beginning with the basic features of the model basin and of its hydrography. They are the background for definitive characteristics of the two main sub-environments of the basin: saline pans and evaporite shoals.

Definition of a salina-type basin

Grabau (1920, p. 123) defined a marine salina as 'salt lake or salina near the sea, the water of which is supplied from the sea, not by overwash or canals, as in the case of salt pan, but by underground passages such as seepage through the sand or other material which separates it from the sea'.

The basin model considered for the Carpathian Foredeep basin is that of a salina or salina-type

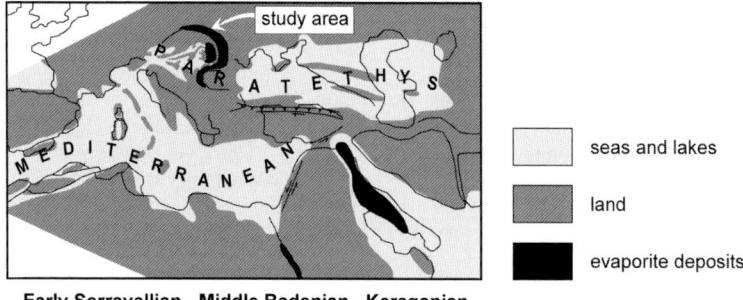

Fig. 1. Palaeogeography of Paratethys and Mediterranean in the Middle Miocene and location of the evaporite basin studied (after Rögl 1998).

Fig. 2. Present distribution of the Badenian evaporite basins in the Carpathian region (after Garlicki 1979; Khrushchov & Petrichenko 1979, modified).

basin (Bąbel 2004a, b) because its principal features fit to the marine salina defined by Grabau (1920), avoiding the mixture of definitions employed in many later studies. The basin model employed here is devoted mainly to shallow water subaqueous deposition. The most important factor in this model salina basin is that its water level is drawdown below world sea-level – as in typical coastal salinas (Fig. 3). The model belongs to the wide group of related models known under various names (for reviews see Sonnenfeld 1984; Dronkert 1985; Bąbel 2004b). Specific features are extracted from these previous models to define the basin.

The salina basin model assumes that: (1) the evaporite basin is a depression separated from the ocean by some topographical barriers which eliminate the open water connections between the basin and the sea; (2) evaporation is the main reason for water deficit, and for the depressed basinal water level lowering (known as evaporite drawdown), and for its chemical sedimentation; (3) the sea water enters the basin mainly by subsurface seepage, although surface inflows are also possible.

Additionally, it can be assumed that (4) the salina basin is a system of interconnected or temporarily disconnected subbasins or saline pans.

There are many possible variants of a salina basin depending on various parameters, like the shape of the basin, stage of its evolution, and even the chemistry of seawater supplied to the basin, all of which could change through time. The model considered in this paper is applied to the basin at the stage of gypsum deposition from seawater, the same in composition or not too much different from the recent seawater.

Main features of a salina-type basin

The main feature in the basin is its lowered water level, which determines the development of high accommodation and a possibility of the rapid and large water level fluctuations within the basin. The accommodation is limited by mean sea-level which is the upper limit for generating any evaporites from seawater brines ('equilibrium level' of Logan 1987). The resultant spatial accommodation

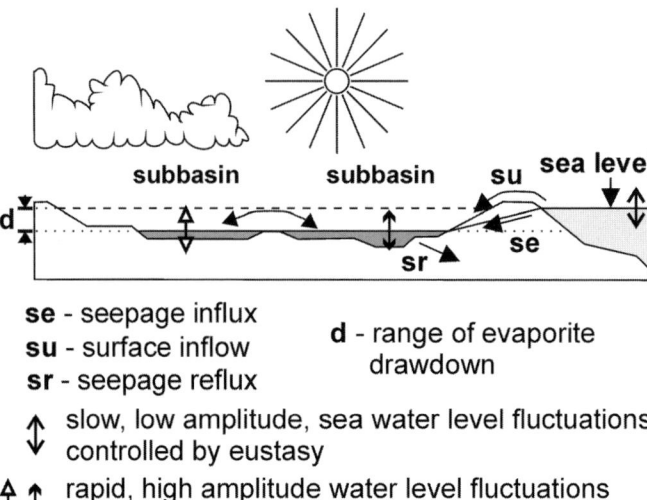

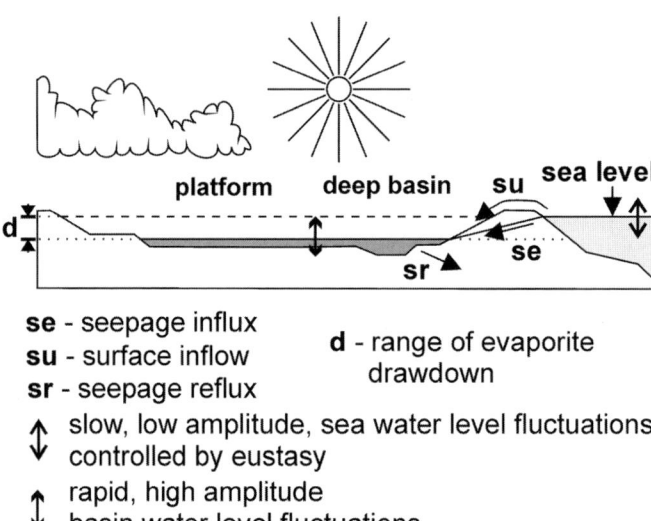

Fig. 3. Schemes showing basic features of a salina-type evaporite basin. Both types of the basin (**a** and **b**) can be considered as evolving one into the other in time. Type (b) corresponds to basinal highstand, type (a) to lowstand (see Bąbel 2004a, fig. 3).

makes deposition of thick sequences pedogenically formed evaporites in the salina basin possible (see Lowenstein *et al.* 2003). The water level fluctuations are the most important for the depositional dynamics, sequences and architecture within the basin. In a salina basin they are dependent on particular local parameters, like regional climate, seepage inflow and outflow rates, tectonics and sediment supply (see Logan 1987; Strecker *et al.* 1999; Ilgar & Nemec 2005). As in modern coastal salinas, the water level fluctuations are similar to those observed in saline lakes. The intrinsic feature of all the salina-type and salt lake basins is that their water level is unstable and always fluctuates, rising and falling over various time periods that are commonly irregular (Langbein 1961; Jauzein & Hubert 1984; Williams 1996; Comín et al. 1999). Additionally, a salina-type basin can be rapidly flooded by marine water, and pass through an abrupt evaporite drawdown after disconnection from the sea.

In most recent salt lakes the annual rise and fall of the water level is variable (Anati 1997) and always fluctuates seasonally, annually, and over longer time periods. This fluctuation is dictated mainly by climatic changes (Comín et al. 1999). Such dramatic fluctuations are not observed in freshwater lakes that usually show relatively stable water levels (Langbein 1961; Scholz et al. 1998). The depositional environment of a salina-type basin is very similar to those of salt lakes or closed basins, although some specific differences are also present. The architecture of depositional sequences in a shallow-water salina basin is similar to the so-called underfilled lake basins (Caroll & Bohacs 1999; Bohacs et al. 2000). Variable water level fluctuations can produce both shallowing- and deepening-upwards sequences, as well as deepening- and brining-upwards sequences. The latter sequences appear to be barely possible in lagoonal evaporite basins with the water level coinciding with the sea-level (Bąbel 2004b).

Among many possible environments of the salina basin, two main ones are discussed below: perennial and ephemeral saline pans (or subbasins), and evaporite shoals (as defined in Bąbel 2004a), which are particularly important for analysis of the Badenian evaporites (Fig. 4).

Hydrographical classification of saline pans

Saline pans are, for practical purposes, lakes showing more or less persistent water stratification which can be permanent, seasonal or temporary depending on many parameters like climate, size and shape of the pan (fetch), depth and salinity. Following limnological and hydrographical terminology the pans can be subdivided on the basis of the frequency of the periods of complete mixing per year (Fig. 5; Lewis 1983) into: (1) meromictic

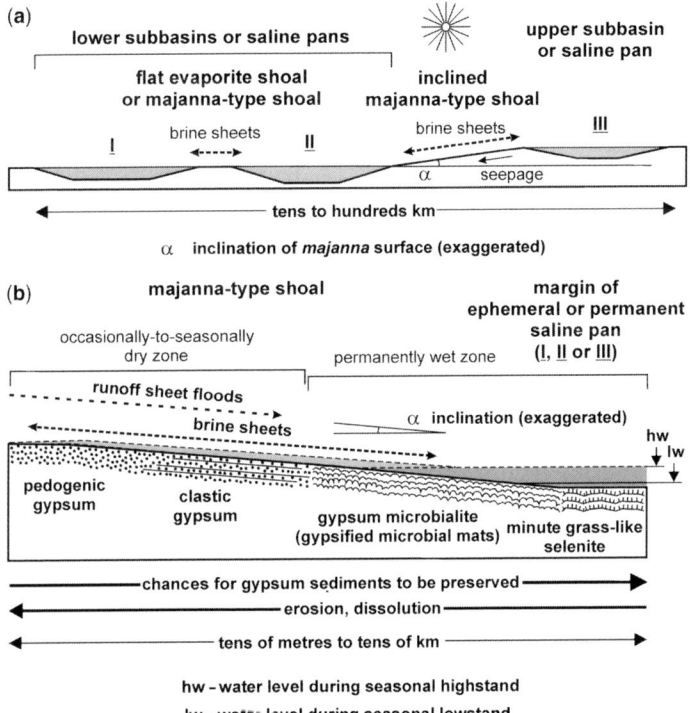

Fig. 4. Schemes showing basic shallow-water depositional environments in a salina-type basin (Fig. 3; partly after Friedman & Krumbein 1985; Logan & Brown 1986; Logan 1987). (**a**) Saline pans or subbasins and majanna-type evaporite shoals; (**b**) depositional environment of the majanna-type shoal (a). Vertical sizes and angles exaggerated. Detailed explanations in the text.

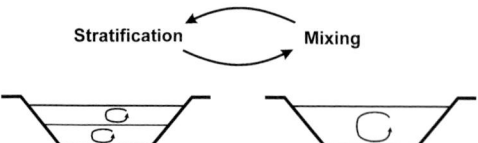

Fig. 5. Simplified scheme showing basic stratification-mixing cycle in a closed basin.

pans, which are permanently stratified (see Bąbel & Bogucki 2007, for more details); and (2) holomictic pans, which undergo complete mixing up to the bottom during the annual cycle (Hutchinson 1975). The latter ones can be subdivided into subordinate types depending on the frequency of complete mixing periods per year; from monomictic pans which show alternately stratification and mixing once a year, through discontinuous polymictic pans which mix more than once a year at intervals of days to weeks, to continuous polymictic pans which mix daily or without interruption (Lewis 1983, 2000).

The limnological classifications adopted here are mainly applied to freshwater lakes and are based on the seasonal thermal variations observed in lakes, namely on the frequency of periods per year when conditions of isothermia are reached in the whole water column due to complete vertical mixing of waters during circulation or de-stratification period (e.g. Ambrosetti et al. 2003). In the saline lakes or pans only the homogenization of all the water properties (temperature, salinity and density, where the first two parameters basically determine the third one, e.g. Anati 1997; and the water chemistry) is indicative of the full mixing (Anati & Stiller 1991), and the mixing processes are more complex than in freshwater systems. Therefore the classification adopted here is a further simplification of classification by Lewis (1983), set up for the purpose of closed saline basin modelling and study. The term 'pycnocline' is used to designate the interface of the two water bodies density stratified on the basis of salinity and/or temperature (similar to the approach adopted by Del Don et al. 2001).

Stratification-mixing phenomena and their cyclicity are crucial for evaporite deposition in the saline pans. It is known from study of many saline lakes and coastal salinas that precipitation of evaporite minerals is strictly linked to particular zones within stratified water masses separated by a pycnocline and to particular periods during the stratification-mixing cycle (Krumbein & Cohen 1977; Niemi et al. 1997; Kirkland 2003; Torfstein et al. 2005; and references in Bąbel & Bogucki 2007). In this outline the author focuses his consideration of evaporite deposition in saline pans on the large bottom-grown crystals which are one of the most characteristic subaqueous Ca-sulphate evaporite facies.

Selenite deposition in saline pans

Of the many varieties of subaqueous precipitates, one of the most common are crusts or layers of *in situ* grown crystals. It is well known that they grow from well-mixed supersaturated brines and, most commonly, in very shallow brine, up to a few metres deep (e.g. Smoot & Lowenstein 1991). Such deposits are typically composed of vertically elongated, large crystals, which in a case of gypsum are known as 'selenite'. The <5 m depth limit for such selenite deposition was suggested by Schreiber (1978), based on observations of many recent environments and ancient deposits, and it was later tentatively extended down to 10–20 m, but without any firm example from the geological record (Schröder et al. 2003, p. 884; Bąbel 2004a) (perhaps gypsum crusts crystallizing at depth of 18 m in the hypersaline Gotomeer basin, island of Bonaire, Netherlands Antilles, can be such an example; Kobluk & Crawford 1990). The reasons for such a depth limit are largely related to the hydrography of stratified brines.

Complete brine mixing (i.e. de-stratification) associated with evaporite concentration of calcium sulphate is the most effective for the accelerated growth of selenite crusts at the bottom of saline pans (Fig. 6, centre; Warren 2006). During stratification periods the selenite growth at the bottom, below the pycnocline (see Fig. 6, top), cannot be promoted directly by evaporation. The waters below the pycnocline cannot evaporate because they are separated from the atmosphere (Sloss 1969). In the deep brine below the pycnocline anoxia can develop, because oxygen is utilized by decomposition of settling dead plankton and other organic debris, and consequently sulphate ions necessary for gypsum precipitation can be removed by sulphate-reducing bacteria. Therefore the selenite crystal growth is hardly possible at the bottom of saline pans during long-term stratification periods, although the crystals can grow below the pycnocline due to temperature changes in the saturated solution or some other effects, like the 'remnant supersaturation' of Stiller et al. (1997, p. 182).

The bottom-growth selenites cannot grow by evaporation of surface brines at the bottom of the permanently stratified meromictic basins (commonly deep and highly saline) although they can grow in the upper water mass over the permanent pycnocline (see Bąbel 2004a; Bąbel & Bogucki 2007). The selenites grow at the bottom of shallow to moderately deep holomictic saline

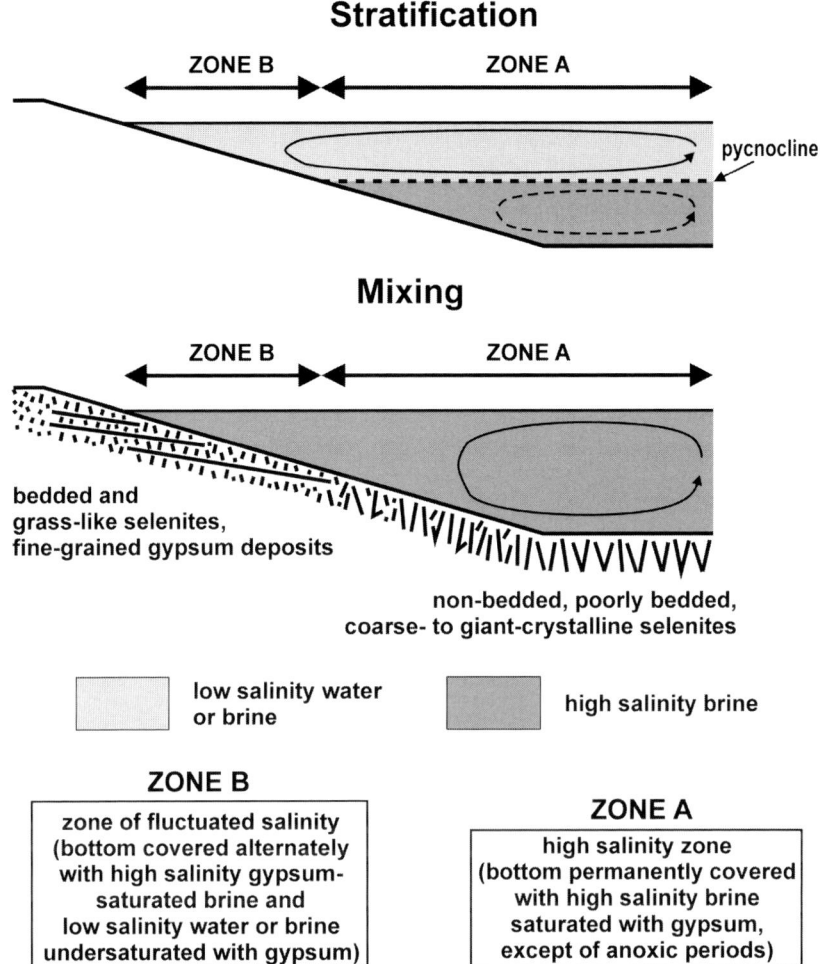

Fig. 6. Depositional model for selenite deposition in a saline pan (Fig. 4a) within a salina-type basin (Fig. 3). Detailed explanations in the text.

pans: monomictic or polymictic, depending on frequency of mixing periods per year, and particularly in those which do not develop the intensive anoxia during stratification periods. Such pans should be well oxygenated and calcium sulphate-saturated during mixing. The deep brine is oxygenated not only by diffusion from air and mixing but by photosynthetic activity of cyanobacteria as well (Horne & Goldman 1994). This activity however decreases with the water depth and the water transparency (e.g. Cloern et al. 1983). The transparency of brines is commonly very low due to planktonic blooms shading the bottom and limiting the growth of phototrophic benthic cyanobacteria.

The more frequently the holomictic pan is mixed the more favourable is its bottom for the growth of the gypsum crystals (assuming that the brine is saturated with calcium sulphate at least during the mixing periods). Consequently polymictic, calcium sulphate-saturated pans appear to be the most favourable for abundant selenite deposition. It is usual that the shallower the pan the more likely is its polymictic state; the shallowest pans are always easily, nearly continuously, mixed by wind, down to the bottom.

The shallowness of polymictic pans implies frequent emersions and associated processes, and hence the interruptions in the crystal growth and destruction of the crystal beds. Indeed the selenite deposits in most modern shallow polymictic basins are composed of small crystals, a maximum of several centimetres long, and commonly contain clastic gypsum as well. The larger, well-developed selenite crystals, decimetres in size, composing thicker selenite crusts, are found only in a few sub-fossil settings where they grew under more permanent cover of relatively deep brine (Warren 1982b). Just such coarse- and giant-crystalline selenite beds are very common in some ancient evaporites. Also in these ancient selenite deposits the larger sizes of crystals appear to coincide with rarity or lack of any evident dissolution or abrasion features. The ancient coarse crystalline selenites apparently grew upwards from the bottom in crystallographic continuity, i.e. syntaxially, for periods of time long enough for creation of decimetres-to-metres thick palisade-like or grass-like beds, without traces of any nucleation or settling of new gypsum crystal seeds on the depositional surface. It seems that the growth of such selenites requires somewhat deeper brine. The environment of coarse-crystalline selenites is interpreted in the following paragraphs.

The optimal place for coarse-crystalline selenite growth is below the seasonal (temporary) pycnocline of monomictic (to polymictic) saline pans in a salina-type basin, at a depth of no more than a few metres (zone A in Fig. 6; Bąbel 2004a). In this zone the brine can be calcium sulphate-saturated nearly all the time, during both stratification and mixing. This practically eliminates crystal dissolution and creates a very stable environment favouring the growth of large crystals. Many conditions in this zone appear to be optimal for the growth of large crystals. These conditions include long-term, persistent and relatively low supersaturation staying within the metastable zone, which eliminates any spontaneous nucleation (Mersmann 1996; Ulrich & Strege 2002). The rises in Ca-sulphate concentration are not so rapid and large as they potentially are in the zone of upper brine mass above the pycnocline (Fig. 6, top), and hence they are not able to promote spontaneous nucleation and to produce a great quantity of new gypsum crystal seeds. On the other hand drops in concentration are too small to promote abundant gypsum dissolution. The crystals at the bottom are nearly always in contact with the gypsum saturated brine. Such crystals can grow syntaxially by deposition of the new portions of gypsum in crystallographic continuity on the pre-existing gypsum crystal faces. Long-term, continuous, syntaxial growth leads to creation of extremely large crystals. Such a growth is not substantially disturbed (mechanically) by deposition of clastic particles or growth of thick microbial mats which could cover the crystal surfaces. Clastic deposition and erosion is uncommon at the bottom of monomictic pans during stratification periods because their bottom is below the seasonal pycnocline and therefore is below the wave base (Sonnenfeld 1984). Because of the greater depth, a higher salinity level (staying within gypsum saturation field in marine brine) and very low illumination caused by possible shading by phytoplankton during eutrophication events, any thick cohesive and impermeable cyanobacterial mats are not developed there (Bąbel 2004a). Such mats commonly cover the gypsum deposits in the shallower and better-illuminated zones above the pycnocline (zone B in Fig. 6) and are a very effective barrier for the syntaxial gypsum growth (Geisler-Cussey 1997, p. 23). Below the pycnocline crystal apices are in contact with basinal brine, even when covered by some biofilms, because the biofilms are very permeable (see Gerdes et al. 2000, p. 204).

It is very likely that in the monomictic saline pans, in the model salina basin, the mixing period coincides with the seasonal lowstand of water level. This is true for many shallow (less than 6 m deep) coastal salinas and salt lakes such as Solar Lake (Eckstein 1970; Cohen et al. 1977), Lake Hayward (Coshell & Rosen 1994; Rosen et al. 1996), salinas of the York Peninsula (Warren 2006), and even for about 2 m deep Red Pond and Green Pond in Arizona (Cole et al. 1967). The beginning of the mixing usually starts in late summer or autumn, being slowed in relation to the climatic cycle. In all these basins (except of those in Arizona) precipitation of evaporite salts (gypsum) coincides with the period of mixing. In some deep saline monomictic basins like the Dead Sea (c. 320 m deep) and Mono Lake (45 m deep) the mixing also starts in late autumn, at the beginning of the highstand, but continues during the advancing winter highstand (Anati 1997; Melack & Jellison 1998). In these deep basins evaporite precipitation mechanisms are more complex and salts can be deposited both during mixing and stratification periods.

The main requirement for the gypsum deposition in this model is that the bottom brines (in zone A in Fig. 6) must be saturated with gypsum both during stratification and mixing. The pans thus should be in their 'Ca-sulphate productive' stage (similar to 'self-precipitating basins' distinguished by Valyashko 1952, 1972; see also Borchert & Muir 1964, pp. 65–67). In reality great disturbances in gypsum productivity can take place which depends on the levels of gypsum concentration within the lower and upper water bodies which influence the level of concentration during mixing. In some lakes containing gypsum saturated brine, after mixing, development of undersaturation of the whole water body with respect to gypsum is possible (depending on temperature, oxygenation etc.; see Last & Schweyen 1983, table 4).

The model presented here for selenite deposition in a saline pan is in agreement with the interpretation by Warren (1982b, 2006), who suggested a monomictic salina for coarse-crystalline selenite growth basing on the Holocene and modern examples from some Australian coastal salinas. Warren indicted two of them, the Deep Lake and Lake Inneston (Marion Lake complex, York Peninsula) where such a growth is currently taking place. However study of these lakes is not complete, but apparently is very similar to the Solar Lake, which is the best hydrographical analog for the model presented (Cohen et al. 1977). The ancient coarse-crystalline selenites commonly show millimetre-scale regular growth zoning which was interpreted as the reflection of an annual, i.e. monomictic, cycle of precipitation (Warren 1982b, 2006; Petrichenko et al. 1997).

Evaporite shoals in a salina-type basin

Gentle, flat-bottomed topography and shallowness of some salina-type basins imply that, between the drying saline pans containing ephemeral or permanent brine bodies, vast semi-emerged shoals will develop (Fig. 4a). Features of these shoals can be to some extent represented in the modern coastal salinas and the best example of such shoals are the flats described under the name 'majanna' (a term introduced by Logan in 1984, see Logan & Brown 1986, p. 6) from the MacLeod basin in Australia (Logan 1987). The evaporite shoals in the model basin are described as majanna-type shoals, although their predicted features come from many salina settings.

The specific feature of shoals in a salina basin appears to be the flow of 'brine sheets' (Last & Schweyen 1983), defined by Logan & Brown (1986, p. 143) as 'bodies of free brine, a few centimetres deep, that are not topographically confined' (Fig. 4a & b). Brine sheets can form as an outflow from seep areas on the slopes of saline pans or as returning outflow from saline pans when the brine level is rising. The former brine sheets, as known from MacLeod basin, are commonly permanent, and the latter ones temporary. Brine sheets are commonly driven by wind in a similar way as 'roving' freshwater sheets in some playa lakes (Torgersen 1984). Sometimes, pushed by wind, they actually can flow upslope as detached migratory brine bodies over large distances attaining a level 2 m higher than the original site, as recorded in Lake Tyrrel in Australia (Teller et al. 1982, p. 169). In the MacLeod basin brine sheets, pushed by persistent, winds are held upslope for days to weeks (Logan 1987, p. 64). Such migratory brine sheets were observed on the drying Chott el Djerid flat in Tunisia (Bryant et al. 1994) and in the drying Kara-Bogaz-Gol in Turkmenistan (Lepeshkov et al. 1981, pp. 186–187; Levine 1998).

The majanna from the MacLeod salina is an evaporite flat, lying below sea-level, controlled by seepage inflow and evaporite outflow, with a suppressed brine table and common migratory brine sheets outflowing from saline pans and permanent brine sheets inflowing into them (Logan 1987). The surfaces of majanna flats are parallel to the inclined brine tables and are permanently wet, when the brine table coincides with, or is close to, the surface, or dry, when the brine table is well below the surface.

Giant flats similar to the Australian majanna were recorded in the area of Kara-Bogaz-Gol, which gradually dried in the early 1980s, when the brine depth fell to 2–5 cm on the surface of a few thousands of square kilometres (Terziev et al. 1986). Similar permanently wet but smaller flats surround coastal salinas in the Red Sea coast (Friedman & Krumbein 1985). They are commonly covered with living microbial mats creating specific 'perennial surface brine' biofacies (Gerdes et al. 2000).

Many salt flats in the continental playa environment show features resembling the majanna when they are covered with halite crystallizing from seeping or rising capillary brine. Such salt flats occur only in places where the groundwater table coincides with the playa surface, or is at a depth less than the depth of the capillary rise (Ullman 1995; Wood et al. 2005). These efflorescent salts accumulate on the surface and are easily dissolved during any inflow of fresh water or even when the humidity is elevated. Unlike in such salt-flats, brine sheets are permanent elements on the majanna flats (Logan 1987). There is a significant difference between the majanna in a salina basin, and the salt flat from continental salt lakes or playa lakes, where the former is commonly flooded by high salinity brine sheets (because the salina basin is supplied with seawater or seawater brine) whereas the latter is flooded mainly by freshwater or brackish water sheet floods (because continental lakes are supplied with meteoric water). Run-off sheet floods also inundate the majannas after rains and the dry majannas are subjected to deflation and pedogenic processes (Fig. 4b). The majanna surface is, however, dominated by in situ crystallization of salts from brine sheets, clastic deposition or re-deposition of earlier precipitates by the subsequent brine sheets and freshwater sheet floods, and microbialite deposition controlled by cyanobacterial mats inhabiting submerged and permanently wet areas (Fig. 4b; Logan 1987).

The depositional environment of majanna-type shoals in a model salina basin is peculiar. Astronomical tides are insignificant in the closed basins and such tides can be excluded in a salina basin. The shoals are flooded seasonally, following the annual cycle of water level fluctuations which can be up

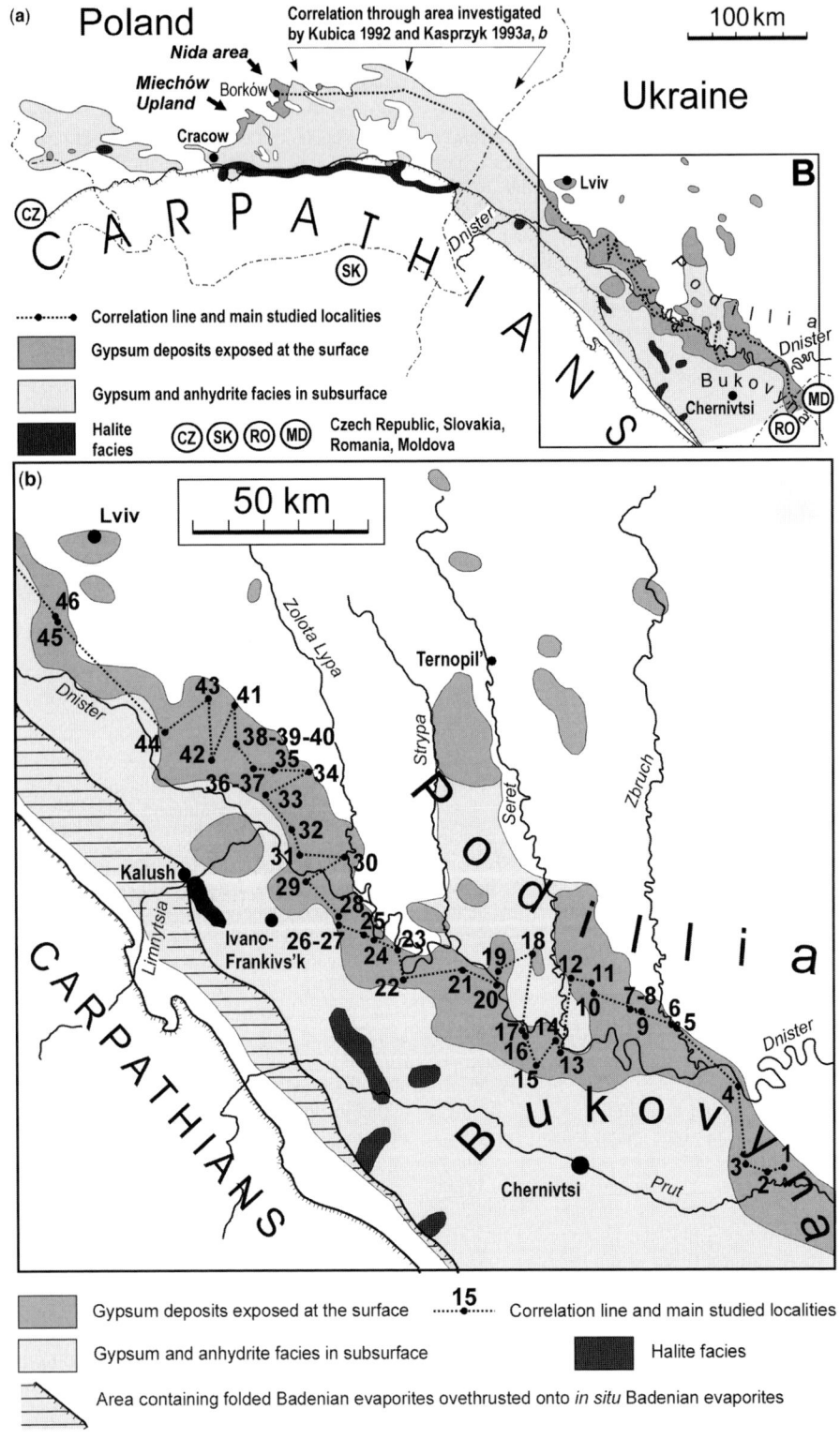

to a few decimetres in amplitude (see Fig. 4b; Logan 1987, fig. 52; Bąbel 2004a, Table 1). Flat shoals forming barriers between saline pans can show specific sedimentation controlled by seepage and brine sheet flow, like on the MacLeod majannas. Adjacent saline pans can always show slight differences in water levels and the shoals between such pans can be gently inclined towards the lower pan. The brine sheets can then flow through such shoals predominantly into the lower pan just as in majanna flats of the MacLeod basin (Fig. 4a).

The majanna-type shoals are similar to some salt flats or inland sabkhas surrounding saline lakes, but differ from typical coastal marine sabkhas, which are flooded only rarely, during extremely high tides by seawater, with the salinity around 35‰. The majanna-type shoals are frequently flooded by brine from local saline pans commonly supplied with brine from interiors of salina basin. This brine can be saturated with respect to gypsum, halite or some other higher salts, and show variable salinity up to the very high values (e.g. c. 300‰ for halite brine). The majanna flat is commonly wet (and some parts of it can be always wet; fig. 4b; Tyler et al. 2006), whereas the sabkhas are traditionally considered as dry (Kinsman 1969), with the groundwater table 0.5–1 m below the surface (Wood et al. 2005). The majanna flat is dominantly formed by *in situ* precipitation at the surface (commonly on or within the surface of microbial mats) directly from the moving brine sheets, and by clastic evaporite deposition determined by sheet flow. Aragonite, gypsum and halite are known as the direct surface precipitates from such brine sheets (Logan & Brown 1986; Bryant et al. 1994). The typical sabkhas are dominated by *in situ* crystallization of evaporite salts from subsurface brines within some host sediments below the surface. The majanna-type shoal can display such features only in its emerged parts. In both environments, majanna and sabkha, the surficial salts can be removed by dissolution caused by intense but infrequent rains (Wood et al. 2005), but in majanna this process can be minimized because of the elevated position of the brine table and the presence of a protective, poorly permeable cover of microbial mats.

The majanna-type shoals in a salina basin usually show large accommodation. The majanna deposits can be accreted together with the rapidly rising water or brine level in saline pans (and brine table levels in dry parts of the shoals) without shallowing-upwards features typical of the coastal sabkha cycles. These extremely shallow water deposits can be well preserved in the sedimentary record, and can form thick sequences, in the same manner as in some saline lakes occupying deep depressions. It is clearly documented that such lakes can preserve even pedogenic evaporite deposits including gypsum efflorescent crusts (Smoot & Castens-Seidell 1994; Li et al. 1996; Bobst et al. 2001, fig. 5). Pedogenic evaporite deposits are probably overlooked facies in many basins (Lowenstein et al. 2003). Owing to large accommodation, the majanna-type shoal deposits of the salina-type basins can develop the aggradational architecture whereas the marine sabkha typically develop the progradational architecture (Warren & Kendall 1985; Alsharhan & Kendall 1994).

The Badenian evaporite basin as a salina-type basin

This part of the paper presents an attempt of the application of the concepts for the evaporite and selenite deposition in the giant 'marine salina' (Grabau 1920) outlined above, for sedimentological interpretation of the Badenian gypsum deposits in Carpathian Foredeep. Depositional features, facies and their architecture are interpreted from the point of view that the basin was a salina (Peryt 1996, 2001; Cendón et al. 2004), and the attention is paid to specific selected features which may be indicative of such a basin type.

Badenian gypsum facies and lithosomes

The Badenian gypsum deposits can be roughly subdivided into two main units: the lower autochthonous unit, dominated by primary selenite and gypsum microbialite deposits, and the upper allochthonous unit, containing mainly clastic

Fig. 7. Distribution of the Badenian (Middle Miocene) evaporites in the northern Carpathian Foredeep (Fig. 2) and correlation line of the gypsum sections studied. (**a**) General sketch map of the study area; (**b**) map of the studied area in western Ukraine (localities 2–46) and northern Moldova (locality 1). Maps after Atlas Geologiczny Galicyi (1885–1914), Garlicki (1979), Panov & Plotnikov (1999) and other sources. Numbered localities: 1, Kriva (Moldova); 2, Mamalyha; 3, Stal'nivtsi; 4, Anadoly; 5, Zavallia; 6, Kudryntsi; 7, Skov'iatyn S; 8, Skov'iatyn N; 9, Kryvche, Krystalna cave; 10, Optymistychna cave; 11, Ozerna cave; 12, Verteba cave; 13, Tovtry; 14, Chun'kiv; 15, Verenchanka; 16, Zveniachyn; 17, Kostryzhivka; 18, Holovchyntsi; 19, Nahoriany; 20, Ustechko; 21, Repuzhyntsi; 22, Harasymiv; 23, Isakiv; 24, Odaiv; 25, Vikniany; 26, Palahychi S; 27, Palahychi N; 28, Oleshiv; 29, Hannusivka; 30, Krasiiv; 31, Lany; 32, Mezhyhirtsi; 33, Kasova Hora; 34, Naberezhne; 35, Podillia; 36, Ozeriany; 37, Kuropatnyky; 38, Obelnytsia; 39, Velyka Holda; 40, Luchyntsi; 41, Rohatyn; 42, Kolokolyn; 43, Pidkamin'; 44, Zahirochko; 45, Pisky; 46, Schyrets'.

gypsum (Peryt 1996; Kasprzyk & Ortí 1998; Rosell et al. 1998). More precisely, the evaporites can be subdivided into several lithosomes or lithostratigraphical units (designated by capital letters: Fig. 8) and the boundary between the autochthonous and allochthonous units is roughly at the base of the 'clastic' unit E destinguished by Kubica (1992). The selenite deposits occur as laterally extensive lithosomes A, C-D, F, SV and SH; the remaining units are dominated by gypsum microbialites and clastic gypsum (Fig. 8). An event-stratigraphical approach based on well-defined marker beds proved that gypsum deposition across the entire basin margin was coeval (Figs 8 & 9; Peryt 2001, 2006; Bąbel 2005a). The architecture of these gypsum lithosomes or lithostratigraphical units reflects dominantly aggradational style of deposition typical of the salina type basins. The onlap pattern of successive gypsum units in some areas (Peryt 2001; Bąbel 2005b) can be a product of 'autocyclic' (Warren 2006, pp. 350–351) or 'autogenetic' transgression (Rouchy & Caruso 2006, p. 47) characteristic of the post-drawdown phase in such basins.

Of particular significance among the Badenian gypsum facies are selenite (coarse-crystalline) and gypsum microbialite (fine-grained) facies. The former facies are considered as deposited in several different types of saline pans, the latter as evaporite shoals and the majanna-type shoals in particular, appearing within a framework of the same salina-type basin (Figs 3, 4 & 9). Microcrystalline and clastic gypsum-dominated facies present in the upper part of the section actually may represent several different environments, but also some saline or brackish pans in which selenites were not able to crystallize.

Selenite facies

Similar to the Messinian selenitic gypsum, two main varieties of the Badenian selenite deposits can be distinguished (excluding some other, more complex varieties of selenites): coarse crystalline selenites forming thicker layers with no or poorly discernible bedding planes, and distinctly thin-bedded, grass-like selenites. Similar to the Messinian, 'as selenite grain-size decreases bedding becomes more distinct' (Hardie & Eugster 1971, p. 191). These two varieties are interpreted as representing two different types of saline pans within a salina basin; namely a deep-brine and a shallow-brine pan (Bąbel 2004a).

The names 'deep-brine' and 'shallow-brine' are used to designate two specific types of saline pans. The 'deep-brine' pan is characterized by a thick zone of brine below a pycnocline. Its pycnocline can fluctuate vertically over a relatively wide range of space but never (or only exceptionally) drops down to the bottom of the pan. The 'shallow-brine' pan shows a thinner zone of brine below a pycnocline and the pycnocline commonly drops to the bottom of the pan (i.e. the whole pan is diluted to the bottom). Such drops or lowstands of the pycnocline are thought to control the growth of selenite at the bottom (due to possible dissolution of crystal tops by less saline water occupying the upper zones of the pan) and are one of the reasons for the bottom of the shallow-brine pan is flat. By contrast to this pan type, the deep-brine pan (below the pycnocline) shows undulating or uneven bottom morphology because the selenites in this zone are sheltered from dissolution. Both pans create separate environments, and show characteristic gypsum facies (see Bąbel 2004a); however they can evolve from one into the other, and this evolution can be linked to the changes in the water depth. The deep-brine pan can be treated as moderately deep and the shallow-brine pan as shallow, however there is no fixed water depth defining these two types of pans (they can show comparable depth) and the shallow-brine pan can be even deeper then the deep-brine pan, depending on local environmental features such as size (fetch), salinity and climate (particularly windiness). Both pans show a depth below the depth limit for selenite deposition and their hydrography is the same as in the earlier described general model of the selenite pan (Fig. 6). Further details and discussion of the both types of the pans are given in Bąbel (2004a, pp. 239–242).

Coarse-crystalline (thick-bedded) selenites as deposits of deep-brine pans

General features Selenites comprising units A, C-D and SV commonly show well-developed palisade structure and coarse to giant crystal sizes (Figs 8 & 10). They appear in thick sequences without well-developed horizontal bedding planes, or the planes are rare or very poorly visible. The Badenian sequences are relatively thin (rarely over 10 m thick; Fig. 8) in comparison with the Messinian equivalents (up to 30 m thick in Sicily; Hardie & Eugster 1971).

Dissolution surfaces Horizontal bedding is very rare in unit A (giant crystal intergrowths; Figs 8 & 10a) considered as representing a slightly deeper environment than the C-D and SV units (sabre-gypsum facies) where the bedding features are more common and crystal sizes smaller (Figs 8, 10b & c). The sole pronounced bedding planes present in unit A are synsedimentary dissolution surfaces. They appear as horizontal

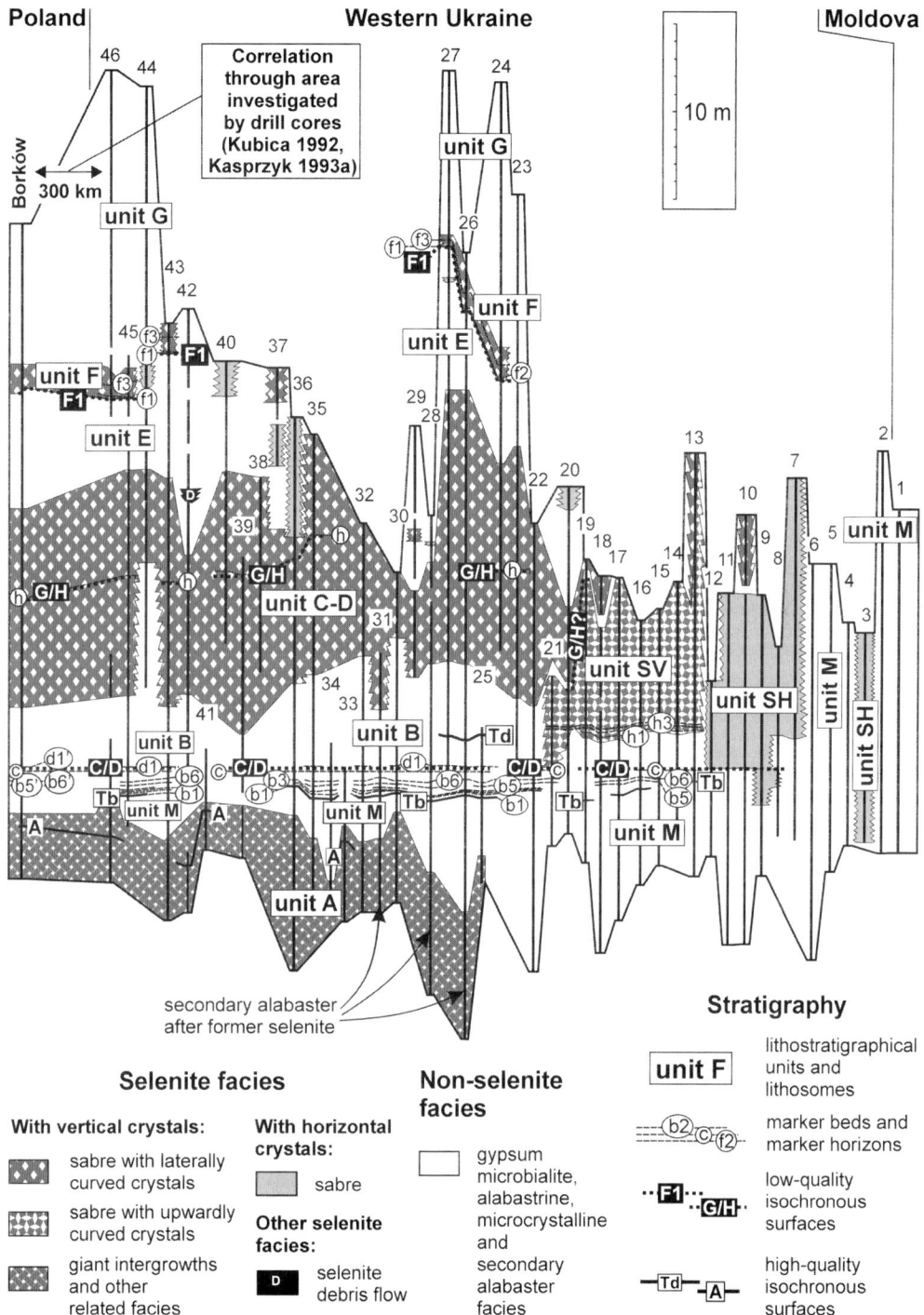

Fig. 8. Scheme showing stratigraphic and facies relations in representative sections of the Badenian gypsum deposits along the correlation line shown in Figure 7. Stratigraphy after Kubica (1992) and Bąbel (2005a); see also Bąbel & Bogucki, 2007, figure 4. A list of the numbered sections is given in Figure 7. Complete documentation and details are shown in Bąbel (2005a), and at www.geo.uw.edu.pl/agp/table/appendixes/55-1/.

discontinuities or fissures cross-cutting the bodies of crystal, (Fig. 10a). Only very rarely are the discontinuities covered with fine-grained gypsum, some clay or carbonates. Even more rarely is the generation of new cycles of grass-like gypsum crystals developed on such surfaces. The dissolution surfaces are flat or slightly irregular, with the shape fitted to the protruding but flattened (truncated) crystal apices. The unusual feature is that such dissolution surfaces, visible as discontinuities or fissures, commonly disappear laterally and are difficult to recognise and correlate even within one quarry (Fig. 10a).

Careful measurement of the sections showed that only the rare and major dissolution surfaces correlate over distances of up to a few kilometres. Correlation over larger distances was not detected in this study although the pronounced long-distance correlation was found within growth zones of the giant crystals showing internal zig-zag shapes characteristic of the crystal apices not truncated by dissolution (Warren 1982b; Bąbel 2005a, On-line Appendix, figs 4 & 5). In vertical sections the dissolution surfaces occur more commonly in the lithological transitional zones to the shallow water gypsum deposits represented by units B and M (Figs 8 & 9). Traced laterally the dissolution surfaces appear more commonly in the areas closer to the margin of the basin or on apparent basinal swells, where the thickness of unit A is apparently reduced. In such areas dissolution surfaces are more densely distributed, even over a distance of a few centimetres one from another.

Interpretation of dissolution events in deep-brine pans The dissolution surfaces, described above, differ from their modern analogues observed in selenites from saltwork pans, that are maintained at depths less than 1 m (Ortí et al. 1984; Geisler-Cussey 1986, 1997), and from coastal salinas in Australia (Warren 1982b). They occur in vertical section as single surfaces that are rarely present within the coarse-crystalline selenite bed present in the modern environments and are not associated with any millimetre-scale horizontal internal lamination within the crystals as recorded in Australia (Warren 1982b, figs 9 & 15b). This suggests that dissolution events were rarer in the Badenian basin and possibly the accommodation was larger in this basin than in the modern analogues. Substantial accommodation is a feature of salina-type basins.

The selenites are interpreted as deposited in moderately deep, permanent saline pans within a salina-type basin (Figs 3a, 4 & 6) or on the vast platforms within such a basin when it is deep enough to be persistently meromictic (Fig. 3b; Bąbel & Bogucki 2007). The term 'deep-brine' is employed for such moderately deep pans but does not imply substantial depths (Fig. 11b). Large topographical differences of bottom-level and thicker brine layers both below and above the pycnocline are characteristic of the deep-brine pan, which develops when a large volume of brine fills in the differentiated relief (this appears to be particularly common during the initial stages of evaporite deposition; as exemplified by unit A in the Badenian basin; Figs 8 & 9). The pan exists in a salina-type basin and therefore:

(1) its water level can fluctuate readily, and the amplitude and rate of these fluctuations are high, as in most saline lakes (where the annual water level fluctuations are up to a few tens of centimetres in amplitude, and the mean water level always is rising and falling throughout the ensuing time interval); and
(2) a stratigraphical record is nearly complete (erosion is minimal or absent because of the high accommodation and the aggradational style of deposition typical of salina basins). The deep-brine selenites do not record small amplitude water and pycnocline level changes (Fig. 11). Deposition of selenite beds is connected with the pycnocline highstands.

In the deep-brine pan the selenites growing in shallower portions are subjected to dissolution by more dilute surficial brines during falls of the average pycnocline level (designated EPI-MIX in Fig. 11). The selenites from below the average pycnocline level grow without interruption (except for some longer meromictic periods), disturbed only by some rare major drops of the pycnocline. These deepest zones are favourable for the deposition of the continuous uninterrupted sequences of coarse-crystalline selenites, which do not display macroscopically visible bedding and are lacking any dissolution features. The bedding planes that are present in coarsely crystalline selenites are commonly marked by dissolution surfaces. It is evident that, because of the uneven basin floor, such bedding planes in the selenites from the deep-brine pans may not show well-developed correlation across the basin (Fig. 11b).

Two scenarios, shown in Figure 12, are suggested for the events producing dissolution surfaces in the deep-brine pan in the salina-type basin, both related to common water level fluctuations typical of such a basin (Figs 3 & 11). The growth of coarsely crystalline selenites is assumed to take place in the zone below the average pycnocline in the pan (Figs 6, 11b & 12, bottom and top). As in the general model (Figs 5 & 6) the brine column in the pan passes through repeated stratification-mixing cycles, and it is assumed that stratification is

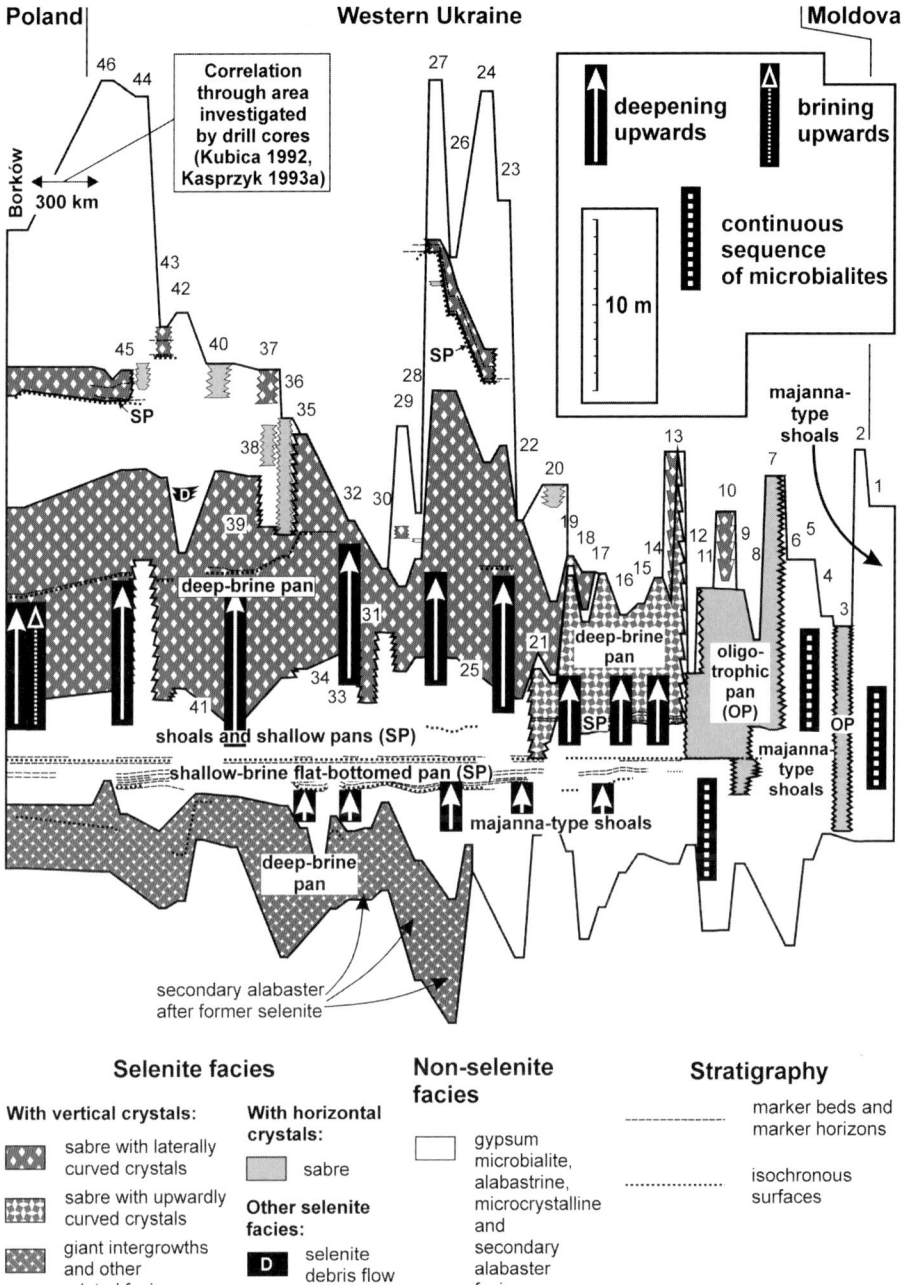

Fig. 9. Environmental interpretation of the selected facies, location of the deepening-upwards and brining-upwards sequences, and continuous microbialite sequences, in the representative sections of the Badenian gypsum deposits along the correlation line shown in Figure 7. Brining-upwards sequence at Borków in Poland (left) was recognized by the rise of Sr content in the selenite crystals from about 600 ppm Sr at the bottom of the sequence, to 1600 ppm Sr at the top (Rosell *et al.* 1998, fig. 9). The same rising trend in Sr content in selenite crystals was documented in this interval of the section in five other sections (from boreholes in the northern margin of the basin in Poland, not shown in Fig. 7, see Rosell *et al.* 1998) and supported by the same trend in Sr content in bulk rock samples in about 20 other sections from this area (Kasprzyk 1989, 1993b, 1994). For more details see Figure 8 and www.geo.uw.edu.pl/agp/table/appendixes/55-1/.

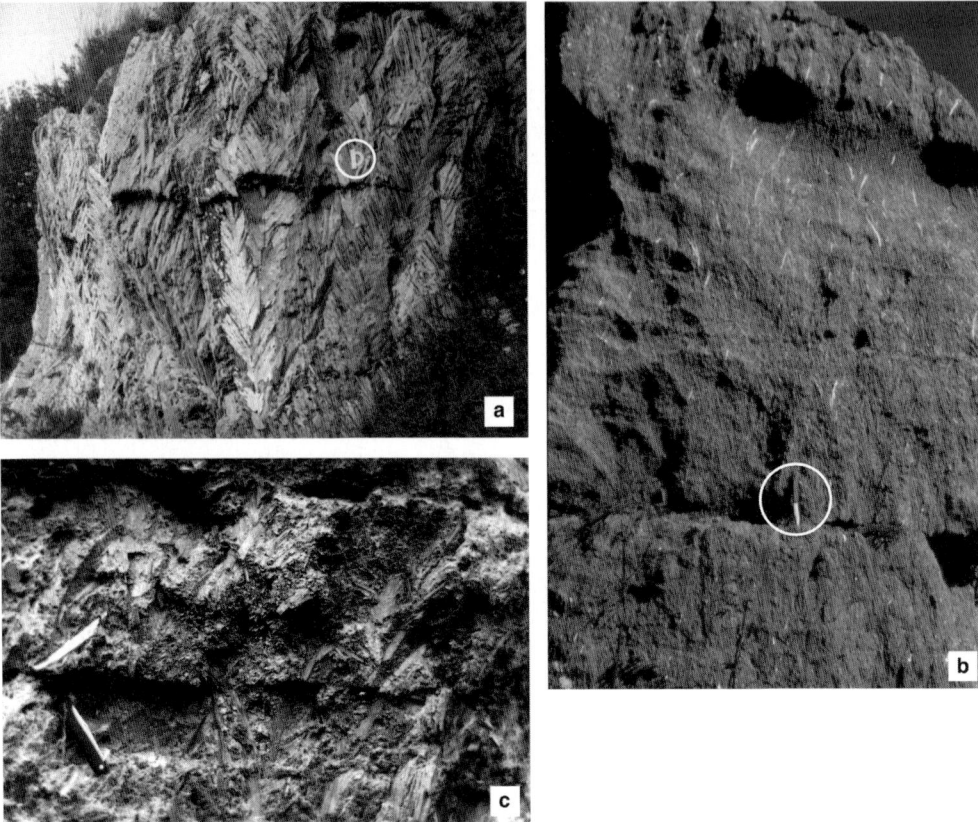

Fig. 10. Synsedimentary dissolution surfaces in coarse-crystalline selenites. (**a**) The giant gypsum intergrowths, unit A; Marzęcin, Nida river valley, Poland; the protractor (circled) is 10 cm long; (**b**) the sabre gypsum facies, unit SV, Nahoriany, locality 19 in Figure 17b, Ukraine; the knife (circled) is 25 cm long; (**c**) the sabre gypsum facies, unit C-D, bed g in local stratigraphy, Skorocice, Nida river valley, Poland; the pocket knife is 14 cm long.

associated with the highstand and the mixing with the lowstand (Figs 12, bottom and top, & 13, top left), as recorded in Red Pond and Green Pond in Arizona, Solar Lake, Lake Hayward and some salinas of the York Peninsula (Cole *et al.* 1967; Eckstein 1970; Cohen *et al.* 1977; Coshell & Rosen 1994; Rosen *et al.* 1996; Warren 2006). The growing crystals develop well-terminated crystal apices during the growth below the lowest pycnocline level in the pan, much like the zig-zag selenites described by Warren (1982*b*). These apices can be dissolved and truncated by the contact with the EPI-MIX water (or brine) undersaturated with gypsum lying above the pycnocline, which can take place without actual emersion (Figs 6, 11b & 12, centre). This contact is made possible by a lowering of the pycnocline level which can be realized in two possible ways:

(1) by shallowing of the pan and drop of water level associated with the pycnocline level drop (Fig. 12, centre left); or

(2) by inflow of the large volume of less saline diluted waters to the pan causing the rise of water level during highstand and associated drop in the pycnocline level (Fig. 12, centre right).

In both cases it is assumed that during stratification periods gypsum dissolution processes take place in the zone above the average pycnocline (zone B in Fig. 6; Fig. 11b), and these processes dominate over the possible gypsum nucleation and growth in this zone during the mixing periods, as predicted by the features of the deep-brine model (Bąbel 2004*a*, pp. 227, 240). During the lowering of the pycnocline, the dissolution surfaces can become covered with microbial mats in the shallow water, possibly due to better illumination (Fig. 12, centre left), unlike the the dissolution surfaces produced in the deeper water (Fig. 12, centre right). Thick, cohesive and impermeable microbial mats covering the crystals can also become a

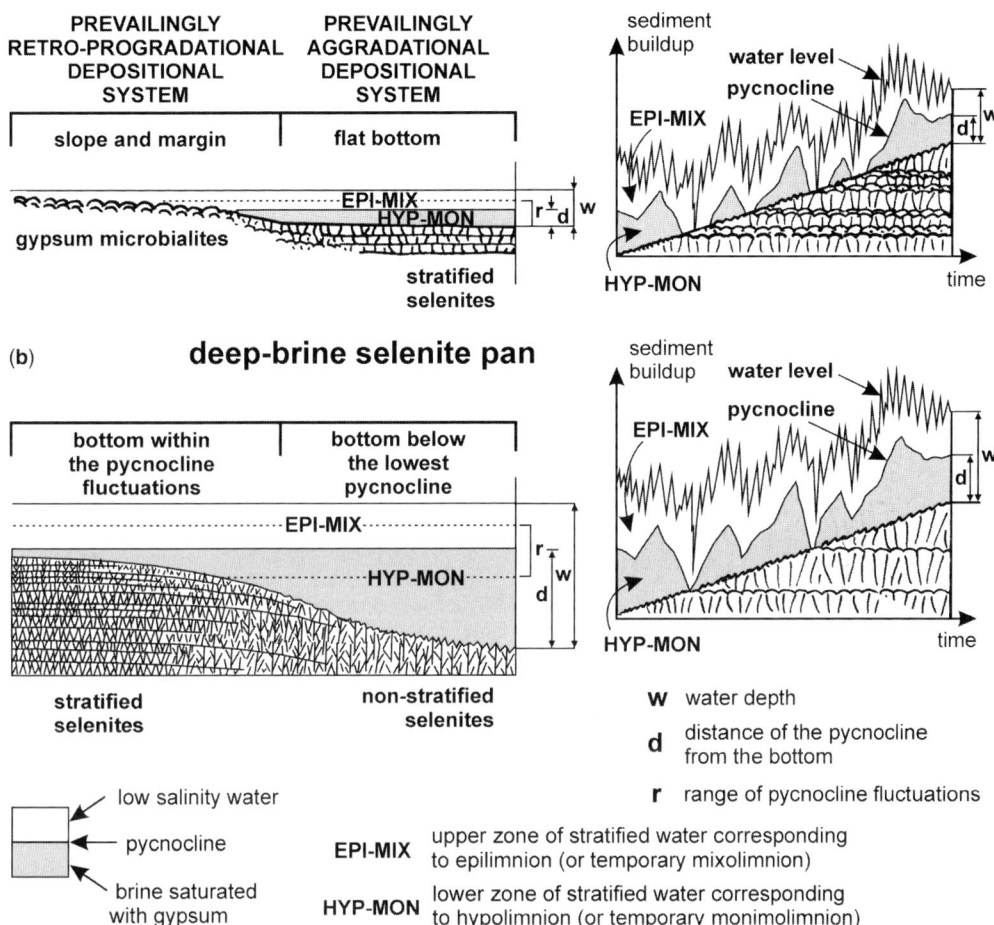

Fig. 11. Depositional models of the shallow-brine flat-bottomed selenite pan (**a**) and deep-brine pan (**b**) and their stratigraphical record (after Bąbel 2004a, modified). The basic concept of the selenite pan is outlined in Figure 6. Detailed explanations in the text.

barrier for the syntaxial growth of these crystals, and, consequently, create a substrate on which a new generation of crystals will grow when conditions for selenite growth return (Figs 6, 11b, 12, top left and centre).

The dissolution events recorded in the deep-brine selenite pans can be longer than one year. In particular, the dissolution events that are related to larger inflows of fresh water to the pan (Fig. 12, centre right) could last longer than one year and could be associated with some short-time meromixis in the pan. It is well known from the observations in the monomictic Dead Sea, Mono Lake and other saline lakes, that large rainstorms have promoted meromixis that lasted a few years (Romero & Melack 1996; Melack & Jellison 1998; Niemi et al. 1997). The inflow floods of run-off water into the Badenian basin is demonstrated by the presence of numerous clay intercalations with floral remains and sedimentary structures suggestive of the sheet flood deposition. Therefore it is highly probable that the Badenian deep-brine pans passed through some short-time meromictic periods promoted by rains or other major influx of less saline water. Some zig-zag growth zones in selenite crystals from unit A were interpreted as recording such meromictic periods promoted by inflow of meteoric waters (Bąbel 2005a, p. 16); however the correlative

124   M. BĄBEL

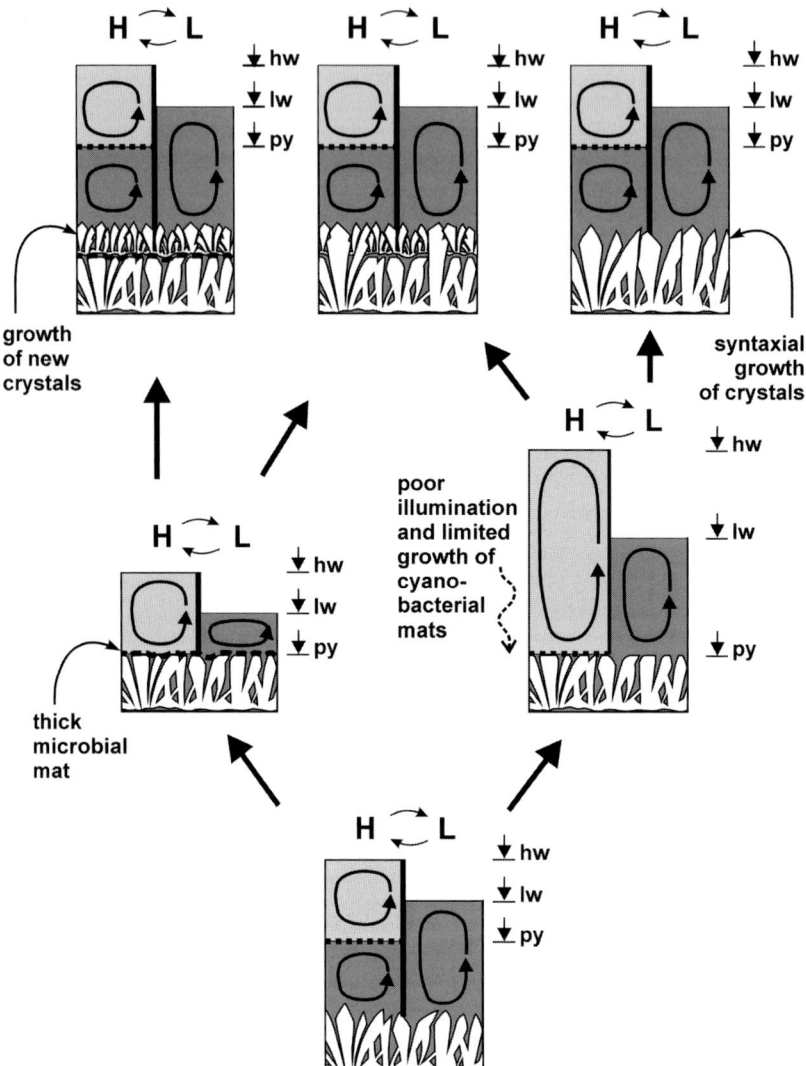

Fig. 12. Scenarios for subaqueous dissolution events in deep-brine selenite pans (general pan models are presented in Figs 6 & 11b) related to pycnocline lowstand associated with moderate shallowing (left), and with relative deepening associated with dilution of the upper brine mass and a pycnocline lowstand (right). Explanations in the text and Figure 13.

dissolution surfaces produced by such waters in the shallower zones are not recognized.

Grass-like (thin-bedded) selenites as deposits of shallow-brine pans

General features Selenite beds in the grass-like facies are laterally continuous and intercalated with fine-grained gypsum. Individual beds of both selenite and fine-grained gypsum are usually less than 20 cm thick (Figs 14 & 15). The fine-grained gypsum intercalations commonly show crenulated laminations and domal structures, typical of gypsum microbialite or gypsified microbial mat deposits. Fine-grained gypsum is also represented by a more or less homogeneous alabastrine rock variety, in places with scattered larger gypsum crystals and brecciated or nodular structures, which was interpreted as *in situ* crystallization from brine sheets (Peryt 1996) or as pedogenic deposits (Kasprzyk 1993a, pp. 49–51; Bąbel 2005a). Unlaminated clay is intercalated in these deposits.

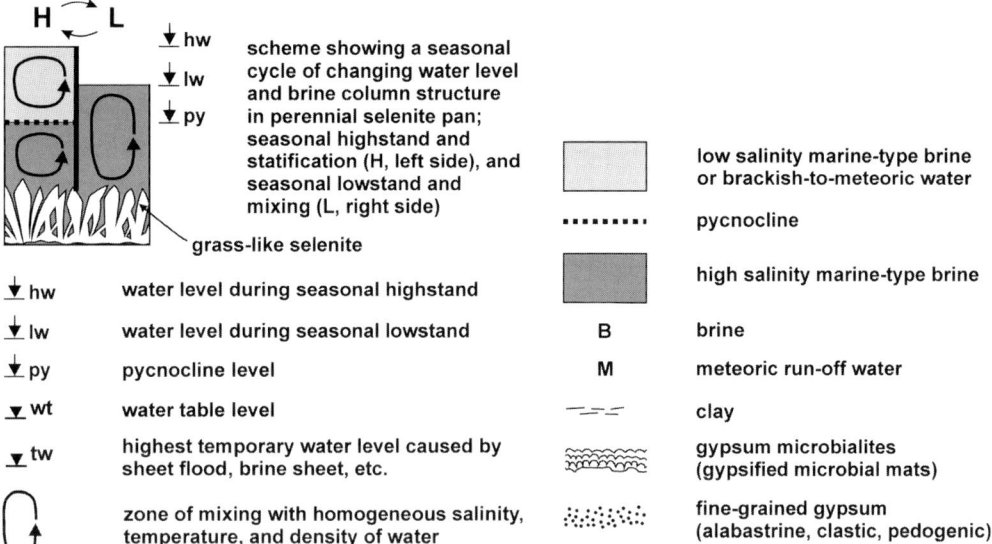

Fig. 13. Key to Figures 12, 18 & 19.

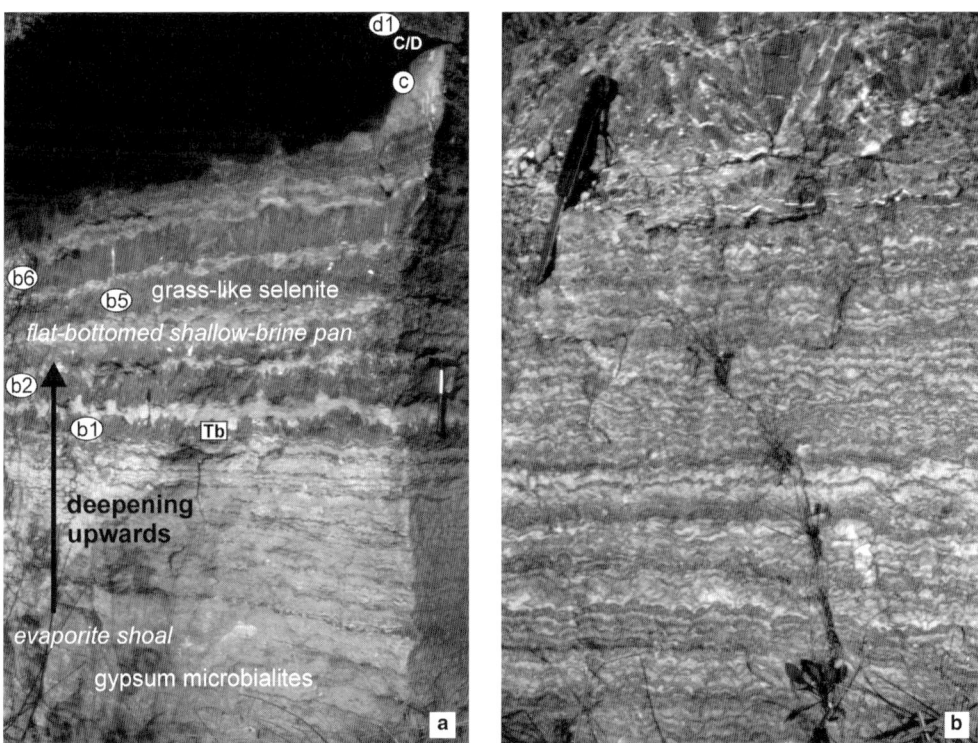

Fig. 14. Deepening upwards transition from gypsum microbialites of unit M (majanna-type shoal facies) to grass-like selenites of unit B (shallow-brine selenite pan facies) (see Fig. 9). (**a**) Microbialites and sequences of marker beds (described in Figs 8 & 16), Mezhyhirtsi, Ukraine (locality 32); (**b**) thick sequence of gypsified microbial mats below the selenite marker bed b1 at Pisky, Ukraine (locality 45); the knife is about 25 cm long.

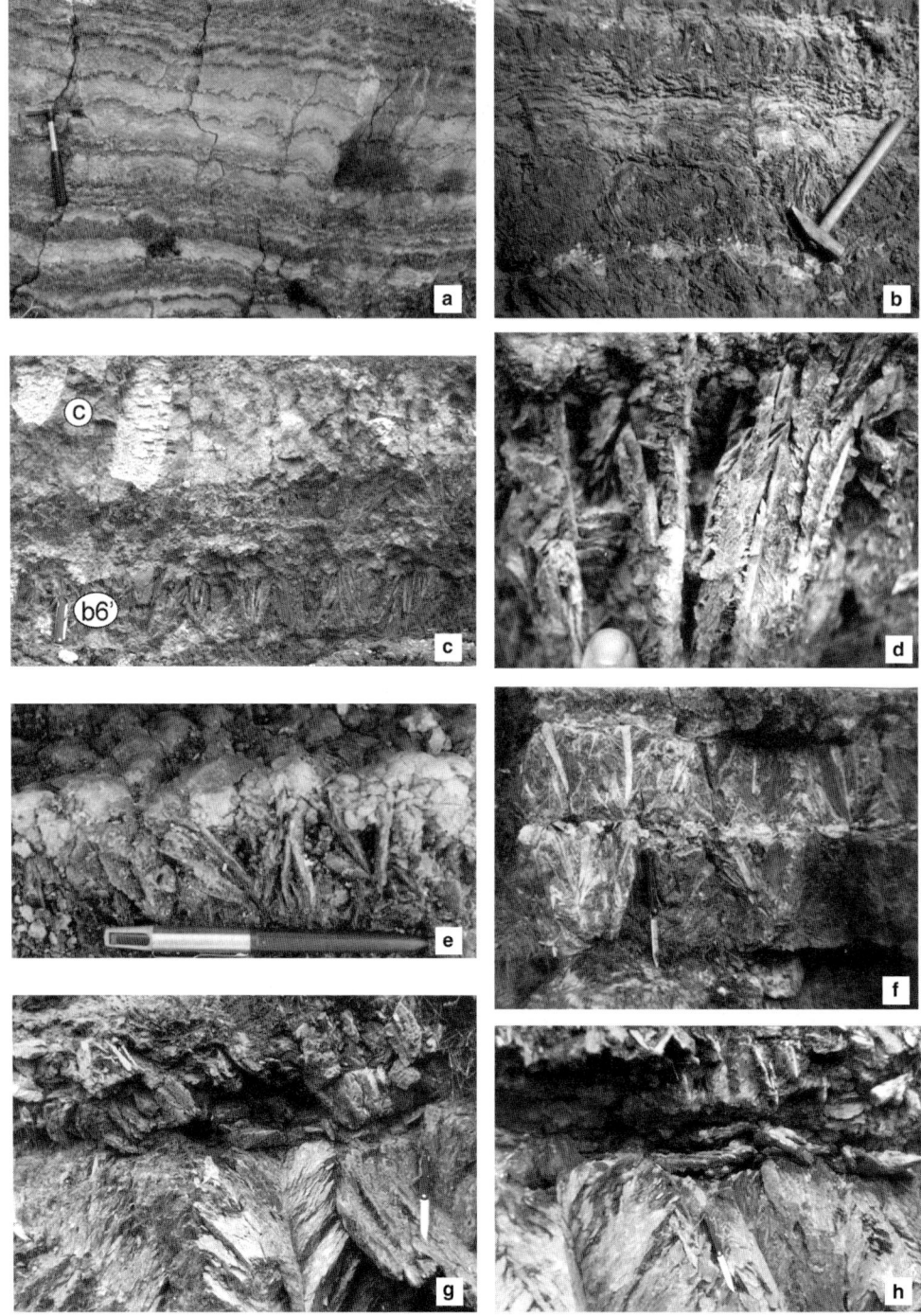

Shallow-brine pan The grass-like facies is interpreted as deposits of the shallow-brine pan (Bąbel 2004a). The shallow-brine pan setting seems to have had an extremely flat bottom (with infilled levelling of the bottom irregularities) and much thinner brine zones separated by a pycnocline than the deep-brine pan (Fig. 11a & b). Similarly, as in the deep pan the water level in a shallow pan fluctuates with time (through the years) frequently, with a high amplitude and rate (note that both pans in Fig. 11 show the same rhythm of fluctuations). It is assumed, following some earlier concepts put forward by Lowe *et al.* (1997) and Verschuren (1999), that the pycnocline tends to fluctuate together with the water level fluctuations, although it is generally more stable, because this allows analysis of the stratigraphical record of this selenite facies (Bąbel 2004a) as in the model presented by Strasser *et al.* (1999) for shallow carbonate platforms.

The shallow-brine selenites record much smaller amplitude changes in both water and pycnocline than the deep-brine pans (Fig. 11a & b). In both pans deposition of selenite beds is connected with the pycnocline highstands. It is evident that, because of the uneven bottom, the bedding planes in the selenites from the deep-brine pans will not develop good correlation across the basin (Fig. 11b). By contrast the selenite beds from the vast flat-bottomed shallow-brine pans are ideal for long-distance correlation (Fig. 11a). The shallow-brine pans develop on the very gentle bottom topography and they appear to be typical of the shallowing events in evaporite basins. These pans may be considered as later stages of the evolution of the deep-brine pans, namely the effect of the sedimentary fill of such pans and associated shallowing (see Figs 8 & 9).

Long-distance correlation Some sequences, made up of several thin selenite beds intercalated with fine-grained gypsum, reveal correlation over distances of tens of kilometres in the Badenian basin. This is similar to some beds present in the Messinian of Sicily (Schreiber *et al.* 1976). The most widespread of these is the sequence of six selenite beds, designated b1–b6, that correlate along nearly the whole northern margin of the basin (Figs 8, 14, 15c, d, f, 16; Bąbel 2005a, On-line Appendix, figs 6a and 8). This sequence was interpreted as deposits within an evolving giant shallow-brine pan (Fig. 11a). The other similar sequences show bed-by-bed correlation over smaller distances, evidently due to smaller sizes of the shallow-brine pans.

Isochronous to near-isochronous deposition of selenite beds in the shallow-brine pan

Each selenite bed in the sequence (Fig. 16) originated due to a stable period, under a long-term highstand of a pycnocline in the shallow-brine flat-bottomed pan (Fig. 11a) and can be treated as isochronous or near-isochronous stratigraphical unit (i.e. an event or marker bed; Bąbel 2005a). The growth of selenite crystals took place below the pycnocline and required a permanent presence of a 'shelter' of calcium sulphate saturated-to-oversaturated brine both during stratification and mixing periods (Figs 5 & 6). The interpretation of associated depositional events taking place in the interval of the section containing the b1–b6 marker beds is shown in Figure 17 and discussed in the next section (see Bąbel 2005a, for more details).

The growth of selenite crystal tops in shallow-brine pans was controlled by a position of the pycnocline (Figs 11a & 17). The crystals could not grow above the pycnocline level and their tops presumably tended to be limited to the ideal horizontal surface in the same way as upper surfaces of many selenite crystal beds in the recent saltwork pans (Ortí *et al.* 1984; Geisler-Cussey 1986, 1997) and in the Messinian deposits of Sicily (Schreiber *et al.* 1976). The other modern example of such horizontally truncated selenite tops can be found in salinas of York Peninsula, in Australia, in selenite beds from the topmost part of the shallowing-upwards gypsum sequences formed during gradual sediment infilling of pans that originally were several meters deep (see Warren 1982b; Warren & Kendall 1985; and photo by J. K. Warren in fig. 8.22 in Schreiber *et al.* 1986).

Fig. 15. Sedimentary features of the grass-like selenite beds. (**a, b**) Grass-like selenites intercalated with gypsified microbial mats; note thin selenite layers coating the domal microbialite structures; unit B above marker bed c; Lokutki near Tlumach, S of locality 26, Ukraine (a) and Gacki quarry, Nida river valley, Poland (b). (**c–e**) Selenite beds with empty fenestral pores between the grass-like crystals, below gypsified microbial mats; (d) details of (c) marker beds b6' and c are shown in (c), bed c represents gypsified microbial mat deposits; unit B, Borków quarry, Nida river valley, Poland; the pocketknife in (c), bottom-left, is 14 cm long. (**f**) Selenite beds with eroded and corroded top surfaces covered with fine-grained gypsum; unit B, near marker bed b5–b6, Oleshiv, locality 28, Ukraine. (**g, h**) Selenite regolith originated from selenite beds destroyed during emersion; note preserved vertical position of crystals within the selenite debris in (g), the regolith covers the apices of the giant gypsum intergrowths flattened by dissolution; unit A and B, Sielec Rządowy, Nida river valley, Poland; the knife in (f), (g) and (h) is about 25 cm long.

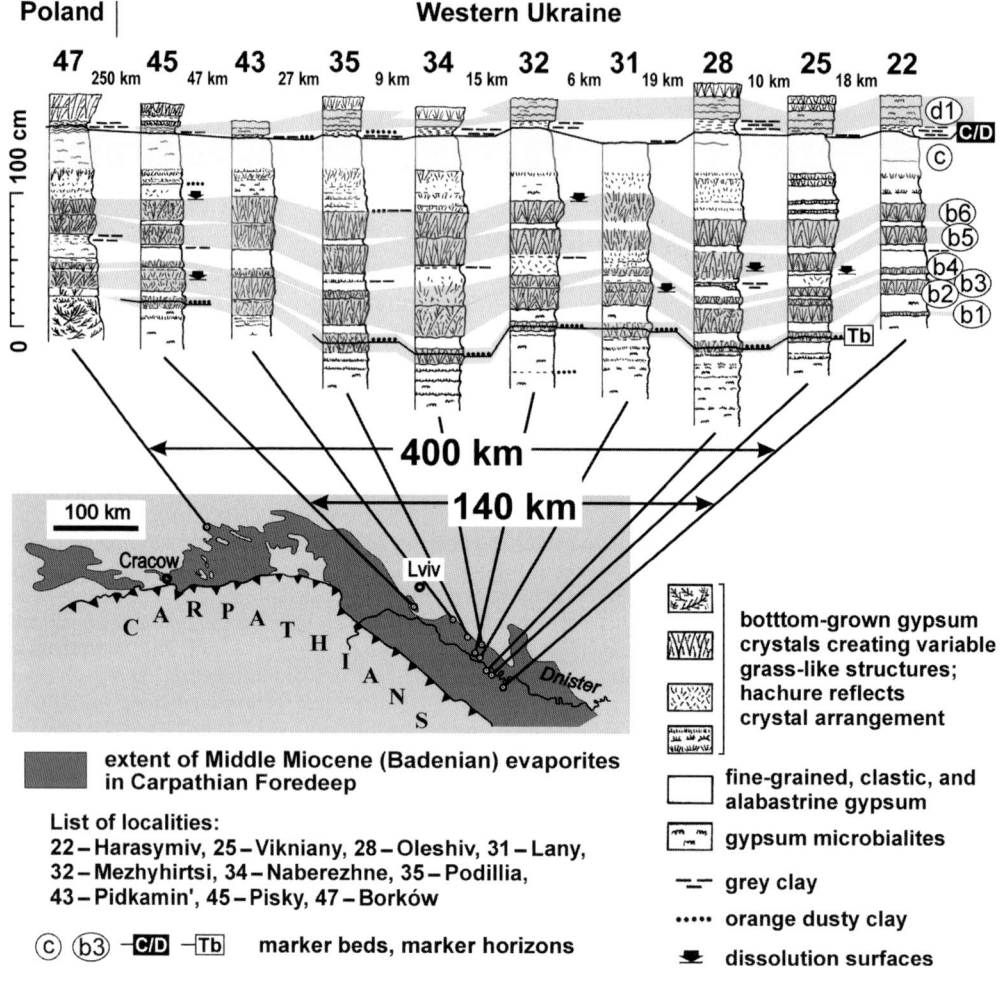

Fig. 16. Long-distance correlation of selenite sequences (marker beds b1–b6, Fig. 8) in the Badenian (Middle Miocene) evaporite basin in the northern Carpathian Foredeep in Ukraine (localities 22–45) and presumed correlation with marker beds b5' and b6' in Poland (locality 47). Complete documentation and details are shown in Bąbel 2005a, and at www.geo.uw.edu.pl/agp/table/appendixes/55-1/.

The other example of similar evaporite growth styles can be observed in trona deposits at the bottom of Lake Magadi in Kenya, where 'the tops of the crystals are roughly at the same elevation throughout the whole lake, which is about 20 km long' (Eugster & Hardie 1978, p. 34). Clear trona crystals from this lake form 2–5 cm thick beds, separated by dark crystals enriched in windblown dust related to seasonal exposure of the bottom of the lake (Eugster 1980). The crystals appear to correlate across the whole lake (see photo by H. P. Eugster in Smoot & Lowenstein 1991, fig. 3.8d). This type of trona deposits was considered by Eugster (1980, p. 215) as product of 'a saline lake at its peak of productivity'. During exceptionally dry seasons the brine level drops slightly below the surface of these grass-like trona crystals (not more than a few centimetres; Warren 2006, pp. 260, 865) causing the formation of expansion polygons typical of a dry or nearly dry bottom. Similar features suggesting that the brine level dropped below the crystal tops in selenite beds are not noted in the Badenian selenite sequences, presumably because fluctuating water or brine level dropped only sporadically and only just below the sediment surface which was nearly always covered with brine or wet (as discussed in the next section).

Thus the considered selenite beds can be used not only for isochronous correlation, but also as one of the most ideal datum surfaces in ancient

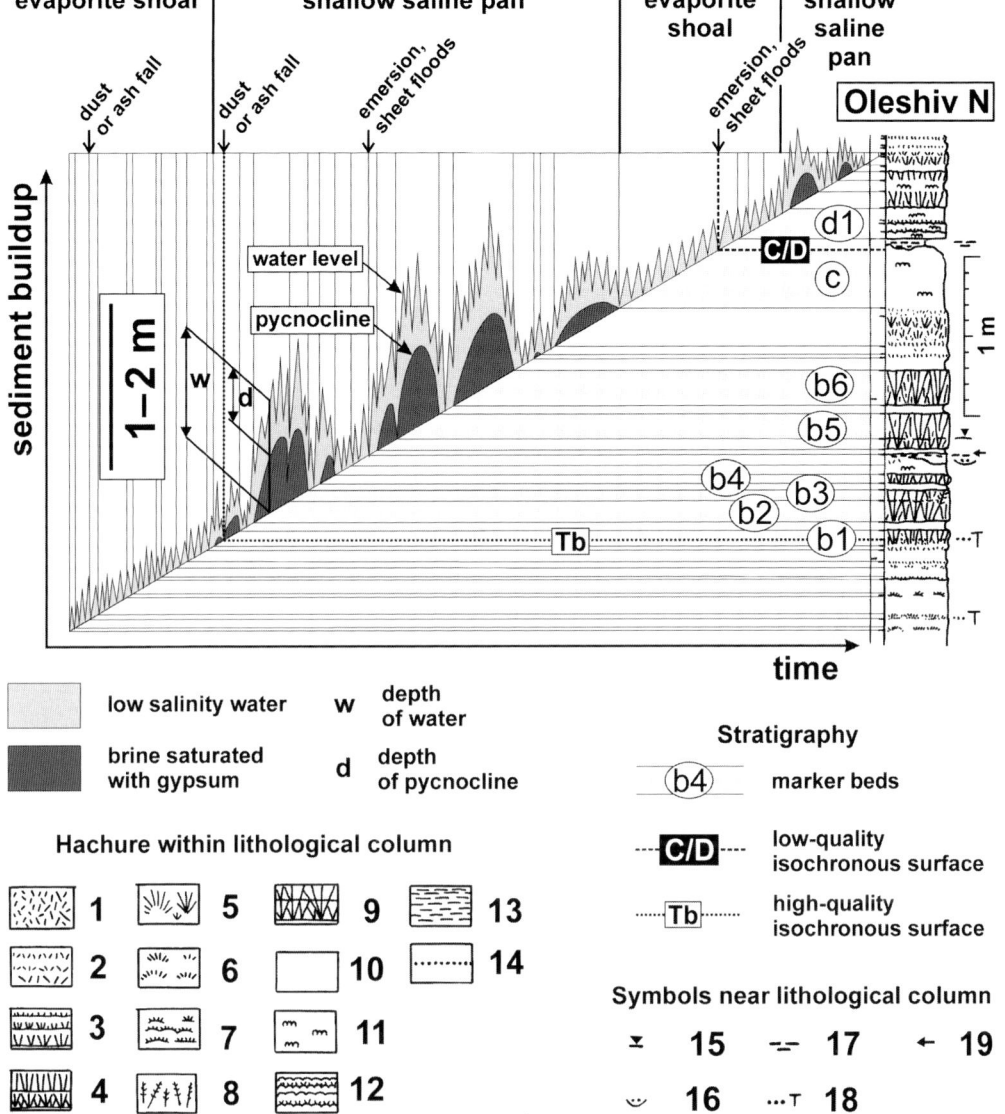

Fig. 17. Scheme showing interpretation of depositional events around marker beds b1–b6, c and d1 in the exemplary gypsum section at Oleshiv, Ukraine (locality 28, Figs 7, 8 & 16). Selenite marker beds are interpreted as response to fluctuations of the average pycnocline level in a shallow flat-bottom saline pan. 1–9, Selenite components (hachure reflects arrangement, sizes, and shapes of crystals): 1, randomly scattered straight rod-like crystals – short (left) and long (right); 2, rows of rod-like crystals creating grass-like structures without sharp common base – short crystals (top), long crystals (bottom); 3, rows of rod-like crystals showing grass-like structures and with flat common base – short crystals (top), long crystals (bottom); 4, straight grass-like crystals showing palisade-parallel (top) and palisade-radial structure (bottom); 5, radial or fan-like aggregates of straight crystals grown from common centres (right) and overgrowing fine-grained gypsum dome (left); 6, rows of small grass-like crystals without common sharp base grown on fine-grained gypsum domes; 7, rows of small grass-like crystals with sharp common base grown on fine-grained gypsum domes; 8, aggregates of crystals resembling palm tree leaves; 9, flat horizontal synsedimentary dissolution surface within radial aggregates of the grass-like crystals showing syntaxial growth over this surface; 10, fine-grained homogeneous gypsum (alabaster when massive and white), or other non-differentiated fine-grained or microcrystalline gypsum; 11, fine-grained (alabaster-like) gypsum with traces of gypsified crenulated microbial mats; 12, small grass-like crystals covering surfaces of gypsified crenulated microbial mats and intercalating gypsum microbialite deposits; 13, clay and gypsum-clay deposit; 14, orange dusty clay; 15, synsedimentary dissolution surface; 16, small channel-filled with sugar-like clastic gypsum; 17, clay intercalation; 18, orange dusty clay interpreted as aeolian dust or pyroclastic ash deposit; 19, horizontal discontinuity surface significant for correlation.

geological record, traceable over distances of hundreds of kilometres (Figs 8 & 16).

Non-selenite deposition in a shallow-brine pan

The deposits and the structures present between the grass-like selenite beds are critical for the interpretation of the evolving depositional environment of the shallow-brine pans.

The main requirement for non-selenite deposition in such pans is considered the 'exposure' of the bottom of the pan from the shelter of calcium sulphate saturated brine (HYP-MON in Fig. 11); or simply lack of a zone A at the bottom of the pan (Fig. 6). This can be realized in two general ways, by dilution and by shallowing, that in both cases can lead to another type of hydrography without the persistent brine cover at the bottom. Sedimentary features suggest that, in the case of many studied selenite sequences, the shallowing events were responsible for intercalated fine-grained gypsum deposition.

These scenarios assume the significant fluctuations of water and pycnocline level in the saline pan, as expected in any salina-type basin without open water connections with the sea (Fig. 3). The shallowing events begin with the development of the dissolution surface and truncation of crystal apices when they are removed from the shelter of saturated brine to the exposure with the more diluted brines above the pycnocline (as during formation of dissolution surfaces in a deep-brine pan; Fig. 12, centre left). Indeed such truncated crystal tops are very common in the Badenian selenite sequences (e.g. Fig. 15c, d & f). Less frequently, the crystal apices protrude from the top surface of selenite beds (Figs 14a & 15a & b), suggesting that a prolonged dissolution stage was omitted, probably because the pycnocline drop was very rapid (Fig. 18, right). In such cases, concentric coatings of laminated gypsum microbialites drape the protruding crystal apices, indicating that the microbial mats developed directly on the pre-existing well-terminated crystal tops (Figs 14a & 15a & b; see also Hardie & Eugster 1971, fig. 19; Peryt 1996, fig. 11e), as in the scenario in Figure 18, right.

It seems that during the water level fluctuations in the shallow-brine pan the water or brine level did not drop below the sediment surface. This can be demonstrated by a lack of any dissolution features on the sides of long crystal faces creating vertical intracrystalline pores within some grass-like selenite layers (Fig. 15c, d & e). No karst forms, infilling or related internal deposits were found at the bottom of these pores. Some intracrystalline pores show features typical of fenestral structures and are covered along their upper surfaces by fine-grained gypsum that apparently is a product of *in situ* gypsification of the microbial mat spread across and between truncated selenite apices (Figs 15c, d & e & 18). The presumed existence of such a mat and its gypsification additionally supports the view that the depositional flat surface created by selenite apices was at least wet. The water level could have dropped to the sediment surface repeatedly during deposition of fine-grained gypsum intercalated with the selenite beds. Such lowering of the water level and associated episodic semi-emersion could be associated with the truncation of the crystal apices (mainly by run-off water sheet floods; Sanz *et al.* 1994), deposition of clastic and pedogenic gypsum as well as deposition of alabastrine gypsum from brine sheets (Logan 1987; Figs 15f & 18 left). Again, these presumed emersion episodes were not associated with significant deep erosion of the grass-like selenite crystals, apparently because the water or brine table level did not drop below the level of selenite apices in the study area. It seems that the water level rose concomitantly with the accretion of gypum beds in these pans, which occurs in salina-type basins characterized by high accommodation.

Nevertheless, even in a salina basin, the periods of emersion, with depression of water level below the sediment surface, and associated weathering and erosion should be recorded in some way. Indeed, in the southern Nida area and the Miechów Upland in Poland layers of residual selenite debris were recognized. This debris represents the long-term emersion of selenite beds associated with the partial or complete disintegration of exposed gypsum (Fig. 15g & h; Ortí *et al.* 1984, figs 16.4, 16.5 & 16.6). The scenario of events leading to deposition of this selenite lag or regolith (Fig. 19) is nearly the same as suggested for the similar Messinian selenites by Garrison *et al.* (1978, fig. 30) and Schreiber (1978, p. 65), although the Badenian debris contains clay in the matrix between the crystal clasts.

The Badenian selenite debris deposits occur at the same interval of the section as the marker beds b1–b6 originated in the giant shallow-brine pan (Fig. 16). However general facies architecture and sedimentological data suggest that the debris represents reworking not from this shallow-brine pan, but rather some associated flat margin of the deep-brine pan in which the giant gypsum intergrowths were crystallised (unit A; Figs 8 & 9), namely the result of long-term emersion and destruction of marginal selenite beds. This style of reworking is recorded in recently exposed to weathering selenite beds crystallized on the slopes of the Marion Lake in Australia (Schreiber 1978, p. 65; references in Bąbel 2005*b*).

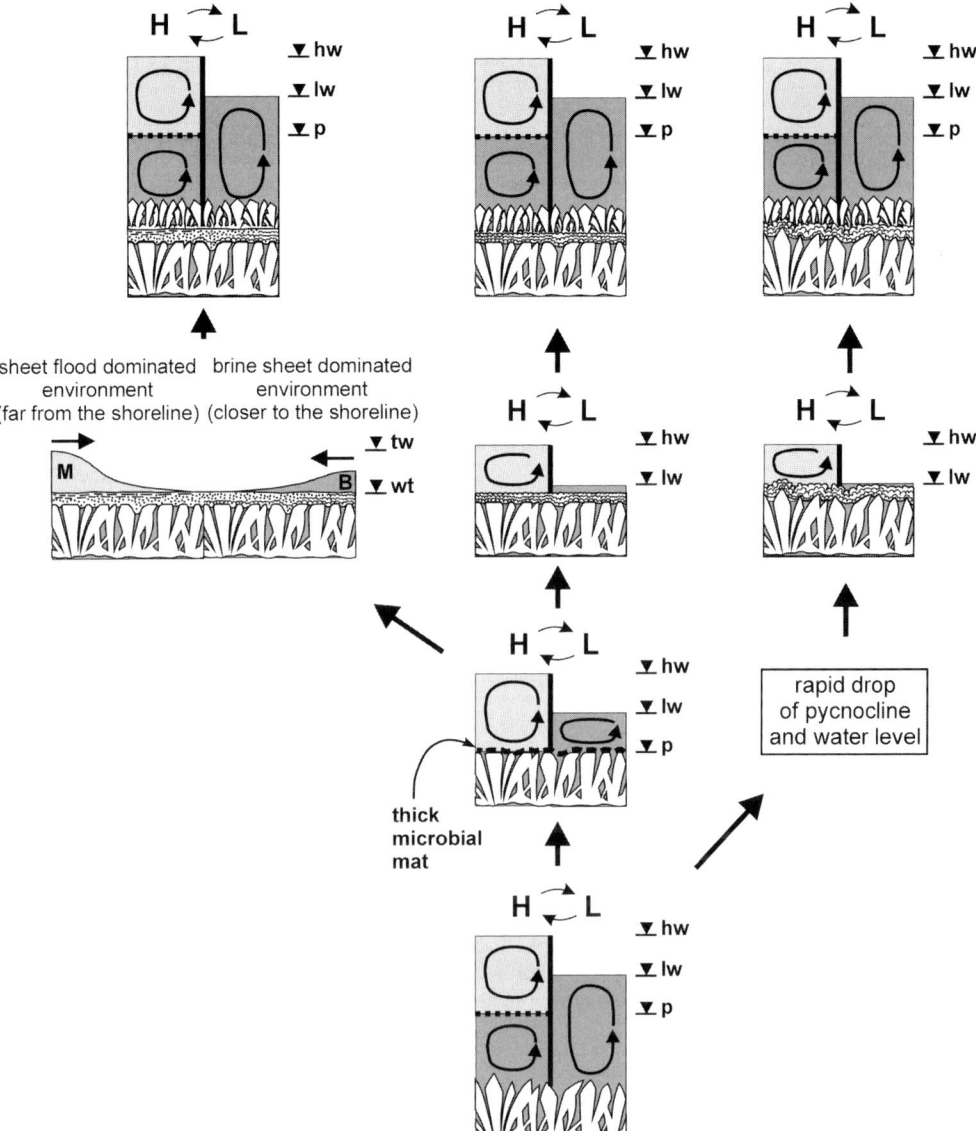

Fig. 18. Scenarios of shallowing events in a selenite pan with transition into majanna-type shoal environment (centre right; see Figs 4, 6 & 11a); further explanations in the text and Figure 13.

Remarks on the concept of a shallow-brine pan and its modern analogs The shallow-brine pan model represents an intermediate form of development between deep perennial saline basin (a deep-brine pan; Fig. 11b) and ephemeral (seasonally drying) lakes. In some aspects the modelled pan is similar to shallow-water selenite salinas, intermittently supplied with seawater, described by Warren (1982b, 2006).

In a manner similar to the shallow-brine pans the ephemeral, seasonally drying saline lakes or playa lakes and continental salt pans are characterized by a flat bottom. They show a well-developed annual cycle of water level fluctuations and pronounced dry phase (e.g. Arakel 1988). During a dry phase a lake bottom is completely dry and exposed to weathering. A seasonal, more or less significant, lowering of the groundwater or brine table facilitates this weathering. Ephemeral saline lakes are described in many models (Valyashko 1952, 1972; Langbein 1961; Handford 1982; Lowenstein & Hardie 1985; Bryant *et al.* 1994).

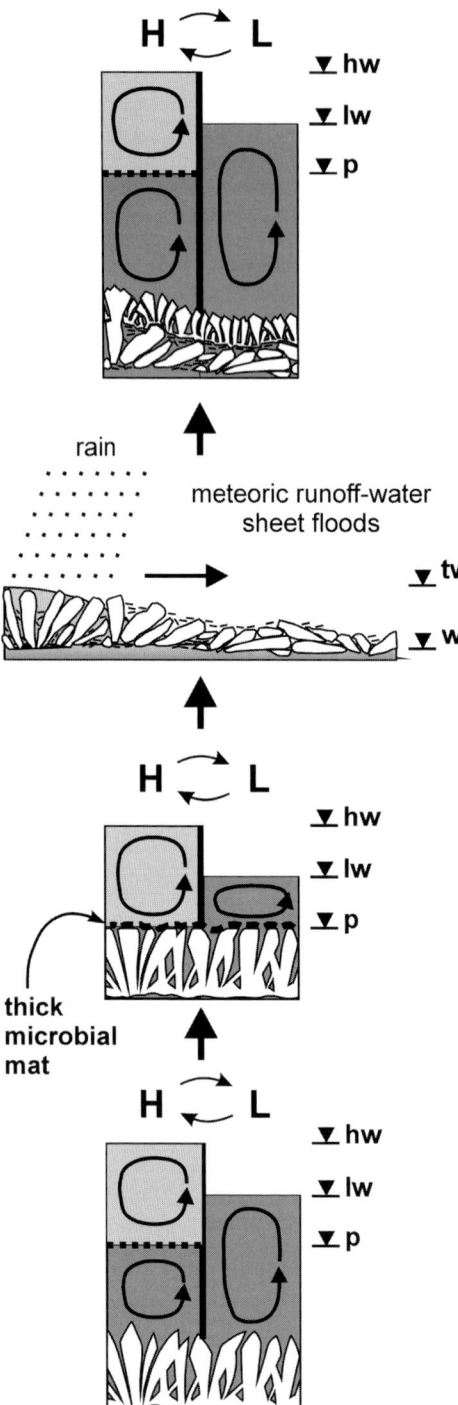

Fig. 19. Scenario of a long-term emersion event with weathering phase and creation of a selenite regolith or lag in a selenite pan (see Figs 6 & 11a); further explanations in the text and Figure 13.

The main difference between the shallow-brine pan and ephemeral lake is that the shallow-brine pan reflects the longer periods of water level fluctuations. Only during the water level lowstand is there a phase of seasonal drying, i.e. temporarily the pan is in an ephemeral lake state, although it is not obligatory for this model. Unlike the typical ephemeral lakes, the shallow-brine pans in the Badenian basin only dried episodically and in such periods the water or brine table level did not fall far below the sediment surface, similar to the brine covering the trona crystals at the bottom of Lake Magadi (Eugster & Hardie 1978; Eugster 1980; Warren 2006). It seems that this Badenian subsurface brine was not able to evaporate during seasonal lowstands.

The shallow-brine pan is particularly similar to saline pan model developed by Bryant et al. (1994) on the basis of the Chott el Djerid studies. These authors supplemented the scenario of events in ephemeral saline pan by Lowenstein & Hardie (1985) and distinguished an additional, 'moving' brine pool stage, i.e. brine sheet stage, appearing before the final complete drying of the pan (see Figs 4b & 18 left). Just such a phase of brine sheet deposition is apparently very important in many shallow-brine pans.

The shallow-brine pan shows seasonal water level fluctuations similar to those from ephemeral lakes (indicated by the scenario of Lowenstein & Hardie (1985), and shown by 'hw' and 'lw' in Figs 12, 13, 18 & 19) but these fluctuations are not as important in the depositional record as some prolonged water and pycnocline level fluctuations that are basically responsible for creation of the selenite sequences and their intercalated deposits (from bottom to the top in Figs 12, 18 & 19, and Figs 11a & 17). This basic fluctuation pattern recorded by selenite sequences in the Badenian basin appears to be some intrinsic feature of the salina basin similar to the long-term water level fluctuations observed recently in saline lakes around the world (e.g. Williams 1996; Comín et al. 1999).

It is possible that during the non-selenite phase of deposition in the shallow brine-pan its hydrography approaches that of the ephemeral saline pans. Owing to shallow water conditions only the temporary short-term pycnoclines were present in the pan and they were unable to maintain a brine shelter at the bottom (facies zone A; Fig. 6) supporting the continuous growth of selenite crystals (although some thin beds or clusters of grass-like selenites are found within fine-grained gypsum intercalations and can be related to some short-term presence of such a brine shelter).

The extreme seasonal fluctuations in salinity are characteristic for all ephemeral saline lakes. Their salinity level changes from the entirely freshwater during the run-off of flood water on to the dry floor of the pan, up to halite saturation during the final drying. Such drastic salinity fluctuations are possible because the ephemeral salt lakes are supplied only or mainly with the meteoric water. It was documented by Campbell (1995) that shallow salt lakes continuously supplied with some saline water show more stable salinity characterized by low-amplitude seasonal fluctuations, opposite to those in common ephemeral saline lakes. The modelled pan discussed here is by definition more or less continuously supplied with marine brine produced from seawater flowing down to the bottom of a salina basin (Fig. 3), or some brine flowing in from the intermediate subbasins (Fig. 4).

Selenite-gypsum microbialite deposits of an oligotrophic pan

Unit SH is composed of peculiar gypsum deposits composed of masses of selenite crystals with predominantly horizontal orientation distributed within fine-grained microbialite gypsum, commonly showing the crenulated lamination characteristic of gypsified microbial (cyanobacterial) mats (Figs 8 & 20; Peryt 2001, fig. 16). These selenites create very large domal structures and microbialites are commonly incorporated within these structures. Some domes are entirely comprised of compact masses of selenite crystals and reach several metres in size.

These deposits were interpreted as having accumulated in some smaller saline pans surrounded by vast majanna-type shoals covered with microbial mats (represented by unit M described below; Figs 4, 8 & 9). Because the pan was cut off from terrestrial run-off waters it was poorly supplied with nutrients and was of the oligotrophic-type (Horne & Goldman 1994), which permitted the abundant growth of cyanobacterial mats even in the deep brine. Photosynthetic activity of such benthic cyanobacterial communities contributed to the oxygenation of the deep brine during stratification periods (Cloern et al. 1983) in a similar way to Lake Hayward (Burke & Knott 1997). Accretion and gypsification of microbial mats was concurrent with the growth of selenite crystals distributed within the microbial mats (Fig. 20a & b). Partial analogs of this peculiar but quite widespread Badenian facies, can be found in Lake Inneston in Australia (Warren 1982b), the Gotomeer basin in Netherlands Antilles (Kobluk & Crawford 1990) and the Solar Lake (Krumbein & Cohen 1977; Krumbein et al. 1977; Friedman & Krumbein 1985) as discussed by Bąbel (2005b).

Channel structures and gypsum microbialite deposition

Unit M is composed of gypsum microbialites interpreted as deposits of the majanna-type shoals (Figs 4, 8, 9 & 14). The microbialites show flat but crenulated lamination that originated from *in situ* gypsification of thin microbial (cyanobacterial) mats with encrustation by fine gypsum crystals (Fig. 18, centre). The laminae are 1–2 mm thick and some of them are covered with minute grass-like gypsum crystals typical of subaqueous syntaxial growth from supersaturated brine directly at the bottom. The laminated structure is very well preserved (Fig. 14b) which suggests that gypsification was periodic, 'mat by mat', i.e. after gypsification of the mat, suppressed by precipitated gypsum, the next living mat was developed on the gypsum substrate. After some period of growth the new mat was subsequently gypsified. Such deposits probably represented permanently wet, semi-emerged and seasonally flooded parts of majanna-type shoals (see Figs 4b, 14a & 15a, b), similar to wet flats covered with living microbial mats creating 'perennial surface brine' biofacies in the coastal salinas of the Red Sea (Gerdes et al. 2000). During seasonal highstand the shoals were flooded by water which was low-to-brackish salinity and permitted the intensive growth of microbial mats (see Logan 1987, fig. 52b & c). Salinity rise during ongoing lowstand, in a dry season, led to gypsification of the mat, in a similar way to the mat observed by Eckstein (1970, fig. 6) on the margin of the Solar Lake, and also similar to the mode of gypsification of microbial mats suggested by Rouchy & Monty (1981, p. 178). The encrustation with gypsum occurring alternately with the microbial mat growth, could also take place during irregular time intervals, dictated by some variable fluctuations of water level and brine sheet flows in the broad boundary zone between the saline pan and the wet majanna-type shoal (Figs 4b & 18, centre). In some instances, in deeper, more saline and permanent brine bodies, minute grass-like crystals (selenite) develop directly on the gypsified mat. Such microbialite gypsum deposits apparently were preserved in sufficiently deep depressions showing much higher accommodation than any known recent coastal salina (e.g. the Solar Lake or MacLeod basin), i.e. when the water and brine levels in the saline pans and majanna-type shoals were rising slowly but permanently.

These microbialites are intercalated and pass laterally into more homogeneous alabastrine facies, sometimes with horizons of scattered larger gypsum crystals. This alabastrine facies was

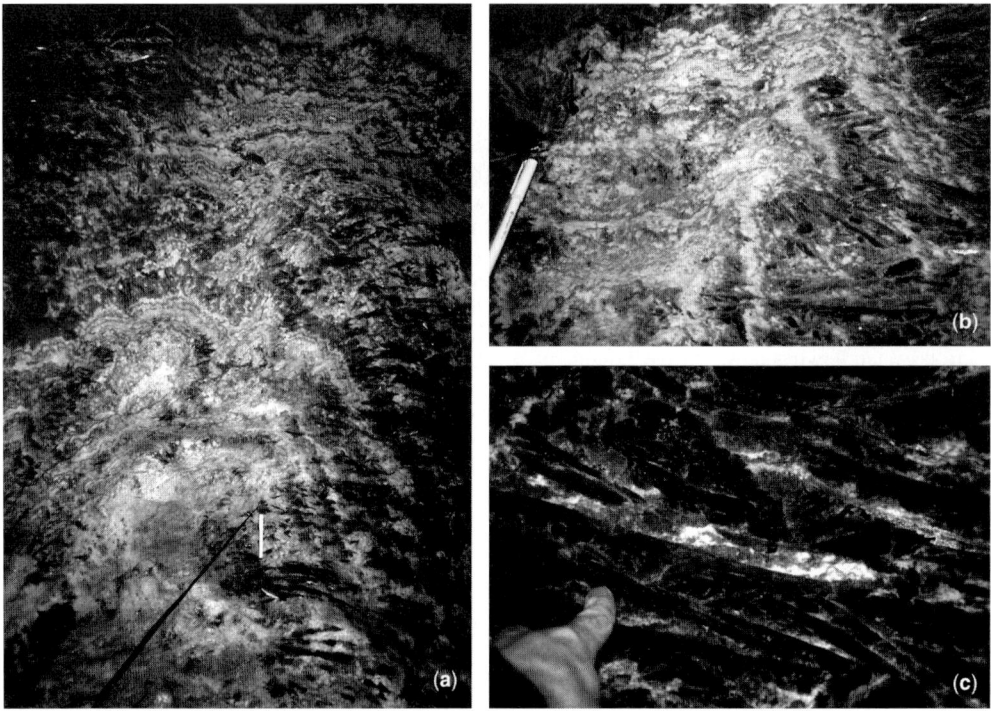

Fig. 20. Selenite–gypsum microbialite deposits representing the oligotrophic-type pan facies, note predominantly horizontal orientation of sabre gypsum crystals; transitional area between units SH and M, walls of Krystalna cave, Ukraine (locality 9, Figs 7–9). (**a**) Complex gypsum microbialite dome with selenite crystals grown on slopes, pen as a scale; (**b**) mushroom-shaped gypsum microbialite dome with selenite crystals grown on slopes; (**c**) horizontally grown selenite crystals typical of unit SH, with fine-grained gypsum showing microbialite laminations in between the crystals.

interpreted as *in situ* precipitates from brine sheets or some pedogenic deposits developed by dissolution and re-precipitation of gypsum on more elevated shoals during phases of temporary emersion (see Figs 4b & 18; Logan 1987; Aigner & Bachmann 1989; Magee 1991; Bryant *et al.* 1994; Peryt 1996; Bąbel 2005a). However, except for one level within the 20 m thick section, and the top surface of the gypsum section (Peryt 2001, 2006), no other remarkable erosion surface suggesting long-term emersion was recognized and the microbialite sections are conspicuously continuous (Peryt 1996, fig. 11d). The continuous vertical sequences of laminated gypsum microbialite deposits in unit M attain several metres in thickness (Figs 8, 9 & 14).

Specific channel structures are plentiful in such microbialite sequences. They are commonly seen as packages of laminated clastic gypsum forming small vertically elongated bodies (fig. 21; Peryt 1996, fig. 12; Bąbel 2005b, plate 3, fig. 2). They resemble infillings of karst cavities or neptunian dykes; however the contact with the surrounding microbialite gypsum is not sharp but indicative of the concurrent deposition (Fig. 21b). Small gypsum microbialite domes are commonly accreted on some clastic laminae within the channels (Peryt 1996, fig. 13).

The laminated gypsum was deposited in shallow (<5 cm deep) flat-bottomed channels present on the majanna-type shoals. The bottoms of the channels do not show scouring features, suggesting the erosion of the substrate. Features produced by lateral migration of channel beds are almost entirely absent (Fig. 21). The channels were evidently infilled by clastic gypsum together with accretion of gypsum microbialites at its banks. Vertically elongated channel infillings suggest long-term existence of the channel beds in one portion of the evaporite shoal. The accretion of the channel structure took place together with the rising water level that presumably coincided with the sediment surface on the shoal (Fig. 22). Such an uninterrupted continuous rise appears to be barely possible in environments directly controlled by seawater level (due to low accommodation); however it is expected and perhaps is indicative of the salina basin where a drawdown of water level

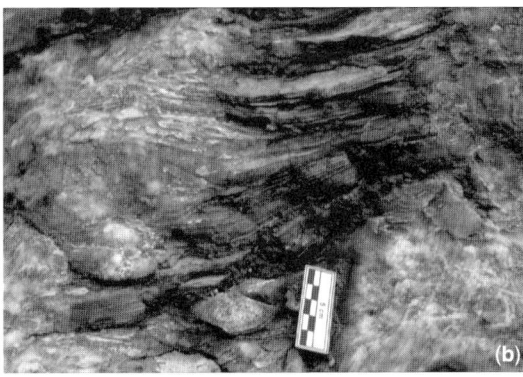

Fig. 21. Laminated gypsum in channels within gypsum microbialite deposits. (**a**, **b**) Majanna-type shoal facies, unit M, Ozerna cave, Ukraine (locality 11, Figs 7–9); scale is in centimetres in (b).

is very unstable (Fig. 3). The water or brine levels in the salina basin can easily rise together with the sediment accumulation as in the closed lake basins (Einsele & Hinderer 1997; Lowenstein *et al.* 2003). It is remarkable that one of the thickest continuous sequences of accreted microbial mat deposits (1.25 m thick) was found at the margin of the Solar Lake salina (Krumbein *et al.* 1977), and not on the marine tidal flat.

Similar channels are known from the zone of permanently flowing brine sheets on the majanna flats in the MacLeod basin (Logan 1987, fig. 53b, p. 64) and the drainage zones on the surface of Al-Khiran sabkha in Kuwait (Gunatilaka & Shearman 1988; Gunatilaka 1990). Warren (1982*a*) described similar forms under the name 'algal channels' in the Deep Lake selenite salina in Australia.

Architecture of Badenian selenite and microbialite facies and sequences

The recognition of the deep- and shallow-brine facies in the Badenian gypsum deposits allows analysis of the sedimentary record in terms of shallowing-upwards and deepening-upwards sequences (e.g. Fig. 14a). Both types of sequences are present in the Badenian evaporites (Kasprzyk 1993*a*; Bąbel 2004*b*). In particular deepening-upwards sequences are relatively common and at least one of them apparently also is a brining-upwards one (Figs 8 & 9; Peryt 2001, fig. 10) that can be indicative of the salina basin with a deeply drawn-down water level. Very thick continuous microbialite sequences are also present and they reflect a brine table level rise that kept pace with the gypsum deposition on the majanna-type shoals (Figs 9, 21 & 22), which is also characteristic of such a basin.

It was demonstrated that marker beds b1–b6 and c (Figs 8 & 16) represent isochronous units and datum levels, i.e. they were nearly ideally horizontal at the time of their deposition. A rather complicated pattern of the architecture of gypsum facies arises from this scheme (Fig. 9), although the aggradational style of deposition is clearly visible. The striking features are:

(1) complete dominance of shallow-water majanna-type shoal facies that occupied the whole thickness of the sections on the eastern end of the basin (near the Ukraine–Moldova border) and testify to a relatively continuous water level rise concordant with gypsum deposition (Fig. 22);

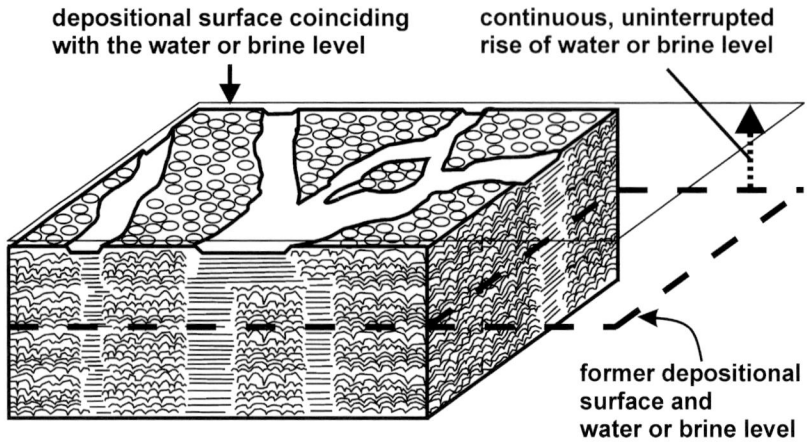

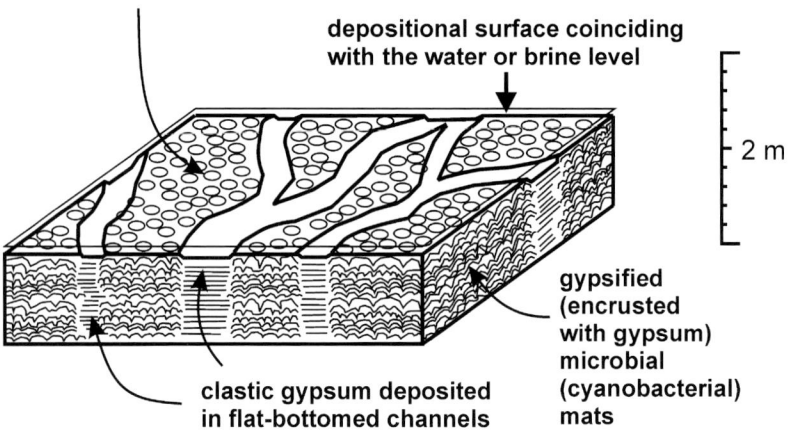

Fig. 22. Scheme showing mode of vertical accretion of gypsum microbialites concurrently with clastic gypsum deposition in shallow channels, controlled by a gradually rising water level on the permanently wet zone of the majanna-type shoal (see Figs 4 & 21).

(2) the uneven thickness of gypsum sections and particular gypsum lithosomes (especially the selenite lithosomes; Fig. 8, see also Kubica 1992); and

(3) poor correlation or expression of deepening-upwards and shallowing-upwards sequences present in the western area of the basin (see also Kasprzyk 1993a, b) in the eastern majanna-dominated area.

The most striking feature is that the deep-brine selenites of unit A seem to occur at the same hypsometric level that the microbialite deposits of the majanna-type shoals (unit M) present to the east (Figs 8 & 9; Peryt 2001, fig. 10). This latter feature, if correctly interpreted, appears to be paradoxical because it means that a deeper water environment, several metres deep, existed at the western side of the basin at the same time when at the same hypsometric level as the basin bottom on the western area a vast semi-emerged shoal, without any significant cover of brine, could exist on the east.

All the listed features are possible in a lagoon-type basin, and can be explained by variable relief in the basin, different rates of sedimentation and subsidence (or uplift) in particular subbasins, differential compaction, as well as some rapid tectonic movements during gypsum deposition (deposition in synsedimentary grabens or half-grabens). However a simpler and more consistent explanation appears when a salina basin concept is applied for the Badenian gypsum basin. In particular, the latter paradoxical features, different water levels and different bottom levels in particular subbasins, shoals and saline pans, are just possible and expected in a salina basin (Fig. 3). These disparate features can be explained by majanna-type shoals inclined parallel to the brine table levels spread between saline pans occupying different topographic levels and having different water levels. As these areas evolved in time and space they developed these seemingly dissimilar features (Fig. 4; Bąbel 2005b). Additionally, the peculiarities of depositional architecture can be explained by specific influence of tectonics on the water level and sedimentation in a salina basin.

Tectonic control of water level changes and deposition in a salina basin

When we assume that the Badenian basin was really a salina, we must take into consideration the possible influence of tectonics on the water level fluctuations in this basin and on the depositional architecture. This could be much greater than in the case where it was a lagoon-type basin with one water level always coinciding with the sea-level.

Subsidence, including rapid subsidence of tectonic blocks, can create accommodation both in salina and lagoonal basins; however its possible influence on the water level and the water depth is drastically different in the two basin types. In a lagoonal basin the water will be always supplemented by inflow from the sea, which can lead to deepening, whereas in a salina basin the water body, that always shows the same volume at any given time, will subside together with the basinal bottom and water depth can remain unchanged. Tectonic movements, however, may influence the water level in a salina basin to a greater degree than in a lagoon basin when the basin floor is tilted and one area is uplifted and the adjacent area subsides (Strecker et al. 1999; Ilgar & Nemec 2005). In such a case the water can flow from the uplifted to the subsident area and the water levels will be exchange directions in both areas (assuming an absence or limitation of sedimentation). Similar effects will take place when the rate of subsidence is irregularly distributed across a basin floor. The water level can rise in a salina basin due to massive transport of sediments into the basin, for example due to giant slumps which can displace some of the water volume onto the shoals (Ilgar & Nemec 2005). By contrast even very abundant and rapid sediment supply cannot promote any rise in water level in a lagoon basin where the water level coincides with the sea-level and is always controlled by global eustasy.

Tectonic movements, and especially the tilting of the basin floor, is expected in an active foreland basin system, such as the Carpathian Foredeep (DeCelles & Giles 1996; Catuneanu 2004; Oszczypko 2006a, b). Kubica (1992, pp. 31–33) assumed that the increased thickness of gypsum deposits in the northern margin of the basin, in comparison with its central area around the Rzeszów Island, is partly a result of increased rate of subsidence in the north, and many other authors accepted the tectonic activity in the foredeep during gypsum accumulation. If these interpretations are correct, then the possible influence of the tectonic movements on changes in the local water level within a salina basin should be taken into account, as indicted by Kwiatkowski (1972, pp. 69, 92).

Thus a relatively continuous presence of majanna-type shoals on the eastern side of the basin contrasting with occurrence of giant saline pans with fluctuating water levels in the western side (Fig. 9) can be explained in two ways: (1) these two areas were hypsometrically different and showed different water levels in the salina basin (Fig. 4a); and/or (2) vertical tectonic movements were irregularly distributed across the basin floor and were responsible not only for local differences in accommodation but also contributed to different style of water level changes in these adjacent areas.

In sum, analysis of the architecture of the salina-type basins in terms of sequences controlled by water level changes appear to be much more complicated than lagoon basins controlled by one common water level coinciding with sea-level.

Conclusions

Ancient evaporite basins occupying the depressions supplied with seawater through seepage and occasional surface inflows, without open water connections with the sea, show specific features in the depositional record. Such distinctions can help in the recognition of these basins (called here salina or salina-type basins) and they may be discerned from the other basin types, and particularly from the lagoonal evaporite basins permanently connected with the sea. The environments of the salina basin were very similar to those of the modern saline lakes, and based on these lake environments, and of the environments of marine coastal salinas (and the MacLeod basin in Australia in particular), an integrated group of conceptual models was built for facies analysis of such a basin, at the stage of gypsum crystallization. Both permanently subaqueous and semi-emerged evnironments were modelled, i.e. saline pans (subbasins) and evaporite shoals in between and around them (they were termed majanna-type sholas by analogy to the specifcic majanna environment in the MacLeod basin). The models developed here were applied and tested in the Badenian selenite-dominated evaporite basin of the northern Carpathian Foredeep.

Many specific features of these models, i.e. the predicted features of the salina-type basins, were recognized in the depositional record of the Badenian gypsum basin. Some depositional features, including the architecture of gypsum facies, seem more readily explained and understood when it is assumed or accepted that:

(1) Water level fluctuations in the basin were more frequent, more rapid, and larger in scale than normal sea-level changes (i.e. the basin showed its own pattern of water level fluctuations, more similar to those in modern saline lakes than in the present seas or lagoons).

(2) The Badenian subaqueous environments (and their depositional record or gypsum facies) can best be described according to the limnological terminology, and characterized by the presence of stratification-mixing cycles typical of monomictic and polymictic lakes and, consequently, by the presence of seasonal to temporary stratification of brine column (i.e. the pycnocline).

(3) The fluctuations in the level of the pycnocline influenced the course of selenite deposition, which predominatly took place below the seasonal (or temporary) pycnocline in saline pans, and mainly during the mixing periods.

(4) Long-term drops of an avarage pycnocline level (pycnocline lowstands) were associated with development of dissolution surfaces and deposition of fine-grained gypsum (commonly fine-grained gypsum microbialites). Rises (highstands) of the pycnocline were associated with crystallization of the grass-like selenite beds. Consequently, such selenite beds, when laterally continuous and arranged into depositional bundles showing some constant pattern of thickness, were assumed to record the pycnocline fluctuations in the particular saline pans.

(5) Some thin and well-recognizable grass-like selenite beds record the pycnocline highstands in the whole gypsum subbasin. These were used as event or marker beds for basin-scale isochronous (or near isochronous) correlation over distances of tens to hundreds of kilometres. Additionally, such beds are treated as ideal datum surfaces, because the growth of gypsum crystal apices was determined by the horizontal position of the pycnocline.

(6) Thick continuous sequences of gypsum microbialite deposits (several metres thick), containing specific vertically accreted stationary channel structures, were deposited on semi-emerged permanently-wet majanna-type shoals, periodically flooded by gypsum-saturated brine sheets inflowing from the adjacent saline pans. The long-term continuous accretion of gypsum microbialites on the shoals apparently kept pace with the gradual uninterrupted water level rise in the basin, a case barely possible in environments directly controlled by a sea-level, but characteristic of the salina basin with the high accommodation potential.

(7) One of the depositional sequences in the gypsum basin is both deepening-upwards and brining-upwards, which is also a case only barely possible in the lagoonal basins but that appears to be indicative of environments not controlled directly by sea-level and showing a high accommodation.

(8) Poor correlation of shallowing-upwards and deepening-upwards gypsum sequences from different sides of the gypsum basin can be a result of tectonic control of the local water changes in the salina basin (which is specific feature of the lake-type depositional systems), or deposition in two subbasins showing different fluctuations of the water levels.

All these depositional features together consistently suggest that the Badenian gypsum basin (at least during the selenite deposition) was not an evaporite lagoon but a salina basin with a water level drawn down below global sea-level, although salts in the water from which the gypsum crystallized were basically of marine derivation, and could pass through recycling processes within the basin.

The author thanks Andrii Bogucki, University of Lviv, and many his colleagues and students, for their help in preparing and carrying out field expeditions to the Badenian gypsum in Ukraine in the period 1994–2002, that allowed development of the concepts presented in this paper. In particular I thank Andrii Yatsyshyn and Svitlana Vizna for help in the field, and the speleologists Sergei Epifanov, Volodymyr Galaichuk and Igor Malavskii for guidance and assistance in visiting the many gyosum caves in the study area. The author also thanks the co-editors of this volume, B. Charlotte Schreiber and Stefano Lugli, for showing me the Mediterranean saltwork pans and Messinian selenites, and for stimulating discussions, which clarified many ideas presented above. The greatest thanks are offered to Federico Ortí Cabo and B. Charlotte Schreiber for constructive criticism which improved the final version of that paper.

References

AIGNER, T. & BACHMANN, G. H. 1989. Dynamic stratigraphy of evaporite-to-red bed sequence, Gipskeuper (Triassic), southwest German Basin. *Sedimentary Geology*, **62**, 5–25.

ALSHARHAN, A. S. & KENDALL, C. G. S. C. 1994. Depositional setting of the Upper Jurassic Hith Anhydrite of the Arabian Gulf: an analog to Holocene evaporites of the United Arab Emirates and Lake MacLeod of Western Australia. *The American Association of Petroleum Geologists Bulletin*, **78**, 1075–1096.

AMBROSETTI, W., BRABANTI, L. & SALA, N. 2003. Residence time and physical processes in lakes. *Journal of Limnology*, **62**, Suppl. 1, 1–15.

ANATI, D. A. 1997. The hydrography of a hypersaline lake. *In*: NIEMI, T. M., BEN-AVRAHAM, Z. & GAT, J. R. (eds) *The Dead Sea: the Lake and Is Setting*. Oxford Monographs on Geology and Geophysics, **36**. Oxford University Press, New York, 89–103.

ANATI, D. A. & STILLER, M. 1991. The post-1979 thermohaline structure of the Dead Sea and the role of double-diffusive mixing. *Limnology and Oceanography*, **36**, 342–354.

ARAKEL, A. V. 1988. Modern halite sedimentation processes and depositional environments, Hutt Lagoon, Western Australia. *Geodinamica Acta*, **2**, 169–184.

ATLAS GEOLOGICZNY GALICYI, 1885–1914. 99 map sheets 1:75000 with explanations; Zeszyty **1–21**, **23**, **25–27**. Wydawnictwo Komisyi Fizyograficznej Akademii Umiejętności, Kraków [in Polish].

BĄBEL, M. 2004a. Models for evaporite, selenite and gypsum microbialite deposition in ancient saline basins. *Acta Geologica Polonica*, **54**, 219–249.

BĄBEL, M. 2004b. Badenian evaporite basin of the northern Carpathian Foredeep as a drawdown salina basin. *Acta Geologica Polonica*, **54**, 313–337.

BĄBEL, M. 2005a. Event stratigraphy of the Badenian selenite evaporites (Middle Miocene) of the northern Carpathian Foredeep. *Acta Geologica Polonica*, **55**, 9–29. On-line Appendix: www.geo.uw.edu.pl/agp/table/appendixes/55-1/.

BĄBEL, M. 2005b. Selenite-gypsum microbialite facies and sedimentary evolution of the Badenian evaporite basin of the northern Carpathian Foredeep. *Acta Geologica Polonica*, **55**, 187–210.

BĄBEL, M. & BOGUCKI, A. 2007. The Badenian evaporite basin of the northern Carpathian Foredeep as a model of a meromictic selenite basin. *In*: SCHREIBER, B. C., LUGLI, S. & BĄBEL, M. (eds) *Evaporites Through Space and Time*. Geological Society of London, Special Publications, **285**, 219–246.

BOBST, A. L., LOWENSTEIN, T. K., JORDAN, T. E., GODFREY, L. V., KU, T.-L. & LUO, S. 2001. A 106 ka paleoclimate record from drill core of the Salar de Atacama, northern Chile. *Palaeogeography, Palaeoclimatology, Palaeoecology*, **173**, 21–42.

BOHACS, K. M., CARROLL, A. R., NEAL, J. E. & MANKIEWICZ, P. J. 2000. Lake-basin type, source potential, and hydrocarbon character: an integrated sequence-stratigraphic-geochemical framework. *In*: GIERLOWSKI-KORDESCH, E. H. & KELTS, K. R. (eds) *Lake Basins Through Space and Time*. AAPG Studies in Geology, **46**, Tulsa, OK, 3–34.

BORCHERT, H. & MUIR, R. O. 1964. *Salt Deposits. The Origin, Metamorphism and Deformation of Evaporites*. Van Nostrand, London.

BRYANT, R. G., SELLWOOD, B. W., MILLINGTON, A. C. & DRAKE, N. A. 1994. Marine-like potash evaporite formation on a continental playa: case study from Chott el Djerid, southern Tunisia. *Sedimentary Geology*, **90**, 269–291.

BURKE, C. M. & KNOTT, B. 1997. Homeostatic interactions between the benthic microbial communities and the waters of a hypersaline lake, Lake Hayward, Western Australia. *Marine and Freshwater Research*, **48**, 623–631.

CAMPBELL, C. E. 1995. Temporal salinity variation in salt evaporation basins in south-eastern Australia. *International Journal of Salt Lake Research*, **4**, 45–55.

CARROLL, A. R. & BOHACS, K. M. 1999. Stratigraphic classification of ancient lakes: balancing tectonic and climatic controls. *Geology*, **27**, 99–102.

CATUNEANU, O. 2004. Retroarc foreland systems – evolution through time. *Journal of African Earth Sciences*, **38**, 225–242.

CENDÓN, D. I., PERYT, T. M., AYORA, C., PUEYO, J. J. & TABERNER, C. 2004. The importance of recycling processes in the Middle Miocene Badenian evaporite basin (Carpathian foredeep): palaeoenvironmental implications. *Palaeogeography, Palaeoclimatology, Palaeoecology*, **212**, 141–158.

CLOERN, J. E., COLE, B. E. & OREMLAND, R. S. 1983. Seasonal changes in the chemistry and biology of a meromictic lake (Big Soda Lake, Nevada, U.S.A.). *Hydrobiologia*, **105**, 195–206.

COHEN, Y., KRUMBEIN, W. E., GOLDBERG, M. & SHILO, M. 1977. Solar Lake (Sinai). 1. Physical and chemical limnology. *Limnology and Oceanography*, **22**, 597–608.

COLE, G. A., WHITESIDE, M. C. & BROWN, R. J. 1967. Unusual monomixis in two saline Arizona ponds. *Limnology and Oceanography*, **12**, 584–591.

COMÍN, F. A., CABRERA, M. & RODÓ, X. 1999. Saline lakes: integrating ecology into their management future. *Hydrobiologia*, **395–396**, 241–251.

COSHELL, L. & ROSEN, M. R. 1994. Stratigraphy and Holocene history of Lake Hayward, Swan Coastal Plain wetlands, Western Australia. *In*: RENAUT, R. W. & LAST, W. M. (eds) *Sedimentology and Geochemistry of Modern and Ancient Saline Lakes*. SEPM, Special Publication, **50**, Tulsa, OK, 173–188.

DECELLES, P. G. & GILES, K. A. 1996. Foreland basin systems. *Basin Research*, **8**, 105–123.

DEL DON, C., HANSELMANN, K. W., PEDUZZI, R. & BACHOFEN, R. 2001. The meromictic alpine Lake Cadagno: orographical and biogeochemical description. *Aquatic Sciences*, **63**, 70–90.

DRONKERT, H. 1985. *Evaporite Models and Sedimentology of Messinian and Recent Evaporites*. GUA Papers of Geology, Series 1, **24**, Utrecht, 1–283.

ECKSTEIN, Y. 1970. Physicochemical limnology and geology of a meromictic pond on the Red Sea shore. *Limnology and Oceanography*, **15**, 363–372.

EINSELE, G. & HINDERER, M. 1997. Terrestrial sediment yield and the lifetimes of reservoirs, lakes, and larger basins. *Geologische Rundschau*, **86**, 288–310.

EUGSTER, H. P. 1980. Lake Magadi, Kenya, and its precursors. *In*: NISSENBAUM, A. (ed.) *Hypersaline Brines and Evaporitic Environments*. Developments in Sedimentology, **28**, Elsevier, Amsterdam, 195–232.

EUGSTER, H. P. & HARDIE, L. A. 1978. Some further thoughts on the depositional environment of the Solfifera Series of Sicily. *Memorie della Societá Geologica Italiana*, **16** (volume for 1976), 29–38.

FRIEDMAN, G. M. & KRUMBEIN, W. E. (eds) 1985. *Hypersaline Ecosystems; the Gavish Sabkha*. Ecological Studies: Analysis and Synthesis, **53**, Springer, Berlin.

GARLICKI, A. 1979. Sedimentation of Miocene salts in Poland. *Prace Geologiczne, Polska Akademia Nauk, Oddzial w Krakowie*, **119**, 1–67 [in Polish with English summary].

GARRISON, R. E., SCHREIBER, B. C., BERNOULLI, D., FABRICIUS, F. H., KIDD, R. B. & MÉLIÉRÉS, F. 1978. Sedimentary petrology and structures of Messinian evaporitic sediments in the Mediterranean Sea, Leg 42A, Deep Sea Drilling Project. *In*: HSÜ, K., MONTADERT, L. *ET AL*. (eds) *Initial Reports of the Deep Sea Drilling Project*, **42**, part 1, U.S. Government Printing Office, Washington, 571–611.

GEISLER-CUSSEY, D. 1986. Approche sédimentologique et géochimique des mécanismes générateurs de formations évaporitiques actuelles et fossiles. Marais salants de Camargue et du Levant espagnol, Messinien méditerranéen et Trias lorrain. *Sciences de la Terre, Mémoires*, **48**, 1–268. Nancy.

GEISLER-CUSSEY, D. 1997. Modern depositional facies developed in evaporative environments (marine, mixed, and nonmarine). *In*: BUSSON, G. & SCHREIBER, B. C. (eds) *Sedimentary Deposition in Rift and Foreland Basins in France and Spain (Paleogene and Lower Neogene)*. Columbia University Press, New York, 3–42.

GERDES, G., KRUMBEIN, W. E. & NOFFKE, N. 2000. Evaporite microbial sediments. *In*: RIDING, R. E. & AWRAMIK, S. M. (eds) *Microbial Sediments*. Springer, Berlin, 196–208.

GRABAU, A. W. 1920. *Geology of the Non-metallic Mineral Deposits other than Silicates*. **1**. *Principles of Salt Deposition*. McGraw-Hill, New York.

GUNATILAKA, A. 1990. Anhydrite diagenesis in a vegetated sabkha, Al-Khiran, Kuwait, Arabian Gulf. *Sedimentary Geology*, **69**, 95–116.

GUNATILAKA, A. & SHEARMAN, D. J. 1988. Gypsumcarbonate laminites in a recent sabkha, Kuwait. *Carbonates and Evaporites*, **3**, 67–73.

HANDFORD, C. R. 1982. Sedimentology and evaporite genesis in Holocene continental-sabkha playa basin – Bristol Dry Lake, California. *Sedimentology*, **29**, 239–253.

HARDIE, L. A. & EUGSTER, H. P. 1971. The depositional environment of marine evaporites: a case for shallow, clastic accumulation. *Sedimentology*, **16**, 187–220.

HORNE, A. J. & GOLDMAN, C. R. 1994. *Limnology* 2nd edn McGraw-Hill, New York.

HUTCHINSON, G. E. 1975. *A Treatise on Limnology*. **1**(1). *Geography and Physics of Lakes*. Wiley, New York.

ILGAR, A. & NEMEC, W. 2005. Early Miocene lacustrine deposits and sequence stratigraphy of the Ermenek Basin, Central Taurides, Turkey. *Sedimentary Geology*, **173**, 233–275.

JAUZEIN, A. & HUBERT, P. 1984. Les bassins oscillants: un modèle de genése des series salines. *Sciences Géologiques, Bulletin*, **37**, 267–282.

KASPRZYK, A. 1989. Content of strontium in Miocene gypsum rocks of the Staszów area. *Przegląd Geologiczny*, **37**, 201–207 [in Polish with English summary].

KASPRZYK, A. 1993a. Lithofacies and sedimentation of the Badenian (Middle Miocene) gypsum in the northern part of the Carpathian Foredeep, southern Poland. *Annales Societatis Geologorum Poloniae*, **63**, 33–84.

KASPRZYK, A. 1993b. Gypsum facies in the Badenian (Middle Miocene) of southern Poland. *Canadian Journal of Earth Sciences*, **30**, 1799–1814.

KASPRZYK, A. 1994. Distribution of strontium in the Badenian (Middle Miocene) gypsum deposits of the Nida area, southern Poland. *Geological Quarterly*, **38**, 497–512.

KASPRZYK, A. 2005. Diagenetic alteration of Badenian sulphate deposits in the Carpathian Foredeep Basin, Southern Poland: processes and their succession. *Geological Quarterly*, **49**, 305–316.

KASPRZYK, A. & ORTÍ, F. 1998. Palaeogeographic and burial controls on anhydrite genesis: the Badenian basin in the Carpathian Foredeep (southern Poland, western Ukraine). *Sedimentology*, **45**, 889–907.

KHRUSHCHOV, D. P. & PETRICHENKO, O. I. 1979. Evaporite formations of Central Paratethys and conditions of their sedimentation. *Annales Géologiques des Pays Helléniques, Tome hors série*, **2**, 595–612.

KINSMAN, D. J. J. 1969. Modes of formation, sedimentary associations, and diagnostic features of shallow-water and supratidal evaporites. *The American Association of Petroleum Geologists Bulletin*, **53**, 830–840.

KIRKLAND, D. W. 2003. An explanation for the varves of the Castile evaporites (Upper Permian), Texas and New Mexico, USA. *Sedimentology*, **50**, 899–920.

KOBLUK, D. R. & CRAWFORD, D. R. 1990. A modern hypersaline organic mud- and gypsum-dominated basin and associated microbialites. *Palaios*, **5**, 134–148.

KRUMBEIN, W. E. & COHEN, Y. 1977. Primary production, mat formation and lithification: contribution of oxygenic and facultative anoxygenic cyanobacteria. *In*: FLÜGEL, E. (ed.) *Fossil Algae*. Springer, Berlin, 37–56.

KRUMBEIN, W. E., COHEN, Y. & SHILO, M. 1977. Solar Lake (Sinai). 4. Stromatolitic cyanobacterial mats. *Limnology and Oceanography*, **22**, 635–656.

KUBICA, B. 1992. Lithofacial development of the Badenian chemical sediments in the northern part of the Carpathian Foredeep. *Prace Państwowego Instytutu Geologicznego*, **133**, 1–64 [in Polish with English summary].

KWIATKOWSKI, S. 1972. Sedimentation of gypsum in the Miocene of southern Poland. *Prace Muzeum Ziemi*, **19**, 3–94 [in Polish with English summary].

LANGBEIN, W. B. 1961. Salinity and hydrology of closed lakes. *Reprinted in*: NEAL, J. T. (ed.) *Playas and Dried Lakes: Occurrence and Development*. Benchmark

Papers in Geology, **20**. Dowden, Hutchinson & Ross, Stroudsburg, PA, 1975, 93–112.

LAST, W. M. & SCHWEYEN, T. H. 1983. Sedimentology and geochemistry of saline lakes of the Great Plains. *In*: HAMMER, U. T. (ed.) Saline Lakes. Developments in Hydrobiology, **16**. Junk, The Hague, 244–263. [Also: *Hydrobiologia*, **105**, 244–263.]

LEPESHKOV, I. N., BUYNEVICH, D. P., BUYNEVICH, N. A. & SEDELNIKOV, G. S. 1981. *Prospects of Utilization of Salt Resources of the Kara-Bogaz-Gol*. Nauka, Moscow [in Russian].

LEVINE, R. M. 1998. The fall and rise of the Garabogaz Aylagy (Kara Bogaz Gol) lagoon. Conference: *Oil and Environment Security in the Black and Caspian Seas*. The Columbia Caspian Project; www.sipa.columbia.edu/RESOURCES/CASPIAN/enviro.html

LEWIS, W. M. J. 1983. A revised classification of lakes based on mixing. *Canadian Journal of Fisheries and Aquatic Sciences*, **40**, 1779–1787.

LEWIS, W. M. J. 2000. Basis for the protection and management of tropical lakes. *Lakes & Reservoirs: Research and Management*, **5**, 35–48.

LI, J., LOWENSTEIN, T. K., BROWN, C. B., KU, T.-L. & LUO, S. 1996. A 106 ka record of water tables and paleoclimates from salt cores, Death Valley, California. *Palaeogeography, Palaeoclimatology, Palaeoecology*, **123**, 179–203.

LOGAN, B. W. 1987. The MacLeod evaporite basin, western Australia. Holocene environments, sediments and geological evolution. *AAPG Memoir*, **44**. AAPS, Tulsa, OK.

LOGAN, B. & BROWN, R. G. 1986. *Sediments of Shark Bay and MacLeod Basin, Western Australia*. Field Seminar Handbook, Sedimentology Research Group, Department of Geology, University of Western Australia.

LOWE, D. J., GREEN, J. D., NORTHCOTE, T. G. & HALL, K. J. 1997. Holocene fluctuations of a meromictic lake in southern British Columbia. *Quaternary Research*, **48**, 100–113.

LOWENSTEIN, T. K. & HARDIE, L. A. 1985. Criteria for the recognition of salt-pan evaporites. *Sedimentology*, **32**, 627–644.

LOWENSTEIN, T. K., HEIN, M. C., BOBST, A. L., JORDAN, T. E., KU, T. L. & LUO, S. 2003. An assessment of stratigraphic completeness in climate-sensitive closed-basin lake sediments: Salar de Atacama, Chile. *Journal of Sedimentary Research*, **73**, 91–104.

MAGEE, J. W. 1991. Late Quaternary lacustrine, groundwater, aeolian and pedogenic gypsum in the Prungle Lakes, southeastern Australia. *Palaeogeography, Palaeoclimatology, Palaeoecology*, **84**, 3–42.

MELACK, J. M. & JELLISON, R. 1998. Limnological conditions in Mono Lake: contrasting monomixis and meromixis in the 1990s. *Hydrobiologia*, **384**, 21–39.

MERSMANN, A. 1996. Supersaturation and nucleation. *Chemical Engineering Research and Design, Transactions of the Institution of Chemical Engineers, Part A*, **74**, 812–820.

NIEMI, T. M., BEN-AVRAHAM, Z. & GAT, J. R. (eds) 1997. *The Dead Sea: the Lake and Its Setting*. Oxford Monographs on Geology and Geophysics, **36**, Oxford University Press, New York.

ORTÍ CABO, F., PUEYO MUR, J. J., GEISLER-CUSSEY, D. & DULAU, N. 1984. Evaporitic sedimentation in the coastal salinas of Santa Pola (Alicante, Spain). *Revista d'Investigacions Geologiques*, **38–39**, 169–220.

OSZCZYPKO, N. 2006a. Late Jurassic-Miocene evolution of the Outer Carpathian fold-and-thrust belt and its foredeep basin (Western Carpathians, Poland). *Geological Quarterly*, **50**, 169–194.

OSZCZYPKO, N. 2006b. Development of the Polish sector of the Carpathian Foredeep. *Przegląd Geologiczny*, **54**, 396–403 [in Polish with English summary].

PANOV, G. M. & PLOTNIKOV, A. M. 1999. Paleogeographic aspects of accumulation of the Badenian evaporites in the Ukrainian Precarpathians (based on data of analysis of lithofacies and thickness). *Geologiya i Geokhimiya Goryuchikh Kopalyn*, **3**, 19–31 [in Ukrainian].

PERYT, T. M. 1996. Sedimentology of Badenian (Middle Miocene) gypsum in eastern Galicia, Podolia and Bukovina (West Ukraine). *Sedimentology*, **43**, 571–588.

PERYT, T. M. 2001. Gypsum facies transitions in basinal-marginal evaporites: middle Miocene (Badenian) of west Ukraine. *Sedimentology*, **48**, 1103–1119.

PERYT, T. M. 2006. The beginning, development and termination of the Middle Miocene Badenian salinity crisis in Central Paratethys. *Sedimentary Geology*, **188–189**, 379–396.

PETRICHENKO, O. I., PERYT, T. M. & POBEREGSKY, A. V. 1997. Peculiarities of gypsum sedimentation in the Middle Miocene Badenian evaporite basin of Carpathian Foredeep. *Slovak Geological Magazine*, **3**, 91–104.

ROMERO, J. R. & MELACK, J. M. 1996. Sensivity of vertical mixing in a large saline lake to variations in runoff. *Limnology and Oceanography*, **41**, 955–965.

ROSELL, L., ORTÍ, F., KASPRZYK, A., PLAYA, E. & PERYT, T. M. 1998. Strontium geochemistry of Miocene primary gypsum: Messinian of southeastern Spain and Badenian of Poland. *Journal of Sedimentary Research*, **68**, 63–79.

ROSEN, M. R., COSHELL, L., TURNER, J. V. & WOODBURY, R. J. 1996. Hydrochemistry and nutrient cycling in Yalgorup National Park, Western Australia. *Journal of Hydrology*, **185**, 241–274.

ROUCHY, J. M. & CARUSO, A. 2006. The Messinian saliniy crisis in the Mediterranean basin: a reassessment of the data and an integrated scenario. *Sedimentary Geology*, **188–189**, 35–67.

ROUCHY, J. M. & MONTY, C. 1981. Stromatolites and cryptalgal laminites associated with Messinian gypsum of Cyprus. *In*: MONTY, C. (ed.) *Phanerozoic Stromatolites*. Springer, Berlin, 155–180.

RÖGL, F. 1998. Palaeogeographic considerations for Mediterranean and Paratethys Seaways (Oligocene to Miocene). *Annalen des Naturhistorischen Museums in Wien*, **99A** (volume for 1997), 279–310.

SANZ, M. E., ROGRÍGUEZ-ARANDA, J. P., CALVO, J. P. & ORDOÑEZ, S. 1994. Tertiary detrital gypsum in the Madrid Basin, Spain: criteria for interpreting detrital gypsum in continental evaporitic sequences. *In*: RENAUT, R. W. & LAST, W. M. (eds) *Sedimentology and Geochemistry of Modern and Ancient Saline*

Lakes. SEPM, Special Publication, **50**. Tulsa, OK, 217–228.

SCHOLZ, C. A., MOORE, T. C. J., HUTCHINSON, D. R., GOLMSHTOK, A. J., KLITGORD, K. D. & KUROTCHKIN, A. G. 1998. Comparative sequence stratigraphy of low-latitude versus high-latitude lacustrine basins: seismic data examples from the East African and Baikal rifts. *Palaeogeography, Palaeoclimatology, Palaeoecology*, **140**, 401–420.

SCHREIBER, B. C. 1978. Environments of subaqueous gypsum deposition. *In*: DEAN, W. E. & SCHREIBER, B. C. (eds) *Marine Evaporites*. SEPM Short Course, **4**. Oklahoma City, 43–73.

SCHREIBER, B. C., FRIEDMAN, G., DECIMA, A. & SCHREIBER, E. 1976. Depositional environments of Upper Miocene (Messinian) evaporite deposits of the Sicilian Basin. *Sedimentology*, **23**, 729–760.

SCHREIBER, B. C., TUCKER, M. E. & TILL, R. 1986. Arid shorelines and evaporites. *In*: READING, H. G. (ed.) *Sedimentary Environments and Facies*. Blackwell Scientific, Oxford, 189–228.

SCHRÖDER, S., SCHREIBER, B. C., AMTHOR, J. E. & MATTER, A. 2003. A depositional model for the terminal Neoproterozoic-Early Cambrian Ara Group evaporites in south Oman. *Sedimentology*, **50**, 879–898.

SLOSS, L. L. 1969. Evaporite deposition from layered solutions. *The American Association of Petroleum Geologists Bulletin*, **53**, 776–789.

SMOOT, J. P. & CASTENS-SEIDELL, B. 1994. Sedimentary features produced by efflorescent salt crusts, Saline Valley and Death Valley, California. *In*: RENAUT, R. W. & LAST, W. M. (eds) *Sedimentology and Geochemistry of Modern and Ancient Lakes*. SEPM, Special Publication, **50**, Tulsa, OK, 73–90.

SMOOT, J. P. & LOWENSTEIN, T. K. 1991. Depositional environments of non-marine evaporates. *In*: MELVIN, J. L. (ed.) *Evaporites, Petroleum and Mineral Resources*. Developments in Sedimentology, **50**. Elsevier Science, Amsterdam, 189–347.

SONNENFELD, P. 1984. *Brines and Evaporites*. Academic Press, Orlando, FL.

STILLER, M., GAT, J. R. & KAUSHANSKY, P. 1997. Halite precipitation and sediment deposition as measured in sediment traps deployed in the Dead Sea: 1981–1983. *In*: NIEMI, T. M., BEN-AVRAHAM, Z. & GAT, J. R. (eds) *The Dead Sea: the Lake and Its Setting*. Oxford Monographs on Geology and Geophysics, **36**, Oxford University Press, New York, 171–183.

STRASSER, A., PITTET, B. HILLGÄRTNER, H. & PASQUIER, J.-B. 1999. Depositional sequences in shallow carbonate-dominated sedimentary systems: concepts for a high-resolution analysis. *Sedimentary Geology*, **128**, 201–221.

STRECKER, U., STEIDTMANN, J. R. & SMITHSON, S. B. 1999. A conceptual tectonostratigraphic model for seismic facies migration in a fluvio-lacustrine extensional basin. *The American Association of Petroleum Geologists Bulletin*, **83**, 43–61.

TELLER, J. T., BOWLER, J. M. & MACUMBER, P. G. 1982. Modern sedimentation and hydrology in Lake Tyrrell, Victoria. *Journal of the Geological Society of Australia*, **29**, 159–175.

TERZIEV, F. S., GOPTAREV, N. P. & BORTNIK, V. N. 1986. The problem of the Kara-Bogaz-Gol Bay. *Vodnye Resursy*, **2**, 64–71. Nauka, Moscow [in Russian]. [English version: *Water Resources*, **13**, 167–172. Kluwer, Dordrecht.]

TORFSTEIN, A., GAVRIELI, I. & STEIN, M. 2005. The sources and evolution of sulfur in hypersaline Lake Lisan (paleo-Dead Sea). *Earth and Planetary Science Letters*, **236**, 61–77.

TORGERSEN, T. 1984. Wind effect and salt loss in playa lakes. *Journal of Hydrology*, **74**, 137–149.

TYLER, S. W., MUÑOZ, J. F. & WOOD, W. W. 2006. The response to Playa and Sabkha hydraulics and mineralogy to climate forcing. *Ground Water*, **44**, 329–338.

ULLMAN, W. J. 1995. The fate and accumulation of bromide during playa salt deposition: An example from Lake Fromme, South Australia. *Geochimica et Cosmochimica Acta*, **59**, 2175–2186.

ULRICH, J. & STREGE, C. 2002. Some aspects of the importance of metastable zone width and nucleation in industrial crystallizers. *Journal of Crystal Growth*, **237–239**, 2130–2135.

VALYASHKO, M. G. 1952. Halite, its basic varieties occurring in salt lakes, and their structural pecularities. *Trudy Vsesoyuznogo Nauchno-Issledovatel'skogo Instituta Galurgii (VNIIG)* [*Publication of the All-Union Institute of Scientific Research in Halurgy*], **23**, 25–53 [in Russian].

VALYASHKO, M. G. 1972. Playa lakes – a necessary stage in the development of a salt-bearing basin. *In*: RICHTER-BERNBURG, G. (ed.) *Geology of Saline Deposits*. Proceedings of the Hanover Symposium 15–21 May 1968. United Nations Educational, Scientific and Cultural Organization, Paris, 41–51.

VERSCHUREN, D. 1999. Influence of depth and mixing regime on sedimentation in a small, fluctuating tropical soda lake. *Limnology and Oceanography*, **44**, 1103–1113.

WARREN, J. K. 1982a. The hydrological significance of Holocene tepees, stromatolites, and boxwork limestones in coastal salinas in South Australia. *Journal of Sedimentary Petrology*, **52**, 1171–1201.

WARREN, J. K. 1982b. The hydrological setting, occurrence and significance of gypsum in late Quaternary salt lakes in South Australia. *Sedimentology*, **29**, 609–637.

WARREN, J. K. 2006. *Evaporites: Sediments, Resources and Hydrocarbons*. Springer, Berlin.

WARREN, J. K. & KENDALL, C. G. S. C. 1985. Comparison of sequences formed in marine sabkha (subaerial) and salina (subaqueous) settings – modern and ancient. *The American Association of Petroleum Geologists Bulletin*, **69**, 1013–1023.

WOOD, W. W., SANFORD, W. E. & FRAPE, S. K. 2005. Chemical openness and potential for misinterpretation of the solute environment of coastal sabkhat. *Chemical Geology*, **215**, 361–372.

WILLIAMS, W. D. 1996. What future for saline lakes? *Environment*, **38**, 13–20, 38–39.

Experimental evaporation of superficial brines from continental playa–lake systems located in Central Ebro Basin (northeast Spain)

P. L. LOPEZ & J. M. MANDADO

Area de Petrologia y Geoquimica, Departamento de Ciencias de la Tierra, Universidad de Zaragoza, 50.009 Zaragoza, Spain (e-mail: pllopez@unizar.es)

Abstract: Solutions coming from two natural playa–lake saline systems located in Central Ebro Basin (NE Spain) have been evaporated in the laboratory, in order to obtain the precise path of chemical evolution followed until high concentration stages. The lakes belong to two chemically different neutral brines: La Playa (with Na–Cl solutions) and La Salada (Na–Mg–SO$_4$ type). Experimental evaporation has been carried out at 25 °C until total dryness, and samples collected along the experiment have been analysed for their major components. Application of geochemical modelling techniques allowed calculation of the saturation indexes for the main saline minerals using the PHRQPITZ program, which incorporates Pitzer's model. The mineral precipitation sequence for La Playa brines following the saturation data is: gypsum, halite, thenardite and epsomite. Brines reach saturation almost simultaneously with respect to both halite and thenardite, but halite precipitates more massively and hence the solid samples collected after total desiccation were composed mainly of halite. In the case of La Salada brines, the order of precipitation is somewhat different and is as follows: gypsum, mirabilite, thenardite and bloedite. Solid samples here consisted of bloedite and thenardite, this latter formed after subaquatic dehydration of mirabilite when brines attained a peritectic point. The evolution of saturation indexes is in good agreement with mineral determinations carried out on the solid experimental samples, and it allowed us to interpret the evaporative evolution of both La Playa and La Salada brines.

Classical studies on saline rocks have been carried out by using petrographical, sedimentological and geochemical methodologies. Genetic interpretations of sedimentary environments for the formation of such geological materials have been made starting from knowledge of present-day evaporitic systems. Primary precipitation processes of the more common saline minerals (gypsum and halite) are well known, but great discrepancies have been observed in the case of more soluble salts, mainly because of their high reactivity in synsedimentary and/or early diagenesis stages.

In the past few decades, the progress of geochemical modelling in highly concentrated solutions [with the development of several programs ('codes') based on Pitzer's model (Pitzer 1973)] allowed the realization of a more accurate description of saline crystallization sequences observed in several present-day saline systems. This knowledge may be applied to interpretation of saline geological records.

In this work, two active playa–lake systems have been chosen. These natural systems have been sampled by several investigators, and their general features are known (Pueyo 1978; Mingarro et al. 1981). Both playa–lake systems have superficial brines belonging to the neutral type described in the Eugster & Hardie (1978) classification scheme, but with different ionic ratios.

The system called 'La Salada' (or 'La Salada de Mediana') is located in an endorheic area covered by Cenozoic sedimentary rocks (mainly gypsum and marly lutites) and Quaternary sediments (river gravels and infill of flat bottom valleys). Brines in this system are of the Na–Mg–SO$_4$ chemical type, and the main saline minerals identified are mirabilite, bloedite, thenardite and gypsum (Pueyo 1978; Lopez 2004).

On the other hand, the 'La Playa' saline system is located in an ample endorheic area developed on carbonate and lutitic Paleogene rocks. Brines in this case belong to the Na–Cl type, and the main saline minerals identified were halite and gypsum (Pueyo 1978).

The geographical location of both saline systems is shown in Figure 1. In this study, experimental evaporation of brines collected in these chosen systems has been carried out to total dryness. The main goal is to determine the sequence of primary saline minerals that precipitate in both systems at high concentration stages.

Methodology

Most geochemical studies on modern saline systems consist of periodic collection of liquid

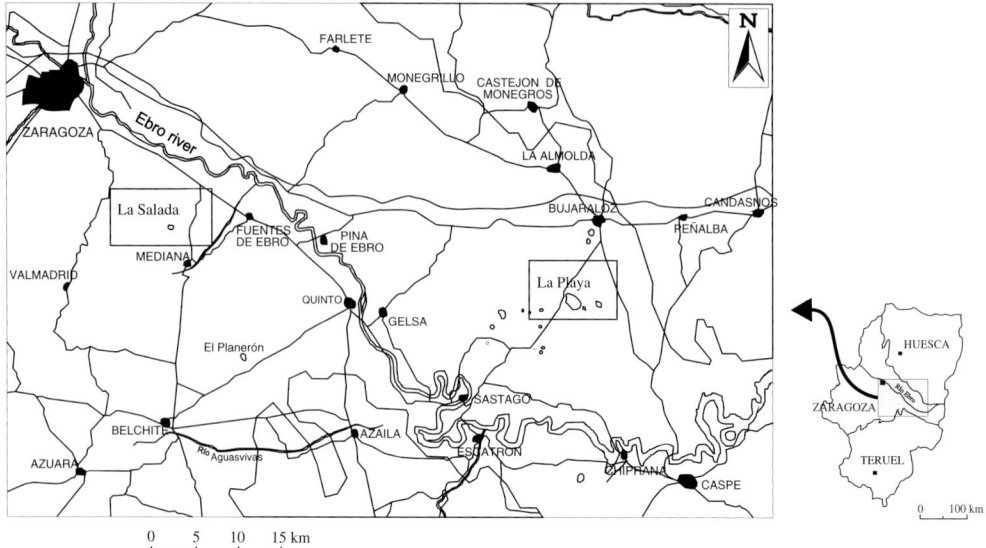

Fig. 1. Location of La Playa and La Salada playa–lake systems, and their position in the Aragon region of NE Spain.

and solid samples in the field resulting in a discontinuous record of the evolution of the system. Furthermore, such a sampling methodology includes the effect of all active processes in the systems: evaporative concentration, daily temperature variations and reactions between brines and solid materials (salts previously precipitated and clastic bottom sediments). Finally, sampling of inner areas in playa–lake systems becomes progressively more difficult as brine concentration increases, for which reason samples corresponding to high concentration stages are normally very scarce.

Here, experimental evaporation has been performed under isothermal conditions. A temperature value of 25 °C has been chosen, thus assuring a high degree of accuracy of thermodynamic data. Moreover, such a temperature value is very close to field mean temperature for seasons in which the two studied saline systems evolve until total dryness (May to July; Lopez 2004). The experimental device consisted of an open (to air) cylindrical container partially submerged in a thermic bath, in which 25 l of brine evaporated until total dryness, in closed system conditions with respect to minerals precipitating within the experiment.

Natural brines collected were first passed through a 5 μm paper filter in order to remove clastic and organic particles. In samples chosen from the evaporation experiment, analyses of Ca, Mg, Na, K (atomic absorption and emission spectrometry), SO_4 (colorimetry), Cl (potentiometry) and carbonate alkalinity (potentiometric volumetry) were made. After total dryness was achieved, salts precipitated were collected and analysed by means of XRD (X-ray diffractometry) and SEM-EDX (scanning electronic microscopy combined with energy dispersive X-ray spectrometry). This experimental stage allowed us to undertake a progressive sampling of brines along the evaporative concentration process until the most advanced stages of system evolution.

Application of geochemical modelling allowed us to estimate several parameters of brines: ionic strength, water activity, ionic activity coefficients and mineral saturation indexes. The geochemical code program used was PHRQPITZ (Plummer et al. 1988), which incorporates the chemical model developed by Harvie and collaborators (Harvie & Weare 1980; Harvie et al. 1984). This chemical model is based on Pitzer's equations (Pitzer 1973) and is at present considered the most accurate model for brine geochemical treatment (Krumgalz 1996).

Results

Chemical data of analysed solutions sampled along the experimental device have been compiled in Table 1 (La Playa brines) and Table 2 (La Salada brines). These analytical data and the results obtained by means of geochemical modelling have been plotted against a concentration factor (CF). This parameter was calculated as the ratio between the potassium concentration in each sample and that of the most dilute solution

Table 1. Chemical data of La Playa experimental brines, in mmol/l. Alkalinity has been expressed as mmol/l of HCO_3^-, density in g/ml, and ionic strength in molality

Sample	pH	Density	Ionic strength	Water activity	Ca^{2+}	Mg^{2+}	Na^+	K^+	Cl^-	SO_4^{2-}	Alkalinity
LP-0	8.35	1.02	0.76	0.99	5.33	112.98	258.37	6.67	301.81	117.42	6.38
LP-1	8.44	1.03	1.00	0.98	21.90	150.17	358.42	9.05	406.17	169.06	9.23
LP-2	8.40	1.04	1.49	0.97	29.21	202.84	485.43	12.30	572.59	230.06	13.10
LP-3	8.43	1.06	1.92	0.97	24.17	274.06	651.59	16.90	744.64	286.28	16.06
LP-4	8.34	1.08	2.79	0.95	16.02	404.85	962.17	25.52	1105.69	405.99	23.67
LP-5	8.25	1.11	3.98	0.93	10.20	568.60	1294.05	34.29	1675.46	565.27	29.60
LP-6	8.03	1.17	6.17	0.88	5.49	873.89	1918.25	51.10	2569.59	851.55	44.51
LP-7	8.09	1.15	5.46	0.89	6.44	782.14	1683.36	45.19	2312.92	755.78	36.00
LP-8	8.08	1.16	5.80	0.88	5.91	830.69	1779.06	48.44	2434.21	801.58	45.00
LP-9	7.85	1.20	7.71	0.85	4.11	1381.56	2305.38	59.39	2197.27	998.25	n.a
LP-10	7.79	1.26	11.56	0.74	3.61	2032.50	3218.83	79.49	3187.31	1382.77	n.a.
LP-11	7.76	1.34	14.25	0.64	1.52	2804.36	3340.62	118.88	3497.59	1958.09	n.a.
LP-12	7.48	1.37	14.57	0.65	1.19	2315.58	3097.04	132.99	4287.37	2152.82	n.a.
LP-13	7.35	1.39	15.52	0.64	1.37	2626.00	2829.53	191.62	4089.92	2383.92	n.a.

n.a., not analysed because of insufficient volume of sample.

Table 2. Chemical data of La Salada experimental brines, in $mmol/l$. Alkalinity has been expressed as $mmol/l$ of HCO_3^-, density in g/ml, and ionic strength in molality

Sample	pH	Density	Ionic strength	Water activity	Ca^{2+}	Mg^{2+}	Na^+	K^+	Cl^-	SO_4^{2-}	Alkalinity
LS-0	8.08	1.14	4.29	0.95	14.47	376.47	1739.96	10.90	211.55	1185.72	8.38
LS-1	8.13	1.16	4.72	0.95	12.99	432.01	1974.86	11.43	234.11	1281.49	9.44
LS-2	8.11	1.17	5.39	0.94	12.79	478.09	2244.55	12.51	265.14	1458.46	10.44
LS-3	8.18	1.23	7.51	0.91	8.28	630.73	2979.69	13.86	359.63	2025.82	13.56
LS-4	8.01	1.28	9.59	0.89	8.30	786.67	3597.37	17.72	456.94	2547.37	16.52
LS-5	7.92	1.31	11.39	0.85	7.66	1017.90	4210.71	23.10	644.52	2828.44	21.87
LS-6	7.82	1.33	12.11	0.84	4.97	1184.53	4110.66	24.76	739.01	2976.26	24.90
LS-7	7.74	1.35	12.91	0.82	4.72	1330.18	4519.55	27.75	829.27	3052.26	27.86
LS-8	7.70	1.35	14.24	0.80	1.85	1570.46	3971.46	34.99	1011.20	3308.35	33.27
LS-9	7.78	1.35	12.61	0.82	1.32	1351.16	4389.06	36.91	922.35	2986.67	35.08
LS-10	7.91	1.34	12.54	0.82	0.67	1264.35	4325.11	40.62	1030.94	2957.53	38.09
LS-11	7.90	1.36	11.66	0.83	0.97	1148.32	4337.29	44.81	1146.59	2812.83	40.98

coming from the natural system. Potassium was chosen here because of its conservative behaviour under these conditions until the last stages of very high concentration. In fact, cationic exchange with clayey sediments is here excluded as brines have been previously filtered, and potassium salts are expected to reach saturation very close to total brine desiccation.

La Playa brines

This evaporation experiment has allowed us to collect samples with a concentration factor near 30 with respect to initial experimental solution (which is also, in this case, the most dilute solution collected in La Playa natural system). Ionic strength has been calculated for the PHRQPITZ geochemical code and it will be considered here as the expression of the global concentration of the brine. Evolution of such a parameter indicates that La Playa brines should reach values near 15 molal in the most advanced stages of evolution (see Fig. 2), but the increase is less marked for the most concentrated solutions. This concentration level is extremely high (seawater has a mean ionic strength of 0.6 m), and in natural playa–lake systems it is very complicated to collect samples with ionic strength values over 10 m.

Another parameter estimated by means of the PHRQPITZ code is water activity. In Figure 2 it can be observed that water activity descends very slowly along the evaporation curve until the final samples (with values near 0.70).

Ionic evolution has been plotted separately for anions and cations (Fig. 3). Scatter diagrams show conservative behaviour for sulphate until the last sample. Chloride concentration evolves linearly until brines reach a concentration factor near 10; beyond this point the slope becomes reduced. Bicarbonate content increases gradually with the concentration factor but it has a concentration level much lower than sulphate and chloride.

Sodium and magnesium show a parallel behaviour, with an initial conservative path and a marked descent in final stages. Change in the slope of magnesium takes place for a concentration factor slightly higher than that of sodium. On the other hand, calcium and potassium contents are much lower and the observed behaviour is clearly different. Potassium evolution is conservative, while calcium shows an initial rise, which soon turns into a descending path.

Saturation indexes (SI) calculated for the main saline minerals are represented in Figure 4. An error band around SI = 0 was considered and it includes both the analytical imprecision (unavoidable when dilution is necessary prior to chemical analysis, as in the case of brines) and the thermodynamic uncertainties (Jenne et al. 1980). This band has a width of 0.40 units of the saturation index for carbonates and 0.20 units of SI for the rest of saline minerals of interest.

Aragonite saturation indexes lie on the supersaturation field from the initial stage. When brines attain a concentration factor about 8, aragonite SI values start to descend until they reach the lower part of saturation band. Significant problems in interpreting the geochemical behaviour of carbonated system in concentrated brines are known (Lazar et al. 1983) but have not yet been solved. Carbonate minerals are very scarce (almost absent) in the natural saline deposits of La Playa and La Salada. Therefore, a suitable definition of the carbonate system will not be a decisive subject in saline systems of the neutral chemical type as occurs in the case of alkaline saline lakes, where chemical evolution of brines is commonly closely controlled by the carbonate system.

Brines begin subsaturated with respect to gypsum, but saturation is reached beginning with the second sample and after an initial phase of slight supersaturation. Equilibrium with gypsum holds until last brine sampled.

Halite and thenardite (anhydrous sodic sulphate) saturation indexes show an almost identical behaviour, starting from marked subsaturation conditions and reaching saturation for a CF near 18. From this point, brines are in equilibrium with respect to both saline minerals until the last samples are collected. Epsomite (heptahydrated magnesium sulphate) saturation indexes evolve in a similar manner, but equilibrium conditions have been reached for a slightly higher concentration

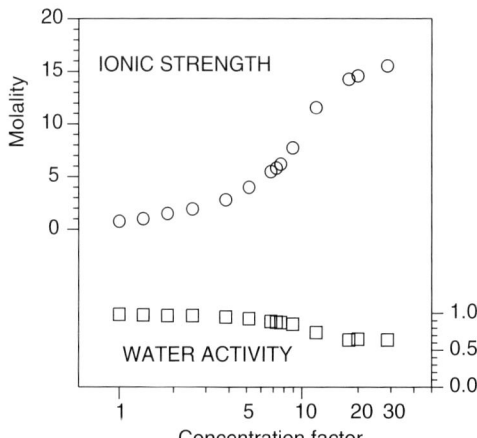

Fig. 2. Evolution of ionic strength and water activity for La Playa experimental brines plotted against the concentration factor.

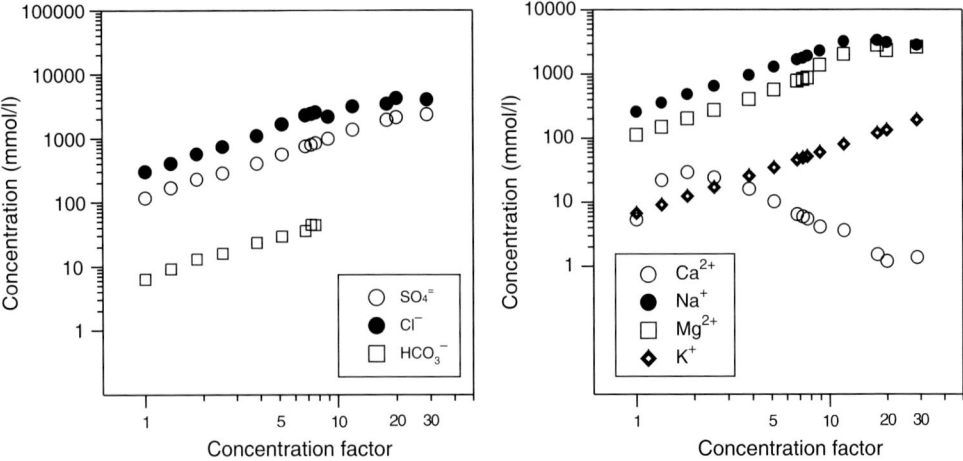

Fig. 3. Anionic and cationic concentration for La Playa experimental brines represented vs the concentration factor.

factor. Finally, mirabilite (decahydrated sodium sulphate) starts from subsaturated conditions and it almost reaches equilibrium for a concentration factor near 18, but beyond this point the path descends and mirabilite subsaturation increases in the last stage of the experiment.

Other saline minerals have not been plotted because they do not seem to have an active participation in the evolution of the system.

The mineral precipitation sequence established after geochemical modelling results is as follows: gypsum, halite, thenardite and then epsomite. The solid salt deposit collected after total desiccation of La Playa brines was analysed. Results were obtained by means of XRD and SEM-EDX, and indicate that the saline deposits collected consist mainly of halite, with gypsum and thenardite as accessories. Morphological aspects of the precipitated minerals can be observed in Figure 8 (d–f). Halite precipitated as millimetric cubic crystals (Fig. 8e & f). Both thenardite and gypsum also showed euhedral habit crystals but were much smaller in size (less than 10 μm) than those of halite (Fig. 8d).

The effective saline precipitation along the pathway of experimental evaporation has a clear influence on the chemical evolution of solutions, according to chemical divide rule (Hardie & Eugster 1970). Thus, gypsum precipitation produces a marked change in the calcium concentration path, although the effect is almost imperceptible in that of sulphate because of the large difference in concentration of both ions. On the other hand, chloride shows a change in slope when the concentration factor reaches about 10; it can be observed that halite attains a saturation field around this same point, thus influencing the chemical evolution of chloride. The sodium line does not show noticeable changes at this point of evolution, but when a CF of about 18 is reached it is marked because of the precipitation of thenardite. Finally, the evolution of the magnesium concentration shows a marked change in slope that corresponds to the epsomite equilibrium at a CF near 20.

La Salada brines

Samples collected in this case reached concentration factor values up to c. 30 with respect to the most dilute solution coming from the natural system (sample 'A' in table 3.3 after Lopez 2004).

Ionic strength values estimated for these brines increase until a maximum of 15 molal when the concentration factor reaches about 21 (Fig. 5). After this point ionic strength values start to descend, following this trend until the last sample collected. In the same figure it can be observed how water activity values descend smoothly from the early stages of concentration until the CF almost reaches 21; beyond this point, water activity stabilizes around 0.80.

The plot for ionic evolution (Fig. 6) permits verification that both bicarbonate and potassium show a conservative behaviour. Sulphate displays changes in the slope, first at a CF near 9 and later when this parameter is about 21.

The linear evolution of the chloride ion reduces its increasing slope in the final stages of the experiment. The path of sodium is very similar to that of sulphate but at a slightly higher concentration level. Magnesium content increases until a CF near 21, but after this point its trend

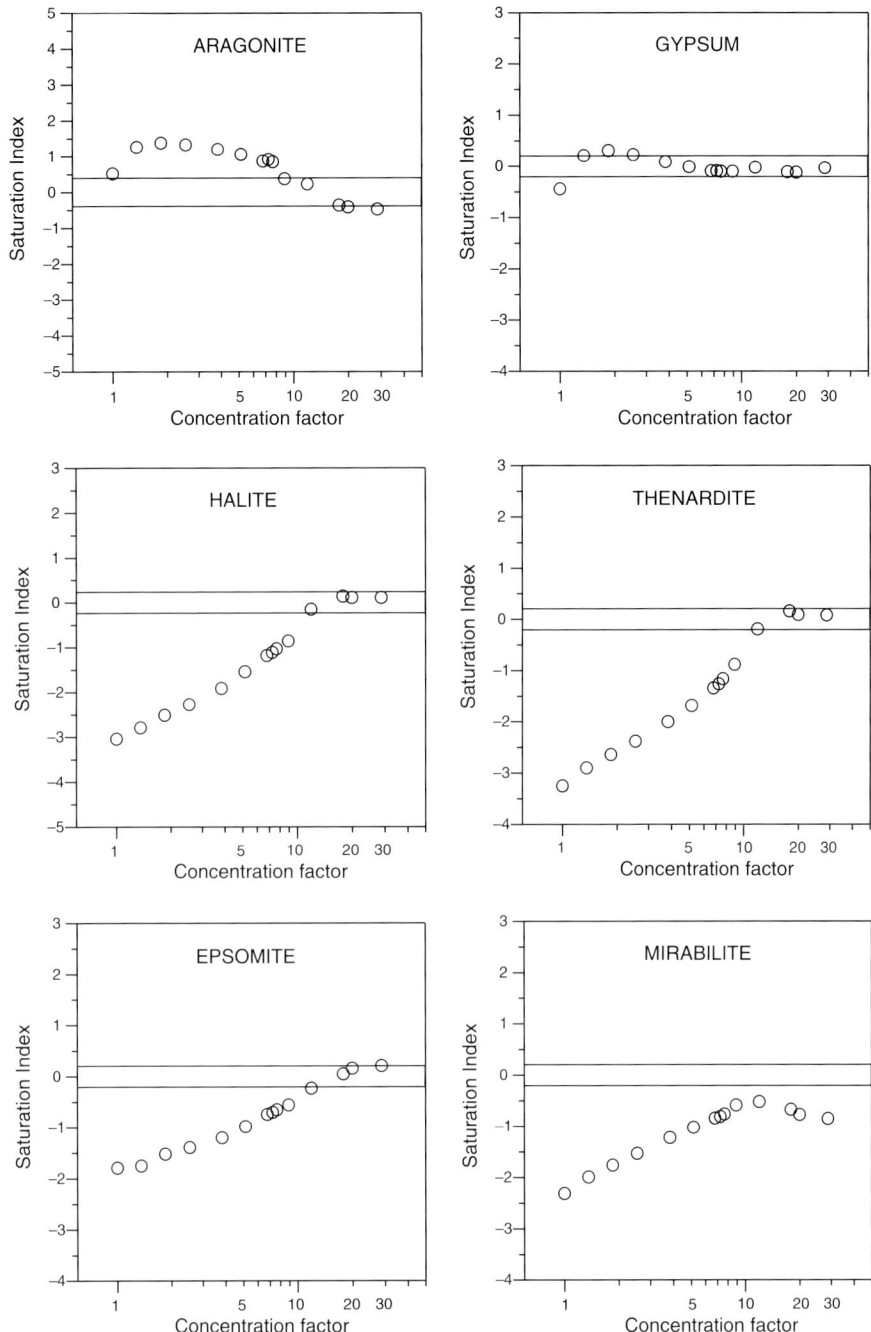

Fig. 4. Evolution of indexes of main mineral saturation in the La Playa evaporation experiment.

starts to descend until the last samples. Finally, calcium concentration shows a descending behaviour from the initial stages, but with a marked increase in the final phase.

Saturation indexes calculated for La Salada brines are represented in Figure 7. Solutions are in equilibrium with respect to aragonite from the start until the most advanced stages, where a

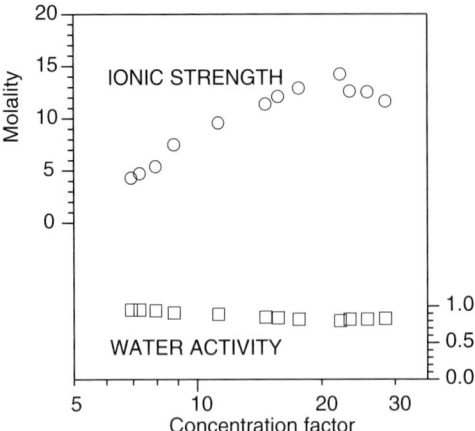

Fig. 5. Evolution of the ionic strength and water activity for the La Salada experimental brines plotted against the concentration factor.

slightly undersaturated state is achieved. Gypsum evolution is very similar to that of aragonite, though the degree of undersaturation is most marked in the final stages for that particular calcium sulphate phase.

Brines begin undersaturated with respect to mirabilite, but saturation is reached for a concentration factor near 9. Thenardite and bloedite (tetrahydrated sodium–magnesium sulphate) evolve within an undersaturated field, reaching equilibrium when CF is about 18 (thenardite) and 21 (bloedite), this latter after an initial phase of slight supersaturation. Finally, halite begins from undersaturated conditions and it approaches equilibrium until CF near 21, after this point saturation indexes remain unchanged until the final stage.

The precipitation sequence in this case can be established as follows: gypsum, mirabilite, thenardite and bloedite. XRD analysis of saline deposits collected after total brine dryness indicates that both thenardite and bloedite are the most abundant minerals precipitated, with gypsum as an accessory. Several morphological features of precipitates can be observed in Figure 8 (a–c). Bloedite appear mainly as radial aggregates of tabular crystals around 400 μm in length (Fig. 8a), whereas thenardite precipitated mainly as isolated crystals with similar size to that of bloedite but also as needle-shaped crystals (Fig. 8a & c). In other samples, both bloedite and thenardite precipitated as smaller crystals (about 20 μm in size) without noticeable morphologic features (Fig. 8b).

Application of chemical divide rule indicates that calcium concentration in La Salada brines already have been influenced from the initial experimental stage, showing a progressively more marked descending path. Precipitation of mirabilite produces a change in slope of both sulphate and sodium when the concentration factor is about 9. Bloedite reaches saturation for a CF near 21, thus affecting the path of sodium, magnesium and sulphate. It is important to indicate that thenardite precipitation does not seem to affect the chemical evolution of system.

Discussion

The results obtained in the experimental stage allowed establishment of the evaporative saline precipitation sequence for two types of neutral

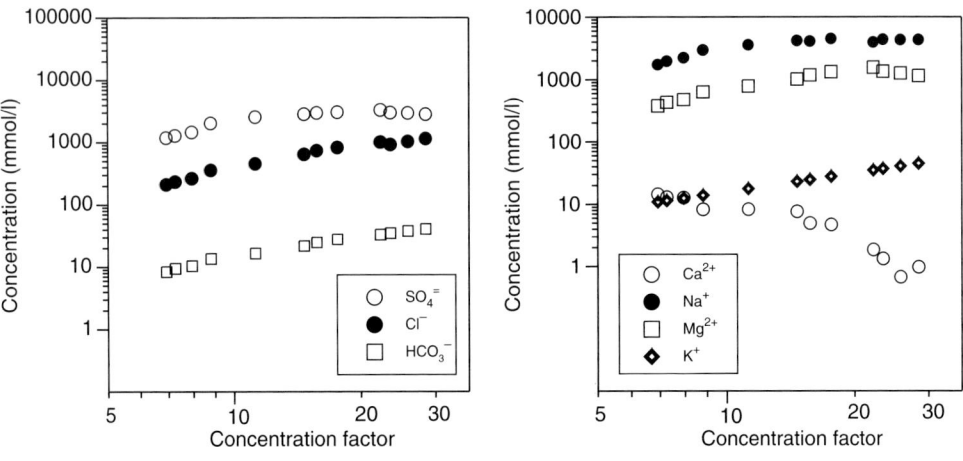

Fig. 6. Anionic and cationic concentration for La Salada experimental solutions represented against concentration factor.

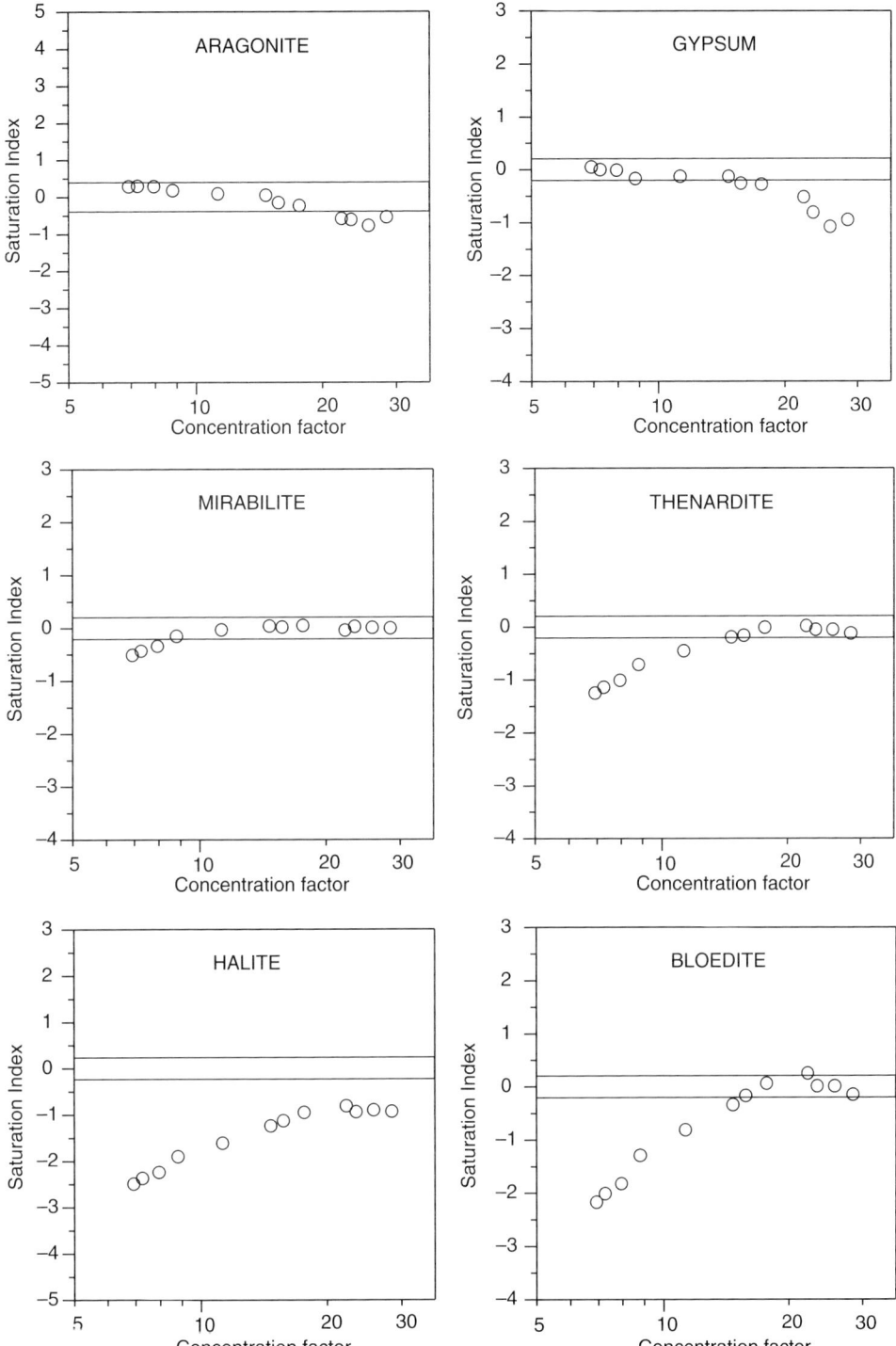

Fig. 7. Evolution of main mineral saturation indexes in the La Salada evaporation experiment.

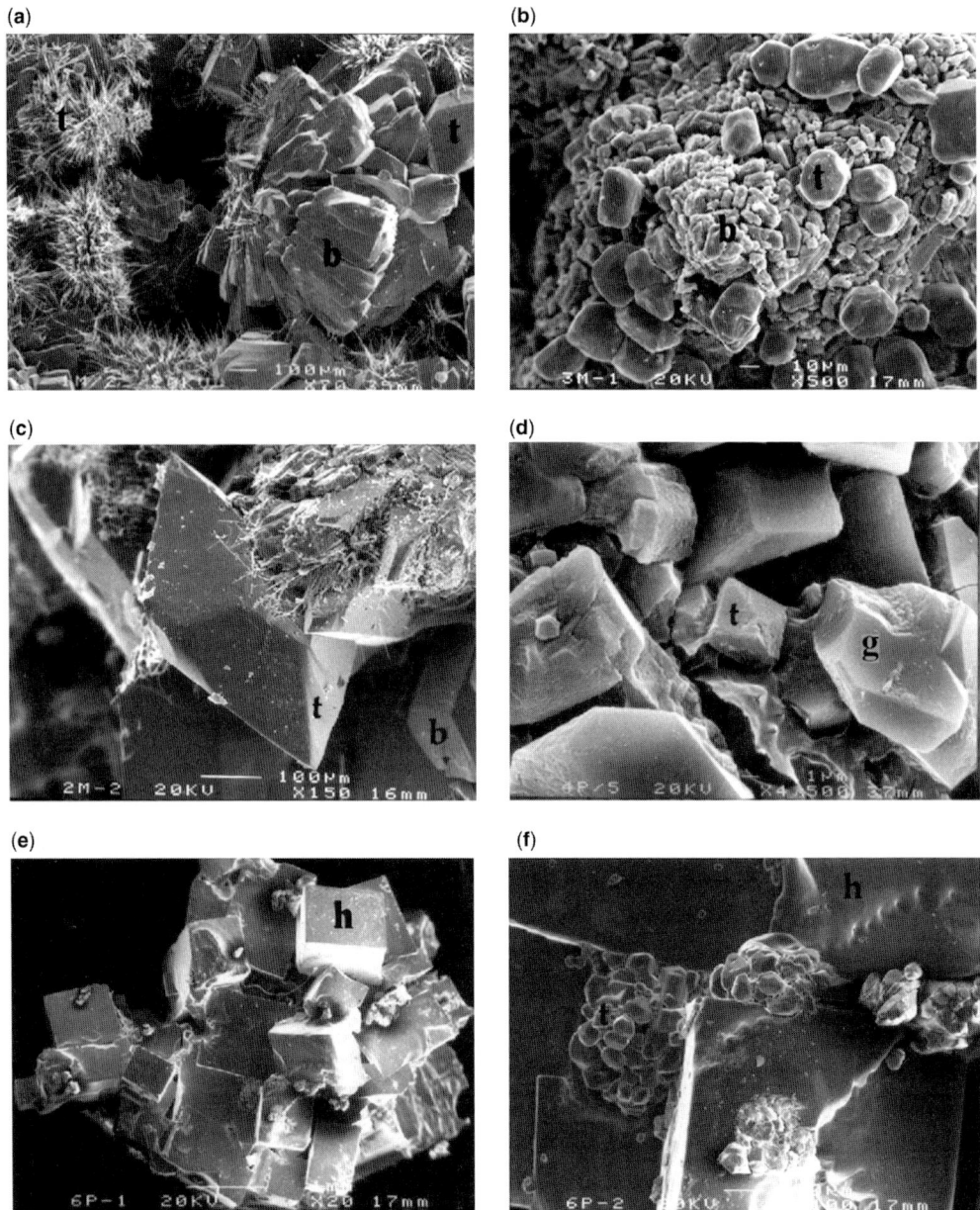

Fig. 8. SEM images showing detailed features of saline minerals collected in La Salada (**a–c**) and La Playa (**d–f**) evaporation experiment (b, bloedite; t, thenardite; g, gypsum; h, halite).

brines but with different ionic ratios. The experimental device, employed in this study, also enabled us to collect brine samples with extremely high concentrations (near 15 molal in both cases), whose sampling is very complicated in natural systems.

The evolving path of experimental evaporation registered in La Salada seems to indicate that brines have reached an invariant point (Risacher & Clement 2001) for a concentration factor near 21, where solutions are simultaneously saturated with respect to mirabilite, thenardite and bloedite.

Under these conditions, water activity values remain fixed by the mineral reactions, and global solution concentration decreases as evaporation advances in the final stages of evolution (Fig. 5).

On the other hand, La Playa brines do not show such an evolution, as solution ionic strength increases until the last collected samples, which are saturated with respect to gypsum, halite, thenardite and epsomite.

All these results permit a check on how the initial ionic ratio controls both the effective precipitation and the order of appearance of saline minerals under identical thermal conditions. Thus, when solutions have high Na:Mg and Cl:SO_4 ratios, as in the La Playa saline system, brines reach equilibrium with respect to epsomite at high concentration stages, whereas bloedite is the main magnesium salt in La Salada brines (in which Mg and SO_4 proportions are clearly higher) and solutions do not attain saturation with epsomite.

Equilibrium of sodium sulphate salts is also affected by solution chemistry. La Playa brines attain equilibrium only with respect to thenardite through the experiment, whereas in the La Salada system both phases mirabilite and thenardite reach saturation in two different stages of evaporative evolution. After reaching the peritectic point, La Salada brines remain saturated with both sodium sulphate phases until the last solution sampled. Geochemical interpretation indicates that in such equilibrium conditions mirabilite dissolves and thenardite precipitates, both processes taking place simultaneously. In this case, brine chemical evolution thus remains stopped at an invariant point until mirabilite fully dissolves. Mirabilite/thenardite transformation reaction is well known in present-day saline systems (Pueyo 1978) as a spontaneous dehydration process that occurs in subaerial conditions. Results obtained here indicate that the mirabilite/thenardite transformation process also takes place in subaquatic conditions without having an apparent effect on the chemical evolution of the solutions.

The magnitude of mass mineral precipitation is another important feature in both systems. In La Playa brines, halite and thenardite reach equilibrium at very close points of evolution, but halite precipitation takes place more massively than in the case of thenardite. Differences in mineral precipitation kinetics and the common ion effect (both phases share the cation) markedly limit the amount of thenardite that precipitates along the evolution of the evaporative system, as analysis of the solids showed (Fig. 8f, see the interstitial location of thenardite between much larger halite cubes).

On the other hand, results of mineralogical analysis carried out on La Salada solids indicate quantitatively similar proportions between precipitated sodium sulphate (thenardite) and bloedite.

Finally, it is important to notice the absence of sulphate minerals with an intermediate degree of hydration both in experimental and natural solid samples collected at La Salada and La Playa systems.

Conclusions

The genetic interpretation of saline deposits in geological record has been carried out historically starting from both the knowledge of present-day systems and the performance of laboratory experiments in controlled conditions. It has benefited in the last few decades from the advances in geochemical modelling in highly concentrated solutions. The results presented here indicate that present-day saline deposits generated by evaporation of brines with high Na:Mg and Cl:SO_4 ratios (and under similar climatic conditions to that of Central Ebro basin) are mainly composed of halite, but they may locally present sodium sulphate in which the primary precipitation has been produced as thenardite.

On the other hand, deposits formed after brines similar in chemistry to that of La Salada (with higher Mg and SO_4 content) are composed of bloedite and sodium sulphate, but in this latter case it depends upon a system that has evolved to total dryness, in which case all the sodium sulphate will be thenardite formed by dehydration of mirabilite in subaquatic and/or subaerial conditions (in areas exposed, early on, to air). In all evolving intermediate situations mirabilite and thenardite can appear together in the saline sediments formed.

All these results are in very good agreement with the mineralogical data coming from natural systems. The experimental device employed here has allowed us to determine both the order of effective mineral precipitation, in chosen saline systems, and the differences between amounts of precipitated minerals, especially in the case of La Playa brines. This type of information is very difficult to obtain starting from only natural samples, where an unknown number of processes (including short-range polythermic ones) is superimposed upon that of evaporative brine concentration.

References

EUGSTER, H. P. & HARDIE, L. A. 1978. Saline lakes. *In*: LERMAN, A. (ed.) *Physics and Chemistry of Lakes*. Springer, Berlin, 237–293.

HARDIE, L. A. & EUGSTER, H. P. 1970. The evolution of closed-basin brines. *Mineralogical Society of America, Special Papers*, **3**, 273–290.

HARVIE, C. E. & WEARE, J. H. 1980. The prediction of mineral solubilities in natural waters: the Na–K–Mg–Ca–Cl–SO$_4$–H$_2$O system from zero to high concentration at 25°C. *Geochimica et Cosmochimica Acta*, **44**, 981–997.

HARVIE, C. E., MOLLER, N. & WEARE, J. H. 1984. The prediction of mineral solubilities in natural waters: the Na–K–Mg–Ca–H–Cl–SO$_4$–OH–HCO$_3$–CO$_3$–CO$_2$–H$_2$O system to high ionic strengths at 25 °C. *Geochimica et Cosmochimica Acta*, **48**, 723–751.

JENNE, E. A., BALL, J. W., BURCHARD, J. M., VIVIT, D. W. & BARKS, J. H. 1980. Geochemical modelling: apparent solubility controls on Ba, Zn, Cd, Pb and F in waters of the Missouri Tri State mining area. *In*: HEMPHILL, D. D. (ed.) *Trace Substances in Environmental Health*, **XIV**, 353–361. University Missouri, Columbia, MO.

KRUMGALZ, B. 1996. Some aspects of physical chemistry of natural hypersaline waters. *Recent Research Developments in Solution Chemistry*, **1**, 9–28.

LAZAR, B., STARINSKY, A., KATZ, A. & SASS, E. 1983. The carbonate system in hypersaline solutions: alkalinity and CaCO$_3$ solubility of evaporated sea water. *Limnology and Oceanography*, **28**, 978–986.

LOPEZ, P. L. 2004. *Estudio geoquimico de la sedimentación salina actual en un sistema playa–lake estacional de tipo sulfatado-sodico: La Salada de Mediana (Zaragoza, España)*. Ph.D Thesis. Servicio de Publicaciones, Universidad de Zaragoza (Spain).

MINGARRO, F., ORDOÑEZ, S., LOPEZ DE AZCONA, M. C. & GARCIA DEL CURA, M. A. 1981. Sedimentoquimica de las lagunas de Los Monegros y su entorno geologico. *Boletin Geologico y Minero*, **92**, 171–195.

PITZER, K. S. 1973. Thermodynamics of electrolytes. I. Theoretical basis and general equations. *Journal of Physical Chemistry*, **77**, 268–277.

PLUMMER, L. N., PARKHURST, D. L., FLEMING, G. W. & DUNKLE, S. A. 1988. A computer program incorporating Pitzer's equations for calculation of geochemical reactions in brines. *United States Geological Survey, Water Resources Investigations Report*, **88–4153**.

PUEYO, J. J. 1978. La precipitacion evaporitica actual en las lagunas saladas del area: Bujaraloz, Sastago, Caspe, Alcaniz y Calanda (provincias de Zaragoza y Teruel). *Revista del Instituto de Investigaciones Geologicas*, **33**, 5–56.

RISACHER, F. & CLEMENT, A. 2001. A computer program for the simulation of evaporation of natural waters to high concentration. *Computers & Geosciences*, **27**, 191–201.

Genesis and evolution of 'pseudocarniole': preliminary observations from the Susa Valley (Western Alps)

W. ALBERTO, F. CARRARO, M. GIARDINO & D. TIRANTI

University of Torino, Earth Sciences Department, Via Valperga Caluso, 35.
I-10125, Torino, Italy (e-mail: walter.alberto@unito.it)

Abstract: In the Alpine geological literature some particular carbonate rocks having a vuggy appearance, associated with evaporite rocks, have a controversial origin and diversified, uneven, nomenclature. In this paper they are provisionally called 'pseudocarniole'. An excellent exposure along the Susa Valley (Italian Western Alps) permitted a relatively full analysis of a group of such vuggy rocks *in situ*. Five main lithologies have been identified with respect to their nature, shape, dimensions and organization of the clasts, also according to the matrix and the composition of the cements.

Utilizing the interpretation of geological and geomorphological field data that cover these deposits, detailed stratigraphical descriptions and laboratory microanalyses (optical microscope and cathode-luminescence), a genetic interpretative model is proposed. The main formational process is dissolution, starting from solution of gypsum and anhydrite; then the processes affecting carbonate 'parental residual rocks' are considered. Thereafter a general instability of the rock masses is introduced with a consequent gravitational collapse. Other general processes are usually associated: tectonics, gravity and Plio-Quaternary karst phenomena. Finally, a particularly young age for the formation of the pseudocarniole deposits is noted, based on palynological data, Plio-Quaternary.

This paper deals with carbonate rocks having a vuggy appearance, commonly grey to yellowish coloured, widely distributed in the Western Alps in association with evaporitic Triassic rocks, here provisionally called 'pseudocarniole'. Formerly (Zaccagna *et al.* 1911a, b), these rocks were interpreted as 'true carniole', in that they are normally associated with gypsum formations. More recently they have been distinguished from these last and interpreted as 'carbonate breccias' of unknown age and/or as 'residual breccias' of Pliocene–Upper Holocene age (Polino *et al.* 2002).

In the present study these rocks have been described, from both lithological and structural points of view, in comparison with other similar alpine rocks (Rauhwacken, carniole, cornieules, cargneules) and Apenninic rocks (Calcare Cavernoso, in: Ciarapica 1985; Ciarapica *et al.* 1987; Gandin *et al.* 2000) of controversial origin and diversified nomenclature.

Throughout the geological literature it is possible to recognize the different interpretative models for the genesis and age of these much-discussed carbonate breccias and conglomerates in the Alps. Schematically the hypothesis for the origins of these rocks can be gathered into three main groups:

(1) 'sedimentary origin' (from late Neogene) due to their unconformity with respect to Paleogene structural elements and/or because they also rework late Eocene or Oligocene sediments in (Grandjacquet & Haccard 1973, 1975; Patacca *et al.* 1973; Federici & Raggi 1974);
(2) 'direct tectonic origin' (e.g. cataclastites) or of 'indirect tectonic origin', e.g. reworked tectonic clasts (Leine 1971; Masson 1972; Debelmas *et al.* 1980; Fudral 1998);
(3) breccias derived from alteration or weathering of carbonate rocks, i.e. processes following sedimentation and tectonics (Bruckner 1941; Ellenberger 1954), also from recent genesis, for instance climate-controlled processes related to alternate advances and retreat of glaciers (Debenedetti & Turi 1975).

Lastly, other studies offer an intermediary view on the matter, pointing out a 'mixed' origin for the much-discussed carbonate breccias and conglomerates: 'sedimentary and tectonic' (Wiesender 1971) or 'tectonic and alteration' (Warrak 1974; Müller 1982; Richards & Vearncombe 1984) because the last process commonly overlaps the tectonic brecciation (also see sedimentation, tectonics and weathering; in Warren 1999).

Recently, the complexity and the variety of processes related to these diverse rock formations has also been recognized as being dependent on local factors. This way it is possible to have many different breccia types, depending on texture, structure, geometry of the rock body and relationships with surrounding rocks (Jeanbourquin 1988; Amieux &

Jeanbourquin 1989; Giardino 1995; Schaad 1995; Carraro *et al.* 2002; Polino *et al.* 2002).

In the Susa valley (Fig. 1) detailed macro- and microscopical analyses of this group of rocks have been conducted (Alberto 2004; Alberto *et al.* 2005). A complex but organic framework resulted from these analyses: all the studied pseudocarniole are carbonate rocks (breccias, conglomerates, calcarenites, travertine crusts) of vuggy appearance (often enhanced by centimetric carbonate crusts) similar to that of the above-mentioned 'true carniole', i.e. from Triassic evaporites (Zaccagna *et al.* 1911*a*, *b*).

For this group of rocks a genetic model and a relative classification scheme are here proposed: the provisional name of 'pseudocarniole' indicates the similarity of appearance but the different origin and age with respect to the 'true carniole' and also to the deformed sulphate evaporites (Passeri 1975; Malavieille & Ritz 1989; Nury & Schreiber 1997; Schreiber & Helman 2005).

In the Susa Valley the pseudocarniole are typically associated with Triassic successions of limestones and dolomites (Fig. 2a–c), being generally located above gypsum/anhydrite rock masses (Fig. 3) and/or along tectonic contacts. The Susa valley pseudocarniole genesis is here interpreted as a consequence of the relevant volume reduction induced by dissolution of gypsum/anhydrite rock masses: the complex story of pseudocarniole development seems to start with hydrothermal fluid ascension and consequent dissolution. As consequence of this, the overlying dolomite and carbonate rocks are deeply transformed by means of gravity collapse; hydrothermal fluid activities and associated dissolution can both be followed by later karst phenomena. The sector is characterized by a significant degree of reworking of the original carbonate rocks so that their earlier structures and textures are completely obliterated. Thus new rocks are created, necessitating the introduction of the new name, pseudocarniole.

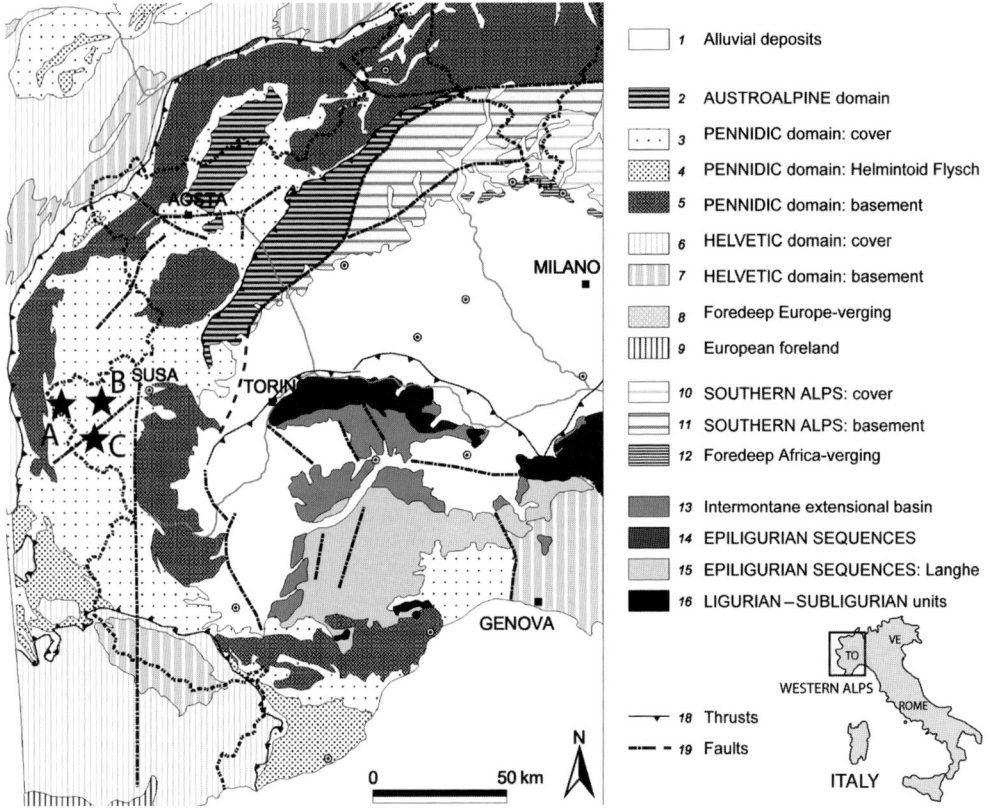

Fig. 1. Structural geology of the Western Alps and locations of case studies in the Susa Valley: ★A, Bardonecchia (E 315.556 m, N 4.990.889); ★B, Seguret (E 328.920, N 4994.109); ★C, Thuras Valley (E 329.651, N 4.975.035). Geographic coordinates UTM 32 ED 50.

Fig. 2. Geomorphological context of the pseudocarniole and geometrical relationships between the pseudocarniole and other lithological units (pc, pseudocarniole; gy, gypsum; cs, calcschistes; do, dolomites). (**a**) Best outcrop of gypsum–pseudocarniole association near Bardonecchia. (**b**) Steep landforms characterizing the geomorphological landscape of the Val Thuras pseudocarniole. (**c**) Karst caves (Grotta dei Saraceni) into the Seguret Dolomites (middle Triassic); detrital pseudocarniole outcrops at the cave's exit.

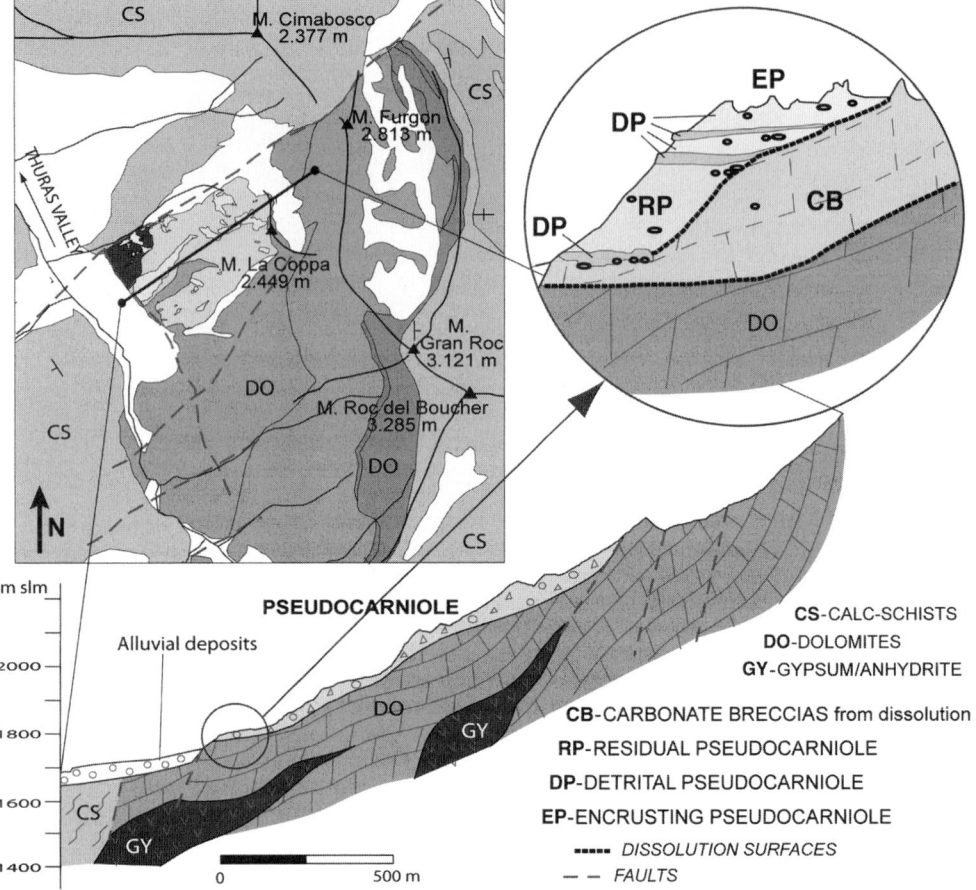

Fig. 3. Thuras Valley case study: geological map and cross-sections.

Other processes are usually associated with dissolution: these are connected to tectonics (usually normal or strike–slip faulting), gravity (deep hollow collapse and slope dynamics) and karst (shallow dissolution and re-precipitation). In consequence of all these different possible formation processes, many types of pseudocarniole have been recognized and described, which constitute rock bodies with different dimensions, geometries and lithofacies associations. Finally, a particularly young age for some facies of pseudocarniole in this area has been suggested: Plio-Quaternary, from preliminary palynological data.

Methodology

The data collection was first developed on the field: detailed geological maps and sections were made, and stratigraphical descriptions, logs and samples were collected. Subsequently, further studies have been carried out in the laboratory, using thin-section optical microanalysis of the samples, some of which were also examined by cathode luminescence (CL).

The pseudocarniole

The main object of the research presented in this paper is the geological data concerning the diverse members of the group pseudocarniole, interpreted and organized in such a way as to create a functional classification, present the evidence and possibly solve some interpretative problems concerning their origin.

These detailed studies have permitted us to specify relationships between different types of pseudocarniole, depending on characteristics and mutual relationships between different carbonate cements found in them. It has also allowed a complete description of the nature of constituent clasts

Table 1. *Hierarchy of pseudocarniole facies and sub-facies*

(A) Carbonate breccias from dissolution		
(B) Residual pseudocarniole from dissolution and collapse		
(C) Detrital pseudocarniole	(C1) Detrital sub-facies without evidence of transport	
	(C2) Detrital sub-facies with evidence of transport	(C2A) Detrital sub-facies, by channelled water transport
		(C2B) Detrital sub-facies, formed by stream water transport
		(C2C) Detrital sub-facies, by stream water from infill of subterranean karst cavities
(D) Encrusting pseudocarniole		
(E) 'Tectonic' pseudocarniole		

present in all the varieties of pseudocarniole, the reconstruction of the sedimentary processes that have determined or conditioned their origin, as well as the influences of the deformation processes on their texture and structure.

The field data collection, supported by laboratory data and results, has resulted in the establishment of five principal facies and to the elaboration of a general genetic model for the pseudocarniole of the Upper Susa Valley. Several related sub-facies also have been recognized, which correspond to several other genetic mechanisms: the complex hierarchy is summarized in Table 1, and detailed descriptions are presented in the following pages.

(A) Carbonate breccias from dissolution

Carbonate breccias from dissolution are constituted by monogenetic breccias (calcareous or dolomitic); the fragments are always angular, do not show any kind of transport or indications of rotation, and have varying dimensions from less than a centimetre to some metres and generally a chaotic distribution. Macroscopically, most of the carbonate breccias show a vuggy fabric and an inhomogeneous greyish colour. Locally, dolomitic remnants of the original clasts are characterized by a poor degree of cohesion, which can decrease until they form a dolomitic powder; generally they show a peculiar boxwork structure, made of angular moulds individuated by prominent calcite walls and selective erosion of incoherent dolomite powder (Fig. 4).

Sometimes they preserve traces of stratification, more or less disguised by the fractures, inherited from the original 'parent rock'. In some cases empty interstitial spaces between fragments have been described. The analyses in CL have shown the presence of several phases of cementation that alternate with dissolution and fracturing phases for the genesis of carbonate breccias from dissolution.

Calcite cements are present (sparry, clear or turbid in transmitted light; sometimes zoned yellow, mustard, brown in CL), as well as dolomite cements (sparry, also in pluri-millimetre crystals, generally turbid and not zoned); in some cases within these last phases there are sparse fluorite crystals present.

As discussed below in the proposed interpretative model, this facies is here interpreted as the product of the partial dissolution and alteration of the dolomite and calcareous parent rocks, after massive dissolution of gypsum/anhydrite rocks and consequent volume reduction, thus carbonate breccias are formed: they are genetically connected to the slow re-arrangement of overhanging dolomite and calcareous rocks; the original sedimentary structures are largely preserved (McWhae 1953).

(B) Residual pseudocarniole from dissolution and collapse

Residual pseudocarniole from dissolution and collapse are made up of monogenetic breccias (calcareous or dolomitic), more rarely polygenetic (containing minor calc-schist and quartzite clasts), with angular, not selected fragments; they are always associated with typical yellowish, seldom reddish, matrix locally showing a weak stratification. The residual pseudocarniole also show a vuggy fabric locally characterized by a boxwork structure (Fig. 5).

They generally include lithosomes of carbonate breccias from dissolution with varying dimensions, up to 10 m long. The residual pseudocarniole are bounded from the underlying rocks (carbonate breccias from dissolution, dolomites) by means of unconformities often underlain by accumulations of insoluble products (Fig. 5c).

The analyses of samples using CL also demonstrate the existence of cement phases having

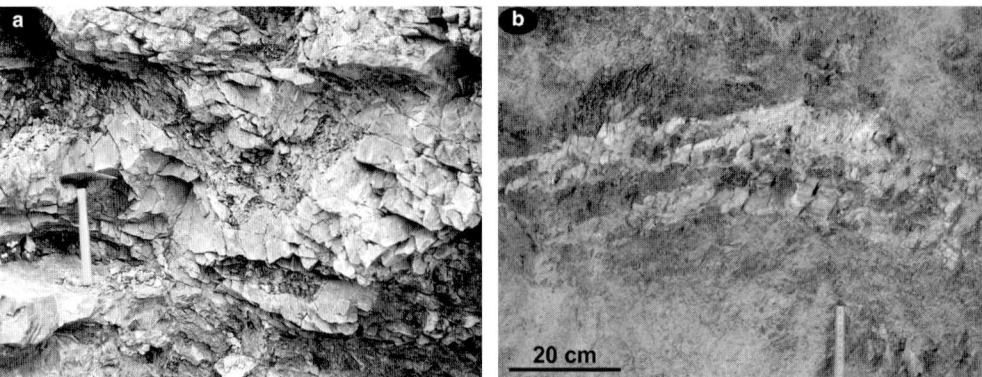

Fig. 4. Carbonatic breccias from dissolution. Note the stratification inherited by the genetical process, the poorly vacuolar detrital aspect and the heterogeneity in the dimensions of the clasts. (**a**) Outcrop view; (**b**) detailed view.

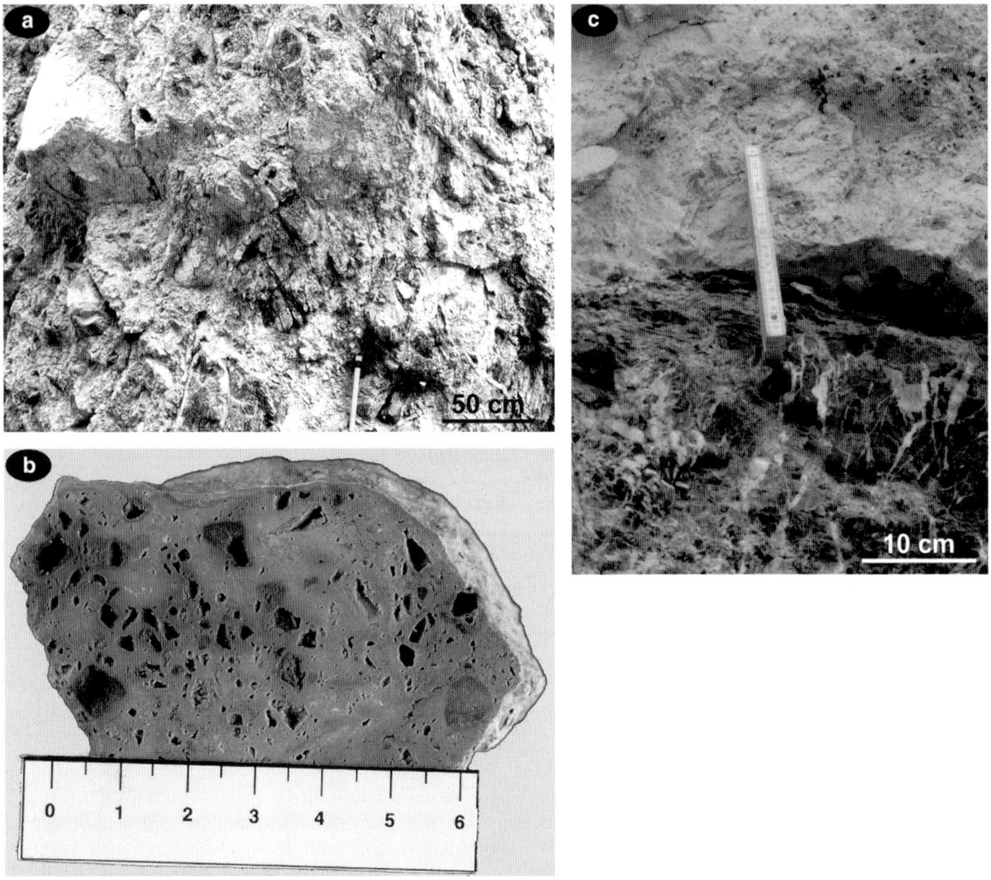

Fig. 5. Residual pseudocarniole from dissolution and collapse. The monogenetic clasts of carbonatic rocks (here dolomites), angular and not sorted, are separated by discontinuous and irregular portions of yellow matrix. (**a**) Outcrop view. (**b**) Sample of polygenetic breccia, note the vacuolar aspect (scale is in centimetres). (**c**) Insoluble product accumulation at the lower boundary of carbonatic breccias from dissolution.

various compositions: several dolomitic phases with large turbid spars (turbid, also in transmitted light), with some neo-formed crystals of quartz and thin dolomite veins, that appear weakly zoned in CL; at least one calcite cement phase has been identified, made up of clear spar, as seen in plane polarized light, but zoned in CL. Like the carbonate breccias from dissolution, the residual pseudocarniole also do not show evidence of 'ductile' deformation while the carbonate 'parent rocks', are generally deformed.

We consider the residual pseudocarniole from dissolution and collapse to originate due to gravity collapse, by complete re-arrangement of rock mass at an advanced stage of dissolution and by accumulation of residual products largely in subsurface environments.

(C) Detrital pseudocarniole

Detrital pseudocarniole are made up either by polygenetic breccias (clast- or matrix-supported, with fragments of varying dimensions and lithology, matrix always present) or by fine sand-silt sediments. They show a vuggy fabric, but are seldom characterized by boxwork structure. Macroscopically the rock colour ranges from yellowish to orange (Fig. 6).

Depending on the textural variability, it is possible to make a further subdivision into a number of sub-facies with different genetic mechanisms.

(C1) Detrital sub-facies without evidence of transport. These are breccias with polygenetic clasts, mainly angular and poorly sorted, with varying dimensions from few centimetres to meters. The texture varies from clast-supported to the rare matrix-supported. The matrix introduces a high variability in grain dimension, from silts to gravelly sands; the grains have the same composition as the clasts. It is possible to observe rare sedimentary structures (e.g. gradation); the cementation by calcite and volume of accumulation vary notably from place to place. This sub-facies can be interpreted as cemented detrital material from the carbonates of the circulating groundwaters.

(C2) Detrital sub-facies with evidence of transport. This sub-facies is represented by breccias or micro-breccias, evolving with the intermediary compositions to conglomerates or microconglomerates; the clasts can be polygenetic or monogenetic. The C2 sub-facies includes matrix-supported, strongly cemented by calcite pseudocarniole characterized by such sedimentary structures as ripples, laminations and well-orientated clasts; the clasts are equal-dimensioned, a few centimetres in size. The very fine and well-sorted carbonate sediments (silts to sands) commonly act as a matrix with respect to the coarsest clasts, or can constitute homo-metric lenses.

All of these characteristics give a non-homogeneous appearance to the C2 sub-facies pseudocarniole. Commonly they constitute the exhumed infill of subterranean karst forms: sediments that are entirely or partially fill-in channels; and spindle-shaped tubes and cavities. Inside this sub-facies a further subdivision has been recognized (three parts):

(C2A) Detrital sub-facies by channelled water transport: the sub-facies is represented by well-sorted carbonate sediments, with grain sizes ranging from clayey silts to coarse sands. The finest-grain deposits, clayey silts to silts, show plane-parallel or parallel-undulating laminations but sedimentary structures formed by current action were not noticed. As the grain dimensions increase, more complex sedimentary structures have been observed: from silts to fine sands, in which low angle laminations and ripples (either asymmetrical or symmetrical) are preserved. The sand size sediments include more marked structures such as festoons and convolute laminations, concave erosion surfaces and evident normal gradations. This sub-facies is the most widely distributed and abundant among the detrital pseudocarniole types. Its fine sediments have a broad distribution: they seem to represent the only preserved record of a long sedimentary phase in the evolution of the pseudocarniole. The C2A sub-facies is significant because fine sediments are associated with almost all the other sub-facies, representing the final local product of their karst reworking. Although rare, the presence of water splash-marks on the strata surfaces suggests genesis within a subterranean karst environment, where the dripping from the ceilings of cavities and into channels was a common event. The different facies geometries and internal structures mark various environments of sedimentation: from one side, diffuse depositional sedimentary structures due to current action (either on the fine isolated deposits or on other constituent of the sub-facies matrix) indicate a high-energy underwater environment sedimentation. On the other hand, the distribution across wide areas seems to represent the deposition by subterranean water sheets. Finally, the C2A sub-facies also fills in fractures and small dissolution cavities in the substratum or in other types of pseudocarniole (Fig. 6c).

(C2B) Detrital sub-facies formed by stream water transport: this is constituted of matrix supported micro breccias, poorly cemented and dense; in a

Fig. 6. Detrital pseudocarniole. (**a**) Outcrop view; note the rounded clasts into the detrital sub-facies by stream water from infill of subterranean karst cavities (C2C). (**b**) Sample of stratified polygenetic breccia of detrital sub-facies formed by stream water transport (C2B), scale in centimetres; the clasts vary from sub-angular to sub-rounded; the hand-sample in the upper portion is composed by fine detrital sub-facies sediments. (**c**) Geometrical relationships detail among different facies of the pseudocarniole: detrital sub-facies by channelled water transport (C2A) that fills in open fractures within the carbonatic breccias from dissolution.

sandy-silt matrix, poorly sorted, polygenetic clasts are found, up to centimetre-size. The sediment is organized in heterogeneous layers of different thicknesses and irregular shapes, even though parallel lamination surfaces are also found. The clasts of small dimensions (up to millimetres in size) are sub-rounded and organized in 'stone-lines', while the larger clasts are angular, have a chaotic disposition and are wrapped by the lamination surfaces. The C2B sub-facies forms rock bodies with an irregular geometry articulated in a series of small lobes. The angle of lamination-dip is entirely concordant with the depositional slope. This sub-facies is thought to have originated from sedimentary deposition in channels and by stream water transport, through localized flow mechanisms. The coarse material comes from the dismantlement of the detrital sub-facies, by the action of channelled waters throughout the karst flow path.

(C2C) Detrital sub-facies, by stream water from infill of subterranean karst cavities: this is composed of poorly sorted and strongly cemented by calcite deposits, matrix- or clast-supported. The polygenetic clasts are heterogeneous in dimension and form: ranging from big blocks, tabular and angular or sub-spherical and rounded, to smoothed pebbles, or equidimensional or prismatic angular clasts inside a silt/sand to sand/gravel matrix. The clast composition is very diverse and includes fragments of bedrock, other pseudocarniole deposits (*residual* and *detrital*) and material from surface formations. Isolated C2C sub-facies rock bodies show characteristic pinnacle morphology. Local development of inverted relief morphology is facilitated by poor mechanical conditions of the rocks caused by the morphology of the karst in which they formed. The C2C sub-facies represents the sediments deposed in funnel-like sockets, probably representing small subterranean karst landforms: these are modelled in the carbonate bedrocks or in the carbonate breccias from dissolution and residual pseudocarniole from dissolution and collapse facies after their exhumation. Accumulation and cementation processes have made these pseudocarniole more resistant to weathering and erosion in comparison to their bedrocks.

(D) Encrusting pseudocarniole

These pseudocarniole show abundant quantities of carbonate polygenetic cement, polychrome (various yellow and white shades), with small angular fragments of different lithological natures. They also show a typical vuggy aspect. This facies is quite diffuse and forms thin layers (thickness up to 30 cm) on exposed surfaces of other types of pseudocarniole, disguising their real aspect and therefore making them appear apparently similar. The encrusting pseudocarniole can also constitute small bodies, up to 50 cm high, at the base of the rock walls (Fig. 7).

The origin of this facies can be compared with that of the detrital travertines (D'Argenio & Ferreri 1987): both are the result of the calcium carbonate precipitation contained in the surface stream waters. The thickness of the mineral crust layers directly depends on the exposure time of the pseudocarniole to meteoric agents.

(E) 'Tectonic' pseudocarniole

These are made up of monogenetic breccias (calcareous or dolomitic), more rarely polygenetic, with angular, not selected fragments; they are always associated to matrix of varied colours; they show a vuggy fabric commonly characterized by a boxwork structure.

The 'tectonic' facies shows characteristic variation in texture, with a typical tabular geometry and distribution, following a sort of parallel band around fault planes and shear zones. Inside the rock it is possible to recognize some deformation markers (boudin lithons, minor shear zones) as well as non-deformed portions (Fig. 8a & b).

In fact, it is possible to observe in thin-section study the presence of hydrothermal minerals that have grown in the matrix pores (hydrothermal albite, quartz and/or ankerite). At outcrop-scale the hydrothermal activity is underlined by a high concentration of hydrothermal products as infill of the fracture systems involving the pseudocarniole formations (hematite crusts, Fe- and Pb-sulfide veins or other mineralogic associations such as albite, hematite, ankerite, quartz (Fig. 8c) and/or dolomite, quartz, chalcedony and fluorite).

All these characteristics suggest that E facies pseudocarniole are formed and evolved in an environment influenced by tectonic activity, either directly or indirectly. Some pseudocarniole portions can be absolutely considered rock shear zones, with a syn-deformation genesis. Moreover, the genesis of some pseudocarniole would seem to be influenced by chemical processes induced by possible hydrothermal fluids flowing along the structural discontinuities.

Interpretative model

The analysis of stratigraphical, sedimentological and geomorphological characteristics of pseudocarniole pointed out different types of relationships

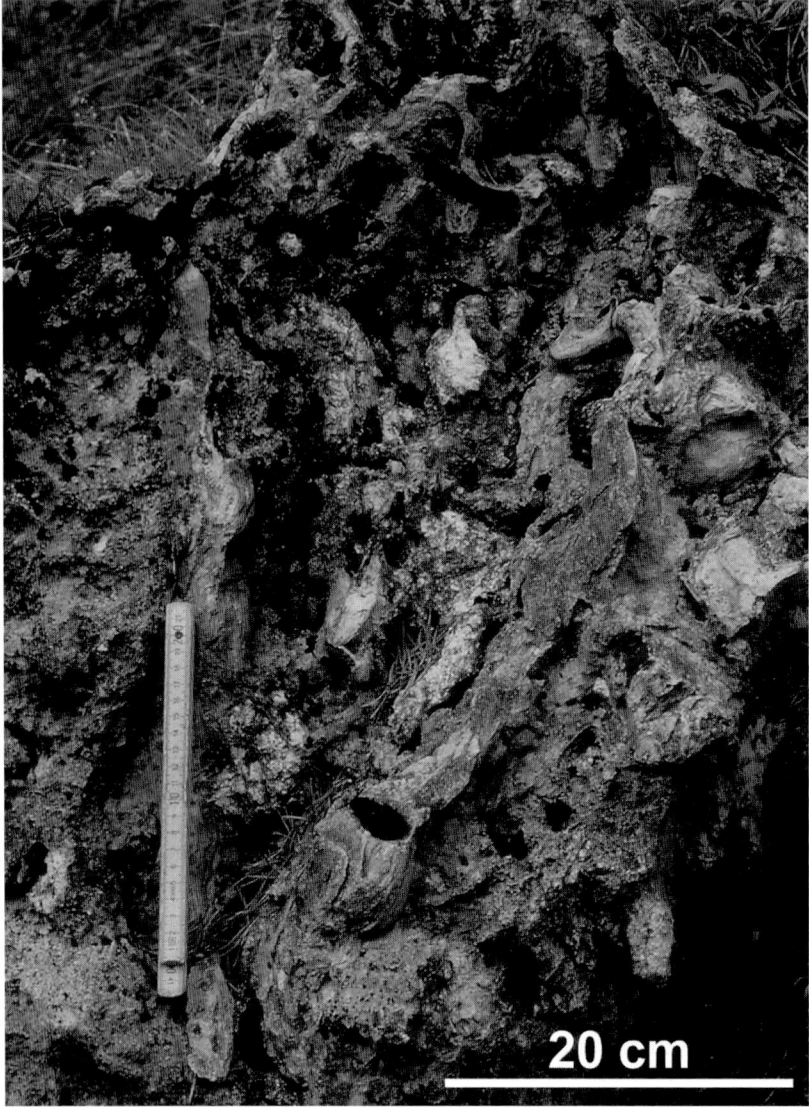

Fig. 7. Encrusting pseudocarniole. Mineral crusts and travertine formation on the aerial exposure surfaces.

between single lithofacies and geological elements and processes:

- with their bedrock (calcareous-dolomitic rock units, i.e. their 'parent rocks');
- with the 'ductile' (plastic) deformation;
- with events of 'brittle' (cataclastic) deformation;
- with fluid flow – ascending (hydrothermal) and descending (karst).

The complex variations of the key features (texture, cementation, corrosion, alteration) observed through different pseudocarniole facies, or inside the same facies, or even in a single sub-facies, testify to a diverse and differential sedimentological/lithological evolution, even in nearby portions of the same rock body.

From a general point of view, considering the chemical and physical modifications responsible for the transformation of carbonate parent rocks into pseudocarniole, the fluids occupy a primary determinant role.

The change of the original gypsum/dolomite/ calcareous rocks into pseudocarniole is carried out through several stages of progressive transformation (Fig. 9) In the first stage, after gypsum/ anhydrite rocks' massive dissolution and

Fig. 8. 'Tectonic' pseudocarniole. (**a**) In the outcrop, local diagenesis of the dolomitic wall along a shear zone; breccia accumulated along some fault planes, altered because of the circulating fluids. (**b**) 'Tectonic pseudocarniole' sample (scale in centimetres); note the partial cementation and the general vacuolar aspect. (**c**) Microscopic view of hematite, ankerite and quartz cements in a tectonic pseudocarniole (thin section, photo taken under crossed polars).

consequent volume reduction, carbonate breccias are formed (carbonate breccias from dissolution): they are genetically connected to the slow re-arrangement of overhanging dolomite and calcareous rocks. The original sedimentary structures are largely preserved.

Subsequently, by means of sulphate-ion rich fluids, induced dissolution extended to include dolomite and calcareous rocks, local reworking of rock masses goes on; an intense leaching of Ca-sulphate by waters could have caused the enrichment in Ca^{2+} and SO_4^{2-} ions in the circulating fluids, resulting in the progressive dissolution of dolomite and the precipitation of calcite associated with dedolomitization (Bischoff et al. 1994; Cañaveras et al. 1996).

Dedolomitization processes induce very important transformation of parent rock and new facies are formed, characterized by progressively stronger internal re-arrangement and by accumulation of residual insoluble products (residual pseudocarniole from dissolution and collapse).

After uplift of the local Alpine structures, and local processes of exhumation and relief formation, a karst landscape forms. The karst flow paths mainly involve the previously formed pseudocarniole. Some rock masses are further reworked by means of transport and sedimentation processes, typical of the hypogenic karst environment, as shown by the close association of sedimentary structures: fill-in channels, spindle-shaped tubes and cavities, laminations and ripples, and water splashmarks on the strata surfaces (detrital pseudocarniole).

Dissolution can be accompanied by other processes (strike–slip or normal faulting, gravitational tectonic) that can act independently, repeatedly and in different moments, with varying intensity. The rocks resulting from these processes are indicated as 'tectonic' pseudocarniole.

The different types of pseudocarniole constitute bodies with different dimensions, geometries and associations, each characterized by key-features left behind by the conditioning processes. Through the pseudocarniole and parent rock outcrops, the Ca-carbonate precipitation from the circulating waters and the surface streams has caused travertine formation with the development of cemented crusts, up to decimetres thick: the 'encrusting pseudocarniole'.

Chronology

Some comments concerning the chronological significance of the Susa Valley pseudocarniole are

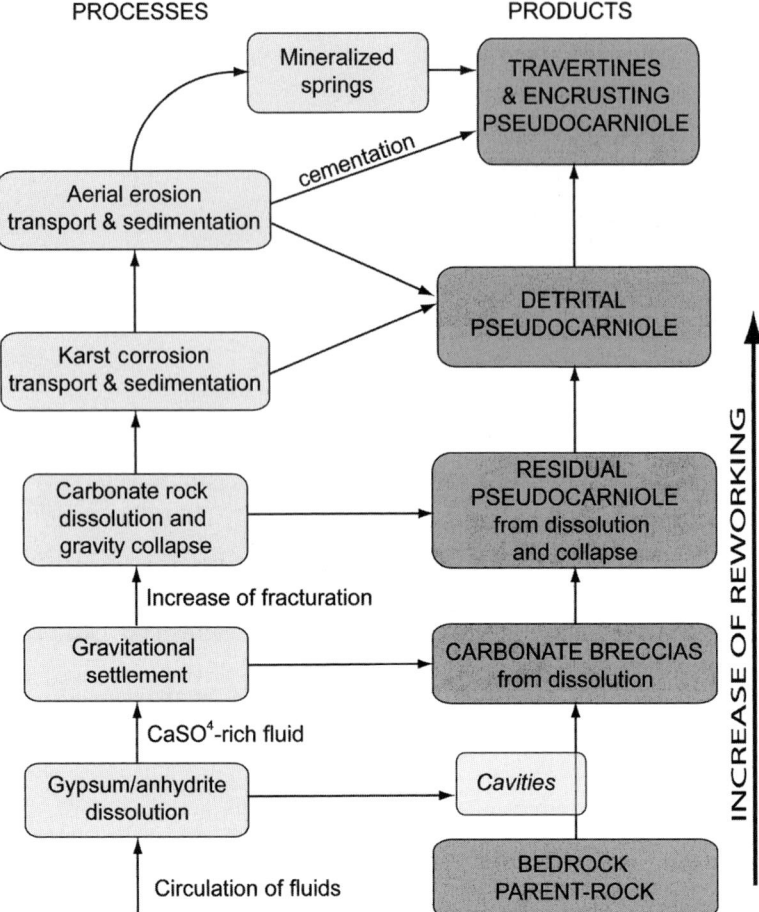

Fig. 9. Outline of possibile processes and products relationships in the proposed model for pseudocarniole genesis.

possible. First of all, the presented data pointed out the objective interpretation problems: at present, there are neither 'time zero' nor significant 'time steps' in the formation process that can be absolutely placed either on climatic and/or tectonic chronologies. Nevertheless, a general framing of the pseudocarniole is possible through indirect dating evidences; for instance the total lack of alpine metamorphism or alpine deformation throughout the pseudocarniole deposits (Alberto 2004; Alberto et al. 2005), the unconformity of pseudocarniole units present on different alpine deformational structures (Polino et al. 2002; Alberto et al. 2005) and the residual pseudocarniole containing clasts of deformed alpine metamorphic rocks (Alberto et al. 2005). The above mentioned evidence represents *post-quem* elements indicating a post-Lower Oligocene age for pseudocarniole.

In this study, some results of more precise dates for pseudocarniole development have been furnished by preliminary pollen analysis (carried out by L. Cattani, Ferrara University) and detailed stratigraphical studies. Collected samples of detrital sub-facies by stream water transport contained pollen of Upper Pliocene–Early Pleistocene arboreal taxa from coastal environments. Moreover, the relationships among the detrital sub-facies by stream waters and natural landforms or man-made structures indicate that the genetic process of pseudocarniole formation took place between the Upper Pliocene and the present. In this way it is possible to ascertain that the genesis of the pseudocarniole does not correspond to the Triassic as the time of formation of 'true' carnioles, but that it might have started much more recently, at the very begining of the recent evolution of the present-day Alpine relief.

Conclusions

After the collection and analysis of the collected data and their comparison with the information from the geological literature, the provisional term of 'pseudocarniole' has been introduced to indicate carbonate breccias and conglomerates of vuggy appearance, in the past interpreted as 'true carniole' of the Susa Valley and assimilated to other alpine rocks of controversial origin and diversified nomenclature.

A classification of the pseudocarniole of the Susa Valley (Western Alps) has been proposed. This is articulated in five major facies, and further sub-facies, the divisions depending on their textural, structural and geometrical characteristics. The genesis of every facies has been inserted in a general evolutional model starting from the Triassic carbonates (particularly dolomitic) and evaporite successions, that are widely distributed in the Western Alps.

The genesis of pseudocarnioles is connected to the interaction and to the superimposition of various processes. The dissolution and alteration of carbonate rocks are induced by a sulphate-rich fluid, which apparently acts as the main mechanism. The dissolution induces a general instability and disaggregation of the bedrock with a consequent gravitational settlement. Other processes are also associated with dissolution; these are connected to tectonics (extensional or strike–slip faults), gravity and karst development.

Because of all these many processes, many types of pseudocarniole have been described, which constitute rock bodies with different dimensions, geometries and lithofacies associations. Finally, a particularly young age for the pseudocarniole facies have been suggested: Plio-Quaternary, based on palynological data.

References

ALBERTO, W. 2004. Le *'pseudocarniole' e il loro contesto geologico e geomorfologico: esempi dalla Val Thuras e da altre località della Valle di Susa*. Ph.D thesis, University of Torino.

ALBERTO, W., CARRARO, F., GIARDINO, M. & TIRANTI, D. 2005. Proposta di classificazione delle 'pseudocarniole' dell'alta Valle di Susa (Alpi Occidentali). *Il Quaternario–Italian Journal of Quaternary Sciences*, **18**, 187–200.

AMIEUX, P. & JEANBOURQUIN, P. 1989. Cathodoluminescence et origine diagénétique tardive des cargneules du massif des Aiguilles Rouges (Valais, Suisse). *Bulletin Société Géologique de France*, **8**, 123–132.

BISCHOFF, J. L., JULIÀ, R., SHANKS, W. C. & ROSENBAUER, R. J. 1994. Karstification without carbonic acid: Bedrock dissolution by gypsum-driven dedolomitization. *Geology*, **22**, 995–998.

BRUCKNER, W. 1941. Uber die Entstehung der Rauhwacken und Zellendolomite. *Eclogae Geologiae Helvetiae*, **34**, 117–134.

CAÑAVERAS, J. C., SANCHEZ-MORAL, S., CALVO, J. P., HOYOS, M. & ORDONEZ, S. 1996. Dedolomites associated with karstification. An example of early dedolomitization in lacustrine sequences from the Tertiary Madrid Basin, Central Spain. *Carbonates and Evaporites*, **II**, 85–103.

CARRARO, F., CADOPPI, P., CASTELLETTO, M., SACCHI, R., BAGGIO, P. & GIRAUD, V. 2002. Foglio 154 'Susa'. Carta Geologica d'Italia, scala 1:50.000. Servizio Geologico d'Italia.

CIARAPICA, G. 1985. Il Trias dell'Unità di Porto Venere e confronti con le coeve successioni apuane e toscane: revisione degli 'strati a R. contorta' Auctt. dell' Appennino Settentrionale. *Memorie della Società Geologica Italiana*, **30**, 135–151.

CIARAPICA, G., CIRILLI, S., PASSERI, L., TRINCIANTI, E. & ZANINETTI, L. 1987. 'Anidriti di Burano' et 'Formation du Monte Cetona' (nouvelle formation), biostratigraphie de deux series-types du Trias superieur dans l'Apennin Septentrional. *Revue de Paléobiologie*, **6**, 341–409.

D'ARGENIO, B. & FERRERI, V. 1987. A brief outline of sedimentary models for Pleistocene travertine accumulation in Southern Italy. *Rendiconti Società Geologica Italiana*, **9**, 167–170.

DEBELMAS, J., GIDON, M. & KERCKHOVE, C. 1980. Idées actuelles sur les cargneules alpines. Livre jubilaire Jacques Flandrin. *Documents des Laboratoires de Géologie de Lyon*, **4**, 195–201.

DEBENEDETTI, A. & TURI, B. 1975. Carniole della Valle d'Aosta studio isotopico ed ipotesi genetica. *Bollettino della Società Geologica Italiana*, **94**, 1883–1894.

ELLENBERGER, F. 1954. Réunion extraordinaire de la Société Géologique de France en Maurienne et Tarentaise (Savoie). *Compte Rendu Sommaire des séances de la Société Géologique de France*, 454–458.

FEDERICI, P. R. & RAGGI, G. 1974. Brecce sedimentarie e rapporti tra le unità tettoniche toscane nel gruppo delle Alpi Apuane. *Bollettino della Società Geologica Italiana*, **92**, 435–452.

FUDRAL, S. 1998. Etude géologique de la suture téthysienne dans les Alpes franco-italiennes nord occidentales de la Doire Ripaire (Italie) à la région de Bourg Saint-Maurice (France). *Géologie Alpine, Mémoire*. **29**, 1–306.

GANDIN, A., GAIMELLO, M. & GUASPARRI, G. 2000. The Calcare Cavernoso of the Montagnola Senese (Siena, Italy): mineralogical–petrographic and petrogenetic features. *Mineralogica et Petrographica Acta*, **XLIII**, 271–289.

GIARDINO, M. 1995. *Analisi di deformazioni superficiali: metodologie di ricerca ed esempi di studio nella media Valle d'Aosta*. Ph.D thesis, University of Torino.

GRANDJACQUET, C. & HACCARD, D. 1973. Mise en évidence de la nature sédimentaire et de l'age néogène de certaines séries de 'cargneules et gypses' des chaines subalpines meridionales; implications structurales. *Comptes Rendus de l'Académie des Sciences, Paris*, **276**, 2369–2372.

GRANDJACQUET, C. & HACCARD, D. 1975. Analyse des sédiments polygéniques néogènes à faciès de cargneules associés à des gypses dans les Alpes du Sud. Extension de ces faciès au pourtour de la Méditerranée occidentale. *Bulletin Société Géologique de France*, **7**, 242–259.

JEANBOURQUIN, P. 1988. Nouvelles observations sur les cornieules en Suisse occidentale. *Eclogae Geologiae Helvetiae*, **81**, 511–538.

LEINE, L. 1971. Rauhwacken und ihre Entstehung. *Geologischen Rundschau*, **60**, 488–524.

MCWHAE, J. R. H. 1953. The Carboniferous breccias of Billefjorden, Vestspitsbergen. *Geological Magazine*, **95**, 287–298.

MALAVIEILLE, J. & RITZ, J. F. 1989. Mylonitic deformation of evaporites in décollements: examples from the Southern Alps, France. *Journal of Structural Geology*, **11**, 583–590.

MASSON, H. 1972. Sur l'origine de la cornieule par fracturation hydraulique. *Eclogae Geologiae Helvetiae*, **65**, 27–41.

MÜLLER, W. H. 1982. Zur Entsthung der Rauhwacke. *Eclogae Geologiae Helvetiae*, **75**, 481–494.

NURY, D. & SCHREIBER, B. C. 1997. The Paleogene Basins of Southern Provence. *In*: BUSSON, G. & SCHREIBER, B. C. (eds) *Sedimentary Deposition in Rift and Foreland Basins in France and Spain (Paleogene and Lower Neogene)*. Columbia University Press, New York, 240–300.

PASSERI, L. 1975. L'ambiente deposizionale della Formazione Evaporitica nel quadro della paleogeografia del Norico Tosco–Umbro–Marchigiano. *Bollettino della Società Geologica Italiana*, **94**, 231–268.

PATACCA, E., RAU, A. & TONGIORGI, M. 1973. Il significato geologico della breccia sedimentaria poligenica al tetto della successione metamorfica dei Monti Pisani. *Atti Società Toscana Scienze Naturali*, **A80**, 126–161.

POLINO, R., DELA PIERRE, F., FIORASO, G., GIARDINO, M. & GATTIGLIO, M. 2002. *Foglio 132–152–153 'Bardonecchia'*. *Carta Geologica d'Italia, scala 1:50.000*. Servizio Geologico d'Italia.

RICHARDS, M. T. & VEARNCOMBE, J. R. 1984. Sur l'origine des faciès de cargneules et de brèches rencontrés dans une coupe du Trias de la Vésubie, Alpes-Maritimes, France. *Géologie Méditerranée*, **11**, 283–286.

SCHAAD, W. 1995. Die Entstehung von Rauhwacken durch die Verkastung von Gips. *Eclogae Geologiae Helvetiae*, **88**, 59–90.

SCHREIBER, B. C. & HELMAN, M. L. 2005. Criteria for distinguishing primary evaporite features from deformation features in sulphate evaporites. *Journal Sedimentary Research*, **75**, 525–533.

WARRACK, M. 1974. The petrology and origin of dedolomitised, veined or brecciated carbonate rocks, the 'cornieules' in the Fréjus region. French Alps. *Journal of the Geological Society, London*, **130**, 229–247.

WARREN, J. K. 1999. *Evaporites. Their Evolution and Economics*. Blackwell Science, Oxford.

WIESENDER, H. 1971. Klassifikation und Entstehung terrigener und karbonatischer Sedimentgesteine. *Mitteilungen Geologische Gesellschaft*, Wien, **64**, 219–236.

ZACCAGNA, D., MATTIROLO, E. & FRANCHI, S. 1911a. *Carta Geologica d'Italia alla scala 1:100.000, Foglio 54–Oulx*. Servizio Geologico d'Italia.

ZACCAGNA, D., MATTIROLO, E. & FRANCHI, S. 1911b. *Carta Geologica d'Italia alla scala 1:100.000, Foglio 66–Cesana Torinese*. Servizio Geologico d'Italia.

Messinian halite and residual facies in the Crotone basin (Calabria, Italy)

S. LUGLI[1], R. DOMINICI[2], M. BARONE[2], E. COSTA[3] & C. CAVOZZI[3]

[1]*Dipartimento di Scienze della Terra, Università degli Studi di Modena e Reggio Emilia, Largo S. Eufemia 19, 41100 Modena, Italy (e-mail: lugli.stefano@unimore.it)*

[2]*Università degli Studi della Calabria, Dipartimento di Scienze della Terra, 87036 Arcavacata di Rende (Cosenza), Italy*

[3]*Dipartimento di Scienze Geologiche, Università di Parma, Parco Area delle Scienze 157/A, 43100, Parma, Italy*

Abstract: The Neogene Crotone basin in eastern Calabria contains extensive Messinian evaporite deposits, including thick gypsarenite and halite. The halite deposit reaches a maximum thickness of *c.* 300 m and in some areas forms relatively small diapirs piercing late Messinian and Pliocene sediments. Halite is strongly modified by folding and recrystallization, but a few primary features are preserved. Four primary halite facies have been recognized: (a) banded halite consisting of folded white and dark bands deposited in a salt pan and/or saline mudflat; (b) white facies, massive halite containing anhydrite nodules, probably formed in a variably desiccating saline lake; (c) clear facies made up of a mosaic of large blocky halite crystals separated by mud, possibly the product of displacive halite growth in a saline mudflat; and (d) breccia facies, a product of dissolution of halite/mudstone/siltstone layers. Residual facies formed from halite dissolution are present as both weld and cap rocks. Weld rocks are thick, undeformed, and composed only of insoluble phases originally included in the salt, whereas cap rocks are thin, strongly sheared and include clasts from the cover rocks.

As one of the less investigated deposits formed during the Messinian Salinity Crisis in the Mediterranean, halite was deposited in the Crotone basin, Calabria (south Italy). The extent of the halite deposit is well known, based on the early works of the 1960s that were produced mostly for mineral extraction purposes (Roda 1964), but no detailed description of the halite facies has ever been produced. Since then, our understanding of the Messinian Salinity Crisis has largely improved and facies analysis of the evaporites provided us with new information on their environmental evolution. Although the significance of the Mediterranean halite deposits in the Salinity Crisis framework is not yet well constrained (Manzi *et al.* 2006), the study of halite facies in Sicily revealed desiccation surfaces (Lugli *et al.* 1999, 2006) and significant constraints on their depositional environments (Lugli & Lowenstein 1997).

This paper documents the halite facies and its dissolution products in the Crotone basin. The scope of this study is to provide new insights on the complexity of the Messinian Salinity Crisis and to provide criteria for the recognition of halite dissolution products that may be useful in tracing elusive weld surfaces.

Geological setting and stratigraphy

The Crotone basin is located on the northeastern margin of the Calabrian arc and represents the filling of the wedge-top basin of the southern Italy foreland basin system (DeCelles & Giles 1996; Critelli 1999; Fig. 1). Sedimentation in this basin began in the Serravallian?–Tortonian with deposition of a shallow-marine arenaceous– conglomerate succession (San Nicola Fm) passing upward to marine argillaceous deposits (Ponda Fm, Tortonian) and organic-rich laminites (Tripoli Fm, Tortonian–Messinian).

During the Messinian the sedimentation in the area was dominated by evaporitic conditions producing sulphate deposits and halite. In the Crotone basin such conditions are recorded mostly by halite (maximum thickness 300 m) and by extensive clastic evaporite deposits, such as a thick gypsarenite body (up to 120 m thick), with minor gypsrudite and limestone breccias.

These evaporite deposits are associated with a sedimentary chaotic complex (SCC), consisting of metric blocks of limestone and alabastrine gypsum included in an argillaceous matrix. A late Messinian succession consisting of gypsarenites, sandstones and pelites with a Lago Mare fauna at

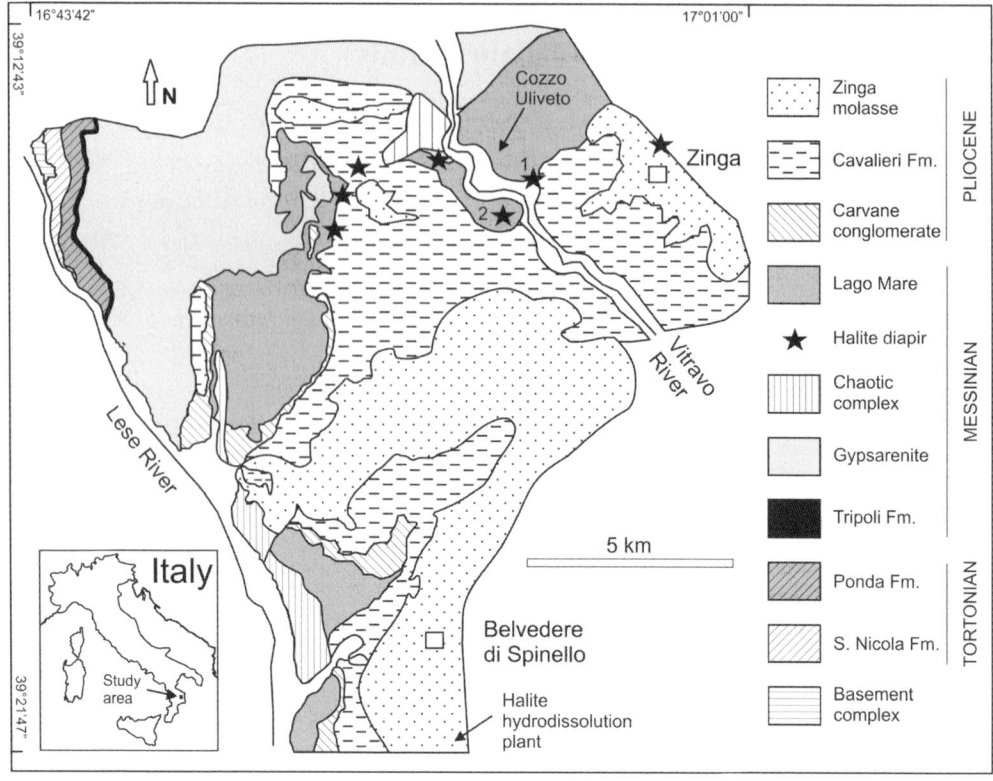

Fig. 1. Schematic geological map of the Zinga-Belvedere di Spinello area of the Crotone basin, eastern Calabria.

the top (Gennari & Iaccarino 2004) onlaps both the chaotic complex and the gypsarenites. An erosional surface separates these latest Messinian deposits from the overlaying fluvial conglomerates (Carvane conglomerates), which are in turn abruptly covered by the off-shore Cavalieri marls (Early Pliocene). The marls grade upward into the shallow-marine Zinga Molasse deposits (Pliocene; Roda 1964; Zecchin et al. 2004).

The original stratigraphic relationships of halite with other evaporite sediments (the main gypsarenite body) in the Crotone basin is unclear at present due to diapiric condition of the salt. Seismic and well data from salt mining areas (Belvedere di Spinello; Roda 1964) show that halite is interposed between organic-rich laminites similar to the Tripoli Fm at the base and latest Messinian gypsarenites, arenites and pelites on the top. A few boreholes show that in some areas halite probably lay on top of the Chaotic Complex. Unfortunately these cores are no longer available for scientific examination. Further investigations and planned new cores will help to reconstruct the stratigraphic position of salt.

The investigated diapirs are located along N 40°- and N 80°-oriented morphostructural highs and pierce the late Messinian and early Pliocene deposits (Fig. 1).

Materials and methods

To describe the halite facies we directly sampled diapiric and wall structures in the Zinga village area (Fig. 1). In addition, we also examined a few remaining samples from cores drilled in the past in the salt mining areas (Belvedere di Spinello, Crotone). Detailed provenance and stratigraphic position of these samples are not known and we considered them just for the purpose of facies description completion.

For thin section preparation, halite samples were cut into $4 \times 5 \times 3$ cm slabs using a diamond saw with a small amount of water to prevent formation of microcracks and significant dissolution artefacts (Schléder & Urai 2005). The slabs were polished dry on grinding paper, mounted on glass plates using epoxy resin and then cut with a wet diamond saw to a thickness of about 5 mm. The sections were then ground down with dry grinding paper to a thickness ranging from less than 1 to about 2 mm, depending on the specific sample. Residual samples were impregnated in epoxy resin (araldite) and then prepared using the standard techniques for thin section production.

Halite facies: description

In diapiric structures halite clearly shows the action of moderate to high plastic deformation and flow: sheared marl clasts, crude imbrication of marl clasts, isoclinal folds, shear folds (Fig. 2) and alignment of halite crystals (Fig. 3). Four distinct original (primary) halite facies were recognized.

Banded facies

This is a folded alternance of centimetre- to decimetre-thick dark and white bands (Fig. 2); the white bands consist of up to 5 mm elongate clear crystals, whereas the dark bands contain variable amounts of non-halite clasts; the clasts are up to 10 cm in size and consist of mudstones, contorted mudstone/anhydrite laminites and graded siliciclastic-carbonate siltstones. The halite crystals are free of primary fluid inclusions (Fig. 3).

White facies

This is a massive halite rock formed by large, equant, blocky crystals (up to 5 cm in size) containing sulphate nodules up to 1 cm in size (Fig. 4); the halite crystals are clear but most of them contain a whitish core rich in fluid inclusions. Some of these cores show halite with chevron-like shapes (marked by fluid inclusions, Fig. 4). Anhydrite nodules and thin films of mud and microcrystalline anhydrite are located at the grain boundaries and within the clear overgrowth zone of the large halite crystals but are not present in the fluid inclusion-rich core. These characteristics suggest that the fluid inclusion-rich milky core represent remnants of primary halite crystals, which underwent rim recrystallization to form a clear overgrowth by grain boundary migration (Schléder & Urai 2005).

Clear facies

This rock is composed of a mosaic of large blocky irregular and elongate crystals (up to 5 cm) separated by thin films of mud (Fig. 5). This facies has been documented only in cores. The zones with thicker mud partings (a few millimetres across) contain small displacive halite cubes and hexagonal shapes filled by microcrystalline anhydrite; some of the large halite crystals have fluid inclusion-rich cores and thick, clear, overgrowths. The clear

Fig. 2. Salt diapir cropping out along the Vitravo valley (*1 in Fig. 1). The diapir pierced the marls of the Pliocene Cavalieri Fm and consists of the banded halite facies. White bands are pure halite, dark bands contain variable proportions of mudstone clasts. Notice the relatively thin cap rock cut by satin spar veins on top of the salt.

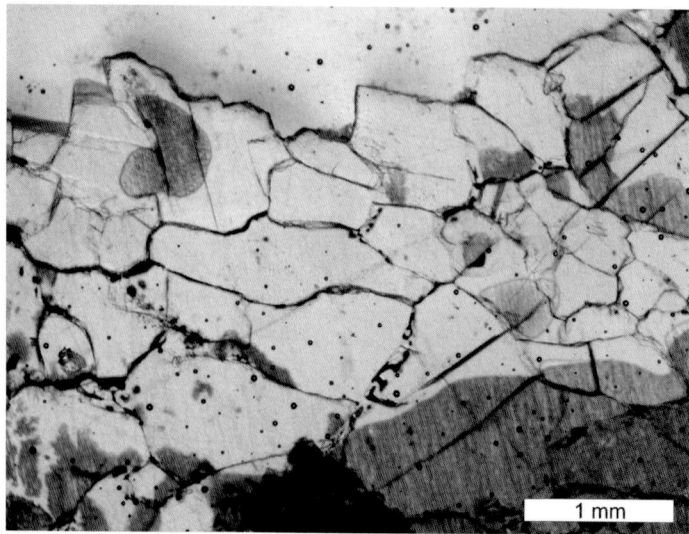

Fig. 3. Photomicrograph showing the white part of the banded halite facies. Crystals are elongated, fluid inclusion-free and meet at triple junctions with angles approaching 120°, suggesting complete recrystallization. Plane polarized light. Salt diapir, ∗ 1 of Figure 1 (see also Fig. 2).

rims commonly include growth bands marked by thin mud sheets, suggesting a displacive overgrowth in the mud after initial mud-free growth (Fig. 6).

Breccia facies

This rock is composed of clear irregular halite crystals (a few centimetres across) cementing decimetric clasts of graded laminites of siliciclastic–carbonate siltstone/mudstone/anhydrite; in the larger clasts the mudstone contains small displacive halite cubes and nodular anhydrite (Fig. 7); this rock has been observed only in cores and probably represents the result of early dissolution collapse of the interbedded halite/mudstone/siltstone layers.

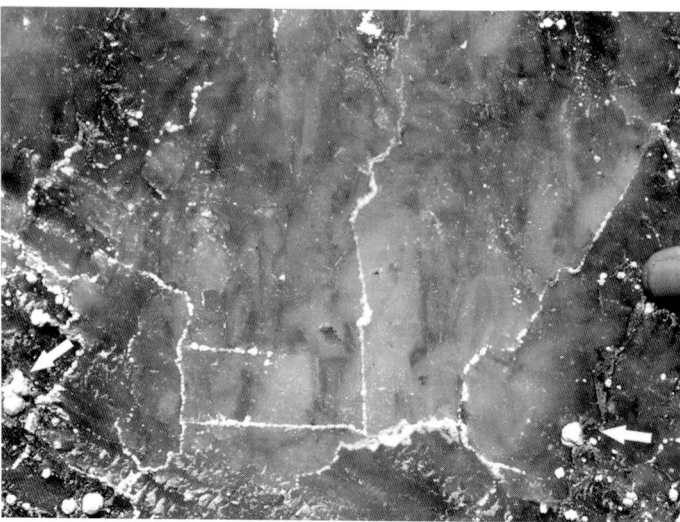

Fig. 4. Outcrop view of the white halite facies (diapir ∗ 2 in Fig. 1); large equant blocky crystals contain a whitish core marked by fluid inclusions. Some of these cores evidence chevron-like shapes. Several anhydrite nodules are present (arrows). Thin white lines crossing the rock are ephemeral salt efflorescence crusts grown along fractures. Finger for scale.

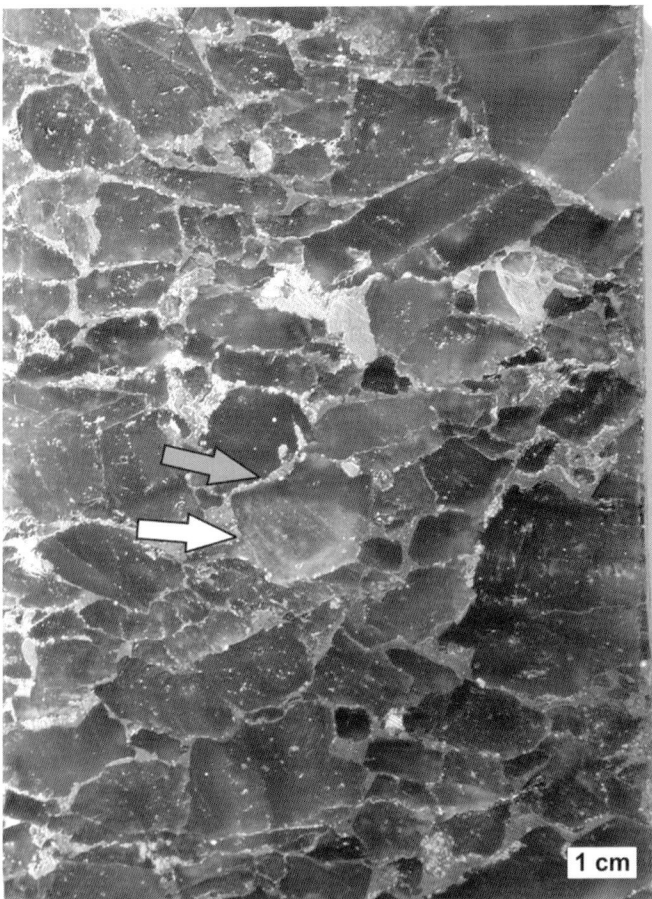

Fig. 5. Clear halite facies. Core slab consisting of a mosaic of large irregular and elongated crystals, separated by mud. Some crystals show a milky core marked by fluid inclusions (white arrow) and a clear overgrowth (gray arrow) (see also Fig. 4). Belvedere di Spinello hydrodissolution plant, stratigraphic position of sample unknown.

Halite facies: interpretation

Although most salt is deformed by flow and recrystallization, a few primary features that may help to reconstruct the original sedimentary setting are still preserved in some of these halite rocks (Shearman 1970; Hardie & Lowenstein 1985; Handford 1991). The original facies were massive halite and halite/mud/anhydrite sequences.

The massive facies (white facies) may represent former primary halite crusts cut by dissolution pits that were then filled by precipitation of clear halite cement. This facies possibly represents deposition in a saline lake which underwent stages of desiccation to form a dry salt pan (Lowenstein & Hardie 1985).

The banded facies can be interpreted as the result of flow, disruption and complete recrystallization of former alternance of pure halite (white bands) and sequences of halite/mudstone/siltstone laminites (dark bands). Deformation and recrystallization completely obliterated the primary features of the halite. The presence of mud and anhydrite laminite intercalations, however, suggests repeated flooding and dilution of the salt basin by undersaturated, muddy, waters. Although no specific sedimentary features are preserved in this type of halite, with the exception of dissolution breccias (breccia facies), the typical halite/mud alternation strongly suggests deposition in a salt pan and a saline mudflat (Lowenstein & Hardie 1985). The presence of these thin beds of graded siliciclastic siltstone may be the product of sheet floods on both/either a salt flat or a mudflat.

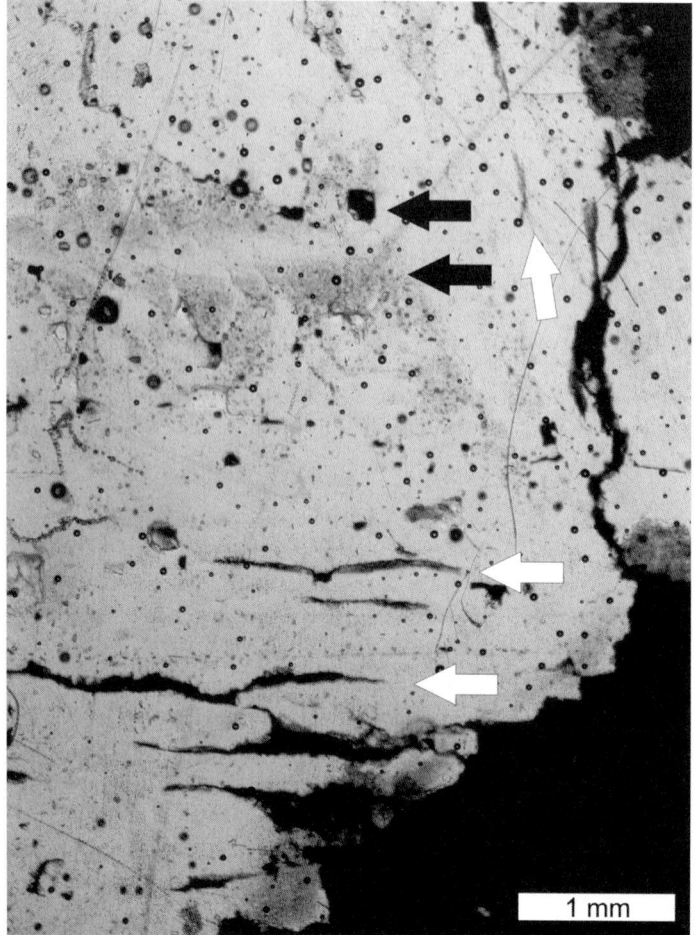

Fig. 6. Clear halite facies of Figure 5 under the optical microscope. The core of the halite crystal contains primary fluid inclusion marking growth bands (black arrows), but the clear further overgrowth is inclusion-free and contain growth bands marked by thin sheets of mud (white arrows). These characteristics suggest a displacive growth into the mud after an initial subaqueous growth. Plane polarized light. Belvedere di Spinello hydrodissolution plant, stratigraphic position of sample unknown.

The clear halite facies probably formed from subaqueous primary crystals (either cumulate or bottom nucleated) that underwent further displacive overgrowth within the mud, together with minor amounts of gypsum, in a saline mudflat.

Residual facies: description

Both weld rocks (Jackson & Cramez 1989) and cap rocks crop out in the studied area allowing for the first time, a direct comparison of their macroscopic and microscopic features. Weld rocks mark removal horizons from which salt flowed, whereas cap rocks represent horizons of salt dissolution at the diapir crest. For this reason, the two horizons differ in position within the sedimentary sequence and in lateral continuity, as well as in their composition.

Cap rocks

In the studied area the halite diapirs pierce the latest Messinian and the Pliocene sediments and are characterized by thin cap rocks (maximum thickness is about 2 m) composed of a silty–marly matrix cemented by prismatic gypsum crystals (Fig. 8) containing anhydrite and gypsum nodules and rosettes and including blocks of sandstone, marls and gypsarenites (Fig. 2). Also, in some places pebbles and cobbles from the Carvane Conglomerate Fm are included within the cap rock matrix. The cap rocks are deformed, sheared and cut by satin-spar veins.

Fig. 7. Photomicrograph of the breccia halite facies showing hollow halite crystals grown displacively into the mud together with gypsum now replaced by anhydrite (hexagonal shape at left). Plane polarized light. Belvedere di Spinello hydrodissolution plant, stratigraphic position of sample unknown.

Weld rocks

A weld horizon is located in between the SE-verging anticline and the northernmost diapirs, at Cozzo Uliveto (Fig. 1). It is represented by a chaotic complex with a maximum visible thickness of about 10 m, lying just below the latest Messinian Lago Mare sediments (Gennari & Iaccarino 2004).

This deposit is composed of prismatic gypsum crystals, anhydrite and gypsum nodules, rosettes and gypsarenite clasts floating in a marl matrix (Fig. 9). The gypsum nodules formed by hydration of anhydrite nodules and are similar to those included in halite (Figs 9 & 4). In contrast to gypsum nodules, which include anhydrite relics, however, prismatic gypsum is relic-free, suggesting

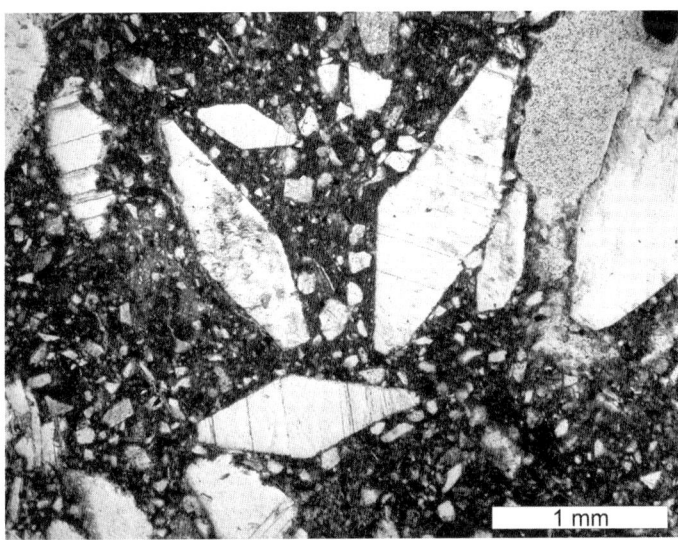

Fig. 8. Photomicrograph of cap-rock residual facies of the diapir of Figure 2. Idiomorphic gypsum crystals into a fine-grained matrix. The crystals are devoid of anhydrite relics suggesting direct displacive growth into the matrix. This facies also contains Pliocene foraminifera. Plane polarized light.

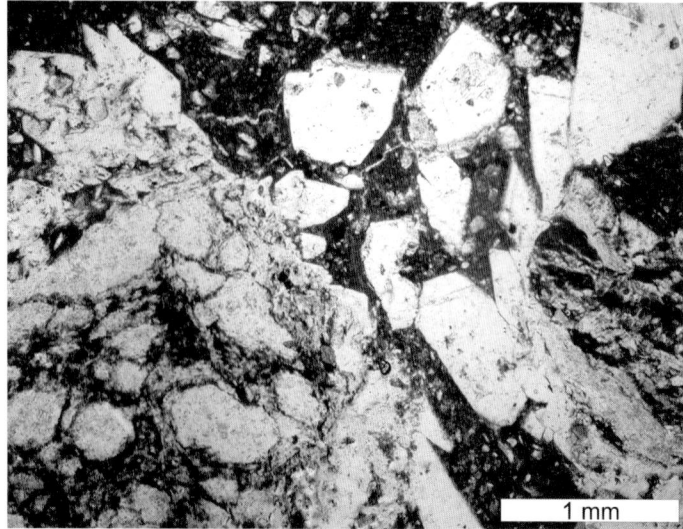

Fig. 9. Photomicrograph of the weld-rock at Cozzo Uliveto. Gypsum nodules (left and right) and euhedra (centre) included into a fine-grained matrix. The nodules appear light grey-coloured due to the presence of numerous anhydrite relics, indicating that they formed by hydration of anhydrite nodules originally included into the vanished halite. The idiomorphic gypsum crystals are devoid of anhydrite relics and grew displacively into the matrix. Plane polarized light.

a pristine displacive growth (*de novo*) into the matrix (Fig. 9).

Residual facies interpretation: salt weld vs cap rock

Welding surfaces may be caused by passive sagging of the salt sedimentary cover, by normal faulting and/or associated rollover synclines. They mark a hiatus lasting the duration of salt deposition. The weld faces are located between the basement and cover rocks (the sub-salt and supra-salt rocks, respectively).

The composition of weld-rocks is dominated by the insoluble phases originally inter-layered in the salt and left behind when the salt flows away. Weld rocks may also contain the lowermost parts of the cover collapsed as result of salt withdrawal. By contrast, cap rocks may contain portions of the cover rocks intruded by diapirs at various stratigraphic levels, which form the insoluble phases together with the insoluble residue originally stratified within the vanished salt.

Weld horizons play an important role in petroleum geology, as they allow hydrocarbon migration from the sub-salt to the cover rocks. They are often easily recognized in calibrated seismic sections, even when they separate geometrically conformable sequences. It is more difficult to recognize weld horizons in the field, where few markers remain after the salt flowed out from under its overlying load.

The main features distinguishing weld from cap rocks in the Crotone basin are their thickness, composition and deformation pattern. Weld rocks are much thicker (possibly more than 10 m in thickness), contain only insoluble phases that were included in the salt, and appear relatively undeformed. In contrast, cap rocks are thin (up to 2 m-thick), include clasts from the latest Messinian and Pliocene cover rocks and are deeply deformed by shearing between the rising diapir and the pierced cover (Fig. 2).

Late gypsum cement

Associated with evaporites in the Crotone Basin, but not related to halite dissolution, are fine-grained deposits that appear similar in general aspect and composition to the residual sediments. These deposits consist of an argillaceous matrix including prismatic and lenticular gypsum crystals. No anhydrite relics are present and no other insoluble residue after halite dissolution could be recognized. Therefore, caution must be used in interpreting fine-grained sediments including sulphate crystals as products of halite dissolution. This is because late displacive prismatic gypsum is relatively common in argillaceous formations surrounding the many Italian gypsum deposits of both Triassic (Lugli

2001) and Messinian age (Testa & Lugli 2000). These fine-grained sediments, cemented by gypsum, are always located in the vicinity of existing sulphate deposits and do not contain residual remnants of halite dissolution such as mudstones and carbonate clasts or anhydrite nodules (and their hydration products). These gypsum crystals are not related to halite dissolution, but are late precipitates from gypsum-saturated groundwater seeping through fine-grained sediments after flowing into sulphate outcrops.

General discussion

The array of salt-related facies that have been recognized in the Crotone basin is the product of depositional features that have been modified at various level of complexity by shearing, flow, recrystallization and, finally, dissolution. The main depositional facies that have been observed point to a saline basin that evolved into a saltpan and a saline mudflat through repeated episodes of flooding and desiccation.

After burial, it seems that the salt began moving very early as a consequence of an early Pliocene extension phase. Salt flowed away from the northernmost area (downbuilding, *sensu* Barton 1933; Jackson *et al.* 1988; Jackson 1997) producing weld rocks and forming diapirs in the adjacent sector further south. In the diapirs the salt flows cause varying degrees of destruction of the primary features of the salt facies so that they became strongly modified with complete recrystallization and loss of primary relics (banded facies). The only possibility to read through and try to reconstruct the primary depositional environment in this case is studying the associated sediments that are intercalated in the salt.

Dissolution in the sub-surface and at surface exposure finally produced the residual cap rock facies that commonly appear very similar to weld rocks. This is because the main component of both weld and cap rocks are the non-salt components originally intercalated in the salt, such as anhydrite nodules and anhydrite/mudstone layers. In the case of weld rocks, the non-salt elements are left behind by the migrating salt, whereas in the case of cap rocks they are accumulated as the insoluble residue of salt dissolution. However, distinguishing the two type of rocks after halite has vanished is possible by considering the different mechanism of formation and the relative stratigraphic position. As a rule, cap rocks also contain clasts deriving from the sedimentary cover that were sheared and pierced by the diapirs, whereas weld rocks are devoid of rock fragments younger than the salt itself.

Conclusions

In this paper we describe and consider the halite facies of the Crotone Basin and their associated residual products. The halite is strongly modified by folding and recrystallization, but some preserved features reveal clues on its origin. Four quasi-primary halite facies were recognized:

(1) banded facies consisting of pure halite and halite including mudstone and anhydrite clasts; this facies was deposited in a salt pan and a saline mudflat;
(2) white facies of massive halite containing anhydrite nodules, probably deposited in a saline lake which underwent phases of desiccation;
(3) clear facies composed of coarse blocky halite separated by thin mud stringers, possibly the product of displacive halite overgrowth in a saline mudflat after initial subaqueous nucleation; and
(4) breccia facies, a dissolution collapse product of halite/mudstone/siltstone layers.

Halite flow and dissolution also produced two residual facies: weld and cap rocks. Criteria for the recognition of weld rocks and cap rocks in the Crotone Basin are thickness, composition and deformation pattern. Weld rocks are thick (possibly more than 10 m), contain only the insoluble phases included in the salt and appear relatively undeformed. In contrast, cap rocks are thin (up to 2 m-thick), include clasts from the cover rocks and are strongly sheared.

This study benefited from the kind cooperation offered by the staff of the Syndial hydrodissolution plant (Belvedere di Spinello): L. Musso, C. Fortugno U. Rizzo. This research was founded by a Cofin grant (E. Costa).

References

BARTON, D. C. 1933. Mechanics of formation of salt domes, with special reference to Gulf Coast salt domes of Texas and Louisiana. *Bulletin of the American Association of Petroleum Geology*, **17**, 1025–1083.

CRITELLI, S. 1999. The interplay of lithospheric flexure and thrust accomodation in forming stratigraphic sequences in the Southern Apennines foreland basin system, Italy. *Accademia Naturale dei Lincei, Rendiconti Lincei Scienze Fisiche e Naturali*, serie IX, **10**, 257–326.

DECELLES, P. G. & GILES, K. A. 1996. Foreland basin system. *Basin Research*, **8**, 105–123.

GENNARI, R. & IACCARINO, S. 2004. An overview of the Messinian 'Lago-Mare' paleontological record. *The Messinian Salinity Crisis Revisited*, 4th International congress 'Environment and Identity in the

Mediterranean 2004', Corte, 19–25 July, 2004. Abstract volume, 43.

HANDFORD, C. R. 1991. Marginal marine halite: sabkhas and salinas. *In*: MELVIN, J. L. (ed.) *Evaporites, Petroleum and Mineral Resources Developments in Sedimentology*, **50**, Elsevier, Amsterdam, 1–66.

HARDIE, L. A., LOWENSTEIN, T. K. & SPENCER, R. J. 1985. The problem of distinguishing between primary and secondary features in evaporites. *In*: SCHREIBER, B. C. & HARNER, H. L. (eds). *Sixth International Symposium on Salt*. Salt Institute, Alexandria VA, **1**, 11–39.

JACKSON, M. P. A. 1997. *Conceptual Breakthroughs in Salt Tectonics: a Historical Review, 1856–1993*. Bureau of Economic Geology, Report on Investigations, **246**.

JACKSON, M. P. A. & CRAMEZ, C. 1989. Seismic recognition of salt welds in salt tectonics regimes. *10th Annual Research Conference Gulf Coast Section, Program and Abstracts*. SEPM Foundation, Houston, TX, 66–80.

JACKSON, M. P. A., TALBOT, C. J. & CORNELIUS, R. R. 1988. *Centrifuge Modeling of the Effects of Aggradation and Progradation on Syndepositional Salt Structures*. Bureau of Economic Geology, Report on Investigations, **173**.

LOWESTEIN, T. M. & HARDIE, L. W. 1985. Criteria for the recognition of salt-pan evaporites. *Sedimentology*, **32**, 627–644.

LUGLI, S. 2001. Timing of post-depositional events in the Burano Formation of the Secchia Valley (Upper Triassic, northern Apennines), clues from gypsum–anhydrite transitions and carbonate metasomatism. *Sedimentary Geology*, **140**(1–2), 107–122.

LUGLI, S. & LOWENSTEIN, T. K. 1997. Paleotemperatures preserved in fluid inclusions in Messinian halite, Realmonte Mine (Agrigento, Italy). *Neogene Mediterranean Paleoceanography*, 28–30 September 1997, Erice. Abstract volume, 44–45.

LUGLI, S., SCHREIBER, B. C. & TRIBERTI, B. 1999. Giant polygons in the Realmonte mine (Agrigento, Sicily): evidence for the desiccation of a Messinian halite basin. *Journal of Sedimentary Research*, **69**, 764–771.

LUGLI, S., DI STEFANO, A., GULFI, V., CORALLO, G., AMENTA, G., MENEGON, S. & MANZI, V. 2006. Halite facies in the Racalmuto mine (Agrigento): further evidence of an exposure surface in the Messinian salt of Sicily. *'The Salinity Crisis Revisited', Interim Colloquium R.C.M.N.S.*, Parma, 7–9 September 2006. Abstract volume.

MANZI, V., LUGLI, S. *ET AL*. 2006. Clastic versus primary precipiteted evaporites in the Messinian Sicilian basins. 'The Messinian salinity crisis revisited', Interim Colloquium R.C.M.N.S., Parmas, (Italy) 7–9 September 2006. Post-meeting excursion. *Acta Naturalia de 'L'Ateneo Parmense'*, **42**(3), 76.

RODA, C. 1964. Distribuzione e facies dei sedimenti neogenici nel Bacino Crotonese. *Geologica Romana*, **3**, 319–366.

SCHLÉDER, Z. & URAI, J. L. 2005. Microstructural evolution of deformation-modified primary halite from the Middle Triassic Röt Formation at Hengelo, The Netherlands. *International Journal of Earth Scince (Geologische Rundschau)*, **94**, 941–955.

SHEARMAN, D. J. 1970. Recent halite rock, Baja California, Mexico. *Institute of Mining and Metallurgy Transactions*, **79**, 155–162.

TESTA, G. & LUGLI, S. 2000. Gypsum-anhydrite transformations in Messinian evaporites of central Tuscany (Italy). *Sedimentary Geology*, **130**, 249–268.

ZECCHIN, M., MASSARI, F., MELLERE, D. & PROSSER, G. 2004. Anatomy and evolution of a Mediterranean-type fault bounded basin: the Lower Pliocene of the northern Crotone Basin (Southern Italy). *Basin Research*, **16**, 117–143.

The Messinian 'Vena del Gesso' evaporites revisited: characterization of isotopic composition and organic matter

S. LUGLI[1], M. A. BASSETTI[2] *, V. MANZI[3], M. BARBIERI[4], A. LONGINELLI[3] & M. ROVERI[3]

[1]*Dipartimento di Scienze della Terra, Università degli Studi di Modena e Reggio Emilia, Largo S. Eufemia 19, 41100 Modena, Italy (e-mail: lugli.stefano@unimore.it)*

[2]*IUEM0-UMR, 6538 Place Copernic, 29280 Plouzanè, France*

[3]*Dipartimento di Scienze Geologiche, Università di Parma, Parco Area delle Scienze 157/A, 43100, Parma, Italy*

[4]*Dipartimento di Scienze della Terra, Università 'La Sapienza', Piazzale Aldo Moro, 00185 Roma, Italy*

**Present Address:* LEGEM, Université de Perpignan, 52 Avenue Paul Alduy, 66860 Perpignan, France*

Abstract: The 'Vena del Gesso' (Gessoso-Solfifera Fm, Messinian) is a 227 m-thick ridge along the western Romagna Apennines (Italy) consisting of up to 16 selenite cycles separated by shales and minor carbonate. The total organic carbon values of these deposits range between 0.087–0.016% (gypsum) and 3% (shales). Organic matter is dominated by black debris associated with continental debris. Algae and dynocysts are rare (<1%). The amount of amorphous organic matter is low but it may reach up to c. 40%. The $^{87}Sr/^{86}Sr$ of gypsum and carbonate vary from 0.708890 to 0.709024, yielding non-oceanic values with several exceptions that plot within error of coeval oceanic values only in the upper part of the section (from the 6° bed). The sulphur isotope composition of gypsum range between $\delta^{34}S = +21.8$ and $+23.7$‰ and may represent precipitation of $\delta^{34}S$-enriched gypsum due to the fractionation effect or recycling of coeval gypsum with contributions from brine-sediment redox variations. The isotope values of carbonates show a large variability ($-6.4 < \delta^{18}O < +6.05$‰; $-14.68 < \delta^{13}C < +2.5$‰), suggesting a complex origin by mixing of marine and non-marine waters with a significant contribution of reduced organic matter. These data point to an evaporite basin dominated by continental waters which received significant phases of marine recharge in the upper part together with a marked facies change. Because seawater recharges and a similar facies change are present in other Messinian sections, it follows that we have new possible geochemical and facies markers to correlate the Lower Evaporites across the Mediterranean.

Isotope stratigraphy represents a fundamental tool in palaeoenvironmental and palaeoclimate reconstruction, especially for evaporite sediments. The integration of geochemical data is imperative in facies interpretation, as the peculiar and restricted setting of evaporite deposition makes the use of geochemical data problematic. This is because brine composition may be the result of a mixture of marine and continental water, and sediment composition may be strongly influenced by recycling of older sediments, biologic activity and diagenesis. These uncertainties may be overcome only with the integration of geochemical data in a detailed stratigraphic and facies framework.

This integration is particularly needed for the Messinian evaporites in the Mediterranean, because their origin is still debated and because many of the available isotope data are scattered and commonly obtained from sections whose stratigraphy is not well constrained in the regional framework. The information provided by such scattered data is very difficult to interpret if its significance is not correctly placed into the salinity crisis framework. This is probably one of the reasons why the debate still exists and even is growing.

This paper illustrates the geochemistry of the Vena del Gesso evaporites (Northern Apennines), which represent the most significant example of detailed facies stratigraphy for the Lower Evaporites in the Mediterranean. The aim is to discuss new possible stratigraphic markers to correlate the elusive evaporite sediments across the Mediterranean during the Messinian salinity crisis.

From: SCHREIBER, B. C., LUGLI, S. & BĄBEL, M. (eds) *Evaporites Through Space and Time.*
Geological Society, London, Special Publications, **285**, 179–190.
DOI: 10.1144/SP285.11   0305-8719/07/$15.00 © The Geological Society of London 2007.

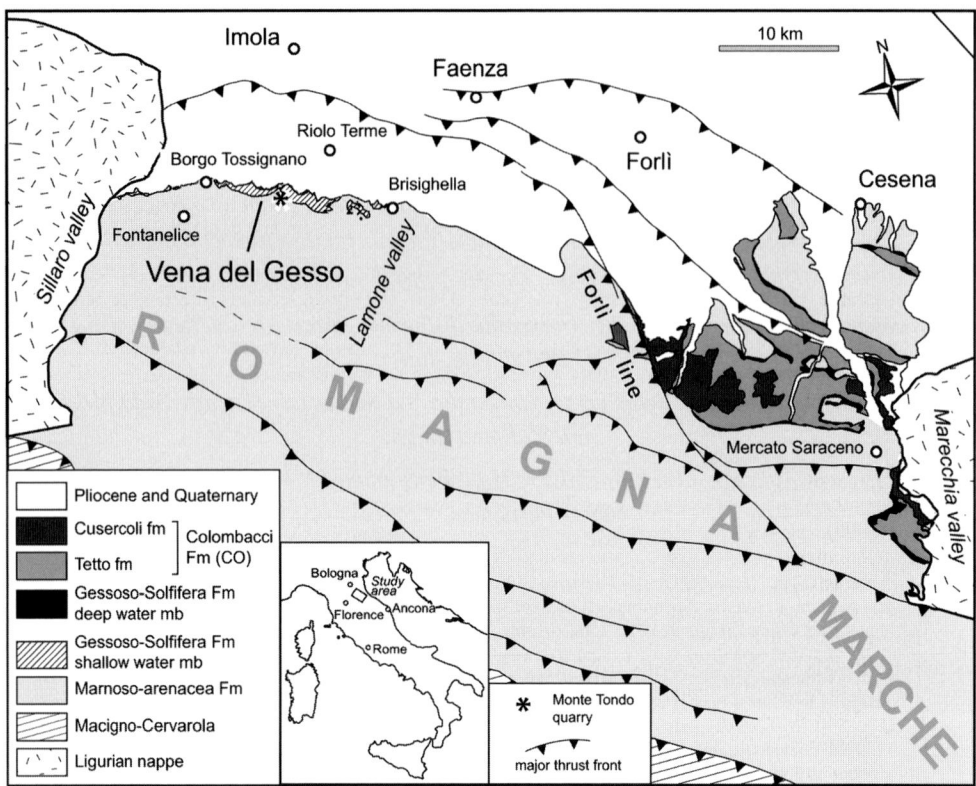

Fig. 1. Schematic geological map of the Romagna Apennines.

Geological setting and stratigraphy

The Romagna Apennines extends from the Sillaro valley to the west and to the Marecchia valley to the east (Fig. 1). This portion of the chain can be subdivided into two minor basement-detached units affected by fold-and-thrust structures (Barchi et al. 1998): (1) a lower and older one composed of Mesozoic to Cenozoic carbonates; and (2) an upper one, up to 3500 m-thick, consisting of a lower Miocene–Quaternary siliciclastic wedge representing the deep-water to continental infill of successive foredeep basins progressively migrating toward the NE, ahead of the advancing Apennines thrust belt (Ricci Lucchi 1986).

According to Roveri et al. 2003, the upper unit of the Romagna Apennines can be subdivided into four formations ranging from Langhian to Pliocene (Vai 1988):

(1) The Marnoso-arenacea Formation (Langhian–Messinian) mainly consists of deep-water siliciclastic turbidites representing the sedimentary infill of the Adriatic foredeep. In the upper part a clayey unit, mainly consisting of slope mudstones ('ghioli di letto') contains minor turbiditic sandstones and chaotic bodies; these deposits are capped by a thin horizon of cyclically interbedded organic-rich laminites and mudstones, informally named 'euxinic shales' (upper Tortonian–lower Messinian). The well-defined biomagnetostratigraphic events recognized within this unit (Vai 1997; Krijsman et al. 1999) allow a detailed correlation with other pre-evaporitic successions of the Mediterranean (Tripoli Formation of Sicily and Spain). The euxinic shales unit, recording the palaeoceanographic changes that affected the whole Mediterranean area before the Messinian Salinity Crisis, contains the Tortonian–Messinian boundary and spans a 1.5 million year time interval.

(2) The Gessoso-solfifera Formation (Messinian) is made up of both primary (Vena del Gesso, Vai & Ricci Lucchi 1977; Table 1) and clastic, resedimented evaporites with interbedded organic-rich shales (eastern Romagna and Marche area), deposited during the evaporitic and post-evaporitic

Table 1. *Facies description of the Vena del Gesso evaporites (from Vai & Ricci Lucchi 1977)*

Facies	Description
F1 Bituminous shale	Organic-rich, laminated marly clay, containing abundant vegetal, fish and insect remains
F2 Calc-gypsum stromatolite and selenite, calcareous breccia and sandstone, flat-lying selenite	Laminated calc-gypsum limestone of algal origin, selenite enclosing algal laminae, mechanically reworked stromatolite clasts, fine-to very coarse sandstone composed of gypsum and carbonate grains, loose selenite crystals lying on their long axis
F3 Massive selenite	Vertical selenite crystals enclosing algal filaments
F4 Banded selenite	Alternation of autochthonus and clastic selenite, with muddy carbonates draping dissolution surfaces
F5 Nodular, lenticular and flaser-bedded gypsum	Clastic selenite with lenses of siltstone, sandstones and micritic carbonate, presence of diagenetic structures
F6 Chaotic gypsum	Mixture of selenite crystal of variable sizes in clayey matrix, clay chips, wood fragments

stages of the Messinian Salinity Crisis (Manzi et al. 2005).

(3) The Colombacci Formation (upper Messinian), consisting of siliciclastic sediments derived from Apenninic sources, was variably deposited in both shallow and deep brackish or freshwater basins developed during the final phase of the Messinian Salinity Crisis (Lago–Mare stage; Bassetti et al. 2004).

(4) The Argille Azzurre Formation (lower Pliocene) is made up of deep marine mudstones recording the return to fully marine conditions.

According to most recent physical stratigraphic frameworks (Roveri et al. 1998, 2001, 2003, 2004, Manzi et al. 2005), the deposition of the Messinian evaporites within the upper unit was controlled by a tectonically related basin topography which began to develop in the upper Tortonian. Primary evaporites (Vena del Gesso) were precipitated in thrust-top semi-closed basins, whereas in the deeper portion of the foredeep only dark euxinic shales were deposited. Additionally, during the post-evaporitic phase clastic evaporites, a large variety of gravity-driven deposits deriving from the dismantlement of the Vena del Gesso primary evaporites, were also deposited in the deep basins.

The 'Vena del Gesso' is a NW–SE elongated relief, approximately 15 km in length, located in the northern Apennines between the Sillaro and the Lamone river valleys. This feature consists of a 227 m-thick succession of 16 cycles of Messinian evaporites. Their deposition was controlled by astronomical precession in a time span ranging from 5.96 to 5.61 ± 0.02 Ma (Krijsmann et al. 1999). The Vena del Gesso evaporites were earlier taken as an example for the formulation of the sedimentary model for the cyclical deposition of primary selenitic gypsum (Vai & Ricci Lucchi 1977, Marabini & Vai 1985). We are currently revising the facies interpretation (Lugli et al. 2005), but in this paper we refer to the original facies description of Vai & Ricci Lucchi (1977; Table 1).

Materials and methods

Detailed sedimentological observations were carried out in the Monte Tondo quarry section and the entire evaporite succession was measured and sampled. A total of 100 samples of gypsum, carbonates and shales have been collected and some of these samples were selected for petrographic and geochemical investigations. The scope of the selection was to provide at least one sample for each facies in the stratigraphic column, following the facies classification proposed by Vai & Ricci Lucchi (1977; Table 1). Because the cycles are thicker and with fewer variations in facies (F3 and F4) in the lower part of the section, samples are more numerous in the upper part, where cycles are thinner but show the complete facies association (F3–F6). For these reasons, the apparent higher frequency of samples in the upper part does not represent a sampling bias.

A total of 54 samples were chosen for strontium isotopic analyses. For gypsum analysis, 100 mg of sample were mixed with 1 g of Na_2CO_3 (strontium-free) and treated with 40 ml of bidistilled water for 6 h at 70 °C to obtain pure $Ca(Sr)CO_3$, which was then dissolved with 2.7 M—HCl. After evaporation to dryness, the resulting $Ca(Sr)Cl_2$ was dissolved in HCl and the solution passed through a cation exchange resin to separate strontium. The strontium was then eluted and analysed for $^{87}Sr/^{86}Sr$ ratios by means of a VG Isomass 54E mass spectrometer. The carbonate samples were dissolved with 0.2 M—HCl, and Sr was separated by cation exchange resin, as previously described for the gypsum samples. For the NBS 987 SrCO$_3$ standard, repeated analyses yielded a mean value of 0.71024 ± 0.00002 (2σ).

The same samples were used to analyse the sulphur isotopic composition: 0.3 g of sulphate by powder was dissolved in distilled water and converted to barium sulphate by adding 0.2 M BaCl$_2$ (~5%) in boiling acid water. The SO$_2$ gas produced by the sulphates was analysed on a Delta C Finnigan Mat mass spectrometer and all data are expressed in the usual notation as δ^{34}S ‰ CDT. The standard deviation for δ^{34}S is ±0.2‰.

Twenty-three bulk carbonate samples were measured for their carbon and oxygen isotopic composition. Measurements were carried out on CO$_2$ obtained by reaction of CaCO$_3$ with 100% H$_3$PO$_4$ at 25 °C and further cryogenic purification in a high vacuum line. The CO$_2$ isotopic values were measured in a Finnigan Delta S mass-spectrometer vs a CO$_2$ working standard obtained from a very pure Carrara marble and calibrated periodically vs NBS-19 and NBS-20 international standards. Because our working standard was systematically calibrated over about 45 years against these two NBS standards which, in turn, were calibrated directly vs PDB-1, we report our isotopic results vs the latter reference standard. The results are reported in the usual δ notation. The standard deviation (2σ) of our measurements is ±0.1‰ for oxygen and ±0.08‰ for carbon.

Total organic carbon (TOC) was determined on 51 samples of both shale and gypsum. A total of 40 samples of both gypsum and shale were also treated with HCl and HF, in order to isolate the residual organic matter for a qualitative visual inspection under the optical microscope (palynofacies analysis).

Results

Organic matter

The TOC values ranged between 0.087 and 0.016% in the selenitic facies, but increased up to 3% in the shales separating the gypsum beds (Table 2).

The accumulation of organic carbon and its preservation in evaporitic basins is favoured by the establishment of marked water stratification together with anoxic bottom conditions: the dense brines from which the evaporites are precipitated inhibit the exchange with the overlaying waters and the organic carbon accumulates at a rapid rate. However, the preservation of accumulated organic matter (principally of algal origin in this restricted basin) depends on the consumption of organic carbon by anaerobic bacteria under anoxic/suboxic conditions: many marine microbes require cations to maintain their osmotic balance and the halophilic marine bacteria communities increase their maximum number activity when the salinity reaches 40‰ (Klinkhammer & Lambert 1989).

The amount of organic matter preserved in the samples appears to be quite low. Despite the fact that we cannot estimate accurately the effect of diagenesis on organic matter preservation in these sediments, the geochemical data (see the next section) strongly suggest that bacterial activity was significant during evaporite deposition.

The composition of organic matter (OM) does not vary greatly throughout the section and is typified by the dominance of black debris (oxidized organic matter) that can reach 90% of the identified material. The content of black debris generally increases upsection with the exception of the 8°, 9° and 13° beds. These elements are associated with other types of continental debris such as well-preserved woody fragments, degraded woody debris (not identifiable) leaf cuticles, spores and pollen. As a whole, the proportion of OM of continental origin can reach up to 99% of the total (Table 2, Fig. 2).

Algae and dynocysts are present in very low amounts (<1%) and are very rare. The amount of amorphous organic matter (AOM) that is generally related to a marine origin is low except for a few samples where it reaches up to 39.5% (shale). The AOM is occasionally well-preserved in the gypsum crystals (37.5%) as well. The total amount of AOM shows a marked increase starting from the 7° bed and the highest contents are found in the 8° and 9° beds, where the black debris reaches its lowest concentration (Table 2, Fig. 2).

These palynofacies data show the predominance of continental organic elements during the sedimentation of the Lower Evaporites of the Vena del Gesso basin with low and localized input of marine elements.

Sr ratio

The strontium isotope ratios (^{87}Sr/^{86}Sr) of the Vena del Gesso gypsum and carbonate vary from 0.708890 to 0.709024 (Table 2) and are in the range of the Messinian Lower Evaporites in the Mediterranean (Müller & Mueller 1991; Flecker & Ellam 1999; Keogh & Butler 1999; Matano et al. 2003; Aharon et al. 1993 measured three samples from the base of the 3° bed of of Vena del Gesso, yielding 0.708904–0.708928). Most of the samples yield non-oceanic Sr isotope ratios with several exceptions that plot within error of coeval oceanic waters (McArthur et al. 2001; Fig. 3).

According to Flecker & Ellam (2006), Sr isotope ratios diverging from coeval oceanic water values indicate that the proportion of oceanic water entering the basin was less than c. 50% and that brines derived mostly from river run-off and rain waters. Values typical of the coeval global ocean thus represent pulses of direct ingression of oceanic water into a restricted marginal basin characterized by brines deriving from continental waters mixed with less than 50% of oceanic water. These pulses of direct oceanic water ingressions appear only in the upper part of the section, starting from the 6° bed, which is the first showing the complete classic facies assemblage (Vai & Ricci Lucchi 1977), and in particular in the 8° and 9° beds (Fig. 3). The oceanic values measured in the upper part of the section appear in every facies of both gypsum and carbonate, with the exception of massive selenite at the base of each cycle, which is characterized by the lowest recorded values (Table 2). This is probably because massive selenite represents the first depositional product on top of the shales that separate each gypsum cycle. The shales contain large amounts of organic matter, mostly from non-marine sources (see previous section), and thus indicate that the basin was flooded by continental waters just before new evaporitic conditions were established with the precipitation of the massive selenite.

Sulphur isotopes of gypsum

The sulphur isotope compositions of gypsum range between $\delta^{34}S = +21.8$ and $+23.7‰$ (Table 2, Fig. 3). These values are in the range of those measured in the Messinian evaporites of the Mediterranean (Ricchivto & McKenzie 1978; Longinelli 1979; Pierre & Rouchy 1990; Lu & Meyers 2003) and are enriched by about 1–3‰ over the expected value assuming the sulphur isotopic composition of dissolved marine sulphate during the Messinian equal or very close to that of modern oceans.

These values may be interpreted in two different ways. If we accept the conclusions by Thode & Monster (1965), confirmed by Raab & Spiro (1991) concerning the fractionation effect which may take place between dissolved sulphate and solid sulphate during the precipitation of gypsum (enrichment of the solid phase by about 2‰), the measured values may be the result of a primary deposition of sulphate. This implies, obviously, that the sulphur isotopic composition of dissolved marine sulphate during Messinian was equal or very close to that of modern oceanic sulphate (about $+21‰$ vs CDT, according to the most reliable measurements, Rees (1978), Rees et al. (1978). According to the evolution curve through time of the $\delta^{34}S$ of oceanic sulphate (Holser 1977), this hypothesis appears reasonable. However, Pierre & Fontes (1978) suggested a reasonable alternative model to explain the ^{34}S-enriched values. During gypsum precipitation, bacterial activity may easily take place below the brine–sediment interface with sulphate reduction followed by sulfide diffusion through the overlying brine and its partial or total re-oxidation at the brine surface layer in a well oxygenated (^{18}O-enriched) environment (evaporated water and dissolved molecular oxygen). A similar hypothesis has been suggested more recently by Lu & Mayers (2003). Both these hypotheses seem to be theoretically acceptable, independent of gypsum age, as similar enriched $\delta^{34}S$ values were measured in modern Mediterranean salinas (Longinelli 1979; Pierre 1982).

According to Cendón et al. (2004) the high $\delta^{34}S$ values (as high as $\delta^{34}S = +23.3‰$) may be the result of recycling of coeval sulphates deposited on the marginal settings. Lower values are generally related to freshwater input or to recycling of ancient evaporites. In Tuscany the recycling of Triassic sulphates ($\delta^{34}S = 14.6‰$) has significantly lowered the $\delta^{34}S$ values of some of the Messinian sulphates ($\delta^{34}S$ ranges from 17.4 to 25.1‰; Dinelli et al. 1999). In the northern Apennines we have no evidence of contributions from recycling of older evaporite deposits, first because the Permian evaporites in the Alps and Upper Triassic evaporites in the Apennines were probably not exposed during the Messinian, second because the resulting $\delta^{34}S$ values should be significantly lower than the measured ones.

The highest $\delta^{34}S$ values are grouped in the 8° and 9° beds and then, going upsection, the oscillating curve shows a decrease upward, a trend which is very similar to that of the Sr ratio curve. Exceptions to this trend are noted the uppermost two beds (15° and 16°), where the curve shows the same general trend as noted in the Sr ratio curve but with higher values and larger amplitude for the $\delta^{34}S$ (Fig. 3).

Oxygen and carbon isotopes of carbonate

The isotope values of carbonates show a large variability $(-6.4 < \delta^{18}O < +6.05‰; -14.68 < \delta^{13}C < +2.5‰$; Table 2, Fig. 3) and are in the range of other Messinian carbonates associated with evaporites in the Mediterranean (Longinelli 1979; Rouchy & Pierre 1990; Lu et al. 2001; Aharon et al. 1993 measured two samples from the base of the 3° bed of Vena del Gesso, yielding $-6.09 < \delta^{18}O < -5,19‰$ and $-12.43 < \delta^{13}C < -7.85‰$).

Oxygen values range from those characteristic of evaporating brines ($\delta^{18}O$ from $+3.34$ to $+6.05‰$) to those characteristic of freshwater ($\delta^{18}O$ from -1.68 to -6.4), whereas negative

Table 2. *Lithology and isotope data of the Vena del Gesso evaporites*

Sample	Distance from base (m)	Bed	Lithology	Facies	$^{87}Sr/^{86}Sr$ gypsum	$^{87}Sr/^{86}Sr$ carbonate	$\delta^{34}S$ (‰ CDT)	$\delta^{13}C$ (‰ PDB-1)	$\delta^{18}O$ (‰ PDB-1)	TOC (%)
MT93	226.5	16	Gypsarenite	F6	0.708914	0.708916	23.1			0.03
MT90	220.3	16	Gypsrudite	F6						0.03
MT94	218.8	16	Gypsrudite	F6	0.708900		22.7			
MT96	218.3	16	Bituminous shale	F1						0.16
MT95	218	16	Selenite	F5	0.708890		22.6			0.01
MT89	217.8	15	Selenite	F5	0.708900		23.3	−0.45	−0.82	0.04
MT88	216	15	Nodular and lenticular selenite	F5	0.708914	0.708930	23.2	−2.32	−2.91	0.04
MT87	210.6	15	Banded selenite	F4	0.708923	0.708940	23.2	−0.06	−1.75	0.03
MT85	207.2	15	Selenite with limestone	F6				−5.48	−5.88	0.02
MT83	206.8	15	Selenite	F5				−5.32	−4.86	0.02
MT82	206	15	Selenite	F5				−5.81	−5.20	0.03
MT81	203.7	15	Massive selenite	F3	0.708899		23.6			0.03
MT80b	203.3	15	Massive selenite with limestone	F3				−3.23	5.12	0.18
MT80	203.3	15	Massive selenite	F3						0.41
MT79	202.5	14	Bituminous shale	F1						0.88
MT77	200.5	14	Nodular and lenticular selenite	F5	0.708954	0.708943	22.5	−3.11	−5.02	0.04
MT76	188.5	14	Banded selenite with limestone	F4	0.708950	0.709001	22.2	−1.96	−3.61	
MT75	187	14	Massive selenite with limestone	F3	0.708893		23.3			
MT74	185.5	13	Nodular and lenticular selenite	F5	0.709024	0.708910				0.06
MT73	185	13	Banded selenite	F4	0.708910	0.708900				0.14
MT72	184.5	13	Massive selenite	F3						0.24
MT70	184	13	Massive selenite	F3	0.708920	0.708974	22.9			
MT100	180.5	12	Selenite	F6	0.708948	0.708944	22.4			
MT99	176.5	12	Nodular and lenticular selenite	F5	0.708921	0.708923	22.2			0.12
MT98	172.8	12	Banded selenite	F4	0.708923		22.8			0.03
MT97	169.3	12	Massive selenite	F3	0.708943		22.6			0.02
MT67	167	11	Nodular and lenticular selenite	F5	0.708893	0.708933	23.0	−0.45	−2.67	0.03
MT66	165.7	11	Nodular and lenticular selenite	F5	0.708951	0.708930	21.8			
MT65	161	11	Banded selenite	F4	0.708900		22.9			
MT64	158	11	Massive selenite	F3	0.708920		22.3			
MT62	157	10	Bituminous shale	F1						0.73
MT61b	155	10	Nodular and lenticular selenite	F5	0.708900	0.708940	22.9			
MT61a	155	10	Nodular and lenticular selenite	F5	0.708971	0.708960	23.1			
MT60	150.5	10	Banded selenite	F4	0.709000		23.2			
MT59	149	10	Massive selenite	F3	0.708895		22.8			
MT58	148.8	9	Bituminous shale	F1						0.48
MT57	148.5	9	Selenite	F6	0.708923		22.8			

Sample			Lithology	Facies						
MT56	147.5	9	Banded selenite with limestone	F4						0.06
MT55	143.5	9	Banded selenite with limestone	F4				−1.63	−4.05	0.24
MT54	143	9	Bituminous shale	F1						1.3
MT53b	141.5	9	Nodular and lenticular selenite	F5	0.708945		23.7			
MT53a	141.5	9	Nodular and lenticular selenite	F5	0.709000	0.709010	22.9	−2.5	−1.68	
MT53	141.5	9	Banded selenite	F4			23.4			
MT52	139	9	Banded selenite with limestone	F4	0.709018	0.708979	22.9			0.05
MT51	137	9	Massive selenite	F3	0.708934		23.0			0.04
MT50	136.5	8	Bituminous shale	F1						0.96
MT49	136	8	Selenite	F5						0.07
MT48b	135.5	8	Nodular and lenticular selenite	F5	0.708920		23.6			
MT48a	135.5	8	Nodular and lenticular selenite	F5	0.708940	0.708940	22.7			
MT48	135.5	8	Nodular and lenticular selenite	F5			23.7			
MT47	134	8	Banded selenite with limestone	F4	0.708965	0.709014	23.3	−0.8	−4.39	0.06
MT46	130.5	8	Massive selenite	F3	0.708931		23.4			0.04
MT39	130.3	8	Bituminous shale	F1						1.38
MT37	129	7	Nodular and lenticular selenite	F5						0.04
MT36	126.5	7	Banded selenite with limestone	F4				−1.04	5.76	0.06
MT35	120.5	7	Massive selenite	F3						
MT34	119.7	7	Stromatolite	F2				1.77	−3.64	0.11
MT31	118.0	7	Banded selenite	F4						0.07
MT29	107.8	6	Nodular and lenticular selenite	F5	0.708900		23.3			
MT27b	104.3	6	Nodular and lenticular selenite	F5	0.709020		23.1			
MT27a	104.3	6	Banded selenite	F4	0.708900		22.5			
MT26	102	6	Banded selenite	F4	0.708920		23.3			
MT25	99	6	Massive selenite with limestone	F3				−3.11	3.34	0.05
MT24	95.5	6	Massive selenite	F3	0.708910		22.6			
MT23	94.5	6	Bituminous shale	F1						0.97
MT21	93.5	6	Massive selenite	F3	0.708910		22.7			0.08
MT20	92.5	6	Stromatolite (carbonate)	F2		0.708920		−0.83	4.95	0.14
MT19	92.3	5	Selenite	F6	0.708910		22.6			
MT18	91	5	Massive selenite	F3	0.708910		22.8		6.05	0.1
MT17	75.2	5	Banded selenite with limestone	F4	0.708920		22.6	−4.74		0.05
MT16	69	5	Massive selenite	F3	0.708900		22.4			0.03
MT11	37.8	4	Massive selenite	F3	0.708930		23.0	−6.16	0.07	0.07
MT10	37.3	4	Stromatolite	F2		0.708930		−14.68	−2.85	0.03
MT12	36.5	3	Limestone	F5				−5.77	−6.40	
MT13	36.3	3	Selenite-bearing limestone	F5	0.708900		23.4	−6.67	−4.62	
MT14	36	3	Massive selenite	F3	0.708940		23.2			0.03
MT09	18.2	3	Bituminous shale pocket	F3						3.13
MT05	11.3	3	Massive selenite	F3	0.708930		22.9			
MT04	10	3	Stromatolite (carbonate)	F2		0.708900		1.25	5.02	0.04
MT02	5.5	2	Massive selenite	F3	0.708930		23.6			
MT01	0	1	Massive selenite	F3	0.708920		23.2			

186 S. LUGLI ET AL.

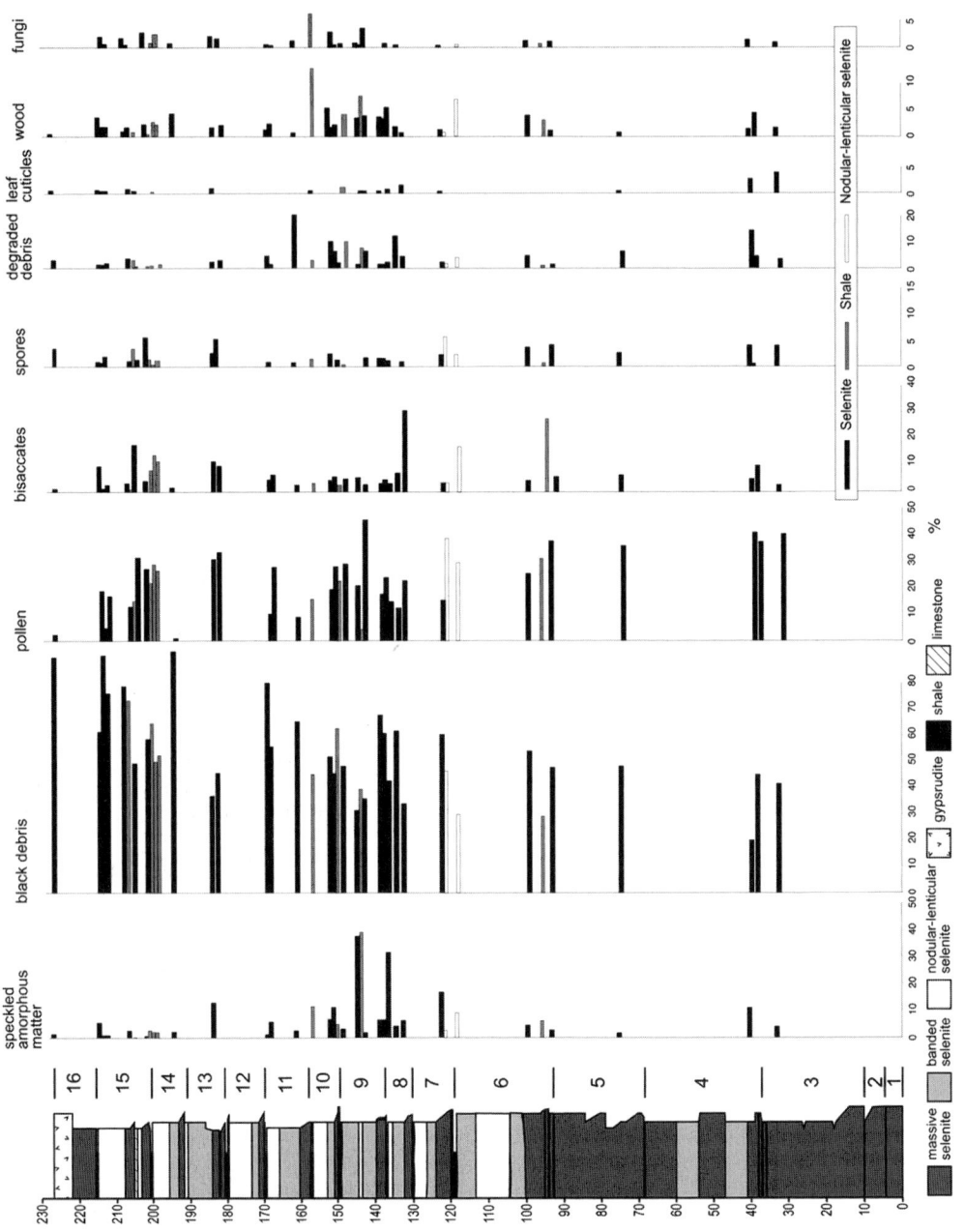

Fig. 2. Organic matter characterization of the Vena del Gesso section.

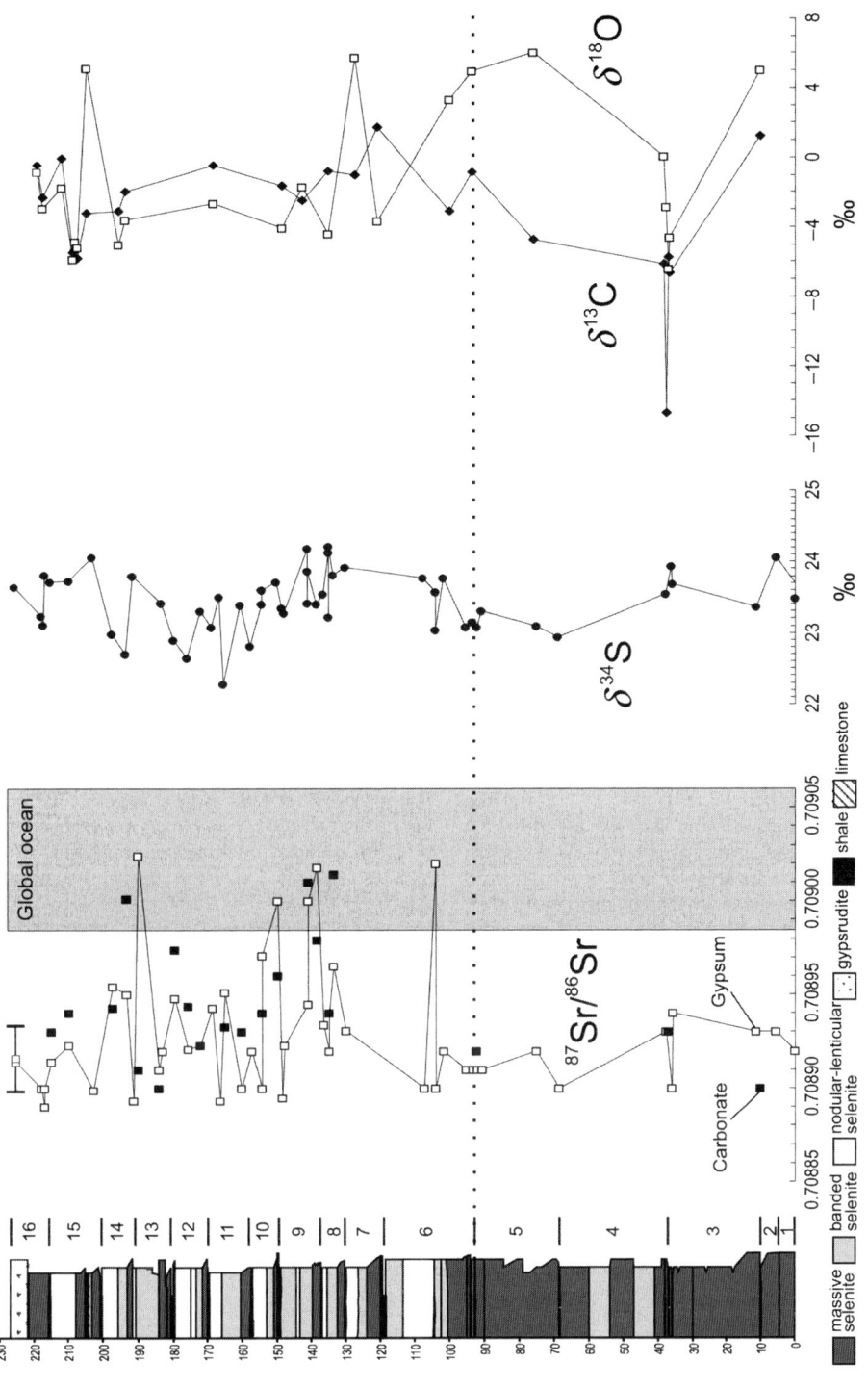

Fig. 3. Isotope geochemistry of the Vena del Gesso section. Global ocean values from McArthur et al. (2001).

carbon data indicate continental water inputs and a possible significant contribution of isotopically light carbon from organic matter reduction. The lowest carbon value ($\delta^{13}C = -14.68$ sample MT 10) is from a laminar stromatolite layer rich in algal filaments (Table 2). These data suggest a complex origin of the carbonates by mixing of marine and non-marine waters with a significant contribution of organic matter reduction, as suggested for the Messinian of the Nijar basin, Spain (Lu et al. 2001).

Discussion

The geochemical data and the organic matter association seem to point to an evaporite basin dominated by continental waters which received significant marine recharges. Seawater inputs have been detected only in the upper part and are concomitant with a marked facies change starting from the 6° bed where nodular, lenticular and flaser bedded gypsum (F5 of Vai & Ricci Lucchi 1977) appear for the first time upsection.

Variations of stable isotopes thorough section suggest that the strongest evaporating conditions during carbonate formation were common in the lower part. A significant contribution of continental water marks the passage from the 3° to the 4° beds. The upper part of the section is dominated by freshwater inputs and only two phases of strong evaporitic conditions are revealed by high $\delta^{18}O$ values in the carbonates from the 7° and 15° beds (Fig. 3).

It is interesting to note that the highest $\delta^{18}O$ values of carbonates that indicate strongly evaporating conditions are not found at the same stratigraphic levels where the associated sulphates show their higher $\delta^{34}S$ values. In particular, the highest $\delta^{18}O$ values do not match the highest Sr isotope ratios of carbonate samples and are also widely distributed in the lower part of the section, where Sr isotopes indicate a prevailing proportion of freshwater over seawater. This apparent contradiction is probably due to the fact that $^{87}Sr/^{86}Sr$ is insensitive to salinity change and evaporation conditions, but is directly controlled by simple mixing of ocean and river water (Flecker et al. 2002). On the other hand, $\delta^{34}S$ and stable isotopes in carbonates are less sensitive, with respect to Sr isotopes, in detecting marine to non-marine influence during evaporite deposition. Moreover, Sr isotopes are more sensitive to restriction of ocean exchange than fauna and lithology when net-evaporation is positive (Flecker & Ellam 2006). The highest recorded $\delta^{34}S$ values may be related to some degree of gypsum recycling from the basin margins and the similar trend of the $\delta^{34}S$ and Sr ratio curves may indicate that these conditions were possibly concomitant with the marine ingressions.

In a marginal basin such as the Vena del Gesso, which records the northernmost environmental conditions during the salinity crisis, the influences of continental waters may have been stronger than in any other Mediterranean setting. These conditions are recorded by the palinology of the shales intercalations which, in contrast with Sicily, suggests a forested environment, the absence of severe dry conditions and the presence of swampy areas (Bertini 2002). On the other hand, significant departures from Sr oceanic values are documented by scattered data in the Lower Evaporites of the Mediterranean (Flecker et al. 2002) and we have observed a similar dramatic facies change in the gypsum sections of Spain, Crete, Tuscany, Calabria and Sicily. If these considerations are correct, then we have new markers that support an attempt of a large scale correlation for the Lower Evaporites. Rates of inflow condition variations (Atlantic exchange) may have triggered the gypsum facies associations across the entire Mediterranean.

Conclusions

This is the first detailed isotope geochemistry and organic matter characterization of a Lower Evaporite section that is well constrained from the stratigraphic point of view. The results indicate that the Vena del Gesso marginal basin was dominated by continental water-derived brines and received repeated pulses of oceanic water in the upper part. The onset of oceanic water recharge coincides with a marked change in facies association producing the nodular, lenticular and flaser bedded gypsum (F5 of Vai & Ricci Lucchi 1977), which appear for the first time starting from the 6° bed.

Because seawater recharges and a similar facies change are present also in other Messinian sections, the implication of these results is that we have new possible geochemical and facies markers to correlate the Lower Evaporite sediments across the Mediterranean. These oceanic signals revealed by geochemistry and facies change also may have been preserved in the succession of the deep Mediterranean, which has not been investigated in detail.

We thank BPB Italia for permission to sample the Monte Tondo quarry. The manuscript has benefited from the constructive review of R. Flecker, C. Pierre and B. C. Schreiber.

References

AHARON, P., GOLDSTEIN, S. L., WHEELER, C. W. & JACOBSON, G. 1993. Sea-level events in South Pacific linked with the Messinian Salinity Crisis. Geology, 21, 771–775.

BARCHI, M., DE FEYTER, A., MAGNANI, M. B., MINELLI, G., PIALLI, G. & SOTERA, B. M. 1998. The structural style of the Umbria-Marche fold and thrust belt. *Memorie della Società Geologica Italiana*, **52**, 557–578.

BASSETTI, M. A., MANZI, V., LUGLI, S, ROVERI, M. A., LONGINELLI, RICCI LUCCHI, F. & BARBIERI, M. 2004. Paleoenvironmental significance of Messinian post-evaporitic lacustrine carbonates in the northern Apennines, Italy. *Sedimentary Geology*, **172**, 1–18.

BERTINI, A. 2002. Palynological evidence of upper Neogene environments in Italy. *Acta Universitatis Carolinae–Geologica*, **46**(4), 15–25.

CENDÓN, D. I., PERYT, T. M., AYORA, C., PUEYO, J. J. & TABERNER, C. 2004. The importance of recycling processes in the Middle Miocene Badenian evaporite basin (Carpathian foredeep): Paleoenvironmental implications. *Palaeogeography, Palaeoclimatology, Palaeoecology*, **212**, 141–158.

DINELLI, E., TESTA, G., CORTECCI, G. & BARBIERI, M. 1999. Stratigraphic and petrographic constraints to trace element and isotope geochemistry of Messinian sulfates of Tuscany. *Memorie della Società Geologica Italiana*, **54**, 61–74.

FLECKER, R. & ELLAM, R. M. 1999. Distinguishing climatic and tectonic signal in the sedimentary successions of marginal basins using Sr isotopes: an example from the Messinian salinity crisis, Eastern Mediterranean. *Journal of the Geological Society of London*, **156**, 847–854.

FLECKER, R. & ELLAM, R. M. 2006. Identifying Late Miocene episodes of connection and isolation in the Mediterranean–Paratethyan realm using Sr isotopes. *Sedimentary Geology*, **188–189**, 189–203.

FLECKER, R., DE VILLIERS, S. & ELLAM, R. M. 2002. Modelling the effect of evaporation on the salinity–$^{87}Sr/^{86}Sr$ relationship in modern and ancient marginal–marine systems: the Mediterranean Messinian Salinity Crisis. *Earth Planetary Science Letters*, **203**, 221–233.

HOLSER, W. T. 1977. Catastrophic chemical events in the history of the ocean. *Nature*, **267**, 403–408.

KEOGH, S. M. & BUTLER, R. W. H. 1999. The Mediterranean water body in the late Messinian: interpreting the record from marginal basins on Sicily. *Journal of the Geological Society of London*, **156**, 837–846.

KLINKHAMMER, G. P. & LAMBERT, C. E. 1989. Preservation of organic matter during salinity excursions. *Nature*, **339**, 271–274.

KRIJSMANN, W., HILGEN, F. J., MARABINI, S. & VAI, G. B. 1999. New paleomagnetic and cyclostratigraphic age constraints on the Messinian of the Northern Apennines (Vena del Gesso Basin, Italy). *Memorie della Società Geologica Italiana*, **54**, 25–33.

LONGINELLI, A. 1979. Isotope geochemistry of some Messinian evaporites; paleoenvironmental implications. *In*: CITA, M. B. & WRIGHT, R. (eds) *Geodynamic and Biodynamic Effects of the Messinian Salinity Crisis in the Mediterranean*. Paleogeography, Paleoclimatology, Paleoecology, **29**, 95–123.

LU, F. H. & MEYERS, W. 2003. Sr, S, and SO_4 isotopes and depositional environments of the upper Miocene evaporites, Spain. *Journal of Sedimentary Research*, **73**, 444–450.

LU, F. H., MEYERS, W. J. & SCHOONEN, M. A. A. 2001. $\delta^{34}S$ and $\delta^{18}O$ (SO_4) fractionation modeling and environmental significance. *Geochimica et Cosmochimica Acta*, **65**, 3081–3092.

LUGLI, S., MANZI, V., ROVERI, M. & SCHREIBER, B. C. 2005. The Messinian Lower Evaporites in the Mediterranean: a new facies model. FIST Forum Geoitalia 2005, 22–23 September 2005, Spoleto (italy), Abstract book, 2.

MCARTHUR, J. M., HOWARTH, R. J. & BAILEY, T. R. 2001. Strontium isotope stratigraphy: LOWESS version 3: Best fit to the marine Sr-isotope curve for 0–509 Ma and accompanying look-up table for deriving numerical age. *Journal of Geology*, **109**, 155–170.

MANZI, V., LUGLI, S., RICCI LUCCHI, F. & ROVERI, M. 2005. Deep-water clastic evaporites deposition in the Messinian Adriatic foredeep (northern Apennines, Italy): did the Mediterranean ever dry out? *Sedimentology*, **52**, 875–902.

MARABINI, S. & VAI, G. B. 1985. Analisi di facies e macrotettonica della Vena del Gesso in Romagna. *Bollettino della Società Geologica Italiana*, **104**, 21–42.

MATANO, F., BARBIERI, M., DI NOCERA, S. & TORRE, M. 2003. Stratigraphy and strontium geochemistry of Messinian evaporite-bearing successions of the southern Apennines foredeep, Italy: implications for the Mediterranean 'salinity crisis' and regional palaeogeography. *Palaeogeography, Palaeoclimatology, Palaeoecology*, **217**, 87–114.

MÜLLER, D. W. & MUELLER, P. A. 1991. Origin and age of the Mediterranean Messinian evaporites: implications from Sr isotopes. *Earth Planetary Science Letters*, **107**, 1–12.

PIERRE, C. 1982. *Teneurs en isotopes stables (^{18}O, ^{2}H, ^{13}C, ^{34}S) et conditions de genèse des évaporites marines: application à quelques milieux actuels et au Messinien de Méditerranée*. Unpublished doctoral thesis. Université Paris-Sud Orsay.

PIERRE, C. & FONTES, J. C. 1978. Isotope composition of Messinian sediments from the Mediterranean Sea as indicators of paleoenvironments and diagenesis. *Initial Reports of the Deep Sea Drilling Project*, **42**, 635–650.

PIERRE, C. & ROUCHY, J. M. 1990. Sedimentary and diagenetic evolution of Messinian evaporites in the Tyrrhenian Sea (ODP Leg 107, Sites 652, 653 and 654): petrographic, mineralogical, and stable isotope records. *In*: KASTENS, K. A. & MASCLE, J. (eds) *Proceedings of the Ocean Drilling program, Scientific Results*, **107**, 187–201.

RAAB, M. & SPIRO, B. 1991. Sulfur isotopic variations during seawater evaporation with fractional crystallization. *Chemical Geology*, **86**, 323–333.

REES, C. E. 1978. Sulphur isotope measurements using SO_2 and SF_6. *Geochimica et Cosmochimica Acta*, **42**, 383–389.

REES, C. E., JENKINS, W. J. & MONSTER, J. 1978. The sulfur isotopic composition of ocean water sulfate. *Geochimica et Cosmochimica Acta*, **42**, 377–381.

RICCI LUCCHI, F. 1986. The foreland basin system of the Northern Apennines and related clastic wedges: a preliminary outline. *Giornale di Geologia*, **48**(1–2), 165–186.

RICCHIUTO, T. & MCKENZIE, J. 1978. Stable isotope investigation of Messinian sulfate samples from DSDP Leg 42A, Eastern Mediterranean Sea. *In*: HSÜ, K. J. *ET AL*. (eds) *Initial Reports of Deep Sea Drilling Project*, **42**, 657–660.

ROUCHY, J.-M. & PIERRE, C. 1990. Sedimentary and diagenetic evolution of Messinian evaporites in the Thyrrenian Sea (ODP Leg 107, sites 652, 653, and 654); petrographic, mineralogical, and stable isotope records. *Proceedings of the Ocean Drilling Program, Thyrrenian Sea*, **107**, 187–210.

ROVERI, M., BASSETTI, M. A. & RICCI LUCCHI, F. 2001. The Mediterranean Messinian Salinity Crisis: an Apennine foredeep perspective. *Sedimentary Geology*, **140**, 201–214.

ROVERI, M., LANDUZZI, A., BASSETTI, M. A., LUGLI, S., MANZI, V., RICCI LUCCHI, F. & VAI, G. B. 2004. The record of Messinian events in the Northern Apennines foredeep basins. B19 Field trip guidebook. 32nd International Geological Congress, 20–28 August 2004.

ROVERI, M., MANZI, V., BASSETTI, M. A., MERINI, M. & RICCI LUCCHI, F. 1998. Stratigraphy of the Messinian post-evaporitic stage in eastern-Romagna (northern Apennines, Italy). *Giornale di Geologia*, **60**, 119–142.

ROVERI, M., MANZI, V., RICCI LUCCHI, F. & ROGLEDI, S. 2003. Sedimentary and tectonic evolution of the Vena del Gesso basin (Northern Apennine, Italy): Implications for the onset of the Messinian salinity crisis. *Geological Society of America Bulletin*, **115**, 387–405.

THODE, H. G. & MONSTER, J. 1965. Sulfur isotope geochemistry of petroleum, evaporites and ancient seas. *American Association of Petroleum Geologists, Memoir* **4**, 367–377.

VAI, G. B. 1988. A field trip guide to the Romagna Apennine geology – the Lamone Valley. *In*: DE GIULI, C. & VAI, G. B. (eds) *Fossil Vertebrates in the Lamone Valley, Romagna apennines*. Field Trip Guidebook, International Workshop: Continental Faunas at the Mio-Pliocene Boundary, Faenza, 28–31 March 1988, Litografica Faenza, Faenza, Italy, 7–37.

VAI, G. B. 1997. Cyclostratigraphy estimate of the Messinian stage duration. *In*: MONTANARI, A., ODIN, G. S. & COCCIONI, R. (eds) *Miocene Stratigraphy: an integrated approach*. Elsevier Science, Amsterdam, 463–476.

VAI, G. B. & RICCI LUCCHI, F. 1977. Algal crusts, autochtonous and clastic gypsum in a cannibalistic evaporite basin; a case history from the Messinian of Northern Apennine. *Sedimentology*, **24**, 211–244.

The 'Evaporiti di Monte Castello' deposits of the Messinian Southern Apennines foreland basin (Irpinia–Daunia Mountains, Southern Italy): stratigraphic evolution and geological context

F. MATANO

Dipartimento di Scienze della Terra, Università Federico II, Largo San Marcellino 10, 80138 Napoli, Italy (e-mail: matano@unina.it)

Abstract: An evaporitic limestone–gypsum succession, belonging to the Evaporiti di Monte Castello Formation, is recognized in the late Messinian (Upper Miocene) of the Irpinia–Daunia Mountains in the Southern Apennines arc (Southern Italy). This unit is formed by diatomaceous marls, massive and laminated evaporitic limestones, and by primary and clastic gypsum. Detailed stratigraphical, sedimentological and strontium geochemistry data has permitted reconstruction of stratigraphic and facies relations of gypsum deposits, depositional environment and basin evolution. Genetically related gypsum lithofacies can be grouped into two facies associations. The autochthonous gypsum facies association consists of shallow water selenitic, acicular and laminated gypsum and is characterized by the absence of high-energy sedimentary structures. The redeposited clastic gypsum facies association consists of shallow- to deeper-water fine-grained laminated gypsum, gypsarenites, pebbly gypsarenites and gypsrudites, showing common features of resedimented deposits. Nodular structures occurring in the laminated gypsum lithofacies seem to be mostly related to late diagenetic processes. The sedimentary evolution during the evaporative phase was characterized by a gradual increase in salinity until gypsum precipitated; then the sedimentary conditions in the basin were characterized by almost homogenous salinity conditions, and influenced by events of gypsum reworking and resedimentation probably related to flooding episodes and local tectonic activity. The gypsum was deposited from mainly marine brines, based on their Sr isotopic compositions. This sedimentary series is an equivalent of the Lower Evaporites of other parts of the Mediterranean. The Messinian Monte Castello Formation evaporites represent an uncommon type of evaporitic succession, probably developed in a extensional setting in a basin located along the Apulian foreland ramp, in contrast with the northern Apennines and Sicilian basins (e.g. Vena del Gesso, and Caltanisetta basins), which are considered to be thrust-top basins of the Apennine–Maghrebian foreland basin system.

An important episode of evaporitic deposition during the Messinian stage, related to the so-called 'salinity crisis' (Selli 1960; Hsü *et al.* 1973a, b), further complicated the very complex Late Miocene geological context of the Mediterranean region (Fig. 1), characterized by the advanced stage of collisional coupling between the African and the Eurasian plates (Cavazza & Wezel 2003). During the Messinian salinity crisis, the Mediterranean basins episodically desiccated and large volumes of evaporitic deposits precipitated on the floor of deep marine basins (Hsü *et al.* 1973a, 1977; Kastens & Mascle 1990), as well as on their shallower, marginal portions. As a matter of fact, Messinian evaporite deposits crop out in a considerable number of localities around the Mediterranean Sea, such as Sicily (Decima & Wezel 1973; Schreiber *et al.* 1976; Butler *et al.* 1995, 1999), Southern Apennines (Di Nocera *et al.* 1975, 1981; Dazzaro *et al.* 1988; Matano *et al.* 2005), Northern Apennines (Vai & Ricci Lucchi 1977; Roveri *et al.* 2001, 2003), Southest Spain (Michalzik 1996; Riding *et al.* 1998, 1999; Playà *et al.* 2000), Northern Africa, Crete, Cyprus and Southern Turkey (Rouchy 1982).

Messinian evaporitic deposition occurred in a series of smaller discrete basins, which were different in dimensions and form from the large pre-Messinian Mediterranean basins, characterized by open marine conditions. The initiation of the salinity crisis resulted mainly from tectonic processes, which progressively restricted and partly isolated the Mediterranean Sea from the Atlantic Ocean (Hodell *et al.* 2001; Vidal *et al.* 2002; Duggen *et al.* 2003; Krijgsman *et al.* 2004). High-resolution cyclostratigraphic studies show that the onset of the Messinian salinity crisis is dated at 5.96 ± 0.02 Ma and is synchronous over the entire Mediterranean basin (Krijgsman *et al.* 1999). Some data are in contrast with this result, such as those referred to the Sicilian Maghrebides foreland basin, where the beginning of evaporitic deposition is considered to be diachronous over a period of at least 0.8 Ma by Butler *et al.* (1999).

It seems that each evaporitic basin had its own tectonic and hydrologic history and hence a

From: SCHREIBER, B. C., LUGLI, S. & BĄBEL, M. (eds) *Evaporites Through Space and Time.*
Geological Society, London, Special Publications, **285**, 191–218.
DOI: 10.1144/SP285.12   0305-8719/07/$15.00 © The Geological Society of London 2007.

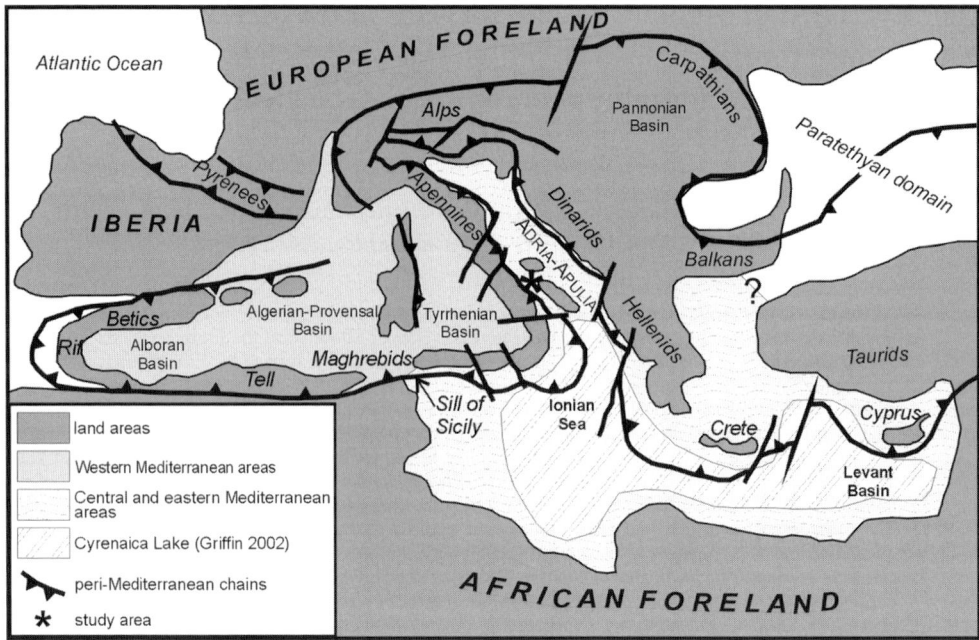

Fig. 1. Paleogeographic and tectonic reconstruction of the Mediterranean domains during Late Messinian (modified after Matano et al. 2005). Land areas and main thrusts distribution are from Cipollari et al. (1999a, b) and Ziegler (1999). The Mediterranean basin subdivision is from Riding et al. (1999, and references therein), Roveri et al. (2001) and Griffin (2002). Western Mediterranean areas are formed by basins characterized by deep or complete desiccation, followed by marine reflooding with scarce evidence of brackish Lago Mare onditions; central and eastern Mediterranean areas are formed by basins characterized by partial desiccation, with deep portions of the basin never desiccated, followed by widespread brackish Lago Mare conditions.

different history of evaporite deposition; as a consequence, a full understanding of the Messinian evaporites must await resolution of the tectonic history of the Mediterranean basin and its component sub-basins (Hardie & Lowenstein 2004). The Messinian was also a period of widespread and short-lived tectonic activity along the contractional fronts from Sicily and Apennines to Crete and Cyprus (Fig. 1). The so-called 'intra-Messinian tectonic phase' caused thrusting, development of widespread unconformities and deposition of syntectonic coarse-grained sediments and reworked evaporites (Decima & Wezel 1973; Elter et al. 1975; Di Nocera et al. 1976; Torre et al. 1988; Patacca & Scandone 1989; Butler et al. 1995; Cavazza & De Celles 1998; Roveri et al. 1998, 2001, 2003, 2004).

Two major Messinian sedimentary cycles separated by a deeply incised erosion surface are usually recognized in the Mediterranean basins. The classic Messinian composite stratigraphic succession exposed in Sicily starts with alternations of open marine marls and sapropels, laminated diatomites ('Tripoli Formation') and evaporitic limestones ('Calcare di Base'), and passes into gypsum and halite ('Lower Evaporites'). The 'Upper Evaporites' overlie an erosional unconformity and comprise gypsum and marls, and are in turn overlain by the Lago-Mare deposits of the Arenazzolo Formation (Ogniben 1957; Decima & Wezel 1973).

In northern Apennines the Upper Evaporites are not present and two mainly terrigenous unconformity-bounded units form the post-evaporitic Messinian deposits (Roveri et al. 1998, 2001). The 'Lower Evaporites' (Gessoso-Solfifera Formation) are formed by mainly shallow-water facies in the Romagna and Emilia sectors (Vena del Gesso basin, Vai & Ricci Lucchi 1977) and by resedimented evaporites in the Marche and Abruzzi sectors of nothern Apennines (Roveri et al. 2003). These evaporitic sequences have been referred to wedge-top and foredeep depocenters of the Apennine foreland basin system (Roveri et al. 1998, 2001, 2003, 2004).

In southern Apennines there are two main areas of Messinian evaporites outcrops (Fig. 2). In the Sannio–western Irpinia area the evaporitic deposits have been referred to two different units separated

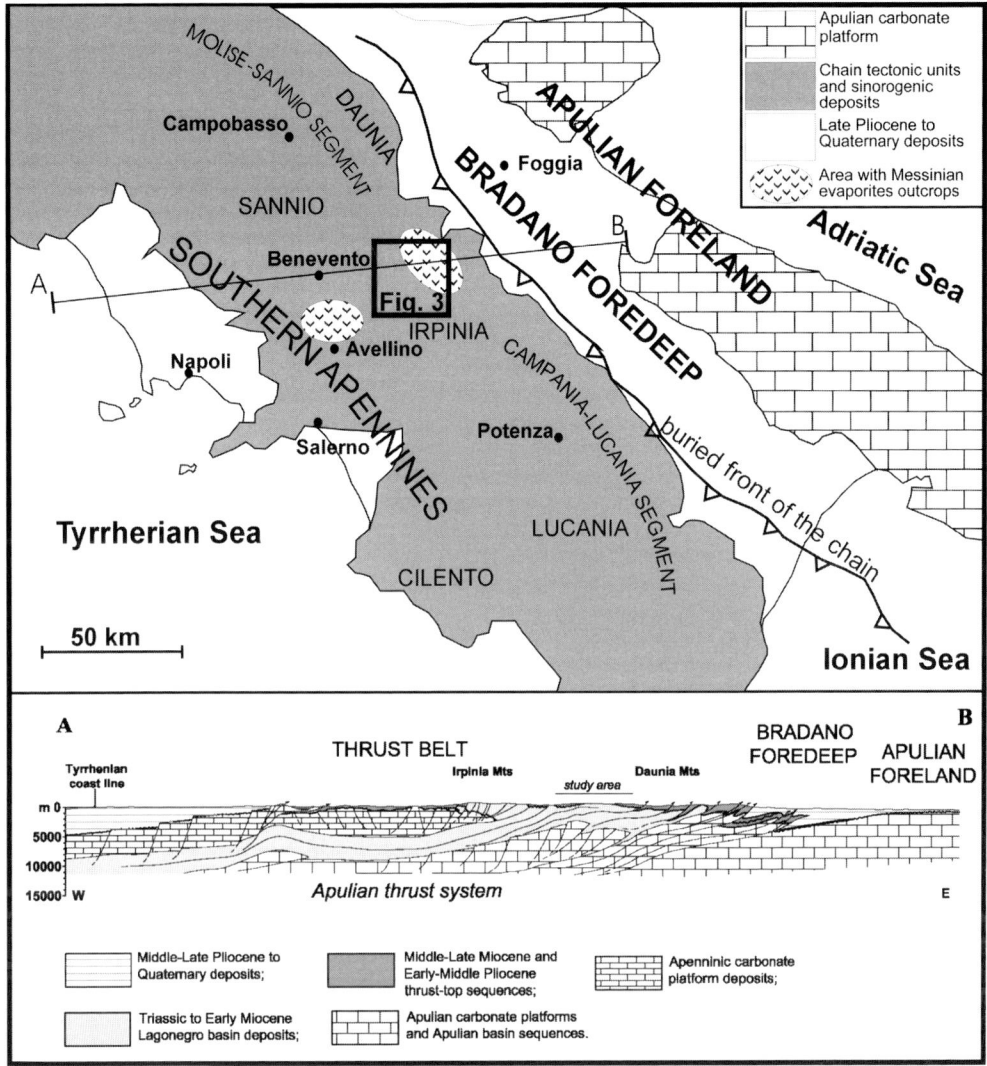

Fig. 2. Tectonic sketch-map of the Southern Apennines, with a simplified geological section modified after Mostardini & Merlini (1986).

by an angular unconformity (Di Nocera et al. 1981), such as the Late Tortonian–Late Messinian p.p. Villamaina Unit and the Late Messinian p.p.– Early Pliocene Altavilla Unit. In the eastern Irpinia–Daunia area, only the Evaporiti di Monte Castello Formation (Crostella & Vezzani 1964) crops out. It is formed by evaporitic limestones, selenites, laminated and nodular gypsum, and clastic gypsum, and is referred to the 'Lower Evaporites' (Dazzaro et al. 1988; Matano et al. 2005).

The aim of this paper is to provide a stratigraphic study of the Evaporiti di Monte Castello Formation, interpreting facies and stratigraphic relationships of the evaporitic deposits, and to frame them into a physical stratigraphic scheme able to help in the reconstruction of the tectono-sedimentary evolution of the basin. A detailed mapping at 1:10,000 scale was carried out in the study area (Basso et al. 2002) and stratigraphic sections of all the Messinian evaporite outcrops were measured to different scales; gypsum facies analyses were performed in all the best-exposed outcrops. The gypsum lithofacies (selenitic, acicular, laminated, nodular and coarse clastic gypsum) are distinguished on the basis of stratal geometries and sedimentary structures. Available geochemical data (Dazzaro et al. 1988; Matano et al. 2005) have been discussed to help in the interpretation of sedimentary facies and in the definition of palaeoenvironmental settings.

Geological setting

The southern Apennines fold-and-thrust belt (Fig. 2) developed from the latest Cretaceous to Early Pleistocene at the W-dipping subduction–collisional boundary between the Apulia/Adria continental microplate and the European continental plate (D'Argenio et al. 1975; Channel et al. 1979; Mostardini & Merlini 1986; Doglioni 1991; Doglioni et al. 1996; Roure et al. 1991; Elter et al. 2003). The Apennines chain comprises a stack of Adria-verging rootless thrust sheets, which overthrusted a buried foreland thrust belt, known as the Apulian thrust system (Mostardini & Merlini 1986; Lentini et al. 1990; Roure et al. 1991). They are bounded by a complex system of frontal arcs, which overlie the terrigenous deposits of the Pliocene–Pleistocene Bradano foredeep and the western flexured margin of the foreland, represented by the Apulian Mesozoic–Cenozoic carbonate platform (Fig. 2).

The chain developed through the deformation of two major palaeogeographic domains: an internal oceanic domain represented by Late Jurassic to Early Miocene Liguride and Sicilide units, and an external domain represented by Triassic to Miocene sedimentary units derived from the deformation of the continental Apulia/Adria passive margin (Mostardini & Merlini 1986; Elter et al. 2003). The Apulia continental margin consists of carbonate platforms and pelagic basins (D'Argenio et al. 1975; Mostardini & Merlini 1986; Sgrosso 1998), which were gradually covered since the Miocene to Early Pleistocene by the deep-marine to continental foreland clastic wedges related to the progressive flexure of the Adria lithosphere beneath the advancing Apenninic thrust belt (Patacca et al. 1990; Sgrosso 1998; Critelli 1999). The progressive eastward migration of the outer Apenninic front is the main constraint to the large-scale evolution of the southern Apennines thrust-belt (Parotto & Praturlon 2004). A significant shift of the foreland basin system depozones toward E–NE occurred starting from the Late Tortonian to Early–Middle Pleistocene, after the occurrence of rifting and the formation of oceanic crust in the southern Tyrrhenian Sea (Patacca et al. 1990, 1993). During the Late Messinian, evaporitic to non-marine sedimentary basins are widespread both along the Tyrrhenian border of the chain and in the foreland basin for the occurrence of the salinity crisis at the Mediterranean scale.

In the study area, located in the Irpinia–Daunia sector along the eastern margin of the Southern Apennines (Fig. 2), Middle Triassic to Upper Miocene shallow- to deep-marine carbonate and pelagic basin successions and Middle Miocene to Middle Pliocene foreland clastic successions crop out (Basso et al. 2002; Di Nocera et al. 2002; Improta et al. 2003). The Frigento, Fortore, Daunia and 'Vallone del Toro' tectonic units have been distinguished in the study area (Fig. 3); they are strongly deformed and thrusted eastward upon the buried Apulian thrust system (Roure et al. 1991; Matano & Di Nocera 2001; Basso et al. 2002). The latest Messinian Altavilla (Basso et al. 2002) and the Pliocene Ariano (Ciarcia et al. 2003) super-synthems are formed by thrust-top basin fillings located on the thrust-sheets of the Frigento, Fortore, Daunia and 'Vallone del Toro' tectonic units (Fig. 3). The age of tectonic deformation is different for the four tectonic units. As a matter of fact, during the Messinian salinity crisis, the Frigento and Fortore basinal units were already deformed and stacked in the orogenic wedge, while Daunia and 'Vallone del Toro' units were located in the not yet deformed foreland domain (Pescatore et al. 2000; Basso et al. 2002; Di Nocera et al. 2002; Matano et al. 2005).

Messinian units in the study area

The studied evaporitic deposits are stratigraphically related to the Daunia unit (Santo & Senatore 1988; Basso et al. 2001), which is made up of upper Oligocene–Burdigalian calcarenites, marls and clays ('Monte Sidone' Formation), Langhian–Lower Messinian calcareous–marly turbidites and hemipelagites ('Faeto Flysch' Formation), Middle Tortonian–Lower Messinian hemipelagic clayey marls ('Toppo Capuana' Formation), and locally Upper Messinian evaporitic deposits ('Evaporiti di Monte Castello' Formation).

The Early Messinian upper portions of Faeto Flysch and Toppo Capuana formations were deposited in basins located along the outer undeformed sector of the Late Miocene southern Apennine foreland basin (Pescatore & Senatore 1986), whose eastern margin was represented by the Apulian carbonate platform (Fig. 2). Only in the Late Messinian, after the deposition of the Evaporiti di Monte Castello Formation, was the Daunia unit deformed and stacked in the orogenic wedge (Matano et al. 2005), as testified by the latest Messinian post-evaporitic thrust-top units ('Torrente Fiumarella' unit and Anzano Molasse), which unconformably overlie the Daunia unit formations (Figs 3 & 4).

The Torrente Fiumarella unit includes lacustrine and alluvial conglomerates, quartzose sandstones containing abundant carbonate detritus, marly–silty

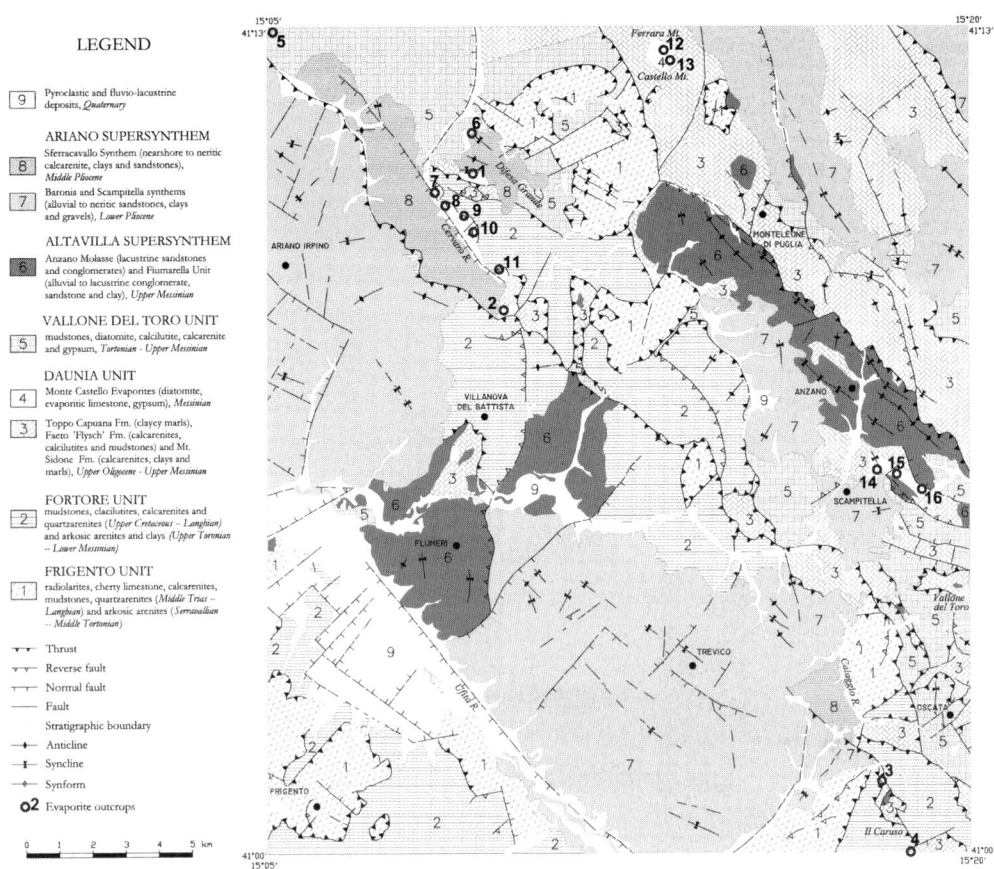

Fig. 3. Geological map of the Irpinia–southern Daunia Mt studied sector of southern Apennines (modified after Matano *et al.* 2005). The area location is shown in Figure 2.

clays and shales and reworked clastic gypsum. The Anzano Molasse Formation includes thick-bedded deltaic to turbiditic conglomerates and quartzo-feldspathic sandstones passing upward to thin-bedded turbidite sandstones and marly–clayey siltstones. Their sedimentary lithofacies, stacking patterns and petrography have been described in detail in Matano (2002), Matano *et al.* (2005) and Barone *et al.* (2006). Fossil associations in the post-evaporitic units are formed by rare freshwater ostracods, like *Cyprideis torosa*, *Ilyocipris gibba*, *Candona* sp. and *Loxoconcha* sp., and by Miocene reworked planktonic foraminifera (*Globigerina multiloba*, etc.) and calcareous nannoplankton (*Amaurolithus amplificus*, etc.) species (Basso *et al.* 2001, 2002; Matano 2002). The post-evaporitic units can be referred to a mainly fresh-water lacustrine basin with an overall deepening trend, receiving initially a great amount of coarse detritus from surrounding relief; then the basin shows a deeper-water phase characterized by finer-grained gravitative deposits (Matano *et al.* 2005; Barone *et al.* 2006).

All the Messinian intervals are characterized by the presence of some discontinuous decimetre-thick volcaniclastic layers (Fig. 4), which have been described in pre-evaporitic (Toppo Capuana Formation), evaporitic (Evaporiti di Monte Castello) and post-evaporitic (Anzano Molasse) units (Matano *et al.* 2005). In northern and central Apennines similar volcaniclastic layers have been used as key beds for basin-wide correlations (Vai 1997; Roveri *et al.* 2003, 2004); in the study area absolute age data are not available, even if the compositional data suggest a homogeneous felsic (rhyolitic to rhyodacitic) composition (Di Girolamo *et al.* 1986; Barone *et al.* 2006). New analyses are in progress in order to obtain absolute age and more complete compositional data, which could permit their use as key beds for correlations.

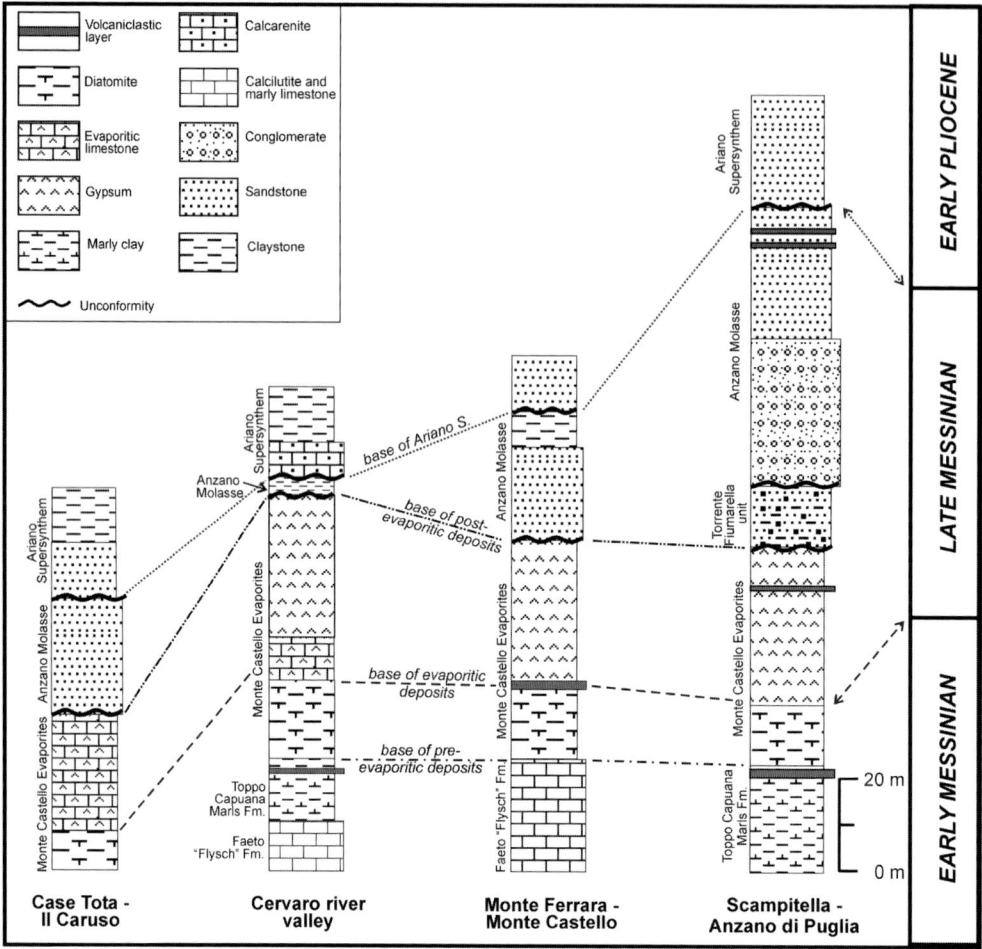

Fig. 4. Representative stratigraphic columns of the late Messinian successions cropping out in the Irpinia–southern Daunia Mts.

Stratigraphy of the Evaporiti di Monte Castello Formation

The Evaporiti di Monte Castello Formation comprises diatomaceous and euxinic marls, massive and laminated evaporitic limestone and primary and clastic gypsum. The evaporites crop out in an area of about 250 km² with small discontinuous exposures (Fig. 3); their thickness varies from 30 to 180 m with an average thickness of about 100 m. In the study area 16 sections have been studied in detail (Fig. 5), mainly located along the eastern slope of the Cervaro river, near Monte Castello and the city of Scampitella (Fig. 3).

The Evaporiti di Monte Castello Formation conformably overlies the Early Messinian portion of the Toppo Capuana and the Faeto Flysch formations, which belong to the Daunia tectonic unit, and is unconformably capped by the Late Messinian post-evaporitic clastic deposits of the Anzano Molasse (Figs 4 & 5). The Evaporiti di Monte Castello Formation comprises at its base thin-bedded whitish diatomites and organic-rich shales, dark marls and clays, whose thickness range from 0 to 40 m (logs 3, 8, 9, 12, 13 & 14; Fig. 5).

In the Il Caruso section (log 3; Fig. 5) the 20 m thick pre-evaporitic succession is formed by silty to sandy dark grey to blackish organic-rich laminated marls beds, 1–5 cm thick, rich of fish scales and skeletal remains with intercalations of whitish diatomites (Fig. 6). In the Contrada Ciccotti and Cervaro Mancone quarries sections (logs 8 and 9; Fig. 5) the 40 m-thick succession is formed at its

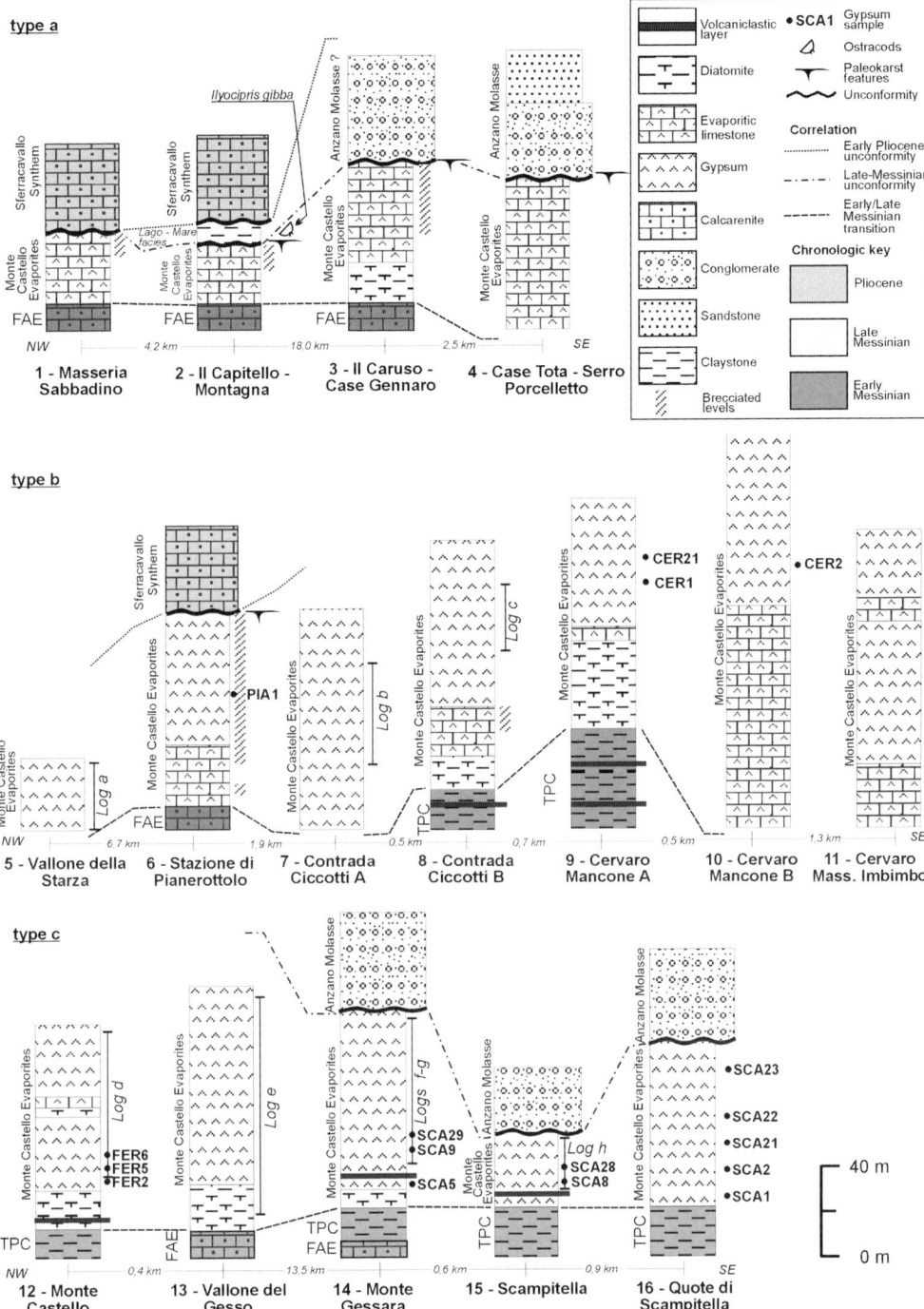

Fig. 5. Stratigraphical columns of the main outcrops of the Evaporiti di Monte Castello Formation. The location of the outcrops is shown in Figure 3.

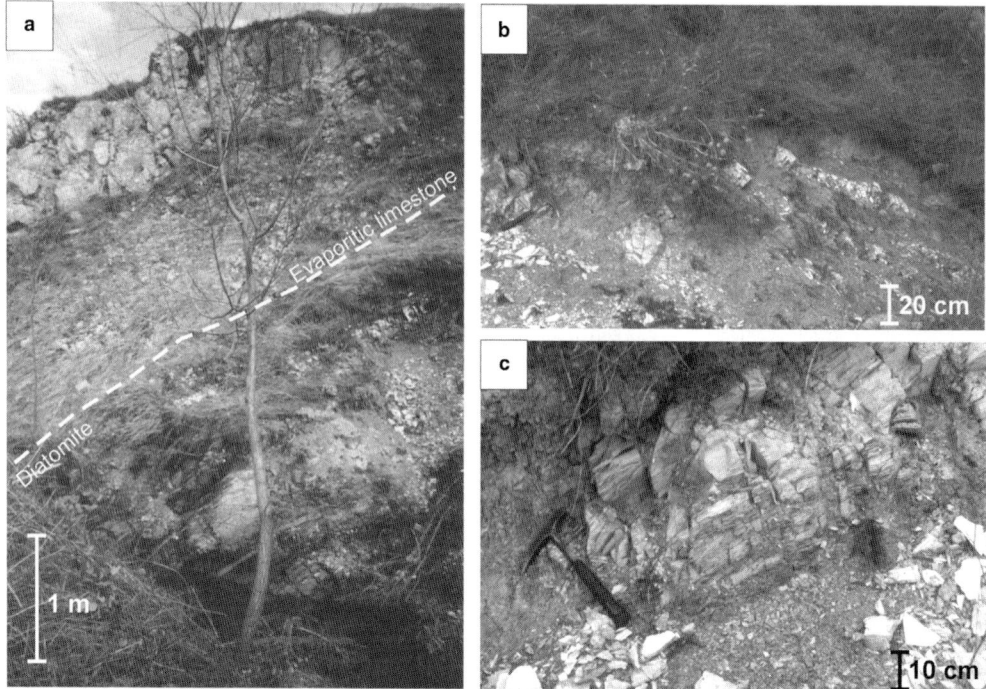

Fig. 6. Pre-evaporitic lithofacies in the 'Il Caruso' section (log 3, Fig. 5). (**a**) Transition between pre-evaporitic interval and evaporitic limestone. (**b** & **c**) Details of the dark grey, organic-rich, laminated marl beds with intercalations of white to pale tan diatomites.

base by light grey silty to marly clays with intercalations of whitish cineritic and marly beds, passing to greenish grey silty marls, rich of fish scales and skeletal remains with very thin beds of whitish diatomites. At the base of the Monte Gessara section (log 14; Fig. 5) about 5 m laminitic dark grey organich-rich marly clays have been recognized. Marl and clay alternations passing to diatomites are also present in the Vallone del Gesso (log 13) and Monte Castello (log 12) sections (Fig. 5).

The environment of deposition of the diatomitic and euxinic facies was probably a pelagic, more or less anoxic, starved basin (Dazzaro et al. 1988; Matano et al. 2005), as in other similar successions (Decima & Wezel 1971; Vai & Ricci Lucchi 1977; Bellanca et al. 2001; Blanc-Valleron et al. 2002).

The transition between pre-evaporitic and evaporitic intervals of the Evaporiti di Monte Castello Formation is characterized by a stratal conformity marked by a sharp change in lithology (Fig. 6a) from euxinic pelitic facies into evaporites (limestone and gypsum).

In some outcrops (logs 1–4; Fig. 5) the evaporitic limestones, up to 65 m thick, represent the whole evaporitic succession, even if more frequently the gypsum may follow (logs 6 & 8–11; Fig. 5) or is locally interbedded (logs 11 and 12; Fig. 5) with evaporitic limestone. In the Monte Castello and Scampitella areas the transition beetween pre-evaporitic deposits and gypsum is characterized by the absence of evaporitic limestone (logs 12–16; Fig. 5).

The limestones are usually massive or laminated, and have been referred to intertidal to supratidal settings (Decima & Wezel 1971; Dazzaro et al. 1988). The massive limestones are very abundant and are formed by whitish micrite and dolomicrite of mudstone/wackstone texture with millimetric, laminar gypsum and sulphur crystals, and sponge spicules. The laminated limestones are formed by whitish dolomicrite of mudstone/packstone texture marked by well-developed porosity (birdseye and fenestral). Some layers with laterally linked hemispheroidal (LLH) stromatolitic structures are also present (Dazzaro et al. 1988).

The limestones are commonly brecciated (Fig. 7), such as in 'Il Capitello' and 'Il Caruso' sections (logs 2 & 3; Fig. 5), where the brecciated levels are present in the upper part of the carbonate sequence, just under the contact with the post-evaporitic Anzano Molasse deposits. In the 'Il Capitello' section, a 2 m thick layer of breccias

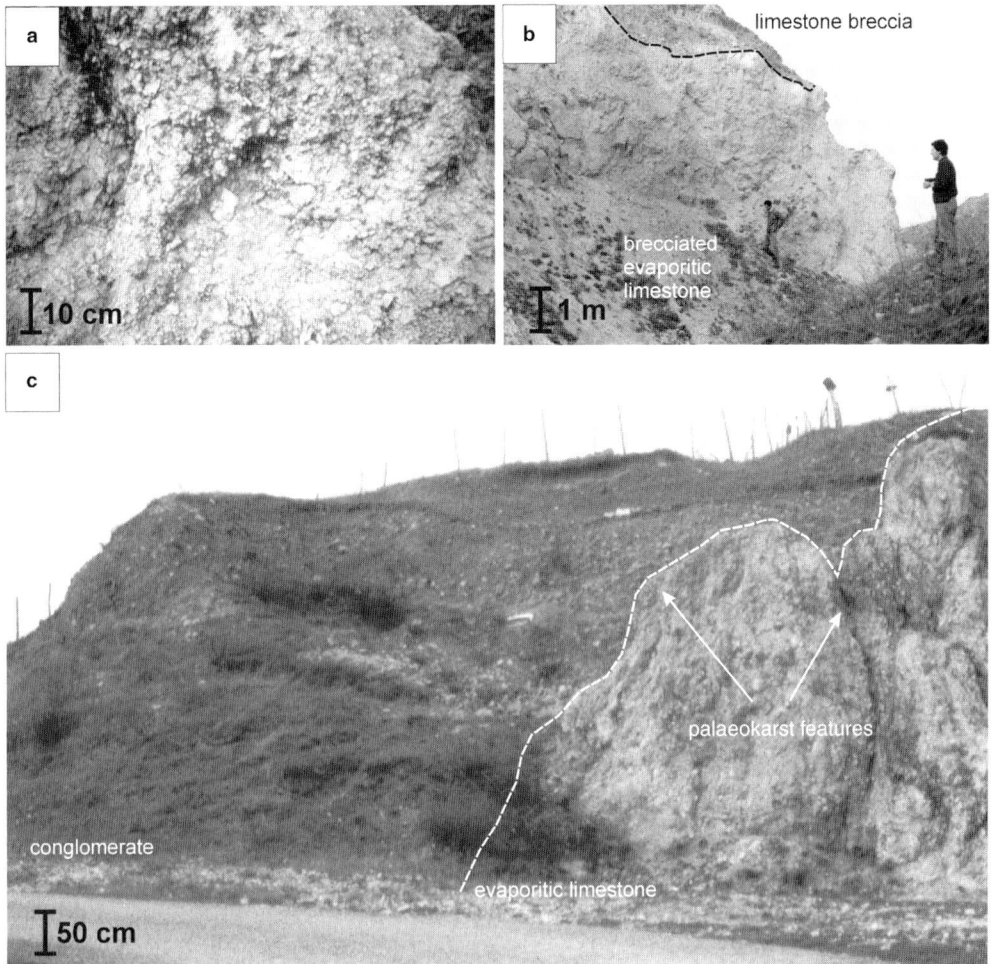

Fig. 7. Evaporitic limestone lithofacies. (**a**) Brecciated limestones in the 'Il Caruso' section (log 3, Fig. 5). (**b**) Brecciated limestones in the 'Il Capitello' section (log 2, Fig. 5) passing upward to a breccia layer. (**c**) Erosional unconformity surface between Late Messinian conglomerates and evaporitic limestone, marked by palaeokarst features in the limestone in the upper part of the 'Il Caruso' section (log 3, Fig. 5).

with limestone pebbles, showing only very local transportation, is present in the uppermost portion of the limestone succession (Fig. 7b). In the 'Il Caruso' and 'Il Capitello' sections the erosional unconformity surface is marked by well-developed desiccation cracks in the limestone (Fig. 7b & c). The cracks are locally filled by reddish silty clays and are characterized by subvertical oxidation bands in the fractured rock. The limestones are sometimes also brecciated at the transition with the gypsum, such as in the Stazione di Pianerottolo and Contrada Ciccotti B sections (logs 6 & 7; Fig. 5). The presence of gypsum crystals and *in situ* brecciated layers in evaporative limestones could suggest partial dissolution or bacterial breakdown of gypsum and replacement by carbonate (Schröder *et al.* 2003) or, more probably, solution collapse of interbedded carbonate and gypsum (Vai & Ricci Lucchi 1977).

In the Evaporiti di Monte Castello Formation the boundary between the evaporitic limestone and the overlying gypsum lithofacies is usually sharp and characterized by stratal conformity (logs 6 & 8–11; Fig. 5). In Monte Castello and Scampitella areas (logs 12–16; Fig. 5) evaporitic limestones are absent and gypsum conformably overlies pre-evaporitic diatomitic and euxinic marls or directly Toppo Capuana Formation clayey marls.

The gypsum can be found as primary crystalline facies (mainly selenites), nodular gypsum and reworked coarse to fine clastic gypsum with a thickness up to 80 m. The gypsum is rarely brecciated, such as in the Stazione di Pianerottolo section (log 6; Fig. 5).

Both evaporitic limestones and gypsum of the Evaporiti di Monte Castello Formation are unconformably capped by the post-evaporitic clastic deposits of the Anzano Molasse and Torrente Fiumarella units (Figs 4 & 5); the angular unconformity surface is erosional and is sometimes marked by desiccation cracks and brecciated levels in both gypsum and evaporitic limestone (logs 1–4 & 6; Fig. 5), suggesting possible subaerial exposure and erosion (Matano et al. 2005).

Stratigraphical relations

The detailed study and the correlation of the evaporites sections across the study area (Fig. 5), approximately along NW–SE direction, which probably corresponds to the basin axis, allow the recognition of stratigraphic and lithofacies relations. Three main types of evaporite successions (Fig. 5) are found at different locations in the study area (Fig. 3): (a) the first type consists entirely of evaporitic limestones (logs 1–4); (b) the second type is composed of evaporitic limestones passing upward to gypsum (logs 5–11); and (c) the third type is formed entirely of gypsum (logs 12–16). All the three different types of successions are locally characterized by the presence of diatomaceous and euxinic marls at their base. The differences among the evaporitic successions could be related to their position in the basin and to its morpho-structural evolution.

Type 'a' sections (Fig. 5) are formed only by 30–70 m-thick evaporitic limestones, which are commonly brecciated; they are located along the hills near the Cervaro River, in the northern sector of the study area, and southward the Calaggio River, in the southern sector of the study area (Fig. 3). The limestone-only sections are always truncated at the top by an erosive unconformity, so it is not possible to reject that their highest portions could have been eroded and that the eroded portion could be formed by gypsum.

Type 'b' sections (Fig. 5), formed by both limestone and gypsum up to 180 m-thick, are located in the northwestern sector of the study area, mainly along the Cervaro River (Fig. 3). In some cases their different features seem to be related only to present-day outcrop conditions; for example some gypsum-only sections (logs 5 & 7) have the base and top missing.

Type 'c' sections (Fig. 5), which do not contain evaporitic limestones, are 40–100 m in thickness and are located in the northeastern sector of the study area, near Monte Castello and Scampitella (Fig. 3).

Gypsum lithofacies

Various types of gypsum lithofacies have been found within the outcrops studied (Figs 5 & 8). They can be attributed to two broad types of gypsum (Peryt 1996), autochthonous gypsum, mainly including selenites, acicular and laminated gypsum, and allochthonous gypsum, including fine-grained laminated gypsum, gypsarenites, pebbly gypsarenites and gypsrudites. Secondary gypsum typically occurs as scattered nodules or layers in the laminated gypsum facies (nodular-laminated gypsum). Recognition and classification of gypsum facies and facies groups have been done mainly following the nomenclature of Vai & Ricci Lucchi (1977), Michalzik (1996) and Peryt (1996, 2000) and for resedimented evaporites Manzi et al. (2005).

Selenitic gypsum

Description Selenitic gypsum is exposed as massive beds of crystalline gypsum, up to more than 3 m thick each (logs a–c & h; Fig. 8). The best exposures of this facies are from the 'Vallone della Starza' (log a) and the Contrada Ciccotti A (log b) outcrops (Fig. 8). Beds of large twinned crystals form the main type of selenite (Fig. 9). Individual beds consist of gypsum crystals, mostly shallow-tail twins normally arranged vertically with respect to the depositional surface, with the crystal size ranging from 10 to 50 cm, and with scarce detrital infillings. Many selenitic crystals, in particular the larger ones, are very limpid without visible impurities. Algal filaments are present in some beds (Dazzaro et al. 1988). Grass-like gypsum beds (Michalzik 1996) are sometimes present and are formed by 5–10 cm thick massive beds of smaller vertically arranged gypsum (selenitic) twins.

Interpretation The Messinian selenitic gypsum is generally accepted as the product of primary, bottom-nucleated growth in a shallow-subaqueous environment (Schreiber & Decima 1976; Schreiber et al. 1976; Vai & Ricci Lucchi 1977; Michalzik 1996), in which brine concentration was just enough to allow gypsum precipitation (Ciarapica & Passeri 1980). Bottom-nucleated selenite deposition is usually referred to shallow (<1–20 m) hypersaline pools or salinas (Schreiber et al. 1982), or also to a marginal lagoonal environment with stratified waters up to a few metres deep (Kasprzyk 2003).

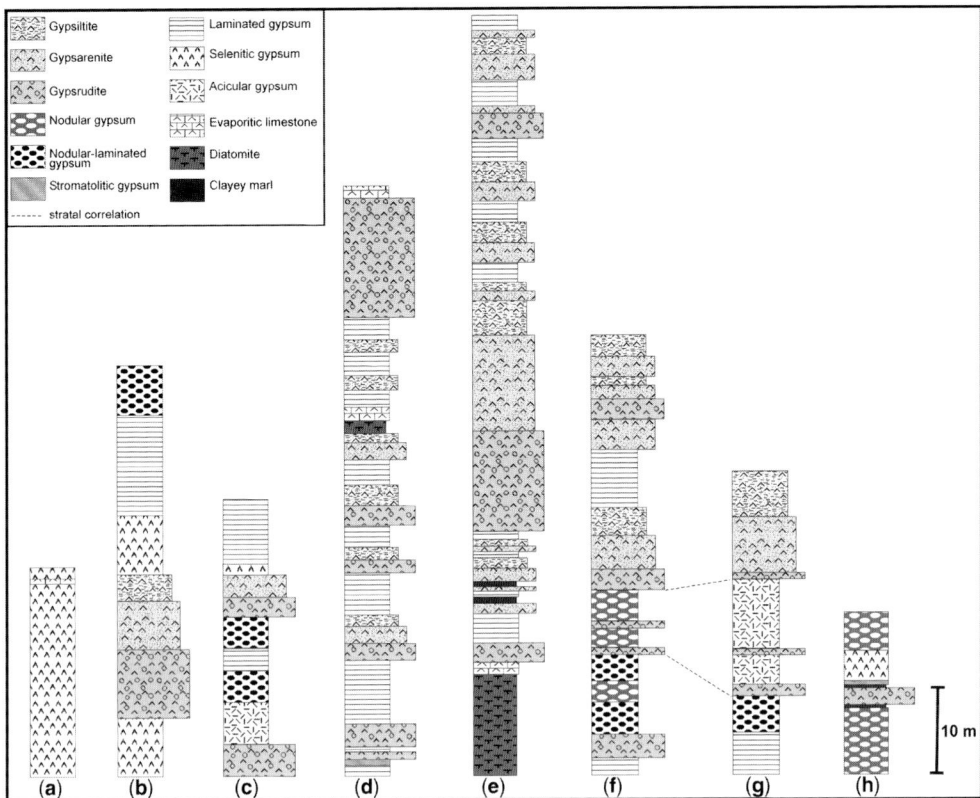

Fig. 8. Selected stratigraphical logs of the main gypsum outcrops. Logs a, d, e & g have been reinterpreted and redrawn by Dazzaro *et al.* (1988). Log names: (**a**) Vallone della Starza, (**b**) Contrada Ciccotti A, (**c**) Contrada Ciccotti B, (**d**) Monte Castello, (**e**) Vallone del Gesso, (**f** & **g**) Monte Gessara and (**h**) Scampitella. The position of the logs is shown in the stratigraphical columns of Figure 5.

There is some evidence for an early diagenetic origin of some structures, such as some large selenite crystals (Michalzik 1996). Alternatively, selenitic gypsum was considered to be a diagenetic secondary product (Ogniben 1957) or is referred to deposition in deep water (a few hundreds of metres) settings (Kirkland *et al.* 2000). The selenitic facies of Evaporiti di Monte Castello Formation shows evidence for deposition in subtidal shallow water setting (Dazzaro *et al.* 1988).

Acicular gypsum

Description Beds, 1–3 m thick, of elongated, up to 5 mm long, crystals form the acicular gypsum (logs c & h; Fig. 8). The gypsum fabric is relatively random. The acicular gypsum layers may show irregular lamination and are often interbedded with laminated, locally cross-bedded and rippled, gypsarenite lenses (Fig. 10). Some layers of granular gypsum are interbedded within acicular gypsum, gypsum granular crystals of arenitic size but without fine-grained matrix.

This gypsum lithofacies is very abundant in the Contrada Ciccotti B and Monte Gessara quarries (logs c & h; Fig. 8).

Interpretation Gypsum precipitates as small needles in hypersaline water bodies, characterized by elevated salinity (300–325 g/l) near halite saturation (Rosell *et al.* 1998). The small prismatic gypsum crystals form near the top of the water column and at boundaries between layers in stratified water body and accumulate at the bottom of the basin as cumulates (Schreiber & El Tabakh 2000).

Laminated gypsum

Description Thick beds of stratified laminated gypsum are one of the most remarkable features in outcrops of the studied evaporitic deposits (logs b–g; Fig. 8). They usually consist of evenly

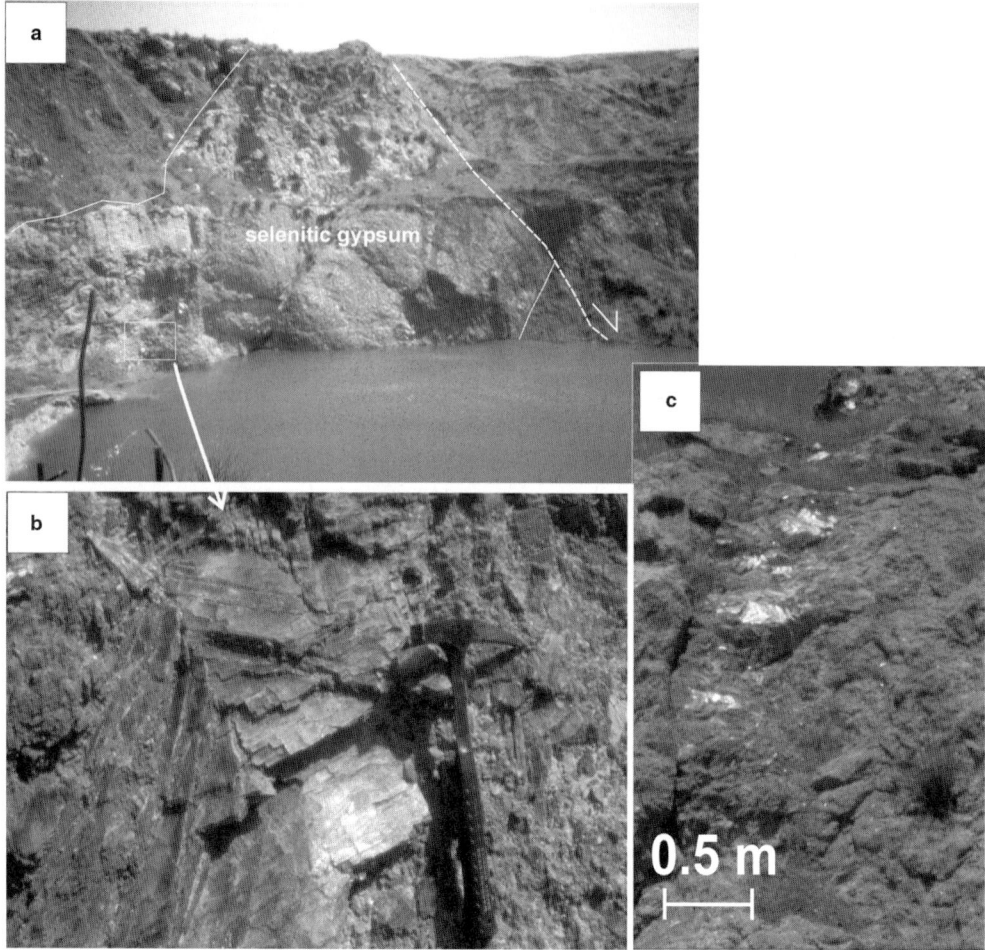

Fig. 9. Selenitic gypsum lithofacies. (a) General view of Contrada Ciccotti A quarry (log 7, Fig. 5). (b) Large gypsum twin crystal. (c) Bed of shallow-tail twin crystals arranged vertically with respect to the bedding surface.

laminated (millimetre-scale) layers, whose colouring ranges from white to very dark grey (Fig. 11a). The lamination is mainly parallel and sometimes undulating; it may also show enterolithic or crinkly. The beds are up to 20 cm thick and form bed sets up to 10 m thick, displaying a certain lateral continuity. Laminated gypsum can be of both clastic and microcrystalline nature.

Reworked fine- to very fine-grained crystals, crystal fragments and gypsum or carbonate clasts form the fine-grained laminated gypsum (similar to facies R6-R7, Manzi et al. 2005); planar and wavy parallel lamination is present. Medium- to fine-grained graded gypsarenites, gypsiltites and clays interbeds are also present.

Very small acicular or granular gypsum crystal layers alternated with micrite layers can sometimes form the laminated gypsum. Stromatolitic gypsum layers (microbialites of Bąbel 2005), up to 40 cm thick, are present in logs d & h (Fig. 8). They are formed by gypsum alternating with organic-rich laminae with domal and crinkled lamination (Dazzaro et al. 1988).

Interpretation Laminated sulphates could represent deeper water depositions (below a few tens of metres; Schröder et al. 2003). The evenly laminated fine-grained gypsum with thin intercalations of graded gypsarenites represents very fine-grained clastic reworked gypsum, related to low-density turbiditic currents and to gypsum-saturated density flows associated with hemipelagic sedimentation (Manzi et al. 2005), suggesting a somewhat deeper and more distal subaqueous environment

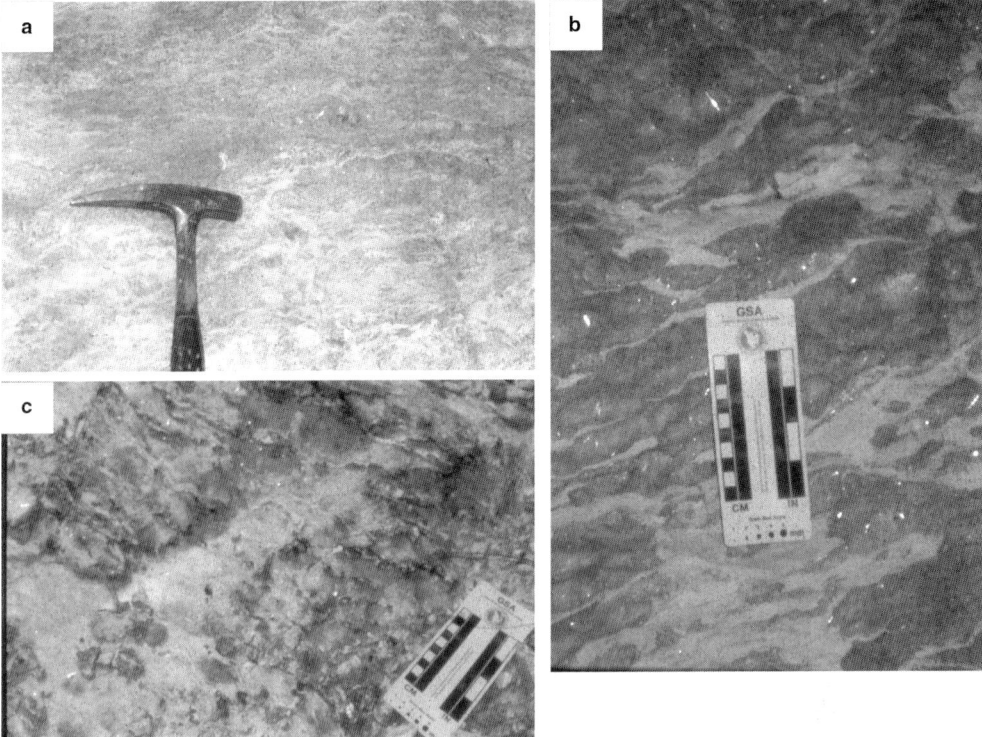

Fig. 10. Acicular gypsum lithofacies. (**a**) Irregular lamination in acicular gypsum. Monte Gessara quarry (log g, Fig. 8). (**b** & **c**) Gypsarenite lenses interbedded within acicular gypsum in Monte Gessara (b) and in Contrada Ciccotti B (c) quarries.

of deposition, such as outer platform and slope with density stratified waters (Kasprzyk 2003) or intrabasinal highs (Manzi *et al.* 2005). The enterolithic folding may be attributed to *in situ* soft sediment deformation (Michalzik 1996) or to diagenetic transformations such as rehydration (Schreiber & El Tabakh 2000).

The laminated gypsum strata characterized by a microcrystalline texture can be generally attributed to chemical precipitation in shallow or very shallow subaqueous settings (Schreiber *et al.* 1976; Playà *et al.* 2000). The alternation of gypsum and micrite laminae suggests alternating salinity in a shallow gypsum salina (Schröder *et al.* 2003) with episodes of dilution (carbonate) and concentration (gypsum). The stromatolitic gypsum growth was probably favoured during prolonged periods of lowered salinity (Peryt 1996).

Nodular gypsum

Description The nodular gypsum is commonly associated with laminated gypsum; individual scattered white nodules, aggregated or packed nodules and nodular layers form it (Fig. 11b–d). Nodules display large variations in size (millimetres to centimetres) and shape, which is occasionally elliptical and spheroidal but usually irregular (Fig. 11b–d). The nodules present peloidal, elongated and grumous (clotted: 'structure grumeleuse') structures; the adjacent gypsum laminae show features of displacive growth in the host sediment. The nodular-laminated lithofacies occurs as an irregular alternation of a set of usually dark gypsum laminae and layers of isolated or coalescent whitish nodules (up to 3 cm), forming massive beds a few decimetres to 5 m thick. This lithofacies is very abundant in the Contrada Ciccotti A and B outcrops and in the quarries near Scampitella (logs b, d, f, g & h; Fig. 8).

Interpretation The nodular gypsum lithofacies is usually considered to have been formed diagenetically by rehydration of anhydrite back to gypsum. Michalzik (1996), Playà *et al.* (2000) and Dazzaro *et al.* (1988) have reported evidences for a sabkha origin in a supratidal environment of these gypsum nodules. The association with subaqueous

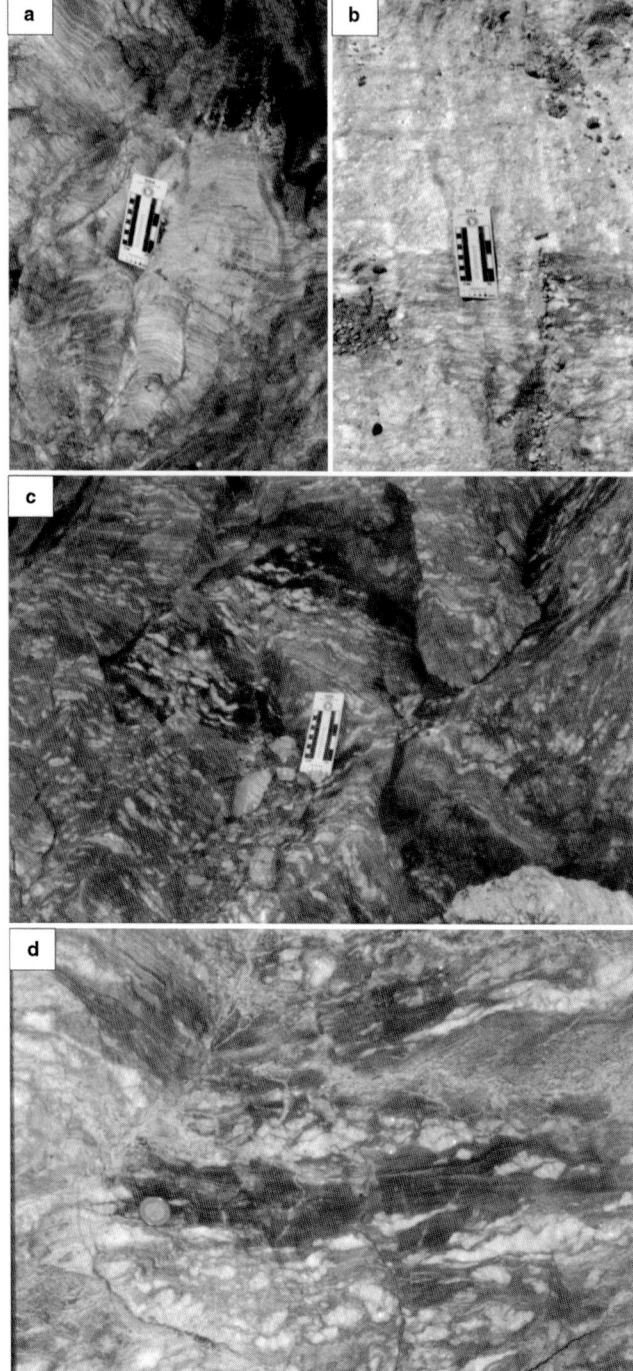

Fig. 11. Laminated and nodular gypsum lithofacies. (a) Even, thinly laminated gypsum (millimetre-scale), with mainly planar lamination. Monte Gessara quarry (log g, Fig. 8). (b) Nodular bedded gypsum with horizontally arranged whitish nodules changing in size from 1–2 cm at the bottom to 5–6 cm at the top of the photo. Contrada Ciccotti A. (c) Laminated gypsum with layers of isolated whitish, 1–2 cm thick gypsum nodules. Monte Gessara quarry (log g, Fig. 8). (d) Laminated gypsum with layers of coalescent whitish 2–3 cm thick gypsum nodules and gypsiferous silty lenses. Monte Gessara quarry (log g, Fig. 8).

sediments (laminated and reworked gypsum) and the observed deformation in adjacent gypsum laminae of surrounding sediment produced by their growth indicate a late diagenetic origin of these nodules linked with partial anhydritization of gypsum (Ciarapica & Passeri 1980; Peryt 1996) and related to interstitial growth of displacive sulphate within laminated gypsum lithofacies.

Resedimented clastic gypsum

Description Thick beds of coarse clastic gypsum constitute a considerable part of the evaporite succession (logs b–h; Fig. 8); they are made up of gypsrudites, pebbly gypsarenites, gypsarenites and subordinated gypsiltites. A general feature at the metric scale is the presence of fining-upward sequences of allochthonous clastic gypsum formed by a basal gypsruditic or pebbly gypsarenitic beds followed by massive or graded gypsarenite passing to gypsiltite and sometimes to fine-grained laminated gypsum. The gypsum detritus mainly derived from erosion and reworking of coeval exposed crystalline and clastic gypsum beds.

The gypsrudites (similar to facies R1, Manzi *et al.* 2005) are formed mainly of beds up to 10 m thick of disorganized, usually matrix-supported, crudely stratified gypsrudites, characterized by gypsum pebbles chaotically embedded in a sandy–silty gypsiferous, and shaly matrix (Fig. 12a & b); pebbles are subrounded to angular, up to 50 cm in size, and include selenite twins, nodular and laminated gypsum, gypsarenite, evaporitic limestones and clay chips. The gypsrudite beds frequently show erosive basal contacts. Some layers of clast-supported gypsrudites are also present (Fig. 12c).

The pebbly gypsarenites (similar to facies R2, Manzi *et al.* 2005) are formed by well-stratified massive coarse gypsarenites, granular gypsrudites and micro-gypsrudites forming medium to thick beds, which often present basal erosional surfaces. The gypsarenites (similar to facies R3–R5, Manzi *et al.* 2005) are well-stratified, massive or graded, with medium beds with basal erosional surfaces; sometimes they have laminated, cross-bedded or rippled intervals. The gypsarenites are usually coarse-grained and sometimes interbedded with evenly laminated fine-grained clastic gypsum (Fig. 12d), and often follow gypsrudites and pebbly gypsarenites layers.

The gypsiltites (similar to facies R6, Manzi *et al.* 2005) are formed by some decimetres up to 4 m thick graded and plane laminated beds of gypsum siltites and fine gypsarenites; in the Vallone del Gesso section (log e, Fig. 8) the presence of layer rich of oncolites and *Quercus* sp. leaves has been reported by Dazzaro *et al.* (1988), indicative of proximity of emerged lands.

Interpretation Beds of coarse clastic gypsum represent gravity flow deposits derived from intrabasinal reworking. The gypsrudites and the coarse gypsarenites include debris flows and hyperconcentrated (granular) flows, sometimes evolving to lower-density turbiditic flows represented by tabular graded and laminated gypsarenitic strata. They represent proximal deposits, which can be referred to shallow- to deeper-water environments. The gypsiltitic strata can be referred to processes within low-density turbiditic currents (Manzi *et al.* 2005) and could represent more distal deposits.

The occurrence of gypsum turbidites and debris flows does not necessarily indicate a great water depth but suggest the existence of parts of the basin with considerable paleoslope.

Strontium geochemistry of the gypsum

The Sr geochemistry of the studied gypsum has been discussed in Matano *et al.* (2005). The available data (Sr concentrations and the Sr isotope) about Evaporiti di Monte Castello Formation gypsum are plotted in Figure 13 and a brief discussion is reported.

Strontium content

The strontium content of the gypsum is between 1560 and 1900 ppm (Fig. 13) with an average value of 1774.7 ± 118.7. Because during the analytical preparation of the samples the possible contribution of other source of Sr has been minimized (Matano *et al.* 2005), the Sr content values suggest no chemical reworking of the studied gypsum with loss of Sr, and all the analysed gypsum samples can be referred to as primary gypsum.

The various gypsum lithofacies of the Evaporiti di Monte Castello Formation present low differences in the average values of the Sr content: (1) 1820 ± 160 ppm for the acicular gypsum; (2) 1787.5 ± 89.6 ppm for the laminated gypsum; and (3) 1736.7 ± 155 ppm for the entherolithic gypsum. The small differences in the strontium contents among these analysed lithofacies of the Evaporiti di Monte Castello Formation gypsum could reflect deposition under hydrologically stable conditions.

It has been assumed that salinity is the main factor controlling the strontium coprecipitation in the gypsum, so the Sr content can be used as a salinity indicator in the gypsum facies (Rosell *et al.* 1998). The acicular facies present the relatively highest range of Sr values, as in other studied contexts (Rosell *et al.* 1998), where it is

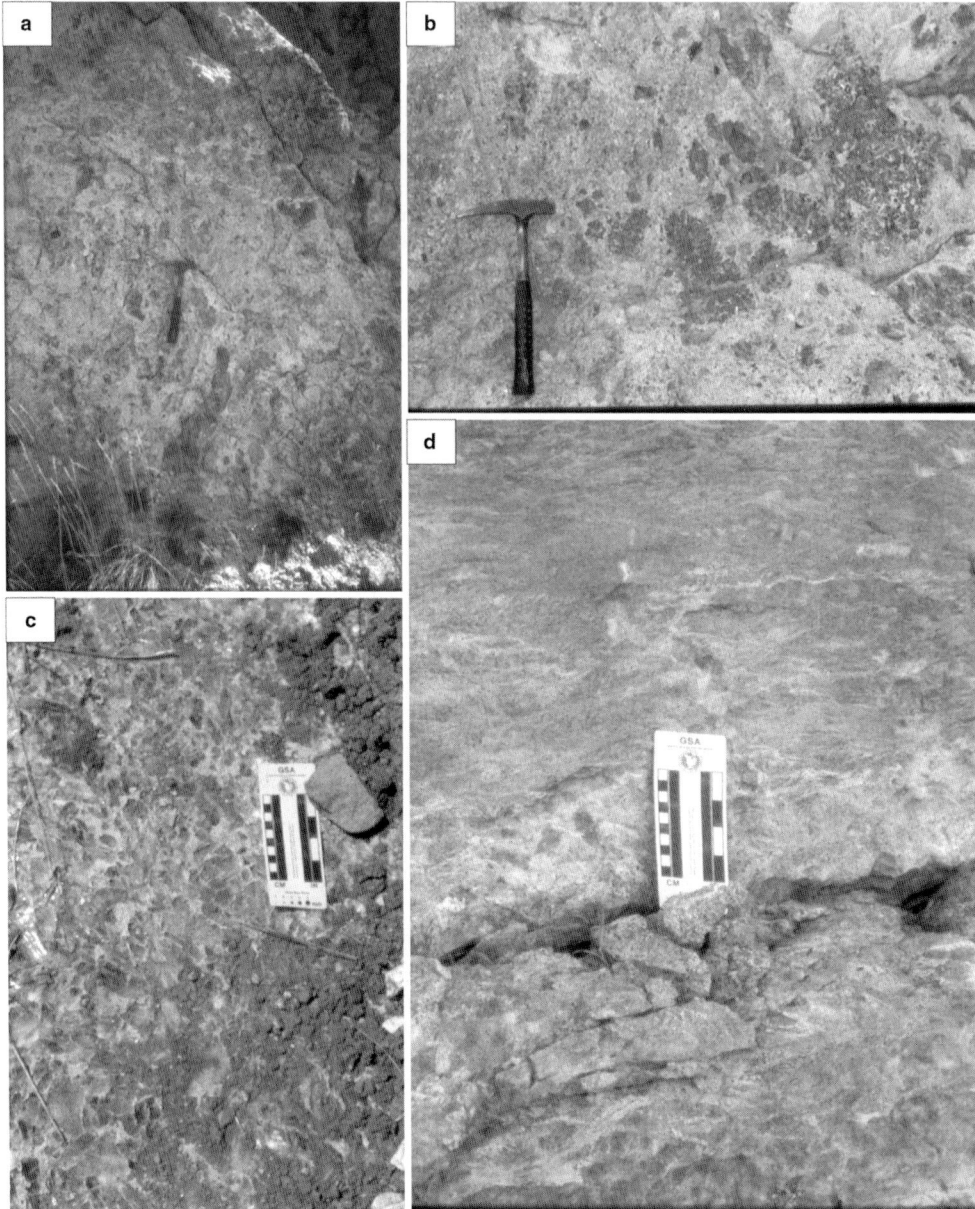

Fig. 12. Coarse clastic gypsum lithofacies. (**a, b**) Disorganized gypsruditles with sandy-silty matrix-supported fabric and angular to subrounded gypsum pebbles. Monte Gessara quarry (log g, Fig. 8). (**c**) Clast-supported gypsrudite with clayey–silty matrix. Contrada Ciccotti A quarry. (**d**) Graded gypsarenite bed with erosive basal contact and a laminated interval in the upper part of the photo. Monte Gessara quarry (log g, Fig. 8).

considered as formed under salinities between 300 and 325 g/l; also the laminated facies present similar even if lower values. The control on salinity required for gypsum precipitation is not exclusively a proportion of fresh/marine water but is forced also by evaporation. The high values of salinity assumed for the Evaporiti di Monte Castello Formation gypsum could have been reached with net evaporation to dominate the local palaeohydrological budget (Flecker *et al.* 2002).

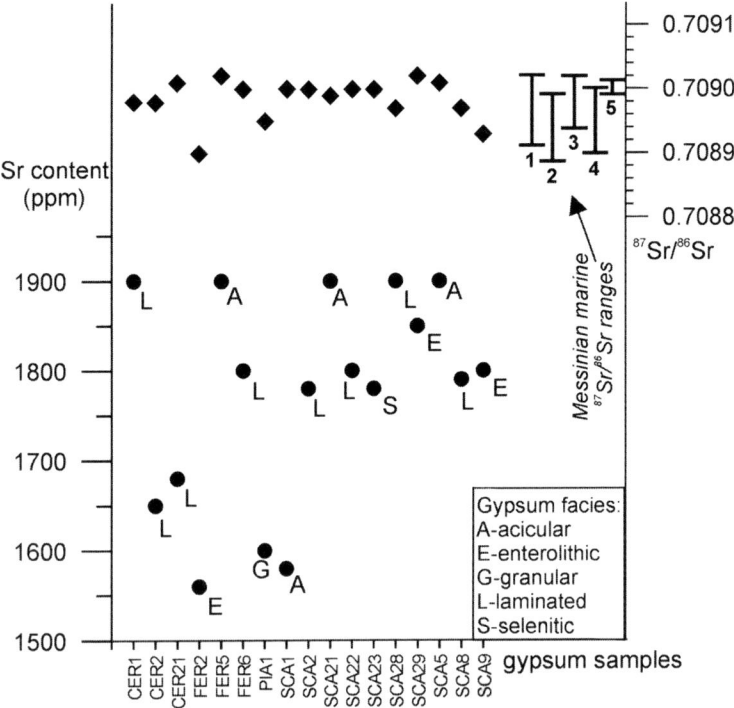

Fig. 13. Sr content (black circles) and $^{87}Sr/^{86}Sr$ isotopic ratio (rhombs) of the gypsum samples collected in the Evaporiti di Monte Castello Formation (modified after Matano et al. 2005). The analytical error for the $^{87}Sr/^{86}Sr$ isotopic ratio is $\pm 2 \times 10^{-5}$ ($2\sigma SD$). The bars of the Messinian marine $^{87}Sr/^{86}Sr$ values range refers to: (1) McKenzie et al. (1988) and Müller et al. (1990); (2) Hodell et al. (1989, 1991) and Müller & Mueller (1991); (3) Keogh & Butler (1999); (4) Howarth & McArthur (1997) and McArthur et al. (2001); (5) Flecker et al. (2002). The gypsum lithofacies of each analysed sample is given. Samples location is shown in Figure 5.

Strontium isotopes

The gypsum shows values ranging from 0.70890 ± 2 ($2\sigma SD$) to 0.70902 ± 2 ($2\sigma SD$) with an average value of 0.70898 ± 3. These values, considering the analytical errors, fall within the range reported for the Messinian seawater (Fig. 13) by most authors (McKenzie et al. 1988; Müller et al. 1990; Hodell et al. 1991; Müller & Mueller 1991; Keogh & Butler 1999) and suggest a major marine origin of the gypsum, even if we are not always in the presence of strictly oceanic values following McArthur et al. (2001) and Flecker et al. (2002). Some contribution of continental freshwater was also suggested by Dazzaro et al. (1988) on the basis of the very low values of Na/Ca ratio and of the very high Ca and Mg content in the evaporite parent brines.

By comparing the $^{87}Sr/^{86}Sr$ data with those reported for upper and lower evaporites (Müller & Mueller 1991; Keogh & Butler 1999; Playà et al. 2000; Flecker et al. 2002), the studied gypsum could be referred to the 'lower gypsum' of the Lower Evaporites in the Mediterranean (Müller & Mueller 1991; Flecker et al. 2002), suggesting that its deposition occurred when the Mediterranean was still connected with the global ocean.

Gypsum facies relations

Eight stratigraphic sections were measured in the gypsum lithofacies (Fig. 8). A general feature of gypsum beds (Figs 5 & 8) is their lithofacies variability at the outcrop scale and their strong lateral discontinuity (Fig. 14) for the intense tectonic deformation that characterizes the study area (Fig. 3). As a matter of fact the outcrops can be walked out along strike only for a distance from 50 to 500 m, and so lateral persistence and thickness uniformity of gypsum strata cannot be observed.

The studied sections show an extensive development of reworked clastic gypsum, varying from 60 to 100% of thickness (Fig. 8, except for log a), whereas autochthonous gypsum usually does not

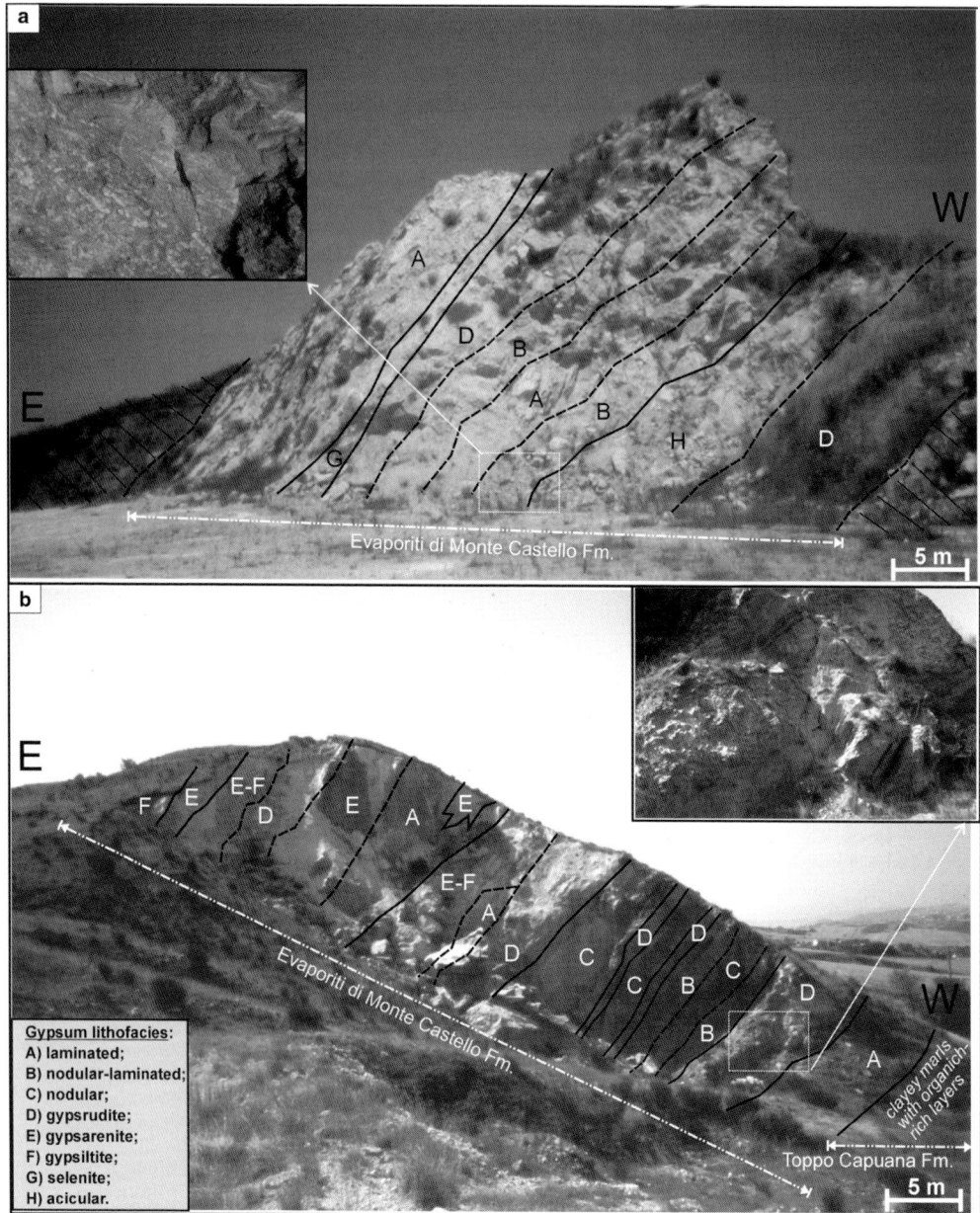

Fig. 14. General views of two type of gypsum sequences: (**a**) Ciccotti B quarry (log 8, Fig. 5; log c, Fig. 8), showing alternating autochthnous gypsum and redeposited clastic gypsum; and (**b**) Monte Gessara quarry (log 14, Fig. 5; log f, Fig. 8), showing redeposited and laminated clastic gypsum organized in fining-upward strata set.

exceed 40% of the total thickness (logs b, c, g & h in Fig. 8) and sometimes is lacking (logs d–f in Fig. 8); only in Vallone della Starza section (log a, Fig. 8) does the selenitic gypsum reach up to almost 100% of the thickness.

Gypsum lithofacies are usually arranged in a general trend characterized by alternation of metric to decimetric horizons of redeposited clastic gypsum and of autochthonous gypsum (logs b, c, g & h in Fig. 8, and Fig. 14a). Sections formed only by reworked gypsum frequently occur (logs d-f in Fig. 8, and Fig. 14b), while selenite-only intervals are less frequent (log a, Fig. 8).

The type-b sections of Fig. 5, containing evaporitic limestone at their base and located to the NW of study area, are usually formed by gypsum lithofacies including both autochthonous gypsum (selenite and acicular gypsum) and clastic gypsum (logs a-c, in Fig. 8). The type-c sections of Fig. 5, which do not contain the evaporitic limestone and are located in the eastern sector of the study area, can be formed both by redeposited clastic gypsum only (logs d-f in Fig. 8) and by alternating autochthonous and allochthonous gypsum (logs g & h in Fig. 8). Examples of the outcrop condition of the two section types are given in Fig. 14, showing facies geometric relations and gypsum bedding in Contrada Ciccotti B quarry (Fig. 14a) and in Monte Gessara quarry (Fig. 14b).

Gypsum facies associations

Genetically related gypsum lithofacies record different depositional environments and can be grouped into two main facies associations: (a) crystalline autochthonous gypsum and (b) redeposited clastic gypsum. Nodular structures in the laminated gypsum lithofacies are mostly of late diagenetic origin.

The crystalline authochthonous facies association consists of selenitic gypsum, acicular gypsum and laminated gypsum, characterized by frequent facies changes, and is based on the absence of high-energy sedimentary structures. The autochthonous gypsum could have been mostly influenced by composition and salinity of brines from which it precipitated (Peryt 1996), as a matter of fact in recent salinas different evaporite facies originate from brine of different salinity, not necessary at different depth (Rosell *et al.* 1998). For example, the highest Sr content mean value (1820 ppm) is found within acicular gypsum variety, which could be referred to salinity between 300 and 325 g/l. Frequent facies changes indicate a dynamic chemical environment occurring in shallow water settings, when even a small additional brine input could influence the established chemical regime (Peryt 1996), while in slightly deeper settings it would be negligible. The presence of bottom-nucleated selenites confirms this interpretation. Some evidence is present of wave action and brine diluition by freshwater inflows, such as detritic laminae, sandy lenses, cross-lamination and variations in Sr geochemistry.

The redeposited clastic facies association consists of fine-grained laminated gypsum, gypsiltites, gypsarenites, pebbly gypsarenites and gypsrudite, showing common features of redeposited sediments that might have formed both in shallow-water and in deeper-water environments. These facies are interpreted as gypsum clastic deposits derived from the reworking and the redeposition of the earlier deposited or coeval shallow-water gypsum deposits. Facies patterns show lateral and vertical facies changes from breccias and matrix-supported gypsrudites to fine-grained laminated gypsum and gypsiltites.

The occurrence of breccia accounts for synsedimentary deformation (Peryt 2000), probably related to the tectonic activity in the southern Apennines flexured Apulian foreland. Also hydrodynamic events may led to a greater or lesser degree of gypsum, forming thick turbidite and debrite layers in more distal basin sectors with a higher local gradient and relatively thinner clastic gypsum layers alternated with autochthonous gypsum in shallower basin sectors. The fining-upward sequences of redeposited clastic gypsum are probably related to single local hydrodynamic events or to local reactivations of the boundary faults of the basin, difficult to correlate between different sectors of the basin.

Depositional setting of the evaporites

As documented in the Central Adriatic Sea and in the Hyblean plateau in Sicily, evaporitic basins located in a foreland ramp setting usually developed in a semi-closed and silled basin, formed as half-graben during the late Miocene. For the studied Monte Castello evaporative basin, a preliminary schematic model of facies relationships and geological evolution during the Messinian is proposed (Fig. 15), among other possibilities.

The overall lithofacies successions (diatomite–carbonate–autochthonous gypsum) suggest deposition under increasingly evaporitic salinity conditions (Schreiber & El Tabakh 2000). A general trend of increasing basin restriction from marine conditions, represented by the Early Messinian Faeto Flysch and Toppo Capuana formations marine deposits at the base of the diatomites, to euxinic settings (organich-rich clays and marls) is the prelude of the salinity crisis. The diatomaceous marl layers could represent episodic increase in bio-siliceous productivity (diatomites) (Blanc-Valleron *et al.* 2002) during the pre-evaporitic deposition of marls and laminites (Fig. 15a). The pre-evaporitic diatomaceous and euxinic marls underlie both evaporitic limestone and gypsum (Fig. 5), and the transition between pre-evaporitic and evaporitic intervals is characterized by a stratal conformity marked by a sharp change in lithology from pelagic euxinic pelitic facies into evaporitic facies.

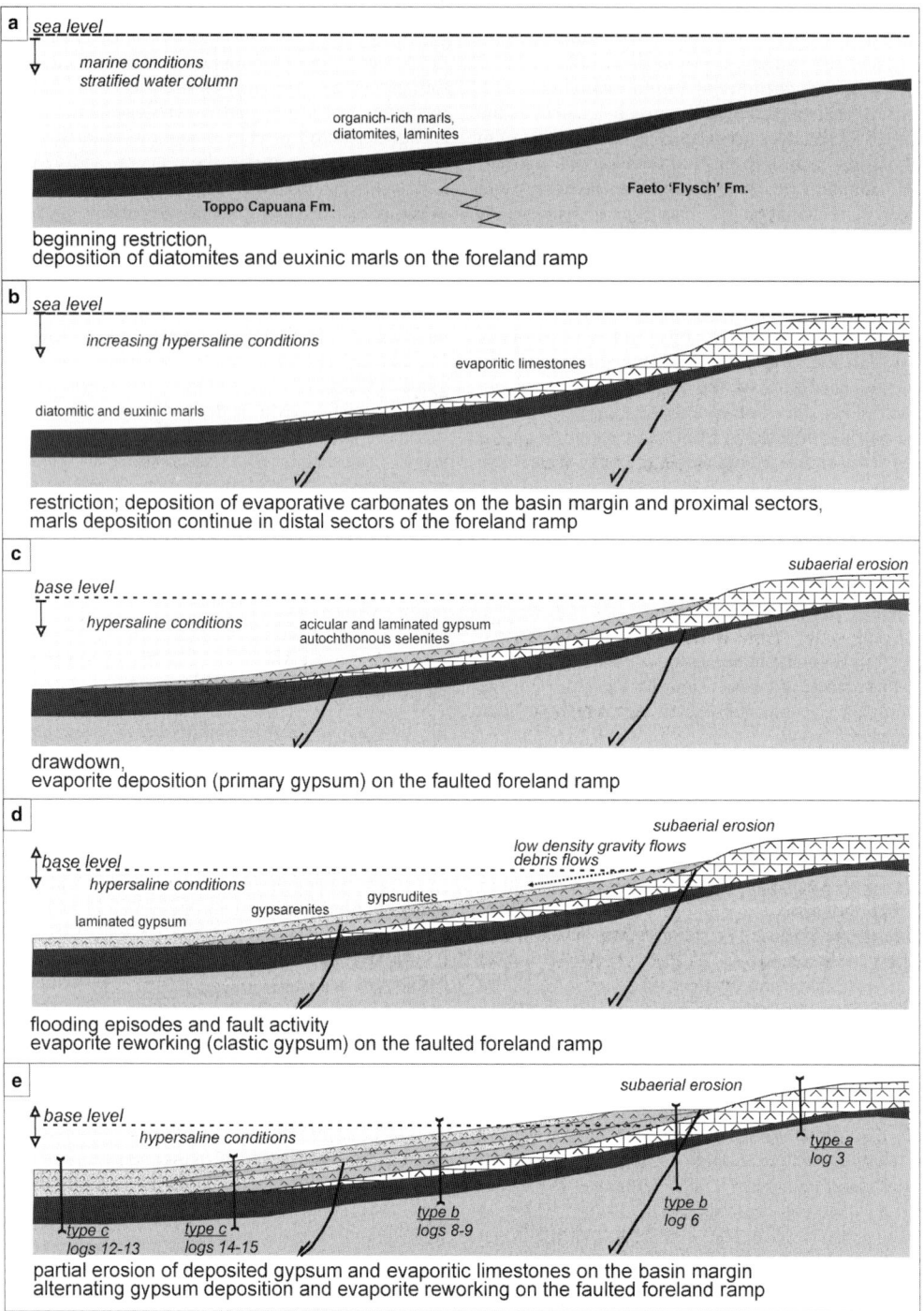

Fig. 15. Sketches showing the Messinian evolution of the Evaporiti di Monte Castello basin from Early (**a**) to Late (**b–e**) Messinian. In (e) reference has been made to logs and types of Figure 5.

The evolution to the evaporitic limestone witnesses the rapid transition to semi-closed basin setting in the early Late Messinian, characterized by increasing hypersaline conditions (Fig. 15b). The evaporative carbonates, referred to intertidal settings (Dazzaro et al. 1988), could have formed before major evaporite deposition, mainly in marginal structural highs of the evaporitic basin, as in Sicily basin (Butler et al. 1995).

The onset of primary gypsum deposits in the Late Messinian could correspond to drawdown with severe lowering of the base level and partial basin desiccation (Fig. 15c). The salinity evolution in the evaporative basin is controlled by climate and by the balance between inflow and outflow of sea water and fresh water (Flecker et al. 2002; Schröder et al. 2003). Seawater must be able to enter the basin throughout most of the evaporative phase in order to provide: (i) a continuous supply for the primary gypsum precipitation, as testified by the gypsum Sr content values, varying from 1550 to 1900 ppm, revealing gypsum deposition under almost stable salinity conditions in the basin, and the gypsum $^{87}Sr/^{86}Sr$ value (average 0.70898 ± 3), showing the marine origin of the evaporites (Matano et al. 2005); (ii) a sulphate-saturated water column in which evaporite resedimentation could have occurred, in order to prevent the dissolution of gypsum clasts during the transport (Manzi et al. 2005).

In the described context the interplay between base-level oscillations, margin erosion and redeposition processes, probably related to fault activity and flooding episodes, produced the observed gypsum facies distribution (Figs 5, 8 & 15d & e). As a matter of fact gypsum facies preserves evidences of variability from shallow-water (mainly crystalline gypsum) to deeper-water (mainly redeposited clastic gypsum) deposition conditions in an extensional evaporative basin located along the Apulian foreland ramp. The gypsum detritus was mainly formed at the expense of the gypsum beds exposed on the flanks of the basin by base-level oscillations and fault-activity (Fig. 15c–e).

Three main types of evaporite successions have been recognized in the study area (Fig. 5), corresponding to different sectors in the evaporative basin: (i) basin margin, (ii) proximal sector and (iii) distal sector (Fig. 15). The differences among the evaporitic successions have interpreted to be related to their position in the basin and to the morpho-structural evolution of the basin, which strongly conditioned the sedimentary evaporitic evolution in the different sectors along the foreland ramp.

In the basin margin the succession is incomplete and is formed only by diatomaceous and euxinic marls passing to evaporitic limestones, often brecciated and truncated at the top by an erosive unconformity (Figs 5a & 15). Limestone brecciation could be related to fault activity controlling the basin margin morpho-structural evolution, related to the faulting of foreland ramp during foreland flexure. The erosive unconformity is the effect of subaerial erosional process, related to base level lowering together with the uplifting and consequent emersion of the basin margin.

In the proximal sector the evaporitic succession is complete (diatomite–carbonate–gypsum), even if sometimes partially eroded at the base and the top (Figs 5b & 15). Local limestone and gypsum brecciation could be related to fault activity. Gypsum lithofacies are arranged in a general trend characterized by alternation of decimetric horizons of reworked allochthonous gypsum, including fine-grained laminated gypsum, gypsarenites, gypsrudites and breccias, and of autochthonous gypsum, including selenite, acicular and laminated gypsum (Fig. 8a–c). The alternation of clastic and primary evaporites is related to the interplay of local tectonics and other factors (regional and local base-level oscillations, flooding events, etc.).

In the distal sector the evaporitic succession is not complete, being formed only by diatomite and gypsum, commonly truncated at the top by an erosive unconformity (Figs 5c & 15). Evaporitic limestones are replaced with diatomaceous and euxinic marls, which may accumulate in these more distal basin sectors while the evaporitic limestones are deposited on the structural highs along the basin margins and proximal sectors (Butler et al. 1995). In the sections not containing basal evaporitic limestones the relationships between primary and reworked gypsum are more complex. As matter of fact in some sections (Fig. 8g & h) we have the same alternations of redeposited clastic and primary evaporites such as the previous type b sections. In other sections the log is formed exclusively by fining upward redeposited clastic gypsum and laminated gypsum (Fig. 8d–f), suggesting a more distal setting.

The proposed model (Fig. 15) takes into account the interplay between base-level oscillations, basin margin erosion and gypsum redeposition. Alternatively, the differences in thickness, facies patterns and stratigraphy can be explained by variable morphology and relief of the evaporative basin, which could be formed within sectors showing different base levels and sedimentation, subsidence and uplift rates (Butler et al. 1995; Bąbel 2005; Manzi et al. 2005). The rates of subsidence and uplift are usually ignored in the gypsum sedimentological analysis, as the gypsum deposition rate is very rapid in comparison with them (Bąbel 2005). High rates of evaporite accumulation rates are considered a major factor in the development of sedimentary sequences containing gypsum (Schröder et al. 2003), so that evaporite accumulation requires a continued supply of accommodation space, which is usually provided by tectonic subsidence. The same tectonic activity could have caused clastic gypsum redeposition processes in the basin.

One of the more consistant characteristics of primary massive selenites all around the Mediterranean is their cyclic organization, which has also a strong chronostratigraphic value (Krijgsman et al. 1999). The occurrence of similar types of depositional cycles in widely separated basins, such as Sicily and Northern Apennines, was likely triggered by external eustatic factors operating in the Mediterranea region (Decima & Wezel 1973; Vai & Ricci Lucchi 1977; Krijgsman et al. 1999). In these basins (Sicily and Northern Apennines) local tectonic movement did not play a major role (Vai & Ricci Lucchi 1977).

No clear cyclic facies organization has been observed in the studied area, even where several flooding episodes probably caused the repeated superposition of shallow-water gypsum and relatively deeper-water gypsum facies associations in the marginal sector and partly in the distal sector of the basin, while in the more distal sector only fining-upward horizons of redeposited clastic gypsum occur. The Monte Castello evaporites were deposited in a tectonically active area in front of the Southern Apenninic orogen and therefore rapid tectonic events (fault activity, tilting of basin margin, slope instability, etc.) could have influenced the pattern of evaporite deposition. Lateral and vertical transition from breccias to laminated fine-grained clastic gypsum suggests the existence of a palaeoslope in the Monte Castello basin with fault-controlled instability of the depositional margins (Fig. 15).

Regional tectonic control on evaporite deposition

The pre-evaporitic, evaporitic and post-evaporitic sequences reflect very different depositional settings, and record the evolution of the Messinian southern Apennine foreland basin.

In the Early Messinian (Fig. 16a), Toppo Capuana clayey marls and Faeto Flysch formations (Daunia unit), which were referred to a marine mainly pelagic environment in the foreland region, evolved into a partially restricted marine setting (pre-evaporitic diatomaceous marls); this trend continued during the 'salinity crisis' of the Mediterranean Sea (Hsü et al. 1973a, b).

The region underwent, during the lower part of the late Messinian, a marked restriction, which resulted in the sedimentation of evaporites (Evaporiti di Monte Castello Formation) in the marginal ramp of the foreland region (Fig. 16b).

The sedimentary evolution during the evaporative phase was characterized by a gradual increase in salinity until gypsum precipitated (Fig. 15). The onset of basin restriction is recorded by the first occurrence of primary gypsum deposits. The gypsum Sr values vary within a narrow range (from 1550 to 1900 ppm), revealing gypsum deposition under almost stable homogeneous salinity conditions in the basin, even if a minor variability is recorded. The sedimentary conditions in the basin were influenced by drastic events of gypsum reworking and resedimentation probably related to flooding events or local tectonic activity (e.g. boundary fault activity). The gypsum was deposited from mainly marine brines, according to the Sr isotopic compositions (Matano et al. 2005). This sedimentation was equivalent to the Lower Evaporites (Müller & Mueller 1991) of the Mediterranean.

The transition from evaporitic interval to post-evaporitic interval of Late Messinian coincides with a strong tectonic phase in the Southern Apennines foredeep, which led to the migration of the depocenters with the formation of an important erosional surface in marginal settings (Fig. 16c), such as in Northern Apennines basins (Roveri et al. 2001).

During the upper part of the Late Messinian ('post-evaporitic interval'), the marginal sector emerged, as a consequence of the intra-Messinian tectonic phase, and was subject to a clastic alluvial-lacustrine deposition (Fig. 16c). The basin received a meteoric water influx during the post-evaporitic clastic deposition, as suggested by the presence of freshwater ostracods in the Torrente Fiumarella and Anzano Molasse units (Matano et al. 2005), which unconformably capped the Evaporiti di Monte Castello Formation.

The occurrence of unconformable siliciclastic deposits (Torrente Fiumarella and Anzano Molasse units) of wedge-top depozone confirms that the Daunia unit was tectonically deformed and stacked in the thrust wedge during late Messinian.

Conclusions

Deposition of evaporites in the Messinian basins of the southern Apennines foredeep is related to 'salinity crisis' events recognized at Mediterranean scale. The stratigraphic and sedimentological data discussed in the paper allow better definition of the geological evolution, the palaeoenvironmental

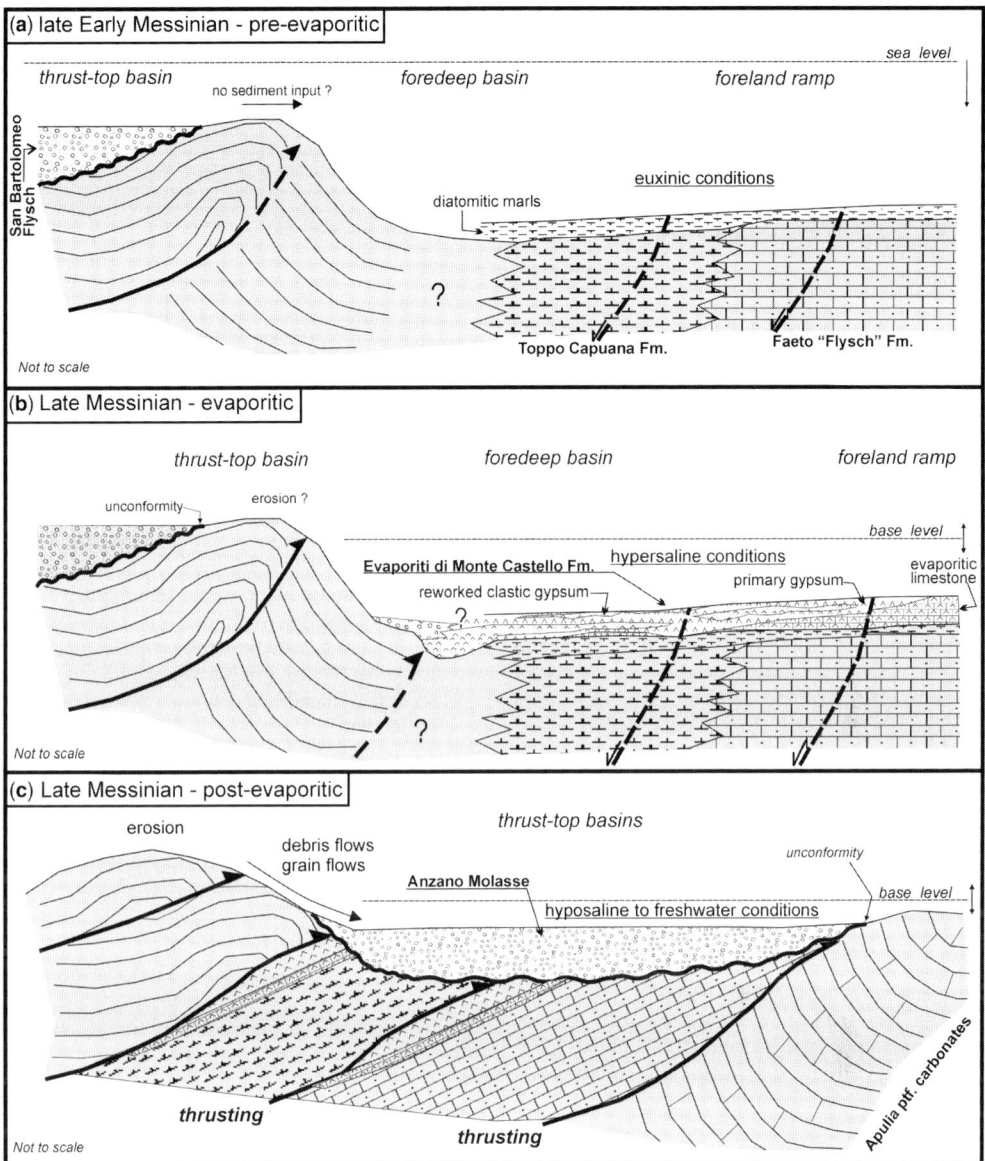

Fig. 16. Tectono-stratigraphic Messinian events in the northern sector of the southern Apennine foreland basin system.

conditions of the late Messinian southern Apennines evaporitic basins and the tectonic controls on the evaporite sedimentation.

The Messinian evaporites of Irpinia–Daunia sector of southern Italy have probably formed within small foreland basin of the yet undeformed sector of the early late Messinian Apenninic foredeep. This tectonic setting (Fig. 17) is very different from the one of the northern Apennines and Sicilian Messinian evaporitic basins (e.g. Vena del Gesso and Caltanisetta basins), which are referred to thrust-top basins located in the wedge-top depozone of the Messinian foreland basin system (Butler et al. 1995; Roveri et al. 2003).

The Messinian evolution in the study area is characterized by the following features (Figs 15 & 16):

(1) Progressive establishment of, more or less anoxic, starved conditions (diatomaceous

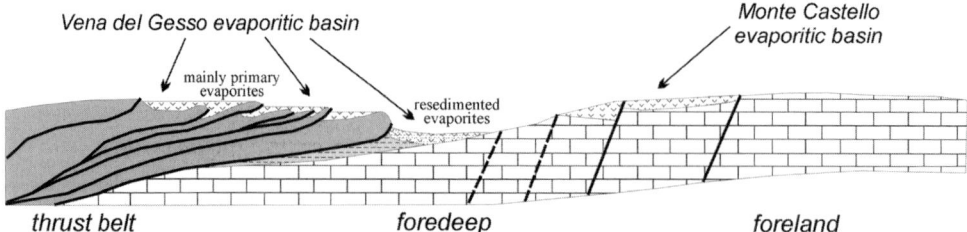

Fig. 17. Tentative section of Late Messinian Apennines thrust belt-foreland system showing the tectonic setting of the evaporitic basins in northern Apennines (Vena del Gesso basin; Roveri *et al.* 2003; Manzi *et al.* 2005) and southern Apennines (Evaporiti di Monte Castello basin).

and euxinic marls) occurred during the pre-evaporitic stage.

(2) The pre-evaporitic and evaporitic continuous sedimentary succession (diatomaceous–carbonate–autochthonous gypsum) suggests a slow buildup of salinity conditions in the basin of Evaporiti di Monte Castello Formation.

(3) Permanent marine connection and minor influence of continental waters characterize the evaporitic interval (Matano *et al.* 2005). Seawater must be able to enter the basin throughout most of the evaporative phase in order to provide supply for the gypsum precipitation and redeposition under almost homogenous salinity conditions, as suggested by Sr content of the gypsum, and from mainly marine brines, as suggested by the Sr isotopic ratio.

(4) Presence of both shallow-water and deeper-water gypsum facies associations, dominated respectively by autochthonous gypsum (laminated, acicular and selenitic lithofacies) and by redeposited clastic gypsum (pebbly gypsarenite, breccia and fine-grained laminated gypsum), with maximum thickness of about 200 m, characterizes the Evaporiti di Monte Castello Formation.

(5) Facies successions observed in outcrops are related to interplay of base-level oscillations, local tectonics, hydrodynamic events and rates of evaporation, which control water depth and basin dynamics. Gypsum facies usually preserve evidences of variability from shallow-water to deeper-water deposition conditions in an extensional evaporative basin. Frequent vertical changes of gypsum facies associations record variations in brine depth (flooding events) and tectonic activity in the basin; minor variations in salinity are recorded by Sr content data and facies variations in autochthonous gypsum.

(6) The absence of a clear cyclic organization in gypsum facies, which is one of the more consolidated characteristics of primary massive selenites in the Mediterranean (Vai & Ricci Lucchi 1977), having also a strong chronostratigraphic value (Krijgsman *et al.* 1999), could be attributable to the actual lack of clear stratigraphic data due to the outcrop conditions and to the strong tectonic deformation in the study area, or the occurrence of some not easily recognized sedimentary or tectonic factors which have obliterated the cycle signals.

(7) An erosional angular unconformity, as a consequence of the intra-Messinian tectonic phase, which cut off evaporitic deposition in the study area, marks the base of the post-evaporitic alluvial and lacustrine deposits where primary evaporites are absent (Basso *et al.* 2001, 2002; Matano 2002; Matano *et al.* 2005).

B.C. Schreiber and an anonymous referee are acknowledged for their reviews that greatly improved this paper. S. Lugli and M. Roveri are gratefully acknowledged for their helpful discussions and suggestions during congress presentations at Florence 2004 and Spoleto 2005. I wish thank also S. Di Nocera, S. Critelli and M. Torre for long discussions concerning Messinian geological problems. The research was supported by funds "Ateneo" 2006–2007 of Federico II University and by the COFIN – PRIN 2004–2005 "Eventi stratigrafici e paleogeografia del Messiniano nell'Appennino campano-molisano" (Resp. Prof. Silvio Di Nocera).

References

BABEL, M. 2005. Selenite–gypsum microbialite facies and sedimentary evolution of the Badenian evaporite basin of the northern Carpathian Foredeep. *Acta Geologica Polonica*, **55**, 187–210.

BARONE, M., CRITELLI, S., DI NOCERA, S., LE PERA, E., MATANO, F. & TORRE, M. 2006. Stratigraphy and detrital modes of the upper Messinian post-evaporitic sandstones of the southern Apennines foreland-basin system, Italy: evidence of foreland-basin evolution during Messinian Mediterranean 'salinity crisis'. *International Geology Review*, **48**, 702–724.

BASSO, C., CIAMPO, G., CIARCIA, S., DI NOCERA, S., MATANO, F. & TORRE, M. 2002. Geologia del settore irpino-dauno dell'Appennino meridionale: unita' meso-cenozoiche e vincoli stratigrafici nell'evoluzione tettonica mio-pliocenica. *Studi Geologici Camerti, nuova serie*, **2**, 7–26.

BASSO, C., DI NOCERA, S., ESPOSITO, P., MATANO, F., RUSSO, B. & TORRE, M. 2001. Stratigrafia delle successioni sedimentarie evaporitiche e post-evaporitiche del Messiniano superiore in Irpinia settentrionale (Appennino meridionale). *Bollettino della Società Geologica Italiana*, **120**, 211–231.

BELLANCA, A., CARUSO, A., FERRUZZA, G., NERI, R., ROUCHY, J. M., SPROVIERI, M. & BLANC-VALLERON, M. M. 2001. Transition from marine to hypersaline conditions in the Messinian Tripoli Formation from the marginal areas of the central Sicilian Basin. *Sedimentary Geology*, **140**, 87–105.

BLANC-VALLERON, M. M., PIERRE, C. ET AL. 2002. Sedimentary, stable isotope and micropaleontological records of paleoceanographic change in the Messinian Tripoli Formation (Sicily, Italy). *Palaeogeography, Palaeoclimatology, Palaeoecology*, **185**, 255–286.

BUTLER, R. W. H., LICKORISH, W. H., GRASSO, M. & RAMBERTI, L. 1995. Tectonics and sequence stratigraphy in Messinian basins, Sicily: Constraints on the initiation and termination of the Mediterranean salinity crisis. *Geological Society of America Bulletin*, **107**, 425–439.

BUTLER, R. W. H., MCCLELLAND, E. & JONES, R. E. 1999. Calibrating the duration and timing of the Messinian salinity crisis in the Mediterranean: linked tectonoclimatic signals in the thrust-top basins of Sicily. *Journal of the Geological Society, London*, **156**, 827–835.

CAVAZZA, W. & DE CELLES, P. G. 1998. Upper Messinian siliciclastic rocks in southeastern Calabria (southern Italy): palaeotectonic and eustatic implications for the evolution of the central Mediterranean region. *Tectonophysics*, **298**, 223–241.

CAVAZZA, W. & WEZEL, F. C. 2003. The Mediterranean region – a geological primer. *Episodes*, **26**, 160–168.

CHANNEL, J. E. T., D'ARGENIO, B. & HORVARTH, H. F. 1979. Adria, the african promontory, in Mesozoic Mediterranean paleogeography. *Earth Science Review*, **15**, 213–292.

CIARAPICA, G. & PASSERI, L. 1980. Contributions on evaporites diagenesis. *Géologie Méditerranéenne*, **7**(1), 43–48.

CIARCIA, S., DI NOCERA, S., MATANO, F. & TORRE, M. 2003. Evoluzione tettono-sedimentaria e paleogeografica dei depocentri 'wedge-top' nell'ambito del 'foreland basin system' pliocenico dell'Appennino meridionale (settore irpino-dauno). *Bollettino della Società Geologica Italiana*, **122**, 117–138.

CIPOLLARI, P., COSENTINO, D. & GLIOZZI, E. 1999*a*. Extension- and compression-related basins in central Italy during the Messinian *Largo-Mare* event. *Tectonophysics*, **315**, 163–185.

CIPOLLARI, P., COSENTINO, D. & PRATURLON, A. 1999*b*. Thrust-top lacustrine-lagoonal basin development in accretionary wedges: late Messinian (Lago-Mare) episode in the central Apennines (Italy). *Palaeogeogr. Palaeoclimatol. Palaeoecol.*, **151**, 149–166.

CRITELLI, S. 1999. The interplay of lithospheric flexure and thrust accomodation in forming stratigraphic sequences in the southern Apennines foreland basin system, Italy. *Rendiconti Accademia Nazionale dei Lincei, Scienze Fisiche e Naturali*, **10**, 257–326.

CROSTELLA, A. & VEZZANI, L. 1964. La geologia dell'Appennino foggiano. *Bollettino della Società Geologica Italiana*, **83**, 121–142.

D'ARGENIO, B., PESCATORE, T. & SCANDONE, P. 1975. Structural pattern of the Campania-Lucania Apennines. *In*: OGNIBEN, L., PAROTTO, M. & PRATURLON, A. (eds) *Structural Model of Italy*. Quaderni de La ricerca scientifica, Rome, **90**, 313–327.

DAZZARO, L., IANNONE, A., MORESI, M., RAPISARDI, L. & ROMEO, M. 1988. Stratigrafia, sedimentologia e geochimica delle successioni messiniane dell'Irpinia al confine con la Puglia. *Memorie della Società Geologica Italiana*, **41**, 841–859.

DECIMA, A. & WEZEL, F. C. 1971. Osservazioni sulle evaporiti messiniane della Sicilia centro-meridionale. *Rivista Mineraria Siciliana*, **22**, 130–132.

DECIMA, A. & WEZEL, F. C. 1973. Late Miocene evaporites of the Central Sicilian Basin. *Initial Report DSDP, Leg 13*, 1234–1240.

DI GIROLAMO, P., NARDI, G., ORTOLANI, F., STANZIONE, D. & TORRE, M. 1986. The petrological and geodynamic significance of late miocene-pliocene tuffites from the Southern Apennines (Italy). *Neues Jahrbuch Mineralogie Abhstract*, **155**(2), 203–218.

DI NOCERA, S., MATANO, F. & TORRE, M. 2002. Le unità 'sannnitiche' Auct. (Appennino centro-meridionale): rassegna delle correnti interpretazioni stratigrafiche e paleogeografiche e nuove ipotesi con l'introduzione dell'Unità di Frigento. *Studi Geologici Camerti, nuova serie*, **1**, 87–102.

DI NOCERA, S., ORTOLANI, F., RUSSO, M. & TORRE, M. 1975. Successioni sedimentarie messiniane e limite Miocene-Pliocene nella Calabria settentrionale. *Bollettino della Società Geologica Italiana*, **93**, 575–607.

DI NOCERA, S., ORTOLANI, F. & TORRE, M. 1976. Fase tettonica messiniana nell'Appennino meridionale. *Bollettino della Società dei Naturalisti in Napoli*, **84**, 1–17.

DI NOCERA, S., ORTOLANI, F., TORRE, M. & RUSSO, B. 1981. Evoluzione sedimentaria e cenni di paleogeografia del Tortoniano-Messiniano dell'Irpinia occidentale. *Bollettino della Società dei Naturalisti in Napoli*, **90**, 131–166.

DOGLIONI, C. 1991. Aproposal for the kinematic modelling of W-dipping subductions – possible applications to the Tyrrhenian–Apennines system. *Terra Nova*, **3**, 423–434.

DOGLIONI, C., HARABAGLIA, P., MARTNELLI, G., MONGELLI, F. & ZITO, G. 1996. A geodynamic model of the southern Apennines accretionary prism. *Terra Nova*, **8**, 540–547.

DUGGEN, S., HOERNLE, K., VAN DEN BOGAARD, P., RÜPKE, L. & MORGAN, J. P. 2003. Deep roots of the Messinian salinity crisis. *Nature*, **422**, 602–606.

ELTER, P., GIGLIA, G., TONGIORGI, M. & TREVISAN, L. 1975. Tensional and compressional areas in the recent (Tortonian to Present) evolution of the Northern Apennines. *Bollettino di Geofisica Teorica e Applicata*, **17**, 1–65.

ELTER, P., GRASSO, M., PAROTTO, M. & VEZZANI, L. 2003. Structural setting of the Apennine–Maghrebian thrust belt. *Episodes*, **26**, 205–211.

FLECKER, R., DE VILLIERS, S. & ELLAM, R. M. 2002. Modelling the effect of evaporation on the salinity–$^{87}Sr/^{86}Sr$ relationship in modern and ancient marginal-marine systems: the Mediterranean Messinian Salinity Crisis. *Earth and Planetary Science Letters*, **203**, 221–233.

GRIFFIN, D. L. 2002. Aridity and humidity: two aspects of the late Miocene climate of North Africa and the Mediterranean. *Palaeogeography, Palaeoclimatology, Palaeoecology*, **182**, 65–91.

HARDIE, L. A. & LOWENSTEIN, T. K. 2004. Did the Mediterranean Sea dry out during the Miocene? A reassessment of the evaporite evidence from DSDP Legs 13 and 42A cores. *Journal of Sedimentary Research*, **74**, 453–461.

HODELL, D. A., CURTIS, J. H., SIERRO, F. J. & RAYMO, M. E. 2001. Correlation of late Miocene to early Pliocene sequences between the Mediterranean and North Atlantic. *Paleoceanography*, **16**, 164–178.

HODELL, D. A., MUELLER, P. A., MCKENZIE, J. A. & MEAD, G. A. 1989. Strontium isotope stratigraphy and geochemistry of the late Neogene ocean. *Earth Plan. Sci. Lett.*, **92**, 165–178.

HODELL, D. A., MUELLER, P. A. & GARRIDO, J. R. 1991. Variations in the strontium isotopic composition of seawater during the Neogene. *Geology*, **19**, 24–27.

HOWARTH, R. J. & MCARTHUR, J. M. 1997. Statistics for strontium isotope stratigraphy: a robust LOWESS fit to the marine Sr-isotope curve for 0 to 206 Ma, with a look-up table for derivation of numerical age. *J. Geol.*, **105**, 441–456.

HSÜ, K., CITA, M. B. & RYAN, W. B. F. 1973a. The origin of Mediterranean evaporites. *Initial Report DSDP, Leg 13*, 1203–1231.

HSÜ, K., RYAN, W. B. F. & CITA, M. B. 1973b. Late Miocene desiccation of the Mediterranean. *Nature*, **242**, 240–244.

HSÜ, K. & MONTADERT, L. ET AL. 1977. History of the Mediterranean salinity crisis. *Nature*, **267**, 399–403.

IMPROTA, L., BONAGURA, M., CAPUANO, P. & IANNACCONE, G. 2003. An integrated geophysical investigation of the upper crust in the epicentral area of the 1980, $M_s = 6.9$, Irpinia earthquake (Southern Italy). *Tectonophysics*, **361**, 139–169.

KASPRZYK, A. 2003. Sedimentological and diagenetic patterns of anhydrite deposits in the Badenian evaporite basin of the Carpathian Foredeep, southern Poland. *Sedimentary Geology*, **158**, 167–194.

KASTENS, K. & MASCLE, J. 1990. The geological evolution of the Tyrrhenien Sea: an introduction to the scientific results of the ODP Leg 107. *Proceedings ODP Scientific Results*, **107**, 3–26.

KEOGH, S. M. & BUTLER, W. H. 1999. The Mediterranean water body in the late Messinian: interpreting the record from marginal basins on Sicily. *Journal of the Geological Society, London*, **156**, 837–846.

KIRKLAND, D. W., DENISON, R. E. & DEAN, W. E. 2000. Parent brine of the Castile Evaporites (Upper Permian), Texas and New Mexico. *Journal of Sedimentary Research*, **70**, 749–761.

KRIJGSMAN, W., GABOARDI, S., HILGEN, F. J., IACCARINO, S., DE KAENEL, E. & VAN DER LAAN, E. 2004. Revised astrochronology for the Ain el Beida section (Atlantic Morocco): no glacio-eustatic control for the onset of the Messinian Salinity Crisis. *Stratigraphy*, **1**, 87–101.

KRIJGSMAN, W., HILGEN, F. J., RAFFI, I., SIERRO, F. J. & WILSON, D. S. 1999. Chronology, causes and progression of the Messinian salinity crisis. *Nature*, **400**(6745), 652–655.

LENTINI, F., CARBONE, S., CATALANO, S. & MONACO, C. 1990. Tettonica a thrust neogenica nella catena appenninico-maghrebide: esempi dalla Lucania e dalla Sicilia. *Studi Geologici Camerti, volume speciale*, 19–26.

MANZI, V., LUGLI, S., RICCI LUCCHI, F. & ROVERI, M. 2005. Deep-water clastic evaporites deposition in the Messinian Adriatic foredeep (northern Apennines, Italy): did the Mediterranean ever dry out? *Sedimentology*, **57**, 875–902.

MATANO, F. 2002. Le Molasse di Anzano nell'evoluzione tettono-sedimentaria messiniana del margine occidentale della microzolla apula nel settore irpino-dauno dell'orogene sud-appenninico. *Memorie della Società Geologica Italiana*, **57**, 209–220.

MATANO, F. & DI NOCERA, S. 2001. Geologia del settore centrale dell'Irpinia (Appennino meridionale): nuovi dati e interpretazioni. *Bollettino della Società Geologica Italiana*, **120**, 3–14.

MATANO, F., BARBIERI, M., DI NOCERA, S. & TORRE, M. 2005. Stratigraphy and strontium geochemistry of Messinian evaporite-bearing successions of the southern Apennines foredeep, Italy: implications for the Mediterranean 'salinity crisis' and regional palaeogeography. *Palaeogeography, Palaeoclimatology, Palaeoecology*, **217**, 87–114.

MCARTHUR, J. M., HOWARTH, R. J. & BAILEY, T. R. 2001. Strontium isotope stratigraphy: LOWESS Version 3: best fit to the marine Sr-isotope curve for 0–509 Ma and accompanying look-up table for deriving numerical age. *Journal of Geology*, **109**, 155–170.

MCKENZIE, J. A., HODELL, D. A., MUELLER, P. A. & MÜLLER, D. W. 1988. Application of strontium isotopes to late Miocene–early Miocene stratigraphy. *Geology*, **16**, 1022–1025.

MICHALZIK, D. 1996. Lithofacies, diagenetic spectra and sedimentary cycles of Messinian (Late Miocene) evaporites in SE Spain. *Sedimentary Geology*, **106**, 203–222.

MOSTARDINI, F. & MERLINI, S. 1986. Appennino centro-meridionale. Sezioni geologiche e proposta di modello strutturale. *Memorie della Società Geologica Italiana*, **35**, 177–202.

MÜLLER, D. W. & MUELLER, P. A. 1991. Origin and age of the Mediterranean Messinian evaporites: implications from Sr isotopes. *Earth and Planetary Science Letters*, **107**, 1–12.

MÜLLER, D. W., MUELLER, P. A. & MCKENZIE, J. A. 1990. Strontium isotopic ratios as fluid tracers in Messinian evaporites of the Tyrrhenian Sea (western Mediterranean Sea). *Proceedings ODP Scientific Results*, **107**, 603–614.

OGNIBEN, L. 1957. Petrografia della Serie Solfifera Siciliana e considerazioni geologiche relative. *Memorie Descrittive della Carta Geologica Italiana*, **8**, 453–763.

PAROTTO, M. & PRATURLON, A. 2004. The southern Apennine arc. In: CRESCENTI, U., D'OFFIZI, S., MERLINO, S. & SACCHI, L. (eds) *Geology of Italy*, Società Geologica Italiana, Rome, 33–58.

PATACCA, E. & SCANDONE, P. 1989. Post-Tortonian mountain building in the Apennines. The role of the passive sinking of a relict lithosperic slab. In: BORIANI, A., BONAFEDE, M., PICCARDO, G. B. & VAI, G. B. (eds) *The Lithosphere in Italy*. Advances in Earth Sciences Research, Italian National Commetee International Lithosphere Program, Mid-term Conference. Proceedings, Rome Atti Convegni Lincei, **80**, 157–176.

PATACCA, E., SARTORI, R. & SCANDONE, P. 1990. Tyrrhenian basin and Apenninic arcs: Kinematic relations since late Tortonian times. *Memorie della Società Geologica Italiana*, **45**, 425–451.

PATACCA, E., SARTORI, R. & SCANDONE, P. 1993. Tyrrhenian basin and Apenninic. Kinematic evolution and related dynamic constraints. In: BOSCHI, E., MANTOVANI, E. & MORELLI, A. (eds) *Recent Evolution and Seismicity of the Mediterranean Region*. Kluwer Academic, Dordrecht, 161–171.

PERYT, T. M. 1996. Sedimentology of Badenian (middle Miocene) gypsum in eastern Galicia, Podolia and Bukovina (West Ukraine). *Sedimentology*, **43**, 571–588.

PERYT, T. M. 2000. Resedimentation of basin centre sulfate deposits: Middle Miocene Badenian of Carpathian foredeep, southern Poland. *Sedimentary Geology*, **134**, 331–342.

PESCATORE, T. & SENATORE, M. R. 1986. A comparison between a present-day (Taranto Gulf) and a Miocene (Irpinian Basin) foredeep of the Southern Apennines (Italy). *Special Publication of International Association of Sedimentology*, **8**, 169–182.

PESCATORE, T., DI NOCERA, S., MATANO, F. & PINTO, F. 2000. L'unità del Fortore nel quadro della geologia del settore orientale dei Monti del Sannio (Appennino meridionale). *Bollettino della Società Geologica Italiana*, **119**, 587–601.

PLAYÀ, E., ORTÌ, F. & ROSELL, L. 2000. Marine to non-marine sediementation in the upper Miocene evaporites of the Eastern Betics, SE Spain: sedimentological and geochemical evidences. *Sedimentary Geology*, **133**, 135–166.

RIDING, R., BRAGA, J. C. & MARTIN, J. M. 1999. Late Miocene Mediterranean desiccation: topography and significance of the 'salinity crisis' erosion surface on-land in southeast Spain. *Sedimentary Geology*, **123**, 1–7.

RIDING, R., BRAGA, J. C., MARTIN, J. M. & SANCHEZ-ALMAZO, I. M. 1998. Mediterranean Messinian salinity crisis: constraints from a coeval marginal basin, Sorbas, southeastern Spain. *Marine Geology*, **146**, 1–20.

ROSELL, L., ORTÌ, F., KASPRZYK, A., PLAYÀ, E. & PERYT, T. M. 1998. Strontium geochemistry of Miocene primary gypsum: Messinian of southeastern Spain and Sicily and Badenian of Poland. *Journal of Sedimentary Research*, **68**, 63–79.

ROUCHY, J. M. 1982. *La genèse des évaporites messiniennes de Méditerranée*. Muséum national Historie naturelle, Memoirs, **50**.

ROURE, F., CASERO, P. & VIALLY, R. 1991. Growth processes and melange formation in the southern Apennines accretionary wedge. *Earth and Planetary Science Letters*, **102**, 395–412.

ROVERI, M., LANDUZZI, A., BASSETTI, M. A., LUGLI, S., MANZI, V., RICCI LUCCHI, F. & VAI, G. B. 2004. The record of Messinian events in the Northern Apennines foredeep basins. In: *B19 Field Trip Guidebook*. 32nd International Geological Congress, Firenze, 20–28 Agosto 2004, APAT, 1–42.

ROVERI, M., BASSETTI, M. A. & RICCI LUCCHI, F. 2001. The Mediterranean Messinian salinity crisis: an Apennine foredeep perspective. *Sedimentary Geology*, **140**, 201–214.

ROVERI, M., MANZI, V., BASSETTI, M. A., MERINI, M. & RICCI LUCCHI, F. 1998. Stratigraphy of the Messinian post-evaporitic stage in eastern-Romagna (northern Apennines, Italy). *Giornale di Geologia*, **60**, 119–142.

ROVERI, M., MANZI, V., RICCI LUCCHI, F. & ROGLEDI, S. 2003. Sedimentary and tectonic evolution of the Vena del Gesso basin (Northern Apennines, Italy): implication for the onset of the Messinian salinity crisis. *Geological Society of America Bulletin*, **115**, 387–405.

SANTO, A. & SENATORE, M. R. 1988. La successione stratigrafica dell'unità dauna a Monte Sidone (Castelluccio Valmaggiore, Foggia). *Memorie della Società Geologica Italiana*, **41**, 431–438.

SCHREIBER, B. C. & DECIMA, A. 1976. Sedimentary facies produced under evaporitic environments: a review. *Memorie della Società Geologica Italiana*, **16**, 111–126.

SCHREIBER, B. C. & EL TABAKH, M. 2000. Deposition and early alteration of evaporites. *Sedimentology*, **47** (supplement 1), 215–238.

SCHREIBER, B. C., FRIEDMAN, G. M., DECIMA, A. & SCHREIBER, E. 1976. Depositional environment of Upper Miocene (Messinian) evaporite deposits of the Sicilian Basin. *Sedimentology*, **23**, 729–760.

SCHREIBER, B. C., ROTH, M. S. & HELMAN, M. L. 1982. Recognition of primary facies characteristics of evaporites and the differentiation of these forms from diagenetic overprints. In: HANDFORD, C. R., LOUCKS, R. G. & DAVIES, G. R. (eds) *Depositional and Diagenetic Spectra of Evaporites – a Core Workshop*. SEPM Core Workshop, **3**, 1–32.

SCHRÖDER, S., SCHREIBER, B. C., AMTHOR, J. E. & MATTER, A. 2003. A depositional model for terminal Neoproterozoic – Early Cambrian Ara Group evaporites in south Oman. *Sedimentology*, **50**, 879–898.

SELLI, R. 1960. Il Messiniano Mayer-Eymar. Proposta di un neostratotipo. *Giornale di Geologia*, **28**, 1–33.

SGROSSO, I. 1998. Possibile evoluzione cinematica miocenica nell'orogene centro-sud appenninico. *Bollettino della Società Geologica Italiana*, **117**, 679–724.

TORRE, M., DI NOCERA, S. & ORTOLANI, F. 1988. Evoluzione post-tortoniana nell'Appennino meridionale. *Memorie della Società Geologica Italiana*, **41**, 47–56.

VAI, G. B. 1997. Cyclostratigraphic estimate of the Messinian stage duration. *In*: MONTANARI, A., ODIN, G. S. & COCCIONI, R. (eds) *Miocene Stratigraphy: an Integrated Approach*, Elsevier Science, Amsterdam, 463–476.

VAI, G. B. & RICCI LUCCHI, F. 1977. Algal crusts, autochthonous and clastic gypsum in a cannibalistic evaporite basin: a case history from the Messinian of Northern Apennines. *Sedimentology*, **24**, 211–244.

VIDAL, L., BICKERT, T., WEFER, G. & ROHL, U. 2002. Late Miocene stable isotope stratigraphy of SE Atlantic ODP Site 1085: relation to Messinian events. *Marine Geology*, **180**, 71–85.

ZIEGLER, P. A. 1999. Evolution of the Arctic-North Atlantic and the Western Tethys. *American Association of Petroleum Geology Memoir*, **43**, 164–196.

The Badenian evaporite basin of the northern Carpathian Foredeep as a model of a meromictic selenite basin

M. BĄBEL[1] & A. BOGUCKI[2]

[1]*Institute of Geology, Warsaw University, Al. Żwirki i Wigury 93, PL-02-089 Warszawa, Poland*
(e-mail: m.babel@uw.edu.pl)

[2]*Faculty of Geography, Ivan Franko National University of Lviv, P. Doroshenka 41, 79000 Lviv, Ukraine*

Abstract: The hydrography and brine flow patterns in the Middle Miocene (Badenian) evaporite basin of the northern Carpathian Foredeep (in Ukraine, Poland and the Czech Republic) are reconstructed based on studies of the peculiar, conformably oriented, bottom-grown gypsum crystals present in the selenite deposits along the basin margin. The crystal apices are turned in a similar horizontal direction that is interpreted as the product of consistent flow of the bottom brines during crystal formation. Similarly the regular millimetre-scale growth zoning in these crystals presumably reflects the annual stratification-mixing pattern in the brine column typical of monomictic basins. In the central, deeper parts of the basin deposition was dominated by Na-chloride, and the selenitic facies are lacking. These central areas are interpreted as being meromictic during the oriented selenite deposition. The permanent pycnocline separated a mixolimnion, at the surface, from an anoxic (euxinic) monimolimnion, at the bottom, where direct evaporative crystallization of gypsum was not possible. The mixolimnion, which extended far onto the shallow margin of the basin, showed only a seasonal (annual) pycnocline and monomictic hydrography. Oriented selenites grew just in this mixolimnetic marginal zone, under predominantly counterclockwise (cyclonic) flow.

In some models and palaeogeographical reconstructions of ancient evaporite basins (Warren 2006), and in a few modern basins (references in Bąbel 2004*a*), the selenite facies (i.e. large bottom-grown gypsum crystals) are distributed along the basin margins and are absent in the deeper basin centres. Such a depth-controlled distribution of selenite crystals is explained by the well-known fact that the shallow-water environment (strictly, some specific aspects of this environment) apparently favours the *in situ* growth of characteristic gypsum crystals. It is also known that the features of brine at the given depth change regularly following the annual hydrographical stratification-mixing cycle of the basin (Bąbel 2007). The crystallization of evaporite minerals and distribution of evaporite facies appear to be closely related to this changing hydrography of the brine column, as is proved by the modern evaporite deposition in some basins (e.g. Valero-Garcés & Kelts 1995; Last *et al.* 2002; Last & Ginn 2005; Warren 2006) and in the drying Dead Sea in particular (Neev & Emery 1967; Niemi *et al.* 1997; Herut *et al.* 1998). Ancient, giant, deep selenite basins have no modern analogs. However, certainly, as in the Dead Sea, some link between the hydrography of the deep brines and distribution of selenite facies in such basins can and should exist. So far the hydrographical and hydrodynamical environment of such basins is poorly recognized and has not been reconstructed with the use of the modern limnological knowledge gained from saline lakes.

The Middle Miocene Badenian evaporite basin of the Carpathian Foredeep is one of such basins with the widespread marginal selenite facies. The main aim of this paper is to reconstruct the brine hydrography in this basin and, based on the actualistic limnological data as a model, to create a new comprehensive hydrographical model of a deep selenite basin to explain more clearly the specific marginal distribution of selenite facies and its lack in the deeper basinal centre. This model also may be valid for many other evaporite basins showing the same lateral differentiation of facies: the selenite or gypsum facies on the basin margin and Na-chloride (halite) facies in the basin centre, where the reason for this differentiation is poorly understand and thus remains controversial (Cohen 1993).

The reconstruction presented in this study is the result of the advanced investigation of the unique sedimentary structure present in some selenite deposits. Here, on the margin of the Badenian basin, there are large areas containing oriented gypsum crystals that grew on the bottom in a concordant manner under influence of an oriented flow of brines that

From: SCHREIBER, B. C., LUGLI, S. & BĄBEL, M. (eds) *Evaporites Through Space and Time.*
Geological Society, London, Special Publications, **285**, 219–246.
DOI: 10.1144/SP285.13 0305-8719/07/$15.00 © The Geological Society of London 2007.

were calcium sulphate oversaturated (Pawlikowski 1982, p. 36; Dronkert 1985, p. 205). These oriented crystals were used as indicators for the brine flow paths and helped reconstruction of the hydrography in this marginal zone of the basin (Bąbel et al. 1999; Bąbel 2002; Bąbel & Becker 2006) while the hydrography in the remaining area remained unknown. This paper supplements these previous investigations. It demonstrates a methodology for brine palaeocurrent analysis and the validity of the previous hydrodynamical and hydrographical reconstruction in a small area of the basin chosen as a case study (located near Halych in Ukraine). This earlier reconstruction, valid for the marginal selenite zone of the basin, is integrated with a new concept concerning the brine hydrography of the adjacent deeper portion of the basin now represented by laminated Ca-sulphate and halite facies, where selenite deposits are lacking. The new model of the basin is developed, in which distribution of chemical facies is shown as a product of a specific stratification and hydrodynamic of the basinal brines.

Geology and palaeogeography of the Badenian evaporite basin

The Carpathian Foredeep basin is the largest evaporite basin developed during the Badenian salinity crisis in the Central Paratethys area (Fig. 1). In the northern portion of this basin, in Ukraine, Poland and the Czech Republic, the Badenian evaporites crop out widely in the northern platform margin of the foredeep but are deeply buried in the foredeep (Fig. 2). The foredeep is mostly filled with the Miocene deposits and is subdivided into the autochthonous zone in the north and the allochthonous zone in the south, in front of the external flysh unit of the Carpathians. The Badenian evaporites are present in both these zones, however only in the autochthonous zone as *in situ* sediments. The evaporites form a relatively continuous horizon of gypsum, anhydrite and halite deposits, up to a few tens of metres thick, in places with thicker clay intercalations. The margin of the basin (i.e. mainly the platform zone) was dominated by gypsum deposition including extensive accumulations of selenites, particularly those selenites with the oriented crystals used for palaeocurrent analysis in this paper. The selenite facies was also deposited on the poorly preserved southern margin of the basin. The central areas of the basin (i.e. mostly the foredeep zone) were dominated by the laminated Ca-sulphate facies, clay and halite and in this presumably deeper zone of the basin the selenite deposition did not take place or is not preserved (Korenevskyi et al. 1977; Garlicki 1979; Połtowicz 1993; Smirnov et al. 1995; Petrichenko et al.

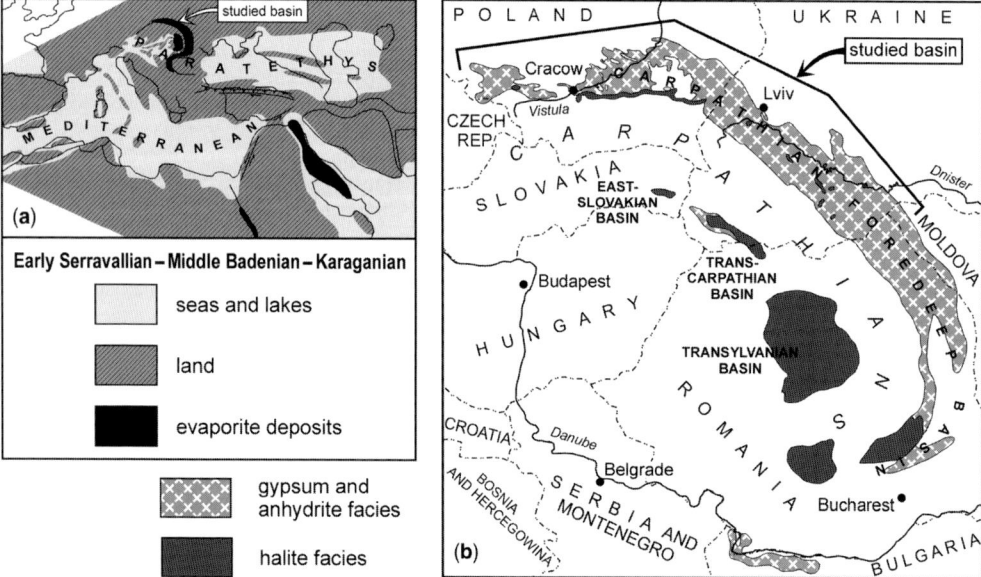

Fig. 1. Maps showing palaeogeography and distribution of Badenian evaporites. (**a**) Palaeogeography of Paratethys and Mediterranean in the Middle Miocene and location of the evaporite basin studied (after Rögl 1999); (**b**) present distribution of the Badenian evaporite basins in the Carpathian region and location of the basin studied (after Garlicki 1979; Khrushchov & Petrichenko 1979; modified).

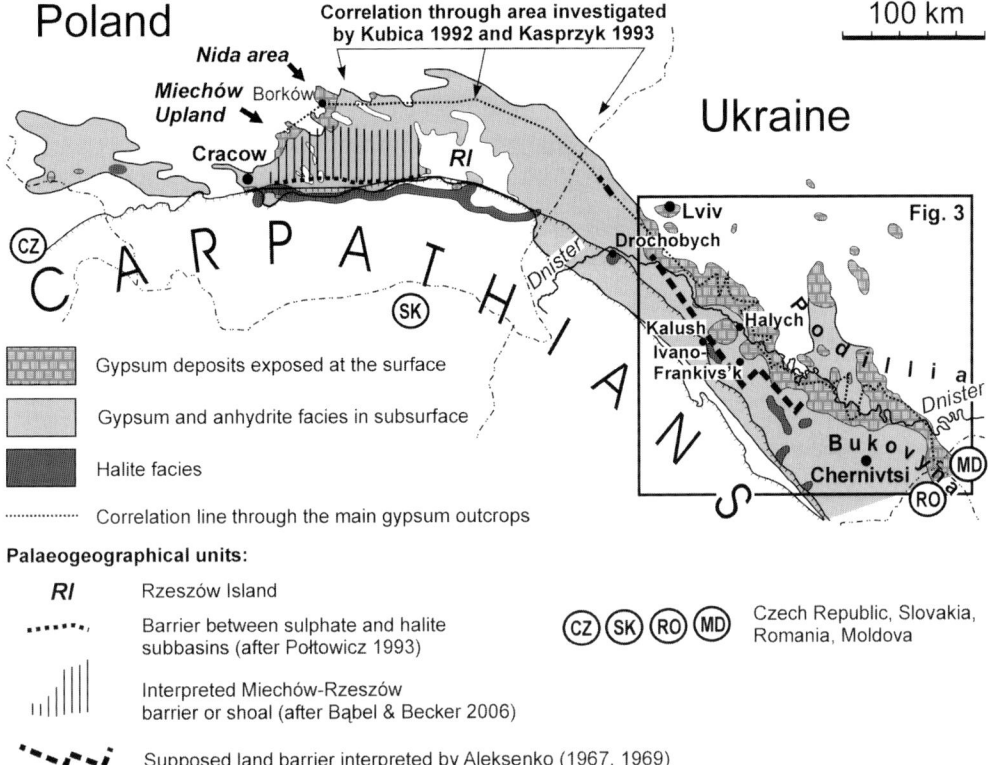

Fig. 2. Distribution of the Badenian (Middle Miocene) evaporites in the northern Carpathian Foredeep (after Atlas Geologiczny Galicyi 1885–1914; Garlicki 1979; Panow & Płotnikow 1996; and other sources).

1997; Kasprzyk & Ortí 1998; Peryt 2000, 2006; Kasprzyk 2003; Cendón et al. 2004; Bąbel & Becker 2006).

The evaporite basin comprised several more or less apparent subbasins. Two halite subbasins occurred along the axis of the foredeep in Poland (Figs 1 & 2; Garlicki 1979; Połtowicz 1993; Ślączka & Oszczypko 2002). Several smaller halite subbasins also are recognized in Ukraine (Fig. 3; Klimov 1974; Petrichenko et al. 1974; Korenevskyi et al. 1977; Panow & Płotnikow 1996). The less pronounced gypsum subbasins were situated in the northern margin of the basin, commonly considered to be part of a giant sulphate platform or 'shelf' (Kasprzyk & Ortí 1998). Presumed shoals and islands apparently lay in between and within the subbasins, and their position could change with time. The largest central area, devoid of evaporite deposits, near Rzeszów (Poland), was interpreted as an island (Rzeszów Island), although evaporites could have been eroded from this area either during or soon after the evaporite deposition (Połtowicz 1993; Oszczypko 2006). The area extending from the Miechów Upland to the Rzeszów Island was presumably a broad longitudinal uplift or barrier separating sulphate and halite subbasins during deposition of the selenite facies (Bąbel 2005b, p. 199; Bąbel & Becker 2006). A similar barrier might also separate the marginal gypsum 'platform' from the southern 'foredeep' subbasin in Ukraine (see Czarnocki 1935, pp. 114–115; Nowak 1938, p. 180). Aleksenko (1967, 1969) recognized a very narrow barrier, running roughly between the platform and the foredeep, based on absence of gypsum in drill cores (Figs 2 & 3). In this area it may be that the lack of gypsum is only an effect of the intersection of the cores taken in the drilled area by numerous oblique post-Badenian faults (Polkunov et al. 1979). The extension of the basin into the Romanian area is poorly recognized, although the evaporite shoal facies with gypsum microbialites are particularly widespread at the border of Ukraine, Moldova and Romania indicating extremely shallow water deposition in this area (Peryt 2001, 2006; Bąbel 2005b).

Stratigraphical correlation of the Badenian evaporites from the platform and particular

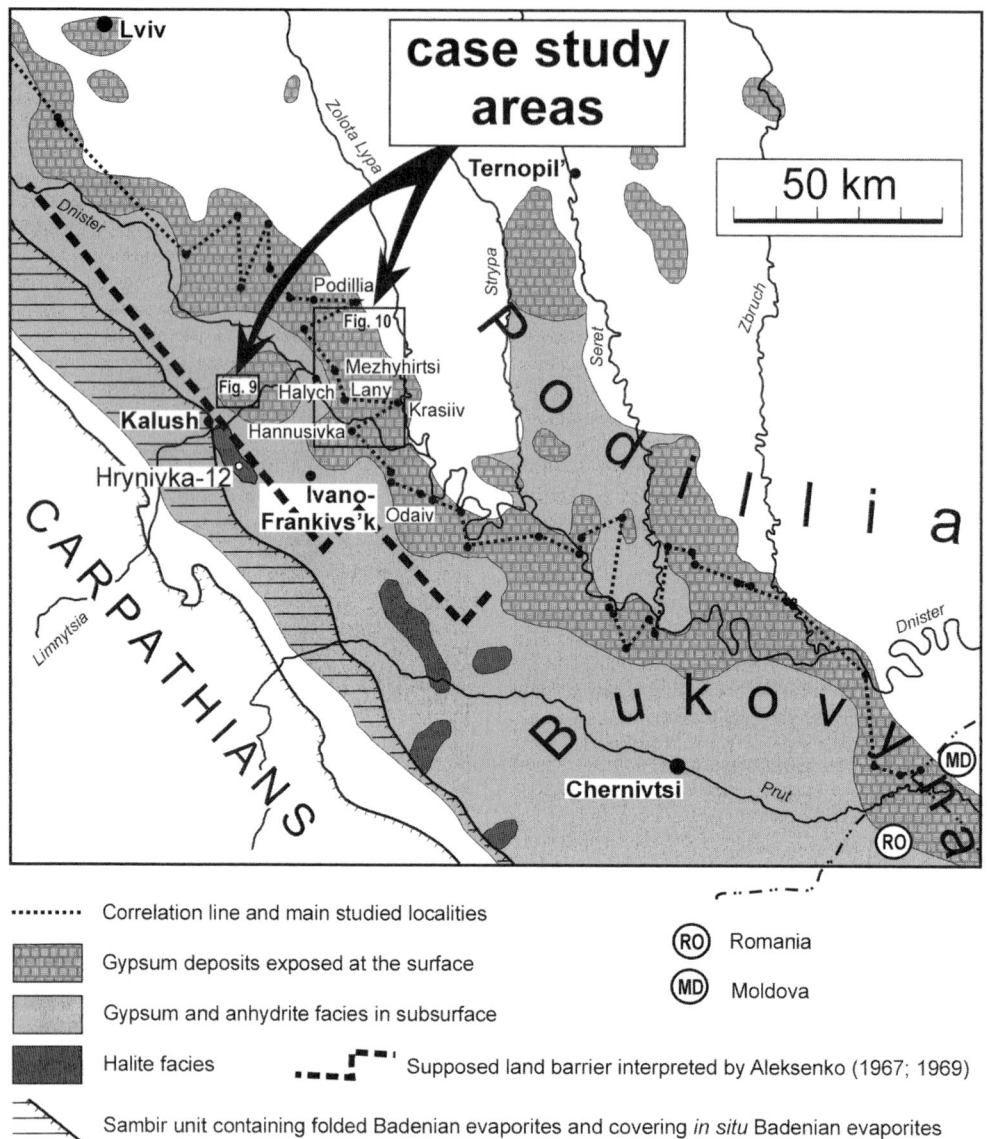

Fig. 3. Location and geology of the study area in the Ukrainian Carpathian Foredeep.

foredeep zones is unclear, although apparently they were deposited within a framework of the same evaporite basin. Event stratigraphical studies, based on a set of thin marker beds, proved that the gypsum deposits in the platform zone are coeval (Peryt 2001; Bąbel 2005a). However the time correlation of these selenite-dominated marginal deposits with the deposits of the remaining parts of the basin, where the facies are different, and selenites or their traces are absent, remains controversial.

The Badenian evaporite basin of the Carpathian Foredeep was probably a depression without open water connections with the sea and with the water level lowered below the sealevel (Peryt 2001, 2006; Cendón et al. 2004; Bąbel 2007). Such a basin can be compared with the modern saline lakes and its hydrography can be described by limnological terminology. Some evidence suggesting that the water level in the basin was below the sealevel can also be found in the studied oriented selenites (Bąbel 2004b, pp. 324–225).

Geology of the area selected for palaeocurrent studies

The area chosen for a case study is situated in Ukrainian Galicia (Halychyna), near Halych on the Dnister River (Fig. 3), the historical centre of Halychyna (from Greek 'hals' = salt). It occupies the basinal platform zone which borders the foredeep zone along the regional Kalush fault. The area extends from the Kalush K-salt open pit mine (Hryniv et al. 2007a, b) on the west, to Zolota Lypa, the left tributary of Dnister, to the east. Kalush area belongs to the allochthonus Sambir zone or unit. This unit is overthrust to the northeast onto the deposits of the autochthonous foredeep and is in direct contact with the platform just at environs of Kalush, where the autochthonous foredeep deposits are completely overlain by it.

On the studied platform area the Miocene deposits lie horizontally, and with an erosional gap, upon the Cretaceous siliceous limestone. Gypsum deposits are underlain by marine clastic and carbonate deposits (most commonly glauconitic marls and sandstones) that are a few metres thick. Locally freshwater-limnic limestones are present at the base of Miocene sequence (e.g. at Lany). The Miocene (Badenian) strata below the gypsum are commonly much reduced in thickness and near Voinyliv (e.g. at Medynia), gypsum deposits lie almost directly on the eroded Cretaceous substrate. Just near Voinyliv and Medynia, Aleksenko (1967, figs 6 & 18; 1969, fig. 2) and Burov (1971) draw the mentioned morphological barrier: a narrow belt of emerged land running in a NW–SE direction in the Badenian evaporite basin (Fig. 3).

Gypsum sections in the studied area generally show the same sequence of lithostratigraphical units and arrangement of facies as in the whole northern margin of the Carpathian Foredeep basin (Kasprzyk 1995; Peryt 2001). The studied oriented selenites occupy the lithostratigraphical unit (or lithosome) C-D. The C-D unit is up to 15 m thick and is present in the entire northern margin of the basin except of its easternmost peripheries where it laterally passes into the other selenite and gypsum microbialite units (Figs 2–4). A thin layer of clastic gypsum–clay deposits in the middle part of unit C-D is traced across nearly the whole basin (e.g. at Ozeriany and Podillia, north of the study area, and at Palahychi and Odaiv, south and SE of the study area; Fig. 3; Bąbel 2005a, On-Line Appendix, fig. 17). This layer is apparently isochronous or nearly so, which suggests that the whole unit C-D was deposited roughly in the same time interval (Peryt 2001). This clastic marker bed was not found in the study area where mainly the lowermost part of unit C-D crops out.

Southwest of the study area, in the foredeep zone, the Badenian evaporites (mostly halite facies and thin-bedded calcium sulphates with clay intercalations) were recognized both within and below the allochthonous Sambir unit (Figs 2 & 3); however the precise time correlation of these evaporites with the gypsum sections from the platform zone is controversial. Available data indicate that no traces of the selenite facies were found in this zone (e.g. Smirnov et al. 1995), although core sections from sulphate evaporites in the vicinity of the halite field near Drohobych show traces of oriented selenite crystals and lithostratigraphical sequence the

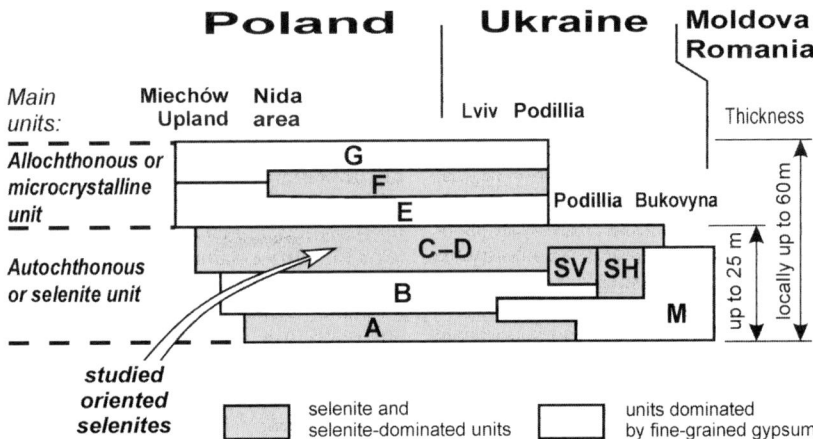

Fig. 4. Lithostratigraphy of the Badenian gypsum evaporites in the northern margin of the Carpathian Foredeep along the correlation line from the Miechów Upland in Poland to Bukovyna in Moldova and Rumania; see Figures 2 & 3 (after Kubica 1992; Bąbel 2005a).

Fig. 5. *Continued.*

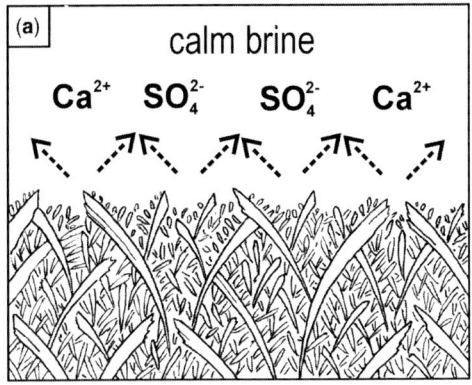

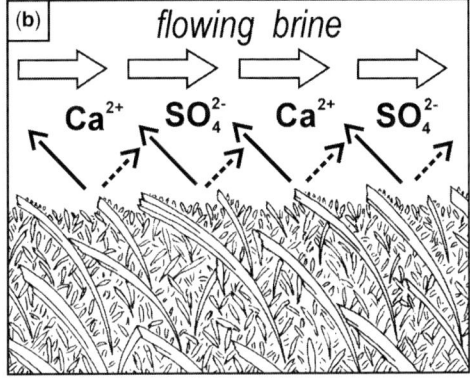

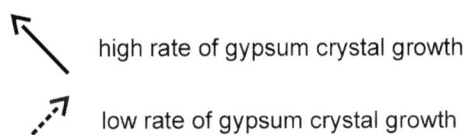

Fig. 6. Scheme showing competitive growth of gypsum crystals at the bottom of evaporite basin: (a) in calm brine; (b) in flowing brine.

thick layer of anhydrite, a 18 m thick layer of rock salt, a 6 m thick layer of anhydrite, a 66 m thick layer of rock salt and a 16 m thick layer of anhydrite (Panow & Płotnikow 1996). This section is similar to the core sections from the five other halite facies fields in the autochthonous foredeep zone in Ukraine, which are recognized as Badenian in age, although most probably only the upper parts of these thick evaporite sequences are age equivalents of the Badenian gypsum on the platform zone (see Figs 2 & 3; Ladyzhenskyi & Antipov 1961, p. 225; Burov et al. 1971, fig. 12; Petryczenko et al. 1994, fig. 2; Petrichenko et al. 1997, fig. 9; Peryt 2006).

Summarizing, the available data suggest that selenite deposition, abundant at the studied edge of the platform zone, did not take place in the adjacent foredeep zone where the halite deposition was commonly associated with clay and took place in some deeper depressions or subbasins.

Method and results of measurements in the environs of Halych

The palaeocurrent analysis in this area of the Carpathian Foredeep basin is based on the directional structures occurring in the whole of unit C-D and formed by selenite crystals with the apices directed in a similar horizontal direction (Fig. 5; Bąbel et al. 1999; Bąbel & Becker 2006, with references). The asymmetric crystal fabric was interpreted as the product of competitive growth of gypsum crystals on the bottom, modified by their more rapid growth in one horizontal direction, towards the inflowing brine (Fig. 6; Dronkert 1977, 1985, p. 205; Pawlikowski 1982; B. C. Schreiber 1982 in Folk et al. 1985; Bąbel 2002; Bąbel & Becker 2006, with references). The largest crystals, with their apices turned horizontally in the same direction (upstream), representing the winners in the competitive growth, are palaeocurrent indicators.

The measurements were collected from intervals of the section usually 2 m thick. In three localities measurements were made separately in the lower and upper part of the strata to reveal the possible change of the brine flow in time (Figs 7 & 8). The azimuths of the apices of the large oriented crystals were measured statistically not only in outcrops of

same as in the platform area (Fig. 2; Kasprzyk 1995; Kasprzyk & Ortí 1998, fig. 3).

Closer to the study area, c. 20 km SSE of Kalush, the halite evaporites were found at about 1.5 km depth in the foredeep area and were recognized as Badenian in age (Korenevskiy et al. 1977, fig. 69). The evaporite section from Hrynivka-12 borehole (Fig. 3), apparently not corrected for dip, comprised (from the bottom to the top): a 12 m

---

Fig. 5. (Continued) Studied oriented selenite deposits: (a, b) selenite beds showing upstream oriented crystals at Lany (a), and Krasiiv (b), Ukraine, compare with Figure 6b; (c) detail of (a) showing aggregates of splited crystals; (d) selenite crystal seen from its side view (at centre), with {010} cleavage surfaces parallel to the picture, and the crystals seen from the side of the curved surface, with {010} cleavage surfaces normal to the picture (at left and right); Lany, Ukraine; (e) variable oriented crystals presumably growing in a calm brine (see a); the wavy bedding surface (at top) is a dissolution surface; Krasiiv, Ukraine; (f) relics of the primary selenite crystals in secondary alabaster deposits, Hannusivka, Ukraine; (g) ghosts of the primary selenite crystals in secondary alabaster deposits, Voinyliv, Ukraine.

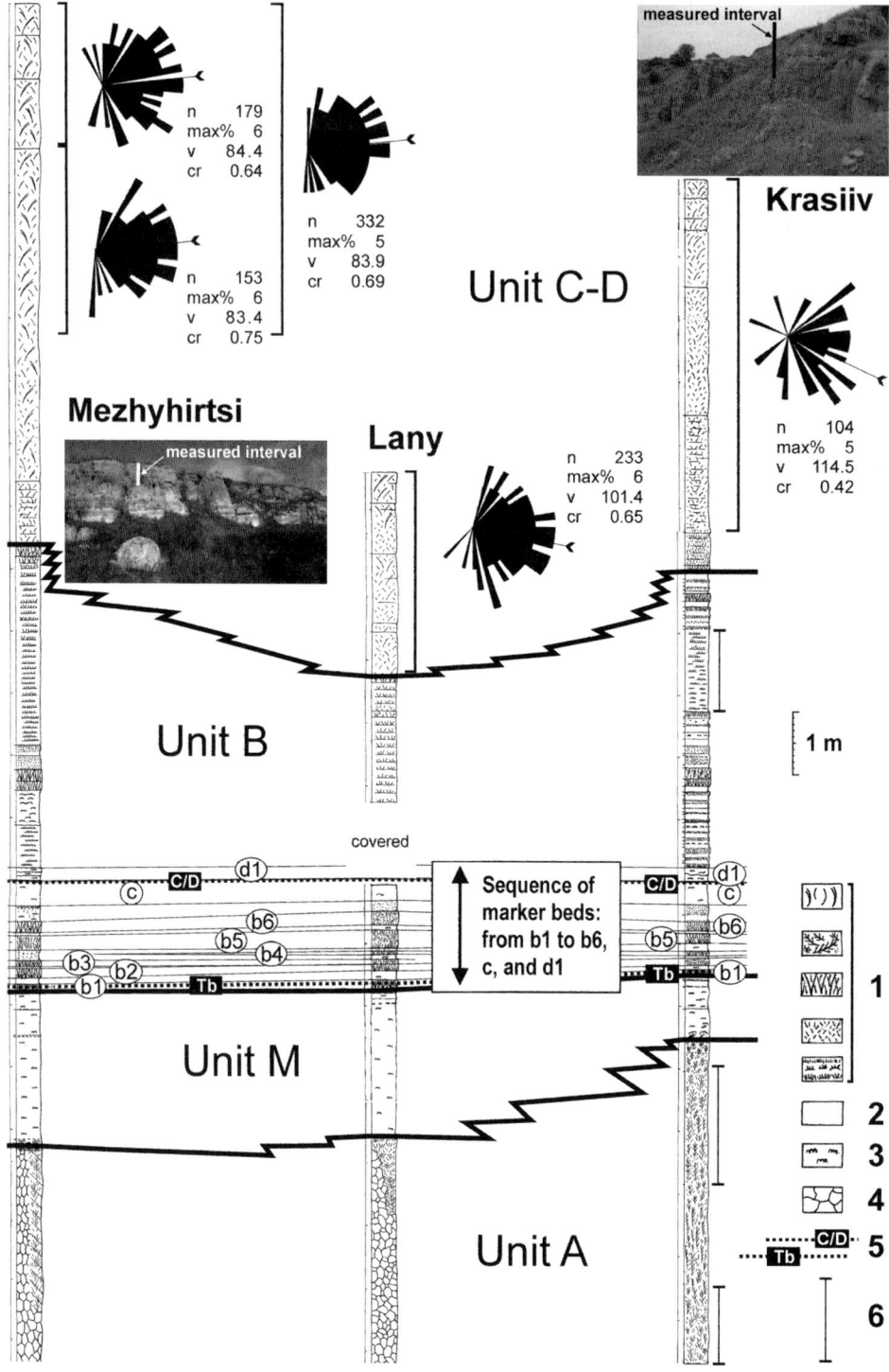

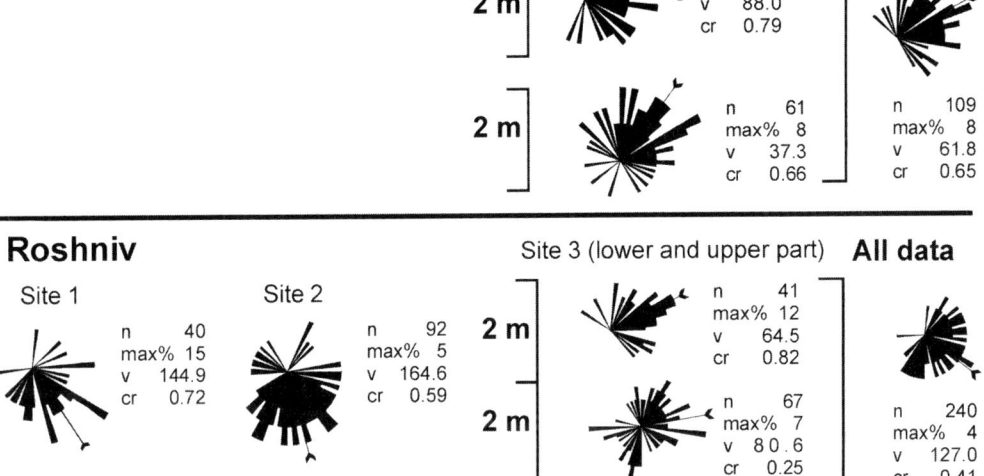

n-number of measurements, %max-maximum percent, v-vector mean (in degrees), cr-consistency ratio

Fig. 8. Rose diagrams showing orientation of apices of curved selenite crystals, unit C-D, at Medukha, Ukraine, and from four different sites at Roshniv, Ukraine (at Roshniv the measurements include relics and ghosts of primary crystals).

primary gypsum facies but also in outcrops of secondary alabaster facies (at Voinyliv, Hannusivka and Roshniv; Fig. 5f & g). In the latter outcrops both relics and ghosts of crystals were measured.

Measurements were subjected to classical statistical analysis used in palaeocurrent studies (Potter & Pettijohn 1977). The statistical data are documented in Figures 7–10, and can be interpreted in the same manner as in the earlier studies (Bąbel et al. 1999; Bąbel 2002; Bąbel & Becker 2006). Mean vectors were recognized for particular groups of measurements and for grouped data and they were interpreted as parallel to the average brine flow.

The measurements made in the lower and upper gypsum layer at Mezhyhirtsy showed nearly the same mean vectors, although the shapes of the rose diagrams and consistency ratio values were slightly different (Fig. 7). The upper layer showed broader span of the azimuths than the lower one. The results suggest that the average flow direction did not change during deposition of the oriented selenites at Mezhyhirtsi although brine flow could be more dispersed in the later stage of gypsum crystallization.

The measurements made in the lower and upper layer at Medukha showed a different orientation of mean vectors, suggesting that the average flow direction changed with time (Fig. 8), and shifted clockwise about 50°. Shapes of rose diagrams and consistency ratio values suggest that the flow directions were more uniform during deposition of the upper layer.

The measurements in the lower and upper layer at Roshniv suggest a similar brine flow direction (Fig. 8). However because of the poor preservation of the primary gypsum crystals, systematicatic data collection throughout the whole bed was not possible; thus at this site the results are not conclusive.

Palaeocurrents at the environs of Halych

The measurements revealed great uniformity of the orientation of the gypsum crystals within the

Fig. 7. (*Continued*) Correlation of studied gypsum sections at Mezhyhirtsy, Lany, and Krasiiv, Ukraine (compare Figs 3 & 4), with location of measured intervals and rose diagrams showing azimuths of selenite crystal apices; top of marker bed c as datum. 1, Botttom-grown gypsum crystals creating variable grass-like structures; hachure reflects arrangement of crystals (for details see Bąbel 2005a, On-Line Appendix); 2, fine-grained homogeneous gypsum; 3, fine-grained gypsum with traces of gypsified crenulated microbial mats; 4, secondary alabaster with homogeneous or nodular structure; 5, isochronous surfaces (after Bąbel 2005a); 6, poorly exposed part of the section. Other explanations: see Figures 8 & 10.

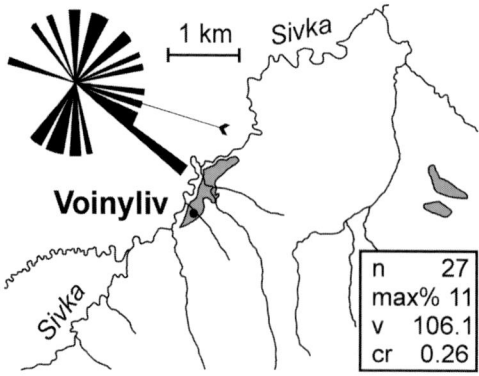

Fig. 9. Rose diagram showing orientation of ghosts of primary curved selenite crystals at Voinyliv, Ukraine (Fig. 3). Badenian gypsum deposits exposed at the surface are shadowed; after Atlas Geologiczny Galicyi (1885–1914). Further explanations see Figures 8 & 10.

studied area. This indicates that the brine in the environs of Halych flowed very uniformly from east to west (Figs 9 & 10). At Hnil'tse, however, the brine flowed from the north to the south apparently because of the presence of some local islands, peninsulas or shoals which disturbed the uniform flow.

The results suggest that during crystallization of the selenite unit C-D the brine flowed *en mass* from east to west in the study area. It apparently formed a broad stream a few tens of kilometres wide moving to the west. The results from Voinyliv (Fig. 9), a few kilometres north of the Kalush fault zone coinciding with the supposed land barrier separating the north from south area of the basin drawn by Aleksenko (1967, 1969), suggests that the brine flowed further west, sub-parallel to this hypothetical barrier (Figs 2 & 3). However the number of measurements is too small and consistency ratio too low for any sure determination of the brine flow direction. If the result is correct the presence of such a barrier in the environs of Kalush cannot be excluded.

The mean vectors from localities east of Halych indicate a swirl or rotary pattern as if the real brine streamlines were arcuate (Fig. 10). Such a pattern is very similar to the large-scale water swirls observed in the modern basins particularly to water swirls formed within separate subbasins where the water tends to flow parallel to the isobaths and shoals (e.g. Csanady 1978). This swirl pattern suggests that the brine, on average, flowed into the study area from SE and then gradually changed the direction of flow into east–west. At Hannusivka the brine flow direction again turned WSW, directly toward the central basin areas (Figs 2 & 10). This brine could flow further into the zone of halite deposition in the environs of Hrynivka (Fig. 3; Korenevskiy *et al.* 1977, fig. 69; Panow & Płotnikow 1996), apparently not being disturbed by any land barrier crossing the basin in this area as proposed by Aleksenko (1967, 1969). The results from Hannusivka are exceptional; it is the only place in the whole basin where the direct brine flow towards the halite facies was detected (see Bąbel *et al.* 1999).

Problem of brine flow into the halite zone

The results from Hannusivka suggest that the calcium sulphate saturated brine flowed over the selenite bottom toward some presumably deeper area in the basin centre where selenite deposition was not recorded. Such deposition should, however, took place there at least from time to time, assuming the downslope flow of brine evidently capable of selenite crystallization.

The nature of the recorded brine currents and hydrography of the selenite area of the basin were originally interpreted by Bąbel & Becker (2006). Their interpretation omitted, however, the transitional zone between the selenite and deeper halite part of the basin and, in particular, the reasons for the disapearance of selenite deposition towards the central halite facies areas, and this requires explanation.

To explain the absence of the selenite facies in the halite areas of the Badenian basin we suggest that these areas were permanently stratified, i.e. they were meromictic subbasins during the time of the oriented selenite crystallization. They were deep enough so that persistent bodies of brine existed, separated from the upper moving and mixing water masses by a permanent pycnocline, at the bottom. These bottom brines remained unmixed with the surface water and the pycnocline acted as a 'virtual bottom'.

During the unit C-D deposition the brine inflowing from the marginal sulphate zone into the halite area was thus moving over the permanent pycnocline (cf. Ahrnsbrak 1974). Inflowing into the halite area these waters lose contact with the bottom and being deflected from it they flowed further, along the pycnocline, in a way similar as the so-called mesopycnal flows known from the studies of turbidity currents (see Last & Schweyen 1983, p. 259; Rimoldi *et al.* 1996; Mulder & Alexander 2001). The inflowing brine could mix partly with the brine below similarly conditions known in the Dead Sea (Stiller *et al.* 1997). In spite of being oversaturated with calcium sulphate, the horizontally flowing brine was not able to promote the growth of selenites simply because it was not in contact with the basin floor.

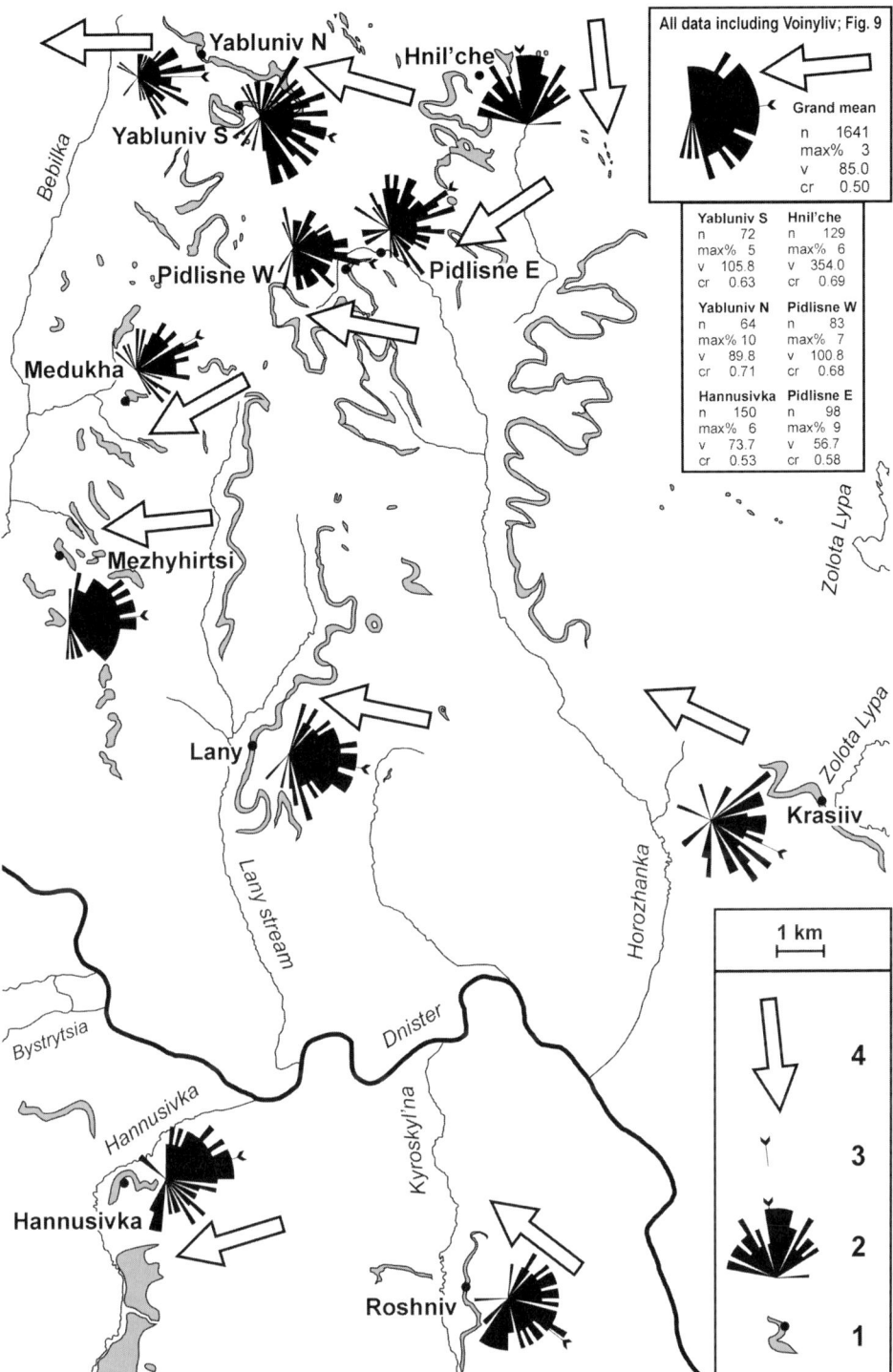

Fig. 10. Palaeocurrents indicated by oriented selenite crystals from unit C-D in area west of Zolota Lipa in Ukraine (Fig. 3); measurements at Hannusivaka and Roshniv include relics and ghosts of primary crystals. 1, Badenian gypsum deposits exposed at the surface (shadowed; after Atlas Geologiczny Galicyi 1885–1914) and locality studied (pointed); 2, rose diagram of the azimuth of the gypsum crystal apices; 3, mean vector of the measured azimuths; 4, interpreted brine flow direction; further explanations in Figure 8.

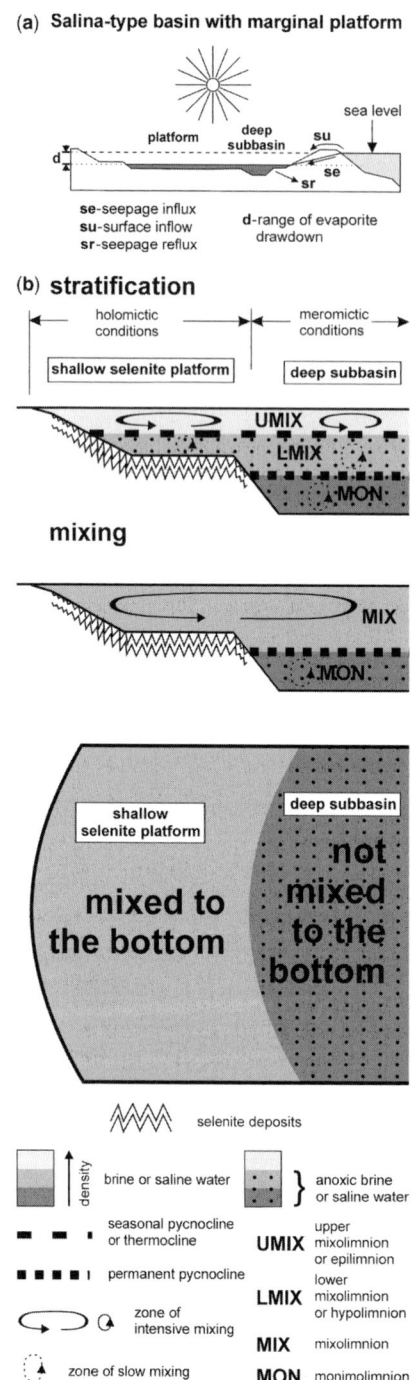

Fig. 11. Model of the Badenian evaporite basin. (**a**) Scheme showing basic features of the basin; (**b**) hydrographical features of the meromictic (selenite) evaporite basin during stratification-mixing cycle (scheme is showing the northern part of the Badenian basin; further explanations in the text).

The concept outlined above is developed below into a more universal hydrographical model of the deep permanently stratified (i.e. meromictic) evaporite basin. The model is presented first, and than it is applied to the geology of the Badenian basin and discussed.

Hydrographical model of the meromictic evaporite basin

To construct an applicable model, and reconstruct the hydrography of the Badenian basin, actualistic data from hydrography of the modern meromictic freshwater and saline lakes were used. They were integrated with the fossil record from the Badenian basin, as well as with the previous hydrographical interpretation of the marginal selenite zone of the basin (Bąbel & Becker 2006). The model is similar to the meromictic selenite basins discussed by Bąbel (2004a, pp. 242–243) and mentioned by Warren (2006, p. 350). A related model was described by Valero-Garcés & Kelts (1995).

According to limnological terminology (Hutchinson 1975; Lewis 1983) in the meromictic basins the deepest water body or zone which is not affected by seasonal vertical mixing is called monimolimnion (Fig. 11). Owing to high concentrations of dissolved ions (or, in some cases, dispersed solids) this deep water is denser than the overlying mixolimnion, defined by the greatest depth reached by mixing of surface waters. The transition zone between mixo- and monimolimnion is defined by a gradient in dissolved solids and is called a chemocline (see Hutchinson 1975, p. 480; Ludlam 1996, p. 116). Because the model concerns evaporite basins with the stratified brine bodies showing pronounced vertical density gradients a chemocline is here described as a permanent pycnocline (as in the Dead Sea and some other basins; Neev & Emery 1967; Miller et al. 1993). As recommended by Hutchinson (1975, p. 537) and Lewis (1983), the mixolimnion can be described according to the terminology of holomictic basins, i.e. basins completely mixed down to the bottom at least once a year. Mixolimnion can be subdivided into the upper warmer epilimnion, metalimnion with the seasonal thermocline, and the lower colder hypolimnion. The terms epi-, meta- and hypolimnion, as originally defined in lakes of temperate zone, in many cases do not fit to the evaporite basins and saline lakes in the arid and tropical zones. In such basins the surface waters are commonly colder that the waters below the seasonal thermocline which shows the temperature gradient reverse than in temperate zone. Therefore the mixolimnion is here described differently than suggested by Hutchinson (1975). The average, relatively thin

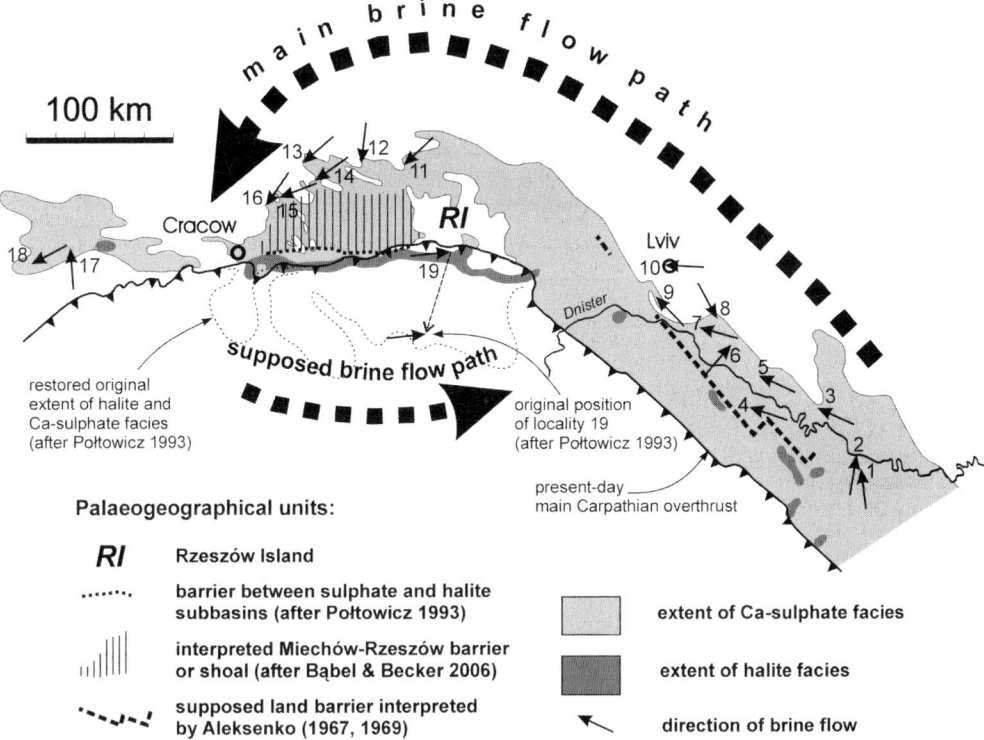

Fig. 12. Mixolimnetic brine palaeocurrents recorded in the Badenian (Middle Miocene) evaporite basin of the northern Carpathian Foredeep reconstructed from orientation of bottom-grown gypsum crystals (after Bąbel & Becker 2006). Localities: 1, Verenchanka; 2, Kostryzhivka; 3, Holovchyntsi; 4, Oleshiv; 5, Krasiiv (see Figs 7 & 10); 6, Podillia; 7, Zahirochko; 8, Pidkamin'; 9, Schyrets'-Pisky; 10, Lviv; 11, Piaseczno; 12, Staszów; 13, Chwałowice; 14, Siesławice; 15, Działoszyce; 16, Kosciejów; 17, Czernica; 18, Kobeřice; 19, Broniakówka.

mixed layer, being an equivalent of epilimnion in temperate lakes, which tends to follow the annual regional temperature changes, is called the upper mixolimnion, the layer below it, defined by the maximum depth of mixing, the lower mixolimnion (Fig. 11). The intermediate zone between the upper and the lower mixolimnion can be described by a seasonal thermocline because temperature is most commonly responsible for seasonal stratification. For simplicity it is assumed that this thermocline roughly coincides with the seasonal pycnocline. In many basins the stratification of mixolimnion is more complex and the upper and the lower mixolimnion can be destinguished only arbitrary or the mixolimnion subdivided into the other pattern of zones.

The model basin contains stratified brines forming three distinct zones which can be described as follows (from the bottom to the top; see the simplified scheme in Fig. 11): (1) monimolimnion enclosing the densest brines showing constant temperature and salinity; (2) pycnocline (chemocline) which shows density gradually decreasing up; and (3) mixolimnion where density, salinity, and, in particular, temperature change seasonally, as well as in smaller time periods. The mixolimnioin is subdivided into lower and upper zone and the average seasonal thermocline (or some average seasonal pycnocline) is the boundary between these zones. The mixolimnion is mixed down to the permanent pycnocline at least once a year and then mixolimnetic waters are fully homogenized.

Modern analogues for the basin model and its mixolimnion

Assumed seasonal stratification-mixing pattern of the mixolimnion (Fig. 11), driven mostly by annual cycle of air temperature changes, is common and characteristic feature of all the modern meromictic basins, both freshwater and saline ones (Table 1). Also in the permanently

Table 1. *Exemplary meromictic basins showing seasonal stratification of mixolimnion (with a few exceptions where the data were lacking) and comparative data for thickness of water zones*

Name and location of the basin	Salinity	Mixolimnion, mixing character	Mixolimnion, structure, depth (m)	Chemocline-pycnocline, depth (m)	Monimolimnion, depth (m)	References	Notes no.
Berkeley pit lake, Montana, USA	FS	Supposedly dimictic	EPI (0–10), THER (10–15), HYP (15–60)	40–60	60 to >250	Gammons & Duaime 2006; references in Boehrer & Schultze 2006	1
Big Soda Lake, Nevada, USA	S	Monomictic	For 1933: EPI (0 to 5–19), THER (5–19), HYP (5–19 to 20). For 1980s: HYP (5–19 to 34.5)	1933: c. 20; In 1980s: c. 34.5	1933: 20–60; In 1980s: 34.5–60	Kimmel et al. 1978; Cloern et al. 1983a, b; Zehr et al. 1987	
Crawford Lake, Ontario, Canada	FS	Dimictic	EPI (0–4), THER (4–8), HYP (8 to 13–16)	13–16	16–24	Dickman 1985	
Deadmoose Lake, Saskatchewan, Canada	S		MIX (0–8)	8	8–48.10	Last & Slezak 1986; Last 1994; Last & Ginn 2005	2
Dead Sea, Israel–Jordan–Autonomous Territory of Palestine	S	Monomictic	UMIX (0 to 15–20), THER (15–25), LMIX (15–25 to 40)	40–100	100 to >350	Neev & Emery 1967; Oren & Shilo 1982; Gertman et al. 2003	3
Faetteville Green Lake, New York, USA	FS	Dimictic	EPI (0–2), THER (2–6), HYP (6–29)	29–40	40–51	Anderson et al. 1985; Thompson et al. 1997; Suits & Wilkin 1998	4
Freefight Lake, Saskatchewan, Canada	S		UMIX (0–3), THER (3), LMIX (3–6)	c. 6	6–24.92	Last 1993, 1994; Last & Ginn 2005	5
Isabela Crater-Lake, Mexico	S	Supposedly monomictic	UMIX (0–2.5), UPYC (2.5–4.5), LMIX (4.5–12)	LPYC (12–13)	13–17.5	Alcocer et al. 1998	6
Lac Pavin, France	FS	Dimictic	EPI (0–10), THER (10–20), HYP (20–60)	60–70	70–92	Aeschbach-Hertig et al. 2002	
Lake Bogoria, Kenya	FS		MIX (0–6)	c. 6	6–11.5	Renaut & Tiercelin 1994	
Lake Cadagno, Switzerland	FS	Dimictic	UMIX (0 to 7–8), THER or UPYC (6–8); LMIX (7–8 to 10–12)	8–15.5	15.5–21	Del Don et al. 2001; Tonolla et al. 2004	17
Lake Goitsche (mine pit lake), Germany	FS	Dimictic	EPI (0–11), THER (c. 11), HYP (11–27)	c. 27	27–31	Boehrer & Schultze 2006	
Lake Fouthiaux (mine pit lake), France	FS		EPI (0–5), THER (5–10), HYP (10 to 20)	20–25	25–37	Denimal et al. 2005	

Lake	Type	Mixing	Stratification	Range 1	Range 2	Reference	#
Lake Kauhako, Moloka'i, Hawaii, USA	S		UMIX (0–1), THER (1–2), LMIX (2 to 3–5)	4–4.5	4.5–248	Donachie et al. 2004	17
Lake La Cruz, Spain	FS	Monomictic	For 1987–1988: EPI (0 to 5–10), THER (5–10 to 10–15), HYP (10–15 to 18–19)	18–19 to 21	21–24	Rodrigo et al. 2001, fig. 1; Romero et al. 2006	
Lake Malawi (Nyassa), Tanzania–Mczambique–Malawi	FS	Monomictic	UMIX (0 to <40) THER (<40 to >80), LMIX (220–230 to >250–300)	c. 250–300	300–706	Eccles 1974; Reynolds et al. 2000; Branchu et al. 2005; Fillippi & Talbot 2005	18
Lake Mary, Wisconsin, USA	S	Dimictic	EPI (0 to 1–4), THER (1–4), HYP (1–4 to 5–9)	5–9 to 11	11–21.5	Weimer & Lee 1973	7
Lake Sonachi, Kenya	S		UMIX (0–1), THER (1–2), LMIX (2 to 3–5)	3–5	5 to c. 8	MacIntyre & Melack 1982; Njugund 1988; Verschuren 1999	8
Lake St Louis (mine pit lake), France	FS		EPI (0–5), THER (5–15), HYP (15 to 20–22)	20–22 to 28	28–60	Denimal et al. 2005	
Lake Tanganyika, Tanzania–Zambia–Zair–Burundi	FS	Presumably monomictic	UMIX (0 to 40–50), THER (40–50), LMIX (40–50 to 100–250)	100–250	250–1471	Reynolds et al. 2000	18
Lake Vähä-Pitkusta, Finland	FS		MIX (0 to 17–23)	17–25	25–35	Hakala et al. 2004; Hakala 2005	17
Lake Zige Tangco, Amdo Caunty, Xizang Autonomous Region, China	S		UMIX (0–8), LMIX (8–22)	21–22	22 to c. 38	Li & Li 2004	9
Mahoney Lake, British Columbia, Canada	S		UMIX (0–1), UPYC (1–2), LMIX (2 to 6–7)	6–10	10–18.3	Northcote & Hall 1983; Overmann et al. 1996; Lowe et al. 1997	10
Medicine Lake, South Dakota, USA	S		MIX (0 to 3.5–4.5)	3.5–4.5	4.5–9	Valero-Garcés & Kelts 1995, table 2; Valero-Garcés et al. 1995	11
Mono Lake, California, USA	S	Monomictic	EPI (0 to 10–20), THER (10–20), HYP (10–20 to 14–28)	14–28	28 to c. 45	Jellison & Melack 1993; Miller et al. 1993, p. 1044	12
Pao-Cachinche reservoir, Venezuela	FS	Monomictic	EPI (0 to 2–4), THER (2–7), HYP (7–10)	PYC and MON (10–20) c. 15	PYC and MON (10–20) 15–21.8	González et al. 2004	13
Piaseczno reservoir (mine pit lake), Poland	FS	Dimictic	EPI (0–4.5), THER (4.5–7.5), HYP (7.5–15)			Wilk-Woźniak & Żurek 2006	

(Continued)

Table 1. *Continued*

Name and location of the basin	Salinity	Mixolimnion, mixing character	Mixolimnion, structure, depth (m)	Chemocline-pycnocline, depth (m)	Monimolimnion, depth (m)	References	Notes no.
Pretoria Salt Pan, South Africa	S		MIX (0–0.55)	c. 0.55–1.00	1.00–2.85	Ashton & Schoeman 1988; Ashton 1991	14
Soap Lake, Grant County, Washington, USA	S		EPI (0 to 6–8), THER (c. 6–8), HYP (6–8 to 17)	c. 17	17–25	Edmondson & Anderson 1965; Walker 1974	
Waldsea Lake, Saskatchewan, Canada	S	Dimictic	UMIX (0–2), THER (2–3), LMIX (2–3 to 6–8)	6–8	6–8 to 14.8	Swanson & Hammer 1983, p. 155; Last et al. 2002	15
West Basin Lake, Victoria, Australia	S		MIX (0 to 4–11)	4 to 10–11	10–11 to 13	Timms 1972; Last & De Deckker 1990	16

S, saline basin; FS, freshwater or slightly saline basin; EPI, epilimnion; THER, thermocline (or metalimnion); HYP, hypolimnion; MIX, mixolimnion; UMIX, upper mixolimnion; LMIX, lower mixolimnion; CHEM, chemocline; PYC, pycnocline; UPYC, upper pycnocline; LPYC, lower pycnocline; MON, monimolimnion.

Notes: 1. 'The mixolimnion has an annual cycle of mixing comparable to a cold monomictic lake, with thermal stratification from spring through autumn and near-complete turnover in winter' (Zehr et al. 1987, p. 783).
2. The mixolimnion exibits seasonal thermal stratification (Last & Slezak 1986, p. 7).
3. Meromictic before 1979, in 1980–1982 and 1992–1995. Brine zones interpreted from figure 32 in Neev & Emery (1967).
4. Interpreted from summer profile (Anderson et al. 1985, fig. 9).
5. 'The mixolimnion is thermally stratified during the ice-free season with a thermocline located at about 3 m depth' (Last 1993, pp. 435–436).
6. Thin 'epilimnion' is produced by rain. Water zones interpreted from description by Alcocer et al. (1998).
7. Thermocline position interpreted from figure 2 in Weimer & Lee (1973).
8. Meromictic during high lake level in 1969–1979; at that period the mixolimnion showed the seasonal thermocline at depth 1–2 m (in January–February 1979; Njuguna 1988, fig. 4); 'the low lake level ($Z_{max} = 5$ m) prevailing since 1985 currently prevents chemical stratification from persisting year-round, with deep wind-driven mixing most likely to occur during the dry-season lowstand in February–March' (Verschuren 1999, p. 115). A shallow daily thermocline is present within mixolimnion.
9. Water zones interpreted from description by Li & Li (2004).
10. Arbitrary interpretation based on spring stratification (Northcote & Hall 1983, p. 192). Depth after Overmann et al. (1996). Mixolimnion shows complex stratification and bias from 'normal' dimictic pattern of mixing due to influence of ice cover.
11. 'Although Medicine Lake possesses a permanent chemocline, the bacterial plate forms in a temporary chemocline at the lower part of the mixolimnion, which evolves during the year in response to changes in amount of rainfall and temperature' (Valero-Garcés et al. 1995, p. 257).
12. Meromictic in 1983–1988 and after 1995.
13. Water zones interpreted from figure 2 in González et al. (2004).
14. Note possibly one of the thinnest natural mixolimnions.
15. Position of the thermocline in spring. Mixolimnion shows complex stratification due to influence od ice cover (see Northcote & Hall 1983, p. 192).
16. 'The mixolimnion stratified during summer and as it cooled in autumn the lake became mesothermal and finally by winter the temperature profile was inverted' (Timms 1972, p. 918).
17. Lack of data on seasonal stratification pattern in mixolimnion.
18. In large lakes (like Malawi, Tanganyika) 'the depth and thickness of seasonal thermoclines are often difficult to determine from temperature profiles alone: solute content (including gases) is generally more reliable guide to vertical integration or to the lack of it' (Reynolds et al. 2000, p. 5).

stratified Black Sea, the shallow seasonal (summer) thermo-, halo- and pycnoclines develop under influence of cyclically (annually) changing air temperature, river flow, precipitation, evaporation, over the permanent deeper thermo-, halo- and pycnoclines, the position of which depends on the intensity of the general cyclonic circulation of surface waters (Titov 2005). Among the saline lakes, the monomictic stratification-mixing pattern of mixolimnion was observed by Neev & Emery (1967) in 1959–1960 in the Dead Sea which showed meromictic hydrography at that time. Neev & Emery (1967) used their own terminology for description of the lake hydrography. According to the terminology accepted in this paper, and by Steinhorn et al. (1979), and Steinhorn (1985), in the Dead Sea in 1959–1960 a permanent pycnocline was at depth 40–100 m (and was less sharp towards the bottom) over deep monimolimnion extending down to over 350 m. Mixolimnion was fully mixed in early winter (December–January) and stratified in the rest of the year, with the most pronounced seasonal thermocline showing gradually lowering tendency with time (from April to September) at depth of 15–25 m (Neev & Emery 1967, fig. 32). Such monomictic pattern of mixing was also recorded in 1980–1981 within much thinner mixolimnion at that time (Oren & Shilo 1982; Anati & Stiller 1991; Anati 1997). Many other meromictic saline lakes show the monomictic mixolimnion, stratified mostly on the temperature basis (Table 1). Because monomixis is typical of the tropical warm zone (Lewis 1983), it is very likely that the mixolimnion of ancient meromictic evaporite basins in the warm and arid zones was also monomictic (as in the Dead Sea). Saline meromictic lakes situated in a temperate and cold zone commonly show dimictic mixolimnion (Table 1).

In some cases and particularly in temperature zone stratification-mixing pattern in the mixolimnion of saline lakes is more complex. Saline lakes seasonally covered with ice commonly show temporary multiple stratification patterns in the mixolimnion (Northcote & Hall 1983; Ludlam 1996; Overmann et al. 1996). Small meromictic saline lakes with a shallow permanent pycnocline (a maximum of a few metres from the surface) appear to not show the seasonal stratification-mixing cycle in the mixolimnion (Kirkland et al. 1983; Ashton & Schoeman 1988), although certainly their mixolimnion exhibits some temporary or daily stratification (MacIntyre & Melack 1982). Any saline lake can develop such a temporary stratification after rain, due to inflow of the fresh waters covering the lake surface.

A stratification pattern similar to those from stratified stage in the model basin (Fig. 11) exists persistently in some deep fiords separated from the sea by a bottom barrier (Ross et al. 1993, fig. 1).

Features of mixolimnion in the Badenian basin

The hydrography of mixolimnion in the Badenian basin was characterized by Bąbel & Becker (2006, figs 8 & 9). The brine was maximum of a few metres deep and the observed regular millimetre-scale growth-zoning of the selenite crystals suggests that such a shallow basin was monomictic (Bąbel & Becker 2006, figs 3f, 9 & 11). The mixolimnion was thus stratified in one, presumably wet, period of the year and fully mixed down to the bottom during the dry period (Fig. 11). It was further assumed that the enhanced crystal growth could take place at the bottom of mixolimnetic zone only during the period of mixing, because only during this period could the bottom brine be fully oxygenated (due to mixing) and highly oversaturated with calcium sulphate (due to increased evaporation). Just at that time the growing selenites obtained the oriented fabric which reflected the *en masse* flow of brine on the whole margin zone of the basin. During the stratification period the growth of selenites was slowed or arrested because the bottom brines were separated from the atmosphere by some seasonal pycnocline and were not subjected to evaporation that is considered as the main driving force of the gypsum crystal growth. Moreover, during and particularly at the end of stratification period, the lower mixolimnion could develop seasonal anoxia (Fig. 11; hypolimnetic anoxia; Horne & Goldman 1994; Bąbel & Becker 2006, fig. 10), and hence the brines could be temporarily stripped of sulphate due to the activity of sulphate-reducing bacteria. The concentration of SO_4^{2-} ions could be too low for the crystallization of gypsum at that time.

As in many modern basins, the stratification of mixolimnion was presumably caused by annually changing environmental factors influencing the surface brines, such as air temperature, precipitation, rate of evaporation or even marine brine inflow to the basin, like in the Solar Lake on Sinai, where the inflow is dictated by the annually fluctuating mean water level of the Red Sea (Cohen et al. 1977). All these factors acting separately or together could influence the surface waters producing a seasonal thermocline and/or pycnocline in the mixolimnion.

Most probably the mixolimnion was seasonally stratified on a thermal basis (Neev & Emery 1967) and in the summer its upper layer was warmer (and lighter) than its lower layer. During winter this stratification could disappear and the whole mixolimnion could show uniform temperature, salinity and density, down to the permanent pycnocline.

The other possibility is that mixolimnion was stratified in the wet season due to increased inflow of some less saline water into the basin (from run-off, rain or rivers) associated with lowered evaporation rate, when the layer of less saline water could appear floating at the more saline brine below. During the dry season the surface layer could partly evaporate and could be fully mix with the underlying brine (more precisely, at the end of the dry and the beginning of the wet season, because of some lag time necessary for generating the full mixing of the brine).

Modern analogues for the Badenian basin and its palaeocurrents

Bąbel & Becker (2006) inferred that the Badenian oriented selenites recorded the mean direction of the brine flow in a given place (Figs 5 & 6), apparently some average value from a relatively long period of time necessary for the growth of gypsum crystals (possibly hundreds of years for a gypsum stratum 1 m thick). The pattern of mean vectors was compared with the average flow pattern observed in lakes which is called the circulation or mean drift, although these terms were commonly attributed to the surface flow (see references in Bąbel & Becker 2006). The mean vectors represent only the flow pattern from the periods of the Ca-sulphate saturation of the bottom brine, because only during such periods could the gypsum crystals grow and could be influenced by the brine current.

Results of all the measurements throughout the entire basin margin, including the study area, strongly suggest counterclockwise (cyclonic) brine flow around the basin shoreline (Fig. 12; Bąbel et al. 1999; Bąbel & Becker 2006). This general reconstruction is in agreement with the observed pattern of water flow (circulation or mean drift) in nearly all the closed and semi-closed basins of the northern hemisphere (Emery & Csanady 1973; references in Bąbel & Becker 2006). The cyclonic circulation was present only in the surface mixolimnetic zone of the basin, down to the permanent pycnocline (Fig. 11). The monimolimnion could show the other unknown pattern of flows and presumably was more stagnant.

Among modern meromictic lakes showing the same circulation pattern as the Badenian basin, the best analogue is the Great Salt Lake (Utah). It is remarkably large and shallow (a water depth fluctuated and attained c. 12.5 m in 1873, and c. 8.5 m in 1963; Cohenour 1968). The lake brine composition is very similar to that of the seawater and approaches salinity over 300‰. Brine density ranges between 1.104 and 1.221. The Great Salt Lake from 1959 to 1991 was meromictic (Great Salt Lake Planning Team 2007, http://www.ffsl.utah.gov/sovlands/GreatSaltLake/resource-fal.PDF Lin 1976). The deepest central parts of the lake, in the southern basin, below about 7–8 m depth, remained unmixed, while above the permanent pycnocline periodic mixing still occurred. The bottom brines below 7–8 m (monimolimnion) remained unmixed even during winter storms (Stephens & Gillespie 1976) when the wind waves can be over 2.7 m high and seiche waves 0.5–1 m high can flood the shores (Cohenour 1968; Lin & Wang 1978). The pattern of beach spits and bars in the Great Salt Lake indicates that the surface waters circulated along the shores predominantly counterclockwise, exactly as the case of the Badenian evaporite basin (Figs 10–12; Cohenour 1968; Butts 1980).

A similar meromixis and circulation as in the Badenian basin was recently documented in the southern Aral Sea. The water in the 25 m deep Chebotarev Bay below 8 m remained unmixed during annual cycle (Friedrich & Oberhänsli 2004). The mixolimnion showed predominantly cyclonic circulation (Zavialov et al. 2003).

The persistent cyclonic circulation of the slightly saline surface waters over the more stagnant and saltier bottom waters is known also from semi-closed basins connected with the ocean like the 34 m-deep Lake Maracaibo, Venezuela (Laval et al. 2005), or the Black Sea (Titov 2005).

The other partial analogue of the Badenian basin is a 4 m deep meromictic salt pond, Lago Pueblo in Venezuela (Sonnenfeld 1984; and references in Bąbel 2004a), where thin gypsum crusts are deposited exclusively in the uppermost 0–1 m deep mixolimnion and gypsum deposits are absent in the anoxic monimolimnion below. Medicine Lake in USA is the next example (Table 1). Thin gypsum crusts, equivalents of selenites, were found there in the littoral zone and gypsum laminites produced by precipitation within the water column in the whole deeper zone (Valero-Garcés & Kelts 1995). Finally, wind-driven cyclonic flow of mixolimnetic brines over the unmixed monimolimnion was recorded in the Pretoria Salt Pan (Table 1; Ashton 1991).

Modelled basin and the Badenian monimolimnion

The monimolimnion of deep meromictic basins is typically anoxic (Fig. 11) and its brine properties are constant or, strictly speaking, change extremely slowly with time (e.g. Anderson et al.

1985). Crystallization of evaporite salts in the monimolimnion is difficult and very complex because the crystallization cannot be directly driven by evaporation (but can be driven by some temperature changes or by salting-out mechanisms; Last & Ginn 2005). In particular, crystallization of calcium sulphate (gypsum and anhydrite) within the monimolimnion is very difficult when it is anoxic. The monimolimnion of sulphate basins is commonly euxinic (showing sulfidic anoxia). It accumulates large amounts of hydrogen sulfide by bacterial reduction of SO_4^{2-} ions if sufficient organic material is available. Crystallization of gypsum in such an environment, showing the lowered concentration or lack of SO_4^{2-} ions, is not or is barely possible (it was recorded, e.g. in Freefight Lake, Canada; Table 1). Gypsum can, however, precipitate in the brine column over the euxinic monimolimnion for several reasons, such as (1) direct evaporation and concentration rise in the surface mixolimnetic brine, (2) 'salting-out' effect during mixing of brines showing different composition and (3) temperature changes of the saturated solution. Such precipitation within the brine column can produce suspension clouds of tiny gypsum crystals and these crystals can be periodically deposited as a rain of particles settling through a brine column and producing laminated deposits (cumulates) at the bottom in the monimolimnion zone, similarly to the meromictic Dead Sea (Neev & Emery 1967; Heim et al. 1997) and many other lakes listed by Valero-Garcés & Kelts (1995, p. 141) including Medicine Lake studied by them (Table 1).

The model suggests that the selenite deposition (in situ growth of large gypsum crystals at the bottom) is restricted to the better oxygenated mixolimnion zone, where the concentration of SO_4^{2-} ions is high at least seasonally (Fig. 11). The depth limit for selenite deposition coincides with the maximum depth of the surface brine mixing. Some authors suggested that the gypsum crystallization in evaporite basins is limited to the photic zone where the photosynthetic activity of cyanobacteria supply the oxygen and thus support the existence of SO_4^{2-} ions (Sonnenfeld 1984). This model relates the limit for selenite deposition to a depth and intensity of physical mixing of surface waters which are oxygeneted both by photosynthesis and direct diffusion from the air (see Babel 2004a). Thus it is rather related to the external environmental and climatic factors, like wind, fetch, etc., which influence the thickness of mixed layer.

A typical feature of monimolimnion is that it contains a fossil water or brine which is older in age than the water in the mixolimnion above (e.g. Walker 1974, p. 215, Steinhorn et al. 1979) The monimolimnetic water is commonly a product of evaporation pre-dating the present stage of the basin, sometimes even the events not directly related to the evaporation (like the brines from dissolution of the Messinian evaporites filling up the deep depressions of the Mediterranean; Cita 2006). It is possible that such fossil monimolimnetic brines can show substantial differences in the chemical composition in comparison with the mixolimnion above (as in the Mediterranean; references in Cita 2006).

The meromictic basin model clearly explains the lack of selenite deposits in the deep central areas of the Badenian basin (Fig. 11). They could not be deposited in the monimolimnion zone because of two main reasons: (1) sulfidic anoxia and associated low concentration or lack of SO_4^{2-} ions (in similar fashion to the model outlined by Parea & Ricci-Lucchi 1972, fig. 5b); and (2) separation from atmosphere and the environment of evaporation – a necessary condition and main reason for the concentration and rise in salinity required for the rapid crystal growth.

Geological data from the Badenian basin can support this interpretation. Laminated, organic, clay and fine-grained clastic deposits typical of the monimolimnion zone (e.g. Anderson et al. 1985) occur in the central zone of the Badenian basin (Kasprzyk & Ortí 1998). Anoxia in the halite depositional zone was suggested by Garlicki (1979). Laminated calcium sulphate deposits are very common in the basin centre. They could be deposited by settling from the water column from some suspended clouds of fine sediment generated by slumps, storms or low-density turbidity currents (Peryt 2000; Kasprzyk 2003). They are mostly clastic deposits derived both from basin margin or slope and from in situ redeposition, as evidenced by common washout surfaces and slump folds. These structures, as well as numerous soft-sediment deformations, indicate that the laminated calcium sulphate deposits remained loose and unlithified at the bottom, although in places they could be partly consolidated or cemented (Peryt 2000). They were rarely subjected to syndepositional cementation by gypsum or anhydrite, which can mean that the bottom was rarely in the long-term contact with the calcium sulphate saturated to oversaturated brine. The brine over the bottom could become undersaturated with calcium sulphate because it was impoverished in SO_4^{2-} ions due to anoxia. There are evidences of the syndepositional and early diagenetic in situ growth of calcium sulphate nodules directly at the bottom and within some laminae (Kasprzyk & Ortí 1998; Kasprzyk 2003), which can suggest that some oxygenated sulphate brines from the surface layers reached the bottom promoting the growth of gypsum or anhydrite crystals (cf. Beyth 1980).

Concepts of brine transport in evaporite basins; a discussion

The concept of the meromictic basin and the data gained from palaeocurrent studies throw some new light into the hydrodynamics of brine both in the Badenian and other similar evaporite basins. This application of hydrodynamics has been commonly understood and described by traditional concepts omitting or ignoring the possibility of persistent brine stratification. Some of these concepts are discussed below and addressed with the new approach.

Coriolis effect and cyclonic flow

The reconstruction presented here is similar to the model of lagoonal evaporite basin considered by Sonnenfeld (1974, 1975, 1984 and other papers) who assumed that Coriolis force influences the stream of marine water inflowing through the narrow strait to the basin and forces the cyclonic flow along the shores of basin. In his evaporite lagoon the marine water makes only one open circle within the basin. It evaporates on its way along the shores and gradually becomes a marine brine that finally flows out of the basin as the bottom current through the same strait (in the similar way as the bottom brine current once flowed the Lynch Strait in the Dead Sea; Neev & Emery 1967, fig. 35).

The brine circulation in the Badenian basin is interpreted differently than in Sonnenfeld's model. In particular, the detected brine flow directions do not reflect the inflow path of the marine water or marine brine to the Badenian evaporite basin (Fig. 1). The brine circulation was similar to those present in most lakes which means that the brine made not an open but a closed circle, flowing more or less continuously along the basin shores and above the permanent pycnocline in the deep basin centre (Figs 11 & 12).

The presented view, if correct, strongly suggests that the oriented selenite facies should be present along the entire path of brine flow and thus, consequently, that such selenites also crystallized on the hypothetical southern Carpathian margin of the basin destroyed by erosion after tectonism (Fig. 12). The presence of such a facies continuity, however, is documented only in one place in Poland (locality 19, Fig. 12) and is unrecognized in Ukraine.

'Saturation shelf' concept

Sonnenfeld (1974, 1975, 1984, 1992) adopted the concept of 'saturation shelf' proposed by Richter-Bernburg (1955) to his evaporite lagoon model and assumed that the flat marginal platforms covered by flowing marine water were such a shelf ('gypsum shelves'; Sonnenfeld 1984, p. 41). 'Flowing over the shallow coastal shelves, the surface waters are warmer and evaporate faster as they encounter dry winds from land and receive more radiation per unit of depth' (Sonnenfeld 1984, p. 41). These 'sunlit shelves act as preconcentrators, as preferred sites of gypsum precipitation' (Sonnenfeld 1992, p. 263). When the flowing waters are 'sufficiently concentrated', they 'veer toward the deeper parts' of the basin (Sonnenfeld 1984, p. 41), commonly being rapidly subsiding grabens (Sonnenfeld 1992). Halite deposition eventually takes place in these depressions, in centre of the basin. The 'saturation shelf' concept was applied for the Messinian and Badenian evaporite basins (Garlicki 1979; Cohen 1993; Połtowicz 1993).

The palaeocurrent analysis in the Badenian basin does not exactly support this concept. Although the brine flow through the shallow marginal sulphate platforms is documented (Fig. 12), any evident traces of the current flowing directly downslope to the halite depositional areas were not detected (Bąbel et al. 1999). Only the data from Hannusivka, documented in this paper, indicate the presence of such a current. Bąbel & Becker (2006) speculated that perhaps such downflowing brine currents were sporadic and too short events to be recorded by the growing gypsum crystals. They also assumed that the downflow of brines was partly blocked by some shoals or islands at the basinward edge of selenite platforms (Fig. 12); however, it is difficult to accept that such a continuous barrier was present around the whole basin.

Downslope brine transport

Downslope flow of brine was not documented by the palaeocurrent studies in the Badenian basin. This is in contrast with many theoretical concepts concerning the brine transport in evaporite basins which assume that density underflows (downflowing) are very common. The brine transport to the basinal depressions is commonly described as a process exclusively driven by evaporation; when produced by evaporation, increasingly dense surface water sinks down and 'displaces' the less dense bottom water (e.g. Schmalz 1969). It is believed that the dense water, which is produced by evaporation on shoals ('saturation shelves'), flows basinwards as a permanent bottom current accompanied by a surface compensation counter-current flowing towards the shoals and carring the lighter waters displaced upward by the downflowing brine (Richter-Bernburg 1955, fig. 2).

These concepts are oversimplified and appear to be unrealistic when compared with the real patterns of brine and water transport in the modern basins.

The physical, 'volume by volume' displacement of bottom water by sinking or inflowing surface brine appears to be nearly impossible in deep evaporite basins. Most if not all the modern saline basins show a nearly permanent pycnocline zone existing and separating the bottom dense brines from the surface lighter waters and this does not permit the surface waters to reach the bottom directly. In the Dead Sea the very dense brine produced in saltworks pans operating in the southern shallow subbasin outflowed as a bottom current through the Lynch Strait to the deep permanently stratified (meromictic) northern basin (Neev & Emery 1967; Steinhorn et al. 1979). This brine did not reach the deepest zone of the basin. It spread along the pycnocline, or along the deep water layers having the same density as the inflowing brine (see Ahrnsbrak 1974), possibly contributing to increased halite crystallization at a depth of 170 m (Stiller et al. 1997). This brine continued its flow northward but not downslope. It flowed along the eastern shore in the typical cyclonic way (Neev & Emery 1967).

A persistent, downslope-flowing bottom current accompanied by a surface countercurrent (Richter-Bernburg 1955) is possible in the case of semi-closed shoals connected with the deep basin through some relatively deep strait, like the mentioned Lynch Strait in the Dead Sea (Neev & Emery 1967); however, this is less probable in case of open-type shoals or platforms. It seems that evaporation which produces highly saline and dense water on the open shoals is too slow a process to generate any rapid, permanent downslope density flow because water movements on the shoals are rather generated, and controlled nearly all the time, by much more effective and stronger wind forces (Walker 1982). Certainly evaporation contributes to the pattern of water movements in every evaporite basins; however in nature many forces act on the shallow water and commonly deflect the flow direction rather parallel to the isobaths and shorelines (e.g. Csanady 1978).

Besides the wind, the downslope outflow of brines from the shoals can also be controlled by inclination of the slope, temperature of brine and horizontal temperature gradients in particular. Cornée et al. (1992) investigated shallow saltwork pans and found that the more saline and warmer gypsum saturated brine was formed on shoals (0–20 cm deep) and stayed there nearly permanently. Its flow down into the depressions (30–40 cm deep) occupied by less saline and cooler brine was absent or very sluggish. Apparently the horizontal salinity gradient persisted because the density difference, caused by salinity difference, was compensated by difference in brine temperature (Sonnenfeld 1984, p. 73; Anati 1998). The density differerence was too low to cause the rapid downflow of brine on a very gently inclined slope. Similar persistent horizontal salinity and temperature gradients exist in the so-called compensation fronts, from 10 m to 100 km wide (typically 3–4 km), in the 100 m thick mixed layer of the open oceans (Rudnick & Ferrari 1999; Rudnick & Martin 2002). Such fronts persist because seawater mixing dissipate the density gradient in the front without obliteration of the coinciding salinity and temperature gradients which compensate each other exactly. Thus it seems that some sufficient temperature difference between the warmer and more saline waters occupying the basin margin, and cooler and less saline water in the basin centre (as in the saltwork pans; Cornée et al. 1992), can indeed inhibit any basinward density flow when the density difference related to salinity difference between such water masses is compensated by this temperature difference. However there is no any data to prove that such a 'compensation' barrier or front was present in the Badenian basin, and on the other hand such a hypothetical front, if ever possible, appears to be only a temporary phenomenon, and downflow of brine to the basinal depressions should take place at least from time to time.

Brine accumulation in meromictic basins

The model presented here suggests a more complex mechanism of brine accumulation in the deep Badenian halite subbasins than formerly assumed (Garlicki 1979; Połtowicz 1993). It was not caused by a downslope flow of dense brine directly to the bottom but rather by its flow into the permanent pycnocline. The brine accumulation could be realized by a series of mixing events at the boundary between mixolimnion and monimolimnion associated with the vertical fluctuations of the permanent pycnocline.

As in the Dead Sea the effective salinity rise in the halite subbasins was probably connected with deeper mixing and thickening of their mixolimnion. Such a deeper mixing in the meromictic Dead Sea was associated with the shallowing which caused the gradual rise in the salinity and density of the mixolimnion up to the level approaching the density of monimolimnion. This destabilizes the permanent pycnocline and permits the deeper mixing of the mixolimnion. The gradual thickening of the more saline and denser mixolimnion was associated with 'erosion' of the monimolimnion which took place rapidly during succeeding five annual mixing periods and culminated with the complete mixing of the entire water column in 1979 (Steinhorn et al. 1979; Steinhorn 1985). The thicker mixolimnion was formed at the expanse of monimolimnion. The salinity and density of the thicker mixolimnion rose mainly due to the mixing with the monimolimnion and additionally due to the evaporite concentration (Jellison & Melack 1993). It is thought that such deep mixing events, sporadically culminating in short holomictic (monomictic) periods, took place in the previous history of the meromictic Dead Sea but they have been stopped. A new thinner mixolimnion and a new shallower permanent pycnocline were developed during some years due to extensive inflow of meteoric waters (Oren & Shilo 1982; Steinhorn 1985; Jellison & Melack 1993, p. 1012). Owing to such flooding the monimolimnion contained, after many years of development, 'fossil' layers of various 'ages' recording some past deep mixing events (Steinhorn 1985, p. 460). The deepest layers contained the oldest, saltiest and densest brine.

Halite crystallization in the Badenian subbasins

The effective salinity rise at the bottom of the Badenian halite subbasins, which could eventually initiate the halite crystallization there, presumably required complete mixing, i.e. temporary or permanent transition of the meromictic hydrography into holomixis. One of the reasons for such a transition is shallowing or evaporite drawdown. The shallowing of the meromictic saline basin can change its hydrography into monomixis, as recently it has happened in Mono Lake (Melack & Jellison 1998) and the Dead Sea (Gertman et al. 2003). Rapid, massive, halite crystallization at the entire bottom of the Dead Sea started with the monomixis in 1979 and continues with the water level drop and gradual salinity rise in this drying lake (Niemi et al. 1997; Herut et al. 1998).

If such a scenario for halite deposition in the deep Badenian subbasins is true it means that the halite deposits in the basin centre could be associated with the shallowing or emersion events on the basin margin and the halite deposition should be proceeded by salinity rise of the mixolimnetic brines. Some geological data support such an interpretation for the period connected with the oriented selenite deposition. Indeed, some geochemical evidence (rise in Sr content in selenite crystals; Bąbel 2007, with references) suggests a gradual salinity increase during deposition of unit C-D, and hence in the presumed mixolimnion in the Badenian basin. However this salinity increase is associated with the deepening rather than shallowing, although gypsum microbialites and traces of erosion are common at the top surface of unit C-D and suggest such a shallowing (Bąbel 2005b, pp. 199–200). Traces of dissolved halite were found directly above unit C-D (Fig. 4; Bąbel 2005b, On-Line Appendix, figs 12a, 13a, 14a & 15b), suggesting the possible halite deposition also in the basin centre just after deposition of unit C-D. This fits to the scenario proposed previously and is in accordance with the former interpretation that the brines from which the halite on the basin margin was crystallized could partly derive from the central subbasins (Bąbel 2005b, p. 202). As in the drying Dead Sea and Mono Lake, where the beginning of the meromictic–holomictic transition was associated with the deepening of the permanent pycnocline (Steinhorn et al. 1979; Anati & Stiller 1991; Jellison & Melack 1993), the Badenian mixolimnion (Fig. 11) could become thicker at the expanse of the monimolimnetic brines and hence enriched in NaCl, necessary for halite crystallization, by mixing with these brines, presumably containing a large amount of NaCl.

Implication for stratigraphy of the Badenian chemical facies

The model presented in Figure 11 is appropriate for the conditions of crystallization of selenite unit C-D, when the gypsum and halite subbasins were well connected, but not necessary for the other units. In particular selenite unit A was deposited when the large morphological barriers present in the basin (see Bąbel 2005b, p. 199; Bąbel & Becker 2006) could make its hydrography different from the model. According to the model, it is unlikely that during unit C-D deposition on the basin margin the halite could crystallize in the monimolimnion zone in the basin centre (but it could be deposited there after crystallization of that selenite unit). If the interpretation is correct then the oriented selenites correlate with some non-halite deposits in the halite facies areas, most probably with some laminated clay or laminated Ca-sulphate deposits. Such laminated deposits are found directly below, over and within the halite evaporites; however, which of these deposits are the age equivalent of the investigated unit C-D is unknown.

It is likely that during unit C-D deposition the monimolimnion contained some fossil brines from some earlier evaporite events or even from dissolution of some older halite deposits (see Cendón et al. 2004; Peryt 2006). Such a view fits the interpretations which assume that the halite deposition at the basin central areas pre-dated the gypsum deposition on the northern margin of the basin (Ladyzhenskyi & Antipov 1961, p. 225;

Burov et al. 1971, fig. 12; Petryczenko et al. 1994, fig. 2; Petrichenko et al. 1997, fig. 9; Peryt 2006).

Conclusions

The distribution of selenite facies along the margin of the Badenian evaporite basin in the northern Carpathian Foredeep, and the lack of selenites in the deeper centre of the basin occupied by halite, laminated Ca-sulphate and clay facies, is interpreted as a result of the meromictic hydrography of the basin. Because the basin was presumably isolated from the sea, its hydrography can be compared with the modern meromictic and saline lakes and described by limnological terminology. There were two main brine bodies separated by a permanent pycnocline in the basin: monimolimnion, in the deepest halite areas, and mixolimnion at the surface, extending onto the selenite basin margin. The monimolimnion was or tended to be euxinic and direct evaporative precipitation of salts in it, and selenite crystallization, in particular, was not possible (because of the complete separation from the atmosphere and scarcity of SO_4^{2-} ions which were mostly decomposed by bacterial reduction). Some clastic calcium sulphate was deposited in the monimolimnion zone, mostly by settling from the brine column. The mixolimnion contained calcium sulphate saturated brine. It was seasonally stratified and mixed down to the permanent pycnocline once a year (i.e. it was monomictic as suggested by regular millimetre-scale growth zoning in the selenite crystals presumably reflecting such stratification-mixing cycle in the brine column typical of monomictic basins). During seasonal stratification of mixolimnion the selenite growth at the bottom was slowed or arrested because the bottom brine was not able to evaporate and presumably developed temporary anoxia. Only during the mixing periods was the brine in the entire mixolimnion zone fully oversaturated with calcium sulphate due to direct evaporation. At that time the selenites could grow rapidly but only in the marginal zone of the basin, not in the deeper halite facies zone below the mixolimnion. Currents were common in the mixolimnion whereas monimolimnion was relatively stagnant because it was separated from the flowing surface waters above by a permanent pycnocline. As in the modern basins, the surface mixolimnetic brines showed a dominantly cyclonic pattern of circulation along the basin shores. This average circulation was recorded on the Badenian basin margin by the selenite crystals growing at the bottom. These crystals show the apices turned in the similar horizontal direction which is interpreted as a result of upstream growth in the moving calcium sulphate oversaturated brine.

The research presented herein was founded by the grant 6 P04D 038 09 to M. B. from the State Committee for Scientific Research (KBN, Poland). The authors are indebted to Vasyl Tycholiz, Svitlana Vizna and Andrii Yatsyshyn for help in the field work. Andrzej Gąsiewicz, Stefano Lugli and B. Charlotte Schreiber are thanked for discussions and critical, constructive comments which improved the text.

References

AESCHBACH-HERTIG, W., HOFER, M., SCHMID, M., KIPFER, R. & IMBODEN, D. M. 2002. The physical structure and dynamics of a deep, meromictic crater lake (Lac Pavin, France). *Hydrobiologia*, **487**, 111–136.

AHRNSBRAK, W. F. 1974. A saline intrusion into Seneca Lake, New York. *Limnology and Oceanography*, **19**, 275–278.

ALCOCER, J., LUGO, A., DEL ROSARIO SÁNCHEZ, M. & ESCOBAR, E. 1998. Isabela Crater-Lake: a Mexican insular saline lake. *Hydrobiologia*, **381**, 1–7.

ALEKSENKO, I. I. 1967. *Sulphur of the Fore-Carpathians*. Izdatelstvo Nedra, Moscow [in Russian].

ALEKSENKO, I. I. 1969. Geological structure of the Fore-Carpathian sulphur-bearing basin and genesis of the native sulphur deposits. In: SOKOLOV, A. S., TRUHACHEVA, A. G. & SHUGYN, A. A. (eds) *Geology of the Native Sulphur Deposits*. Nedra, Moscow, 65–95 [in Russian].

ANATI, D. A. 1997. The hydrography of a hypersaline lake. In: NIEMI, T. M., BEN-AVRAHAM, Z. & GAT, J. R. (eds) *The Dead Sea: the Lake and Its Setting*. Oxford Monographs on Geology and Geophysics, **36**. Oxford University Press, New York, 89–103.

ANATI, D. A. 1998. Dead Sea water trajectories in the T-S space. *Hydrobiologia*, **381**, 43–49.

ANATI, D. A. & STILLER, M. 1991. The post-1979 thermohaline structure of the Dead Sea and the role of double-diffusive mixing. *Limnology and Oceanography*, **36**, 342–354.

ANDERSON, R. Y., DEAN, W. E., BRADBURY, J. P. & LOVE, D. 1985. Meromictic lakes and varved lake sediments in North America. *U.S. Geological Survey Bulletin*, **1607**, 1–18.

ASHTON, P. J. 1991. Horizontal heterogeneity of the Pretoria Salt Pan. *Salinet*, **5**, 37.

ASHTON, P. J. & SCHOEMAN, F. R. 1988. Thermal stratification and the stability of meromixis in the Pretoria Salt Pan, South Africa. In: MELACK, J. M. (ed.) *Saline Lakes*. Developments in Hydrobiology, **44**, 253–265. The Hague. [Also: *Hydrobiologia*, **158**, 253–265.]

ATLAS GEOLOGICZNY GALICYI, 1885–1914. 99 map sheets 1:75000 with explanations; Zeszyty **1–21**, **23**, **25–27**. Wydawnictwo Komisyi Fizyograficznej Akademii Umiejętności, Kraków [in Polish].

BĄBEL, M. 2002. Brine palaeocurrent analysis based on oriented selenite crystals in the Nida Gypsum deposits (Badenian, southern Poland). *Geological Quarterly*, **46**, 49–62.

BĄBEL, M. 2004a. Models for evaporite, selenite and gypsum microbialite deposition in ancient saline basins. *Acta Geologica Polonica*, **54**, 219–249.

BĄBEL, M. 2004b. Badenian evaporite basin of the northern Carpathian Foredeep as a drawdown salina basin. *Acta Geologica Polonica*, **54**, 313–337.

BĄBEL, M. 2005a. Event stratigraphy of the Badenian selenite evaporites (Middle Miocene) of the northern Carpathian Foredeep. *Acta Geologica Polonica*, **55**, 9–29. On-line Appendix: www.geo.uw.edu.pl/agp/table/appendixes/55-1.

BĄBEL, M. 2005b. Selenite-gypsum microbialite facies and sedimentary evolution of the Badenian evaporite basin of the northern Carpathian Foredeep. *Acta Geologica Polonica*, **55**, 187–210.

BĄBEL, M. 2007. Depositional environments of a salina-type evaporite basin recorded in the Badenian gypsum facies in northern Carpathian Foredeep. *In*: SCHREIBER, B. C., LUGLI, S. & BĄBEL, M. (eds) *Evaporites Through Space and Time*. Geological Society of London, Special Publications, **285**, 107–142.

BĄBEL, M. & BECKER, A. 2006. Cyclonic brine flow pattern recorded by oriented gypsum crystals in the Badenian evaporite basin of the northern Carpathian Foredeep. *Journal of Sedimentary Research*, **76**, 996–1011.

BĄBEL, M., BOGUCKI, A., VIZNA, S. & YATSYSHYN, A. 1999. Reconstruction of brine paleocurrents in the Middle Miocene evaporitic basin of the Carpathian Foredeep. *Biuletyn Państwowego Instytutu Geologicznego*, **387**, 12–13.

BEYTH, M. 1980. Recent evolution and present stage of Dead Sea brines. *In*: NISSENBAUM, A. (ed.) *Hypersaline Brines and Evaporitic Environments*. Developments in Sedimentology, **28**. Elsevier, Amsterdam, 155–166.

BOEHRER, B. & SCHULTZE, M. 2006. *On the relevance of meromixis in mine pit lakes*. www.imwa.info/docs/imwa_2006/0200-Boehrer-DE.pdf

BRANCHU, P., BERGONZINI, L., DELVAUX, D., DE BATIST, M., GOLUBEV, V., BENEDETTI, M. & KLERKX, J. 2005. Tectonic, climatic and hydrothermal control on sedimentation and water chemistry of northern Lake Malawi (Nyasa), Tanzania. *Journal of African Earth Sciences*, **43**, 433–446.

BUROV, V. C. 1971. Structure of the Miocene molasses. *In*: GLUSHKO, V. V. & KRUGLOV, S. S. (eds) *Geological Structure and Combustible Resources of the Ukrainian Carpathians. Trudy Ukrainskogo Nauchno-Issledovatelskogo Geologorazvedochnovo Instituta (UkrNIGRI)*, **25**. Nedra, Moscow, 58–70 [in Russian].

BUROV, V. C., GLUSHKO, V. V. & PISHVANOVA, L. S. 1971. Neogene deposits of the Forecarpathian Depression. *In*: GLUSHKO, V. V. & KRUGLOV, S. S. (eds) *Geological Structure and Combustible Resources of the Ukrainian Carpathians. Trudy Ukrainskogo Nauchno-Issledovatelskogo Geologorazvedochnovo Instituta (UkrNIGRI)*, **25**. Nedra, Moscow, 42–54 [in Russian].

BUTTS, D. S. 1980. Factors affecting the concentration of Great Salt Lake brines. *In*: GWYNN, J. W. (ed.) *Great Salt Lake: a Scientific, Historical and Economic Overview*. Utah Geological and Mineral Survey Bulletin, **116**. Artistic Printing, Salt Lake City, 163–167.

CENDÓN, D. I., PERYT, T. M., AYORA, C., PUEYO, J. J. & TABERNER, C. 2004. The importance of recycling processes in the Middle Miocene Badenian evaporite basin (Carpathian Foredeep): palaeoenvironmental implications. *Palaeogeography, Palaeoclimatology, Palaeoecology*, **212**, 141–158.

CITA, M. B. 2006. Exhumation of Messinian evaporites in the deep-sea and creation of deep anoxic brine-filled collapsed basins. *Sedimentary Geology*, **188–189**, 357–378.

CLOERN, J. E., COLE, B. E. & OREMLAND, R. S. 1983a. Seasonal changes in the chemistry and biology of a meromictic lake (Big Soda Lake, Nevada, U.S.A.). *Hydrobiologia*, **105**, 195–206.

CLOERN, J. E., COLE, B. E. & OREMLAND, R. S. 1983b. Autotrophic processes in meromictic Big Soda Lake, Nevada. *Limnology and Oceanography*, **28**, 1049–1061.

COHEN, A. 1993. Halite–clay interplay in the Israeli Messinian. *Sedimentary Geology*, **86**, 211–228.

COHEN, Y., KRUMBEIN, W. E., GOLDBERG, M. & SHILO, M. 1977. Solar Lake (Sinai). 1. Physical and chemical limnology. *Limnology and Oceanography*, **22**, 597–608.

COHENOUR, R. E. 1968. Great Salt Lake. *In*: FAIRBRIDGE, R. W. (ed.) *Encyclopedia of Geomorphology*. Reinhold, New York, 506–517.

CORNÉE, A., DICKMAN, M. & BUSSON, G. 1992. Laminated cyanobacterial mats in sediments of solar salt works: some sedimentological implications. *Sedimentology*, **39**, 599–612.

CSANADY, G. T. 1978. Water circulation and dispersal mechanisms. *In*: LERMAN, A. (ed.) *Lakes – Chemistry, Geology, Physics*. Springer, New York, 21–64.

CZARNOCKI, S. 1935. On the main problems concerning the stratigraphy and palaeogeography of Polish Tortonian. *Sprawozdania Państwowego Instytutu Geologicznego*, **8**, 1–206 [in Polish with German summary].

DEL DON, C., HANSELMANN, K. W., PEDUZZI, R. & BACHOFEN, R. 2001. The meromictic alpine Lake Cadagno: orographical and biogeochemical description. *Aquatic Sciences*, **63**, 70–90.

DENIMAL, S., BERTRAND, C., MUDRY, J., PAQUETTE, Y., HOCHART, M. & STEINMANN, M. 2005. Evolution of the aqueous geochemistry of mine pit lakes – Blanzy-Montceau-les-Mines coal basin (Massif Central, France): origin of sulfate contents; effects of stratification on water quality. *Applied Geochemistry*, **20**, 825–839.

DICKMAN, M. 1985. Seasonal succession and microlamina formation in a meromictic lake displaying varved sediments. *Sedimentology*, **32**, 109–118.

DONACHIE, S. P., HOU, S. ET AL. 2004. The Hawaiian Archipelago: a microbial diversity hotspot. *Microbial Ecology*, **48**, 509–520.

DRONKERT, H. 1977. The evaporites of the Sorbas basin. *Instituto de Investigaciones Geológicas Diputación Provincial Universidad de Barcelona*, **32**, 55–76.

DRONKERT, H. 1985. Evaporite Models and Sedimentology of Messinian and Recent Evaporites. GUA Papers of Geology, Series 1, **24**, 1–283.

ECCLES, D. H. 1974. An outline of the physical limnology of Lake Malawi (Lake Nyasa). *Limnology and Oceanography*, **19**, 730–742.

EDMONDSON, W. T. & ANDERSON, G. C. 1965. Some features of saline lakes in Central Washington. *Limnology and Oceanography*, **10**, Supplement: Alfred C. Redfield 75th Anniversary Volume, 87–96.

EMERY, K. O. & CSANADY, G. T. 1973. Surface circulation of lakes and nearly land-locked seas. *Proceedings of the National Academy of Sciences of the United States of America*, **70**, 93–97.

FILIPPI, M. L. & TALBOT, M. R. 2005. The palaeolimnology of northern Lake Malawi over the last 25 ka based upon the elemental and stable isotopic composition of sedimentary organic matter. *Quaternary Science Reviews*, **24**, 1303–1328.

FOLK, R. L., CHAFETZ, H. S. & TIEZZI, P. A. 1985. Bizarre forms of depositional and diagenetic calcite in hot-spring travertines, Central Italy. *In*: SCHNEIDERMANN, N. & HARRIS, P. M. (eds) *Carbonate Cements*. SEPM, Special Publications, **36**, 349–369.

FRIEDRICH, J. & OBERHÄNSLI, H. 2004. Hydrochemical properties of the Aral Sea water in summer 2002. *Journal of Marine Systems*, **47**, 77–88.

GAMMONS, C. H. & DUAIME, T. 2006. Long term changes in limnology and geochemistry of the Berkeley pit lake, Butte, Montana. *Mine Water and the Environment*, **25**, 76–85.

GARLICKI, A. 1979. Sedimentation of Miocene salts in Poland. *Prace Geologiczne, Polska Akademia Nauk, Oddział w Krakowie*, **119**, 1–67 [in Polish with English summary].

GERTMAN, I., ANATI, D. A., HECHT, A., BISHOP, J. & TSEHTIK, Y. 2003. *The Dead Sea hydrography from 1992 to 1999*; http://marine.ocean.org.il/DS_website/DS_Database.htm

GONZÁLEZ, E. J., ORTAZ, M., PEÑAHERRERA, C. & DE INFANTE, A. 2004. Physical and chemical features of a tropical hypertrophic reservoir permanently stratified. *Hydrobiologia*, **522**, 301–310.

HAKALA, A. 2005. *Paleoenvironmental and paleoclimatic studies on the sediments of Lake Vähä-Pitkusta and observations of meromixis*. Academic Dissertation, Helsinki; http:æthesis.helsinki.fijulkaisutmat-geologovkhakalapaleoenv.pdf

HAKALA, A., SARMAJA-KORJONEN, K. & MIETTINEN, A. 2004. The origin and evolution of Lake Vähä-Pitkusta, SW Finland – a multi-proxy study of a meromictic lake. *Hydrobiologia*, **527**, 85–97.

HEIM, C., NOWACZYK, N., NEGENDANK, J. F. W., LEROY, S. A. G. & BEN-AVRAHAM, Z. 1997. Near East desertification: evidence from the Dead Sea. *Naturwissenschaften*, **84**, 398–401.

HERUT, B., GAVRIELI, I. & HALICZ, L. 1998. Coprecipitation of trace and minor elements in modern authigenic halites from the hypersaline Dead Sea brine. *Geochimica et Cosmochimica Acta*, **62**, 1587–1598.

HORNE, A. J. & GOLDMAN, C. R. 1994. *Limnology*, 2nd edn. McGraw-Hill, New York.

HRYNIV, S., PARAFINIUK, J. & PERYT, T. M. 2007a. Sulphur isotopic composition of K–Mg sulphates of the Miocene evaporites of the Carpathian Foredeep, Ukraine. *In*: SCHREIBER, B. C., LUGLI, S. & BĄBEL, M. (eds) *Evaporites Through Space and Time*. Geological Society of London, Special Publications, **285**, 265–273.

HRYNIV, S., DOLISHNIY, B. V., KHMELEVSKA, O. V., POBEREZHSKYY, A. V. & VOVNYUK, S. V. 2007b. Evaporites of Ukraine: a review. *In*: SCHREIBER, B. C., LUGLI, S. & BĄBEL, M. (eds) *Evaporites Through Space and Time*. Geological Society of London, Special Publications, **285**, 309–334.

HUTCHINSON, G. E. 1975. *A Treatise on Limnology, Vol. 1(1) Geography and Physics of Lakes*. Wiley, New York, 1–540, S1–S10, B1–B66, I1–I37.

JELLISON, R. & MELACK, J. M. 1993. Meromixis in hypersaline Mono Lake, California. 1. Stratification and vertical mixing during the onset, persistence, and breakdown of meromixis. *Limnology and Oceanography*, **38**, 1008–1019.

KASPRZYK, A. 1995. Correlation of sulfate deposits of the Carpathian Foredeep at the boundary of Poland and Ukraine. *Geological Quarterly*, **39**, 95–108.

KASPRZYK, A. 2003. Sedimentological and diagenetic patterns of anhydrite deposits in the Badenian evaporite basin of the Carpathian Foredeep, southern Poland. *Sedimentary Geology*, **158**, 167–194.

KASPRZYK, A. & ORTÍ, F. 1998. Palaeogeographic and burial controls on anhydrite genesis: the Badenian basin in the Carpathian Foredeep (southern Poland, western Ukraine). *Sedimentology*, **45**, 889–907.

KHRUSHCHOV, D. P. & PETRICHENKO, O. I. 1979. Evaporite formations of Central Paratethys and conditions of their sedimentation. *Annales Géologiques des Pays Helléniques, Tome hors série*, **2**, 595–612.

KIMMEL, B. L., GESBERG, R. M., PAULSON, L. J., AXLER, R. P. & GOLDMAN, C. R. 1978. Recent changes in the meromictic status of Big Soda Lake, Nevada. *Limnology and Oceanography*, **23**, 1021–1025.

KIRKLAND, D. W., BRADBURY, J. P. & DEAN, W. E. 1983. The heliothermic lake – a direct method of collecting and storing solar energy. *Archiv für Hydrobiologie, Supplementband (Monographische Beiträge)*, **65**, 1–60.

KLIMOV, M. A. 1974. Potassium-bearing saliferous formations of the Forecarpathians and prospects for searching the potassium salt deposits. *In*: KITYK, V. I. (ed.) *Geology and Minerals of the Saliferous Layers*. Naukova Dumka, Kiev, 146–156 [in Russian].

KORENEVSKYI, S. M., ZAHAROVA, V. M. & SHAMAHOV, V. A. 1977. Miocene saliferous formations of Carpathian Foreland. *Trudy Vsesojuznovo Nauchno-Issledovatelnovo Geologicheskovo Instituta (VSEGEI)*, Novaya Ser., **271**, 146–248 [in Russian].

KUBICA, B. 1992. Lithofacies development of the Badenian chemical sediments in the northern part of the Carpathian Foredeep. *Prace Państwowego Instytutu Geologicznego*, **133**, 1–64 [in Polish with English summary].

LADYZHENSKYI, N. P. & ANTIPOV, V. I. 1961. *Geological Structure and Gas- and Oil-bearing of the Soviet Forecarpathians*. Gostoptechyzdat, Moscow [in Russian].

LAST, W. M. 1993. Geolimnology of Freefight Lake: an unusual hypersaline lake in the northern Great Plains of western Canada. *Sedimentology*, **40**, 431–448.

LAST, W. M. 1994. Deep-water evaporite mineral formation in lakes of western Canada. *In*: RENAUT, R. W. & LAST, W. M. (eds) *Sedimentology and*

Geochemistry of Modern and Ancient Lakes. SEPM, Special Publications, **50**, 51–59.

LAST, W. M. & DE DECKKER, P. 1990. Modern and Holocene carbonate sedimentology of two saline volcanic maar lakes, southern Australia. *Sedimentology*, **37**, 967–981.

LAST, W. M. & GINN, F. M. 2005. Saline systems of the Great Plains of western Canada: an overview of the limnogeology and paleolimnology. *Saline Systems*, **1**:10; http://www.salinesystems.org/content/1/1/10.

LAST, W. M. & SCHWEYEN, T. H. 1983. Sedimentology and geochemistry of saline lakes of the Great Plains. *In*: HAMMER, U. T. (ed.) *Saline Lakes*. Developments in Hydrobiology, **16**. Junk, Hague, 245–263. [Also: *Hydrobiologia*, **105**, 245–263.]

LAST, W. M. & SLEZAK, L. A. 1986. Paleohydrology, sedimentology, and geochemistry of two meromictic saline lakes in southern Saskatchewan. *Géographie Physique et Quaternaire*, **40**, 5–15.

LAST, W. M., DELEQIAT, J., GREENGRASS, K. & SUKHAN, S. 2002. Re-examination of the recent history of meromictic Waldsea Lake, Saskatchewan, Canada. *Sedimentary Geology*, **148**, 147–160.

LAVAL, B. E., IMBERGER, J. & FINDIKAKIS, A. N. 2005. Dynamics of a large tropical lake: Lake Maracaibo. *Aquatic Sciences*, **67**, 337–349.

LEWIS, W. M. J. 1983. A revised classification of lakes based on mixing. *Canadian Journal of Fisheries and Aquatic Sciences*, **40**, 1779–1787.

LI, S. & LI, W. 2004. Hydrochemistry in meromictic Lake Zige Tangco, Central Tibetan Plateau. *Asian Journal of Water, Environment and Pollution*, **1**, 1–4.

LIN, A. 1976. The meromictic Great Salt Lake. *Journal of Great Lakes Research*, **2**, 374–383.

LIN, A. & WANG, P. 1978. Wind tides of the Great Salt Lake. *Utah Geology*, **5**, 17–25.

LOWE, D. J., GREEN, J. D., NORTHCOTE, T. G. & HALL, K. J. 1997. Holocene fluctuations of a meromictic lake in southern British Columbia. *Quaternary Research*, **48**, 100–113.

LUDLAM, S. D. 1996. The comparative limnology of high arctic, coastal, meromictic lakes. *Journal of Paleolimnology*, **16**, 111–131.

MACINTYRE, S. & MELACK, J. M. 1982. Meromixis in an equatorial African soda lake. *Limnology and Oceanography*, **27**, 595–609.

MELACK, J. M. & JELLISON, R. 1998. Limnological conditions in Mono Lake: contrasting monomixis and meromixis in the 1990s. *Hydrobiologia*, **384**, 21–39.

MILLER, G. M., JELLISON, R., OREMLAND, R. S. & CULBERTSON, C. W. 1993. Meromixis in hypersaline Mono Lake, California. 3. Biogeochemical response to stratification and overturn. *Limnology and Oceanography*, **38**, 1040–1051.

MULDER, T. & ALEXANDER, J. 2001. The physical character of subaqueous sedimentary density flows and their deposits. *Sedimentology*, **48**, 269–299.

NEEV, D. & EMERY, K. O. 1967. *The Dead Sea. Depositional Processes and Environments of Evaporites.* Monson Press, Jerusalem.

NIEMI, T. M., BEN-AVRAHAM, Z. & GAT, J. R. (eds) 1997. *The Dead Sea: the Lake and Its Setting.* Oxford Monographs on Geology and Geophysics, **36**. Oxford University Press, New York.

NJUGUNA, S. G. 1988. Nutrient–phytoplankton relationships in a tropical meromictic soda lake. *In*: MELACK, J. M. (ed.) *Saline Lakes*. Developments in Hydrobiology, **44**. Junk, The Hague, 15–28. [Also: *Hydrobiologia*, **158**, 15–28.]

NORTHCOTE, T. G. & HALL, K. J. 1983. Limnological contrast and anomalies in two adjacent saline lakes. *In*: HAMMER, U. T. (ed.) *Saline Lakes*. Developments in Hydrobiology **16**, Junk, Hague, 179–194. [Also: *Hydrobiologia*, **105**, 179–194.]

NOWAK, J. 1938. Dnister and the Tortonian gypsum. *Rocznik Polskiego Towarzystwa Geologicznego*, **14**, 155–194 [in Polish with German summary].

OREN, A. & SHILO, M. 1982. Population dynamics of *Dunaliella parva* in the Dead Sea. *Limnology and Oceanography*, **217**, 201–211.

OSZCZYPKO, N. 2006. Late Jurassic-Miocene evolution of the Outer Carpathian fold-and-thrust belt and its foredeep basin (Western Carpathians, Poland). *Geological Quarterly*, **50**, 169–194.

OVERMANN, J., BEATTY, J. T. & HALL, K. J. 1996. Purple sulfur bacteria control the growth of aerobic heterotrophic bacterioplankton in a meromictic salt lake. *Applied and Environmental Microbiology*, **62**, 3251–3258.

PANOW, G. M. & PŁOTNIKOW, A. M. 1996. Badenian evaporites of the Ukrainian part of Carpathian Foredeep: lithofacies and thickness. *Przegląd Geologiczny*, **44**, 1024–1028 [in Polish].

PAREA, G. C. & RICCI-LUCCHI, F. 1972. Resedimented evaporites in the Periadriatic Trough (Upper Miocene, Italy). *Israel Journal of Earth-Sciences*, **21**, 125–141.

PAWLIKOWSKI, M. 1982. Mineralogical and petrographical study of alteration products of the Miocene gypsum rocks in the Wydrza sulphur deposit. *Polska Akademia Nauk, Oddział w Krakowie, Prace Mineralogiczne*, **72**, 1–60 [in Polish with English summary].

PERYT, T. M. 2000. Resedimentation of basin centre sulfate deposits: Middle Miocene Badenian of Carpathian Foredeep, southern Poland. *Sedimentary Geology*, **134**, 331–342.

PERYT, T. M. 2001. Gypsum facies transitions in basin–marginal evaporites: middle Miocene (Badenian) of west Ukraine. *Sedimentology*, **48**, 1103–1119.

PERYT, T. M. 2006. The beginning, development and termination of the Middle Miocene Badenian salinity crisis in Central Paratethys. *Sedimentary Geology*, **188–189**, 379–396.

PETRICHENKO, O. I., KOVALEVYTS, V. M. & TSALYI, V. N. 1974. Geochemical environment of salt formation in the Tortonian evaporite basin of the north-west Forecarpathian. *Geologiia i Geochimiia Goriuchich Iskopaiemych*, **41**, 74–80 [in Russian].

PETRICHENKO, O. I., PERYT, T. M. & POBEREGSKY, A. V. 1997. Pecularities of gypsum sedimentation in the Middle Miocene Badenian evaporite basin of Carpathian Foredeep. *Slovak Geological Magazine*, **3**, 91–104.

PETRYCZENKO, O. I., PANOW, G. M., PERYT, T. M., SEREBRODOLSKI, B. I., POBEREŻSKI, A. W. & KOWALEWICZ, W. M. 1994. Outline of geology of the Miocene evaporite formations of the Ukrainian part of the Carpathian Foredeep. *Przegląd Geologiczny*, **42**, 734–737 [in Polish].

POLKUNOV, V. F., GERASIMOV, L. S. & KOSTROVSKAYA, A. I. 1979. Tectonics. *In*: KITYK, V. I. (ed.) *Structure and Regularities of Distribution of the Sulphur Deposits of the SSSR*. Naukova Dumka, Kiev, 29–42 [in Russian].

POŁTOWICZ, S. 1993. Palinspastic palaeogeography reconstruction of Badenian saline sedimentary basin in Poland. *Zeszyty Naukowe AGH*, **1559**. *Geologia, Kwartalnik*, **19**, 174–178, 203–233 [in Polish with English summary].

POTTER, P. E. & PETTIJOHN, F. J. 1977. *Paleocurrents and Basin Analysis*, 2nd edn. Springer, Berlin.

RENAUT, R. W. & TIERCELIN, J.-J. 1994. Lake Bogoria, Kenya Rift Valley – a sedimentological overview. *In*: RENAUT, R. W. & LAST, W. M. (eds) *Sedimentology and Geochemistry of Modern and Ancient Lakes*. SEPM, Special Publications, **50**, 101–123.

REYNOLDS, C. S., REYNOLDS, S. N., MUNAWAR, I. F. & MUNAWAR, M. 2000. The regulation of phytoplankton population dynamics in the world's largest lakes. *Aquatic Ecosystem Health and Menagement*, **3**, 1–21.

RICHTER-BERNBURG, G. 1955. Über salinare Sedimentation. *Zeitschrift der Deutschen Geologischen Gesellschaft*, **105** (volume for 1953), 593–645.

RIMOLDI, B., ALEXANDER, J. & MORRIS, S. 1996. Experimental turbidity currents entering density-stratified water: analogues for turbidites in Mediterranean hypersaline basins. *Sedimentology*, **43**, 527–540.

RODRIGO, M. A., MIRACLE, M. R. & VICENTE, E. 2001. The meromictic Lake La Cruz (Central Spain). Patterns of stratification. *Aquatic Sciences*, **63**, 406–416.

ROMERO, L., CAMACHO, A., VICENTE, E. & MIRACLE, M. R. 2006. Sedimentation patterns of photosynthetic bacteria based on pigment markers in meromictic Lake La Cruz (Spain): paleolimnological implications. *Journal of Paleolimnology*, **35**, 167–177.

RÖGL, F. 1999. Mediterranean and Paratethys. Facts and hypotheses of an Oligocene to Miocene paleogeography (short overview). *Geologica Carpathica*, **50**, 339–349.

ROSS, A. H., GURNEY, W. S. C., HEATH, M. R., HAY, S. J. & HENDERSON, E. W. 1993. A strategic simulation model of a fjord ecosystem. *Limnology and Oceanography*, **38**, 128–153.

RUDNICK, D. L. & FERRARI, R. 1999. Compensation of horizontal salinity gradients in the ocean mixed layer. *Science*, **283**, 526–529.

RUDNICK, D. L. & MARTIN, J. P. 2002. On the horizontal density ratio in the upper ocean. *Dynamics of Atmospheres and Oceans*, **36**, 3–21.

SCHMALZ, R. F. 1969. Deep-water evaporite deposition: a genetic model. *The American Association of Petroleum Geologists Bulletin*, **53**, 798–823.

ŚLĄCZKA, A. & OSZCZYPKO, N. 2002. Paleogeography of the Badenian salt basin (Carpathian Foredeep, Poland and Ukraine). *Geologica Carpathica, Special Issue*, **53**, Proceedings of 17 Congress of Carpathian-Balkan Geological Association, Bratislava 2002; www.geologicacarpathica.sk/special/S/Slaczka_Oszczypko.pdf

SMIRNOV, S. E., SAMARSKA, O. V., SMOGOLIUK, N. V. & TROFIMOVICH, N. A. 1995. Deep-water Tyrassian gypsum – the member of the continuous Miocene sequence of the Precarpathian. *In*: *Proceedings of the 15 Congress of the Carpatho-Balcan Geological Association, Athens, Greece 1995*. Geological Society of Greece, Special Publications, **4**, 391–396.

SONNENFELD, P. 1974. The Upper Miocene evaporite basins in the Mediterranean Region – a study in paleo-oceanography. *Geologische Rundschau*, **63**, 1133–1172.

SONNENFELD, P. 1975. The significance of Upper Miocene (Messinian) evaporites in the Mediterranean. *Journal of Geology*, **83**, 287–311.

SONNENFELD, P. 1984. *Brines and Evaporites*. Academic Press, Orlando, FL.

SONNENFELD, P. 1992. Genesis of marine evaporites – a summation. *Geologica Carpathica*, **43**, 259–274.

STEINHORN, I. 1985. The disappearance of the long term meromictic stratification of the Dead Sea. *Limnology and Oceanography*, **30**, 451–472.

STEINHORN, I., ASSAF, G. ET AL. 1979. The Dead Sea: deepening of the mixolimnion signifies the overture to overturn of the water column. *Science, New Ser.*, **206**, 55–57.

STEPHENS, D. W. & GILLESPIE, D. M. 1976. Phytoplankton production in the Great Salt Lake, Utah, and a laboratory study of algal response to enrichment. *Limnology and Oceanography*, **21**, 74–87.

STILLER, M., GAT, J. R. & KAUSHANSKY, P. 1997. Halite precipitation and sediment deposition as measured in sediment traps deployed in the Dead Sea: 1981–1983. *In*: NIEMI, T. M., BEN-AVRAHAM, Z. & GAT, J. R. (eds) *The Dead Sea: the Lake and Its Setting*. Oxford Monographs on Geology and Geophysics, **36**. Oxford University Press, New York, 171–183.

SUITS, N. S. & WILKIN, R. T. 1998. Pyrite formation in the water column and sediments of a meromictic lake. *Geology*, **26**, 1999–1102.

SWANSON, S. M. & HAMMER, U. T. 1983. Production of *Cricotopus ornatus* (Meigen) (Diptera: Chironomidae) in Waldsea Lake, Saskatchewan. *In*: HAMMER, U. T. (ed.) *Saline Lakes*. Developments in Hydrobiology, **16**. Junk, The Hague. 155–163. [Also: *Hydrobiologia*, **105**, 155–163.]

THOMPSON, J. B., SCHULTZE-LAM, S., BEVERIDGE, T. J. & DES MARAIS, D. J. 1997. Whiting events: Biogenic origin due to the photosynthetic activity of cyanobacterial picoplankton. *Limnology and Oceanography*, **42**, 133–141.

TIMMS, B. V. 1972. A meromictic lake in Australia. *Limnology and Oceanography*, **17**, 918–922.

TITOV, V. B. 2005. Seasonal variations in thermo-, halo-, and pycnocline in the northeastern part of the Black Sea (based on long-term data). *Water Resources*, **32**, 23–30. [Translation of Russian text from *Vodnye Resoursy* 2005.]

TONOLLA, M., PEDUZZI, S., DEMARTA, A., PEDUZZI, R. & HAHN, D. 2004. Phototropic sulfur and sulfate-reducing bacteria in the chemocline of meromictic Lake Cadagno, Switzerland. *Journal of Limnology*, **63**, 161–170.

VALERO-GARCÉS, B. L. & KELTS, K. R. 1995. A sedimentary facies model for perennial and meromictic saline lakes: Holocene Medicine Lake Basin, South Dakota, USA. *Journal of Paleolimnology*, **14**, 123–149.

VALERO-GARCÉS, B. L., KELTS, K. & ITO, E. 1995. Oxygen and carbon isotope trends and sedimentological evolution of a meromictic and saline lacustrine system: the Holocene Medicine Lake basin, North American Great Plains, USA. *Palaeogeography, Palaeoclimatology, Palaeoecology*, **117**, 253–278.

VERSCHUREN, D. 1999. Influence of depth and mixing regime on sedimentation in a small, fluctuating tropical soda lake. *Limnology and Oceanography*, **44**, 1103–1113.

WALKER, K. F. 1974. The stability of meromictic lakes in central Washington. *Limnology and Oceanography*, **19**, 209–222.

WALKER, T. A. 1982. Lack of evidence for evaporation-driven circulation in the Great Barrier Reef lagoon. *Australian Journal of Marine and Freshwater Research*, **33**, 717–722.

WARREN, J. K. 2006. *Evaporites: Sediments, Resources and Hydrocarbons*. Springer, Berlin.

WEIMER, W. C. & LEE, G. F. 1973. Some considerations of the chemical limnology of meromictic Lake Mary. *Limnology and Oceanography*, **18**, 414–425.

WILK-WOŹNIAK, E. & ŻUREK, R. 2006. Phytoplankton and its relationships with chemical parameters and zooplankton in the meromictic Piaseczno reservoir, Southern Poland. *Aquatic Ecology*, **40**, 165–176 DOI: 10.1007/s10452-005-0781-6.

ZAVIALOV, P. O., KOSTIANOY, A. G., EMELIANOV, S. V., NI, A. A., ISHNIYAZOV, D., KHAN, V. M. & KUDYSHKIN, T. V. 2003. Hydrographic survey in the dying Aral Sea. *Geophysical Research Letters*, **30**, 1659.

ZEHR, J. P., HARVEY, R. W., OREMLAND, R. S., CLOERN, J. E., GEORGE, L. H. & LANE, J. L. 1987. Big Soda Lake (Nevada). 1. Pelagic bacterial heterotrophy and biomass. *Limnology and Oceanography*, **32**, 781–793.

# Sedimentology and geochemistry of the Middle Miocene (Badenian) salt-bearing succession from East Slovakian Basin (Zbudza Formation)

K. BUKOWSKI[1], G. CZAPOWSKI[2], S. KAROLI[3] & M. BĄBEL[4]

[1]*University of Mining and Metallurgy, Al. Mickiewicza 3., PL-30-059 Cracow, Poland*
(e-mail: buk@geolog.geol.agh.edu.pl)

[2]*Polish Geological Institute, Rakowiecka 4, PL-00-975 Warsaw, Poland*
(e-mail: grzegorz.czapowski@pgi.gov.pl)

[3]*Geological Survey of Slovak Republic, Branch Košice, Jasenskeho 9, 04011 Košice, Slovakia*

[4]*Institute of Geology, Warsaw University, Al. Żwirki i Wigury 93, PL-02-089 Warsaw, Poland*
(e-mail: m.babel@uw.edu.pl)

Abstract: The Middle Miocene (Badenian) evaporites have formed within more or less restricted basins, occupying the Carpathian Foredeep and inner depressions in the East Slovakian Basin of the Carpathians. Their deposition coincided with the increased tectonic activity in the Carpathians, evidenced by volcanic eruptions (widespread volcaniclastics) and frequent earthquakes. These phenomena apparently initiated landslides, submarine slumps and turbidity currents which formed deposits characterized by rapid facies changes and relatively limited lateral continuity. The clastic–evaporitic deposition in the Zbudza area (East Slovakian Basin) was controlled by frequent tectonic and seismic phenomena and high continental clastic supply that produced repeated slump sediments with dominant proximal mass flows and distal flows and diverse salt types (salt/clay rhythmites, finer halite–arenites, coarse halite–rudites). These zones are separated by primary salt units (halite) precipitated *in situ* from bottom brines during calm periods. The observed cyclicity (I–V cycles) reflects varied tectonic activity of the basin margins, that mechanically remobilized the sediments from marginal salt pans, flats and adjacent uplifts, and could be correlable with cycles in the Wieliczka Formation from Carpathian Foredeep.

Middle Miocene (Badenian) evaporites have formed within more or less restricted basins, occupying the Carpathian Foredeep (Wieliczka Formation in Poland and Tirass Suite in Ukraine) and in the inner depressions of the Carpathians (e.g. East Slovakian Basin; Zbudza Formation). The deposition coincided with the increased tectonic activity of the Carpathian area, evidenced by volcanic eruptions (widespread volcanoclastics) and common earthquakes (Peryt & Kasprzyk 1992). These phenomena initiated landslides, submarine slumps and turbidity currents, which formed deposits characterized by rapid facies changes and relatively limited areal extent (e.g. Bukowski 1997; Ślączka & Kolasa 1997).

The Badenian evaporitic succession within the East Slovakian Basin (Inner Carpathians, in the area between Košice and Michalovce), known as the Zbudza Formation (>150 m thick, with halite beds up to 75 m thick intercalated with siliciclastics and anhydrite) were studied in detail, enabling a reconstruction of their depositional mechanisms and environment. They are commonly the equivalent of the evaporites of the Carpathian Foredeep.

Geological setting

The studied salt-bearing succession was deposited in the Parathetys seaway. The Parathetys was an enclosed sea that existed from Oligocene to Middle Miocene times and consisted of an interrelated chain of basins of diverse tectonic origin. These basins were covered most of the time by the same water mass, which shared a common aquatic biota, and was a link between the Mediterranean and the Indo-Pacific areas (Bàldi *et al.* 2002).

Both the Carpathian Foredeep and adjacent marginal basins located in the Inner Carpathians belong to the Central Parathetys (the area from Bavaria to the Eastern Carpathians). It was developed between the Carpathian orogene to the south and the Palaeozoic Platform of Eastern Europe to the north and the Precambrian East-European Platform to the southeast (Fig. 1). Evaporites of the Middle-Upper Badenian age (the period equivalent to the Upper Langhian-Lower Serravallian (Fig. 2) have accumulated in the foredeep area stretching from Upper Silesia in Poland (Garlicki 1979, 1994*a*, *b*) and on into the Doftana Valley in

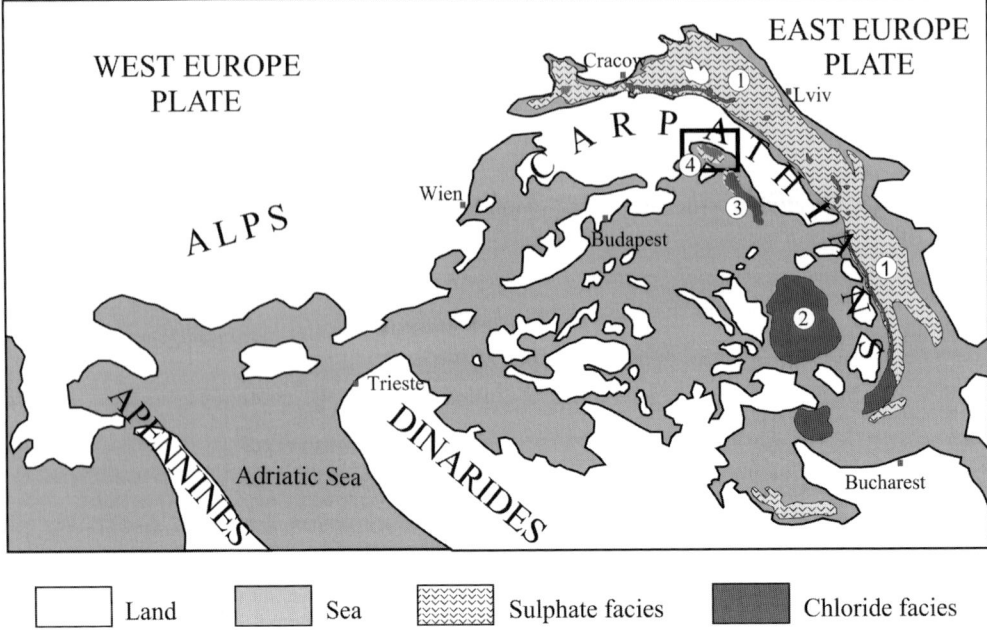

Fig. 1. Distribution of evaporites (chloride and sulphate facies) in the Central Paratethys (after Bąbel 2004 and references therein) during the 'Badenian salinity crisis' projected onto the palaeogeography of Middle to Late Badenian Carpatho-Pannonian Basin. The paleogeographic map was simplified and compiled after Rögl (1998a, b), Studencka et al. (1998) and Popov et al. (2004). 1, Carpathian Foredeep; 2, Transylvanian Basin; 3, Transcarpathian Basin; 4, East Slovakian Basin (study area).

Romania (Crihan 1999; Mǎrunţeanu 1999). Contemporaneous evaporites also were deposited in the Transylvanian Basin in Romania (Balintoni & Petrescu 2002), the Trans-Carpathian Basin of the Ukraine (Kityk et al. 1983) and the East Slovakian Basin (Karoli et al. 1997).

The East Slovakian Basin was the most northern part of the extensive Pannonian Basin and it formed in an intermountain depression having a generally 'horst-and-graben style' of morphology (Vass et al. 2000) that resulted in mainly syndepositional fault activity (Figs 1 & 3). It was created in the Lower Miocene (Eggenburgian stage) along the internal side of the Klippen Belt as a narrow NW–SE oriented depression and that persisted until the Pliocene. The depression infill is mainly marine but toward the top of the section the deposits are replaced by the deltaic and lacustrine deposits of the Sarmatian–Pannonian age (Paratethys equivalent stages to the Serravallian and Tortonian). Their total thickness in the depression centre reaches 8–9 km (Vass et al. 2000). Two evaporite horizons, dominated by chlorides (rock salt with siliciclastic and anhydrite interbeds), occur within the marine succession. The lower horizon of the early Miocene (the Lower Karpatian stage) age is distinguished as the Solna Bania Formation and developed locally in the western part of East Slovakian Basin (Karoli et al. 1997). The upper horizon of the middle Miocene (the Badenian stage) is defined as the Zbudza Formation covering most of the basin area (Vass & Čverčko 1985).

The evaporite series of the Middle Miocene (Badenian) age, known from the Carpathian region and based on the most recent stratigraphic data, were formed at the boundary of nannoplankton zones NN5/NN6 and/or to the lower part of NN6 zone (e.g. Peryt 1997; Chira & Draghici 2002). According to some authors this episode lasted up to the boundary of NN6/NN7 zone (Andreyeva-Grigorovich et al. 2003).

The Zbudza Formation, recognized by Vas & Čverčko (1985), is traditionally included as the upper part of Middle Badenian (Vass et al. 2000) or to the lower part of the Upper Badenian (Gasparikova 1963; Zlinska 2004). Its position in the stratigraphic succession (Fig. 2) corresponds with the uppermost stratigraphic divisions within the Central Paratethys. This relationship is supported by a single radiometric dating from tuffite intercalations found within the salt-bearing series from Bochnia (13.6 ± 0.2 Ma; Dudek et al. 2004).

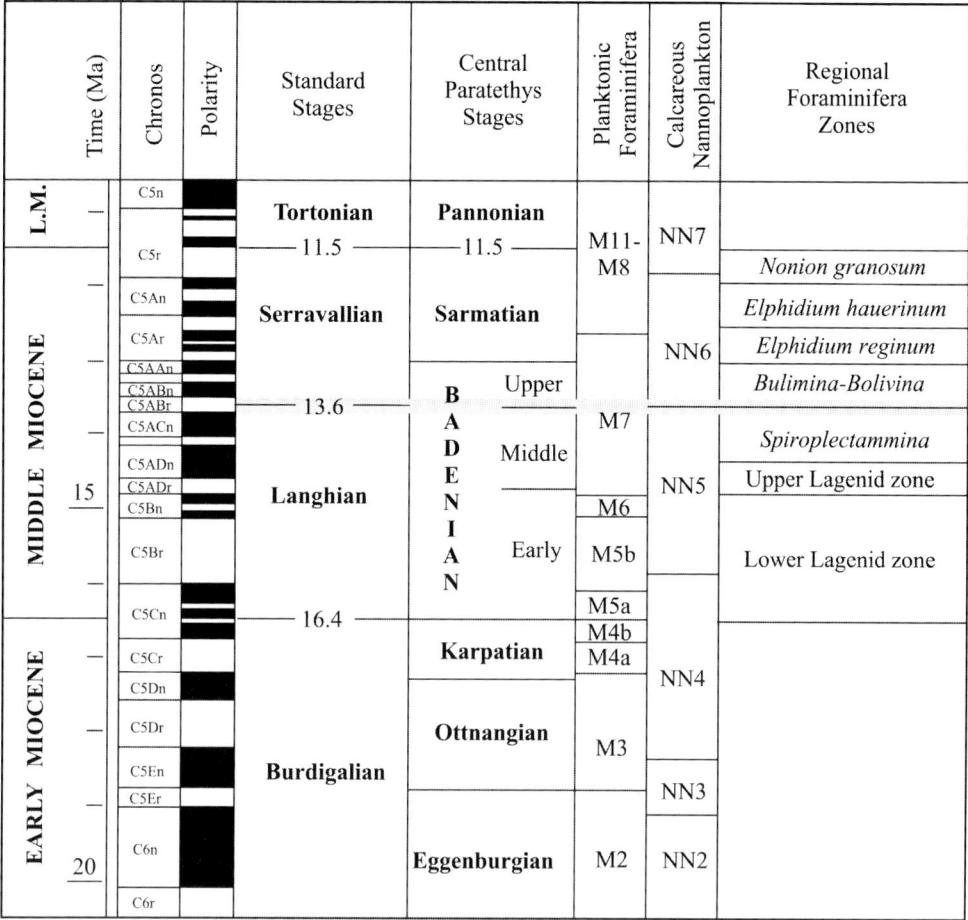

Fig. 2. The modified Miocene geochronology and biostratigraphy (after Harzhauser *et al.* 2003). Note re-calibrated position of the Langhian/Serravallian boundary now located at 13.59 Ma (Foresi *et al.* 2002) based on the LAD of *Sphenolithus heteromorphus*. This event is marked by the NN5/NN6 boundary and corresponds to the Badenian salinity crisis (Peryt 1997; Chira & Draghici 2002; Andreyeva-Grigorovich *et al.* 2003; Krézsek & Filipescu 2005), the age of which was defined by K-A dating of tuffite from Wieliczka Fm (*c.* 13.6 ± 0.2 Ma, after Dudek *et al.* 2004). The shaded area indicates the stratigraphic position of the Zbudza and Wieliczka formations.

Lithology and facies of the clastic-evaporitic deposition in the Zbudza area

Deposits of Zbudza Formation accumulated in local depressions of the eastern part of East Slovakian Basin and they cover an area of *c.* 350 km² (Fig. 3). This evaporite horizon was first noted in several places during the gas-oil exploration in the 1950s, but it proved economically viable only in the Michalovce region (Janáček 1958, 1959, 1960; Slávik 1967a). In the 1990s six new wells were drilled (from P-2 to P-8, Fig. 3) and this salt-bearing series (100–140 m thick) was detected at depths ranging from 440 to 530 m.

Salt-bearing deposits in all studied profiles (Fig. 4) contain pure or clayey halites interbedded with siliciclastics (laminated siltstones and claystones, sandstones and conglomerates) and nodular anhydrites. The chlorides range in appearance from chaotic to stratified (laminated, graded and cross-bedded) and are subdivided into four distinct macroscopic facies:

(A) coarse halite–rudite to halite breccia (Fig. 5a & b);
(B) fine halite–arenites (Fig. 5a);
(C) fine crystalline halites with clay laminae (Fig. 5c);
(D) chevron halite layers (Fig. 5d).

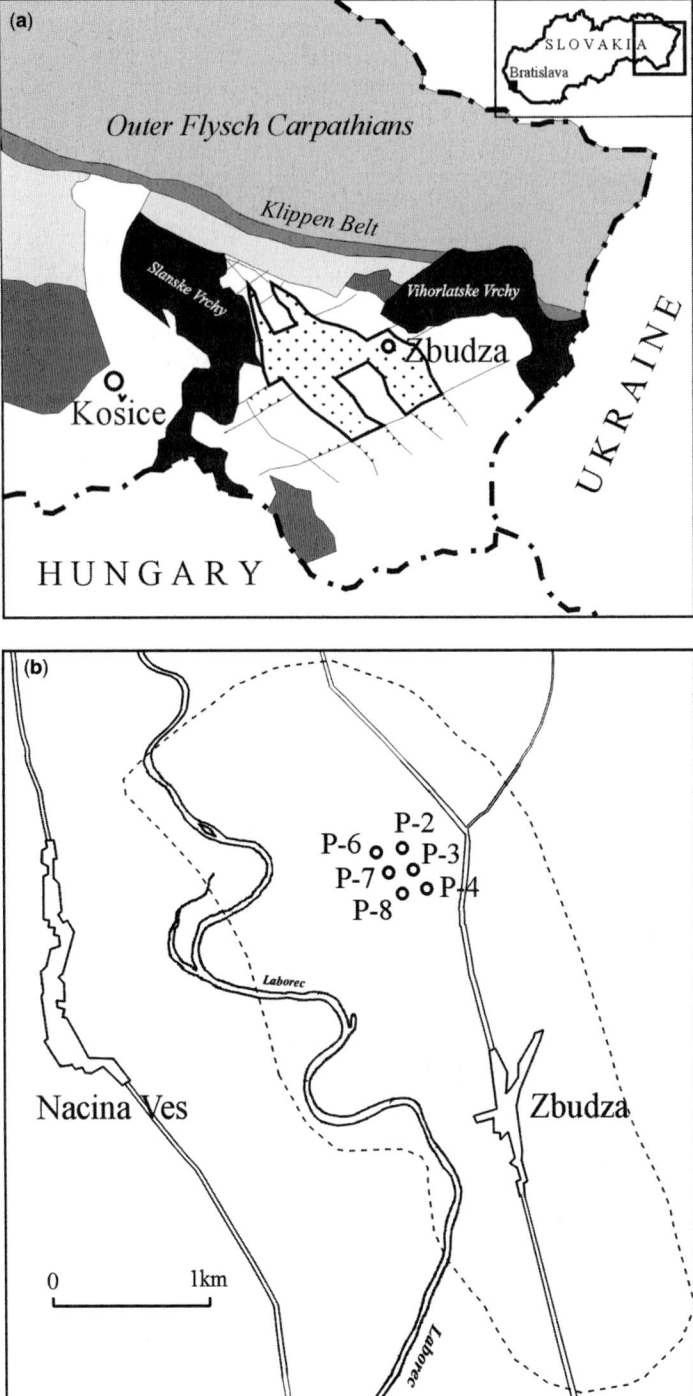

Fig. 3. Location of the study area in this paper. (**a**) Schematic map of East Slovakian Basin with the location of Zbudza Formation (dotted area) (after Karoli *et al.* 1997; Vass *et al.* 2000). (**b**) The Zbudza salt deposit area (shaded and outlined with a dashed line) together with locations of studied boreholes.

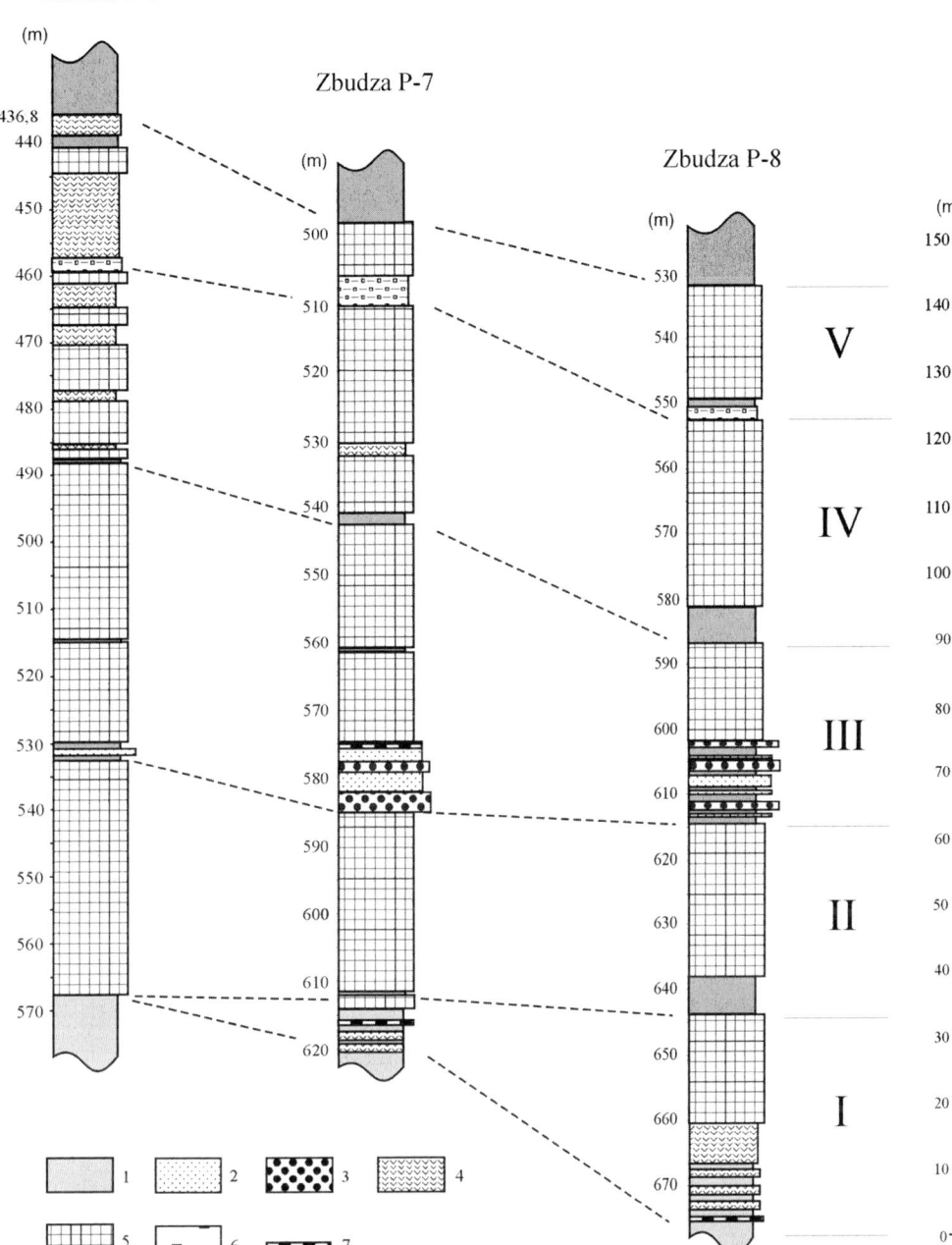

Fig. 4. Schematic lithological profiles of three studied boreholes with correlation of five observed sedimentary cycles (I–V). Legend: 1, claystone and mudstone; 2, sandstone; 3, conglomerate; 4, anhydrite; 5, salt; 6, mixed clay-salt rock (zuber); 7, tuffite.

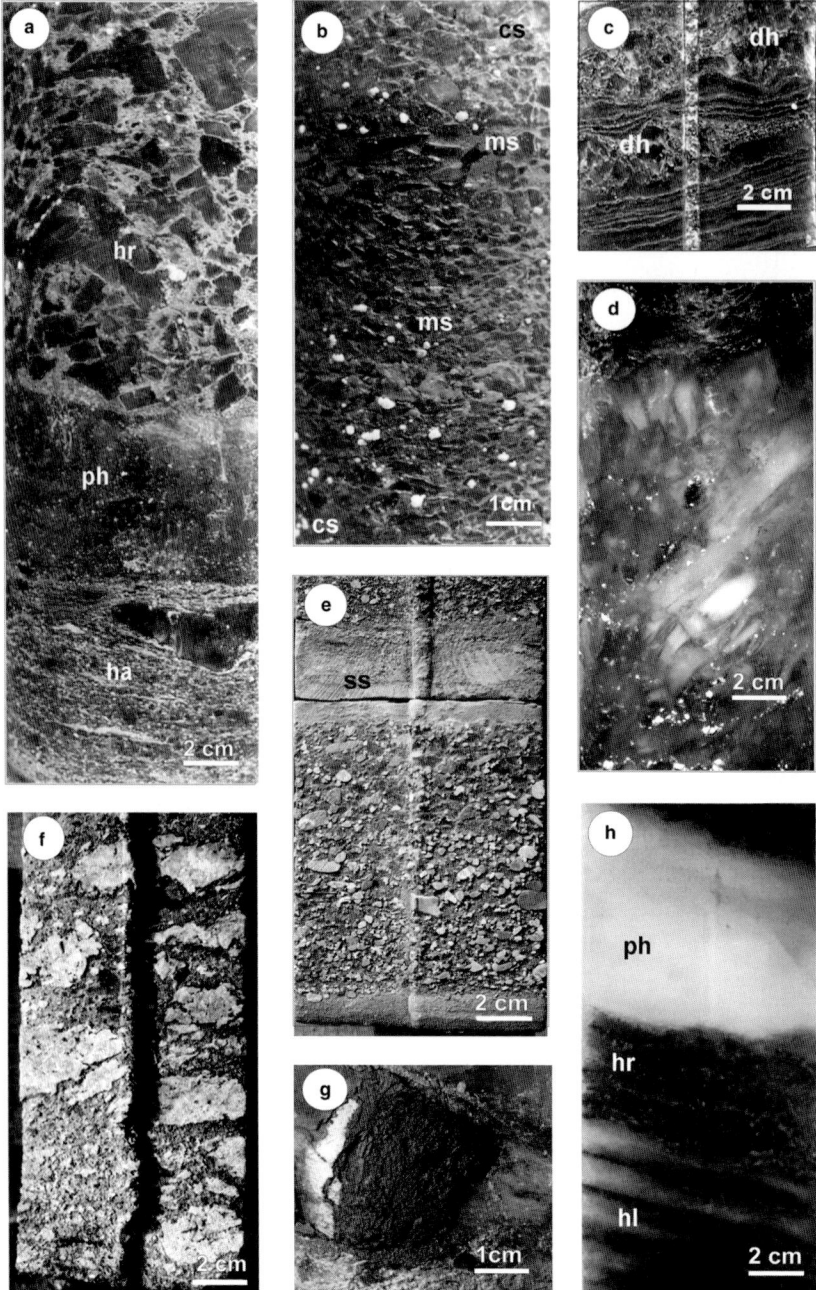

Fig. 5. Photographs of different types of sedimentary structures occurring in the salt-bearing succession. (a) Primary halite (ph) layer separating halite–arenite (ha, facies B) and halite–rudite (hr, facies A). Zbudza Fm P-7 well, depth 567 m. (b) Clast-supported (cs) and matrix-supported (ms) halite–rudite (facies A). Zbudza Fm, P-7 well, depth 537 m. (c) Displacive halite crystals (dh) between clay-rich parallel and wavy laminae (light) within fine halite. Salt–clay rhythmites (facies C), Zbudza Fm, P-6 well, depth 532 m. (d) 'Cloudy' elongate halite crystals (fluid-rich) after chevron halite. Zbudza Fm P-6 well, depth 460 m. (e) Conglomerate and fine, halite-cemented sandstone (ss) interbeds from the Zbudza Fm, P-8 well, depth 613 m. (f) Nodular anhydrite clasts within a halite-clay matrix. Zbudza Fm, P-8 well, depth 603 m. (g) Lignite clast within a halite–rudite. Zbudza Fm, P-2 well, depth 538 m. (h) Tectonically deformed salt – pure recrystallized halite (ph), halite–rudite (hr) and laminated halite (hl). Zbudza Fm, P-8 well depth 647 m.

Facies A: halite–rudites

Description These rocks are composed of salt debris and crystals, mixed with anhydrite nodules and sporadically contain clasts of other rocks, e.g. shales, limestones or sandstones, distributed/located within the fine to pelitic matrix (Fig. 5a & b). Also present are a range of structureless halite–rudites, devoid of clast selection and preferred orientation, as well as varieties having both normal and reversed grading.

Halite particles vary from 0.5 to 5 cm in diameter, but the larger ones (up to several centimetres) are commonly slightly rounded but the finer ones are commonly angular. Grains and clasts bearing relic chevron and hopper structures and of diagenetic cloudy halite are dominant. Very large clasts of transparent halite are most common in rocks with a higher clay content and some have regularly developed corners (primary euhedral halite crystals) evidencing their generation as intrasediment displacive halite. These rocks also contain up to several percent of anhydrite nodules. Terrigenous material (clay matter, quartz grains) content is up about 12% and is concentrated at grains boundaries, producing a stained-glass structure, or in flames with rare subtle lamination.

Locally in the top of the Zbudza Formation profile (Fig. 4) the mixed clay–salt deposits become increasingly common, with a clay matrix of up to 50%, called by miners 'zubers' (such rocks have a halite content from 15 to 85%; Poborski 1959). Zubers contain clasts of halite and nodular anhydrite 4–8 cm in diameter (Fig. 5f) and blocks of halite–rudites and of very large transparent halite, all dispersed within the clay matrix. The high volume, size and shape of these clasts permit these rocks to be classified as a salt breccia.

Common clasts of other sedimentary rocks noticed in halite–rudites include rounded fragments of claystones, marls and sandstones (Fig. 5e). Also present are dispersed and carbonized plant remains, especially abundant in the lower part of Zbudza Formation (Fig. 5g).

Interpretation These halite–rudites were interpreted as proximal mass flow deposits, accumulated in the most proximal part of a subaqueous slump fan from density currents highly overloaded with both pelitic and coarse clastic material (Fig. 10-1).

Facies B: halite–arenites

Description This facies includes fine-grained halite rocks, more homogenous than the halite–rudites and frequently laminated with clay matter. They commonly coexist with the halite–rudites forming rhythmic couplets (cycles) (Fig. 5a), with thickness of each halite-arenite unit up to 10 cm.

Interpretation Halite–arenites represent:

- distal mass flow deposits, formed in the seaward part of subaqueous slump fans from density currents, dominantly transporting fine clastics (Fig. 10-1),
- resedimented deposits produced by any traction currents (wave and gravity), replacing bottom clastic material.

Fine clay laminae occurred within this facies resulting from a suspension drop of pelitic material during the fair-weather periods separating the dynamic current and wave events. Commonly observed coexisting arenite and rudite layers resulted from redeposition by currents (both density and wave) of primary halite crystals and sediment fragments coming both from emerged and eroded coastal salt pans and from local bottom sediments. In some layers, the halite–arenite bed with such a depositional rhythm is overlain by a thin halite layer with *in situ* chevron crystals, indicating the calm period of intensive halite bottom growth.

Facies C: fine crystalline halites with clay laminae

Description This rock variety is characterized by regular rhythms of fine halite interlaminated with clayey matter. Halite layers (up to 10 cm thick), composed of primary precipitated finely crystalline, monomorphic halite, are separated by clay laminae (up to 1 cm), producing a parallel and/or wavy rock lamination. Also frequent interbeds of primary chevron halite are observed in this variety as well as having displacive halite crystals within clay laminae (Fig. 5c).

Interpretation This facies was interpreted as a primary salt deposit, precipitated relatively slowly on a basin bottom (Fig. 10-1) from halite-concentrated bottom brines during calm periods. Interbedded clay laminae represent a pelitic fallout from more diluted top brines.

Facies D: layers with chevron halite crystals

Description These layers are commonly observed as the interbeds within the laminated salts but also sometimes they accompany halite–rudite and halite–arenite cycles. The thickness of homogenous units with chevron halite in the Zbudza Formation profiles reached a thickness of 40 cm and they were topped with a clay lamina (Fig. 5d).

Interpretation This rock variety represents a primary salt deposit, precipitated quickly from highly concentrated brines that probably correspond to a distinct drop of water level (Fig. 10-2). Top clay lamina, covering the chevrons, could reflect: (a) an ensuing water level rise with input of fresh waters and subsequent accumulation of additional pelites (b) an event of density currents, transporting continental material seaward.

Horizontal beds of the varied salt varieties, described above, occur in most of the studied profiles or they are slightly inclined but locally some fairly thick deformed intervals were observed (several metres thick) particularly in the upper part of the Zbudza Formation, evidenced by a higher bed inclination (30–70°) and oblique veins of secondary fibrous salt and gypsum. Salt in these zones became highly recrystallized and both laminated halites and halite–arenites were transformed to a striped variety of salt (a foliation effect) and later to the transparent giant halite (Fig. 5h). This process has apparently developed due to a displacement of the mineral component, e.g. clay matter between the points of higher and lower pressure. Probably in a halite–rudite variety, when subjected to an intensive stress, the clay matrix behaved as a lubricant, transferring the deformation without destroying the primary rock structure.

Cyclicity of the Zbudza Formation

The salt-bearing series of the Zbudza Formation examined in this study may be subdivided into five chloride-bearing sedimentary cycles, separated by beds of siliciclastics and sulphates. The onset of each cycle coincides with a distinct increase in the content of terrigenous siliciclastics (Figs 4 & 6). Cycles were identified and correlated, based on the results of sedimentological and geochemical profiling supplemented with well-log data, which is exactly matched by gamma and neutron–gamma curves of the variable content of clay (a main component of the water-insoluble residues; Fig. 6).

The lowest cycle (cycle I) begins with claystones having nodular clayey anhydrite intercalations, indicating onset of the first stages of evaporite formation. Within these intercalations carbonized floral remains are common (Fig. 5g). Above this facies the laminated halite–arenites and halite–rudites occur. Such a complete succession of the first cycle, up to 30 m thick, is observed only in the P-8 well but in other bore-holes it is reduced in thickness and is represented only by a single bed of halite–rudites, up several metres thick (wells P-2 and P-7), anhydrite (in the P-3 well) or is entirely absent/possibly eroded, e.g. the P-6 well (Fig. 4).

The second overlying cycle (cycle II) begins with a claystone layer with carbonized flora and sandy flame structures (Figs 4 & 6), overlain by a series of halite–arenites and halite–rudites locally replaced by rhythmites containing primary chevron halite crystals. These chlorides are almost pure halite with a very low content of clayey matter and anhydrite (up to several percent). This cycle is about 30 m thick.

The third cycle (cycle III) begins with a series of conglomerates and fine sandstones, 8–12 m thick (Figs 4 & 6), that are commonly crossbedded, forming several sandy and gravelly subcycles (Fig. 5e). Clasts are weakly sorted, composed of sandstones, shales and carbonates as well as common halite and anhydrite clasts. The overlying *c.* 30 m thick series of halite–rudites and halite–arenites there are mechanically redeposited sediments, interbedded with the primary halite precipitates. Subangular halite clasts predominate, together with oval ones are and their size varies from 0.2 to 5 cm. Clast-supported halite–rudite beds accompanied by rhythmites with fine precipitated halite (cumulates) and *in situ* chevron halite layers compose numerous subcycles with a component thickness less than 50 cm.

The fourth cycle (cycle IV) begins with a *c.* 2 m thick claystone bed (Figs 4 & 6) with displacive halite crystals, corresponding to an event of maximum brine refreshment/solution and accumulation of delivered suspended pelitic material. Overlying this basal clay is an 8.5–12.5 m thick series of matrix-supported halite–rudites with decimetre-thick marly intercalations containing sporadic plant remains. This halite–rudite probably represents the product of slide/slump deposits developed along a basin slope (for analogues see Roveri *et al.* 2003). Clay content in this cycle is significantly higher, up to about 20–30% (Fig. 6). Upward the rudites are overlain by a nodular anhydrite bed 2.5 m thick, produced by the next basin brine refreshment. The upper part of the fourth cycle is built of subcycles consisting of rhythmites and halite–rudites but it has been significantly deformed (beds are inclined at an angle of *c.* 40°) and the primary salt structure was apparently destroyed due to later recrystallization (Fig. 5h).

The uppermost fifth cycle (cycle V) begins with several metres thick bed of clayey anhydrite (Fig. 6), suggesting a new brine refreshment. The sulphates are overlain by 3 m thick layer of mixed salt–clay rock (called 'zuber') with chaotically dispersed halite and nodular anhydrite clasts 4–8 cm in diameter (Fig. 5f) and blocks of halite–rudite and secondary, displacive, very large halite crystals within a clay matrix. Salt deposits are present above the zuber series (halite–rudites and rhythmites) and they apparently represent altered slope

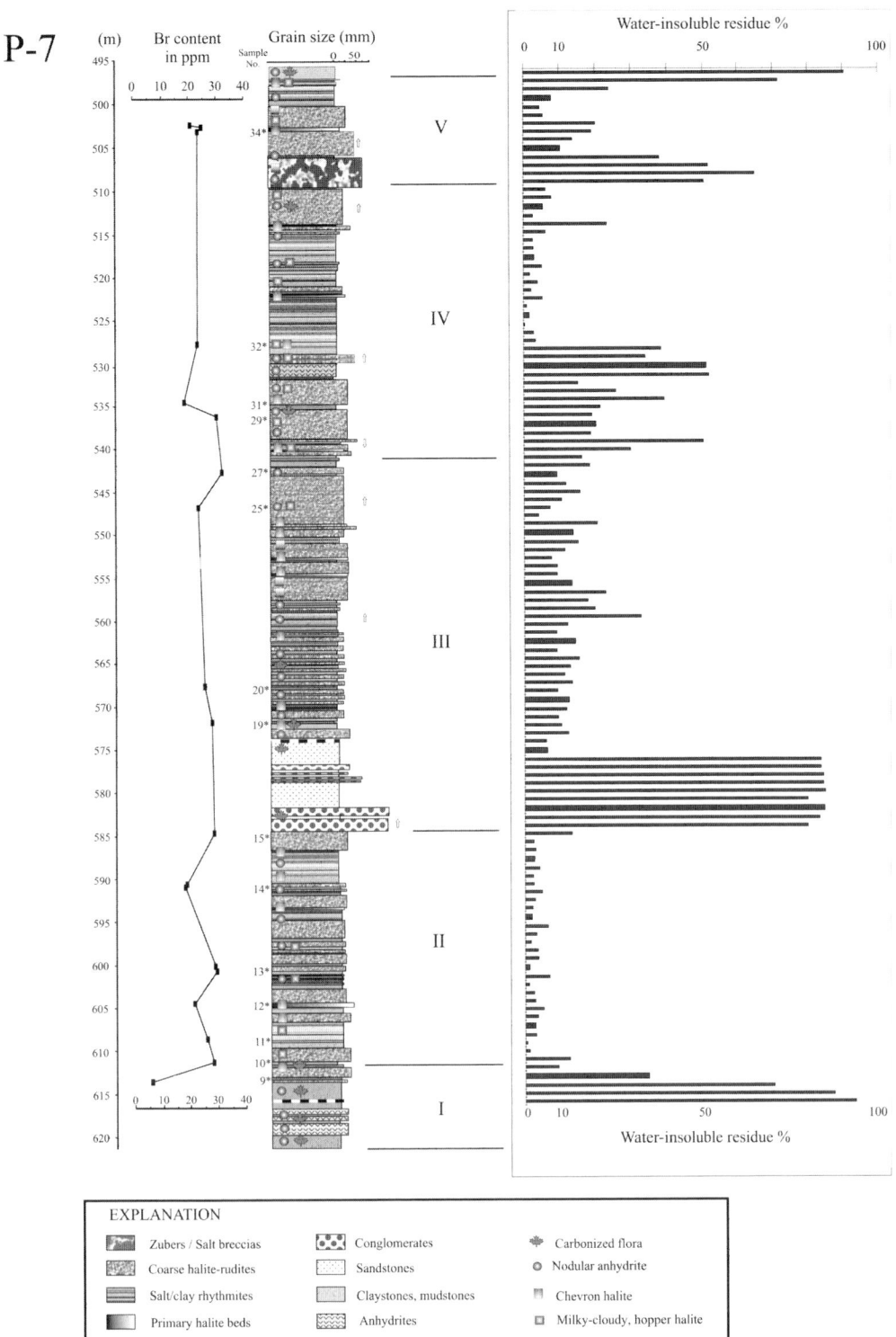

Fig. 6. Distribution of bromine and water-insoluble residue and lithological succession of deposits of the Zbudza Formation in P-7 profile. The lowest part of each cycle (I–V) is marked by a distinct increase in terrigenous material.

slide sediments and are the products of calm chemical precipitation (sections P-3, P-7 and P-8 – Figs 4 & 6) or (see P-6 section in Fig. 4) they are mainly sulphates that resulted from brine refreshment.

The Zbudza Formation, in the studied sections, is overlain by open marine siltstones, called the Lastomir Formation that apparently marks the late Badenian ingression of open marine water into the East Slovakian Basin (Vass et al. 2000).

Geochemistry

Methodology

Geochemical analyses were carried out on 289 samples taken from wells P-6 and P-7. Applying 'classical' methods (e.g. argentometric analysis of chlorine concentration) the content of such main components as NaCl, Ca, Mg, K, SO_4, HCO_3 water-insoluble residue (e.g. in the P-7 section, Fig. 6) were determined in 255 samples but the content of these main components and trace elements were defined in 34 point samples using ICP and INNA methods (Baedecker 1987).

Bromine content was analysed (INNA method) in 34 point samples of pure halite (without the insolubles) taken from wells P-6 and P-7 (Table 1). Oxygen and sulphur isotope content of the anhydrite was determined in 10 point samples taken from wells P-6 and P-7 (Table 2, Fig. 7). Both $\delta^{34}S$ and $\delta^{18}O$ were analysed by a dual inlet and triple collector mass spectrometer analysing SO_2 and CO_2 gases, respectively. SO_2 was extracted by the method developed in the Lublin laboratory (Hałas & Szaran 1999), whereas CO_2 was prepared by the method described by Mizutani (1971). The reproducibility of both the delta analyses was about 0.16‰ (Table 2).

The chemical composition of solutions from individual inclusions has been determined by ultra-microchemical (or glass capillary) analyses described by Petrichenko (1973), which makes it possible to analyse a solution volume in the range $0.001–0.0001$ μm^3 and extract the solution from inclusions of 40–50 μm across and greater. The solution from individual inclusions was extracted by special conical capillary with interior diameter of 3–5 μm at the tip of the cone. Each element was determined directly in the capillary by standard chemical analysis. This method could define the content of ions K^+, Mg^{2+}, Ca^{2+} and SO_4^{2-} in inclusion brines so several tens of such determinations were carried out for the primary inclusions and a dozen or so on the secondary ones. Samples for such studies were prepared from chevron halite crystals taken from wells P-2, P-3, P-6, P-7 and P-8 (Galamay et al. 2004) and results were compared with earlier published data (Galamay & Karoli 1997; Kovalevich & Petrichenko 1997).

Results

Determination of the content of the main components in channel samples taken from each meter of core demonstrated that salts of greatest purity are located in the lower part of Zbudza Formation (in the first cycle). They are composed of nearly pure rock salt (NaCl more than 95%) with a low content of insolubles (clay matter and sulphates). The content of terrigenous material increases upward the Zbudza Formation, resulting in the occurrence of more and more clayey salts.

The studied rock salts of Zbudza Formation are characterized by a variable content of three main components: NaCl, $CaSO_4$ and water-insoluble residues (Fig. 6). Relations of these components remain similar throughout the formation, c. 77% NaCl, 20–22% water-insoluble residue and 1–3% $CaSO_4$. The ranges of potassium and magnesium contents are very similar in salts from both analysed cores and are relatively low. Concentrations of potassium and magnesium in salts from the Zbudza Formation apparently are related to the admixture of terrigenous material, i.e. clay content in the analysed salts. Potassium content varies from 0.01 to 0.49%, and the Mg content is between 0.01 and 0.77% (Table 1).

Concentrations of these elements as well as aluminium (0.01–1.53%) and iron (0.01–0.61%)

Table 1. NaCl, Br, K, Mg content in rock salt rocks of the Zbudza Formation

Element	Well	Number of samples	Range	Mean
NaCl (%)	P-6	127	8.19–98.19	77.13
	P-7	121	14.44–98.54	78.31
Br (ppm)	P-6	15	16–36	26.9
	P-7	19	6–35	23.9
K (%)	P-6	15	0.01–0.49	0.08
	P-7	19	0.01–0.49	0.12
Mg (%)	P-6	15	0.01–0.63	0.15
	P-7	19	0.01–0.77	0.20

are slightly higher in samples taken from the P-7 borehole; however the difference resulted from a greater admixture of terrigenous material in the studied well. Strontium content varied from 5 to 5338 ppm, and is closely related to the amount of dispersed anhydrite (the average Sr content is higher in P-6 core material, 680 ppm in comparison to 498 ppm for samples from P-7 well).

The content of other minor and trace elements in the studied deposits are commonly low and undifferentiated. Several trace elements are connected with non-chloride minerals occurred in the salts, for example, sulphates, carbonates and clay minerals.

Bromine content in Zbudza Formation salts varies between 5.8 and 36 ppm. Similarly low values of bromine concentrations were established earlier in this area from halite sampled in borehole Ep-2 (Galamay & Karoli 1997) although through the whole interval the bromine content was somewhat variable: 13–89 ppm. The average bromine concentration in core material from P-6 and P-7 boreholes (24 and 27 ppm) is close to the average content of this element in zuber rocks from both the Wieliczka and Bochnia salt mines. Such low bromine concentrations in the salts indicate that bromine content in their parent brines was not characteristic for brines of purely marine origin. The bromine distribution in salt profile P-7 is illustrated in Figure 6.

The narrow range of the bromine content is reasonable evidence of small variety of parent brines concentrations during salt sedimentation period. It is very likely that the salts in this study were precipitated from secondary brines that formed due to dissolution of earlier salts deposited in other places. Marine waters with an influx of meteoritic waters probably were the main cause of dissolution, though a particularly low bromine content (5.8 ppm), determined in the sample taken from the bottom of salt profile, indicates that at the onset of salt sedimentation waters of non-marine origin also contained dissolved earlier primary salts.

The oxygen and sulphur isotope compositions of analysed anhydrite samples show a wide scatter of values from 11.16 to 13.15‰ SMOW for oxygen and from 18.28 to 24.45‰ CDT for sulphur (Table 2; Fig. 7). Average values are respectively 12.34 ± 0.47 and 22.22 ± 0.87‰. Overall, these results correspond well to the isotopic composition of Badenian anhydrites in the Carpathian Foredeep and associated halite in Wieliczka and Bochnia salt deposits (Claypool et al. 1980; Parafiniuk & Hałas 1997; Bukowski & Szaran 1997; Peryt et al. 1998, 2002; Kasprzyk 2003). Similar $\delta^{18}O$ and $\delta^{34}S$ were also documented for the Badenian primary gypsum deposits in the Carpathian Foredeep in Poland, Czech Republic and Ukraine (e.g. Parafiniuk et al. 1994; Hałas et al. 1996; Kasprzyk 1997; Peryt et al. 1997; 2002; Peryt 2001).

Data on the chemical composition of the brines in primary inclusions in the chevron halite crystals from the Zbudza Formation indicate that the brines of this basin were of the Na–K–Mg–SO$_4$ type (Galamy et al. 2004). The ratios of the K^+, Mg^{2+} and SO_4^{2-} ions were almost analogous to the ratios in the Badenian seawater, which differed from the contemporaneous seawater (McCaffrey et al. 1987) in terms of the reduced content of Mg^{2+} ions by approximately 20% and SO_4^{2-} ions by approximately 40% (Kovalevich et al. 1998; Zimmermann 2000; Poberezhskyy & Kovalevich 2001). Such lower sulphate values, typical for other Miocene basins of the Carpathian area (Fig. 8) as well, might be due to the chemical evolution of the Phanerozoic ocean (Kovalevich et al. 1998; Horita et al. 2002) or may be due to inflow of considerable amounts of continental fresh waters and terrigenous materials into the basin (Galamay et al. 1997; Cendón et al. 2004).

Table 2. *Isotopic composition of anhydrite in the Zbudza Formation*

Sample	Depth (m)	Anhydrite type	$\delta^{18}O‰_{SMOW}$	$\delta^{34}S‰_{CDT}$
P-6/27a	469.9	Nodular anhydrite with halite pore fill (50%)	12.40	22.10
P-6/27b	469.9	and halite clasts	12.34	22.21
P-6/28	461.1	Massive nodular anhydrite	13.15	22.92
P-6/30	457.0	Porous nodular anhydrite	12.34	22.08
P-6/31	437.0	Anhydritic conglomerate with clay matrix	12.44	22.92
P-6/35	436.7	Anhydritic conglomerate with clay matrix	12.94	24.45
P-7/31	533.5	Halite-rudite with a matrix of nodular anhydrite	12.58	22.95
P-7/8a	612.6	Nodular anhydrite with clay laminae/flame structures and fine clasts of halite and sedimentary rocks	12.47	22.08
P-7/8b	612.6	Br content in halite: very low probable recycling?	11.16	22.19
P-7/2	619.5	Finely laminated anhydrite	11.17	18.28

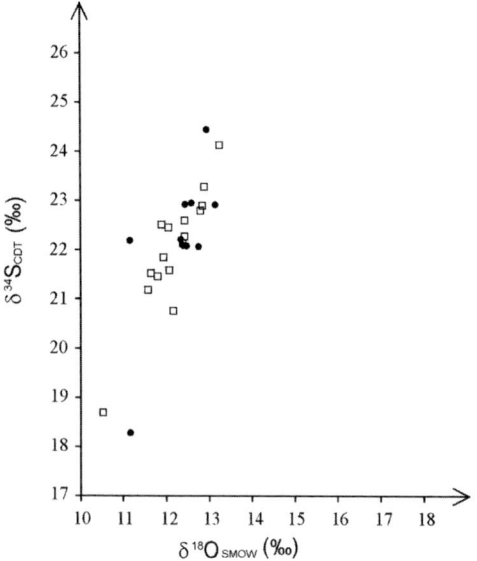

- Zbudza Fm (P6-P7) – East Slovakian Basin
- Wieliczka, Bochnia – Carpathian Foredeep (Bukowski & Szaran 1997)

Fig. 7. Plot of ^{34}S and ^{18}O values measured from the anhydrite in the Zbudza salt deposit. Isotopic compositions for Badenian anhydrite from Wieliczka and Bochnia salt mines in the Carpathian Foredeep, Poland, are also shown (data from Bukowski & Szaran 1997).

Genesis of Zbudza Evaporitic Formation

The redeposition processes generally played a very important role during the deposition of evaporites and, while resedimented gypsum and anhydrite occurrences are well known (e.g. Hardie & Eugster 1971; Parea & Ricci-Lucchi 1972; Schreiber et al. 1976; Schlager & Bolz 1977; Rouchy et al. 1995; Warren 1999; Peryt 2000; Roveri et al. 2001; Manzi et al. 2005), resedimented halite deposits are less commonly noted (e.g. Schreiber 1986; Ślączka & Kolasa 1997). Salt breccia (employed as a morphological term) is more often interpreted as a result of tectonic deformation, additionally the influence of tectonic processes (recrystallization, foliation) is readily causes loss of primary sedimentary structures of halite.

The salt-bearing succession in the Zbudza Formation contains much evidence of sedimentary origin and a considerable part of the studied halite was mechanically redeposited.

1. Both halite clasts (angular and rounded/partly dissolved with zoned, cloudy features) dispersed within the salt–clay matrix and accompanied with clasts (pebbles) of marls, sandstones and claystones and fine-medium clastic quartz (Fig. 9a), so they all were deposited by sedimentary process.
2. In vertical section common salt to siliciclastic units/sets are visible with normal and reversed grading, with erosional lower contacts and an upward increase in the content of siliciclastics and clay (Fig. 9b).
3. Salt/halite rudites–psamites units are interbedded with layers (up to several metres thick) of sandstones with gravels and claystone, having normal grading, erosionally scoured lower contacts, lignite fragments and

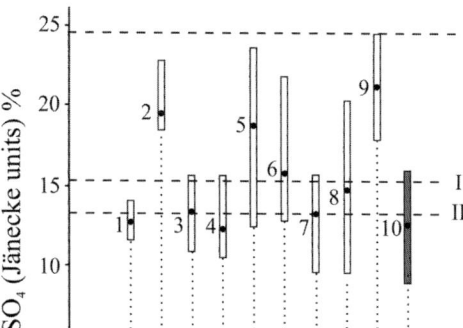

Fig. 8. Content of the SO$_4$ ion in primary fluid inclusions of sedimentary halite (given in Jänecke units) as measured from different boreholes, mines and localities in the Badenian evaporites of the Carpathian region (after Galamay et al. 2004). I, Badenian seawater (Galamay et al. 2004); II, Badenian seawater (Zimmermann 2000). Dots indicate locations: 1, Bochnia mine, Carpathian Foredeep, Poland (Kovalevich & Petrichenko 1997; $n = 2$ samples); 2, Łeżkowice Z-1 and Woszczyce IG-1 boreholes, Carpathian Foredeep, Poland (Garcia-Veigas et al. 1997; $n = 6$ samples); 3, Wieliczka mine, Carpathian Foredeep, Poland (Kovalevich & Petrichenko 1997; $n = 5$ samples); 4, Wieliczka mine, Carpathian Foredeep, Poland (Galamay et al. 1997; $n = 20$ samples); 5, Selets-Stupnitsy and Zabolotiv localities, Carpathian Foredeep, Ukraine (Kovalevich & Petrichenko 1997; $n = 20$ samples); 6, Slanic-Prahova mine, Carpathian Foredeep, Romania (Kovalevich & Petrichenko 1997; $n = 5$ samples); 7, Ep-2 and Zb-1 boreholes, East Slovakian Basin (Kovalevich & Petrichenko 1997; $n = 4$ samples); 8, Transylvanian Basin, Romania ($n = 4$ samples); 9, Solotvyne mine, Transcarpathian Basin, Ukraine (Kityk et al. 1983; $n = 7$); 10, all data, East-Slovakian Basin ($n = 32$ samples);

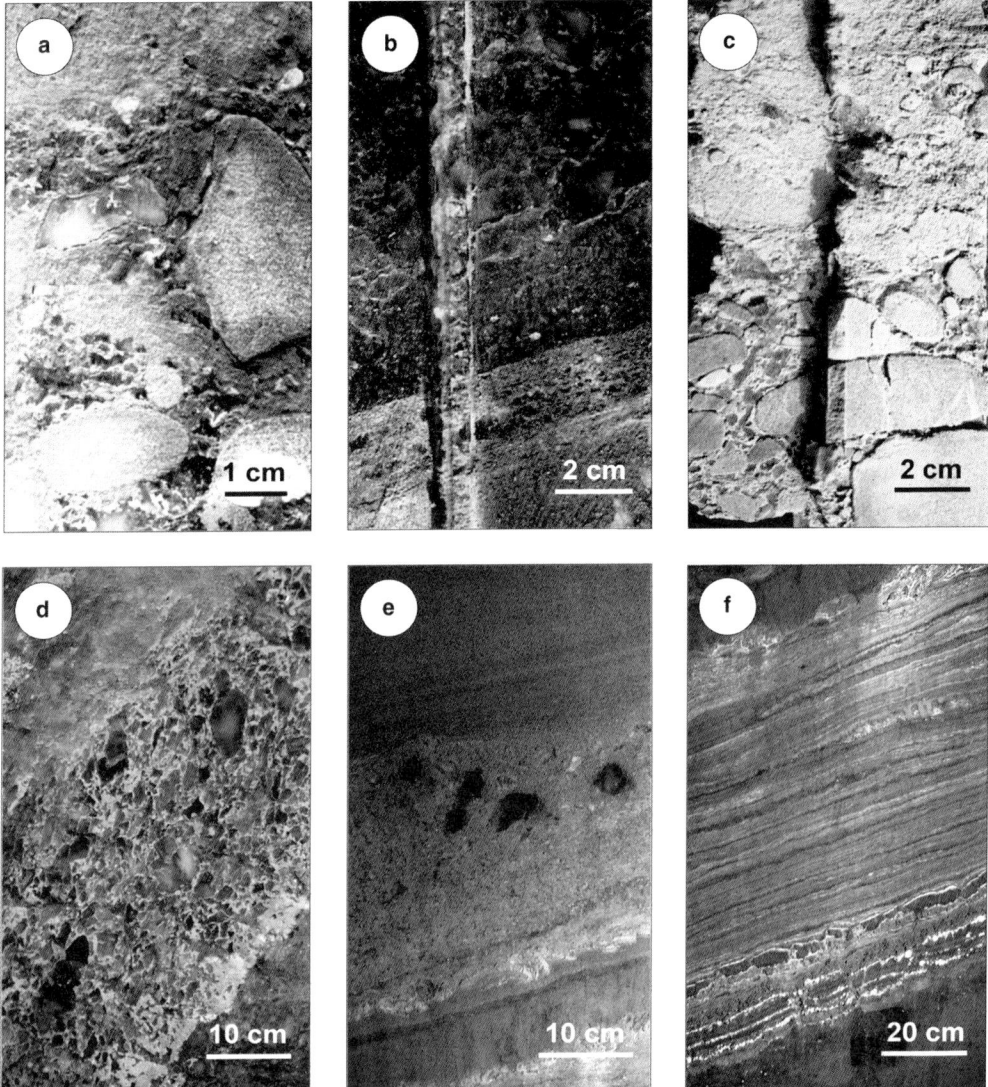

Fig. 9. Photographs of chosen examples being evidence of redeposition within Zbudza Fm. (a) Clasts of 'milky–clouded' halite dispersed within the salt-clay matrix and accompanied with clasts of marls and sandstones, Zbudza Fm, P-8, depth 612 m. (b) Erosional contact between halite with reverse grading and siliciclastic arenite, Zbudza Fm, P-7, depth 584 m. (c) Conglomerate with normal grading, pebbles about 10 cm in diameter, interbeds from Zbudza Fm, P-8, depth 610 m. (d) Layer of matrix-supported halite–rudite and monomictic breccia with large clasts of cloudy and chevron halites. Spiza salt. Gallery Kunegunda, first level in the Wieliczka Salt Mine. (e) Wavy laminated halite–arenite with the large rounded halite monocrystals in the centre. Spiza salt. Gallery Lichtenfeltz in the Wieliczka Salt Mine. (f) Fine laminated Spiza salt with the discontinuous layer of anhydritic siltstone in the lower part. Gallery Lichtenfeltz in the Wieliczka Salt Mine.

x-bedding, evidence of high dynamic transport and the intensive coarse sediment input (Fig. 9c).

4. Tectonically induced features observed in the form of vertical or oblique fractures and veins infilled with secondary clear halite, and as markedly inclined salt-clastic units but typical tectonic breccias are rare and occur locally in the top of the succession.

5. In the studied wells the deposits of the Zbudza Formation and also the over- and underlying sediments are only slightly inclined without features of intensive tectonic displacement

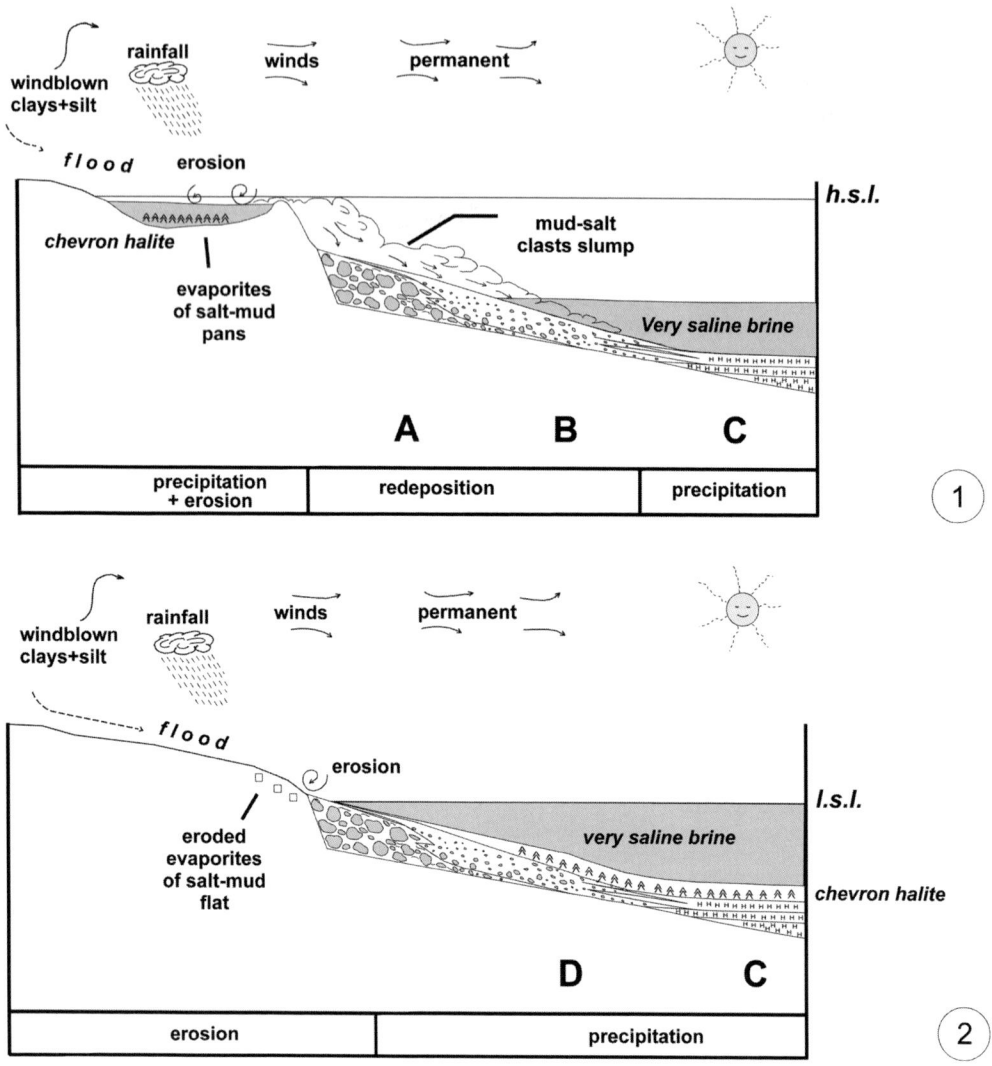

Fig. 10. Deposition model of salt facies as interpreted from deposits that comprise the Zbudza Formation. (1) High water level phase; (2) low water level phase; (A–D) salt facies described in the text.

and rebuilding, the zones of distinct tectonic transformation were found more commonly to the south of the studied profiles and are clearly related to the regional tectonics generated by the Zaluzice and Michalovce fault zones.

The Zbudza evaporites were produced mainly by repeated redeposition/recycling of chlorides precipitated from parent brines of a salt basin of variable/lowered salinity (comparable to brines of a salt basin in the outer Carpathian foredeep). Brine salinity fluctuation was due to periodic high continental influx (siliciclastics and floral input). We found that the mixed clastic–evaporitic deposition in the Zbudza area (East Slovakian Basin), was controlled by frequent tectonic and seismic phenomena and pulses of elevated terrigenous supply, producing the repeated slump sediments with dominant A (proximal mass flow) and B (distal flow) salt types (Fig. 10). These deposits were separated by primary salt units (facies C and D) that precipitated *in situ* from bottom brines during calm periods.

Most of the clastic material composing the evaporitic deposits (redeposited halite–rudites and halite–arenites) and siliciclastics were probably transported from the intensively eroded coastal mud-salt flats framing the basin, but these facies had not been preserved in the sections due to post-depositional tectonics and erosion. Submarine fans and slope slumps, with the clastic salts as the main component and only partly transformed by diagenetic and tectonic processes, were described in the Spiza Salt Series from the Wieliczka salt deposit in the Polish Carpathian Foredeep (Ślączka & Kolasa 1997). Features analogous to the previously described salt types, as described above, were noticed in the galleries of the Wieliczka mine (Fig. 9d–f). In the Wieliczka salt deposit a lateral gradual decrease in the quantity of halite clasts and crystals, the replacement of halite–rudites by finer halite–arenites (Fig. 9e), as well as the more common presence of finely laminated rock beds (Fig. 9e) were noted in the east, west and north directions outside the deposit centre. This suggests a lateral transition from the axial to the distal fan portions.

The repeated cyclicity (cycles I–V) probably reflects varied tectonic activity of the basin margins, that mechanically remobilized the sediments from marginal salt pans, flats and adjacent uplifts, and could be correlated with cycles in the Wieliczka Formation from Carpathian Foredeep.

Analysis of redeposited material and facies succession in the well profiles located nearby the Zbudza salt deposit (Slávik 1967b) showed that the area of basinal salt facies was framed by the sulphate zone, producing a series of anhydrite–siliciclastic layers several metres thick.

Conclusions

The Middle Badenian evaporites of the Paratethys area, present at the Langhian/Serravallian boundary were studied in the East Slovakian Basin (an intermountain depression having a horst-and-graben structure, located in the Inner Carpathians). In this area they were developed as an economically valuable salt-bearing series (Zbudza Formation), identified in several exploration wells.

The Zbudza Formation (100–140 m thick) is composed of pure to clayey halite beds up to 75 m thick, interbedded with siliciclastics (laminated siltstones and claystones, sandstones and conglomerates with carbonized flora remains) and nodular anhydrite. The chlorides are of four main diagnostic types: (A) coarse halite–rudites to salt breccia, being deposited by mass flows; (B) halite–arenites, deposited due to current and/or wave mobilization of halite crystals; (C) salt/clay with parallel to wavy rhythmites, interpreted as chemical precipitates from brines of fluctuating salinity; and (D) chevron halite layers, formed *in situ* due to a halite bottom growth.

The sedimentary succession that makes up the Zbudza Formation was subdivided based on lithological, geochemical and log data into five units, recording the cyclicity of evaporites deposition. Several cycles begin with laminated claystone layers (cycles II and IV), locally with nodular anhydrite interbeds (cycle I at the bottom) or with the anhydrite beds (cycle V at the top), reflecting the refreshment of basin brines (input of fresh marine and/or continental waters).

The middle cycle (III) began with coarser siliciclastic beds (sandstones to conglomerates), deposited by slumps of clastic material along the basin margin. These deposits represent subaqueous fan sediments, produced due to intensive erosion of emerged basin margins (Bukowski et al. 2003). Outside the supposed fan axis the sediment thickness and clast size diminish, replaced by a series of fine gravels and siltstones (e.g. P-6 well). The siliciclastics appear to reflect the period of maximum tectonic activity in the entire region, and their character and position within the Zbudza Formation profile is comparable to the sub-salt sandstone series typical of the Wieliczka Formation in the Carpathian Foredeep (Bukowski 1997).

Slope slides and associated mass flows of unsorted material, removed from coastal mud-salt flats, were responsible for deposition of halite–rudites (type A), accumulated in the axial part of formed subaqueous fan. Further redeposition of finer clastics by diluted distal mass flows and wave currents have produced the halite–arenites (type B), composing the distal fan part. Interbeds of salt/clay rhythmites and primary halite beds (type C), produced during the calm periods of dominant chemical precipitation and suspension drop, complete the evaporate succession of the cycle, illustrating a climatic fluctuations development of slope slumps. The high level of remobilization of both bottom and shallow water sediments, interpreted as a result of increased tectonic activity along basin margins, was observed in the third cycle, dominated by coarse clastics.

The youngest, fifth cycle was characterized by the specific succession of matrix-supported halite–rudites and breccias, called the 'zuber' series. This zuber zone apparently resulted from a single mud flow, triggered by tectonic movements on the basin margin. Comparable but thicker zuber rocks are common in the Wieliczka Formation within the Carpathian Foredeep, documented in the salt deposits at Wieliczka and Bochnia (e.g. Kolasa & Ślączka 1985; Czapowski & Bukowski 2002).

Geochemical data indicate an upward increased in the terrigenous content of the salts of the Zbudza Formation, reflecting fluctuating but generally greater clastic supply to the basin, moving up the section. Fluctuating but low, on average, bromine content in halites – (up to 36 ppm), sulphur and oxygen isotopic composition of anhydrite interbeds and chemistry of primary fluid inclusions in halite (especially low SO_4 concentrations found in fluid inclusions) give evidence that the parent brines for the salts were of low concentration, depleted in sulphates and highly diluted by fresh/continental waters. The extremely low bromine (as low as 5.8 ppm) confirms that secondary brines obtained from dissolution of earlier primary salts by marine waters and non-marine/continental waters suggest a common source for deposition of Zbudza evaporites

Lithological and geochemical features of Zbudza salts are comparable with features of salts that make up the age-equivalent Wieliczka Formation in the Carpathian Foredeep, so the mechanisms of their formation and observed cyclicity, clearly controlled by tectonic activity, is the valid causitive agent within both the intra- (Zbudza area) and the extramountain (Carpathian Foredeep) basins.

Authors are very indebted to B. Charlotte Schreiber for help in the improvement of this paper and to Jean Marie Rouchy and Andrzej Ślączka for the helpful reviews and comments. Special thanks are dedicated to Peter Kotulak for his help during the field studies and fruitful discussions and to Alicja Kasprzyk for interpretation of sulphate isotope results. This work is part of the projects 8T12B 01521 and 11.11.140.300 funded by the Polish Ministry of Education and Science.

References

ANDREYEVA-GRIGOROVICH, A. S., OSZCZYPKO, N., SAVITSKAYA, N. A., ŚLĄCZKA, A. & TROFIMOVICH, N. A. 2003. Correlation of Late Badenian salts of the Wieliczka, Bochnia and Kalush areas (Polish and Ukrainian Foredeep). *Annales Societatis Geologorum Poloniae*, **73**, 67–89.

BĄBEL, M. 2004. Badenian evaporite basin of the northern Carpathian Foredeep as a drawdown salina basin. *Acta Geologica Polonica*, **54**, 313–337.

BAEDECKER, P. A. (ed.). 1987. *Methods for Geochemical Analysis*. US Geological Survey Bulletin, **1770**.

BÀLDI, K., BENKOVICS, L. & SZTANÓ, O. 2002. Badenian (Middle Miocene) basin development in SW Hungary: subsidence history based on quantitative paleobathymetry of foraminifera. *International Journal of Earth Sciences (Geol. Rundisch.)*, **91**, 490–504.

BALINTONI, I. & PETRESCU, I. 2002. A hypothesis on the Transylvanian halite genesis. Studia Universitatis Babes-Bolyai, *Geologia*, Special Issue, **1**, 51–61.

BUKOWSKI, K. 1997. Sedimentation of clastic strata associated with Miocene salts in Wieliczka (Southern Poland). *Slovak Geological Magazine*, **3**, 157–164.

BUKOWSKI, K. & SZARAN, J. 1997. Zawartość izotopów tlenu i siarki w anhydrytach z serii solonośnej Wieliczki i Bochni. *Przegląd Geologiczny*, **45**, 816–818 [in Polish].

BUKOWSKI, K., CZAPOWSKI, G., KAROLI, S. & KOTULAK, P. 2003. Badenian evaporitic-clastic succession in the East Slovakian basin (Zbudza Fm., the well P-7). *Mineralia Slovaca*, **35**, 53–55.

CENDÓN, D. I., PERYT, T. M., AYORA, C., PUEYO, J. J. & TABERNER, C. 2004. The importance of recycling processes in the Middle Miocene Badenian evaporite basin (Carpathian foredeep): palaeoenvironmental implications. *Palaeogeography Palaeoclimatology Palaeoecology*, **212**, 141–158.

CHIRA, C. & DRAGHICI, D. 2002. The calcareous nannoplankton from the Badenian salt and gypsum level in Transylvania. Studia Universitatis Babes-Bolyai, *Geologia*, Special Issue **1**, 97–111.

CLAYPOOL, G. E., HOLSER, W. T., KAPLAN, I. R., SAKAI, H. & ZAK, I. 1980. The age curves of sulfur and oxygen isotopes in marine sulfate and their mutual interpretation. *Chemical Geology*, **28**, 199–260.

CRIHAN, M. 1999. The stratigraphy of Middle Miocene deposits from the Sub Carpathians of Muntenia (Romania). *Biuletyn Państwowego Instytutu Geologicznego*, **96**, 27–29.

CZAPOWSKI, G. & BUKOWSKI, K. 2002. Genesis of clayey salt (zuber) facies (Upper Permian and Middle Miocene case studies from Poland). 16th International Sedimentological Congress, Johannesburg, 8–13 July 2002. Abstract volume, 71–72.

DUDEK, K., BUKOWSKI, K. & WIEWIÓRKA, J. 2004. Radiometric dating of Badenian pyroclastic sediments from the Wieliczka-Bochnia area. *VII Ogólnopolska Sesja Naukowa: 'Datowanie minerałów i skał'*, Kraków 18–19 November 2004 [in Polish].

FORESI, L. M., BONOMO, S., CARUSO, A., STEFANO, A., STEFANO, E., SALVATORINI, G. & SPROVIERI, R. 2002. Calcareous plankton high resolution biostratigraphy (Foraminifera and Nannofossils) of the Uppermost Langhian–Lower Serravallian Ras il-Pellegrin section (Malta). *Rivista Italiana di Paleontologia e Stratigraphfia*, **108**, 195–210.

GALAMAY, A. R. & KAROLI, S. 1997. Geochemical peculiarities of Badenian salts from East-Slovakian basin. *Slovak Geological Magazine*, **3**, 187–192.

GALAMAY, A. R., BUKOWSKI, K. & PRZYBYŁO, J. 1997. Chemical composition and origin of brines of the Forecarpathian Badenian evaporite basin: inclusion data from Wieliczka rock salt deposit, Poland. *Slovak Geological Magazine*, **3**, 165–171.

GALAMAY, A. R., BUKOWSKI, K., POBEREZHSKYY, A. V., KAROLI, S. & KOVALEVYCH, V. M. 2004. Origin of the Badenian salts from East Slovakian Basin indicated by the analysis of fluid inclusions. *Annales Societatis Geologorum Poloniae*, **74**, 267–276.

GARCÍA-VEIGAS, J., ROSELL, L. & GARLICKI, A., 1997. Petrology and geochemistry (fluid inclusions) of

Miocene halite rock salts (Badenian, Poland). *Slovak Geological Magazine*, **3**, 181–186.

GARLICKI, A. 1979. Sedymentacja soli mioceńskich w Polsce. *Prace Geologiczne PAN*, **119**, 1–67 [in Polish with English summary].

GARLICKI, A. 1994a. Comparison of salt deposits in Upper Silesia and Wieliczka [southern Poland]. *Przegląd Geologiczny*, **42**, 752–753 [in Polish with English summary].

GARLICKI, A. 1994b. Formalne jednostki litostratygraficzne miocenu – formacja z Wieliczki fm. *Przegląd Geologiczny*, **42**, 26–28 [in Polish].

GAŠPARIKOVÁ, V. 1963. Mikrobiostratigrafickié pomery okolia ložiska Zbudza. *Geologicke Práce, Správy. Ust. D. Štúra*, **29**, 105–110 [in Slovak].

HAŁAS, S. & SZARAN, J. 1999. Low temperature thermal decomposition of sulfates to SO_2 for on-line $^{34}S/^{32}S$ analysis. *Analytical Chemistry*, **71**, 3254–3257.

HAŁAS, S., JASIONOWSKI, M. & PERYT, T. M. 1996. Anomalia izotopowa w badeńskich gipsach Ponidzia. *Przegląd Geologiczny*, **44**, 1054–1056 [in Polish].

HARDIE, L. A. & EUGSTER, H. P. 1971. The depositional environment of marine evaporites; a case for shallow, clastic accumulation. *Sedimentology*, **16**, 187–220.

HARZHAUSER, M., MANDIC, O. & ZUSCHIN, M. 2003. Changes in Paratethyan marine molluscs at the Early/Middle Miocene transition: diversity, palaeogeography and palaeoclimate. *Acta Geologica Polonica*, **53**, 323–339.

HORITA, J., ZIMMERMANN, H. & HOLLAND, H. D. 2002. Chemical evolution of seawater during the Phanerozoic: Implications from the record of marine evaporites. *Geochimica et Cosmochimca Acta*, **66**, 3733–3756.

JANÁČEK, J. 1958. Nové ložisko soli na východnom Slovensku. *Geologicke Práce, Zprávy*, **14**, 63–72 [in Slovak].

JANÁČEK, J. 1959. Solne ložisko Michalovce. Manuskript. Archív Solivary a.s. [in Slovak].

JANÁČEK, J. 1960. Geologické poměry solného ložiska u Michalovcu na východnim Slovensku. *Geologicke Práce, Správy*, **20**, 151–175 [in Slovak].

KAROLI, S., JANOČKO, J., KOTUĽÁK, P. & VERDON, P. 1997. Sedimentology of Karpatian evaporites in the East-Slovakian Basin (Slovakia). *Slovak Geological Magazine*, **3**, 201–211.

KASPRZYK, A. 1997. Oxygen and sulfur isotope composition of Badenian (Middle Miocene) gypsum deposits in southern Poland: a preliminary study. *Geological Quarterly*, **41**, 53–60.

KASPRZYK, A. 2003. Sedimentological and diagenetic patterns of anhydrite deposits in the Badenian evaporite basin of the Carpathian Foredeep, southern Poland. *Sedimentary Geology*, **158**, 167–194.

KITYK, V. I., BOKUN, A. N., PANOV, G. M., SLIVKO, E. P. & SHAIDETSKA, V. S. 1983. *Evaporite Formations of Ukraine: Transcarpatian Trough*. Naukova Dumka, Kiev [in Russian].

KOLASA, K. & ŚLĄCZKA, A. 1985. Sedimentary Salt megabreccia exposed in the Wieliczka Mine (Fore-Carpathian Depression). *Acta Geologica Polonica*, **35**, 221–230.

KOVALEVICH, V. M. & PETRICHENKO, O. I. 1997. Chemical composition of brines in Miocene evaporite basins of the Carpathian region. *Slovak Geological Magazine*, **3**, 173–180.

KOVALEVICH, V. M., PERYT, T. M. & PETRICHENKO, O. I. 1998. Secular variation in seawater chemistry during the Phanerozoic as indicated by brine inclusions in halite. *Journal of Geology*, **106**, 695–712.

KRÉZSEK, CS. & FILIPESCU, S. 2005. Middle to Late Miocene sequence stratigraphy of the Transylvanian Basin (Romania). *Tectonophisics*, **410**, 437–463.

MANZI, V., LUGLI, S., RICCI LUCCHI, F. & ROVERI, M. 2005. Deep-water clastic evaporites deposition in the Messinian Adriatic foredeep (northern Apennines, Italy): did the Mediterranean ever dry out? *Sedimentology*, **52**, 875–902.

MĂRUNȚEANU, M. 1999. Litho- and biostratigraphy (calcareous nannoplankton) of the Miocene deposits from the Outer Moldavides. *Geologica Carpathica*, **50**, 313–324.

MCCAFFREY, M. A., LAZAR, B. & HOLLAND, H. D. 1987. The evaporation path of seawater and the coprecipitation of Br^- and K^+ with halite. *Journal of Sedimentary Petrology*, **57**, 928–937.

MIZUTANI, Y. 1971. An improvement in the carbon-reduction method for the oxygen isotopic analysis of sulphates. *Geochemical Journal*, **5**, 69–77.

PARAFINIUK, J. & HAŁAS, S. 1997. Sulfur- and oxygen-isotope composition as the genetic indicator for celestite from the Miocene evaporites of the Carpathian Foredeep. *Slovak Geological Magazine*, **3**, 131–134.

PARAFINIUK, J., KOWALSKI, W. & HAŁAS, S. 1994. Stable isotope geochemistry and genesis of the Polish native sulfur deposits – a review. *Geological Quarterly*, **38**, 473–496.

PAREA, G. C. & RICCI-LUCCHI, F. 1972. Resedimented evaporites in the Preadriatic Trough (Upper Miocene, Italy). *Israel Journal Earth Sciences*, Special Issue, **21**, 125–141.

PERYT, D. 1997. Calcareous nannoplankton stratigraphy of the Middle Miocene in the Gliwice area (Upper Silesia, Poland). *Bulletin of the Polish Academy of Sciences, Earth Siences*, **45**, 119–131.

PERYT, T. M. 2000. Resedimentation of basin centre sulphate deposits: Middle Miocene Badenian of Carpathian Foredeep, southern Poland. *Sedimentary Geology*, **134**, 331–342.

PERYT, T. M. 2001. Gypsum facies transitions in basin-marginal evaporites: middle Miocene (Badenian) of West Ukraine. *Sedimentology*, **48**, 1103–1119.

PERYT, T. M. & KASPRZYK, A. 1992. Earthquake-induced resedimentation in the Badenian (middle Miocene) gypsum of southern Poland. *Sedimentology*, **39**, 235–249.

PERYT, T. M., HAŁAS, S., KAROLI, S. & PERYT, D. 1997. Zapis izotopowy zmian środowiskowych podczas depozycji gipsów badeńskich w Kobeřicach koło Opawy. *Przegląd Geologiczny*, **45**, 807–810 [in Polish].

PERYT, T. M., PERYT, D., SZARAN, J., HAŁAS, S. & JASIONOWSKI, M. 1998. O poziomie anhydrytowym badenu w otworze wiertniczym Ryszkowa Wola 7 k. Jarosławia (SE Polska). *Biuletyn Państwowego Instytutu Geologicznego*, **379**, 61–78 [in Polish].

PERYT, T. M., SZARAN, J. *ET AL*. 2002. S and O isotopic composition of the Badenian (Middle Miocene) sulphates in the Carpathian Foredeep. *Geologica Carpathica*, **56**, 391–398.

PETRICHENKO, O. I. 1973. *Metody doslidzhennya vkluchen' v mineralah galogennykh porid.* Naukova Dumka, Kiev [in Ukrainian].

POBEREZHSKYY, A. V. & KOVALEVYCH, V. M. 2001. Chemical composition of seawater in the Cenozoic. *Geology and Geochemistry of Combustible Minerals*, **2**, 90–109 [in Ukrainian].

POBORSKI, J. 1959. Skały solne na tle ogólnej klasyfikacji skał. *Zeszyty Naukowe AGH*, **3**, 73–80 [in Polish].

POPOV, S. V., RÖGL, F., ROZANOV, A. Y., STEININGER, FRITZ, F., SHCHERBA, I. G. & KOVAC, M. 2004. Lithological–Paleogeographic Maps of Paratethys 10 maps Late Eocene to Pliocene. *Courier Forschungsinstitut Senckenberg*. E. Schweizerbart'sche Verlagsbuchhandlung Science, Stuttgart, maps 1–10.

RÖGL, F. 1998a. Palaeogeographic considerations for Mediterranean and Paratethys Seaways (Oligocene to Miocene). *Annalen des Naturhistorischen Museums in Wien*, **99A**, 279–310.

RÖGL, F. 1998b. Das Werden der Zentralen Paratethys im Tertiär. *In*: SCHULTZ, O. (ed.) (1988b) *Tertiärfossilien Österreichs*. Goldschneck-Verlag, 159 [in German].

ROUCHY, J. M., PIERRE, C. & SOMMER, F. 1995. Deep-water resedimentation of anhydrite and gypsum deposits in the Middle Miocene (Belayim Formation) of the Red Sea, Egypt. *Sedimentology*, **42**, 267–282.

ROVERI, M., BASSETTI, M. A. & RICCI LUCCHI, F. 2001. The Mediterranean Messinian salinity crisis: an Apennine foredeep perspective. *Sedimentary Geology*, **140**, 201–214.

ROVERI, M., MANZI, V., RICCI LUCCHI, F. & ROGLEDI, S. 2003. Sedimentary and tectonic evolution of the Vena del Gesso Basin (Northern Apennines, Italy); implications for the onset of the Messinian salinity crisis. *Geological Society of America Bulletin*, **115**, 387–405.

SCHLAGER, W. & BOLZ, H. 1977. Clastic accumulation of sulphate evaporites in deep water. *Journal of Sedimentary Petrology*, **4**, 600–609.

SCHREIBER, B. C. 1986. Arid shorelines and evaporites. *In*: READING, H. G. (ed.) *Sedimentary Environments and Facies*, 2nd edn. Blackwell Scientific, Oxford.

SCHREIBER, B. C., FRIEDMAN, G. M., DECIMA, A. & SCHREIBER, E. 1976. Depositional environments of Upper Miocene (Messinian) evaporite deposits of the Sicilian Basin. *Sedimentology*, **23**, 729–760.

SLÁVIK, J. 1967a. Salt deposits in the East-Slovakian Miocene. *Sbornik Geologických Véd*, **9**, 129–149 [in Slovak].

SLÁVIK, J. 1967b. Gips und Anhydrit aus den Salzführenden Formationen des Miozäns der Ostslowakaei. *Geologický Sbornik*, **18**, 1, 65–77 [in German].

ŚLĄCZKA, A. & KOLASA, K. 1997. Resedimented salt in the Northern Carpathians Foredeep (Wieliczka, Poland). *Slovak Geological Magazine*, **3**, 135–155.

STUDENCKA, B., GONTSHAROVA, I. A. & POPOV, S. 1998. The bivalve faunas as a basis for reconstruction of the Middle Miocene history of the Paratethys. *Acta Geologica Polonica*, **48**, 285–342.

VASS, D. & ČVERČKO, J. 1985. Litostratigrafické jadnotky neogénu východoslovenskej nížiny. *Geologicke Práce, Správy*, **82**, 111–116 [in Slovak].

VASS, D., ELEČKO, M., JANOČKO, J., KAROLI, S., PERESZLENYI, M., SLÁVIK, J. & KALIČIAK, M. 2000. Paleogeography of the East-Slovakian Basin. *Slovak Geological Magazine*, **6**, 377–407.

WARREN, J. 1999. *Evaporites: Their Evolution and Economics*. Blackwell Science, Oxford, 422.

ZIMMERMANN, H. 2000. Tertiary seawater chemistry – implications from primary fluid inclusions in marine halite. *American Journal of Science*, **300**, 3–45.

ZLINSKA, A. 2004. *Zhodnotenie mikrofauny z vrtu P-3 (Zbudza, 505–624, 3 m podložie soli)*. Štatny Geologický Ústav Dionýza Štúra, Bratislava [in Slovak].

Sulphur isotopic composition of K–Mg sulphates of the Miocene evaporites of the Carpathian Foredeep, Ukraine

S. HRYNIV[1], J. PARAFINIUK[2] & T. M. PERYT[3]

[1]*Institute of Geology and Geochemistry of Combustible Minerals, National Academy of Sciences of Ukraine, Naukova 3a, 79053 Lviv, Ukraine (e-mail: Sophia_Hryniv@ukr.net)*

[2]*Institute of Geochemistry, Mineralogy and Petrology, University of Warsaw, Żwirki i Wigury 93, 02-089 Warszawa, Poland*

[3]*Państwowy Instytut Geologiczny, Rakowiecka 4, 00-975 Warszawa, Poland*

Abstract: Miocene evaporites of the Carpathian Foredeep host an interesting sulphates group of potash deposits, including about 20 sulphates minerals. The study of sulphur isotopic composition of 10 of the sulphates minerals from the Kalush–Holyh and Stebnyk potash deposits shows that only the basal Ca-sulphates (anhydrite) from the Kalush–Holyn potash deposits has $\delta^{34}S$ values typical of Neogene marine evaporites (+21.0‰). Potash minerals related to the deposits (polyhalite, anhydrite, kainite, langbeinite and kieserite) show $\delta^{34}S$ values from +15.28‰ to +17.54‰, and the weathering zone minerals (picromerite, leonite, bloedite, syngenite and gypsum) show values ranging from +14.73‰ to +18.22‰. The recorded depletion of sulphur isotopic composition of the salt minerals of potash deposits (and their weathering zone) was probably caused by one or more of the following isotope fractionation factors: bacterial reduction of sulphate, effect of crystallization and inflow of surface waters containing sulphates enriched in light sulphur isotopes due to pyrite oxidation. Accordingly, the observed sulphur isotopic composition of minerals from these potash deposits demonstrates the depletion of the original brines and continual inflow of new (concentrated) seawater. Similar sulphur isotopic composition of minerals from the potash deposits and their weathering zone points out that there was no significant sulphur isotope fractionation during weathering and in this case inflow of surface water has an insignificant influence on sulphur isotopic composition.

Miocene K–Mg sulphates in the Ukrainian part of the Carpathian Foredeep basin have been mined since the mid-nineteenth century (Korenevskiy & Donchenko 1963, with references therein) and their economic importance has led to intensive mineralogical, petrographical and geochemical studies aiming to explain the composition and origin of these potash deposits. Present concepts on the origin of K–Mg salts strongly stress a very shallow water, marginal marine sedimentary environment, in the decimetre range of brine depth, and such a concept was earlier applied to the Miocene potash in the Carpathian foredeep basin of the Ukraine (e.g. Valyashko 1962). However, on the basis of physicochemical considerations, many authors considered that Lower Miocene potash salts were formed in a deeper setting (e.g. Kovalevich 1978; Petrichenko 1988; Peryt & Kovalevich 1997).

During the extensive study of the Miocene K–Mg sulphates in the Ukrainian portion of the Carpathian Foredeep (see Hryniv *et al.* 2007), particular attention was paid to the K–Mg sulphates of the Stebnyk and Kalush–Holyn deposits (Figs 1–4). Despite having a somewhat different stratigraphical position, both deposits show the same mineralogical characteristics (e.g. Kovalevich 1978; Griniw 1994). K/Ar dating of the potassium–magnesium sulphate minerals from both deposits provided evidence that they have recrystallized at roughly the same time, and this recrystallization was apparently the response to three major tectonic events that affected the area (Wójtowicz *et al.* 2003). However, despite the intensive geochemical research, no isotope geochemical study for those deposits has been reported and the aim of this paper is to fill that gap in our knowledge. It should be mentioned that, as noted by Raab & Spiro (1991), data on late evaporative facies are rare. Their experimental data showed that primary, unaltered, highly evaporated minerals can have *c.* 2‰ depletion value, relative to the first gypsum, by the simple mechanism of a continual evaporation process (Raab & Spiro 1991). Strauss (1997) commented that there is isotopic depletion in ^{34}S (by as much as 4‰) of late-stage sulphate, deposited within the halite and potash–magnesia facies.

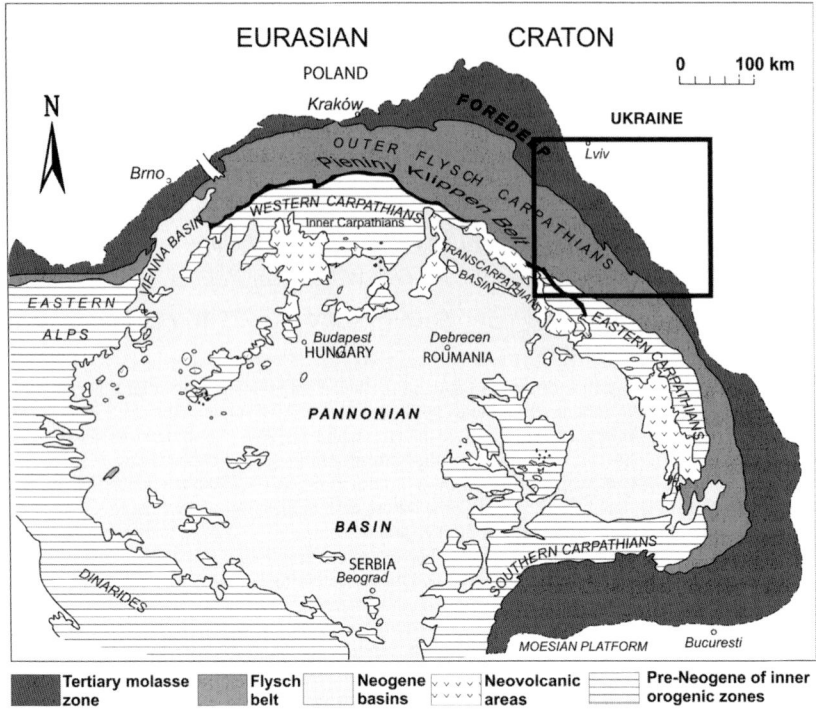

Fig. 1. Geological map of the Carpathian–Pannonian basin system (after Kovač *et al.* 1998) showing the location of Figure 2 (rectangle).

Geological setting

The Carpathian Foredeep basin was formed in the early Miocene, northeast of the overthrusting Carpathian nappes. The basin is filled with Miocene deposits (mostly siliciclastic and minor evaporites) that are more than 3000 m thick in the area adjacent to the Carpathians (Oszczypko *et al.* 2006). In the Ukrainian part of the foredeep, three tectonic zones are distinguished: the outer (Bilche-Volytsa), a central (Sambir Nappe thrust over the foreland) and an inner (Boryslav-Pokuttya Nappe, thrust over the Sambir Nappe and the underlying foreland). The Stebnyk potash deposit is located in the latter zone, near the front of the overthrust Carpathian nappes, and the Kalush potash deposit is located in the central zone (Fig. 2). Because of this geological location, the stratigraphic section is most likely a repeated section (Fig. 3).

The Stebnyk potash deposit is part of the Vorotyshcha suite (Fig. 4). In this paper, instead of *formation*, the term 'suite' is applied, which is more consistent with its poorly defined and complex stratigraphy. The Vorotyshcha suite is composed of three members: a lower salt-bearing, a middle terrigenous and an upper salt-bearing one, with a total thickness more than 2000 m (Petryczenko *et al.* 1994), but new interpretation assumes that upper and lower salt-bearing members belong to the same, Vorotyshcha suite, and that the terrigenous member separating them is an olistostrom (Kulchitsky 1986; Koriń 1994). The occurrence of multiple potash layers and the great thickness of the suite are the result of intensive folding and overthrust tectonics and the normal total thickness of potash deposits in Stebnyk is actually 10–125 m (Dzhinoridze 1980; Koriń 1994; Peryt & Kovalevich 1997). There is only one potash complex (langbeinite–kainite) in the middle part of the suite (Koriń 1994). The complex consists of beds of kainite, langbeinite and kainite–langbeinite (with sylvite and kieserite). Polyhalite beds (5–15 cm thick) occur at the base and the top of the potash complex (Koriń 1994). The mineralogical composition of individual parts of the mined horizons varies considerably (see Kovalevich 1978), but the bulk chemical composition (as far as Mg, K_2 and SO_4 are concerned) is constant. The actual, unfolded thickness of the potash complex is probably less than 15 m. The micropalaeontological data indicate the Early Miocene Eggenburgian age of the Vorotyshcha suite (Andreyeva-Grigorovich *et al.* 1997; Fig. 4). The Vorotyshcha suite is overlain by the Stebnyk suite, a molasse deposit with

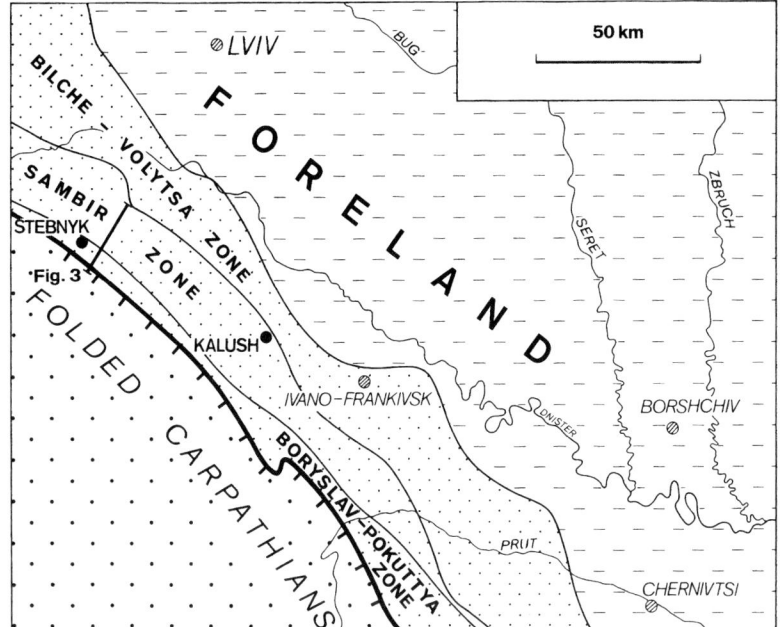

Fig. 2. The main tectonic units of West Ukraine (after Vul *et al.* 1998). The Carpathian Foredeep area is indicated by the fine stripple; adjacent Folded Carpathians and Transcarpathian Depression (SW of the Carpathian Foredeep) are indicated by the coarse stripple, and the Carpathian Foreland area (NE of the Carpathian Foredeep) is indicated by dashed lines.

graded bedding formed from relatively shallow subaqueous turbidite currents as well as from subaerial debris flows (Kulchitskiy 1986, p. 14).

The lithostratigraphical and chronostratigraphical position of evaporite deposits in the Kalush–Holyn deposit is subject to continuing controversy (Fig. 4; see Wójtowicz *et al.* 2003, for review). The potash-bearing sequence consists of interbedded salt claystones, breccias, potash and rock salt, up to 500 m thick (Petryczenko *et al.* 1994).

The present structure of the Kalush–Holyn deposit is complex but it seems that during sedimentation in the evaporite basin only two beds of potash salts were formed: the lower chloride and upper sulphate (Dzhinoridze 1973; Griniw 1994). The unfolded thickness of this sulphate potash complex is 38.5 m (Griniv 1985). The major rock-forming minerals of the potash deposits are halite, langbeinite and kainite. Kieserite, polyhalite, anhydrite, sylvite and carnallite are present in

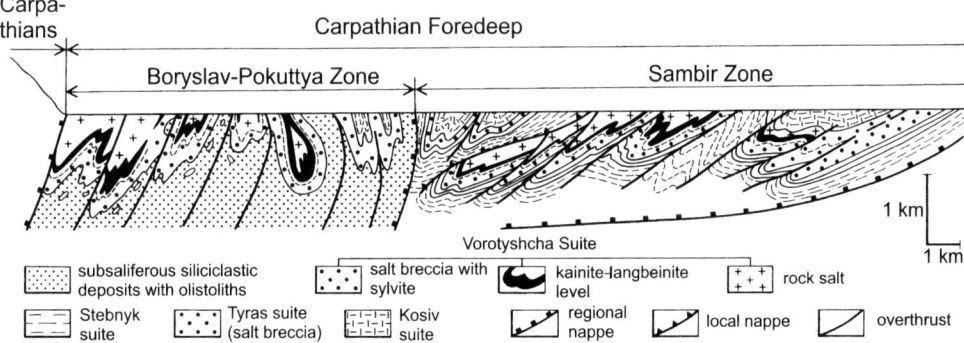

Fig. 3. SW–NE geological cross-section through the inner and central zones of the Carpathian Foredeep (after Koriń 1994); the cross-section line is shown in Figure 2.

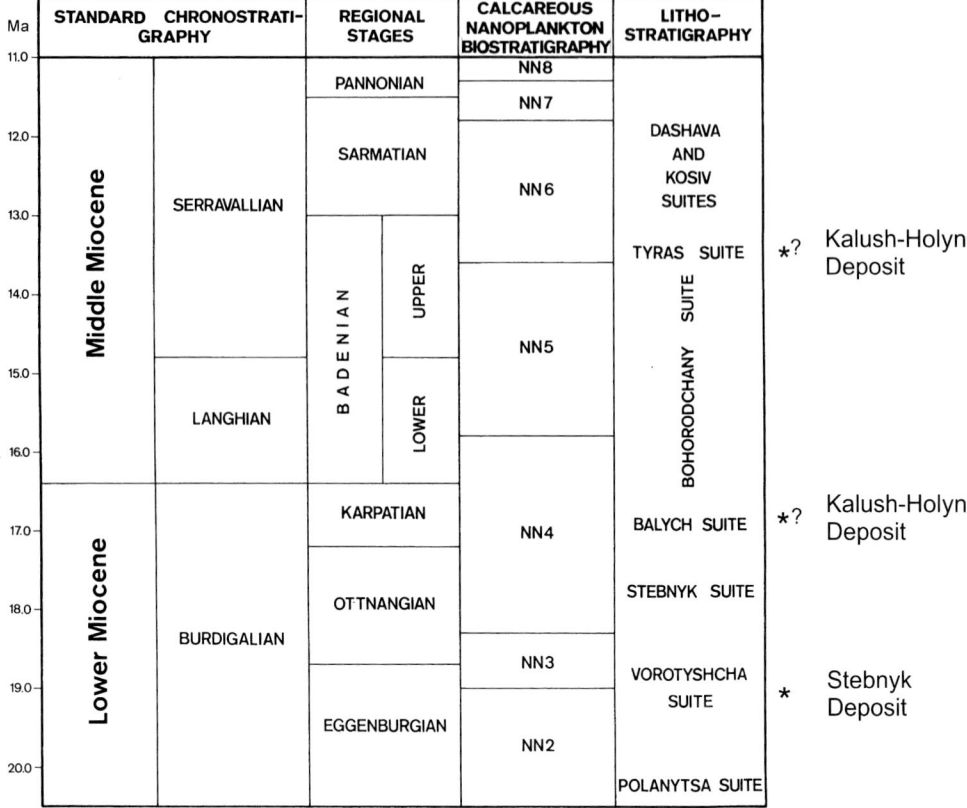

Fig. 4. Stratigraphic position of the Kalush–Holyn Deposit (two interpretations marked by asterisks) and the Stebnyk Deposit in the Miocene of the Ukrainian part of the Carpathian Foredeep. Chronostratigraphy after Rögl (1996), Vakarcs et al. (1998) and Steininger & Wessely (2000); lithostratigraphy after Petryczenko et al. (1994).

smaller quantities. Both the tectonic structure and the composition of potash rocks of the Kalush–Holyn and the Stebnyk deposits show similarities. Evaporite deposits in the Kalush–Holyn area are included within the Dombrovo suite (Dzhinoridze 1973) and are counterparts of the Badenian Tyras suite (Dzhinoridze 1980). The Badenian age of this deposit is supported by the study of calcareous nannoplankton assemblages, suggesting that the evaporites in the Kalush deposit belong to the NN6 Zone (Andreyeva-Grigorovich et al. 2003), similar to the Badenian evaporites of Poland (Peryt 1997), but geochemical arguments and radiometric data suggest that the Kalush evaporites can be Karpatian in age (see Wójtowicz et al. 2003, for discussion).

Mineralogical characteristics

Primary deposited minerals (Table 1) in the Miocene salt basins were halite, epsomite, sylvine, hexahydrite and possibly kainite. During deposition of the potash, the brines were at the sylvite stage of concentration as indicated by a limited occurrence of carnallite, chemical composition of inclusion fluids and the bromine content in halite (Kovalevich 1978; Petrichenko 1988). During early diagenesis, metastable minerals reacted among themselves and with brine and thus new minerals formed: stable (kainite) or metastable (tetra- or pentahydrate of Mg sulphate). There is a deficit of calcium during the potash stage and therefore the polyhalite can be formed only when the calcium is delivered by the (concentrated) seawater. Although the polyhalite crystallization is possible, as indicated by experimental data, it is more probable that the polyhalite was formed due to chemical reaction of the precipitated gypsum with potassium- and magnesium-bearing brines (Valyashko 1962). Under burial conditions, due to the temperature increase, dehydration and recrystallization, the formation of new minerals took place. The most important changes of composition, structure and texture of

Table 1. *Minerals discussed in this paper and their chemical composition*

Anhydrite	$CaSO_4$
Bloedite	$Na_2Mg(SO_4)_2 \cdot 4H_2O$
Gypsum	$CaSO_4 \cdot H_2O$
Halite	$NaCl$
Kainite	$KMg(SO_4)Cl \cdot 3H_2O$
Kieserite	$MgSO_4$
Langbeinite	$K_2Mg_2(SO_4)_3$
Leonite	$K_2Mg(SO_4)_2 \cdot 4H_2O$
Picromerite	$K_2Mg(SO_4)_2 \cdot 6H_2O$
Polyhalite	$K_2Ca_2Mg(SO_4)_4 \cdot 2H_2O$
Sylvite	KCl
Syngentite	$K_2Ca(SO_4)_2 \cdot H_2O$

evaporites of the Carpathian Foredeep are related to overthrusting of Carpathian nappes and the resulting regional overheating that resulted in the major recrystallization phase of the rocks (cf. Wójtowicz et al. 2003).

There also exists another possible model of deposition of evaporites in the Carpathian Foredeep, one that assumes a possibility of precipitation of kainite and langbeinite from brines due to the density- and temperature-stratification of the water (the bottom brines had temperature >45 °C according to Petrichenko 1988, pp. 103–106).

During the uplift of this area, when evaporites became subject to dissolution by surface waters, the clay–gypsum cap began to form. In its lower part, at the boundary with potash rocks, a discontinuous bed and lenses occur that are composed of weathering zone minerals (a schoenite cap, Lobanova 1956).

Materials (sampling)

For the sulphur isotope study we sampled the potash zone and the adjacent polyhalite–anhydrite rock (from mine and open-pit surfaces) occurring at the boundary between the potash zone and salt breccia (seven samples from the Stebnyk deposit and 13 samples from the Kalush–Holyn deposit). In addition, two samples of anhydrite from the Kalush–Holyn deposit were taken (basal anhydrite and anhydrite from the salt breccia with carnallite) and seven samples of minerals from the lower part of the cap rock of the Dombrovo quarry. In total 29 samples of 10 minerals have been analysed: anhydrite, four minerals of the potash zone and five minerals from the cap rock.

Anhydrite forms a basal bed (0.8 m thick in the sampled area) of the Kalush salt sequence (such a basal anhydrite is lacking in the Stebnyk deposit), and it is commonly present within the polyhalite–anhydrite beds (0.01–0.3 m thick) that occur at the contact of the potash zone and salt breccia: anhydrite adjoins this breccia, and polyhalite adjoins the potash zone. One sample of pure gypsum replacing polyhalite from the polyhalite–anhydrite bed was taken and also one sample of anhydrite from the salt breccia with carnallite; anhydrite forms nodules in the breccia. Polyhalite was taken from three varieties of polyhalite rocks: in clayey polyhalite rock from the potash zone, where it forms beds rich in clay (25%) that are typically 5–10 cm thick and grey in colour; in the red variety of pure polyhalite (with beds 1–3 cm thick), in complexes up to 2 m thick, which are characteristic for places where the potash zone disappears laterally; and in polyhalite in the polyhalite–anhydrite beds. In four cases both anhydrite and polyhalite were sampled from the same bed (samples 84 and 95, 89 and 94, 85 and 97, and 88 and 92). Kainite rock is the most common potash rock in the Ukrainian part of the Carpathian Foredeep. It is bedded, consisting of alternation of pure kainite (yellow in colour) and clayey kainite (yellow-grey in colour); the samples were taken from the pure variety of kainite rock. Langbeinite occurs in the form of langbeinite rock that contains small admixtures of halite, polyhalite and clay material (Lobanova 1956). Kieserite was taken from the kainite–langbeinite rock that also contains sylvite and water insoluble residue; we selected pure kieserite.

From the middle part of the cap rock of the Dombrovo quarry we selected samples of monomineralic crystals of gypsum and syngenite, and from the lower part of the cap rock leonite, picromerite and bloedite.

Methods

Each sample was powdered in the agate mortar, then c. 100 mg of the powder was dissolved in distilled hot water. After complete dissolution of sulphates, the solution was filtered and $BaSO_4$ was precipitated by adding $BaCl_2$ solution acidified with HCl. The precipitate was washed by distilled water until the chlorides had been removed, and dried. $BaSO_4$ powder was additionally purified by heating in an electric oven at a temperature of about 500 °C. Pure barium sulphate was then treated with chemical procedure in a vacuum line to obtain SO_2 gas for isotope analysis. We employed this method where SO_2 was extracted by quantitative reduction of $BaSO_4$ using $NaPO_3$ (Halas & Wolacewicz, 1981; Halas & Szaran, 2001). The sulphur isotopic composition was analysed on a dual inlet and triple collector mass spectrometer (reconstructed MI 1305) in the Institute of Physics of the Maria Curie–Sklodowska University in Lublin. Precision of the spectrometer measurements was 0.05–0.08‰ and reproducibility of the chemical procedure about 0.15‰. Obtained results as a conventional delta notation were normalized to the CDT by analysis of the NBS-127 standard.

Results

Results of our analysis are shown in Table 2. The range of $\delta^{34}S$ values for anhydrite, polyhalite and gypsum from the Stebnyk deposits is relatively small (15.28–16.73‰), and both the minimum and maximum values characterize the polyhalite beds within the potash zone. When one polyhalite–anhydrite bed was examined, anhydrite showed a lower $\delta^{34}S$ value (15.75‰) than polyhalite (16.60‰); in turn, when gypsum replaces polyhalite, gypsum shows lower a $\delta^{34}S$ value (15.99‰) than anhydrite (16.11‰).

In the Kalush–Holyn deposit, basal anhydrite shows the highest $\delta^{34}S$ value recorded in the studied sample set (21.00‰). Polyhalite shows a small range of $\delta^{34}S$ values (17.18–17.54‰) and in two pairs of samples that were taken from the same anhydrite–polyhalite bed the $\delta^{34}S$ values for polyhalite are greater than for anhydrite (by 1.59 and 1.60‰, respectively; see Table 2). The $\delta^{34}S$ values for kainite, langbeinite and kieserite as well as red polyhalite are within the range 15.93–17.38‰ (average 16.53‰).

The range of $\delta^{34}S$ values for minerals from the cap rock is 14.73–18.22‰ (average 16.44‰); gypsum and syngenite that were taken from the middle part of the cap rock show smaller $\delta^{34}S$ values (14.73–15.96‰) compared with minerals from the lower part of the cap rock (the $\delta^{34}S$ values of which are 16.57–18.22‰).

Discussion and interpretation

The comparison of isotopic composition of sulphate minerals that originated during various phases of formation of the potash deposits in the Carpathian Foredeep basin indicates that the processes of transformation and recrystallization did not lead to a significant isotopic fractionation of sulphur.

Discussion

The $\delta^{34}S$ values of basal anhydrite indicate the marine provenance of parent water (cf. Paytan et al. 1998). The polyhalite–anhydrite beds occurring at the transition zone between the potash zone and the salt breccia zone were formed during diagenesis. The study of pore waters in the Stebnyk deposit indicated that the brines in the potash zone are buried sulphate brines (that were characteristic for the phase of potash deposition) and the brines in breccia zone are Ca chloride brines (Valyashko et al. 1974). The latter brines are modified due to interaction of buried sulphate brines with clay minerals that supplied Ca ions to the brines, and this led to the brine losing the sulphate ion (Valyashko 1962). The polyhalite–anhydrite beds were formed at the geochemical barrier between those two types of brines, and both anhydrite and polyhalite are newly formed minerals (de novo). Of those two minerals, polyhalite was the first to crystallize and it consumed available potassium and magnesium ions, so that further crystallization of the brine led to precipitation of anhydrite. Accordingly, the $\delta^{34}S$ values recorded in anhydrite from polyhalite–anhydrite beds are lower than they are in polyhalite due to the effect of fractionation during crystallization.

The $\delta^{34}S$ values for the minerals from the cap rock vary within a slightly wider range. However, gypsum and syngenite that are related to the middle part of the cap rock show lower values than those characterizing the potash zone. In turn, minerals from the lower part of the cap rock show slightly higher values (17.25‰ on average) compared with the potash zone. The differentiation can be related to the inflow of surface water and the crystallization of new minerals in more open conditions.

Interpretation

These potash deposits are regarded as a lowstand deposit (Peryt & Kovalevich 1997). The bromine content in halite is very close to that characteristic of marine salt (Bilonizhka 1970, 1975; Kovalevich 1978) and the composition of brines inferred from study of fluid inclusions also suggests a dominantly marine origin for these brines (Kowalewicz 1994). The envisaged scenario presented by Peryt & Kovalevich (1997) assumes that the Lower Miocene salts originated by evaporation of mostly marine parent waters that were syndepositionally recycled. During periods of reduced supply of detrital material by inflow, brines in the basin could reach sylvite-saturation level and potash chlorides and sulphates were deposited at that time. Inflow of new brines concentrated in Na and Cl served as a source for halite beds within the potash complex. These beds were formed by mixing of the new brines with bottom brines concentrated to K and Mg salt precipitation. Then the evaporation resulted in decline of the upper level and increase of the bottom brine density, which in turn led to potash precipitation. The multiple complex of such processes led to the origin of the potash complex composed of interbedded potash

Table 2. $\delta^{34}S$ values of sulphate minerals from the Kalush–Holyn and Stebnyk potash deposits of the Carpathian Foredeep (asterisks in the 'Remarks' column indicate that samples have been taken from the same polyhalite–anhydrite bed)

Sample	Sample description	Locality	$\delta^{34}S(‰)$	Remarks
Stebnyk Deposit				
70	Clayey polyhalite rock	Stebnyk mine	16.73	
77	Red polyhalite rock	Stebnyk mine	15.28	
83	Polyhalite from polyhalite–anhydrite beds	Stebnyk mine	16.48	
84	Polyhalite from polyhalite–anhydrite beds	Stebnyk mine	16.60	*
89	Gypsum replacing polyhalite from polyhalite–anhydrite beds	Stebnyk mine	15.99	**
94	Anhydrite from polyhalite–anhydrite beds	Stebnyk mine	16.11	**
95	Anhydrite from polyhalite–anhydrite beds	Stebnyk mine	15.75	*
Kalush–Holyn Deposit				
74	Red polyhalite rock	Holyn mine	17.18	
85	Polyhalite from polyhalite–anhydrite beds	Dombrovo quarry	17.54	***
88	Polyhalite from polyhalite–anhydrite beds	Sivka–Kalushska mine	17.36	****
92	Anhydrite from polyhalite–anhydrite beds	Sivka–Kalushska mine	15.77	****
93	Anhydrite from polyhalite–anhydrite beds	Holyn mine	16.95	
97	Anhydrite from polyhalite–anhydrite beds	Dombrovo quarry	15.94	***
101	Anhydrite with carnallite from salt breccia	Holyn mine	17.88	
102	Basal anhydrite	Sivka–Kalushska mine	21.00	
111	Kainite	Sivka–Kalushska mine	15.93	
112	Kainite	Sivka–Kalushska mine	16.18	
113	Kainite	Holyn mine	17.38	
118	Langbeinite	Dombrovo quarry	15.95	
121	Langbeinite	Dombrovo quarry	16.27	
141	Langbeinite	Dombrovo quarry	17.17	
122	Kieserite	Dombrovo quarry	16.15	
123	Leonite	Dombrovo quarry	16.57	
124	Picromerite	Dombrovo quarry	18.22	
128	Bloedite	Dombrovo quarry	16.77	
129	Bloedite	Dombrovo quarry	17.43	
134	Syngenite	Dombrovo quarry	14.73	
136	Gypsum	Dombrovo quarry	15.41	
137	Gypsum	Dombrovo quarry	15.96	

and halite beds contaminated with terrigenous material (Peryt & Kovalevich 1997).

It seems that 4–5‰ difference between the isotopic composition of sulphur in the K–Mg minerals and the basal anhydrite is too large for the simple evolution of brine composition during the normal evaporation of the Carpathian Foredeep Basin according to a Rayleigh-type effect. A more probable process of isotopic fractionation of sulphur that resulted in such a difference could be bacterial reduction of sulphates in brines, which is the most effective process of isotope fractionation. During deposition of potash minerals the brines were stratified and sulphate-reducing bacteria probably were living at the boundary of brines ('bacterial plate', see Sonnenfeld 1984, p. 105). Isotopically light H_2S produced by bacteria may have been sequestered in the upper bed of brine where it was oxidized to sulphates. Part of the light H_2S probably escaped. This led to the heavier isotopic composition of the lower bed of brine; part of that brine underwent outflow/leakage. When the complete evaporation of the upper brine layer occurred, the stratified system disappeared and only homogenized brine body existed, the precipitation of potash minerals occurred.

This precipitation of potash took place from the brine body that was isotopically lighter because of the uptake of lighter sulphur isotopes that previously dominated the upper brine layer. The process of brine depletion in $\delta^{34}S$ due to bacterial reduction was probably superimposed on fractionation related to crystallization of K–Mg sulphates. It is difficult to estimate the volume of the isotope fractionation related to the more general crystallization of sulphate minerals in the Carpathian Foredeep basin due to their origin in an open system in a dynamic environment with multiple inflows of seawater and the probable outflow of concentrated brines. Raab and Spiro (1991) concluded that, during the precipitation of primary minerals of the Mg– and K–Mg sulphate facies, the fractionation factor for the minerals was different than for the minerals precipitated for the gypsum facies. However, their model is based on an assumption of a finite quantity of sulphate ions. In the case of the Miocene Carpathian Foredeep basin, due to the continual inflow of seawater, the volume of incoming sulphate ions was infinite.

Another mechanism that potentially influenced the recorded ^{34}S depletion in potash salts was the delivery of isotopically lighter sulphates from the adjacent land areas during the deposition of the salts in the Stebnyk basin. More light sulphur isotopic composition recorded in the Stebnyk deposit can be explained by more intensive inflow of surface waters from the Carpathian nappes and/or by the oxidation of a part of pyrite contained in bituminous clays of the Menilite suite that occur in olistoliths of the Vorotyshcha suite (that contain a considerable pyrite content, usually 2.0–4.5%; Lazarenko et al. 1962).

Conclusions

Isotopic composition of sulphur in the K–Mg sulphate evaporites occurring in the Lower and (?)Middle Miocene of the Carpathian Foredeep basin shows the overprint of isotopic fractionation processes related to precipitation and sulphate reduction resulting in the minerals of potash deposits that show isotopic depletion in ^{34}S by 3.5–5.7‰ and the minerals of the weathering zone by 2.8–6.2‰. The inflow of surface waters from the Carpathians and olistoliths of bituminous shales (Menilite suite) supplied lighter sulphate, formed due to pyrite oxidation, into the Vorotyshcha evaporite basin, resulting in the depletion in ^{34}S by 0.3–1.9‰ in the polyhalites of the Stebnyk deposit compared with similar varieties of that mineral from the Kalush–Holyn deposit.

References

ANDREYEVA-GRIGOROVICH, A. S., KULCHYTSKY, Y. O. ET AL. 1997. Regional stratigraphic scheme of Neogene formations of the Central Paratethys in the Ukraine. *Geologica Carpathica*, **48**, 123–136.

ANDREYEVA-GRIGOROVICH, A. S., OSZCZYPKO, N., SAVITSKAYA, N. A., ŚLĄCZKA, A. & TROFIMOVICH, N. A. 2003. Correlation of Late Badenian salts of the Wieliczka, Bochnia and Kalush areas (Polish and Ukrainian Carpathian Foredeep). *Annales Societatis Geologorum Poloniae*, **73**, 67–89.

BILONIZHKA, P. M. 1970. The content of bromine in salt minerals of potash deposits of the Forecarpathians. *Voprosy Mineralogii Osadochnykh Obrazovaniy*, **8**, 125–133 [in Russian].

BILONIZHKA, P. M. 1975. Use of Br/Cl coefficient for the explanation of genesis of salt rocks (on example of Forecarpathians). *Geologia i Geokhimia Goryuchikh Iskopayemykh*, **45**, 55–62 [in Russian].

DZHINORIDZE, N. M. 1973. Tertiary potassium basins. *In*: RAYEVSKIY, V. I. & FIVEG, M. P. (eds) *Potash Salt Deposits of the USSR*. Nedra, Leningrad, 183–234 [in Russian].

DZHINORIDZE, N. M. 1980. Carpathian potassium-bearing region. *In*: DZHINORIDZE, N. M., GEMP, S. D., GORBOV, A. F. & RAYEVSKIY, V. I. (eds) *Regularities of Distribution and Exploration Criteria for Potassium Salts in the USSR*. Izdatelstvo 'Mecniereba', Tbilisi, 73–159 [in Russian].

GRINIV, S. P. 1985. On the problem of comparison of potassium salt deposits. *In*: KITYK, V. I. (ed.) *Evaporites of Ukraine*, Naukova Dumka, Kiev, 44–50 [in Russian].

GRINIW, S. P. 1994. Composition and lithostratigraphic correlation of salts of the Kalush–Holyn deposit

(Miocene, Ukrainian Fore-Carpathians). *Przegląd Geologiczny*, **42**, 748–750 [in Polish].

HALAS, S. & SZARAN, J. 2001. Improved thermal decomposition of sulfates to SO_2 and mass spectromeric determination of IAEA SO-5, IAEA SO-6 and NBS-127 sulfate standards. *Rapid Communications in Mass Spectrometry*, **15**, 1618–1620.

HALAS, S. & WOLACEWICZ, W. 1981. Direct extraction of sulfur dioxide from sulfates for isotopic analysis. *Analytical Chemistry*, **53**, 686–689.

HRYNIV, S. P., DOLISHNIY, B. V., KHMELEVSKA, O. V., POBEREZHSKYY, A. V. & VOVNYUK, S. V. 2007. Evaporites of Ukraine: a review. *In*: SCHREIBER, B. C., LUGLI, S. & BĄBEL, M. (eds) *Evaporites through Space and Time*. Geological Society of London.

KORENEVSKIY, S. M. & DONCHENKO, K. B. 1963. Geology and conditions of formation of potash deposits of the Soviet Forecarpathians. *Trudy VSEGEI*, **99**, 3–152 [in Russian].

KORIŃ, S. S. 1994. Geology of the Miocene salt-bearing formations of the Ukrainian Fore-Carpathians. *Przegląd Geologiczny*, **42**, 744–747 [in Polish].

KOVÁČ, M., NAGYMAROSY, A. *ET AL*. 1998. Palinspastic reconstruction of the Carpathian-Pannonian region during the Miocene. *In*: RAKÚS, M. (ed.) *Geodynamic Development of the Western Carpathians*. Dionýz Štúr, Bratislava, 189–217.

KOVALEVICH, V. M. 1978. *Physico-chemical Conditions of Salt Formation in the Stebnik Potassium Deposit*. Naukova Dumka, Kiev [in Russian].

KOWALEWICZ, W. M. 1994. Condition of origin of the Miocene salts of the Ukranian. *Przegląd Geologiczny*, **48**, 738–743 [in Polish].

KULCHITSKIY, A. YA. 1986. *Geology and Conditions of Formation of Salt-bearing Molasses of Neogene Carpathian Foredeep and Transcarpathian Depression*. Institute of Geology and Geochemistry of Compostible Minerals, Lvov [in Russian].

LAZARENKO, E. K., GABINET, M. P. & SLIVKO, O. P. 1962. *Mineralogy of Sedimentary Deposits of the Forecarpathians*. Vydavnitstvo Lvivskogo Universytetu, Lviv [in Ukrainian].

LOBANOVA, V. V. 1956. Questions of petrography of potassium deposits of the eastern Forecarpathians. *Trudy Vsesoyuznogo Instituta Galurgii*, **32**, 164–214 [in Russian].

OSZCZYPKO, N., KRZYWIEC, P., POPADYUK, I. & PERYT, T. 2006. Carpathian Foredeep Basin (Poland and Ukraine) – its sedimentary, structural and geodynamic evolution. *American Association of Petroleum Geologists Memoirs*, **84**, 293–350.

PAYTAN, A., KASTNER, M., CAMPBELL, D. & THIEMENS, M. H. 1998. Sulfur isotopic composition of Cenozoic seawater sulfate. *Science*, **282**, 1459–1462.

PERYT, D. 1997. Calcareous nannoplankton stratigraphy of the Middle Miocene in the Gliwice area (Upper Silesia, Poland). *Bulletin of the Polish Academy of Sciences, Earth Sciences*, **45**, 119–131.

PERYT, T. M. & KOVALEVICH, V. M. 1997. Association of redeposited salt breccias and potash evaporites in the Lower Miocene of Stebnyk (Carpathian Foredeep, West Ukraine). *Journal of Sedimentary Research*, **67**, 913–922.

PETRICHENKO, O. Yo. 1988. *Physico-chemical Conditions of Sedimentation in Ancient Salt-bearing Basins*. Naukova Dumka, Kiev [in Russian].

PETRYCZENKO, O. I., PANOW, G. M., PERYT, T. M., SREBRODOLSKI, B. I., POBEREŹSKI, A. W. & KOWALEWICZ, W. M. 1994. Outline of geology of the Miocene evaporite formations of the Ukrainian part of the Carpathian Foredeep. *Przegląd Geologiczny*, **42**, 734–737 [in Polish].

RAAB, M. & SPIRO, B. 1991. Sulfur isotopic variations during seawater evaporation with fractional crystallization. *Chemical Geology*, **86**, 323–333.

RÖGL, F. 1996. Stratigraphic correlation of the Paratethys Oligocene and Miocene. *Mitteilungen der Gesellschaft der Geologie- und Bergbaustudenten in Österreich*, **41**, 65–73.

SONNENFELD, P. 1984. *Brines and Evaporites*. Academic Press, Orlando, FL.

STRAUSS, H. 1997. The isotopic composition of sedimentary sulfur through time. *Palaeogeography, Palaeoclimatology, Palaeoecology*, **132**, 97–118.

STEININGER, F. F. & WESSELY, G. 2000. From the Tethyan Ocean to the Paratethys Sea: Oligocene to Neogene stratigraphy, paleogeography and paleobiogeography of the circum-Mediterranean region and the Oligocene to Neogene basin evolution in Austria. *Mitteilungen der Österreichischen Geologischen Gesellschaft*, **92**, 95–116.

VAKARCS, G., HARDENBOL, J., ABREU, V. S., VAIL, P. R., VÁRNAI, P. & TARI, G. 1998. Oligocene-Middle Miocene depositional sequences of the Central Paratethys and their correlation with regional stages. *Society of Economic Paleontologists and Mineralogists Special Publications*, **60**, 209–231.

VALYASHKO, M. G. 1962. *Geochemical Principles of Origin of Potassium Salt Deposits*. Izd. Moskovskogo Universiteta, Moskva [in Russian].

VALYASHKO, M. G., BOGASHOVA, L. G., BORISENKO, V. I., SADYKOV, L. Z. & VOLKOVA, N. N. 1974. Formation of chemical composition of pore solutions of salt-bearing clays of the Stebnik Deposit of potash salts. *In*: KITYK, V. I. (ed.) *Geology and Mineral Deposits of Saliferous Series*. Naukova Dumka, Kiev, 183–190 [in Russian].

VUL, M. Y., DENEGA, B. I., KRUPSKY, Y. Z., NIMETS, M. V., SVYRYDENKO, V. G. & FEDYSHYN, V. O. (eds) 1998. *Atlas of Oil and Gas Fields of Ukraine in Six Volumes*. Vol. IV: Western Oil-and-Gas-Bearing Region. Vydavnitsvo 'Tsentr Evropy', Lviv.

WÓJTOWICZ, A., HRYNIV, S. P., PERYT, T. M., BUBNIAK, A., BUBNIAK, I. & BILONIZHKA, P. M. 2003. K/Ar dating of the Miocene potash salts of the Carpathian Foredeep (West Ukraine): application to dating of tectonic events. *Geologica Carpathica*, **54**, 243–249.

Generation of primary sylvite: the fluid inclusion data from the Upper Permian (Zechstein) evaporites, SW Poland

S. V. VOVNYUK[1] & G. CZAPOWSKI[2]

[1]*Institute of Geology and Geochemistry of Combustible Minerals of National Academy of Sciences of Ukraine, Naukova str 3a, Lviv, Ukraine (e-mail: greatboludo@yandex.ru)*

[2]*Polish Geological Institute, Rakowiecka str. 4, Warsaw, Poland*

Abstract: The chemical composition of primary inclusions in sedimentary halite from the Polish Zechstein salt deposits (Upper Permian), which represent deposition in salt lagoon to salt pan-salina conditions of two evaporitic units of the Stassfurt and Leine cyclothems, contains evidence that the included brines belong to the Na–K–Mg–Cl–SO$_4$ (SO$_4$-rich) type. They differ from the brines typical for other Zechstein salts by having an elevated potassium content, which produced primary sylvite (daughter crystals in inclusions), caused by redeposition of locally generated potassium salts during inflow of fresh seawater into the sedimentary basin. The homogenization temperature of primary inclusions in halite with sylvite daughter crystals reflects the increased temperature (53–60 °C) of basin brines, caused by a greenhouse effect due to density stratification of the brines. Isotopic composition of the anhydrite (sulphur: 9.4–13.30‰ Canyon Diablo Troilite (CDT); oxygen: 9.44–10.4‰ Standard Middle Ocean Water (SMOW), as well as the bromine content in halite (49–77 ppm) indicated that these salts had a marine origin and were deposited at an elevated temperature, from concentrated seawater with a high potassium content, that favoured precipitation of primary sylvite.

Fluid inclusions in primary halite record the chemical composition of basin brines; however the primary origin of sylvite, present in these deposits, still remains controversial. Syndepositional origin of sylvite from most of ancient evaporitic basins is doubtful because of the lack of primary structures in sylvite crystals. The presence of such fluid inclusion structures (chevrons) in halite serves as evidence that the crystals have grown on the floor of the basin. These structures are known only in halite. Commonly crystals of sylvite contain high-pressure gas inclusions, the presence of which is unambiguous evidence for sylvite crystal growth (or recrystallization) during a postsedimentary (diagenetic) stage of the development of a salt deposit. The incongruent dissolution of carnallite has been suggested as the possible mechanism of sylvite formation (Borchert & Muir 1964; Braitsch 1971; Sonnenfeld 1984). On the other hand, Valyashko (1962), after a series of experiments in artificial solutions, showed that primary deposition of sylvite is possible in the case of a metastable pathway of evaporite crystallization. Lowenstein & Spencer (1990) reported apparent primary formation of sylvite in three deposits (Oligocene Rhine Graben, Alsace, France; Permian Salado Formation, New Mexico, USA; and the Devonian Prairie Formation, Saskatchewan, Canada). In Lowenstein's study, fluid inclusions in primary halite crystals contain daughter crystals of sylvite. Halite in these deposits crystallized from relatively hot brines, the cooling of which permits supersaturation with respect to KCl (Lowenstein & Spencer 1990).

Inclusions containing daughter crystals of sylvite within halite from the Polish Zechstein have been noted and studied. These inclusions form primary sedimentary structures. As a rule, these structures have the shape of chevrons or sometimes appear as cubic hopper crystals. Geochemical and fluid inclusion studies suggested a primary sylvite formation in the deposit. Dissolution of earlier formed potassium salts is the most probable origin of an elevated KCl content within the basin. Similar inclusions have also been noted in halite from the German Zechstein and the Devonian Starobinsk deposits of the Pripyat of Belarus (Petrychenko 1977) and the Permian Upper Kama Potash Deposit, Uralian Foredeep, Russia (unpublished author data). Taking into consideration the data from all these deposits, we can conclude that sylvite crystallization took place by cooling during deposition from brines, supersaturated with respect to KCl, and that this was the general mechanism of sylvite formation in most ancient potash-bearing deposits. During diagenesis many deposits of sylvite also may undergo significant recrystallization.

Geological setting

The Zechstein (Upper Permian) evaporites are present across more than half of Poland (Fig. 1),

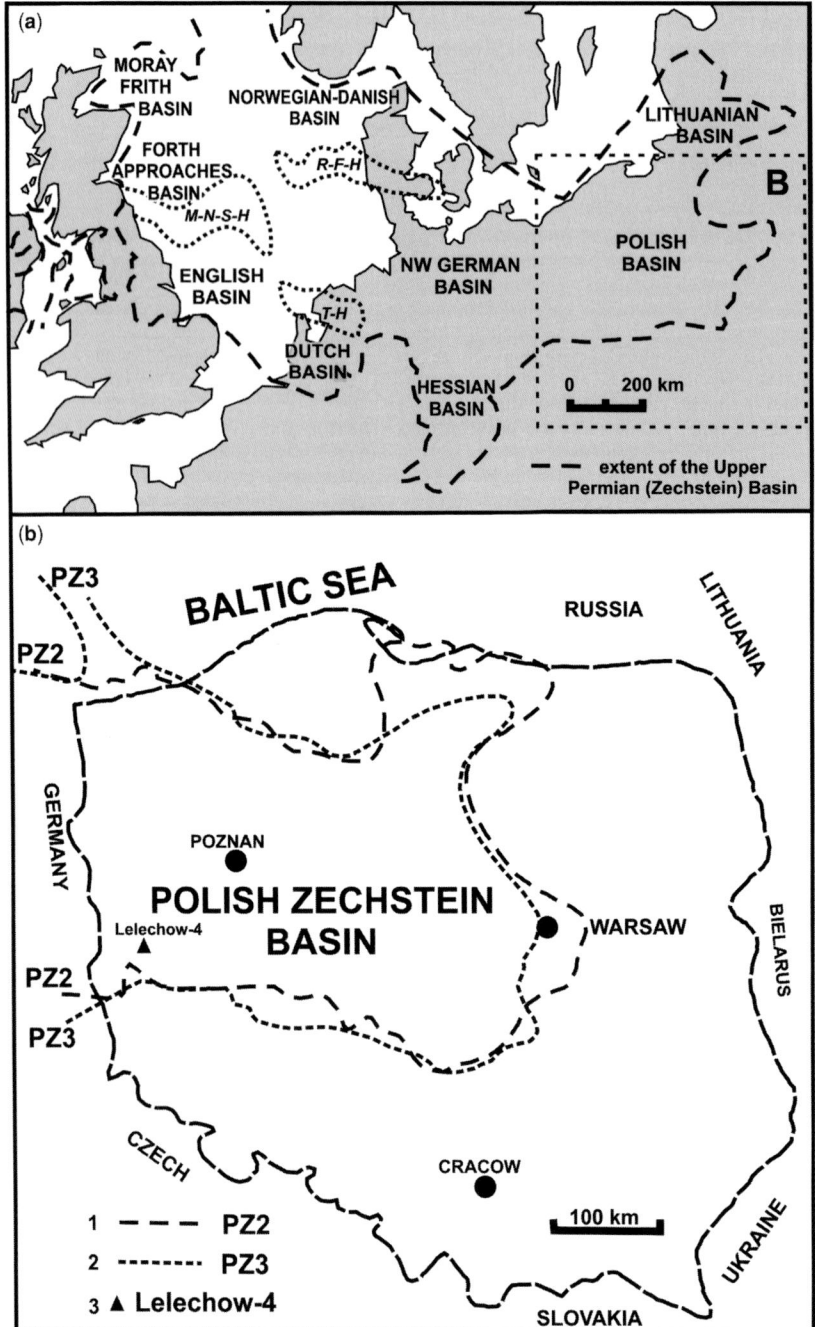

Fig. 1. Location of studied section with Zechstein (Upper Permian) evaporite units in Poland (**b**), belonging to the Zechstein Basin of Europe (**a**; basin contour after Smith 1980). 1, Present limit of PZ2 (with Older Halite unit) deposits; 2, present extent of PZ2 (with Younger Halite unit) deposits; 3, studied borehole Lelechow-4; M-N-S-H, Mid North Sea High; R-F-H, Ringkøbing-Fyn High; T-H, Texel High.

within the Polish sub-basin portion of the giant Zechstein Basin that stretched across much of Europe (Kiersnowski et al. 1995). It is made up of four major stratigraphic units, composed of sulphates, halite and potassium salts, separated by carbonates and siliciclastics, which together comprise four cyclothems: PZ1–PZ4 (Wagner et al. 1981; Wagner 1987), and are equivalent to the German stratigraphic cycles Z1 (Werra) to Z7 (Möln: Wagner & Peryt 1997).

Salts of the PZ2 cyclothem are divided into two rock salt members: the Older Halite (Na2; Czapowski et al. 1991; Czapowski 1995), and the locally developed Screening Older Halite (Na2r; Wagner 1994), separated by the potash salt member – Older Potash (K2; Wagner et al. 1981). The Older Halite member was initially deposited in relatively deep water conditions in the basin centre and in more shallow lagoons and salt pans of its margin, but that environmental pattern was subjected to successive shallowing phases and eventually separated into numerous salt pans and salinas during accumulation of the Older Potash and Screening Older Halite units (Czapowski 1991; Czapowski et al. 1991).

The rock salt member of PZ3 cyclothem is called the Younger Halite (Na3), and is subdivided by the potash salt member – Younger Potash (K3; Wagner 1994; Wagner et al. 1981) – into two units: the Lower Younger Halite (Na2a) and the Upper Younger Halite (Na3b), with thickness ranging from a few up to 240 m (Czapowski 1993). Locally the uppermost part of Upper Younger Halite contains dispersed clay matter and it is distinguished as a separate unit, the Younger Clayey Halite (Na3t). Salts were deposited initially in the relatively shallow open basin, framed by lagoons and salt pans along its margins, but due to a successive brine level drop (and/or sediment infill) salinas, salt pans and shallow lagoons predominated during accumulation of potash and upper rock salt members (Czapowski 1991, 1993). Studied halite samples of both Na2 and Na3 rock salt members were taken from the borehole Lelechow-4, located in the SW margin of the Polish Zechstein Basin (Fig. 1).

The Older Halite (Na2) member is sharply underlain by the laminated to nodular anhydrite of the Basal Anhydrite (A2) member and overlain by nodular anhydrite of the Screening Anhydrite member (A2r) topped with the claystone bed of the Grey Pelite (T3). Rock salt is semitransparent to opaque, red, pink and orange in the lower part but grey and orange in the upper portion, it contains dispersed clay matter and fine aggregates of anhydrite and carnallite. The halite is heteromorphic, with crystal size from 3 to 20 mm, cloudy (having partial, inclusion-rich zonation), and hopper crystals are relatively frequent and locally occur in clusters as 3 cm 'nests' of secondary giant halite as well as displaying solution/erosion scours lined with sulphates and clay matter. In the whole member there are sections observed as horizontal to gently undulating fine anhydrite laminae, present in groupings from 1 to 10 cm in thickness. This salt unit apparently was deposited in the shallow salt lagoon and pan environments as defined by the salt facies criteria in Table 1 (Czapowski 1995). The bromine content in equivalent salts of pan facies from nearby wells varies from 119 ppm to 160 ppm (Kovalevych et al. 2000).

The Older Potash (K2) member, absent in the studied well, is present in many boreholes located just to the north of this study area and consists of halite + anhydrite + polyhalite and halite + sylvite interbeds decimetres to metres thick (Podemski 1972).

The next salt bed in the borehole Lelechow-4 is the lower unit of the Younger Halite (Na3) member, called the Lower Younger Halite (Na3a) unit (22.5 m thick). It is underlain with a 15 m thick Main Anhydrite (A3) member and is overlain by Younger Potash (K3) member, composed of halite + anhydrite + polyhalite and sylvite + kieserite + polyhalite beds (Podemski 1972). These apparently originated in the shallow lagoon to salina environments. The bottom part of the Na3a unit is made up of beige to pink, semitransparent, almost monomorphic halite, with an average crystal size 2–3 mm and common fine parallel laminae of anhydrite. This part of Younger Halite member was deposited in the shallow open salt basin but features of the upper portion one include heteromorphic and secondary giant halites, cloudy crystals indicative of accumulation in a progressively shallowing lagoonal conditions. Bromine content in equivalent salts of shallow salt lagoons, analysed in nearby wells, varies from 45 to 60 ppm, suggesting frequent dissolution/erosion events in less saline waters (Kovalevych et al. 2000). Also the shallow lagoon to salt pan conditions, with erosion of Older Potash sediments at the beginning, is evidenced by inclusion composition data in a nearby well (Kovalevych et al. 2000), characterizing deposition of the Upper Younger Halite (Na3b) member, topping the evaporitic succession of the PZ3 cyclothem. Discussed units are presented in the profile of rocks of the Lower Red Pelite (T4a) and Youngest Halite (Na4a) members (Fig. 2).

Fluid inclusions

Primary sedimentary structures were revealed in halite from six samples (Fig. 2). The fluid inclusions that form these structures are of negative crystal

Table 1. *Features of fossil salt facies from the Upper Permian (Zechstein) rock salt deposits in Poland (after Czapowski 1995)*

Facies	Lithology	Structures	Br content (ppm)
Deep salt basin (I)	Common primary banded halite (C) and salt rhythmites; dominant salt sequences: A + C, A + B + C; sulphates as laminae; sporadic K–Mg salt	Common 'internal lamination' in primary banded halite (LC) and fine sulphate lamination, rare hopper, raft halite and halite intraclasts	30–250, sporadically >300
Shallow salt basin (II)	Rare C-halite, sporadic secondary giant halite (D) and salt rhythmites; dominant salt sequences: A + B, A + C, B + C; sulphates as laminae and nodules; sporadic terrigenous material (T)	Rare LC and fine sulphate lamination; common hopper halite, halite intraclasts and erosion boundaries; rare raft and chevron halite	40–200
Deep salt lagoon (IIIA)	Frequent C-halite, rare salt rhythmites; dominant salt sequences: B + C, A + B; sulphates as laminae; locally K–Mg salt; sporadic T	Frequent LC and fine sulphate lamination; common hopper and raft halite; rare halite intraclasts, chevron halite and erosion boundaries	20–330
Shallow salt lagoon (IIIB)	Rare C-halite, D-halite and salt rhythmites; dominant salt sequences: B + A, B + C; sulphates as laminae and nodules; common T	Rare LC, fine sulphate lamination, raft and hopper halite; frequent halite intraclasts, chevron halite and erosion boundaries	40–100
Salt pan (IV)	Absent C-halite and salt rhythmites, common D-halite; dominant salt sequences: A + B, B + D; sulphates as laminae and nodules; common T; sporadic K–Mg salt	Structures as in facies IIIB, absent LC	10–80 (sporadic 260)
Salina (IVs)	Lithology as in IIIB and IV facies, common K–Mg salt	Structures as in facies IV	>300

A, Monomorphic halite; B, polymorphic halite.

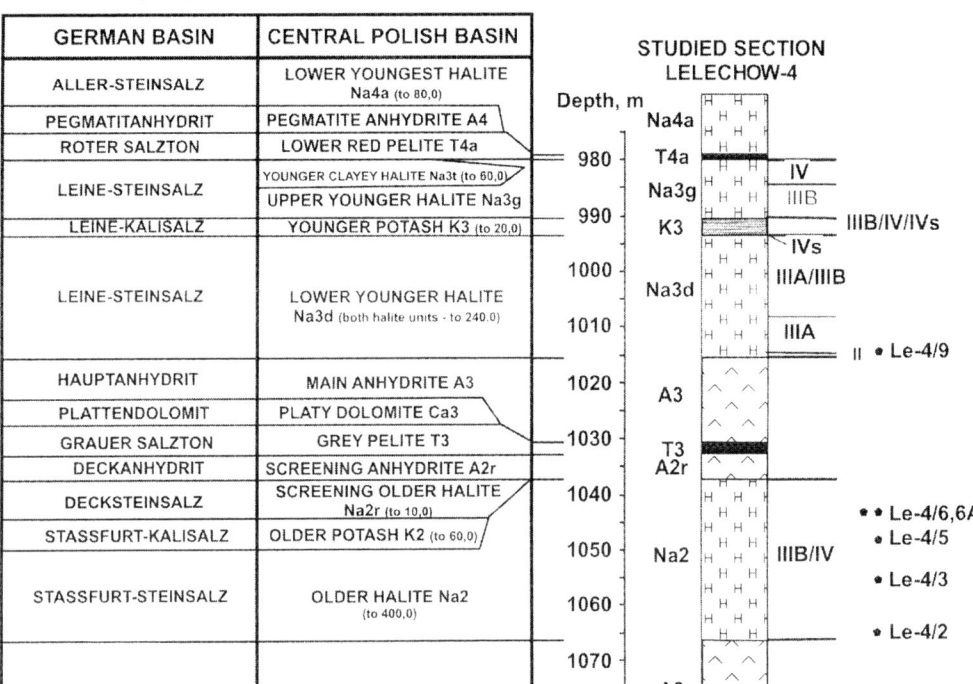

Fig. 2. Simplified profile of studied evaporite units in the Lelechow-4 borehole, shown together with the corresponding part of general stratigraphic succession of the German and the Central Polish Zechstein basins (stratigraphic units after Wagner 1987, 1994). 1, Claystone; 2, anhydrite; 3, rock salt (halite); 4, K–Mg salt (potash); 5, samples in which primary fluid inclusions with daughter crystals of sylvite have been revealed; salt facies: II, open shallow basin; IIIA, deep lagoon; IIIB, shallow lagoon; IV, salt pan; IVs, salina; Na2, A2, lithostratigraphic units (see text); (22,5), unit thickness (in metres).

(cubic) shape and contain daughter crystals of sylvite (Fig. 3) which make up 2–2.5% of inclusion volume.

Sedimentary structures (chevrons) are present in samples Le-4/2, 3, 5, 6, 6A outlined by fluid inclusions with size from a few micrometres to 80 μm (in Le-4/2 sample to 150 μm). In the Le-4/2 sample were also revealed rare cubic hopper crystals and solid inclusions of sylvite in halite. Chevron structures in halite from sample Le-4/9 are formed by rather small (up to 20 μm) two-phase fluid inclusions.

Methods

The chemical composition of brines from individual fluid inclusions was determined by means of an ultramicrochemical (UMCA) method (Petrychenko 1973). This method allows determination of the content of K^+, Mg^{2+}, Ca^{2+} and SO_4^{2-} ions in brines. The analytical error is 15–23% for K and Mg and 37–43% for Ca and SO_4. In order to improve quality of our results we ran a few analyses of each ion. The minimal inclusion size necessary for analyses is 40 μm. Na and Cl ion content within the limits of the method precision could be worked out from the tables (McCaffrey et al. 1987).

The data for potassium content obtained by the UMCA method do not include the amount of this ion in daughter crystals, which were precipitated from the inclusion brine and thus should be included in the total. The quantity of potassium in daughter crystal of sylvite was estimated by comparison of optically determined volumes of

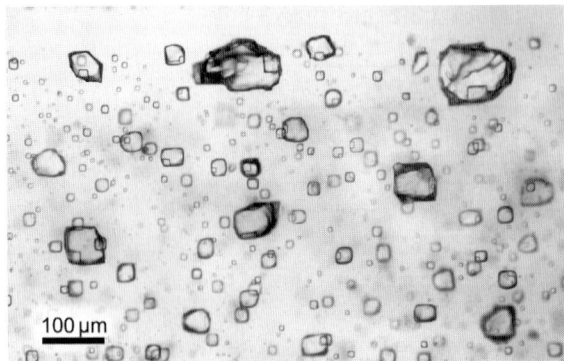

Fig. 3. Fragment of chevron structure in halite, outlined by fluid inclusions with daughter crystals of sylvite. Sample Le-4/2.

inclusion and crystal. This was possible because both negative crystal-shaped inclusions in halite and sylvite daughter crystals are close to cubic shape (Fig. 3). Under the microscope we measured two out of three edges of the cube; the third was taken as an average between the measured ones. Total potassium content was calculated by summing up amounts of K from inclusion brine and sylvite daughter crystals.

The homogenization temperatures (the end of sylvite daughter crystal dissolution in a heated halite plate) were measured under the microscope in cleavage fragments of chevron halite. A modified heat-chamber designed by Kalyuzhny has been used in this study (Kalyuzhny 1982). This camera allows measurement of the temperature with an accuracy of ± 1 °C. The heating velocity utilized in this study was 0.2 °C/min. In each sample seven to 20 inclusions were investigated.

Br content was determined by means of X-ray fluorescent spectrometry. Halite samples were pretreated in order to eliminate Br-enriched fluid inclusion brines. Artificial standards were made by adding solid NaBr to halite samples in which Br content was below the sensitivity of measurement using this method. Measurements were made using a Phillips PW2400 spectrometer. Analytical error was 3%.

The isotopic composition of sulphate sulphur and oxygen was studied in the laboratory of Barcelona University. The samples were dissolved in boiling water and then filtered to remove insoluble residue (mainly silicates and carbonates). Dissolved SO_4^{2-} was precipitated with excess $BaCl_2$ at pH values 1–2, which prevented simultaneous $BaCO_3$ precipitation. $BaSO_4$ was then filtered and dried at 120 °C. $\delta^{18}O$ values were determined from CO, which was obtained by heating the admixture of $BaSO_4$ and graphite. $\delta^{34}S$ values were determined from SO_2, which was obtained by heating the admixture of $BaSO_4$ and V_2O_5. The analytical error was 0.3% for $\delta^{34}S$ and 0.4% for $\delta^{18}O$.

Results

The major ion content is given in Table 2. The potassium content determined with UMCA method varies from 36 to 44 g l^{-1} but it was worked out that the total content of this ion (at an increased homogenization temperature) in the included brine was 20–25 g l^{-1} higher and thus reached 56–70 g l^{-1}.

Homogenization temperatures of the studied fluid inclusions are between 50 and 135 °C (Table 3). Most of the data fall within the range 53–60 °C; in individual plates temperatures higher than 80 °C were also registered. Br content values fell within range 49–77 ppm (Table 4).

Table 2. *Chemical composition of inclusion brines and Janecke units*

Sample	Chemical composition, g l^{-1} (average of 2–4 analyses)			Janecke units		
	K$^+$	Mg^{2+}	SO$_4^{2-}$	2K	Mg	SO$_4$
Le-4/2	44	72	55	14	72	14
Le-4/5	45.2	42.5	12.4	24	71	5
Le-4/6A	35.7	47.15	3	19	80	1
mSW	3.9	12.6	17.6			
mSW*	27.1	85.9	115			

mSW, modern seawater concentrated to the beginning of halite precipitation (McCaffrey et al. 1987); mSW*, modern seawater concentrated to the beginning of sylvite precipitation (McCaffrey et al. 1987).

Table 3. Homogenization temperatures of inclusions in chevron halite

Sample	T (°C)
Le-4/2	59; 59; 59; 59; 60; 60; 60; 60; 60; 60; 61; 61; 61; 61 (60)
Le-4/3	plate 1: 57; 58; 60; 60; 60; 60; 61 (59) 82; 84; 85; 85; 85; 90; 92; 104; 106; 135
	plate 2: 51; 52; 52; 53; 53; 54; 55; 56; 57; 57; 57; (54) 89; 96; 98; 105
Le-4/5	50; 53; 54; 54; 55; (53) 82; 94; 109
Le-4/6	51; 52; 52; 52; 52; 53; 54; 54; 54; 54 (53)
Le-4/9	56; 56; 57; 57; 58; 58; 58; 58; 59; 59 (58)

In brackets is shown the average of results considered as corresponding to basin brine temperatures (see text).

The results of a study of oxygen isotopic composition were between 9.44 and 10.41‰ SMOW; that of sulphate sulphur was between 9.42 and 13.30‰ CDT (Table 5).

Table 4. Bromine content in halite

Sample	Br (ppm)
Le-4/2	77
Le-4/10 A	79
Le-4/15	49

Table 5. Isotopic composition of sulphate sulphur and oxygen in studied salts

Sample	$\delta^{34}S$ (‰ CDT)	$\delta^{18}O$ (‰ SMOW)
Le-4/2	9.44	9.42
Le-4/10a	10.41	13.30
Le-4/15	9.52	12.70

Interpretation and discussion

Zechstein salts are undoubtedly of marine origin (Braitsch 1971). This notion is confirmed by numerous fluid inclusion studies (Kovalevych et al. 2000; Vovnyuk et al. 2004). The results of our study of chemical composition of brines from individual primary inclusions in halite from borehole Lelechow-4 (Table 2) also confirm the marine origin of studied salts and are in good agreement with the existing notion about the Na–K–Mg–Cl–SO$_4$ (SO$_4$-rich) brine type of Zechstein evaporite basin (Kovalevych et al. 2000; Vovnyuk et al. 2004). The SO$_4$/Mg ratio of the studied inclusion brines is lower than that of concentrated modern seawater (McCaffrey et al. 1987) and is typical for Late Permian seawater (Kovalevych et al. 1998, 2002). Furthermore, in studied brines an anomalously high potassium content was detected. The potassium content calculated from UMCA analyses is 36–44 g l^{-1}. Taking into account the additional K from daughter crystals we can estimate that real potassium content in inclusion brines before sylvite precipitation was approximately 20–25 g l^{-1} higher and reached 56–70 g l^{-1}. This is more than twice the maximum possible concentration of this ion at saturated modern seawater (McCaffrey et al. 1987). Such high K content could be reached due to redeposition of earlier formed sylvinite deposits and increased temperature of basin brines.

Most of the homogenization temperatures (the end of sylvite daughter crystal dissolution in a slowly heated plate of chevron halite) fall within narrow range of 50–62 °C. These data reflect the temperature of salt precipitation (Fig. 4). Such temperatures could be reached due to a basinal greenhouse effect within a density-stratified brine column. Thus, the lower, bottom layer of basin brines was heated to this temperature during crystal growth. The brine stratification is confirmed by the occurrence of cubic hopper crystals. Such crystals precipitate when brines of different saturation are mixed, which apparently happens in stratified basin brines (Raup 1970). The individual inclusions show significantly higher homogenization temperatures (up to 135 °C). These results do not reflect the actual temperature in the sedimentary basin; extremely high temperature values and a wide range of data allow us to interpret them as the result of sporadic trapping of sylvite crystals by growing halite.

Salts in studied area precipitated from brines with potassium concentration about 70 g l^{-1} and with temperature about 50–60 °C. Cooling of these brines made it possible to reach supersaturation with respect to K, which favoured primary sylvite precipitation. The points of brine composition of all three studied samples on a Janecke diagram for a six-component system, Na–K–Mg–Ca–SO$_4$–Cl–H$_2$O at 25 °C (Eugster et al. 1980), are within the sylvite stability field, which could be considered as reasonable evidence for sylvite precipitation (Fig. 5). This mechanism was described by Lowenstein (Lowenstein & Spencer 1990) as major for primary formation of sylvite in three ancient basins (Oligocene Rhine Graben, Alsace, France; Permian Salado Formation, New

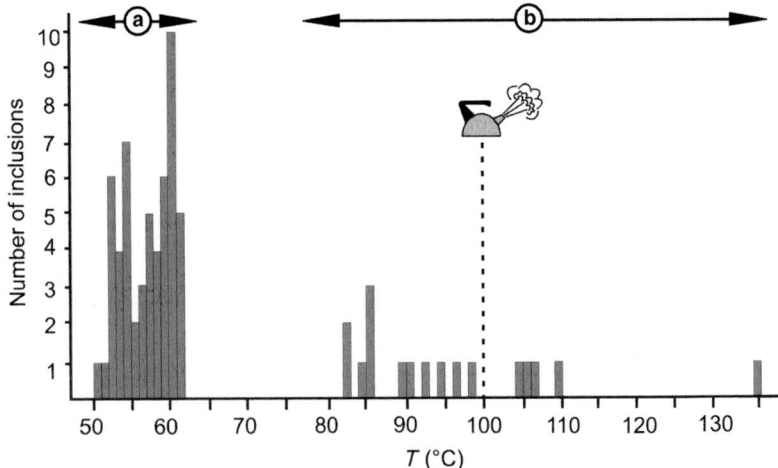

Fig. 4. Histogram of homogenization temperature data (as shown in Table 3). (**a**) Data that reflect the real temperature of basin brines during precipitation of studied halite; (**b**) data considered by us as a result of sporadic trapping of sylvite crystals by growing halite.

Mexico, USA; and the Devonian Prairie Formation, Saskatchewan, Canada). Halite in these deposits crystallized from relatively hot brines (at temperatures 39–71 °C, the cooling of which favoured primary sylvite precipitation. Primary halite from these formations also contains fluid inclusions with daughter crystals of sylvite. The presence of such inclusions in halite are considered as evidence that conditions for sylvite precipitation from basin brines existed in the basin. Such inclusions are also described in halite from other two deposits (German Zechstein and Prypiat Trough; Devonian Starobinsk Deposit, Pripyat, Belarus; Petrychenko 1977). Recently inclusions with daughter crystals of sylvite have also been discovered in primary halite from Lower Permian salts of Uralian Foredeep (Upper Kama Potash Deposit, Russia; unpublished author data). Taking into consideration the data from seven deposits of different ages we can conclude that sylvite crystallization from brines, supersaturated with respect to KCl by cooling, was the general mechanism of sylvite formation in most of ancient potash-bearing deposits. During diagenesis of the deposit formation sylvite underwent significant recrystallization.

The bromine content in these studied samples (Table 3) is typical for salts of marine origin. However, taking into account the high stage of brine concentration, the Br content could exceed 160 ppm (Bilonizhka 1975) or even reach 270 ppm (Valyashko 1962), so we can consider the values of this component, obtained in this study, as quite low. To our opinion this evidences dissolution and redeposition of earlier formed salts. The dissolution had taken place after inflows of fresh portions of less concentrated seawater into this sedimentary basin. The role of non-marine waters in this process was apparently insignificant, which is confirmed not only by Br content but also by sulphate sulphur and oxygen isotopic composition (Table 5), which is typical for Permian marine sulphate (Claypool et al. 1980; Strauss 1997).

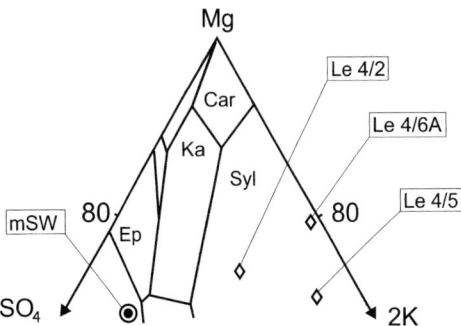

Fig. 5. Chemical composition of individual inclusion brines (as shown in Table 2), plotted on Janecke diagram (Eugster et al. 1980) for a six-component (Na–K–Mg–SO$_4$–Cl–H$_2$O) system at 25 °C. Stability fields: Car, carnallite; Ka, kainite; Ep, epsomite; Syl, sylvite. mSW, point of composition of modern seawater, concentrated to halite precipitation (McCaffrey et al. 1980).

Conclusions

The chemical composition of primary inclusions in sedimentary halite from the Polish Zechstein belongs to the Na–K–Mg–Cl–SO$_4$ (SO$_4$-rich) type and differs from the composition typical for other Zechstein salts by an elevated potassium content. The high K content in basin brines was caused by redeposition of potassium salts during inflow of fresh seawater into the sedimentary basin. The homogenization temperature of primary inclusions in halite with sylvite daughter crystals reflects the increased temperature (53–60 °C) of basin brines, caused by a greenhouse effect due to density stratification of the brines. After study of sulphur isotopic composition of the anhydrite as well as the bromine content in halite, the conclusion has been drawn that these salts had a marine origin and that there was an insignificant role of non-marine waters.

Finally we conclude that the Zechstein salts studied here were deposited with major contributions due to intrabasinal redeposition, at an elevated temperature, from concentrated seawater with a high potassium content. These conditions favoured precipitation of primary sylvite.

Together with the consideration of data from several other deposits, we can assume that primary sylvite precipitation from brines supersaturated by cooling was the typical mechanism of sylvite formation in many ancient potash-bearing basins.

The authors are indebted to Professor B. C. Schreiber for the detailed and patient revision of the paper, helpful and kindly remarks and suggestions due which this work was finalized. Many thanks are due to Professor Volodymyr Kovalevych for his scientific support during the study.

References

BILONIZHKA, P. M. 1975. The use of bromine vs. chlorine coefficient for evaporite genesis clarification. *Geologiya i Geokhimiya Goriuchikh Iskopayemykh*, **45**, 55–61 [in Russian].

BORCHERT, H. & MUIR, R. O. 1964. *Salt Deposits: The Origin, Metamorphism and Deformation of Evaporites*. Van Nostrand Reinhold, New York.

BRAITSCH, O. 1971. *Salt Deposits Their Origin and Composition*. Springer, Berlin.

CLAYPOOL, G. E., HOLSER, W. T., KAPLAN, I. R., SAKAI, H. & ZAK, I. 1980. The age curves of sulfur and oxygen isotopes in marine sulfate and their mutual interpretation. *Chemical Geology*, **28**, 199–260.

CZAPOWSKI, G. 1991. Sedimentation of Upper Permian (Zechstein) marine chlorides in Poland. *Program and Abstracts of International Congress of the Permian System of the World*, 60A.

CZAPOWSKI, G. 1993. Facies characteristics and distribution of the Zechstein (Upper Permian) salt deposits of PZ3 (Leine) Cycle in Poland. *Bulletin of Polish Academy of Sciences, Earth Sciences*, **41**, 229–237.

CZAPOWSKI, G. 1995. Salt facies of the Upper Permian. *XIII International Congress on Carboniferous Permian, Guide to Excursion A3*, 85–96.

CZAPOWSKI, G., ANTONOWICZ, L. & PERYT, T. M. 1991. Facies and paleogeography of the Zechstein (Upper Permian) Older Halite (Na2) in Poland. *Bulletin of Polish Academy of Sciences, Earth Sciences*, **38**, 45–55.

EUGSTER, H. R., HARVIE, C. E. & WEARE, J. H. 1980. Mineral equilibria in six component system, Na–K–Mg–Ca–SO_4–Cl–H_2O, at 25 °C. *Geochimica et Cosmochimica Acta*, **44**, 1335–1347.

KALYUZHNY, V. A. 1982. *The Foundations of Teaching about Mineral-forming Fluids*. Naukova dumka, Kyiv [in Russian].

KIERSNOWSKI, H., PAUL, J., PERYT, T. M. & SMITH, D. B. 1995. Facies, paleogeography and sedimentary history of the southern Permian basin in Europe. *In*: SCHOLLE, P. A., PERYT, T. M. & ULMER-SCHOLLE, D. S. (eds), *The Permian of Northern Pangea*, **2**, 119–136.

KOVALEVYCH, V. M., CZAPOWSKI, G., HAŁAS, S. & PERYT, T. M. 2000. Chemical evolution of brines in the Zechstein (Upper Permian) evaporite basins of Poland: fluid inclusion study of halite from the rock salt units Na1 to Na4. *Przegląd Geologiczny*, **48**, 448–454 [in Polish with English summary].

KOVALEVYCH, V. M., PERYT, T. M. & PETRYCHENKO, O. YO. 1998. Secular variation in seawater chemistry during the Phanerozoic as indicated by brine inclusions in halite. *Journal of Geology*, **106**, 695–712.

KOVALEVYCH, V. M., PERYT, T. M., CARMONA, V., SYDOR, D. V., VOVNYUK, S. V. & HAŁAS, S. 2002. Evolution of Permian seawater: evidence from fluid inclusions in halite. *Neues Jahrbuch fur Mineralogie, Abhandlungen*, **178**, 27–62.

LOWENSTEIN, T. K. & SPENCER, R. G. 1990. The syndepositional origin of potash evaporites: petrographic and fluid inclusion evidence. *American Journal of Science*, **290**, 1–42.

McCAFFREY, M. A., LASAR, B. & HOLLAND, H. D. 1987. The evaporation path of seawater and the coprecipitation of Br^- and K^+ with halite. *Journal of Sedimentary Petrology*, **57**, 928–937.

PETRYCHENKO, O. YO. 1973. *Methods of Fluid Inclusion Study in Evaporates*. Naukova Dumka, Kyiv [in Ukrainian].

PETRYCHENKO, O. YO. 1977. *Atlas of Inclusions in Evaporitic Minerals*. Naukova Dumka, Kyiv [in Russian].

PODEMSKI, M. 1972. Zechstein rock and potash salts of the Z2 and Z3 cyclothems in Nowa Sól vicinity. *Biulletin of Polish Geological Institute*, **260**, 5–62 [in Polish with English summary].

RAUP, O. 1970. Brine mixing: an additional mechanism for formation of basin evaporites. *Bulletin AAPG*, **54**, 2246–2259.

SMITH, D. B. 1980. The evolution of the English Zechstein Basin. *Contribution to Sedimentology*, **9**, 7–34.

SONNENFELD, P. 1984. *Brines and Evaporites*. Academic Press, Orlando, FL.

STRAUSS, H. 1997. The isotopic composition of sedimentary sulfur through time. *Palaeogeography, Palaeoclimatology, Palaeoecology*, **132**, 97–118.

VALYASHKO, M. G. 1962. *Principles of Evaporite Deposits Formation*. Izdatelstvo Moskovskogo Universiteta, Moscow [in Russian].

VOVNYUK, S. V., KOVALEVYCH, V. M., CZAPOWSKI, G. & SYDOR, D. V. 2004. Geochemical peculiarities of formation of Upper Permian salts in Eastern part of Central European basin. *Geologiya i Geokhimiya Goriuchykh Kopalyn*, **2**, 119–135 [in Ukrainian with English summary].

WAGNER, R. 1987. Stratigraphy of the Uppermost Zechstein North-Western Poland. *Bulletin of Polish Academy of Sciences, Earth Sciences*, **35**, 265–273.

WAGNER, R. 1994. Stratigraphy and evolution of the Zechstein Basin in the Polish Lowland. *Prace PIG*, **146**, 1–71 [in Polish with English summary].

WAGNER, R. & PERYT, T. M. 1997. Possibility of sequence stratigraphy subdivision of the Zechstein in the Polish Basin. *Geological Quarterly*, **41**, 457–474.

WAGNER, R., PERYT, T. M. & PIĄTKOWSKI, T. S. 1981. The evolution of the Zechstein sedimentary basin in Poland. *Proceeding of the International Symosium Central European Permian*, 69–83.

Evidence of vanished evaporites in Neoarchaean carbonates of South Africa

A. GANDIN[1] & D. T. WRIGHT[2]

[1]*Dipartimento Scienze della Terra, Università di Siena, Via Laterina, 8-I-53100 Siena, Italy (e-mail: gandin@unisi.it)*

[2]*Department of Geology, University of Leicester, University Road, Leicester LE1 7RH, UK (e-mail: davywright2@ntlworld.com)*

Abstract: A detailed sedimentological and petrographic analysis of the Neoarchaean Campbellrand and Malmani carbonates of South Africa provides evidence that collectively indicates the former existence of evaporites and the early replacement of primary sulphate deposits by calcite. Diagenetic disruption of sedimentary structures, solution collapse breccias, cross-cutting relationships, corrosion surfaces, pseudomorphs of displacive nucleation cones, flowage and nodular structures are all characteristic of evaporites and their ductile behaviour. In thin section, rosettes of length-slow chalcedony around the heads of the silicified stromatolites also testify to the former presence of sulphates. Calcite, in the unusual form of herringbone/flamboyant spar in 'cuspate' and nodular facies, may preserve the fold shapes of enterolithic gypsum, picked out by deformed trails of degraded organic matter and dolomitized clusters of filaments. Subvertical columns of compressed organic material in a matrix of herringbone/flamboyant calcite appear to be the result of squeezing and compression between adjacent sulphate nodules. Carbonate replacement of evaporites was largely driven by bacterial sulphate reduction in sapropels and microbialites during early diagenesis. The petrographic evidence indicates that the replacive herringbone/flamboyant calcite precipitated after enterolithic and nodular structures formed, the sulphate having been used up before the enclosed organic material was totally consumed.

Sedimentary structures in carbonates observed by the authors in the first extensive cratonic platform that appeared in the history of the Earth, the Neoarchaean Kaapvaal carbonate platform, are exposed in two sections (Fig. 1) at Kuruman Kop (Kogelbeen and Gamohaan Formations, Campbellrand Subgroup) and at the R36 road-cut near Echo Caves (Frisco Formation, Malmani Subgroup). These structures, as well as those illustrated from other intermediate localities and interpreted as fenestrate microbialites (Sumner 1997, 2000), show macro- and microfacies that have never been reported from the sediments of the carbonate platforms of the Phanerozoic. They are characterized by deformed and contorted fabrics that can best be explained by the presence during early diagenesis of plastic materials that suggest an evaporitic origin of the sediments. Petrographic analysis and comparison with recent, mainly Messinian, analogues, strongly support this interpretation.

Although calcite or barite pseudomorphs after sulphates have been reported from Neoarchean deposits of Australia and Canada (Walker et al. 1977; Buick & Dunlop 1990; Simonson et al. 1993; Aspler & Chiarenzelli 2002), researchers working in Archaean carbonates have argued that, under an anoxic atmosphere, the SO_4^{2-} content of the oceans would be low (e.g Walker & Brimblecombe 1985; Canfield & Teske 1996; Habicht et al. 2002), and therefore large-scale precipitation of sulphates on the first carbonate platforms of the Earth history would be unlikely. However, cyanobacterial biomarkers have been recorded from Neoarchaean sediments (e.g. Brocks et al. 1999), and the vast regional extent and variety of microbialites comprising the bulk of the Campbellrand carbonates represent huge populations of photosynthesising cyanobacteria, which would have pumped significant volumes of oxygen directly into the shallow epeiric seas. Under restricted conditions, sulphates could easily have precipitated. Interestingly, Melezhik et al. (2005) report widespread syn-depositional, marine, anhydrite-containing pseudomorphs after solid calcium sulphates from the 2100 Ma Tulomozero Formation in the SE Fennoscandian Shield, implying that Palaeoproterozoic surface waters already contained significant aqueous sulphate.

Features suggestive of former evaporitic deposits

Many of the depositional and early diagenetic features observed in the Neoarchean Campbellrand/Malmani carbonate beds show fabrics that can be

From: Schreiber, B. C., Lugli, S. & Babel, M. (eds). *Evaporites Through Space and Time.*

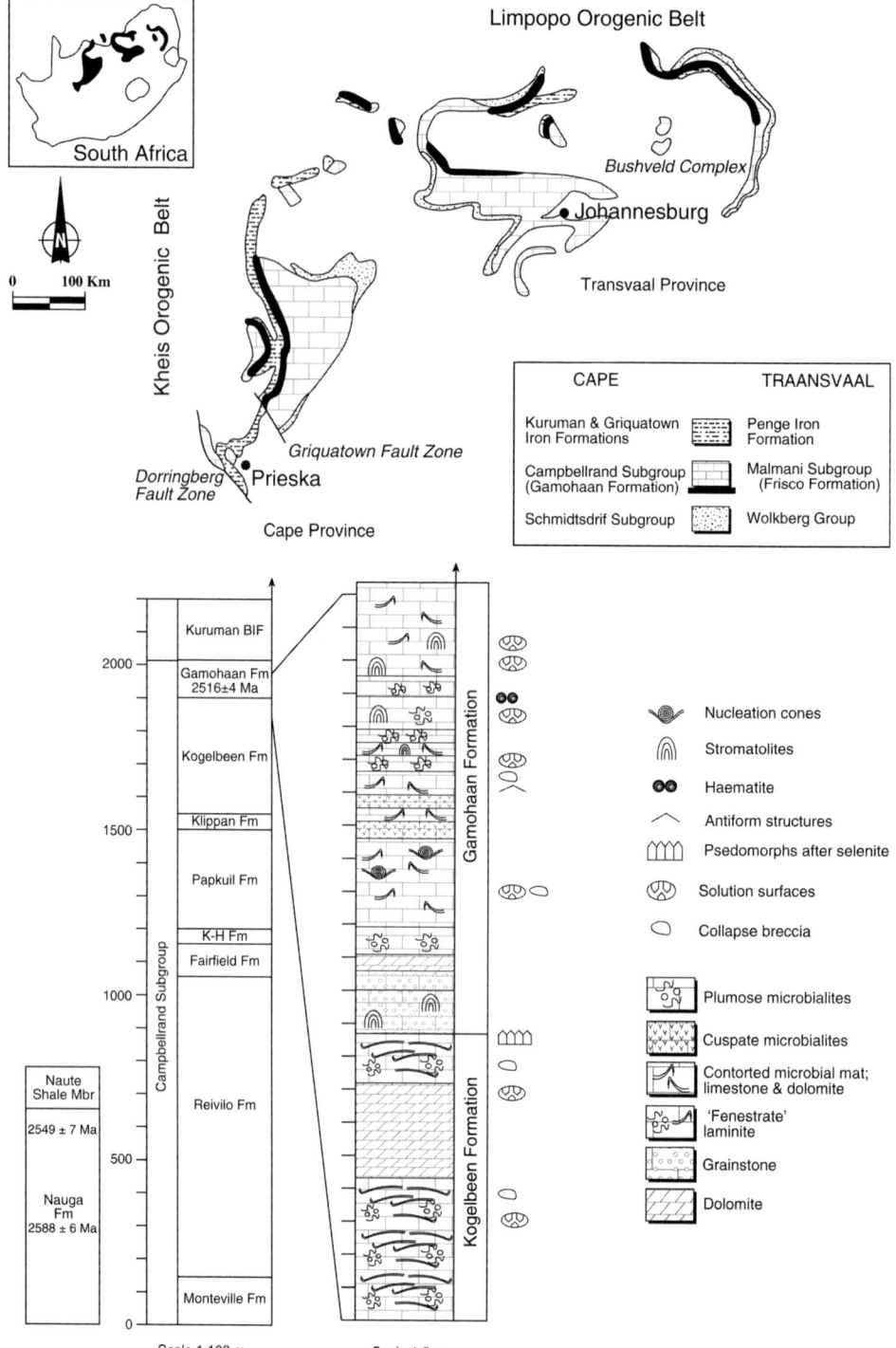

Fig. 1. Location map of the Neoarchaean Campbellrand and Malmani Subgroups of South Africa, together with a simplified stratigraphy of the Gamohaan and Kogelbeen Formations of the Campbellrand Subgroup at Kuruman Kop (after Sumner 1997 and the authors' fieldwork). This is the lateral equivalent of the section exposed on the R36 road-cut near Echo Caves (Frisco Formation; Malmani Subgroup).

better explained as generated by mechanical processes derived by early diagenetic deformation of original evaporitic deposits encasing or interbedded with stromatolites and organic material (Gandin et al. 2005) rather than related to unusual microbialites and associated supposed carbonate sea-floor cements (Sumner 1997, 2000, 2001; Sumner & Grotzinger 2000, 2004). These include solution/collapse breccias, calcite pseudomorphs after halite and probable selenite, and a variety of contorted and nodular structures which can be compared with possible analogues found in well-preserved Triassic and Messinian evaporites of the Mediterranean area. Calcite pseudomorphs after halite have been reported from Neoarchean carbonate or silicified deposits of Canada, Australia and South Africa (Badham & Stanworth 1977; Boulter & Glover 1986; Sumner & Grotzinger 2000).

Direct evidence of the existence of sulphate deposits is subject to interpretation, though according to Usiglio's order of precipitation of evaporitic salts, gypsum should be deposited before halite (Usiglio 1849). However, precipitation of solid sulphate is dependent on Ca^{2+} and SO_4^{2-} concentrations, and it should also be noted that any aqueous sulphate present may have been removed by sulphate-reducing bacteria (SRB), as for example in the dolomitic lakes of the Coorong, South Australia (Wright 1999; Wright & Wacey 2005). Nevertheless observed macro and micro features can be interpreted as primary evaporitic sediments that in an early stage of diagenesis were replaced by calcite.

Depositional components

The Kogelbeen Formation in outcrop comprises over 300 m of limestone, dolomite and chert lithologies, with abundant microbialites, oolites and laminites, and a number of doleritic intrusions (Altermann & Siegfried 1997). It is interpreted as shallow-subtidal to intertidal, with organic-rich and apparently fenestral sediments dominant in the upper part, suggesting a lagoonal facies. Coccoid and filamentous microfossils have been preserved in silicified stromatolites of the Lime Acres Member (Altermann & Schopf 1995).

The c. 2516 Ma (Altermann & Nelson 1998), 110m-thick Gamohaan Formation and its eastern equivalent (Frisco Formation, Malmani Subgroup) exposed on the R36 road-cut near Echo Caves are dominantly sapropelic, with abundant contorted microbial laminae, common dolomitic intercalations and apparently fenestrate limestones (including the 'cuspate' and 'plumose' microbialites of Sumner 2000). Ripple cross-laminated dolarenites occur in the lower part of the Gamohaan Formation, with cycles of grainstones and stromatolites clearly indicating energetic shallow-water conditions. Abundant pyrite nodules occur in the middle and upper parts of the formation.

The dominant depositional facies in the surveyed sections are microbialitic plane parallel laminites and stromatolites (Figs 2a & 5a, b); granular dolomite and small stromatolites are characteristic of the lower part of the Gamohaan Formation. The early diagenetic, contorted laminites and fenestrate facies are commonly associated with grey herringbone calcite and white drusy calcite nodules (Figs 2a, 6a, e, 8e, 9a, 10a, f & 13b).

Laminites

Description Laminites consist of thin, regular layers or bundles of partially degraded organic filaments forming a microbial mat, with less frequent dolomicrosparitic and granular laminae. Preserved micrite is lacking or rarely found as a thin lining of organic filaments. Centimetric, wavy or plane layers (Figs 2a–d, 6a & 14a, b) are commonly found associated with the deformed beds consisting of lenticular or contorted laminar shreds (Figs 2b–d, 3b, c, 7d, e, 11d & 12e, 14a–c) and/or rolled up ribbons (Figs 3a & 9c) enclosed in calcite spar. The contortions generally range from the microscale up to a few centimetres length.

Filamentous laminites consist of microbial filament bundles 10–20 μm thick, embedded in microcrystalline (20–60 μm) brick-like or lenticular hyaline calcite with undulose extinction (Figs 2b–d, & 3b). The filaments, originally tightly disposed, are locally intruded by lenses of undulose calcite that appears to push the filaments upwards (Figs 2b, c & 7d) and displaces and stretches filament bundles (Fig. 12e) in zones of strain. In contrast, bundles appear curled and crushed (Fig. 3a & c) in the zones of compression. The resulting folded, cable-like (Figs 3b & 4e) or contorted structures (Fig. 3b, c) floating in herringbone calcite (Fig. 14a, b), are commonly pervaded by selective dolomitization.

Interpretation The laminated filamentous fabrics, which can be compared to the laminar calcium sulphate facies in the Mediterranean Messinian basins, indicate a subtidal, organic-rich sediment, formed by bacterial films covering the basin floor. Anoxic bottom conditions in density-stratified water, associated with mineralization through organic diagenesis, would favour the preservation of the organic matter.

Dolomitization of the filaments appears to have been an early diagenetic process that occurred before the formation of the enclosing herringbone calcite. The scale and morphology of the contorted

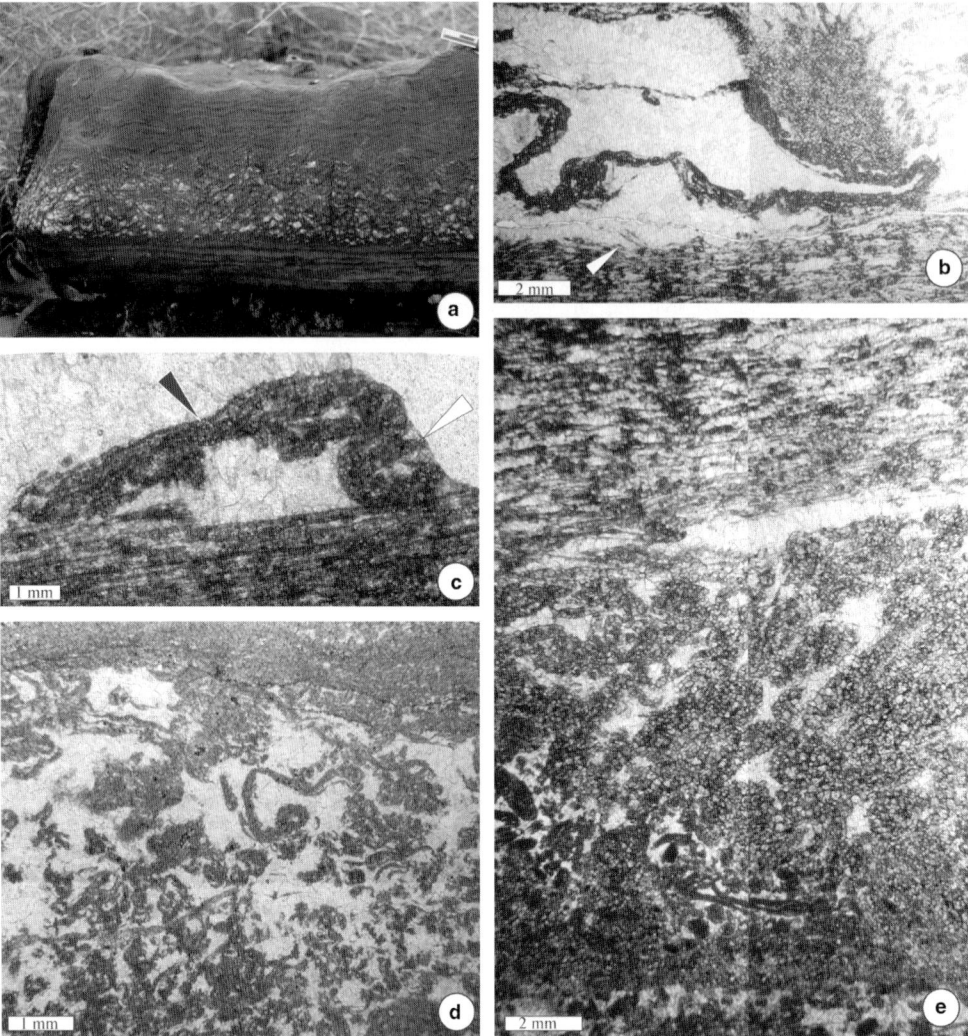

Fig. 2. Depositional and early diagenetic features: (**a**) Outcrop showing calcite nodules growing within a wavy-laminated bed which overlies a planar laminite layer. The growth of the nodules tends to bend the orderly depositional arrangement of the mat (the same relationships can be observed in thin section, see (**b**) (sample R36; Echo Caves section). (**b–e**) Thin section views (photomicrographs) of the transition from a well bedded sediment to a diagenetic nodular fabric. (b) The nodular fabric consists of stacked, half moon domains of herringbone calcite bounded by more or less stretched microbial filaments which laterally converge in deformed bundles. Uplifted single filaments (with arrow) and contorted cable-like dark bundles at the transition with the filamentous mat, suggest that each filament has been pushed upwards by the growth of a ductile precursor to herringbone calcite. Growth of the nodules pushes the filaments upwards and displaces bundles of them that are folded and crushed in the zones of compression and stretched in the zones of strain. The laminite consists of layers of filamentous microbial mat overlying granular facies (b, e). The organic filaments and the poorly sorted, selectively dolomitized ooids/peloids in the lower part of the thin section, are embedded in brick-like, flamboyant calcite that fills the irregular intergranular porosity. The laminated filamentous fabrics are interpreted as representing bacterial films covering the sea floor (sample Rau 5.4; Kuruman Kop section). (**c**) An arched bundle of filaments partially separated from well-preserved, filamentous microbial mat. The bundle appears to have been detached and pushed upward from the planar mat, representing an early stage in the deformational process that led to the development of the nodular fabric of (a). Volume increase of the expanding material, now replaced by herringbone calcite, caused the filaments to curl/fold (white arrow) and locally to break up/be eroded (black arrow) (sample Rau 5.1.a.II; Kuruman Kop section). (**d**) Stromatoclasts floating in flamboyant calcite, partly replaced by a mosaic of neomorphic equant calcite. The uneven distribution of stromatoclasts 'floating' in calcite, and the lack of a fabric characteristic of the grainstone cements (isopachous acicular fringes and later drusy mosaic), suggest that the clasts, rather than forming a sand with grains in contact with each other, were supported in a precursor soft matrix, later replaced by herringbone calcite (sample 1A.VM; Kuruman Kop section). (**e**) Filamentous microbial mat overlying a granular facies (lower part of slide).

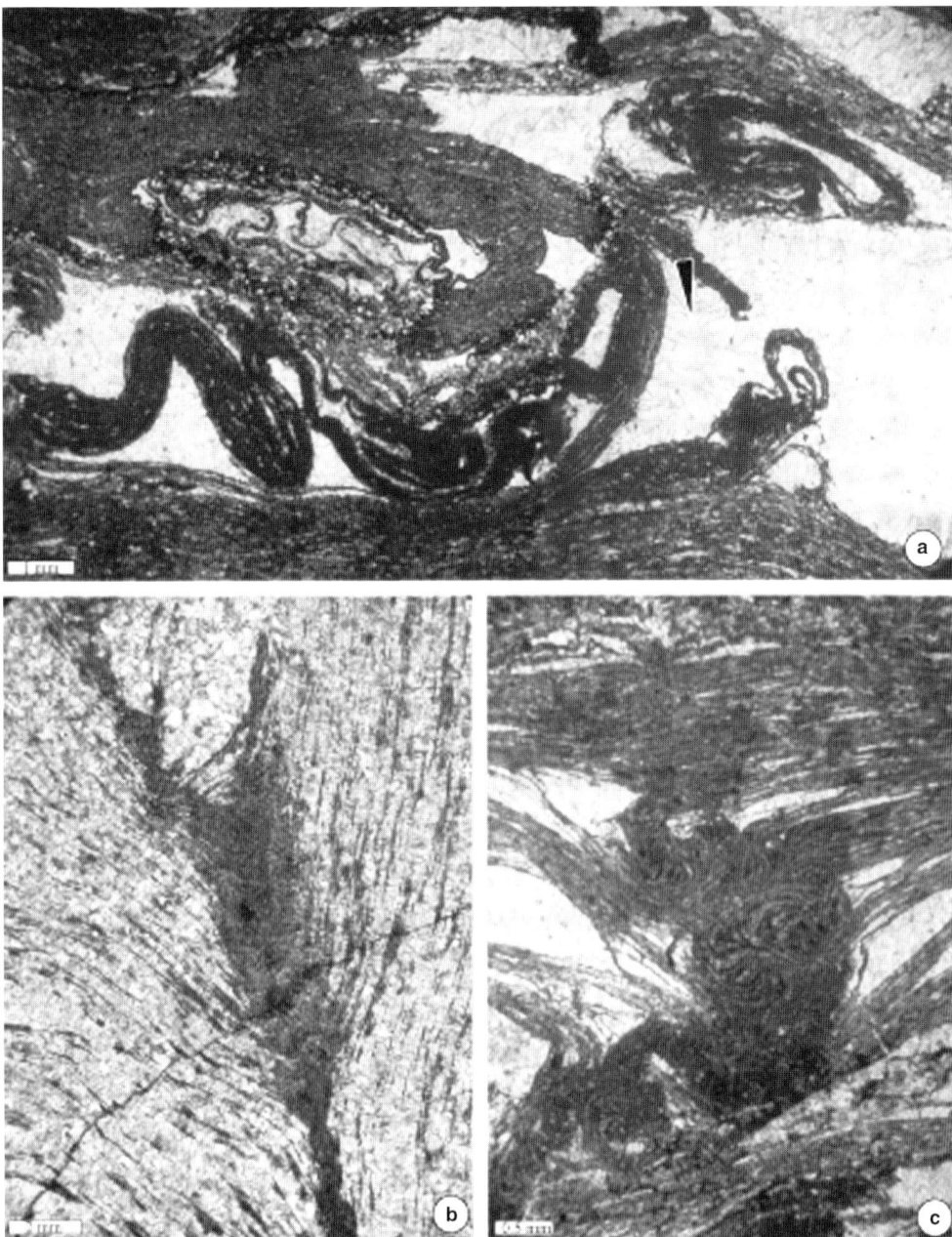

Fig. 3. Unusual structures embedded in herringbone calcite (all seen in thin section; plane light). (**a**) Rolled-up ribbons of filamentous laminites and partially dolomitized bundles. Discrete sets of laminae are in places pushed apart, while elsewhere they are compressed and folded together. The voids formed in the detachment zone and in the loops of the folds are filled by late stage prismatic cement (arrow). The cement grows from the herringbone/flamboyant calcite lining of the filament ribbons. This structure is interpreted as the first stage of plastic deformation driven by the volume increase of the inferred former sulphate mass interbedded in a planar filamentous laminite (sample Rau 5.1C; Kuruman Kop section). (**b**) Flower-like structure encased within the herringbone/flamboyant calcite. The structure is formed by the compression of sets of filaments which are amalgamated in a crinkled 'stem' and dolomitized at their core. The brick-like calcite interspersed with the outer, relict filaments of this peculiar structure does not show any of the features of (and has no affinity with) a marine cement precipitate (sample Rau 5.1; Kuruman Kop section). (**c**) Bundled cyanobacterial mats amalgamate to form a flower-like structure which display an evident plastic behaviour of the filaments, suggesting that the deformation process occurred when the curled bundles were still soft and supported in some way by a 'plastic matrix' probably originally represented by sulphate (sample Rau 5.1C; Kuruman Kop section).

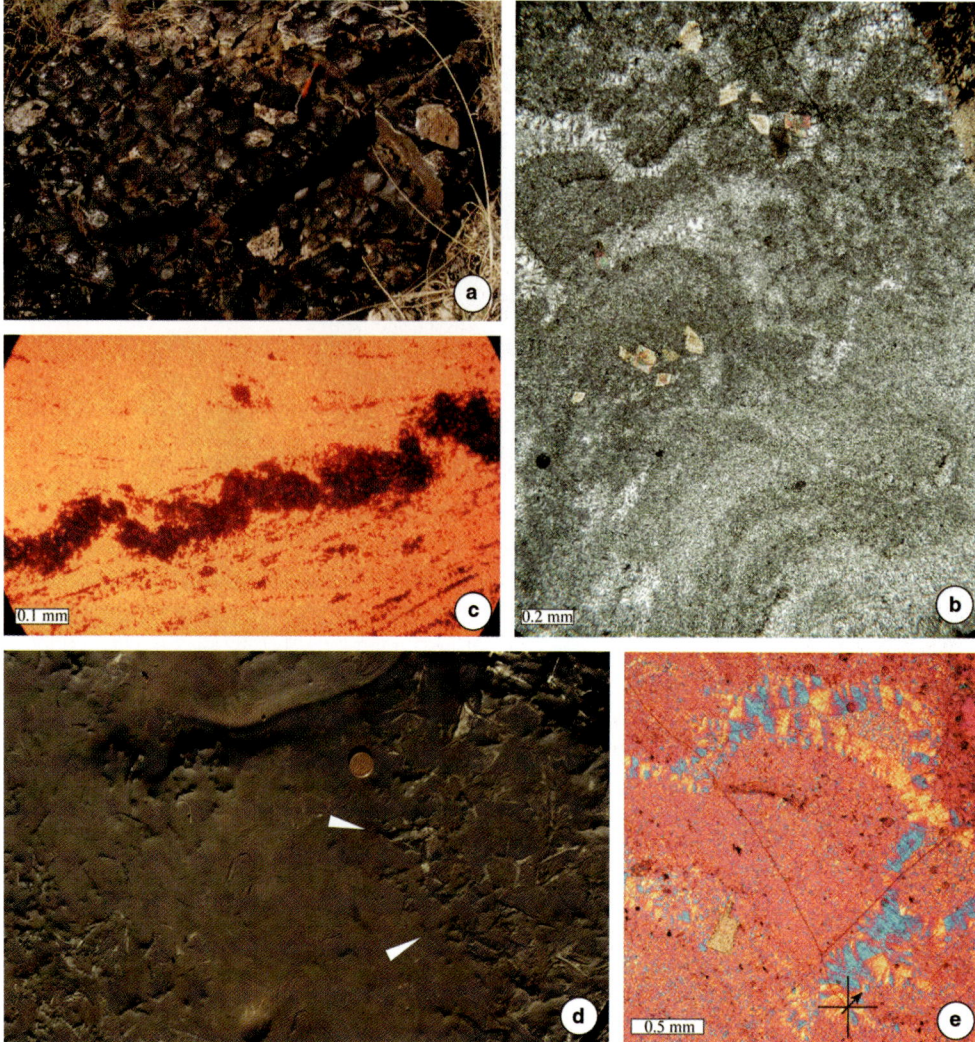

Fig. 4. Silicified and dolomitized microbial sediments. (**a**) Small, silicified stromatolites with a flat, shallow-domed profile. (Kuruman Kop section: outcrop view). (**b**) Thin section photomicrographs of a silicified stromatolite replacing micrite or microcrystalline gypsum mush (cross-polarized light–XPL) (sample Rau 5.2; Kuruman Kop section). (**c**) Dolomitized, cable-like, crinkled 'stem' of a flower-like structure (see Fig. 3b) resulting from amalgamated and folded filaments. Seen under cathodoluminescence is evident that dolomitization is strictly associated with the partially degraded organic filaments (sample Rau 5.1; Kuruman Kop section). (**d**) Moulds with partially hollow interiors (arrows) characteristic of hopper halite crystals form a mesh with other pseudomorphs probably related to gypsum and/or celestite crystals (Kuruman Kop section). The coin in the figure is 25 mm in diameter. (**e**) Thin section photomicrograph showing a detail of (b): the intermound spaces are now filled with spherulitic, length-slow chalcedony. The arrow marks the direction of the c axis parallel to the fibres, indicative of the presence of sulphates in the silicifying fluids: XPL with λ (gypsum) plate.

laminae, from the microscale up to a few centimetres in length, argue for *in situ* deformation caused by expansion and contraction of a ductile material, later replaced by the more brittle crystalline calcite mosaic.

The calcite occurring among the filaments displays a very compact fabric comprising brick shaped crystals with the typical undulose extinction of herring bone calcite. It may represent the replacement of a muddy sediment or of a cement. However no evidence of an earlier cement fabric can be recognized.

Stromatolites

Description Stromatolites varies in size from centimetric (Fig. 4a) to larger (up to 50 cm)

Fig. 5. Stromatolites and autoclastic breccias (outcrop photographs). (**a**) Stromatolites with concentrations of iron oxides, probably resulting from the alteration of early diagenetic, organic-derived pyrite (Kuruman Kop section). (**b**) Finely laminated stromatolites with a very smooth profile. (Kuruman Kop section). (**c**) Autoclastic breccia associated with a diapir-like, domal structure. Contorted and disrupted mat facies appear to have intruded, pierced and displaced overlying sediment, thinning and breaking dolomitized layers which now appear as broken, corroded rafts (Kuruman Kop section). (**d**) Prograding of the brecciation front from left to right: chaotic, contorted sediment encroaches on the rusty red dolomitized sediment and associated grey herringbone calcite (black arrow). Associated, planar and vertical fractures in the undeformed rock have been filled by multiple rims of white diagenetic cement (white arrow) (Kuruman Kop section).

forms (Fig. 5a, b). Some are preserved by silicification (Fig. 4a) with superposed, irregular laminae (Fig. 4b) of dense, microcrystalline former primary sediment (micrite or gypsum mud). Spherulitic length-slow chalcedony is locally associated in the intermound voids (Fig. 4e).

The larger and more developed stromatolites are made from a very regular lamination similar to that of the planar laminites, and commonly exhibit a very smooth profile (Fig. 5a, b). They may be embedded in herringbone calcite and show frequent concentrations of iron oxides, often with yellowish haloes (probably hematite and sulphur after early-diagenetic pyrite) distributed along the laminae (Fig. 5a).

Smaller forms that may be interpreted either as stromatolites or incipient antiform/tepee structures (Fig. 7e) are low-relief domes composed of thin and irregular, often broken laminae of dense microsparite alternated with thicker ones made of stromatoclasts and pyrite crystals enclosed in a fine, irregular mosaic of herringbone calcite.

Interpretation On the whole the stromatolite shapes and sizes suggest a very quiet, shallow-water, low-energy depositional environment, in which the dominant marine chemical precipitate was probably sulphate mush rather than carbonate micritic mud, as suggested by the presence of length-slow chalcedony in the silicified stromatolites, indicating that silicification occurred before the dissolution/disappearance of the primary sulphate.

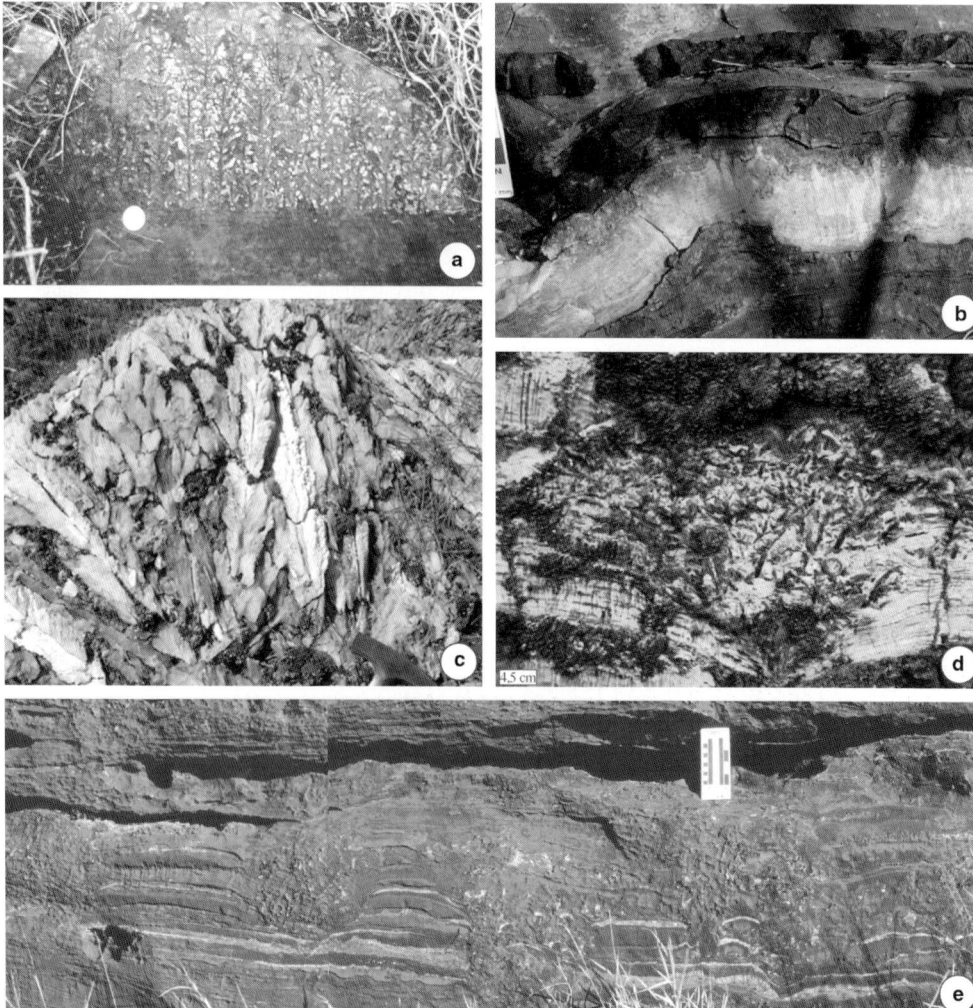

Fig. 6. Macroscopic structures suggesting the former presence of solid sulphate deposits. (**a**) Calcite nodules growing above a planar laminated layer, are disposed in a pattern reminiscent of tickets of 'grass-like', swallow-tail selenite crystals. The white calcite domains (comprising a cortex of herringbone calcite) may be interpreted as replacive after solid sulphate. They resemble nodules grown at the expenses of selenite crystal twins (see for comparison fig. 8.53 in Schreiber 1986) rather than a marine calcite cement (sample R36; Echo Caves section). The coin is 25 mm in diameter. (**b**) A band of white calcite formed by vertically oriented chevron-like crystals interbedded in laminated sediments, interpreted as pseudomorphs after swallow-tail selenite (Kuruman Kop section). (**c**) Vertically oriented swallow-tail crystals of selenite; for comparison with (a) (Montalcinello, Messinian Basin of Radicondoli, Tuscany). (**d**) Gypsum nucleation cone in deflected, gypsum laminites, for comparison with (**e**) (Messinian, Cyprus). (**e**) Cauliflower (cavoli) structures, comparable with the Messinian nucleation cones at Cyprus (d), have displaced laminated, soft-sediments, which are deflected downwards. (Kuruman Kop section).

Granular facies

Description Granular facies locally occur at the base of contorted beds as a grain supported, massive or planar laminated sediment (Fig. 2e), or associated with nodular/contorted structures. The grains are represented by rare allochems: ooids, peloids (Fig. 2e) and more frequent stromatoclasts (Sami & James 1994) comprising microsparitic or dolomicrosparitic thin fragments and shreds derived from the fragmentation of stromatolitic material (Figs 2d, 7d & 14a, c). Most of them are selectively pervaded by dolomicrite rhombs (Fig. 2e), and the coated grains still retain

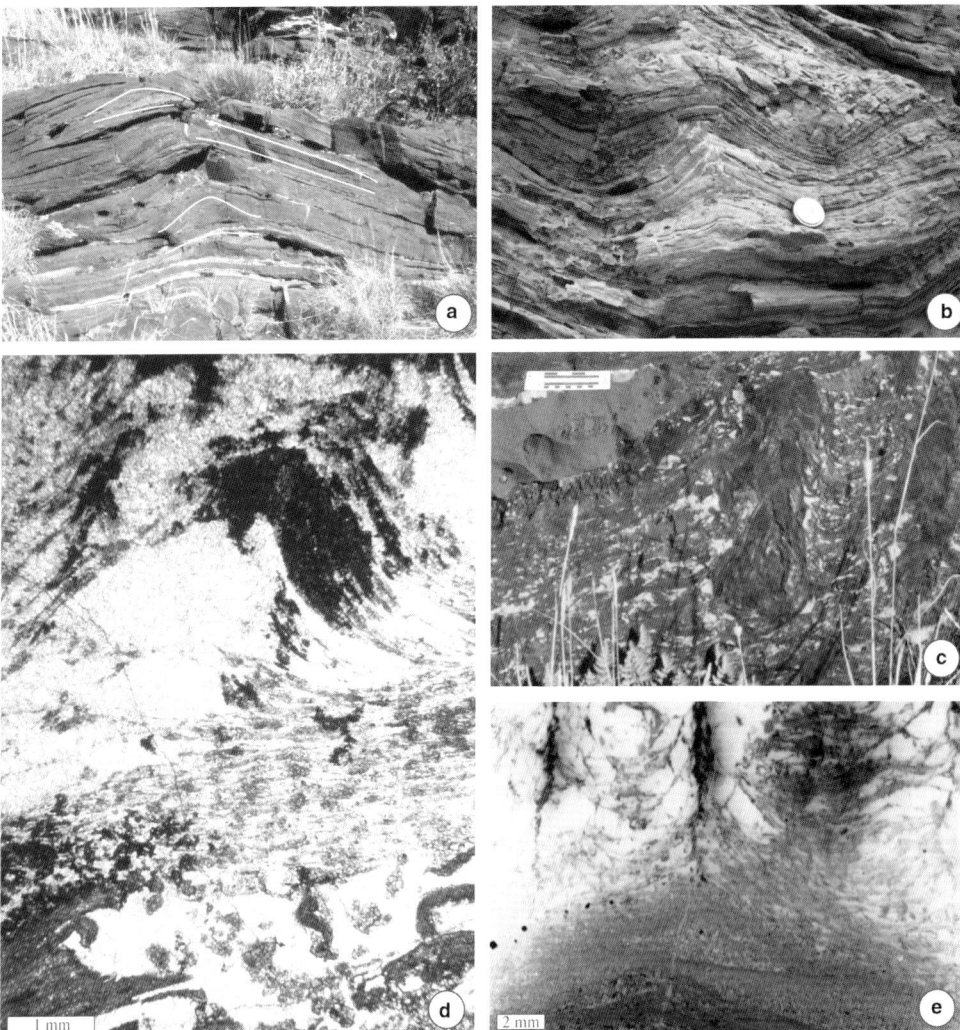

Fig. 7. Deformational antiform structures. (**a**) Antiform structure similar to a tepee (outcrop photograph): apparently originated in shallow burial conditions, as attested by the characteristic accumulation in the hinge zones (marked by white outline) of rusty-coloured, dolomitized material which evidently was deposited in between the grey beds and migrated towards the top during the deformation, and before lithification (Kuruman Kop section). (**b**) Diagenetic, tepee-like fold observed in carbonate beds (outcrop photograph): associated with spectacular intrastratal collapse breccias and calcite pseudomorphs after gypsum and anhydrite (not in the picture), suggests the upward-running flow, in very shallow burial conditions, of interbedded, now vanished, sulphates; for comparison with (a) (Middle Triassic, Campumari section, southern Sardinia). (**c**) Diapir-like, cuspate structure interspersed within contorted and disrupted mat facies, appears to have expanded and displaced the overlying sediment, causing the flexible filamentous laminae to form columns while the more brittle, dolomitized layers are left floating as broken, corroded rafts (Kuruman Kop section; outcrop photograph). (**d, e**) Small-scale antiform structures, seen in thin section, apparently caused by the upward-running flow of plastic material, resulting in a nodular structure now of herringbone calcite. In the zones of strain, the swelling up of the nodules stretched and destroyed filaments, locally leaving relict stromatoclasts while in the zones of compression, filaments and laminae were crushed, and now appear as dark dolomitic material (d). Pyrite crystals (e) probably of early diagenetic origin, are aligned along the inclined shoulder of the antiform structure (samples Rau 24B and R36; Echo Caves).

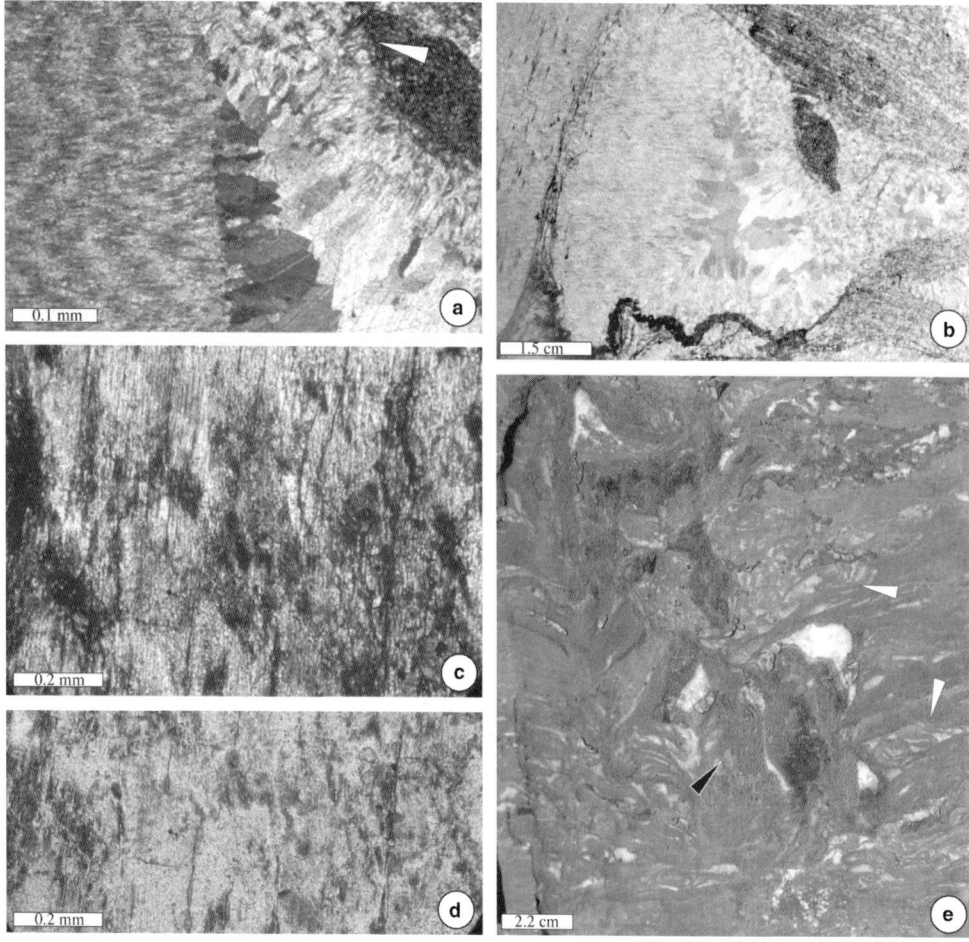

Fig. 8. Neoarchean herringbone calcite compared with the fabric of Messinian selenite crystals. (a, b) Photomicrographs showing well ordered zebra-like mosaic of herringbone calcite, associated with randomly disposed rosettes of flamboyant calcite (arrow), form the cortex of a nodule with a central cavity filled by a true geodic cement. The prismatic and equant spar, which evidently fills a primary structural cavity, is in sharp contact with herringbone calcite (a, XPL; b, PL) (sample Rau 5.1; Kuruman Kop section). (c, d) Photomicrographs of crystal aggregate texture characterized by sweeping extinction of Messinian selenite crystals (c, XPL; d, PL) (La Spicchiaiola, Volterra Basin, Tuscany). (e) Cut section showing folded rinds of herringbone calcite (black arrow) connected with partially folded/deformed filament laminites (white arrows) similar to those of Figure 3a and 9c. The hinge cavities of herringbone calcite folds are infilled with geodic cements of prismatic and equant spar (white irregular patches). (Kuruman Kop section).

relicts of calcite, consistent with a calcite origin for the ooids. They form a grain-supported fabric in which the intragranular spaces are filled by inequigranular calcite spar with undulose extinction, which does not show any of the typical features of cement either of marine or phreatic-meteoric derivation (Figs 2d, e & 7d); rather, it is always in optical continuity with the herringbone calcite of the overlying nodular facies.

Interpretation The allochems and part of the stromatoclasts may have been formed in a marginal site by hydraulic erosion of poorly calcified stromatolites, with the transport of the clasts and other grains (e.g. ooids) into subtidal sapropel–mud deposits.

The unsorted deformed stromatoclasts associated with the nodular and contorted structures are always found as unsupported grains floating in a mosaic of herringbone calcite (Figs 2d & 14c), with no trace of isopachous cements. They clearly derive from *in situ* disruption, evidently resulting from deformation strain of the organic laminites. This very unusual wackestone-like

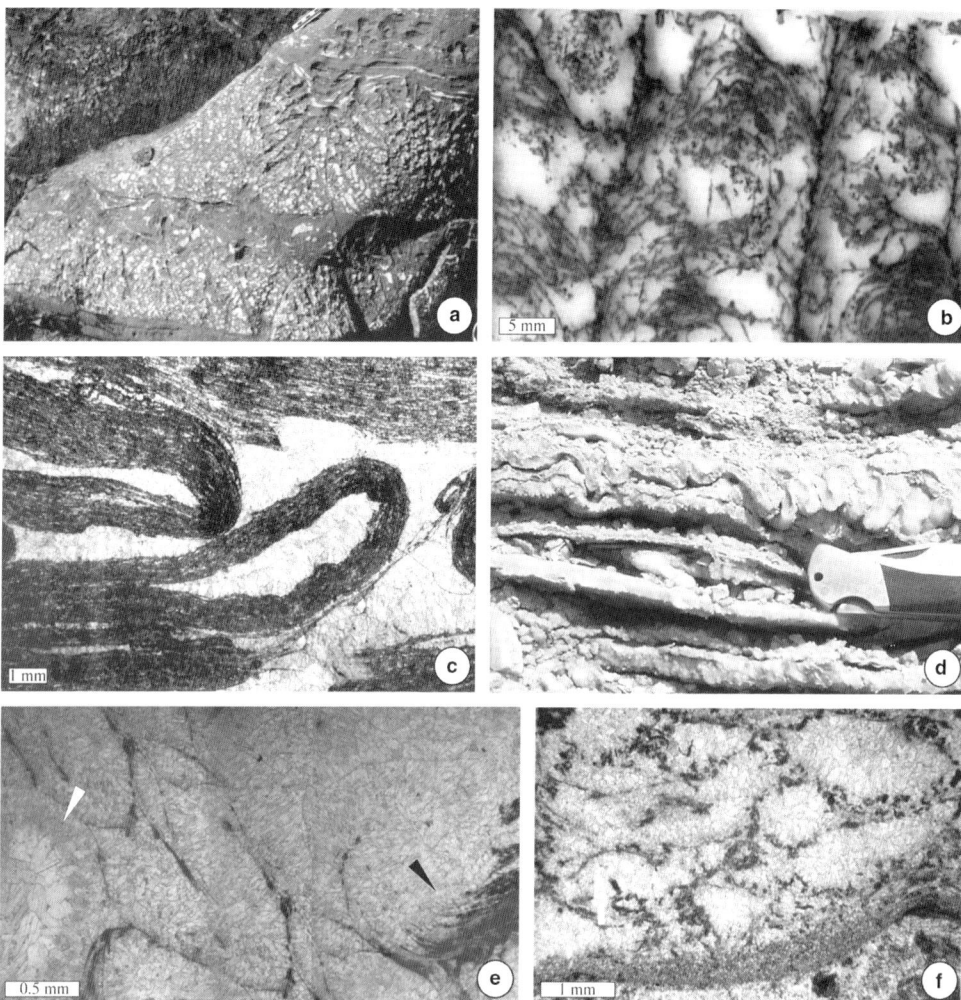

Fig. 9. Enterolithic and chicken-wire structures. (**a**) Outcrop section of large composite ('plumose') nodular structure with radially disposed nodules (compare with Fig. 10d), upward fading in the laminated sediment. Seen in thin section this fabric reveals the presence of assembled enterolithic folds (see Figs 9e & 12d) forming small lobate nodules (Fig. 9b) that in turn tend to coalesce into a chicken-wire fabric (Fig. 9f) (Kuruman Kop section). (**b**) Cut slab showing lobate nodules formed by enterolithic folds, encased in a web of compressed and contorted, relict filaments. This fabric appears to represent an early stage of deformation, when the dolomitization of the organic matter was still incomplete (samples Rau 24C & R36; Echo Caves). (**c**) Folded sets of dark filaments outline enterolithic-like structures in herringbone calcite. They attest to the original presence of a microbial mat displaced and destroyed by the growth of a ductile precursor, interpreted as solid sulphate. The hinge loop of the folds has been filled by diagenetic cement (sample Stop 5.1.B.II; Kuruman Kop section). (**d**) Enterolithic folds in Messinian gypsum. Structural voids at the hinge of the folds are at present empty or filled with residual clayey sediment. (Faltona Quarry, Tuscany). (**e**) Enterolithic folds made of flamboyant, subhedral sucrosic calcite, merge to form hollow subrounded bodies filled with drusy calcite cement (white arrow). Folds and nodules are bounded by relicts of dolomitized filaments (black arrow) which are also concentrated in the tightly folded hinge zones (samples Rau 24 & R36; Echo Caves). (**f**) Photomicrograph of chicken-wire fabric composed of nodules, made from a flamboyant, subhedral sucrosic calcite mosaic, bounded by dolomitized relict clusters of filaments (sample Rau 5.1A; Kuruman Kop section).

fabric can be interpreted by assuming that the original supporting matrix, of which no traces are now preserved, was removed and replaced by the herringbone calcite.

This interpretation is supported by the uneven distribution of the grains and the lack of pore filling cements characteristic of grainstones (isopachous acicular fringes and later drusy mosaic). The

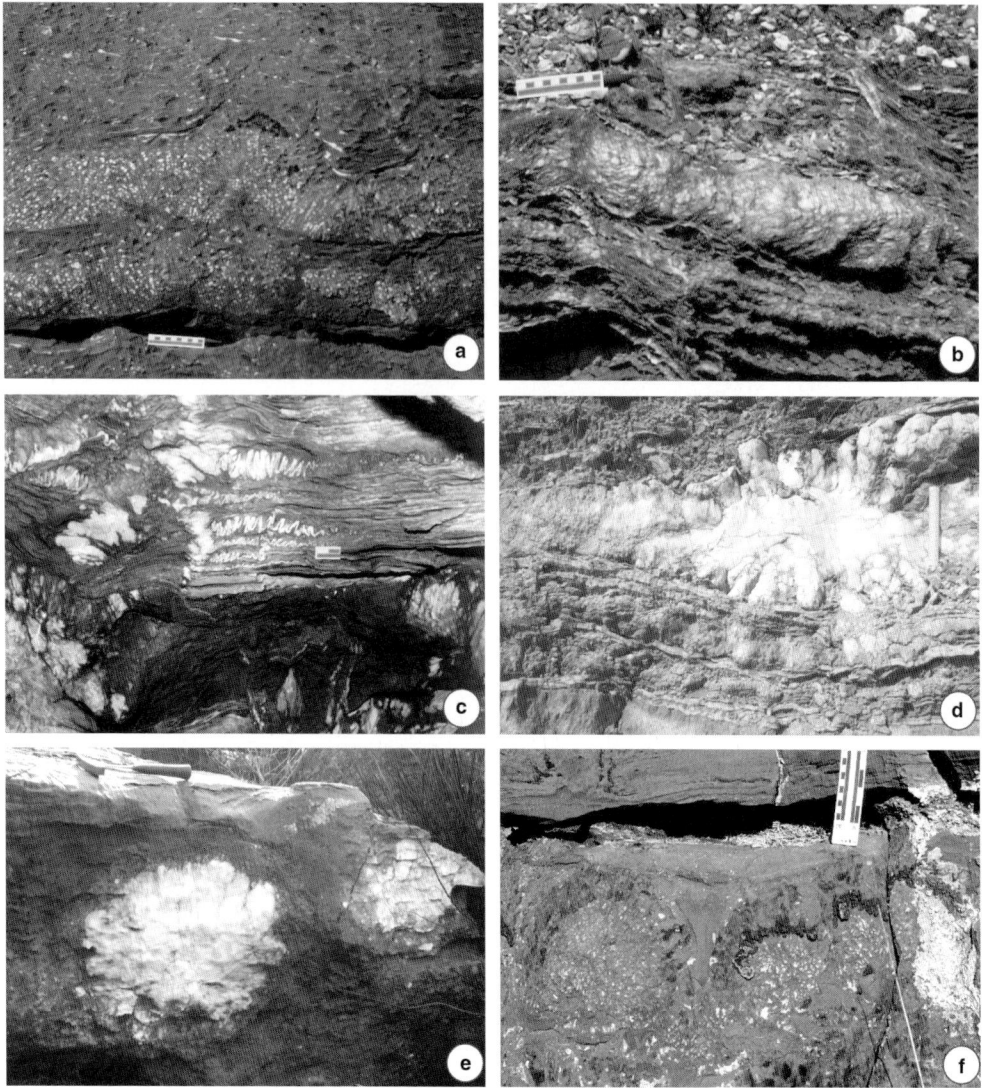

Fig. 10. Evolution of the enterolithic structures in nodular, half-moon and ball-like bodies. (**a, b**) Irregular bands of concentrated calcite nodules interbedded in partially disrupted, contorted laminar sediment (a), (Kuruman Kop section). Similar deformation of the original deposits can be observed in the Messinian gypsum. (b), (Faltona Quarry, Tuscany). (**c, d**) Gypsum nodules derived from the coalescence of enterolithic folds converging to form increasingly larger nodules such as that of (e). A similar trend as well as the incipient radial arrangement of the enterolithic folds can also be seen in the Neoarchean structures of (F) and Fig. 9a (Montalcinello Mine (c) and Faltona Quarry (d). Messinian, Tuscany. The scale bar in 1(c) is 2 cm long). (**e**) Photomicrograph of composite nodule of sucrosic gypsum, for comparison with (f). (Messinian, Ripaiola Quarry, Tuscany). (**f**) Large composite nodules rimmed by concentrations of iron oxides, interbedded with undisturbed, laminated layers (Kuruman Kop section).

embedding herringbone calcite does not therefore represent a cement, but appears to replace a matrix which was not a micritic mud, but a soft, ductile sediment; we suggest that the precursor matrix was sulphate mush.

Calcite mosaics

Description Calcite mosaics are the dominant constituent of the deformed carbonate facies. Herringbone calcite is locally associated with prismatic and equant spar that forms drusy

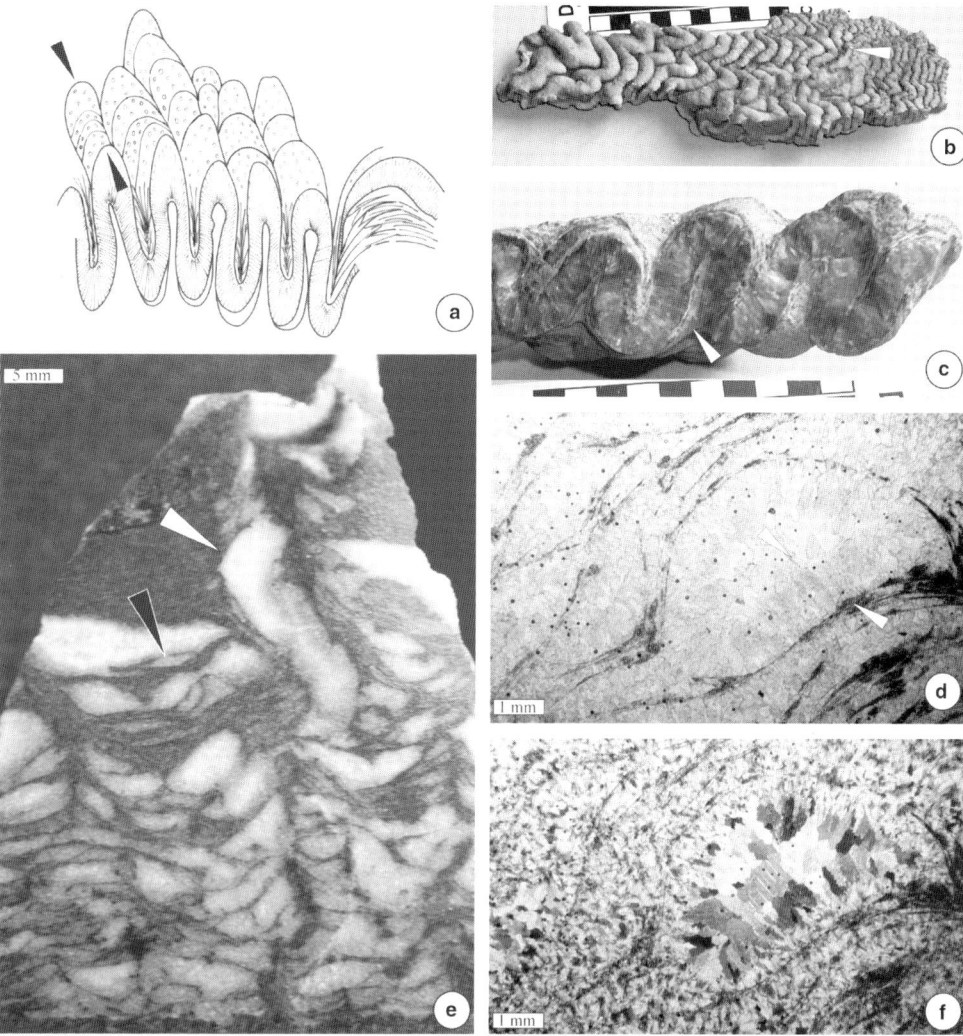

Fig. 11. Evolution of enterolithic folds in solid sulphate (based on Messinian gypsum examples), stage 1. (a) This sketch illustrates the first stage of deformation of a sulphate layer: folds develop as a response to volume increase, resulting from diapiric expansion of evaporites. They are made of fine, palisade-like gypsum crystals disposed normal to the edges of the ribbon (b, c). The ridge of each fold is in turn gently folded (arrows in a, and b) so that a tangential/oblique section of a folded system results in a stacked set of half-moon shaped nodules and contorted veins/funnels. These three-dimensional forms are comparable to those shown in the polished slab of an Archaean sample (e). The nodules are composed of a cortex of herringbone calcite (d) disposed normal to their edges (arrows), which outline a core, originally empty, later filled by geodic cement (f). (b, c) Hand samples from the Faltona Quarry and Montalcinello Mine, Messinian of Tuscany. (d, f) Photomicrographs of parts of sample taken at Stop 5.4A (f, XPL). Kuruman Kop section. (e) Cut sample from Rau 5.1.2; Kuruman Kop section.

infillings of large limpid crystals with normal extinction. Herringbone calcite (consisting of spherulitic calcite aggregates, Sumner & Grotzinger 1996) encloses and appears to intrude rolled up ribbons and bundles of filaments (Figs 2b, c, 3a–c & 7d, e) that locally thin out to single threads (Figs 12e & 14b). Three different, commonly associated, mosaics formed by these spherulitic aggregates of cloudy, inclusion-rich calcite, have been identified, that often merge into each other:

- herringbone calcite *sensu stricto*, made of elongated-lenticular, crenulated crystals arranged in remarkably regular mosaics

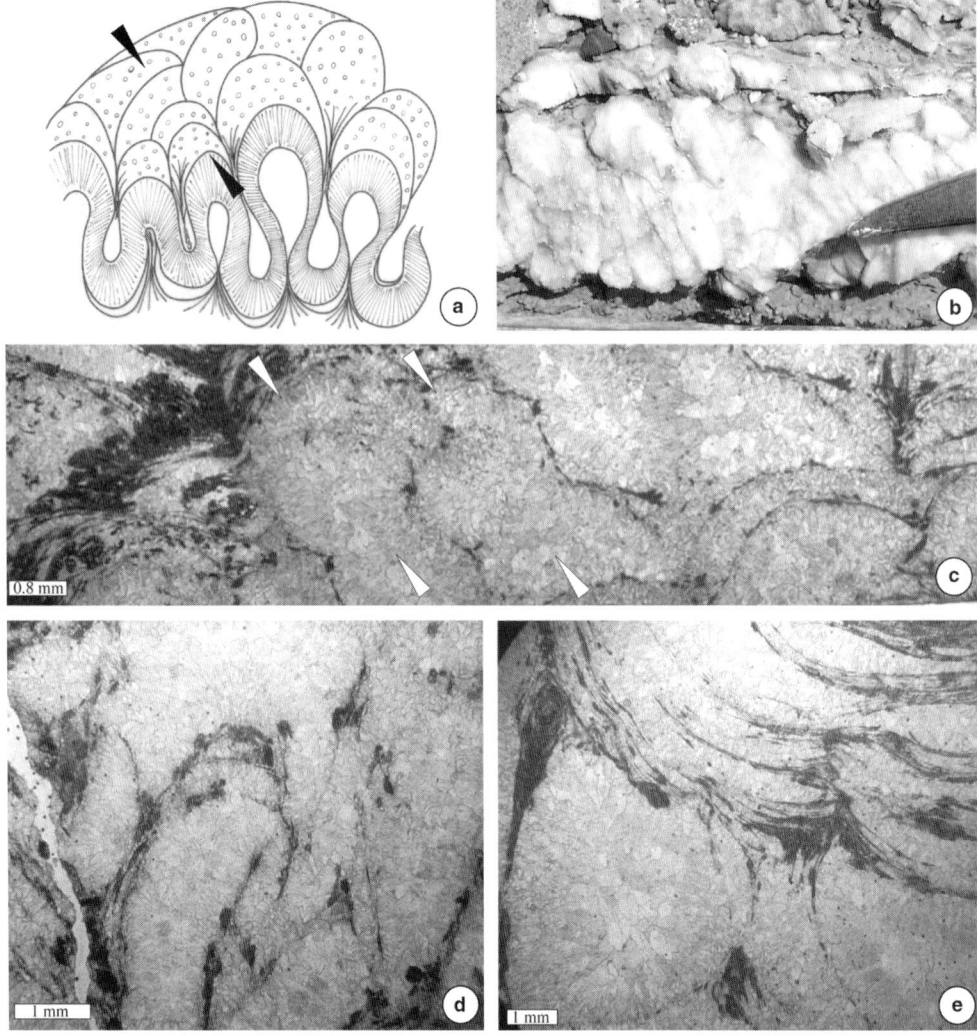

Fig. 12. Evolution of enterolithic folds, stage 2. Demand for space increases: as the folds swell (**a, b**) the microbial filaments are stretched to the point of disappearance (**c–e**) while every other fold is tightly squeezed (d, e) and their hinges or convergence zones entrap crushed remnants of the microbial filaments (c–e). A tangential/oblique section of the system shows chains of well-rounded nodules (c, arrows) associated with half moon ones (e). (a) Sketch of second stage development showing swelling folds. (b) Outcrop photo of sample similar to (a) from Faltona Quarry, Messinian of Tuscany. Note pick tip for scale (at right). (c) Photomicrograph of sample Rau 5.4; Kuruman Kop section. (d) Photomicrograph of samples Rau 24C; & R36 Echo Caves). (e) Photomicrograph of sample Rau 5.1; Kuruman Kop section.

(Figs 3b & 8a, b), composed of alternating, light and dark zebra-like bands of lenticular crystals with undulated optical extinction, disposed with the long axes perpendicular to the banding;
- flamboyant calcite forming a felted fabric made of shorter, irregularly lenticular crystals randomly disposed, or rosettes with undulose extinction (Figs 8a, b, 11d, f & 13c, d);

- subhedral, sucrosic calcite with indented crystal boundaries and undulose extinction (Figs 3a, 7d & 9c, e, f). This forms a coarser mosaic, probably related to incipient later recrystallization of herringbone calcite.

The transitions from herringbone-to-flamboyant-to-herringbone fabric may be sharp or gradual,

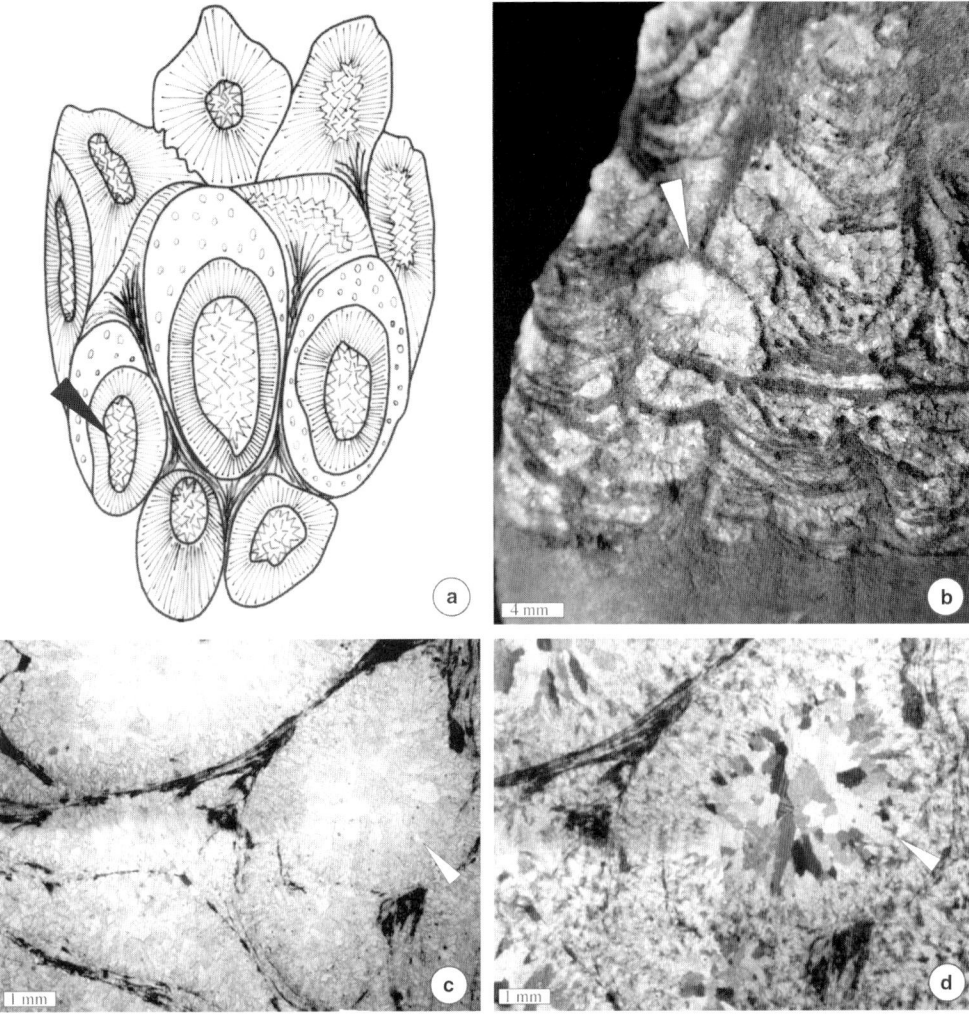

Fig. 13. Evolution of enterolithic folds, stage 3. (**a**) Sketch of the stage 3 folds: the 'necks' are thinned and then the heads severed and isolated to form subspherical nodules. (**b**) Features from (a) are evident even on the weathered surface of an Archaean sample. (**c, d**) Thin sections (photomicrographs) clearly showing the cortex of herringbone calcite with undulose extinction, with a core filled with geodic, prismatic/equant calcite with normal extinction (arrows). Sample from stop Rau 5.4 (d, XPL). Kuruman Kop section.

depending on the angle at which the thin section is cut through the spherulitic aggregates. This arrangement suggests that herringbone and flamboyant fabrics can be interpreted as two aspects of the same felted fabric. The tight, well-ordered mosaic (Figs 3b & 8a, b) appears where the section is cut parallel to the main orientation of the spherulitic aggregates while, where obliquely cut, the elongated crystals appear as randomly disposed flamboyant calcite (Fig. 8a, b).

This arrangement of the calcite crystals, implying an original, three-dimensional sub-spheroidal structure, with the crystals of the herringbone fabric disposed perpendicular to the nodule walls, has been frequently found (Figs 11d–f & 13c, d).

The nodules appear to be made of a cortex of herringbone/flamboyant calcite and an empty core filled with prismatic/equant calcite (Figs 8a, b, 11d, f & 13b, d) that is in sharp contact with the herringbone and flamboyant calcite (Figs 8b, c & 9e).

Interpretation Herringbone/flamboyant calcite has formerly been interpreted as a cement precipitated directly on the sea floor (Sumner & Grotzinger 1996). The only true cement found in these sediments, associated with the carbonates of the

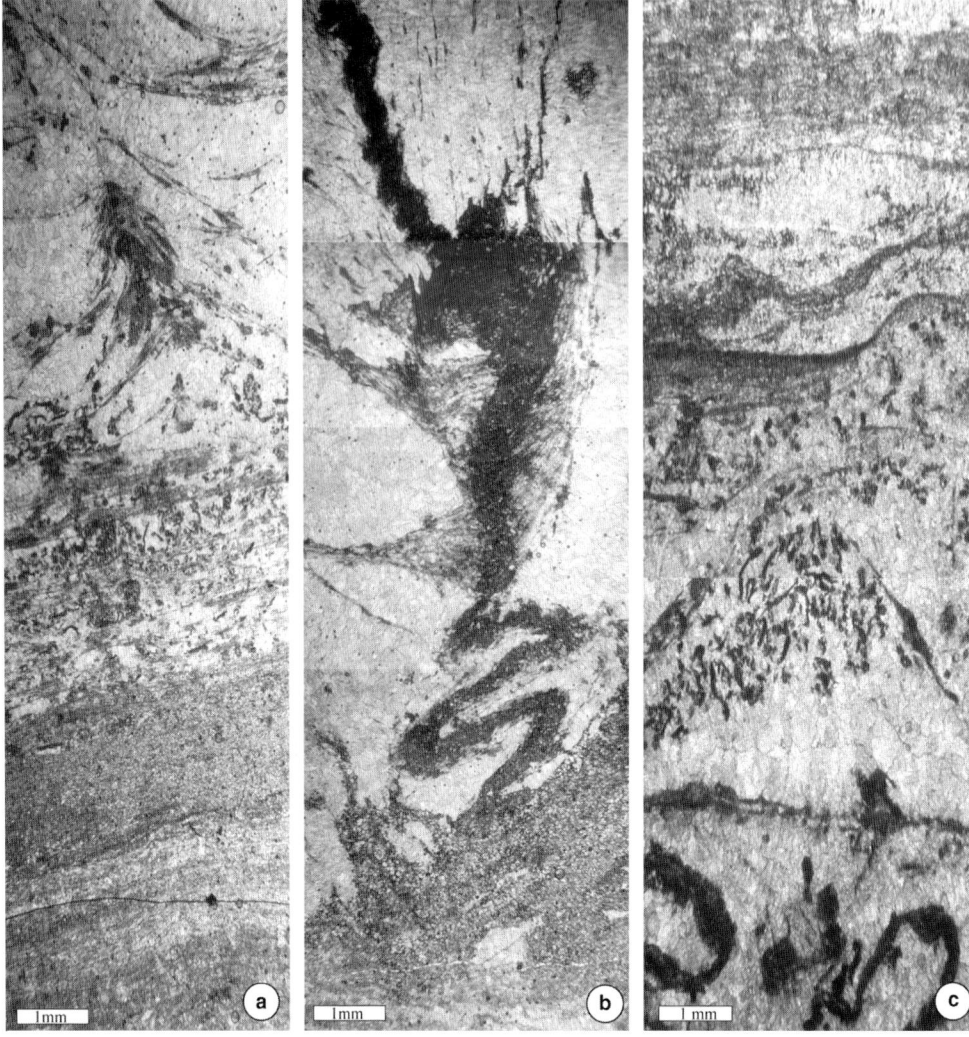

Fig. 14. The fate of the microbial mats: from well bedded sediment (**a**) to antiform/diapiric structures (**b**), to contorted compressed, elongate bundles, and stromatoclasts floating in herringbone calcite (**c**). The genesis of such peculiar fabrics can be related to the volume increase of a ductile precursor sediment, leading to enterolithic folds and the formation of nodules which displacively disrupts the well bedded sediment, and pushes apart single or bunched filaments (a). In the zones of compression between enterolithic fold 'limbs', and at the contact of nodules, laminae are folded, crushed and dolomitized (a, b) and in the zones of strain are stretched and fragmented to form stromatoclasts (a, c). This process ends with the total effacement of the depositional signature. (a) Thin section: sample Stop 5.1.A.II; Kuruman Kop section. (b) Thin section: sample rau 5.1.2; Kuruman Kop section. (c) Thin section: sample Stop 5.1.A; Kuruman Kop section.

Gamohaan and Frisco formations, is the prismatic/equant calcite, which is commonly interpreted as a non-marine, late-diagenetic cement that is found in evident sharp contact, associated with herringbone/flamboyant crystalline rinds (Figs 8b & 13b–d). Careful observation of the crystalline mosaics of the herringbone/flamboyant calcite, and their relationships with unmistakably sedimentary components and with true associated cement, shows that the herringbone/flamboyant calcite represents a sediment rather than a cement. A cement with this kind of crystal fabric has not been described previously, neither should it be mistaken for the distinctive crystal fabric of radiaxial calcite (Sumner &

Grotzinger 1996) formed by large, solid prismatic crystals characterized by distally divergent subcrystals, cross-cut by convergent optic axes (Kendall 1985).

Thus, the herringbone/flamboyant calcite represents a precursor sediment, which acted as a support for the later growth of the cement. The precursor was soft sediment able to migrate within the microbial mat, displacing and deflecting (Figs 2b, c, 3a, c & 7d) the filaments and gradually growing in half-moon shaped (Figs 3c, 11d, 12c, e & 13b, c) and then subspherical bodies/nodules (Figs 8b, 12e & 13b, c).

The fabric of the herringbone/flamboyant mosaics, comprising crystal aggregates with sweeping extinction, displays morphological and mineralogical features that can be compared with the micro-flamboyant quartz described by Milliken (1979) and Arbey (1980) as replacive of sulphate (gypsum) deposits. The shape and arrangement of the nodules are reminiscent of core of originally hollow nodules [equivalent to West's (1979) 'spherical vugs'] that have been later filled by true geodic cement.

Moreover, since marine cements with this kind of undulose extinction, tight, felt fabric and thin lenticular component crystals, have never been described in the Phanerozoic and recent carbonates, and since micrite, representing the lime mud, is commonly missing, the precursor ductile sediment in which the filaments and grains were embedded is likely to have been a gypsum-mush. An alternative is anhydrite which, according to Hardie (1967) and Cody & Hull (1980), can precipitate instead of gypsum directly from more concentrated waters at temperatures lower than 50 °C, in a subaqueous environment, rich in organic matter. Subaqueous sulphate associated with organic material in a sediment associated with minor carbonate can also be found in present day marginal marine salinas, sabkha and crystallizer ponds and salterns (e.g. Krumbein et al. 1977; Taberner et al. 1997; Canfield et al. 2004).

Early diagenetic features

Early diagenetic processes can be recognized from the presence of selective dolomitization, moulds of evaporitic minerals, and a variety of deformational structures (Wright & Altermann 2000; Gandin et al. 2005), commonly occurring in planar beds associated with laminated or stromatolitic layers, and comprising antiformal structures, autoclastic solution/collapse breccias, cuspate and nodular 'fenestrate' fabrics. Collapse breccias are found associated with many of the other soft-sediment deformational structures.

Dolomitization

Dolomitization selectively affected grains and microbial filaments. Stromatoclasts, ribbons, laminae and vertical 'stems'/'supports' are commonly the site of concentration of dark/rusty, organic-rich, microdolomite rhombohedra (Figs 4c, 7d & 14a–c). Dolomitization of the grains appears to have been an early diagenetic process that occurred within the precursor matrix, before the formation of the enclosing herringbone calcite, probably linked to bacterial sulphate reduction associated with degradation of cyanobacterial filaments (Wright 1997; Wright & Altermann 2000). This is supported by the frequent occurrence of iron oxides and pyrite (e.g. Fig. 7e) of likely early diagenetic origin.

Pseudomorphs after evaporitic minerals and nucleation cones

Description In addition to the already reported moulds of halite cubes (Simonson et al. 1993; Sumner & Grotzinger 2000, 2004) occasionally exposed on bed surfaces (Fig. 3d), vertically oriented calcite crystals that resemble selenite crystals (Fig. 6c) may occur either as thickets of 'grass-like' swallow-tail selenite overlying laminated sediments (Fig. 6a) or as bands of white calcite with a chevron-like structure (Fig. 6b).

Cauliform-like (cavoli) structures growing displacively in laminated or thinly bedded sediments (Fig. 6e) are comparable with Messinian nucleation cones (Fig. 6d), also called cauliflower clusters or 'sea-floor fans' (Dronkert 1976, 1985; Schreiber et al. 1976), consisting of crystals oriented normal and oblique to the bedding plane, and underneath their base previous existing lamination is bent and deflected while overlying laminae may also curve downwards around the cone top.

Interpretation The moulds with partially hollow interiors characteristic of hopper halite crystals, and associated probable gypsum and/or celestite pseudomorphs, found on the surface of a bed intercalated with other layers bearing structures evocative of sulphate deposition, suggest general evaporitic conditions and a normal evolution of the evaporative deposition of salts. In modern evaporitic sediments, chevron or swallow-tail selenite crystals typically form *in situ* on the floor or in the upper part of the bottom sediment of sabkha lagoons or of pre-evaporative salt pans, through crystallization from pore-waters and displacement of the bottom sediments. They are frequently found in the Messinian (Testa & Lugli 2000, and references therein), within finely laminated gypsum.

The cone structure which has punctured and shouldered aside adjacent underlying and overlying layers before they were lithified, and thus postdates their deposition, clearly indicates diagenetic growth and displacement within the original sediment. Brown, thin sandy beds and laminae are clearly disrupted by the growth structure, being deflected downwards and thinned as they approach the cone. The structure thus disrupts the adjacent laminae, and depresses, compresses and truncates those at and near the cone's base – clear evidence that it postdates them. Darker patches can be seen near the base of the cone, indicative of the soft brown sandy sediment being disaggregated and dispersed by the cone, with some of that sediment being incorporated within the base. The possibilities for explaining the behaviour during growth of this structure are limited, but nucleation cones (or cauliform structures) in evaporites are known to form by expansive and displacive diagenetic growth within soft sediment. It is therefore interpreted as a formerly evaporitic nucleation cone that displaced the surrounding sediment during differential expansion of gypsum/anhydrite during its precipitation in the sediment, and which was later replaced by carbonate.

Antiformal structures

Description Antiformal structures are observed in the upper part of the Kuruman Kop section, where they appear as incompletely developed antiform folds, fading upwards into flat undisturbed layers, lacking expansion crests and internal sediment (Fig. 7a). The fold hinge is characterized by the accumulation of yellowish, dolomitized material that evidently was deposited in between the grey beds and migrated towards the top during the deformational event. The antiform folds overlie a bed composed of large composite, plumose nodules interpreted in this paper as pseudomorphs of sulphate nodules. At the microscopic scale, small tepee-like structures are made from microsparitic or granular laminae which show a deformation growing upward and ending with the complete disruption of the upper layers (Figs 2b, c & 7d, e).

Interpretation The macro and micro antiform structures are different in shape and origin from true tepees, which are characterized by raised and buckled edges with geopetal infilling of vadose silt and commonly are bounded at the top by an erosional discontinuity. They are typical of the subaerial part of arid peritidal environments (Kendall & Warren 1987). The regularity of the Neoarchean deformed beds suggests a subtidal depositional setting. However, similar tepee-like features have been described in calcrete soils (Kendall & Warren 1987) and observed in a Triassic succession of southern Sardinia (Fig. 7b), where abundant calcite pseudomorphs of gypsum and anhydrite in the carbonate beds and spectacular intrastratal collapse breccias, document and confirm the original presence of now vanished, primary evaporite intercalations (Cocozza & Gandin 1976). The process of formation of the Triassic as well as the analogous Neoarchean antiform folds can be interpreted as diagenetic structures formed in near surface or very shallow burial conditions, probably induced by the upward-running flow of interbedded sulphates exerting on the carbonate beds pressures sufficient to overcome the lithostatic weight of the overlying sediments (Gammon 2004).

Collapse breccias

Description intrastratal collapse and solution breccias are commonly associated with the diapir-like cones and with planar and vertical fractures filled with multiple rims of white diagenetic cement (Fig. 5c, d). They are composed of matrix-supported sub-rounded or angular clasts of dolomite bordered by parallel, folded, stretched and buckled, sparry veins. The fine-grained, contorted and disrupted mat facies, studded with irregularly shaped sublinear calcite blebs, appear to have intruded, pierced and displaced overlying sediment, thinning and breaking dolomitized layers which now appear as broken, corroded rafts (see Alberto *et al.* 2007).

Interpretation The genesis of observed collapse breccias can be best explained with the gradual flow of a plastic matrix and the brittle behaviour of the formerly interbedded dolomitic layers. Similar chaotic breccias are considered as the result of diagenetic processes which led to the complete dissolution or liquefaction of the original salt and gypsum matrix and the collapse of the fragments of the brittle, carbonate intercalations (Anadón *et al.* 1992; Gandin *et al.* 2000; Kendall 2001; Rouchy *et al.* 2001). Locally in the Kuruman Kop section, the solution collapse breccias have been cemented, in proximity to a fault, by abundant white blocky spar and saddle dolomite. The occurrence of saddle dolomite (Gregg & Sibley 1984) and of slightly twinned calcite (types I and II of Burkhard 1993) suggests a rather high temperature (about 200 °C) of the localized, fault-related circulating fluids.

Columnar structures

Description Columnar structures up to 15 cm high, interspersed within contorted and disrupted mat facies and separated by convex-down laminae (Figs 7c & 8e), appear to have pushed upwards at various points into overlying flat laminae, so that the flexible, partially degraded organic laminae are now draped between the columns and the more brittle, dolomitized layers are left floating as broken, corroded rafts. Similar structures occur also at the microscopic scale, associated with the nodular fabric (Figs 7d, e & 14a, c).

Interpretation This structure provides evidence for the early diagenetic folding of originally ductile, filamentous flat layers with the formation of displacive cones which started to rise, displacing the overlying sediment. The development of these structures, and the folding and subsequent fracturing of the encasing/encased brittle layers, may be related to diapiric migration accompanying volume increase through rehydration of evaporites. Similar diagenetic disruption of the primary arrangement of the depositional layers may also have been caused by inverted density gradients within alternate sulphate, organic and carbonate beds as well as by the growth of clusters of selenite crystals.

Nodular 'fenestrate' fabric

Description Complex structures formed by variably-sized nodules of white calcite associated with grey, herringbone calcite (Fig. 8e), arranged in more or less dense concentrations (Fig. 2a) forming irregular bands (Fig. 10a) or large composite nodular structures ('plumose', Figs 9a & 10f). The microfacies of samples taken from these apparently different structures are characterized by assemblages of white crystalline domains composed of one or more concentric bands of radially disposed spherulitic aggregates or rosette-like bunches of herringbone/flamboyant calcite. Their shapes, illustrated in Figures 3a, 7d, e, 8b, e, 9b, c, e, f, 11d–f, 12c–e & 13b–d, vary from subrounded/nodular to biconvex to half-moon, to loop-shaped (see also fig. 4a in Sumner & Grotzinger 2000). These calcite domains are encased and bounded in a web of compressed and contorted ribbons, bundles, clusters and single threads of more or less dolomitized filaments. They appear to correspond to subrounded nodules made from herringbone/flamboyant (Figs 8b, e, 9b, 11d & 13c, d) or subhedral sucrosic (Figs 3a, 9e, f & 12c–e) calcite.

The herringbone/flamboyant calcite forms a cortex of needle-like acicular crystals disposed normal to the edge of the nodule, which outlines a core filled by geodic cement (Figs 3a, 8b, 9e, 11d, f, 12e & 13c, d). The nodules often occur in stacked sets of half-moon shapes (Figs 3c, 11d, e, 12d, e & 13b, c), or chains of irregularly sub-rounded domains (Fig. 12c), or display structures that can be related to enterolithic folds (Figs 9e & 12d). The hinge zone of these folds can be tightly folded so that only relics of the included organic filaments are preserved (Figs 9e & 12d). Alternatively, less tight folding would have created structural voids formed by the fold loops, that represent the only cavities that may have formed within such a plastic material, and that were filled by geodic prismatic and equant cements (Figs 3a, 9e, 8b, 11d, f, 12e & 13c, d). This fabric, reminiscent of the chicken-wire fabric (Figs 2a, 7d, e, 9a, b, f, 10f & 12c), is bounded by the remains of organic filaments or by thicker, dark, crinkly, cable-like (Figs 2b, 3b, 4c, 7e, 8b, 9b & 14b) structures (the 'supports' of Sumner 1997; Sumner & Grotzinger 2004) formed by tightly compressed organic filaments which at the end open in flower-like bundles of closely pressed and sometimes curled filaments (Figs 3b, c & 14b). Folded laminae of partially degraded organic material (Figs 2b, c, 3a, c, 7d, e, 9c & 14a–c), and isolated organic filaments and fragments (Figs 2d, 3b, c, 7d & 14a, c) locally appear to float in the calcite mosaic. The laminae forming the contorted mats, mainly represented by preserved remains of bundled cyanobacterial sheaths (Figs 2c & 3c), but also by dolomicrosparitic ribbons (Figs 2b, 3a, c & 9c), show no signs of compaction suggesting that mineralization occurred before any compaction could take place, but after deformation during an early diagenetic stage. The organic filaments display a plastic behaviour (Figs 2b–d, 3a–c, 7d, e, 9c, 11e & 14a–c), which demonstrates that when they were deformed, they were still soft (Klein *et al.* 1987). Uplifted single filaments and contorted cable-like dark bundles at the transition with the filamentous mat, suggest that each filament has been pushed upwards by the growth of a ductile precursor of herringbone calcite. Displacement and digestion of the parent primary sediment results from the growth of gypsum nodules, the growth of the nodules pushes the filaments upwards and displaces bundles of them that are folded and crushed in the zones of compression and stretched in the zones of strain.

Interpretation The deformation affecting the organic structures can be explained as derived from the radial pressure exerted by the growth of evaporite nodules, during the coalescence of enterolithic folds, and the consequent compaction of the elastic organic filaments squeezed between them.

The relationships between the laminite shreds and the calcite mosaics are very complex, however, and demonstrate that the volume increase of the calcite precursor resulted in the formation of enterolithic folds (Figs 9c, e & 10a) evolving into nodules (Figs 2a, 9a, b, f & 10a, f) and in the meantime in pushing upwards (Figs 2b, 3b, c, 7d, e, & 9a,

b) and laterally (Figs 3a & 8e) sets of laminites that were folded and crushed.

This behaviour can be related to a plastic sulphate deposit and to the pressure produced by its growth as a consequence of a probable change from anhydrite to gypsum. A concurrent factor to deformation of the laminites and the movement of the sulphate mass, may have also been the gas escaped from the decay processes (Klein et al. 1987) of the organic component which grew within it.

The origin of the 'dirty' dolomite may be derived from early diagenetic degradation of the cyanobacteria sheaths involved in the deformational process that caused the planar depositional organic laminae to swell, curl and break. This process was probably driven by volume increase of the sulphate matrix and consequent nodular growth, which pushed aside and compressed insoluble organic matter between adjacent nodules or the limbs of enterolithic folds.

Dolomite probably formed in association with progressive degradation of microbial sheaths located along the pressure/solution seams separating anhydrite nodules (El Tabakh et al. 1998; Schreiber & El Tabakh 2000). Cathodoluminescence confirms this (Fig. 4c) as the recalcitrant Mg-rich sheaths were the last element of the cyanobacteria to be degraded, when sulphate was largely exhausted.

Discussion

The complex structures recognized in the Neoarchaean Campbellrand and Malmani carbonates in this study, unreported from Phanerozoic calcareous platforms, can be related to deformation processes affecting a soft, ductile material that may be envisaged to have been a sulphate deposit. Contorted cuspate and nodular forms have been described as 'fenestrate' fabrics and interpreted as peculiar microbial build-ups produced by organic communities supported only by precipitated carbonate under the form of herringbone calcite (Sumner 1997; Sumner & Grotzinger 2004).

They actually show structural characteristics that suggest soft-sediment deformational processes generated by volume variations, probably induced by crystal growth.

Most of these features appear strikingly similar to the nodular structures resulting from the evolution of enterolithic folds of diagenetic anhydrite growing in the modern sabkhas (Hardie & Eugster 1971) and to the diagenetic and tectonic deformational structures that can be observed in the Mediterranean Messinian evaporites (Schreiber et al. 1976; Schreiber 1989; Testa & Lugli 2000), such as nucleation cones/cauliflower clusters (Fig. 6c, d), tepee-like antiforms (Fig. 7a, b), vertical diapiric flows (Fig. 7c–e) and nodular fabric (Figs 2a, 8b, e, 9a, b, f, 10, 11d–f, 12 & 13).

The sedimentological and petrographic analysis of the Neoarchean fabrics suggests that the early diagenetic, deformational behaviour of the precursor sediment of herringbone/flamboyant calcite, is comparable with that resulting in the nodular structures of Messinian gypsum.

Herringbone calcite, described by Sumner & Grotzinger (1996) as a marine cement, is a poorly known crystalline form of calcite so far reported only from Archean and Early/Middle Proterozoic carbonates and probably from Silurian reefal deposits of the Gaspé Basin (Sumner & Grotzinger 1996). However, the herringbone calcite cannot be a true cement, since there is no evidence that it grew from the walls of cavities. It rather appears to be the result of replacement of a primary soft and plastic sediment, with the rheological properties of sulphates. This assumption is also supported by the herringbone crystal aggregate texture, characterized by sweeping extinction (Figs 8a, b, 11f & 13d), which appears very similar to that of the Messinian gypsum selenite crystals (Fig. 8c, d) and to the felted or alabastrine textures of older evaporites (Holliday 1970) recognized also in replacive carbonates (Chafetz 1980) or in silicified evaporites (Arbey 1980).

Field relationships and petrographic analysis clearly show that herringbone calcite was contemporaneously associated with microbialites. Examination of a polished slab of a partially calcified, laminated sapropel permits the observation of herringbone calcite that has progressively replaced organic-rich sediment (Fig. 11e) (W. Altermann, pers. comm. 2004). The calcite occurs as thin stringers or isolated trains of lenses, embedded in the dark laminae and which in other parts of the slab coalesce into thicker bands of herringbone calcite. The polished slab may thus preserve a stage in the progressive replacement of an original sapropel by herringbone calcite as a result of bacterial sulphate reduction of interlayered evaporites. If carried to completion, this process would have formed a bed comprising entirely herringbone calcite, preserving no clue to its former composition, and allowing it to be interpreted (erroneously) as a chemical precipitate directly on the seafloor. Entire beds of herringbone calcite may have sequentially replaced gypsiferous sapropels and/or stratiform microbialites, and were thus not directly precipitated encrustations on the seafloor. Cathodoluminescence of the microbialitic calcite (Fig. 4c) implies an anaerobic, reducing diagenetic environment (Wright & Alterman 2000). The blocky subhedral crystals, which are found in irregular patches and nodules (Figs 3a, 7d, 9c, e, f, 12c, d & 13b) always associated with the herringbone/flamboyant calcite, resemble the cloudy amoeboid texture of sucrosic gypsum in Messinian nodules (Testa & Lugli 2000). On the contrary the prismatic and equant calcites, which have all the characteristics of a cement, precipitated in the rare structural cavities that are commonly found in the middle of the herringbone/flamboyant mosaic (Figs 3a, 8b, e, 9e, 11d, f, 12e & 13c, d). It grows directly from the herringbone/flamboyant calcite with a rather sharp boundary, forming a geodic structure that suggests the previous existence of a structural cavity corresponding to a void in the loop left open at the fold hinge of enterolithic folds (Figs 3a, 8a, b, 11, 12 & 13) that formed in the sulphate. Volume variations involving anhydrite-to-gypsum-to anhydrite changes are commonly interpreted as driven by hydration/dehydration processes (Holliday 1970; Testa & Lugli 2000).

Depositional and diagenetic evolution

The restricted shallow-marine basin was dominated by benthic microbial mats and stromatolites that apparently only in rare cases built their domes in a carbonate regime with the production of micrite. The main precipitate in the Campbell/Malmani carbonate beds appears to have been calcium sulphate, whereas calcite was almost absent, since no micrite matrix has been so far been detected in these sediments (Sumner 2001).

During the early phases of diagenesis the following processes can be detected:

(1) local deformation of the original bedding due to migration of and displacement by sulphates;
(2) probable contemporaneous formation of calcite and dolomite as a consequence of bacterial sulphate reduction, either below the sulphate reducing zone, or after the sulphate has been locally consumed;
(3) further migration and volume increase of solid sulphate, with interbedded organic material, led to subsequent stages of deformation, giving rise to structures such as enterolithic folds, nodular gypsum, contorted laminae, nucleation cones, diagenetic tepees, and collapse breccias–all of which were preserved by progressive carbonate mineralization of the sediments.

Discussion: gypsum to calcite

The absence of preserved evaporites in sediments of Archaean age is not unexpected (e.g. Scholle et al. 1992; Taberner et al. 1997; Hardie 2003). The major metabolic processes capable of mediating ambient water chemistry associated with benthic microbial communities today are oxygenic photosynthesis in the living surface colonies and bacterial sulphate reduction in the underlying anoxic decaying mat. The sedimentological and petrographic data above provide credible evidence for a shallow subaqueous environment with evaporite precipitation associated with microbialitic sediments supporting widespread bacterial sulphate reduction in the Kogelbeen and Gamohaan Formations. It is known that sulphate reduction alongside iron reduction will actively drive calcite precipitation (e.g. Coleman 1985; Hendry 1993; Kowalski et al. 2001–2002). This is likely to have been the case in the Kogelbeen and Gamohaan depositional environments, where craton-wide cyanobacterial colonies oxidized reduced sulphur species in a largely reducing ocean, providing sulphate and supporting bacterial sulphate reduction on a regional scale. This is supported by results from sulphur isotope analyses of individual pyrite nodules from a sample of the Campbellrand carbonates (Gandin et al. 2005) which show a range of -6.2 to $+8.1$ per million CDT, indicating fractionation by sulphate reducing bacteria.

The petrographic evidence indicates that herringbone calcite has developed by replacing sulphate precipitates, after they began to deform and before they completely destroyed the organic material in which they originally were deposited. The shapes of the laminae between the 'supports' or 'hinges' are the result of evaporite expansion, with compression at the margins of enterolithic folds or nodules related to the diapiric migration of the sulphate, deforming originally planar laminae (Figs 2b, c, e, 3a–c & 7d, e). The lateral and upward growth of the folds and nodules also results in their final coalescence to form the chicken-wire fabric (Figs 2b, 7c & 9b, f).

Although the calcite mosaic enclosing the filaments appears very compact and no primary or dissolution porosity has been found, replacement as a process leading to the calcitization of the primary sediment is probable. Calcified evaporites are frequently cited in the literature even if the mechanism of replacement is not understood, but among other interpretations, it may have acted through bacterial sulphate reduction (Kendall 2001; Gandin et al. 2005) or within cyanobacterial sheaths with the production of calcite spherulites (Verrecchia et al. 1995).

Recent laboratory experiments suggest that potential depositional rates, biomediated carbonatogenesis, given time and abundance of bacteria (Castanier et al. 1999), may explain herringbone calcite production.

Conclusions

The sedimentation in the Neoarchean Campbellrand/Malmani platform was at times represented by huge quantities of microbial filaments arranged in stromatolites, detrital laminites and sulphates. The granular sediments comprise diagenetically calcitized allochems but do not show marine cements. Diapiric movements in solid sulphates caused the development of deformational structures. Sulphates were calcified by the activity of sulphate-reducing bacteria. It is likely that replacement of evaporites occurred during early diagenesis and through the early stages of burial associated with bacterial sulphate reduction. The remaining sulphates were partly replaced by chert, partly dissolved by late diagenetic hydrothermal waters, causing the formation of collapse breccias. It is likely that the sulphates were composed of both gypsum and anhydrite formed during repeated cycles of de- and rehydration.

The data resulting from the petrographic analysis of the present carbonate Campbellrand/Malmani beds, suggest that the rarity of preserved evaporites in the Archaean/Proterozoic can be explained by the domination of the environment by microbes, especially cyanobacteria and sulphate reducers, which were able to colonize every available environment unchallenged at the time. The anoxic conditions of the degrading organic material just beneath the growing surfaces of microbial mats and stromatolites, together with evaporitic sulphate in the colonized environment, provided the perfect environment and food source for the SRB. Large-scale sulphate reduction led to large-scale carbonate precipitation and replacement.

We would like to thank C. Taberner for her constructive criticism of the paper, and B. C. Schreiber for critical encouragement; the manuscript has benefited significantly as a result of their diligence, and through the critical work of an anonymous reviewer. We gratefully acknowledge the assistance provided by Barbara Terrosi and Antonella Mancini in the preparation of figures and drawings.

References

ALBERTO, W., CARRARO, F., GIARDINO, M. & TIRANTI, D. 2007. Genesis and evolution of 'pseudocarniole': preliminary observations from the Susa Valley (Western Alps). *In*: SCHREIBER, B. C., LUGLI, S. & BĄBEL, M. (eds) *Evaporites Through Space and Time*. Geological Society of London, Special Publications, **285**, 129–142.

ALTERMANN, W. & SCHOPF, J. W. 1995. Microfossils from the Neoarchean Campbell Group, Griqualand West Sequence of the Transvaal Supergroup, and their paleoenvironmental and evolutionary implications. *Precambrian Research*, **75**, 65–90.

ALTERMANN, W. & NELSON, D. R. 1998. Sedimentation rates, basin analysis and regional correlations of three Neoarchean and Palaeoproterozoic sub-basins of the Kaapvaal craton as inferred from precise U–Pb zircon ages from volcaniclastic sediments. *Sedimentary Geology*, **120**, 225–256.

ALTERMANN, W. & SIEGFRIED, H. P. 1997. Sedimentology and facies development of an Archean shelf–carbonate platform transition in the Kaapvaal Craton, as deduced from a deep borehole at Kathu, South Africa. *Journal of African Earth Science*, **24**, 391–410.

ANADÓN, P., ROSELL, L. & TALBOT, M. R. 1992. Carbonate replacement of lacustrine gypsum deposits in two Néogene continental basins, eastern Spain. *Sedimentary Geology*, **78**, 201–216.

ARBEY, F. 1980. Les formes de la silice et l'identification des évaporites dans les formations silicifiées. *Bulletin Centre Recherche Exploration–Production Elf-Aquitaine*, **4**, 309–365.

ASPLER, L. B. & CHIARENZELLI, J. R. 2002. Mixed siliciclastic–carbonate storm dominated ramp in a rejuvenated Palaeoproterozoic intracratonic basin: Upper Hurwitz Group, Nunavut, Canada. *International Association of Sedimentologists*, Special Publications, **33**, 293–321.

BADHAM, J. P. N. & STANWORTH, C. W. 1977. Evaporites from the Lower Proterozoic of the East Arm, Great Slave Lake. *Nature*, **268**, 516–517.

BOULTER, C. A. & GLOVER, J. E. 1986. Chert with relict hopper moulds from Rocklea Dome, Pilbara Craton, Western Australia: an Archaean halite-bearing evaporite. *Geology*, **14**, 128–131.

BROCKS, J. J., LOGAN, G. A., BUICK, R. & SUMMONS, R. E. 1999. Archean molecular fossils and the early rise of eukaryotes. *Science*, **285**, 1033–1036.

BUICK, R. & DUNLOP, J. S. R. 1990. Evaporitic sediments of early Archaean age from Warrawoona Group, North Pole, Western Australia. *Sedimentology*, **37**, 247–277.

BURKHARD, M. 1993. Calcite twins, their geometry, appearance and significance as stress-strain markers and indicators of tectonic regime: a review. *Journal of Structural Geology*, **15**, 351–368.

CANFIELD, D. E. & TESKE, A. 1996. Late Proterozoic rise in atmospheric oxygen concentration inferred from phylogenetic and sulfur isotope studies. *Nature*, **382**, 127–132.

CANFIELD, D. E., SØRENSEN, K. B. & OREN, A. 2004. Biogeochemistry of a gypsum-encrusted microbial ecosystem. *Geobiology*, **2**, 133–150.

CASTANIER, S., LE MÉTAYER-LEVREL, G. & PERTHUISOT, J.-P. 1999. Ca-carbonates precipitation and limestone genesis – the microbiogeologist point of view. *Sedimentary Geology*, **126**, 9–23.

CHAFETZ, H. S. 1980. Evidence for an arid to semi-arid climate during deposition of the Cambrian System in central Texas, U.S.A. *Palaeogeography, Palaeoclimatology, Palaeoecology*, **30**, 83–95.

COCOZZA, T. & GANDIN, A. 1976. Età e significato ambientale delle facies detritico-carbonatiche dell'Altopiano di Campumari (Sardegna sud-occidentale). *Bollettino della Società Geologica Italiana*, **95**, 1521–1540.

CODY, R. D. & HULL, A. B. 1980. Experimental growth of primary anhydrite at low temperatures and water salinities. *Geology*, **8**, 505–509.

COLEMAN, M. L. 1985. Geochemistry of diagenetic non-silicate minerals: kinetic constraints. *Transactions of the Royal Society of London*, **A315**, 39–56.

DRONKERT, H. 1976. Late Miocene evaporites in the Sorbas basin and adjoining areas. *Memorie della Società Geologica Italiana*, **16**, 341–362.

DRONKERT, H. 1985. *Evaporite models and sedimentology of Messinian and recent evaporites*. PhD thesis. GUA Papers of Geology, Series 1, **24**, Amsterdam.

EL TABAKH, M., SCHREIBER, B. C., UTHA-AROON, C., COSHELL, L. & WARREN, J. K. 1998. Diagenetic origin of Basal Anhydrite in the Cretaceous Maha Sarakham salt: Khorat Plateau, NE Thailand. *Sedimentology*, **45**, 579–594.

GAMMON, P. R. 2004. *Neoproterozoic Cap Carbonate Tepee Structures*. St Catharines 2004 Technical Program; prgammon@yahoo.com.au

GANDIN, A., GIAMELLO, M., GUASPARRI, G., MUGNAINI, S. & SABATINI, G. 2000. The Calcare Cavernoso of the Montagnola Senese (Siena, Italy): mineralogical–petrographic and petrogenetic features. *Mineralogica et Petrographica Acta*, **43**, 271–289.

GANDIN, A., WRIGHT, D. T. & MELEZHIK, V. 2005. Vanished evaporites and carbonate formation in the Neoarchaean Kogelbeen and Gamohaan formations of the Campbellrand Subgroup, South Africa. *Journal of African Earth Sciences*, **41**, 1–23.

GREGG, J. M. & SIBLEY, D. F. 1984. Epigenetic dolomitization and the origin of xenotopic dolomite texture. *Journal of Sedimentary Petrology*, **54**, 908–931.

HABICHT, K. S., GADE, M., THAMDRUP, B., BERG, P. & CANFIELD, D. E. 2002. Calibration of sulfate levels in the Archean Ocean. *Science*, **298**, 2372–2374.

HARDIE, L. A. 1967. The gypsum–anhydrite equilibrium at one atmosphere pressure. *American Mineralogist*, **52**, 171–200.

HARDIE, L. A. 2003. Secular variations in Precambrian seawater chemistry and the timing of Precambrian aragonite seas and calcite seas. *Geology*, **31**, 785–788.

HARDIE, L. A. & EUGSTER, H. P. 1971. The depositional environment of marine evaporites: a case for shallow, clastic accumulation. *Sedimentology*, **16**, 187–220.

HENDRY, J. P. 1993. Calcite cementation during bacterial manganese, iron and sulfate reduction in Jurassic shallow marine carbonates. *Sedimentology*, **40**, 87–106.

HOLLIDAY, D. V. 1970. The petrology of secondary gypsum rocks: a review. *Journal of Sedimentary Petrology*, **40**, 734–744.

KENDALL, A. C. 1985. Radiaxial fibrous calcite: a reappraisal. *In*: SCHNEIDERMANN, N. & HARRIS, P. M. (eds) *Carbonate Cements*. SEPM Special Publication, **36**, 59–77.

KENDALL, A. C. 2001. Late diagenetic calcitization of anhydrite from the Mississippian of Saskatchewan, western Canada. *Sedimentology*, **48**, 29–55.

KENDALL, C. G. ST. C. & WARREN, J. 1987. A review of the origin and setting of tepees and their associated fabrics. *Sedimentology*, **34**, 1007–1027.

KLEIN, C., BEUKES, N. J. & SCHOPF, J. W. 1987. Filamentous microfossils in the early Proterozoic Transvaal Supergroup: their morphology, significance and palaeoenvironmental setting. *Precambrian Research*, **36**, 81–94.

KOWALSKI, W., HOLUB, W., WOLICKA, D., PRZYTOCKA-JUSIAK, M. & BŁASZCZYK, M. 2001–2002. Sulphur balance in anaerobic cultures of microorganisms in medium with phosphogypsum and sodium lactate. *Archiwum Mineralogiczne*, **54**, 33–40.

KRUMBEIN, W. E., COHEN, Y. & SHILO, M. 1977. Solar Lake (Sinai) 4. Stromatolitic cyanobacterial mats. *Limnology and Oceanography*, **22**, 635–656.

MELEZHIK, V. A., FALLICK, A. E., RYCHANCHIK, D. V. & KUZNETSOV, A. B. 2005. Palaeoproterozoic evaporites in Fennoscandia: implications for seawater sulfate, the rise of atmospheric oxygen and local amplification of the $\delta^{13}C$ excursion. *Terra Nova*, **17**, 141–148.

MILLIKEN, K. L. 1979. The silicified anhydrite syndrome–two aspects of silicification history of former evaporite nodules from southern Kentucky and northern Tennessee. *Journal of Sedimentary Petrology*, **49**, 245–256.

ROUCHY, J. M., TABERNER, C. & PERYT, T. M. 2001. Sedimentary and diagenetic transitions between carbonates and evaporites. *Sedimentary Geology*, **140**, 1–8.

SAMI, T. T. & JAMES, N. P. 1994. Peritidal carbonate platform growth and cyclicity in an early Proterozoic foreland basin, Upper Pethei Group, northwestern Canada. *Journal Sedimentary Research*, **64**, 111–131.

SCHREIBER, B. C. 1986. Arid shorelines and evaporites. *In*: READING, H. G. (ed.) *Sedimentary Environments and Facies*. Blackwell Scientific, Oxford, 189–228.

SCHREIBER, B. C. & EL TABAKH, M. 2000. Deposition and early alteration of evaporites. *Sedimentology*, **47** (Supplement), 215–238.

SCHREIBER, B. C., FRIEDMAN, G. M., DECIMA, A. & SCHREIBER, E. 1976. Depositional environments of Upper Miocene (Messinian) evaporite deposits of the Sicilian Basin. *Sedimentology*, **23**, 729–760.

SCHOLLE, P. A., ULMER, D. S. & MELIM, L. A. 1992. Late-stage calcites in the Permian Capitan Formation and its equivalents, Delaware Basin margin, west Texas and New Mexico: evidence for replacement of precursor evaporites. *Sedimentology*, **39**, 207–234.

SIMONSON, B. M., SCHUBEL, K. A. & HASSLER, S. W. 1993. Carbonate sedimentology of the early Precambrian Hamersley Group of Western Australia. *Precambrian Research*, **60**, 287–335.

SUMNER, D. Y. 1997. Late Archean calcite–microbe interactions: two morphologically distinct microbial communities that affected calcite nucleation differently. *Palaios*, **12**, 302–318.

SUMNER, D. Y. 2000. Microbial vs environmental influences on the morphology of Late Archean fenestrate microbialites. *In*: RIDING, R. E. & AWRAMIK, S. M. (eds) *Microbial Sediments*. Springer, Berlin, 307–314.

SUMNER, D. Y. 2001. Decimeter-thick encrustations of calcite and aragonite on the sea floor and implications for Neoarchean and Neoproterozoic ocean chemistry.

In: ALTERMANN, W. & CORCORAN, P. L. (eds) *Precambrian Sedimentary Environments. A Modern Approach to Ancient Depositional Systems*. International Association of Sedimentologists, Special Publications, **33**, 107–120.

SUMNER, D. Y. & GROTZINGER, J. P. 1996. Herringbone calcite. Petrography and environmental significance. *Journal of Sedimentary Research*, **66**, 419–429.

SUMNER, D. Y. & GROTZINGER, J. P. 2000. *Late Archean Aragonite Precipitation: Petrography, Facies Associations, and Environmental Significance in Carbonate Sedimentation and Diagenesis in the Evolving Precambrian World*. SEPM Special Publications, **67**, 123–144.

SUMNER, D. Y. & GROTZINGER, J. P. 2004. Implications for Neoarchean ocean chemistry from primary carbonate mineralogy of the Campbellrand-Malmani Platform, South Africa. *Sedimentology*, **51**, 1273–1299.

TABERNER, C., ROUCHY, J. M., RUSSELL, M. & GRIMALT, J. O. 1997. Timing of BSR and carbonate replacement after evaporites in organic-rich deposits. Lorca basin, Messinian, SE Spain. *American Association of Petroleum Geologists*, Annual Meeting, Abstracts, 114.

TESTA, G. & LUGLI, S. 2000. Gypsum–anhydrite transformations in Messinian evaporites of central Tuscany (Italy). *Sedimentary Geology*, **130**, 249–268.

USIGLIO, J. 1849. Analyse de l'eau de la Méditerranée sur les côtes de France. *Annales Chimie et Physique*, Série 3, **27**, 92–107, 172–191.

VERRECCHIA, E. P., FREYTET, P., VERRECCHIA, K. & DUMONT, J.-L. 1995. Spherulites in calcrete laminar crusts: biogenic CaCO$_3$ precipitation as a major contributor to crust formation. *Journal of Sedimentary Research*, **65**, 690–700.

WALKER, J. C. G. & BRIMBLECOMBE, P. 1985. Iron and sulfur in the pre-biologic ocean. *Precambrian Research*, **28**, 205–222.

WALKER, R. N., MUIR, M. D., DIVER, W. L., WILLIAMS, N. & WILKINS, N. 1977. Evidence of major sulfate evaporitic deposits in the Proterozoic McArthur Group, Northern Territory, Australia. *Nature*, **265**, 526–529.

WEST, I. 1979. Review of evaporite diagenesis in the Purbeck Formation of southern England. Symposium on West European Jurassic Sedimentation–"Sedimentation Jurassique W. European", A.S.F. (Association of French Sedimentologists), Special Publication, **1**, 407–416.

WRIGHT, D. T. 1997. An organogenic origin for widespread dolomite in the Cambrian Eilean Dubh Formation, north western Scotland. *Journal of Sedimentary Research*, **67**, 54–64.

WRIGHT, D. T. 1999. The role of sulphate-reducing bacteria and cyanobacteria in dolomite formation in distal ephemeral lakes of the Coorong region, South Australia. *Sedimentary Geology*, **126**, 147–157.

WRIGHT, D. T. & ALTERMANN, W. 2000. Microfacies development in Late Archean stromatolites and oolites of the Ghaap Group of South Africa. *In*: INSALACO, E., SKELTON, P. W. & PALMER, T. J. (eds) *Carbonate Platform Systems: Components, Interactions*. Geological Society, London, Special Publications, **179**, 51–70.

WRIGHT, D. T. & WACEY, D. 2005. Dolomite precipitation in experiments using sulphate reducing bacterial populations in simulated lake and pore waters from distal ephemeral lakes, Coorong region, South Australia. *Sedimentology*, **52**, 987–1008.

Evaporites of Ukraine: a review

S. P. HRYNIV, B. V. DOLISHNIY, O. V. KHMELEVSKA,
A. V. POBEREZHSKYY & S. V. VOVNYUK

*Institute of Geology and Geochemistry of Combustible Minerals, National
Academy of Sciences of Ukraine, Naukova 3A, 79053 Lviv, Ukraine
(e-mail: Sophia_Hryniv@ukr.net, igggk@ah.ipm.lviv.ua)*

Abstract: The results of geological and lithological–geochemical investigations of the Devonian, Permian, Jurassic and Miocene evaporite deposits of Ukraine are presented in review. The main regions of evaporite distribution are the Dnipro–Donets depression, Carpathian (Forecarpathians, Transcarpathians) and Foredobrogean regions. The data on tectonics and stratigraphy are presented and information on lithology, the mineralogical and geochemical study of gypsum, anhydrite, rock and potash salts are summarized. The rich mineral composition of the Miocene evaporites in the Carpathian Foredeep (more than 20 salt minerals) is demonstrated, and the unique superimposed hydrothermal mineralization in the rock salt of salt domes from the Dnipro–Donets depression is presented (containing about 40 high- and mid-temperature hydrothermal minerals). In particular, the results of brine inclusion studies in evaporite minerals suggest that seawater was the main source of most of the salts. The brines in both the Miocene and Permian evaporite basins are classified as the Na–K–Mg–Cl–SO_4 (SO_4-rich) chemical type and the Jurassic and Devonian belong to the Na–K–Mg–Ca–Cl (Ca-rich) type. Temperature of solutions during halite precipitation shifted from 25 to 43 °C, while during the stage of potash salt sedimentation it apparently increased to 40–83 °C.

Ukraine possesses Devonian, Permian, Jurassic and Neogene evaporite deposits located in several areas: the Dnipro–Donets depression (D, P), the Foredobrogean trough and western margin of the East European platform (J), the Carpathian Foredeep, the Transcarpathian trough and the Crimea (N) (Fig. 1). Some of the deposits are widespread (e.g. the Devonian evaporites in the Dnipro–Donets depression are distributed over area of about 40,000 km^2) and also are quite thick (Zharkov 1984); another, the Miocene gypsum of Kerch peninsula, has a more restricted distribution and is relatively thin. Small evaporite deposits of Silurian age present in Volyno–Podoliya (Markovskiy *et al.* 1979) and those of Jurassic age, present in the Crimea (Borisenko *et al.* 1974), are not well studied and are not reviewed here.

Evaporites of Ukraine are represented by gypsum, anhydrite and both rock salt and potash salts. In the Carpathian Foredeep saline deposits are characterized by a very rich mineral composition – potash deposits contain about 20 potassium–magnesium, mostly sulphate, minerals. In the rock salt of salt domes of the Dnipro–Donets depression a unique superimposed hydrothermal mineralization has developed.

A brine inclusion study in sedimentary halite provided the data and stimulated the understanding of geochemical principles of marine salt formation evolution (Petrichenko 1988) – changes from SO_4-rich, Ca-poor chemical type to the SO_4-poor, Ca-rich chemical type that have been outlined by Petrichenko and are tied in with an influx of subsurface $CaCl_2$-rich brines into the evaporite basins during orogenesis. Such changes of brine inclusion composition apparently were connected with the evolution of seawater during the Phanerozoic (Kovalevich 1988, 1990). In the Phanerozoic seawater chemistry oscillated between the Na–K–Mg–Ca–Cl and Na–K–Mg–Cl–SO_4 types. From the Cambrian to Carboniferous and from the Jurassic to middle Paleogene the Na–K–Mg–Ca–Cl type prevailed, while in Permian and Triassic as well as starting in the middle Paleogene until the Recent the ocean was of the Na–K–Mg–Cl–SO_4 type (Kovalevich 1988; Kovalevich *et al.* 1998). A similar series of research studies concerning secular changes in the chemical composition of seawater based on the results of brine inclusions study is also noted in the western geological literature (Lowenstein *et al.* 2001, 2003; Horita *et al.* 2002), though the problem of secular changes in the chemical composition of seawater during the Phanerozoic from the results of another investigation (chemical composition of potash salts, mineral composition of carbonates) has been considered and is still widely discussed (e.g. Holland 1972, 1984, 2003; Holland *et al.* 1986; Hardie 1996). Besides being of general scientific interest, the observations pertaining to secular changes in

From: SCHREIBER, B. C., LUGLI, S. & BĄBEL, M. (eds) *Evaporites Through Space and Time*.
Geological Society, London, Special Publications, **285**, 309–334.
DOI: 10.1144/SP285.18 0305-8719/07/$15.00 © The Geological Society of London 2007.

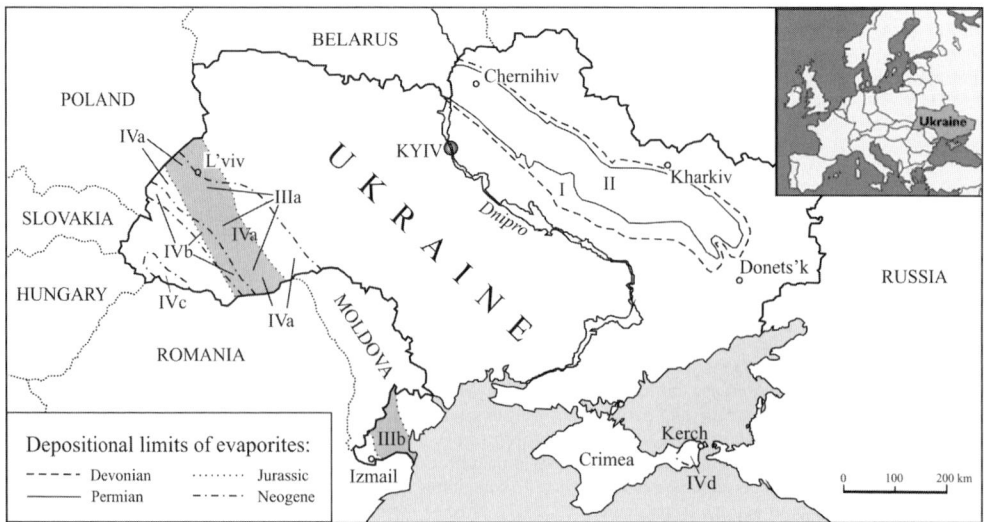

Fig. 1. Location map of evaporite deposits in Ukraine (after Kruglov & Tsypko 1988). I, Devonian evaporites of the Dnipro–Donets depression; II, Permian evaporites of the Dnipro–Donets depression; III, Jurassic evaporites: a, Forecarpathian region; b, Foredobrogean trough; IV, Miocene evaporites: a, b, Carpathian Foredeep (a, Bilche-Volytsya zone and adjacent part of East European platform; b, Sambir and Boryslav-Pokuttya zones); c, Transcarpathian trough; d, Kerch peninsula.

the chemical composition of seawater have a practical application in that they represent a predictor for the determination of the mineral resources connected with evaporites.

In Ukraine these mineral resources include gypsum, rock and potash salts, native sulphur, celestite, brines and mineral–medicinal waters. In most areas having evaporites, oil and gas deposits also are present.

The majority of investigators adhere to the opinion of general principles of classic theory of salt formation that connects the origin of evaporites with evaporation of seawater (Borchert 1959; Valyashko 1962), although alternative points of view have also been presented. These other concepts suggest that salt-saturated hot brines form under the Earth's crust (Sozanskiy 1973) or that exhalation–sedimentary formation of salt deposits is very important (Dzhinoridze 1980).

The results of most of the investigations of evaporites in Ukraine, their geology, composition and origin have been published in Russian and Ukrainian journals (and books) and have not been readily available for foreign scientists. Though recently a good deal of information has been published in English, the goal of this review is to summarize in a single paper the main data on geology, mineralogy, petrography, geochemistry and origin of evaporite deposits in Ukraine. Because this paper is merely a short review, it only permits us to touch on the general aspects of the deposits, but here we outline some of the more significant deposits and their references.

Devonian and Permian evaporites of the Dnipro–Donets depression

Devonian and Permian evaporites are widespread in Ukraine in the Dnipro–Donets depression. This area is located in southeastern part of the East European platform between the margin of the Ukrainian Shield, along the SW, and the Voronezh High to the NE (Fig. 2). In the NW the depression continues along the Prypiat' trough (in Belarus territory), which is separated from the Dnipro–Donets depression by the Chernihiv–Brahin High. To the SE the depression ends in the Donbas Foldbelt.

Geotectonically the depression is a complicated graben with a northwestern extension, in which the hypsometric level of basement is gradually lower in the southeastern direction. The shape and size of Dnipro–Donets depression is controlled by the evolution of marginal faults that have been traced by seismic investigations and penetrate both the granitic and basaltic layers of the Earth's crust (both upper and lower lithosphere) and apparently reach down to the mantle. The length of this depression is about 500 km, and the width ranges from 80 km in the NW part to 120 km to the SE. The thickness of sedimentary cover changes from 3–4 km in the NW to 10–17 km in the SE.

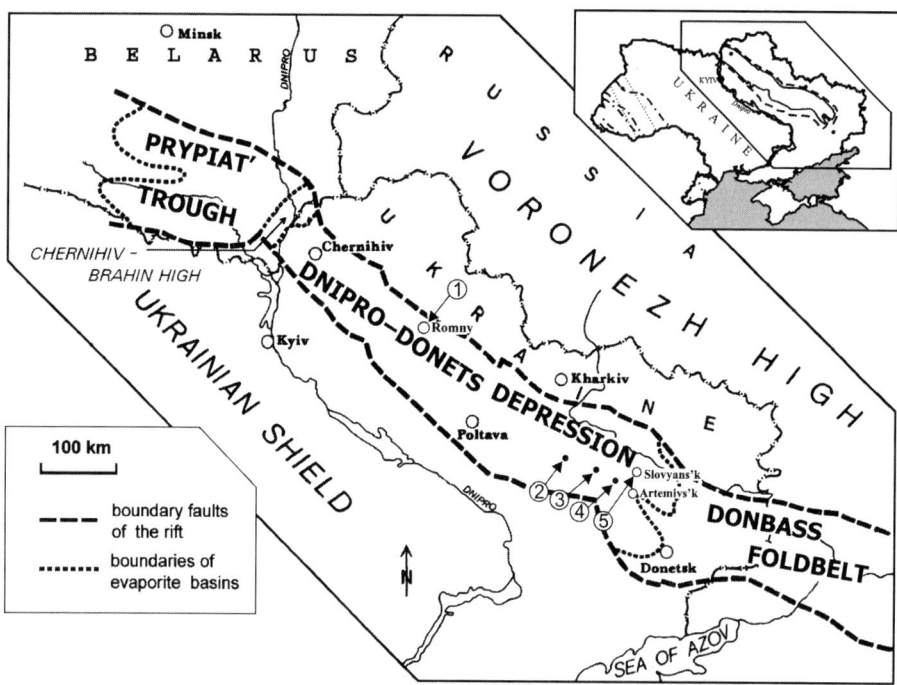

Fig. 2. Location map of the Dnipro–Donets depression (after Ulmishek *et al.* 1994). Salt domes: 1, Romny; 2, Bilyayivka; 3, Berets'k; 4, Bantyshevo; 5, Slovyans'k.

The fill is composed of sedimentary rocks of the Devonian, Carboniferous, Permian, Triassic, Jurassic, Cretaceous, Paleogene, Neogene and Quaternary ages and consists of the terrigenous–carbonate, sulphate and saline deposits, sometimes with layers of tuffaceous rocks. Rock salt (halite) is present mainly in the Upper Devonian (Frasnian and Famennian stages) and in the Lower Permian (lower part).

The sedimentary series in the Dnipro–Donets depression is mainly controlled and determined by basement offsets, which penetrate the sedimentary series, and by disharmonic displacements of the Devonian and Permian salts (Kityk 1970). The basement has a complicated fault-block structure: numerous horsts displaced with vertical amplitude up to 1–2 km. In the sedimentary cover offset along most faults gradually declines from bottom to top and passes into zones of plastic deformation that warped the evaporite sections. These warps lead to disharmonic dislocation of the entire sedimentary sequence.

The plastic deformation of the salt rocks led to the origin of many uplifts in the sediments overlying the salt series. Cores of these uplifts, as a rule, pierce the overlying deposits and sometimes form surface outcrops. In some cases the uplift of overlying rocks is not significant and takes place without true diapirism. The salt domes in the Dnipro–Donets depression form disharmonic folds, which are composed of both undersalt and oversalt structures with salt cores. Diapirism of overlying deposits, formed by salt cores, leads to their destruction and the formation of diapiric and solutional breccias of different thicknesses that vary from place to place.

The first geological investigations of the depression began in the early eighteenth century. These studies were connected mainly with salt springs near the town of Tor (presently called Slovyans'k), which were used as a source of kitchen salt (Novik *et al.* 1960). In 1701 richer springs were found along the Bakhmut River, and this in turn caused the development of an industrial salt centre in the town of Bakhmut (presently called Artemivs'k). In the 1840s the salt-making factories were initiated, and they, in turn, stimulated research for associated fuel resources, e.g. coal.

In 1831 the Petrovsk coal deposit, and in 1870s the rock salt deposit in shallow horizons near towns Slovyans'k and Artemivs'k, were opened. In the 1930s Shatskiy (1931) projected the existence of salt domes near Isachka and Romny on the base of correlation of the lithological peculiarities of rocks from two different parts of the depression. This hypothesis was successfully tested by drilling in the Romny area (1932) and in 1936 in this area oil was discovered in the oversalt rocks. Later, all of the territory of Dnipro–Donets depression was gradually investigated. Salt domes became the object of great interest for geologists as sources for rock salt as well as for oil and gas, sulphur, coal and gypsum. Additionally, some of them were found to contain polymetallic sulfides, fluorite, mercury and others ore manifestations.

Devonian evaporites

The Upper Devonian evaporite deposits in the Dnipro–Donets depression (Fig. 3) are recognized in sediments of both the Frasnian and Famennian stages and are widespread (Vysotskiy *et al.* 1988).

Upper Frasnian In the initial bedded occurrence the Upper Frasnian salt deposits are uncovered in the NW and near the margins of the depression, and in the inner part they are dislocated by processes of salt domes forming. The plastic deformation of the salt rocks led to the origin of many uplifts in the oversalt series, such as brachyanticlines, salt domes and stocks. In these salt uplifts the salt roof may be located at a depth of 40–1000 m, but in zones between the domes it occurs at 3500–4500 m (Vysotskiy *et al.* 1988). The thickness of this salt formation ranges from several hundreds to 1200 m and within the salt uplifts it reaches over 2500 m.

The salt formation is represented by interbedded layers of rock salt (5–20 to 70–100 m thick) and non-saline rocks (2–10 to 20–40 m thick). At the base of the salt formation carbonate, igneous and rarely terrigenous rocks are present (Galabuda 1970). The salt-bearing section in addition to rock salt may contain rocks such as anhydrite, carbonate, terrigenous deposits, volcanic tuff, diabase and basalt.

Sulphate rocks in these deposits are grey, yellowish and white, microgranular, laminated, commonly with breccia-like textures, and sometimes include carbonates and bitumen.

Rock salt is light and dark grey, orange and red, fine- and coarse-grained, sometimes

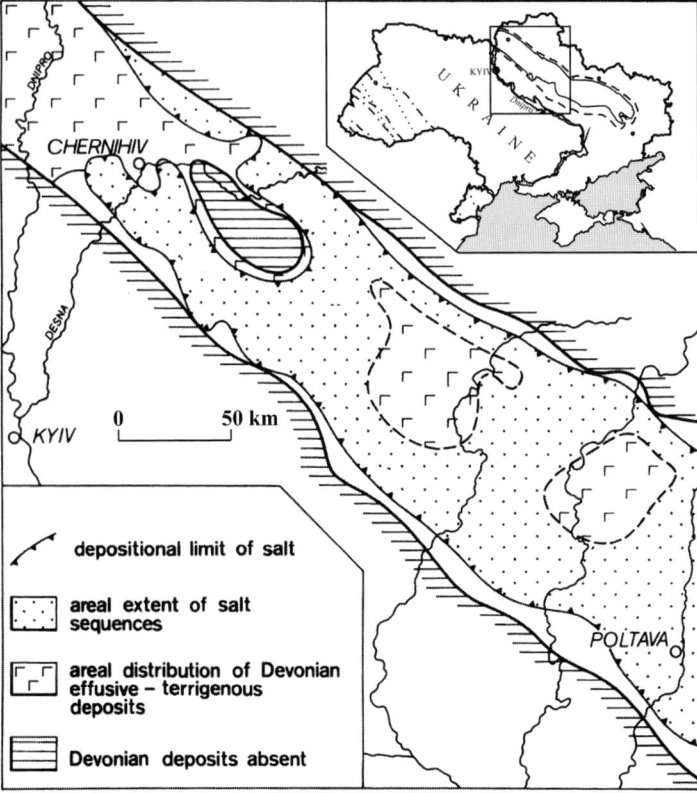

Fig. 3. Scheme of distribution of Devonian rock salt areas in the NW and central parts of the Dnipro–Donets depression (after Zharkov 1984).

mega-crystalline; however, generally isometric grains of 2–10 mm prevail. The texture is massive, laminated, in the cores of salt domes – deformed with flattened grains. The water-insoluble residue is composed of anhydrite, gypsum, calcite, dolomite, ankerite, quartz, feldspar and clays.

In the salt diapirs, as a rule, rock salt contains fragments of layers and blocks of anhydrite, terrigenous, clayey-carbonate and igneous rocks. The latter are made up of diabase and gabbro-diabase. Within the contact with igneous rocks the rock salt is strongly altered, apparently melting (Petrichenko & Shaidetskaya 1977; Petrichenko 1989). The thickness of such melted zones does not exceed several meters.

Potash-bearing salts The most complete data concerning the potash-bearing salts of the Upper Frasnian age were obtained from the Romny salt stock (Korenevskiy *et al.* 1968; Vysotskiy *et al.* 1988). There are two potash-bearing horizons, which are strongly deformed. The initial thickness of the upper one reaches a thickness of 4 m, and the lower is 18 m. The lithological section of the upper horizon is rhythmic. One rhythm is usually composed of three parts: (1) clayey-carbonate or clayey-carbonate–anhydrite rock (0.5–3 cm thick); (2) rock salt (2.5 cm); and (3) sylvinite or rock salt with sylvite (from 1.2 to 8 cm).

The lower horizon is represented by fine- and middle-grained sylvite–halite rock and rock salt with layers of reddish sylvinite. The bromine content in the Frasnian rock salt is 50–150 ppm.

Famennian Salt formation of the Famennian stage is located near the margins of the Dnipro–Donets depression. Its thickness is variable and varies from 0 to 320 m. In the arches of salt structures it is minimal and on the flanks the thickness increases. Salt formation is composed of rock salt interbedded with marls, anhydrites and terrigenous and igneous rocks (Zharkov 1984). The bromine content in the rock salt is 60–130 ppm.

Rock salt in domes is made up of light to dark grey halite with a heterogeneous structure – large salt crystals commonly form separate segregations within a thin- and middle-grained mass. Crystal sizes usually are 2–3 cm, but sometimes reach 5–6 cm. This rock salt contains hydrothermal authigenic minerals, which were first studied in the evaporites by Dolishniy (1976, 2000). The content of water-insoluble residue ranges from less than 1% up to several percent, 1–3% on average.

Minerals are distributed unevenly and form accumulations, where one or other minerals are more common. Such distribution of minerals sometimes causes a spotty colouration in the salt. For example, in borehole cores some intervals of rock salt from the Slovyans'k structure are reddish because of presence of large amounts of thin plates of hematite; in the Bantyshevo dome the salt is green due to the presence of small crystals of tourmaline. Authigenic minerals are typical of high-temperature association (scapolite, albite, muskovite, sphene, microcline, clinozoisite and others), rare minerals (tunisite, dawsonite, juwelite) and some ore minerals (Table 1).

Almost all studied minerals are euhedral crystals, and rounded ones are not present. Some minerals are commonly present as a xenogenic phase within the gas–liquid inclusions in halite. The minerals are present as separate crystals, and in intergrowths as well as monomineralic aggregates, microdruses and crusts. Crystal sizes range from 0.1 to 3 mm, rarely 4–5 mm, on average 1–2 mm.

The origin of superimposed secondary authigenic mineralization in salt domes is related to the broad development of hydrothermal activity in the region. Thermal fluids moved up along the tectonically 'released' zones around salt domes and poured out along microcracks of salt bodies as well as in porous diapiric breccias. The comparison of the authigenic mineral composition in salt from one side and in breccia zones (Kuznietsova *et al.* 1968; Kuznietsova 1970) from another appears to confirm the correctness of such conclusions (Dolishniy 1976, 2000).

The presence of hydrothermal mineralization occurred in three linear zones parallel to the extension of graben: two near the edge and one zone along the central axis. Almost all ore trends are located in breccias of salt domes and are made up of magnesite, galenite, sphalerite, pyrite, pyrotite, chalcopyrite, hematite, ilmenite, titanomagnetite, parisite, fluorite, scapolite, tourmaline, garnet, rutile etc. (Kuznietsova 1971). Based on chemical and spectral analyses the increased content of Co, Ni, Cu, Ba, Mo, Hg and the transition elements, etc. has been determined (Kuznietsova 1971, Khrushchov 1980). Small layers of coal and organic matter also have been noted. In the caprock of the Romny dome there are industrial sulphur deposits. Fluid inclusion investigations in the Slovyans'k dome (Zatsikha *et al.* 1973) showed that hydrothermal processes occurred over a wide scale of temperature, from 500 to 100 °C and lower. Some geochemical peculiarities of the Devonian salts are compared with Permian ones at the end of the Permian evaporites description.

Permian evaporites

The Lower Permian deposits are spread across most of the Dnipro–Donets depression (Fig. 4), but their total volume is smaller than that of the Devonian age. The depth of Permian base ranges from 700

Table 1. *Distribution of minerals in the water-insoluble residues of rock salt from some salt domes in SE part of Dnipro–Donets depression (Dolishniy 1976, 2000)*

Minerals	Saltdome structures			
	Slovyans'k	Bantyshevo	Berets'k	Bilyayivska
Sulphur	3		3	2
Pyrite	2	1	2	2
Pyrrhotite		4		
Molybdenite	4			
Galenite		4		
Sphalerite	4	4		4
Wurtzite				4
Quartz	1	1	1	1
Rutile	4	3		
Anatase		4		
Brookite		4		
Leucoxene		4		
Magnetite		4		
Hematite	3		3	3
Hydrous ferric oxide	3		3	3
Dolomite	1	1	1	1
Calcite	1	1	1	1
Siderite	4			
Ankerite	4			
Rhodochrosite	4			
Dawsonite	3			
Tunisite	2	3		
Anhydrite	1	1	1	1
Barite	3			
Celestite	3			3
Scapolite		3		
Albite	4	3		3
Muscovite	3	2	4	4
Phlogopite		3		
Tourmaline	4	3		4
Microcline		3		
Sphene		4		
Apatite		4		
Fluorite			4	
Juvelite	4			
Anthraxolite	4			3
Clinozoisite		4		
Pumpellyite		4		
Tremolite(?)				

Note: 1, major minerals; 2, minor; 3, rare; 4, single occurrence.

to 5000 m (Vysotskiy *et al.* 1988). The total thickness of the Lower Permian evaporite formation is from 300 m in the NW to 1650 m in the SE part of depression and the Permian deposits are actually exposed at the surface here (Fig. 5). Within the Permian of Ukraine lies the largest rock salt deposit, the Artemivs'k.

Within this depression, the Lower Permian deposits are subdivided up the section to four suites: Kartamysh, Mykytiv, Slovyans'k and Kramators'k. The lowest one (a thickness of 10–1200 m) is represented mainly by terrigenous deposits with dolomite layers. Mykytiv suite (10–300 m) is composed of clay, carbonates, anhydrite and rock salt. This salt is from several to 80 m thick. In the Slovyans'k suite salt thickness increases from 15 m in the NW part of the depression to 900 m in the SE. The uppermost suite, Kramators'k, is 150–900 m thick and consists of 90% salt. Bromine content in rock salt from the three salt-bearing suites ranges from 32 to 85 ppm (Korenevskiy *et al.* 1968; Zharkov 1984).

Also potash salts (of chloride and chloride–sulphate composition) and bischofite are present in some second-order depressions. There are five potash-bearing horizons (up to 30 m each) with

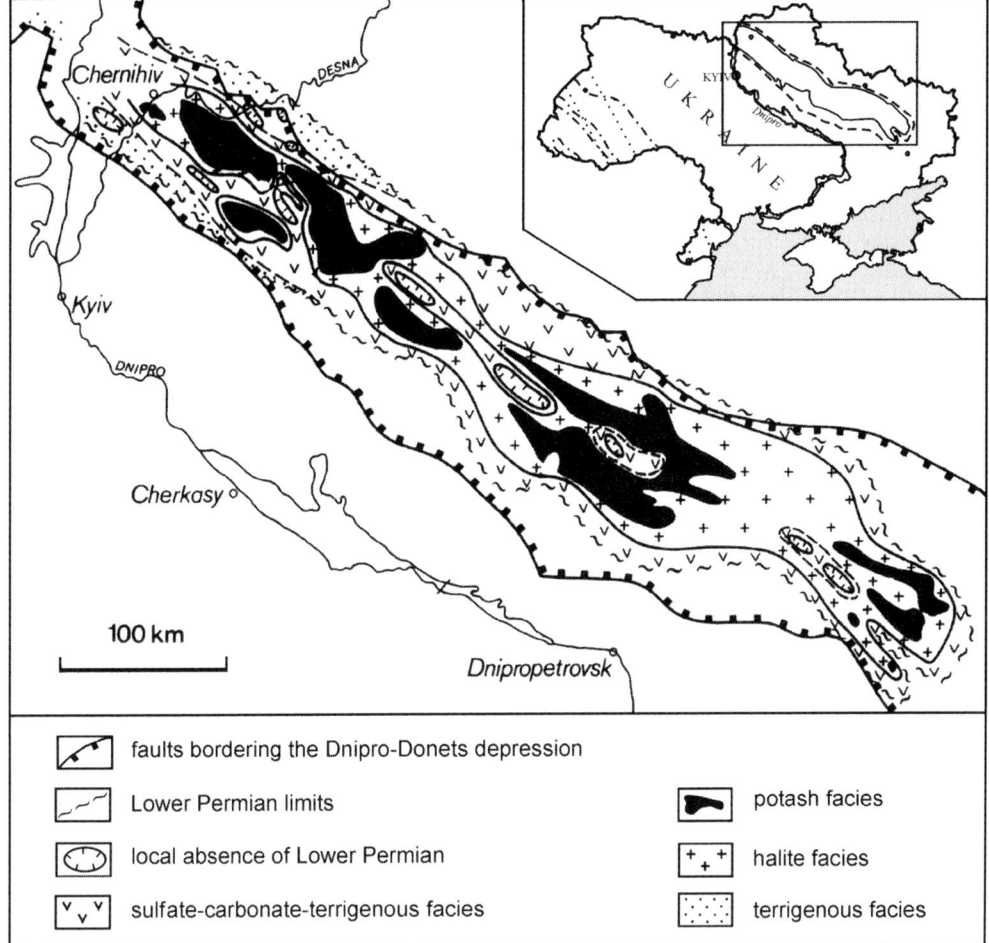

Fig. 4. Location map of the Permian evaporites in the Dnipro–Donets depression (after Zharkov 1984).

bischofite mineralization. Also present are separate layers having very complicated compositions. They are made up of carnallite, carnallite–bischofite, kieserite–carnallite, carnallite–kieserite and carnallite–halite rocks. In the kieserite-bearing rocks sulfoborite and ascharite are present (Khrushchov 1971; Galitskiy 1972; Korenevskiy et al. 1980; Vysotskiy et al. 1988).

The physical–chemical depositional conditions of the Devonian and Permian salt formations (based on data from studies of fluid inclusions in halite) are described in many publications (Petrichenko et al. 1974a; Kovalevich 1985; Shaidetskaya 1990; Kovalevych & Petrychenko 1994; Petrichenko & Shaidetska 1998; Kovalevych et al. 2002; Petrychenko & Peryt 2004).

Based on these investigations, the Devonian salts were formed from solutions of Na–K–Ca–Mg–Cl (Ca-rich) chemical type and the Permian from solutions of Na–K–Mg–Cl–SO_4 (SO_4-rich) type. These differences have permitted stratigraphical differentiation for the twin-salt structures of the Permian and Devonian age in the Dnipro–Donets depression (Petrychenko et al. 1976b), using the ultramicrochemical analysis of brine inclusions in halite (Petrichenko 1973). Based on the fluid inclusion studies the temperature of salt-forming solutions in the Devonian and Permian evaporite basins at the stage of halite precipitation did not exceed 43 °C, elevated up to 60–65 °C in the sylvite stage in the Devonian and up to 78–83 °C on carnallite and 65–77 °C on bischofite stages in the Permian (Petrichenko 1988). The temperature of diagenetic transformations of the rock salt in the Permian deposits was 45–55 °C and that of potash rocks is estimated to have been 75–82 °C.

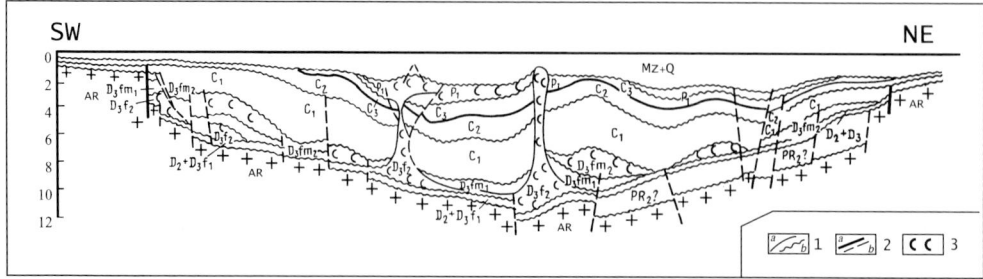

Fig. 5. Schematic geological section across the Dnipro–Donets depression (after Kruglov & Tsypko 1988). 1, boundaries (a, geological, b, of unconformable bedding of rocks); 2, faults (a, marginal, b, other faults); 3, rock salt.

Upper Jurassic evaporites

Upper Jurassic evaporite deposits are broadly spread across the western and southwestern slope of the East European platform in Poland, Ukraine, Moldova and Romania (Fig. 6). In Ukraine they are distributed in the Forecarpathian region (Carpathian Foredeep and the slope of the East European platform) and in the Foredobrogean trough.

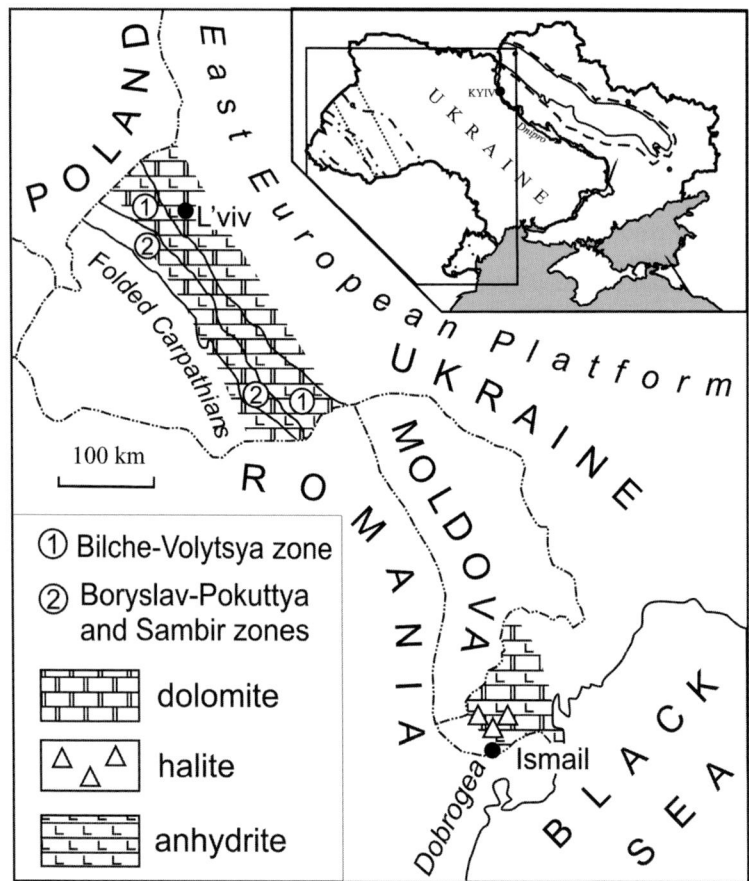

Fig. 6. Schematic map of distribution of Jurassic evaporites, western and southwestern slope of East European platform (after Khmelevskaya 1997). I, Forecarpathian region; II, Foredobrogean trough.

Forecarpathian region

In the territory of the Forecarpathian region of Ukraine Jurassic evaporites are located in the basement of the Bilche–Volytsya zone of the Carpathian Foredeep and the adjacent part of Volyno–Podoliya region (Stryy Jurassic trough). During Jurassic time the Stryy trough was developed on the western and southwestern slope of the East European platform. It has a triangular shape that extended SE toward NW but showed an asymmetric construction, with a NE limb that was gentle and extended and the SW was steep and short, complicated by faults (Glushko 1968).

Jurassic deposits transgressively overlap the eroded and truncated surface of the Paleozoic (Anastasieva 1957) and are represented by all units, Lower, Middle and Upper Jurassic with a total thickness >2500 m. On the Ukrainian territory they are spread as a wide band (60–100 km) from Poland at NW to Romania at SE (Dulub et al. 2003).

Evaporite deposits are recognized in the Rava–Ruska suite of Kimmeridgian age, and their thickness varies from 20 to 250 m (Garetskiy 1985). They are made up of interbedded dolomites, dolomitic limestones, terrigenous rocks (sand- and siltstones), massive and bedded anhydrite, anhydrite breccia and gypsum (Korenevskiy 1982).

Anhydrite is white, light and dark grey, pink, brownish and commonly multicoloured due to admixtures, and the structure is fine-grained, while the texture is massive or irregular. Some anhydrites with thin clayey or carbonate veins have a cellular texture (Khmelevska 1993; Chmielewska 1995). The thickness of individual anhydrite layers in the section ranges from several centimetres to 7 m (Anastasieva 1958).

Anhydrite breccia is composed of oval nodules of dark grey or pink anhydrite cemented by brown or grey fine-grained dolomite or mudstone. Anhydrite nodules range in size from 2–3 mm to 5 cm. Breccias occur in the upper part of sulphate-carbonate suite and are 12–25 m thick. They are intrastratal formations interpreted as possible subaqueous slumps and collapse deposits (Sandler & Vorona 1955).

Gypsum rocks have a restricted distribution and are present as lenses and layers up to 1 m thick, filling cavities and veins in breccia-like carbonate–clayey and clay deposits. The gypsum is white, grey and brownish and commonly contains significant carbonate and clayey admixtures. In some cases the veins and cavities are filled by parallel-fibrous gypsum (satin-spar) (Anastasieva 1958).

The formation of these evaporite deposits apparently took place in lagoons separated from sea only by underwater barriers. Such conditions provided a lateral transition from anhydrite to carbonate rocks (Dulub et al. 1986).

Foredobrogean trough

In the SW Ukraine, near the Black Sea, the Upper Jurassic evaporite deposits accumulated within the Foredobrogean Mesozoic trough, located between Dobrogea and the platform. It is composed of both Triassic and Jurassic deposits. The trough has an asymmetric construction, where the limb adjacent to Dobrogea is steep and that near the platform – sloping (Romanov & Slavin 1970).

The Jurassic rocks are formed into gentle folds with small amplitude. In the central part of trough thickness of Jurassic rocks is greatest, up to 2500 m (Glushko 1968). Evaporites have been recognized in the Kongaz suite of Kimmeridgian age and are composed of carbonate, clastic, sulphate and halite rocks. The total thickness of this suite ranges from 15 to 445 m (Sliusar 1971; Romanov 1976; Korenevskiy 1982; Garetskiy 1985).

The sulphate rocks in the trough are composed of anhydrite, mixed gypsum–anhydrite and gypsum. Near the salt body their thickness reaches 250 m. The anhydrite is white, grey and brown, and usually fine-grained, massive and relatively pure (97–99% calcium sulphate). Among the admixtures are carbonates, celestite and pyrite. Some anhydrite includes the detrital carbonized remains of shells (Khmelevskaya 1990a). As a result of gypsification, mixed gypsum–anhydrite rocks are present and characterized by presence of rosette-like gypsum aggregates within the massive anhydrite.

Gypsum may be fine-grained but there also are coarsely crystalline varieties. Fine-grained gypsum is grey, blue, brownish, multicoloured with massive or irregular patchy textures. Coarsely crystalline gypsum is uncoloured, yellow and grey and forms veins and layers in fine-grained gypsum or anhydrite. Crystals of gypsum are lens-shaped, platy or columnar and may reach 7 cm in size.

Rock salt is located in a restricted area near Izmail (Butkovskiy et al. 1976); its maximal thickness does not exceed 80 m. It is composed of white and grey halite, medium- to coarse-grained, slightly recrystallized, with thin layers and aggregates of anhydrite. The NaCl content ranges from 93 to 98%. The water insoluble portion of the salt consists of authigenic minerals – anhydrite, carbonates (calcite and dolomite), quartz, celestite, fluorite and pyrite and terrigenous ones – quartz, rarely garnet, tourmaline and biotite. Clay minerals include hydromica and chlorite (Khmelevskaya 1990b). Euhedrality is the distinguishing feature for authigenic minerals.

On average the bromine content of the rock salt ranges from 27 to 63 ppm and is 35 ppm. Fluid inclusion investigation showed that rock salt was composed of both sedimentary (primary) and recrystallized halite (Khmelevska 1993; Khmelevskaya 1997). Based on ultramicrochemical analysis of brine inclusions in halite (Petrichenko 1973) it appears that salt-forming solutions were of Na–K–Ca–Mg–Cl composition (Ca-rich type). The maximum concentration determined for the potassium-ion in brine inclusions in halite was 15.5 g l^{-1} and this value was too low to expect any primary potash salt sedimentation (Petrychenko et al. 1976a).

The Upper Jurassic Foredobrogean evaporite basin was formed within the northern continental margin of the Tethys ocean, parallel to the Dobrogea land mass. It was apparently deposited in a lagoon or arm of the sea, which united several isolated basins having different rates of bottom subsidence and concentration of seawater. Only in the most subsident part of the evaporite basin the conditions were favourable for rock salt accumulation.

Miocene evaporites

Miocene evaporite deposits in Ukraine are distributed in the Carpathian region (Carpathian Foredeep and Transcarpathian trough) and in the Crimea on the Kerch peninsula (see Fig. 1). The archeological findings show that salt in the Forecarpathian and Transcarpathian areas has been mined for a long time. In the Transcarpathian rock salt was first quarried in the first to third centuries, and later it was again mined in the second part of the nineteenth century (Kityk et al. 1983).

In the Forecarpathian the salt production was widespread and this geographic area was given the name 'Halychyna' (from the Greek 'galos' for salt). At the end of the nineteenth century the potash deposits were discovered and their exploitation was begun (Korenevskiy & Donchenko 1963).

The Miocene evaporite deposits of the Forecarpathian are potash-bearing and the Stebnyk and Kalush-Holyn' industrial potash fields are located there. At the present time their exploitation is conducted only in the Dombrovo quarry of Kalush-Holyn' deposit. Another 10 deposits are explored here of which the largest one is Markova–Rosilna (Fig. 7).

The results of investigations of these evaporites were published by Polish and Austrian geologists before the second World War as well as in later studies and have been summarized in the following reviews (Byhovier et al. 1941; Lazarenko et al. 1962; Korenevskiy & Donchenko 1963; Dolenko 1975).

Carpathian Foredeep

For this complicated region the peculiarities of evaporites are described separately for sulphate (gypsum–anhydrite) deposits and for the rock salt of the Bilche–Volytsya zone and potash-bearing deposits of the Boryslav–Pokuttya and Sambir zone.

The Carpathian Foredeep in Ukraine territory extends from NW to SE as a band of 25–60 km wide and about 300 km long (see Fig. 7) and is located along the boundary of the Folded Carpathians and East European platform. It occurs on folded flysch as well as on the platform basement and is filled by thick chemogenic–terrigenous formation of the Miocene molasse. In the SW it is bounded by the Carpathian overthrust and to the NE by a system of faults in which the basement rocks have a step-like subsidence towards the Carpathians.

The Carpathian Foredeep is subdivided on the Boryslav–Pokuttya, Sambir and Bilche–Volytsya zones (Vul et al. 1998). In the Boryslav–Pokuttya and Sambir zone the folded flysch deposits of the Cretaceous and Palaeogene age are conformably overlapped by the complex of the lower and upper molasse. In the Bilche–Volytsya zone, only the upper molasse is present and it lies on the Mesozoic platform basement with a sharp break and an angular unconformity. Within the platform portion, the Miocene deposits are present at 10–200 m depths below the surface, and in the foredeep they subsided under the overthrust of the Sambir zone and are at depths of 1200–2200 m. The Boryslav–Pokuttya zone has been thrust to the NE and forms a sheet of 25 km amplitude. The Sambir zone is stripped from its basement and thrust onto the Badenian–Sarmatian deposits of the Bilche–Volytsya zone; near the Kalush–Holyn' deposit the thrust amplitude is 8–12 km (Korenevskiy & Donchenko 1963).

The correlation of different stratigraphic schemes for the Neogene areas of the Eastern and Western Paratethys in Western regions of Ukraine can be seen (Fig. 8) in the review by Vyalov (1980), and the stratigraphic scheme of the Miocene molasse of the Carpathian Foredeep is shown in Figure 9. The number and stratigraphic position of evaporite deposits in the Miocene section have not been fully explored and remain uncertain. The number and stratigraphic position of evaporite deposits in the Miocene section have not been fully explored. It remains uncertain whether there are two (Dzhinoridze 1980), three (Klimov 1977; Khrushchov 1980) or four (Korenevskiy et al. 1977) evaporite formations. According to different researchers, Markova–Rosilna and Stebnyk deposites could belong to the Lower- and Upper-Vorotyshcha subsuites respectively (Klimov 1972; Korenevskiy et al. 1977), or to the undivided Vorotyshcha suite (Dzhinoridze 1980), while Kalush–Holyn' and a number of other deposits are related to the Upper-Stebnyk

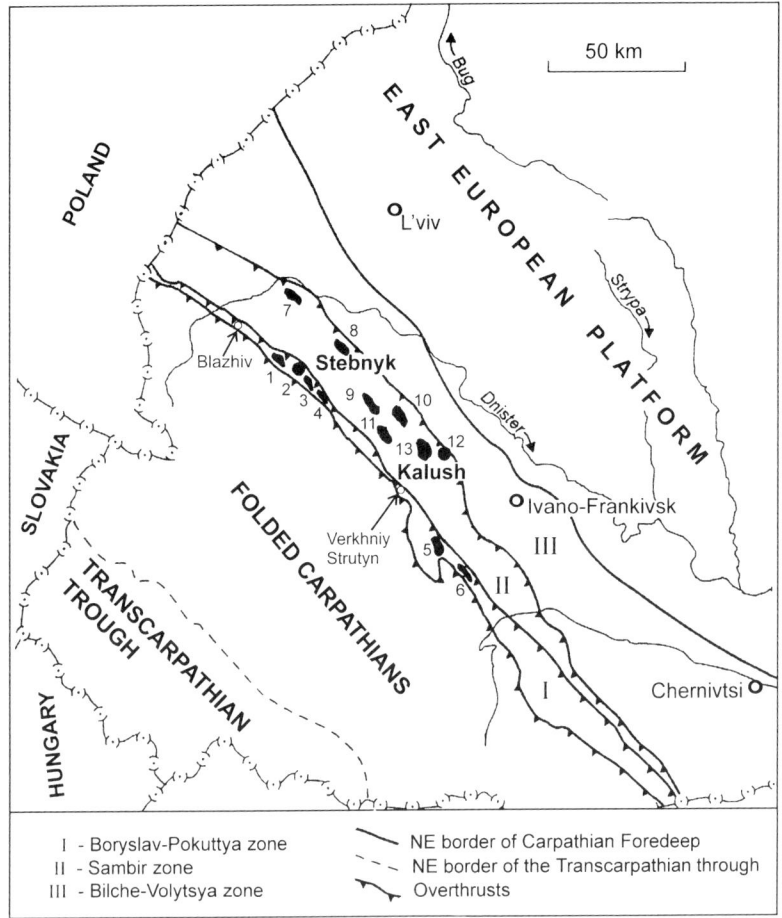

Fig. 7. Location map of potash-bearing deposits in the Carpathian Foredeep. I, Boryslav-Pokuttya zone: 1, Boryslav; 2, Stebnyk; 3, Dobrohostiv; 4, Ulychne; 5, Rosilna; 6, Markova. II, Sambir zone: 7, Bilyna-Velyka; 8, Nezhukhiv; 9, Morshyn; 10, Tura Velyka; 11, Trostyanets'; 12, Kalush; 13, Holyn'. III, Bilche–Volytsya zone.

subsuite (Klimov 1972; Korenevskiy et al. 1977; Khrushchov 1980) or to its facial analogue, the Lower-Balytska one (Skvortsova 1964; Klimov 1977), or to the Tyras suite (Dzhinoridze 1980). In this paper we accept that evaporite deposits belong to the Vorotyshcha (Eggenburgian) and Tyras (Badenian) suites (Dzhinoridze 1980).

In our opinion the most certain view concerns the Badenian age of Kalush–Holyn' deposits. This was first put forward by Polish geologists (Kuźniar 1939) and later supported by the studies of foraminifera microfauna (Dzhinoridze 1976) and nannoplankton (Andreyeva-Grigorovich & Kulchytskiy 1985; Andreyeva-Grigorovich et al. 2003). Salt-bearing deposits from Bilyna Velyka and Nezhukhiv, Morshyn, Tura Velyka, and Trostyanets' (see Fig. 7) are considered to be coeval with Kalush–Holyn' deposit (Korenevskiy et al. 1977).

The deposits of the Lower-Vorotyshcha subsuite form a band in Boryslav–Pokuttya zone and extend along the Carpathian thrust. They are composed of saline clays with intercalated strata of sandstones and siltstones, anhydrite and rock salt. In some areas (Boryslav, Markova–Rosilna potash deposits, Dobrogostiv, Ulychne fields; see Fig. 7) they are represented by salt-bearing breccia with layers of potash salts. Their thickness varies from 100 to 900 m.

The deposits of the Upper-Vorotyshcha subsuite are spread only in zone from Blazhiv to Verkhniy Strutyn, within the Stebnyk deposit and they are composed of a thick sequence (700–1000 m) of saline clays and salt-bearing breccia with layers of

Western Paratethys, regional stage		Transcarpathian trough, suites	Forecarpathian deep, suites	SW margin of East Europian platform	
Old scheme	New scheme				
Sarmatian S. Str.	Sarmatian S. Str.	Lukove	Dashava	Upper Volyn' horizon	
		Dorobratove		Lower Volyn' horizon	
Upper Tortonian	Badenian	Darolyn	Kosiv	Holohory substage	Buhliv beds
		Neresnytsya Tyachiv Shandrove			Vyshgorod beds
					Ternopil' beds
		Solotvyna			Pidhirtsi beds
Lower Tortonian		Tereblya	Tyras		Tyras gypsum
		Talabor	Bohorodchany	Opillya substage	Kryvchytsi horizon Roztocha horizon Narayiv beds
		Novoselytsya			Baranove beds
Upper Helvetian	Karpatian		Balych	Berezhany beds	
Lower Helvetian	Ottnangian		Stebnyk	Nahoryany beds	
Burdigalian	Eggenburgian	Burkalo	Vorotyshcha		
Aquitanian	Egerian	Nehrove			
Chattian			Polyanytsya		

Fig. 8. Regional stages scheme for Miocene of Western Ukraine (after Vyalov 1980).

potash salts (Korenevskiy *et al.* 1977). The Upper-Vorotyshcha subsuite is separated from the Lower Vorotyshcha by clays, sandstones, gritstones and conglomerates of Zagorsk suite that is 100–400 m thick. In recent papers the Zagorsk suite is described as an olistostrome horizon, and the Vorotyshcha suite is not subdivided into the Upper and Lower parts (Dzhinoridze 1980; Kulchytskiy 1986; Korin' & Mosora 1990).

The Tyras suite in Sambir zone of the Foredeep is potash-bearing (Dzhinoridze 1980) but in the Bilche-Volytsya zone it is composed mainly of gypsum and anhydrite. In the Sambir zone the Kalush–Holyn' potash deposit and Bilyna Velyka, Nezhukhiv, Morshyn, Trostyanets' and Tura Velyka all belong to the Tyras suite (Dzhinoridze 1980). It consists of salt-bearing breccia, rock and potash salts. Their thickness in Kalush syncline reaches 300–350 m and in Holyn' *c.* 600–800m. In the Bilche–Volytsya zone the Tyras suite is composed of gypsum, anhydrite and rock salt with terrigenous (clay and sandstone) and carbonate (limestone and marl) rocks; its thickness ranges from 20 to 50 m, sometimes up to 70 m. In Kolomyya depression where the rock salt is known among the anhydrites, the thickness increases to 246 m, and in the Bratkivtsi area up to 347 m (Korenevskiy *et al.* 1977).

Gypsum-anhydrite deposits and rock salt of the Bilche–Volytsya zone

Evaporite deposits of Tyras suite in Bilche–Volytsya zone of the Carpathian Foredeep (Fig. 10) contain large industrial resources of native sulphur and are represented by three complexes: carbonate, sulphate and chloride.

The carbonate complex is composed of sulphate–carbonate, commonly celestite-bearing rocks, secondary metasomatic and sulphur-bearing limestones. The results obtained due to the petrographical and mineralogical investigations of this complex are in agreement with the idea that proposes a metasomatic origin for the limestones from this sulphur-bearing series and the epigenetic origin for celestite, with wide development of superimposed mineralization and intensive leaching of the rocks. Geochemical investigations of the composition of intracrystal solutions and of brine inclusions in calcite from limestones showed an elevated content of calcium hydrocarbonate and potassium chloride in comparison to seawater.

The sulphate complex on 95% consists of chemogenic deposits represented by gypsum, gypsum–anhydrite and anhydrite rocks. It is subdivided into the lower autochthone (*in situ*) and upper allochthone (reworked, clastic) parts. The lower part

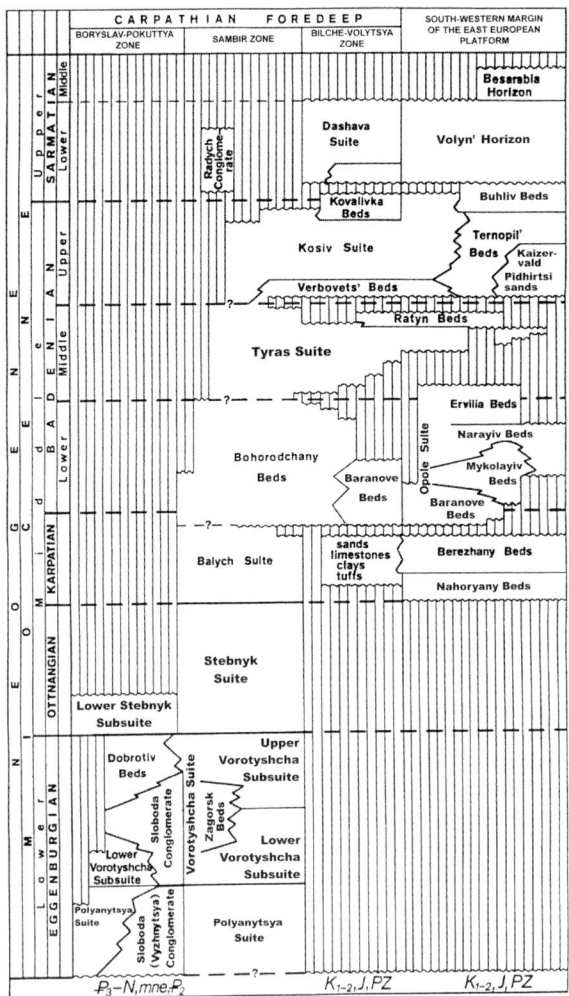

Fig. 9. Stratigraphic scheme of Miocene in the Carpathian Foredeep (after Petrichenko *et al.* 1994).

is locally distributed in tectonic depressions within the Ukraine territory. Within the autochthonic sulphate sequence there are three section types (Peryt 2001), two of which are present in Ukraine. The first one is composed of stromatolitic gypsum and is typical for the NE margin of the basin (up to 15 km wide). The second type of gypsum section forms a zone (40 km wide) near the central part of basin and is composed of stromatolitic gypsum in its lower part and of saber-like gypsum in the upper part (Poberezhskyy *et al.* 2002; Peryt *et al.* 2004*a*). The correlation of these facies zones is based on recognized key horizons in the primary sedimentary gypsum. The typical key horizon is 20–40 cm thick and is composed of wavy layers of fine-grained gypsum and it is present across a very broad area of the Bilche–Volytsya zone.

Another important correlative marker is the contact between the autochthonic and allochthonic (clastic) gypsum and this evidences a significant alteration in the history of basin evolution (Peryt 1996). Correlation between the facies zones evidences the contemporaneous origin of coarse-crystalline and stromatolitic gypsum. It is considered that the near-shore facies associations were formed under sabkha conditions (Peryt 1996; Bąbel 2004, 2005*a*, *b*), and during the whole time of existence of the basin only two periods of elevated brine concentration are noted (Kasprzyk 1999).

The question of the correlation of the sulphate and chloride sedimentation in the region remains unresolved, particularly concerning possible genetic connections between these strictly chemogenic deposits and the time of their sedimentation

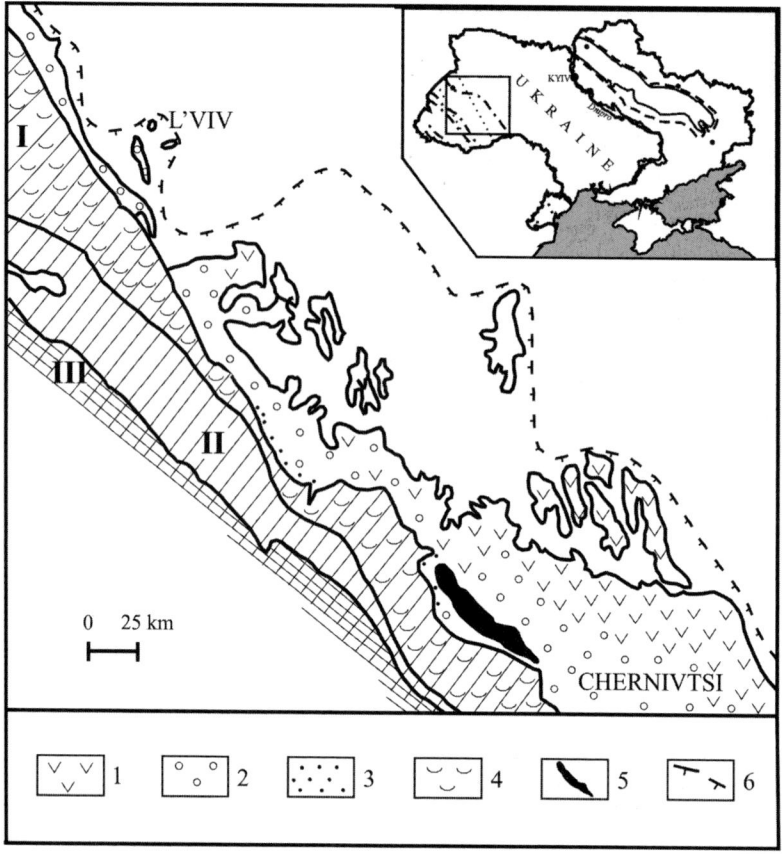

Fig. 10. Location scheme of sulphate and chloride complexes of Tyras suite in Bilche–Volytsya zone (SE part) of the Carpathian Foredeep and adjacent part of East European platform (after Kityk *et al.* 1979). I, Bilche–Volytsya zone; II, Boryslav–Pokuttya and Sambir zones; III, Folded Carpathians. 1, sedimentary gypsum; 2, epigenetic gypsum; 3, gypsum–anhydrites; 4, anhydrites; 5, rock salt; 6, NE boundary of distribution of sulphate complex.

(Panow & Płotnikow 1996; Petrichenko *et al.* 1994; Wyszyński 1939). It has been assumed, but remains uncertain, that the gypsum sequences correspond only to the upper part of evaporite complex, the origin of which was related to a mixed continental-marine evaporite basin at the end of the Badenian time (Petrychenko *et al.* 1997; Poberezhskyy 2000).

Anhydrite deposits are massive or layered rocks, and commonly with intercalations of clay lamina. $CaSO_4$ content is about 94–96% and different admixtures and clays compose 4–6% of the total. Gypsum–anhydrite mixes apparently resulted from incomplete dehydration of gypsum (or incomplete hydration of secondary anhydrite) and sometimes are marked by inherited lithologic structures, relics of coarse-crystalline sabre-like gypsum crystals for example.

Petrographical study of coarse-crystalline gypsum showed that, during its dehydration, bassanite originated first, and then anhydrite. Bassanite forms as crystals of prismatic, needle, plate and rod-like shapes. It is typical that many fluid inclusions (less than 1–3 μm) occur in the gypsum around bassanite crystals. These inclusions are formed from water released during gypsum–bassanite alteration. (Petrichenko & Poberezhskyy in Matkovskyy *et al.* 2003). The water-insoluble residue from single crystals of coarse-crystalline gypsum is composed of chlorite, montmorillonite, mica, kaolinite, euhedral quartz, dolomite and calcite.

The chloride complex is composed of rock salt containing both primary and recrystallized halite. Three varieties of primary halite are distinguished: pyramidal hopper crystals, chevrons and cubic hopper crystals. This last structure was first described by Raup (1970) and demonstrated the mixing of differentiated brines. The bromine content in these salt deposits varies from 20 to 60 ppm (Petrichenko *et al.* 1974*b*). The water-insoluble residue consists of authigenic anhydrite, gypsum, bassanite, quartz, dolomite, magnesite and calcite. The clay fraction contains hydromica and chlorite.

Potash salts of the Forecarpathian region

Potash-bearing deposits of the Carpathian Foredeep generally are composed of kainite, langbeinite and kainite–langbeinite rocks; to a lesser extent they have sylvinite beds; other salts are found as admixtures and form rare accumulations of potash rocks.

The successions of potassic deposits are best studied in the Stebnyk and Kalush–Holyn' areas. The potassic salts occur repeatedly in the section and this repetition formerly was considered to have formed as the result of sporadic refilling, concentration and repeated sedimentation of different salts in small lagoons. In the Stebnyk deposit there are 14 distinct layers (lenses) (Kudryavtsev 1971; Korenevskiy et al. 1977); in the Kalush–Holyn' deposit such lenses are united into four potash horizons for the Kalush syncline; and in the Holyn' area they may variously be divided into nine (Korenevskiy & Donchenko 1963) or into six separate deposits (Korenevskiy et al. 1977). In the papers of Dzhinoridze and Korin' an idea has been proposed concerning the thrust–slice constructs of these evaporite deposits. The number of thrust–slices determines the thickness of the salt-bearing sequence and the number of potash lenses. Thus, using that proposal the multistage and lens-like bedding of potash layers resulted not from numerous repetition of sedimentation but from thrust-fold construction (Dzhinoridze 1980), and this point of view is clearly reflected in the section of Stebnyk deposit (Fig. 11). The true unfolded thickness of salts in Stebnyk deposit is estimated to have been 100–125 m (Korin' 1992). Before the processes of thrust tectonics in the Forecarpathian region only two strata of potash salts were recognized, the lower one of chloride composition (sylvinite) and the upper of sulphate (kainite–langbeinite) composition (Dzhinoridze 1980; Korin' 1992). This point of view is also supported by lithological investigations; the comparison of lithological sections of some potash layers shows their identity (Hryniv 1985).

Based on their compositions the Vorotyshcha and Tyras potash-bearing deposits of the Carpathian Foredeep, which are best studied in Kalush–Holyn' and Stebnyk deposits, have many comparable features such as sulphate composition of salts, significant content of clays, and also that the enclosing rock for potash salts is a clayey, salt-bearing breccia. This breccia is composed of clay and siltstone detritus cemented by halite that is thus far considered as a rock of sedimentary origin (Polikarpov 1974; Peryt & Kovalevych 1997), but may be the result of crushing of the rocks due to migrating brine and gases during late diagenesis and catagenesis, and more importantly, during repeated periods of strain during overthrust formation (Khodkova 1980). In the Kalush–Holyn' deposit the salt-bearing sequence is somewhat different and begins with a basal anhydrite.

The industrial potash deposits are composed of kainite, langbeinite, kainite–langbeinite, sylvinite, polyhalite and carnallite rocks (Lobanova 1956) with layers of rock salt or interbedded clays and

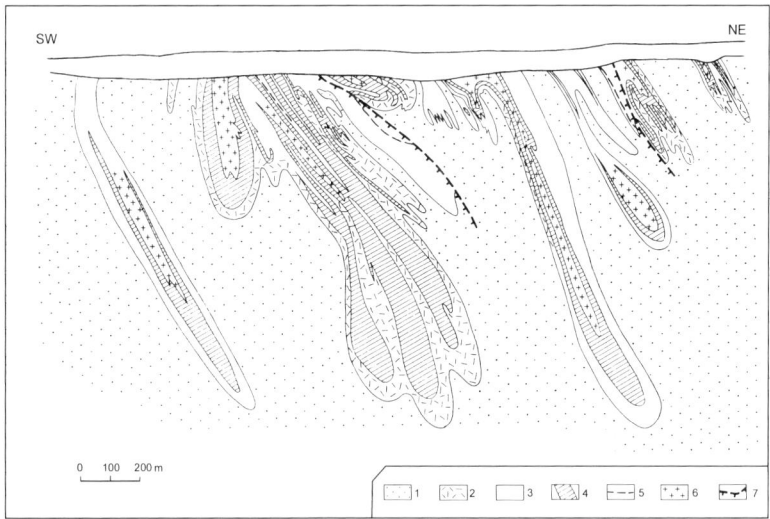

Fig. 11. Geological cross section through the Stebnyk potash-bearing deposits (after Korin' 1992). 1, Terrigenous deposits below evaporites; 2, salt breccia with sylvite; 3, salt breccia; 4, kainite–langbeinite bed; 5, kainite admixture; 6, rock salt; 7, overthrust zones.

rock salt. In the areas of salt-bearing breccia the polyhalite–anhydrite layer occurs along the contact with the potash salts beds.

Kainite rocks are yellow, grey-yellow, mainly fine-grained, commonly clayey, bedded, with the thickness of the layers being 5–10 cm.

Langbeinite rocks are pink, massive, crystalline-grained with a small clayey admixture. In the opinion of Lobanova (1956) kainite and langbeinite are primary minerals precipitated in the evaporite basin. However, Valyashko (1962) and Khodkova (1971) considered them as a result of reactions of metastable minerals (epsomite, hexahydrite, kainite) and burial solutions during early diagenesis.

Kainite–langbeinite rocks contain a mixture of kainite, langbeinite, sylvite, kieserite, halite and water-insoluble residue in approximately equal amounts, 10–15%, and have a breccia-like texture and are considered to be the breccia of sulphate minerals cemented by chloride minerals (Khodkova 1971).

Polyhalite rocks are found in several varieties. The grey clayey polyhalite rock occurs in potash rocks as persistent layers of 5–10 cm thickness. Red polyhalite rock forms layers of 1–3 cm to 1.5–3 m thick in areas where potash deposits pinch out. Two component polyhalite–anhydrite layers of 1–30 cm thick occur in a salt-bearing breccia near the contact with potash salts. Though the crystallization of polyhalite from solution is possible and has been tested experimentally, the gypsum crystallization and its transformation into polyhalite during a stage of early diagenesis are considered as a more reliable possibility for grey polyhalite (Valyashko 1962). Polyhalite–anhydrite layers were formed *de novo* at sulphate geochemical barrier due to the reaction of burial sedimentary sulphate solutions from potash deposits and metamorphosed calcium chloride solutions from salt-bearing breccia (Hryniv 1990).

Carnallite rocks are uneven in thickness and extension beds, which occur in the places of tectonic dislocation of the salt sequence. In the Forecarpathian region the origin of carnallite is connected with altered burial sedimentary brines (Povsten & Povsten 1980).

In potash-bearing deposits and salt-bearing breccia closely related to them, there are occurrences of laterally persistent layers of rock salt (5–10 cm thick, rarely 20 cm). These may have been formed in an evaporite basin as primary precipitates due to salting out, i.e. mixing of solutions of different concentration (Kovalevich 1977; Hryniv 1982). There is, however, another view on their origin, that they are layer-like bodies that were formed during diagenesis (Khodkova 1971).

Potash-bearing deposits of the Carpathian Foredeep contain more than 20 salt minerals. The main rock-forming minerals are halite, sylvite, kainite, langbeinite, kieserite and polyhalite, rarely leonite or carnallite, rare minerals include bassanite and kalistrontite (the last one being the third finding in the world). Anhydrite in salt-bearing breccia is a common admixture, and it occurs rarely in potash rocks. A separate group is formed by the hypergene minerals of the weathering zone (from the gypsum-clayey cap): schoenite (picromerite), epsomite, hexahydrite, glaserite, astrakhanite (bloedite), loeweite, vanthoffite, glauberite, mirabilite, tenardite, syngenite (kaluschite) and gypsum. The results of mineralogical investigations and the vast literature on evaporite minerals are summarized in the monograph *Minerals of Ukrainian Carpathians* (Matkovskyy *et al.* 2003).

In early publications the main minerals were considered to be sedimentary (Korobtsova 1955; Lobanova 1956); in later works their origin was recognized to be connected with reactions between unstable sedimentary minerals and the brines from stages of both early and late diagenesis (Valyashko 1962; Khodkova 1971; Gemp 1979; Kovalevich 1982). A sedimentary origin of kainite and one variety of langbeinite, however, is not excluded (Petrichenko 1988). The origin and transformation of salt minerals is also connected to the thermodynamic metamorphism stage of the region (Gemp 1979; Dzhinoridze 1980), which took place during the overthrust-forming processes in the Carpathians and Carpathian Foredeep. The Radiometric age (K–Ar) of langbeinite from Kalush–Holyn and Stebnyk deposits ranges from 13.63 to 14.65 Ma and shows the regional heating and recrystallization of potash-bearing deposits during overthrust-forming (Wojtowicz *et al.* 2003).

The pelitic fraction of the water insoluble residue of salts is represented by hydromica and chlorite with admixture of quartz and carbonates; in some cases feldspar is present (Yarzhemskaya 1954; Bilonizhka *et al.* 1966). Based on data of absolute ages the hydromica is mainly terrigenous (reworked) (Bilonizhka & Kostin 1977). Kaolinite is present in subsalt deposits and disappears in salt-bearing sequence, and this apparently is a result of its transformation under the aggressive environment of saline solutions (Bilonizhka 1991). In the pelitic fraction from the clayey gypsum cap, except for hydromica and chlorite, the mixed-layered phase hydromica-montmorillonite is also present. This is evidence of the degradation of part of hydromica during leaching of evaporites by surface water over long periods of time (Oliyovych *et al.* 2004).

In the potash rocks from Lower Vorotyshcha subsuite the mineral chambersit (a manganese boracite) has been found in the Stebnyk and Markova–Rosilna deposits. In the heavy fraction of water-insoluble residues from potash-bearing salts it is associated with anhydrite, barite, breinerite and rarely dolomite (Slivko 1979). In salt breccia from Upper Vorotyshcha subsuite luneburgite has been found (Matkovskyy et al. 2003).

The marine genesis of the Miocene salts is supported by the results of a study of the trace elements: Li, Rb and Cs (Slivko & Petrichenko 1967), Rb, Tl and Br (Malikova 1967), B, Br and J (Bilonizhka 1972). By isotope data, $\delta^{34}S$ of anhydrites from Vorotyshcha and Tyras suites is in the typical range for the Neogene sulphates of marine origin (Kovalevych & Vityk 1995).

The bromine content in the halite (397 analyses) (Kovalevich 1979) has been studied for all saline deposits of the Carpathian Foredeep. In halite from those deposits without potash salts the bromine values range from 10 to 100 ppm (on average 56 ppm); in halite from salt breccia with potash salts the bromine content is 30–230 ppm (on average 120 ppm), and in halite from potash beds it is 70–300 ppm (on average 170 ppm).

In the minerals from the potash deposits the bromine content ranges are: in kainite 800–2300 ppm; in sylvite 1410–2660 ppm; in carnallite 1520–2450 ppm (Bilonizhka 1970).

The brines of Vorotyshcha and Tyras salt-forming basins (based on investigation of brine inclusions from sedimentary halite) belong to a Na–K–Mg–Cl–SO$_4$ (SO$_4$-rich) chemical type.

Seawater was a main source of salts and based on the chemical composition it was close to the modern seawater and but differed somewhat, based on a lower content of the sulphate-ion (Kovalevich 1978; Kovalevich & Petrichenko 1997; Galamay 1999; Poberezhskyy & Kovalevych 2001).

The temperature of solution in the Miocene basins of this region at the stage of halite sedimentation was about 25 °C (Kovalevich 1978; Galamay 2003), and at the potash salt stage of sedimentation these solutions were stratified and due to greenhouse effect their lower portion heated up to 40–60 °C. At the stage of catagenesis the temperature rose to 70 °C and such elevated values have a regional character (Kovalevich 1978).

Transcarpathian trough

The Transcarpathian trough is a sag structure filled with a thick sequence of the Neogene molasse (Fig. 12). It formed at the boundary of the folded Eastern Carpathians and the Hungarian (or Pannonian) median mass and was separated from them by deep faults of the Transcarpathian (Laz'ko & Rezvoy 1962) or Utiosovo–Marmaroshsky (Slavin & Khain 1965) and Pripannonsky (Merlich & Spitkovskaya 1965). The trough extends from NW to SE and is about 150 km long and 20–30 km wide.

Crossfaults (anticarpathian extension faults) caused the separation of the trough into second-order

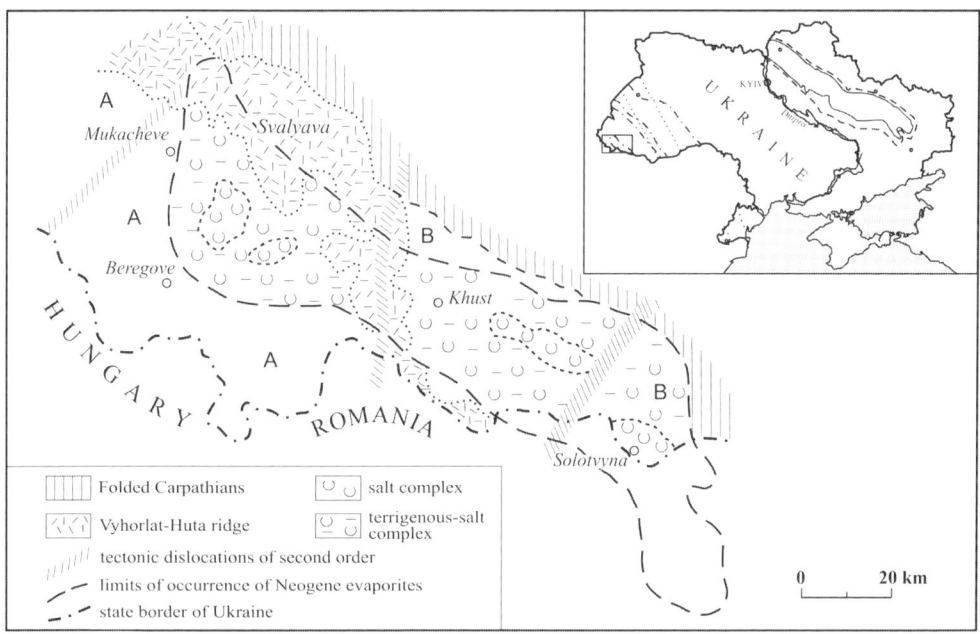

Fig. 12. Location map of Miocene evaporites in Transcarpathian trough (after Kityk et al. 1983). (A) Mukacheve depression; (B) Solotvyna depression.

structural–tectonic units and the major units are the Solotvyna and Mukacheve depressions, divided by the Khust fault (Kityk *et al.* 1983).

Neogene deposits in the Transcarpathian trough lie with an angular and stratigraphic unconformity upon Paleozoic, Mesozoic and Paleogene formations. In the Miocene section evaporite deposits are present in the Tereblya suite. They are in turn overlapped by rocks of the Upper Miocene, Pliocene and Pleistocene.

Disharmonic dislocations which are typical for the salt-bearing deposits of Transcarpathian trough result in the development of diapiric salt bodies with the degree of deformation of overlying deposits from insignificant rise to entire penetration and outcrop. In the central part of Solotvyna depression there are large short, broad, anticlinal domes (diapirs) with open piercement-type cores. In the Mukacheve depression these disharmonic dislocations are less marked, and the salts do not penetrate the overlying deposits (Kityk *et al.* 1983).

Evaporite deposits of Tereblya suite are composed of rock salt with an admixture of clayey material and anhydrite, saline and argillaceous clays, and rare tuffs.

The rock salt is pure, usually white, but sometimes grey with layers of clays and gypsum–anhydrites. In areas of outcrop it is overlapped by grey viscous clays (named 'palah') which are the residue of rock salt weathering. The most widespread rock salt in the Solotvyna deposit is pure with a NaCl content of 93.67–99.34% and a water insoluble residue of 0.8–2.0%. The content of bromine is typical for a marine-sourced water at the halite stage (100–110 ppm on average), but in some areas its content is distinctly lower, *c.* 10ppm, which suggests that the halite is recrystallized (Kityk *et al.* 1983).

The clayey minerals of salt-bearing deposits in Solotvyna are represented by hydromica and Mg–Fe chlorite with admixture of dolomite, calcite and anhydrite (Bilonizhka 1979; Kityk *et al.* 1983). Rock salt in the Svalyava region underlies the sheet of Pliocene volcanic rocks of the Vyhorlat–Huta ridge and its pelitic fraction contains the mixed layered chlorite–Mg–montmorillonite close to corrensite composition (Kityk *et al.* 1983).

The brines that had concentrated during halite sedimentation in Transcarpathian salt basin (based on data from fluid inclusions from the sedimentary halite) belong to the Na–K–Mg–Cl–SO$_4$ (SO$_4$-rich) chemical type and by ratio of K and Mg were close to the modern seawater saturated to the halite stage. The halogenesis had not reached a high stage of concentration and finished in the halite range (Shaidetskaya 1997).

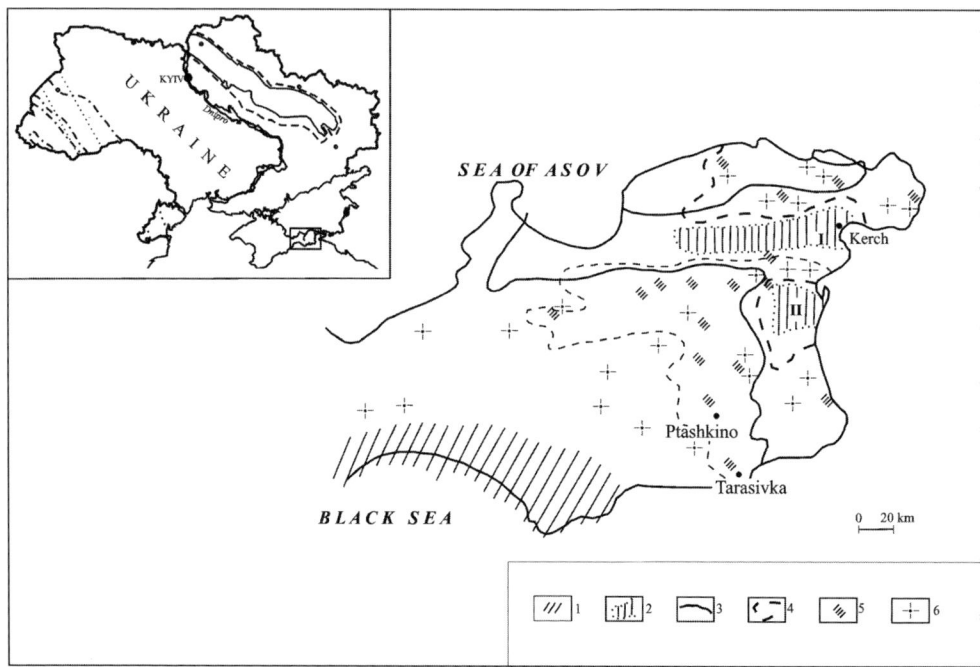

Fig. 13. Paleogeographic scheme of Kerch peninsula in Chokrakian time (after Peryt *et al.* 2004*b*). 1, Land; 2, relatively deep-water part of Chokrakian sea; 3, shore line of late Tarkhanian sea; 4, assumed boundary of area of late Chokrakian gypsum accumulation; 5, areas where late Chokrakian gypsum or metasomatic limestones were found; 6, mud volcanoes active in Chokrakian.

Kerch peninsula (Crimea)

The presence of evaporites in the Miocene deposits of Kerch peninsula (Fig. 13) is well known and mentioned in many papers devoted to the geology and mineral resources of this region. Evaporites are represented by gypsum and are present in all stages of the Middle and Upper Miocene. The largest gypsum deposits are located near Ptashkino (Elkedgi–Eli) and Tarasivka (Tchekur–Koyash). Research on sulphur as an industrial mineral largely governed the development of evaporite investigation in the area of the Kerch peninsula.

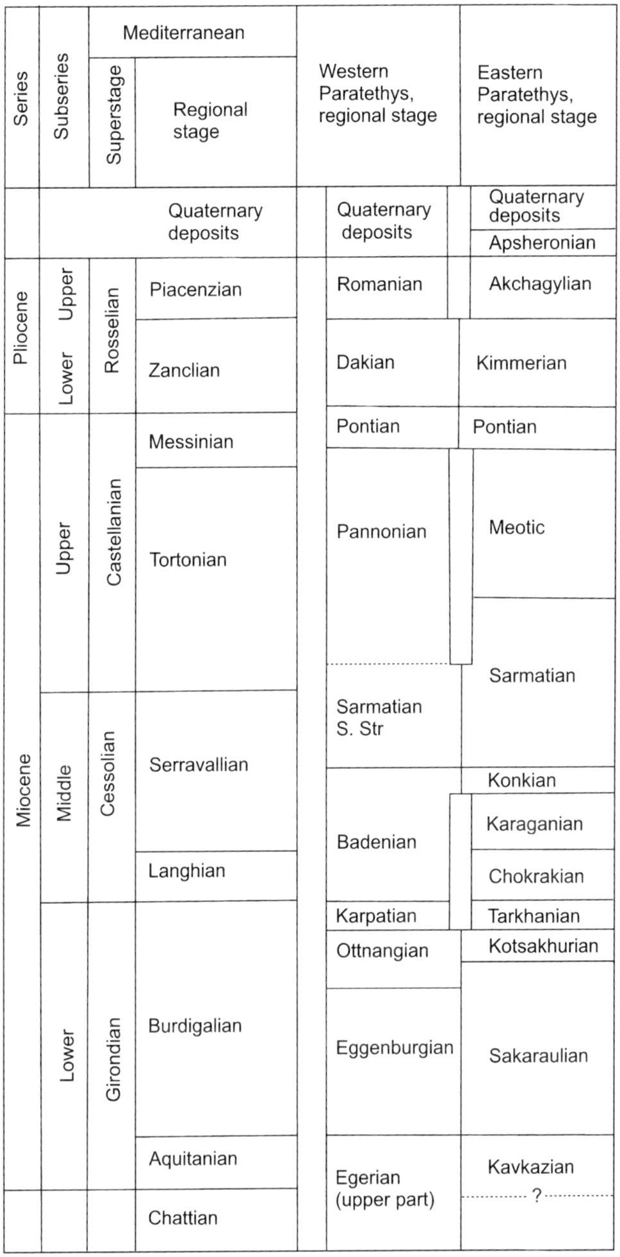

Series	Subseries	Superstage	Mediterranean Regional stage	Western Paratethys, regional stage	Eastern Paratethys, regional stage
			Quaternary deposits	Quaternary deposits	Quaternary deposits
Pliocene	Upper	Rosselian	Piacenzian	Romanian	Apsheronian
					Akchagylian
	Lower		Zanclian	Dakian	Kimmerian
Miocene	Upper	Castellanian	Messinian	Pontian	Pontian
			Tortonian	Pannonian	Meotic
					Sarmatian
	Middle	Cessolian	Serravallian	Sarmatian S. Str	
				Badenian	Konkian
					Karaganian
			Langhian		Chokrakian
				Karpatian	Tarkhanian
				Ottnangian	Kotsakhurian
	Lower	Girondian	Burdigalian	Eggenburgian	Sakaraulian
			Aquitanian	Egerian (upper part)	Kavkazian ...?...
			Chattian		

Fig. 14. Regional stages correlation scheme of Eastern Paratethys, Western Paratethys and Mediterranean (after Nevesskaya *et al.* 1984).

The Miocene evaporites in this region are not uniformly distributed. They are absent in the southwestern plain and in the majority of the uplifts. In the eastern part of the peninsula they overlap the Maikop Oligocene to Lower Miocene clays without a visible break. To the west they are separated by an erosion surface and the Tarkhanian stage is missing from the section. At last case the angular unconformity, sometimes significant, is noticed between Maikop (Oligocene) and Miocene layers.

Gypsum as a rule forms small lenses and layers up to several centimetres thick and is only present in the lower part of the Middle Miocene. The age of these layers has been determined as Chokrakian (Arkhangelskiy et al. 1930), Upper-Chokrakian (Ishchenko & Kisieliov 1967; Pavlenko et al. 1974), Karaganian (Shniukov et al. 1971; Nevesskaya et al. 1984) or Chokrakian–Karaganian (Muratov 1960). It is most likely that a Chokrakian age is correct for these evaporites (Peryt et al. 2004b; Kityk et al. 1979). If this age assignment is correct then the gypsum-bearing deposits of Kerch peninsula are coeval to the Tyras gypsum-bearing sequence of the Forecarpathian region and this in turn supports the synchronous conditions of gypsum sedimentation in the basins of the Eastern and Western Paratethys.

In the Ptashkino and Tarasivka deposits, gypsum crops out as a strata about 0.9 m thick and the thickness increases toward the east into the subsident area up to 6.25 m (Sobolevskiy 1970). Gypsum occurs in two layers. The lower one composed of fine-grained gypsum is accompanied by clayey and clay–carbonate material, but the upper layer is mainly composed of coarsely crystalline gypsum.

Based on the morphology of crystals and internal structure the coarse-crystalline gypsum of Chokrakian horizon is similar to that of the Tyras suite of the Forecarpathian (Fig. 14). The Kerch gypsum contains more solid inclusions and is less transparent and is enriched by biogenic material. The crystallization of coarse-crystalline gypsum took place in conditions within the field of gypsum stability. Possibly that was connected with lower temperature (30 °C) and water concentration in the Chokrakian basin in comparison to Tyras basin (Kulchetskaya 1987; Peryt et al. 2004b).

Conclusions

Evaporite deposits are widespread in Ukraine and are located in the Dnipro–Donets depression (Devonian, Permian), western and southwestern slope of the East European platform (Jurssic), Carpathian Foredeep, Transcarpathian trough and the Crimea (Neogene). The majority of investigators (including the authors of this review) consider them to be mostly of marine origin. The largest evaporite deposits of Devonian and Permian age are present in the Dnipro–Donets depression. They are presented mainly by rock salt.

Upper Devonian (Frasnian and Famennian) evaporites are represented by anhydrites, rock and potash salts (sylvinite). Bedded salts, as well as dislocated salts (brachianticlines, salt domes and stocks), are well known. Rock salt in some domes contains rich superimposed hydrothermal mineralization. In it water insoluble residues, there are about 40 high- and middle-temperature minerals such as scapolite, albite, muskovite, sphene, microcline, clinozoisite and others, and rare minerals – tunisite, dawsonite, juwelite. Bromine content in rock salt of Frasnian stage is 50–150 ppm and of the Famennian stage 60–130 ppm. These salts were formed from solutions of Na–K–Ca–Mg–Cl (Ca-rich) chemical type. The temperature of these solutions during the stage of halite precipitation did not exceed 43 °C, and was elevated to 60–65 °C on sylvite stage.

Lower Permian evaporites in the Dnipro–Donets depression are composed of anhydrite, rock salt, potash and magnesium salts (composed of carnallite, kieserite and bishofite). Rock salt also forms salt domes but the halite has less thicknesses than Devonian deposits. In some cases Devonian salts pierced the Permian ones creating so-called twin-salt structures (Petrychenko et al. 1976b). The stratigraphical differentiation in such structures is possible only with the help of the ultra-microchemical analysis of brine inclusions in halite (Petrichenko 1973). Permian salts were formed from solutions of Na–K–Mg–Cl–SO$_4$ (SO$_4$-rich) chemical type while the Devonian salts were Na–K–Ca–Mg–Cl (Ca-rich). The bromine content in halite of Permian deposits was 32–85 ppm. The temperature of salt-forming solutions on stage of halite precipitation did not exceed 43 °C, but was elevated to 78–83 °C for carnallite and to 65–77 °C in the bischofite stages. The temperature of diagenetic transformations of rock salt in the Permian deposits was 45–55 °C and of potash rocks, 75–82 °C.

Upper Jurassic evaporite deposits in Ukraine are distributed within western and southwestern slopes of the East European platform in the Forecarpathian and Foredobrogean regions. Throughout most of the territory they are represented by anhydrites and gypsum and only in a restricted area, near Izmail, they contain rock salt. The bromine content in this rock salt is 27–63 ppm. Salt was formed from solutions of Na–K–Ca–Mg–Cl (Ca-rich) chemical type and at temperatures of less than 40 °C. The maximal determined concentration of potassium-ion in halite was 15.5 g l^{-1} and this value was too low to expect potash salt sedimentation.

Neogene (Miocene) evaporite deposits in Ukraine are distributed in the Carpathian Foredeep, and Transcarpathian trough and in Crimea (Kerch peninsula). In the Carpathian Foredeep they are located in the Bilche–Volytsya, Sambir and Boryslav–Pokuttya zones. In the Bilche–Volytsya zone they are represented by gypsum, anhydrites and the rock salt of the Tyras suite (Badenian) and large industrial resources of native sulphur are connected with them. Potash salts of the Carpathian Foredeep are related to Vorotyshcha suite (Eggenburgian), which are located in the Boryslav–Pokuttya zone and to the Tyras suite (Badenian) in the Sambir zone. They have many similar features, such as large number of potash lenses in the section, which results from the folded-thrust construction of the evaporite deposits; most of these salts are sulphates with a high clay content. The main potash rocks are of kainite, langbeinite and kainite–langbeinite composition. In the potash deposits and in their weathering zone there are more than 20 salt minerals. Bromine content in halite of the Carpathian Foredeep for deposits without potash salts ranges from 10 to 100 ppm (on average 56 ppm); that in halite from salt breccia with potash salts ranges fom 30 to 230 ppm (on average 120 ppm); and that in halite from potash beds ranges from 70 to 300 ppm (on average 170 ppm). In the minerals from the potash deposits bromine content ranges are: in kainite 800–2300 ppm; in sylvite 1410–2660 ppm; and in carnallite 1520–2450 ppm.

The brines of Vorotyshcha and Tyras salt-forming basins (based on data from brine inclusions in an investigation of sedimentary halite) belong to the Na–K–Mg–Cl–SO$_4$ (SO$_4$-rich) chemical type. The temperature of halite formation in the Miocene basins of Forecarpathian region was about 25 °C at the halite stage. During the stage of potash salts sedimentation these solutions became highly stratified and due to a greenhouse effect the bottom (formative) layer of basin brines was heated up to 40–60 °C. During a later stage of burial and catagenesis the temperature rose to 70 °C and such elevated values had a clear regional distribution.

In the Transcarpathian area the Miocene evaporites of the Tereblya suite (Badenian) are represented by rock salt, which occurs in a bedded series or is present in salt domes. Bromine content in these rock salts is typical of the marine-sourced halite stage of evaporation and is 100–110 ppm, but in some areas its content is distinctly lower (as low as 10 ppm), which indicates the recrystallization of halite. The brines during halite sedimentation in the Transcarpathian salt-forming basin (based on data of fluid inclusions from sedimentary halite) belong to the Na–K–Mg–Cl–SO$_4$ (SO$_4$-rich) chemical type. Here the salt formation (halogenesis) had not reached the highly evaporated stages (potassic salts) and apparently ended within the halite stage.

In the Kerch peninsula Miocene evaporites deposits of the Chokrakian stage are represented only by small lenses and layers of gypsum. The gypsum-bearing deposits of Kerch peninsula are coeval to the gypsum-bearing sequence of the Forecarpathian region.

The data presented above illustrate that evaporite deposits of Ukraine have been studied comprehensively with geological, lithological, mineralogical, petrographic and geochemical methods. Large-scale sampling and thorough investigation of inclusions in primary sedimentary halite deposits demonstrate that the brine inclusions in Devonian and Jurassic evaporites of Ukraine were formed from solutions of Na–K–Ca–Mg–Cl (Ca-rich) chemical type and that Permian and Neogene evolved from a Na–K–Mg–Cl–SO$_4$ (SO$_4$-rich) water type. This, in turn, is in good agreement with the well-documented secular variations of seawater chemistry during the Phanerozoic.

We thank O. Petrychenko and V. Kovalevych for critical comments and suggestions, and B. C. Schreiber and M. Bąbel for their remarks on an earlier version of this paper.

References

ANASTASIEVA, O. M. 1957. Some data about the Upper Jurassic sections of Volyno-Podoliya plate. *Geologicheskiy sbornik L'vovskogo geologicheskogo obshchestva*, **4**, 155–162 [in Russian].

ANASTASIEVA, O. M. 1958. Upper Juassic anhydrites and gypsums in the south-western margin of Russian platform. *Voprosy mineralogii osadochnyh obrazovaniy*, **5**, 47–55 [in Russian].

ANDREYEVA-GRIGOROVICH, A. S. & KULCHYTSKIY, A. Ya. 1985. Concerning a problem about the age of the saliferous series of Kalush–Holyn deposit. *In*: KITYK, V. I. (ed.) *Evaporites of Ukraine*. Naukova Dumka, Kiev, 176–181 [in Russian].

ANDREYEVA-GRIGOROVICH, A. S., OSZCZYPKO, N., SAVITSKAYA, N. A., LĄCZKA, A. & TROFIMOVICZ, N. A. 2003. Correlation of Late Badenian salts of the Wieliczka, Bochnia and Kalush areas (Polish and Ukrainian Carpathian Foredeep). *Annales Societatis Geologorum Poloniae*, **73**, 67–89.

ARKHANGELSKIY, A. D., BLOKHIN, A. A., MENNER, V. V., OSIPOV, S. S., SOKOLOV, M. I. & CHEPIKOV, K. R. 1930. Brief description of the oil fields in the Kerch peninsula. *Trudy glavnogo geologorazviedochnogo upravleniya*, **13** [in Russian].

BĄBEL, M. 2004. Badenian evaporite basin of the northern Carpathian Foredeep as a drawdown salina basin. *Acta Geologica Polonica*, **54**, 313–337.

BĄBEL, M. 2005a. Event stratigraphy of the Badenian selenite evaporites (Middle Miocene) of the northern Carpathian Foredeep. *Acta Geologica Polonica*, **55**, 9–29.

BĄBEL, M. 2005b. Selenite-gypsum microbialite facies and sedimentary evolution of the Badenian evaporite basin of the northern Carpathian Foredeep. *Acta Geologica Polonica*, **55**, 187–210.

BILONIZHKA, P. M. 1970. Content of bromine in salt minerals of potash deposits of Forecarpathians. *Voprosy mineralogii osadochnyh obrazovaniy*, **8**, 126–133 [in Russian].

BILONIZHKA, P. M. 1972. *Geochemistry of Boron, Bromine and Iodine in the Forecarpathian Potash Deposits.* L'vivskyy universytet, L'viv [in Russian].

BILONIZHKA, P. M. 1979. About the mineral composition of carbonates and clays of the Solotvyna rock salt deposit (Transcarpathians). *In:* KITYK, V. I. (ed.) *The Questions of Geology and Geochemistry of Saliferous Deposits.* Naukova Dumka, Kiev, 53–61 [in Russian].

BILONIZHKA, P. M. 1991. The transformation of terrigenous clay minerals in the saliferous process. *Mineralogichnyy zbirnyk*, **45**, 2, 51–56 [in Ukrainian].

BILONIZHKA, P. M. & KOSTIN, V. A. 1977. Origin of hydromicas from salt deposits of the Carpathian Foredeep (based on their absolute ages). *In:* KITYK, V. I. (ed.) *Geology and Geochemistry of Salt-bearing Formations of Ukraine.* Naukova Dumka, Kiev, 53–65 [in Russian].

BILONIZHKA, P. M., VYNAR, O. N. & MELNIKOV, V. S. 1966. About the mineral composition of clay from potash salt deposits of the Forecarpathian. *Voprosy mineralogii osadochnyh obrazovaniy*, **7**, 147–158 [in Russian].

BORCHERT, H. 1959. *Ozeane Salzlagerstatten.* Gebr. Borntrager, Berlin.

BORISENKO, L. S., KROPACHEVA, S. K., PIVOVAROV, S. V. & VASILEVSKAYA, A. E. 1974. First discovery of the Upper Jurassic saliferous deposits in the Crimean mountains. *Doklady AN SSSR*, **219**, 933–935 [in Russian].

BUTKOVSKIY, YU. M., TUREVSKAYA, E. S. & BELIAEV, V. U. 1976. Foredobrogea as Upper Juassic evaporite basin. *Izvestiya VUZov. Geologiya i razviedka.* **2**, 25–30 [in Russian].

BYHOVIER, N. A., VOLGDIN, A. G. & MATVIEYEV, A. K. 1941. Geological investigations. *In:* BYHOVIER, N. A. (ed.) *Geology and Mineral Resources of the Western Regions of USSR.* Gosgeoltehizdat, Moskva-Leningrad, 21–40 [in Russian].

CHMIELEWSKA, E. W. 1995. Litologia kimerydzkich ewaporatow poludniowo-zachodniego sklonu platformy wschodnioeuropejskiej. *Przegland Geologiczny*, **43**, 928–930 [in Polish].

DOLENKO, G. N. (ed.) 1975. Ukrainian SSR (western regions). Period 1918–1945. *In:* TIHOMIROV, V. V. (ed.) *Geological Investigations of USSR*, **31**. Naukova Dumka, Kiev, 1–350 [in Russian].

DOLISHNIY, B. V. 1976. Composition of rock salt insoluble residue from Bantyshevskiy salt stock (northwestern Donbass). *Geologiya i geokhimiya goriuchyh kopalyn*, **47**, 27–30 [in Ukrainian].

DOLISHNIY, B. V. 2000. Hydrothermal mineralization of salt domes from south-eastern part of Dnieper–Donets depression. *Geologiya i geokhimiya goriuchyh kopalyn*, **4**, 56–61 [in Ukrainian].

DZHINORIDZE, N. M. 1976. About the age of salt-bearing deposits of Holyn in Forecarpathian. *Doklady AN SSSR.* **227**, 4, 932–935 [in Russian].

DZHINORIDZE, N. M. 1980. Carpathian potassium-bearing region. *In:* DZHYNORIDZE, N. M., GEMP, S. D., GORBOV, A. F. & RAYEVSKIY, V. I. (eds) *Regularities of Distribution and Exploration Criteria for Potassium Salts in the USSR.* Izdatelstvo Metsniereba, Tbilisi, 73–159 [in Russian].

DULUB, V. G., BUROVA, M. I., BUROV, V. S. & VISHNIAKOV, I. B. 1986. *Explanatory Report to the Regional Stratigraphic Scheme of Jurassic Deposits of the Carpathian Foredeep and Volyno-Podoliya Margin of the East European Platform.* Leningrad [in Russian].

DULUB, V. G., ZHABINA, N. M., OGORODNIK, M. E. & SMIRNOV, S. E. 2003. *Explanatory Report for the Regional Stratigraphic Scheme of Jurassic Deposits of Forecarpathian (Stryy Jurassic basin).* L'viv [in Ukrainian].

GALABUDA, N. I. 1970. Geological construction of Frasnian salt formation of Dnieper–Donets depression. *Geologiya i geokhimiya goriuchyh kopalyn*, **50**, 59–68 [in Ukrainian].

GALAMAY, A. R. 1999. The conditions of the Badenian rock salt formation of fore- and intermountain depressions of the Carpathians. *Geologiya i geokhimiya goriuchyh kopalyn*, **1**, 55–66 [in Ukrainian].

GALAMAY, A. R. 2003. The temperature of halite crystallization in the Badenian evaporite basins of the Carpathian region. *Geologiya i geokhimiya goriuchyh kopalyn*, **1**, 130–139 [in Ukrainian].

GALITSKIY, I. V. 1972. Cyclicity of saliferous, deposits of Lower Permian Kramatorska suite of Dnieper-Donets depression. *In:* KALEDA, G. A. (ed.) *Lithology and paleography of Paleozoic deposits of Russian platform.* Nauka, Moscow, 249–255 [in Russian].

GARETSKIY, R. G. 1985. *Sedimentation and paleogeography of the western part of East European platform in Mesozoik.* Nauka i tekhnika, Minsk [in Russian].

GEMP, S. D. 1979. The role of secondary processes in formation of the mineral composition of potash salts in Forecarpathian. *In:* RAYEVSKIY, V. I. (ed.) *Geological and Mineralogical–Petrographic Estimation of the Distribution, Quality of the Ores and Conditions of Exploitation of Potash Deposits.* VNIIG, Leningrad, 73–91 [in Russian].

GLUSHKO, V. V. 1968. *Tectonics and Oil-and-Gas Content of the Carpathians and adjacent Foredeeps.* Nedra, Moscow [in Russian].

HARDIE, L. A. 1996. Secular variations in seawater chemistry: An explanation for the coupled secular variation in the mineralogies of marine limestones and potash evaporites over the past 600 m.y. *Geology*, **24**, 279–283.

HOLLAND, H. D. 1972. The geologic history of sea water an attempt to solve the problem. *Geochimica et Cosmochimica Acta*, **36**, 637–651.

HOLLAND, H. D. 1984. *The Chemical Evolution of the Atmosphere and Oceans.* Princeton University Press, Princeton, NJ.

HOLLAND, H. D. 2003. The geologic history of seawater. *Treatise on Geochemistry*, **6**, 583–625.

HOLLAND, H. D., LAZAR, B. & MCCAFFREY, M. A. 1986. Evolution of the atmosphere and oceans. *Nature* **320**, 2733.

HORITA, I., ZIMMERMANN, H. & HOLLAND, H. D. 2002. Chemical evolution of seawater during the Phanerozoic: implications from the record of marine evaporites. *Geochimica et Cosmochimica Acta*, **66**, 3733–3756.

HRYNIV, S. P. 1982. Rock salt of salting out origin of Kalush–Holyn deposit. *In*: KITYK, V. I. (ed.) *Geology and Geochemistry of Non-Metallic Mineral Resources*. Naukova Dumka, Kiev, 41–47 [in Russian].

HRYNIV, S. P. 1985. To a question of Kalush-Holyn potash salt deposit correlation. *In*: KITYK, V. I. (ed.) *Evaporites of Ukraine*. Naukova Dumka, Kiev, 44–50 [in Russian].

HRYNIV, S. P. 1990. Conditions of anhydrite-polyhalite beds formation of potash salt deposits of Forecarpathian. *In*: MERZLIAKOV, G. A. (ed.) *Conditions of Formation of Potash Salt Deposits*. Nauka, Novosibirsk, 181–189 [in Russian].

ISHCHENKO, D. I. & KISIELIOV, M. V. 1967. To the question about the geological construction and formation of gypsum deposits of Crimea. *Geologichnyi zhurnal*, **27**(1), 68–76 [in Ukrainian].

KASPRZYK, A. 1999. Sedimentary evolution of Badenian (Middle Miocene) gypsum deposits in the northern Carpathian Foredeep. *Geological Quarterly*, **43**, 449–465.

KHMELEVSKA, O. V. 1993. Physico-chemical conditions of salt accumulation in the Upper Jurassic evaporite basin of Foredobrogea. *Geologiya i geochimiya goriuchyh kopalyn*, (**2–3**), 83–83; 98–102 [in Ukrainian].

KHMELEVSKAYA, E. V. 1990a. About the matter composition of rocks of the Upper Jurassic evaporite formation of the south-western slope of East European platform. *In*: PETRICHENKO, O. I. (ed.) *Geology and Geochemistry of Salt-bearing Deposits of Oil-and-gas Content Provinces*. Naukova Dumka, Kiev, 143–152 [in Russian].

KHMELEVSKAYA, E. V. 1990b. Authigenic mineral-impurities in the Upper Jurassic rock salt of Foredobrogea and Middle Asia. *Mineralogicheskiy sbornik*, **44**, 1, 75–77 [in Russian].

KHMELEVSKAYA, E. V. 1997. Upper Jurassic evaporites of the south-western slope of East European platform. *Slovak Geological Magazine*, **3**(3), 213–216.

KHODKOVA, S. V. 1971. Minerals and rocks of the Stebnik potassium salt deposit. *In*: KHODKOV, A. E. (ed.) *Materials on Hydrogeology and Geological Role of Subsurface Waters*. Izdatelstvo Leningradskoho universiteta, Leningrad, 82–91 [in Russian].

KHODKOVA, S. V. 1980. About the nature of salt-bearing breccias of Eastern Forecarpathian. *In*: KITYK, V. I. (ed.) *Lithology and Geochemistry of Salt-bearing Series*. Naukova Dumka, Kiev 69–77 [in Russian].

KHRUSHCHOV, D. P. 1971. *Lithology and Potash-bearing of Salt Deposits of Dnieper–Donets depression*. Naukova Dumka, Kiev [in Ukrainian].

KHRUSHCHOV, D. P. 1980. *Lithology and Geochemistry of Saliferous Formations of the Carpathian Foredeep*. Naukova Dumka, Kiev, [in Russian].

KITYK, V. I. 1970. *Salt Tectonics of Dnieper–Donets Depression*. Naukova Dumka, Kiev [in Russian].

KITYK, V. I., POLKUNOV, V. F. & STEPANENKO, O. T. 1979. *Construction and Regularity of Distribution of Native Sulfur Deposits of the USSR*. Naukova Dumka, Kiev [in Russian].

KITYK, V. I., BOKUN, F. N., PANOV, G. M., SLIVKO, E. P. & SHAIDETSKAYA, V. S. 1983. *Halogenous Formations of Ukraine: Transcarpathian Trough*. Naukova Dumka, Kiev [in Russian].

KLIMOV, M. A. 1972. The manufacturing potash-bearing of the Lower Vorotyshcha suite and the prospects for further potash salts exploration in the Forecarpathian basin. *Trudy Vsesoyuznoho instituta galurgii*, **60**, 148–162 [in Russian].

KLIMOV, M. A. 1977. The prospects of potash-bearing of Kalush salt bearing formations in the Inner zone of the Carpathian Foredeep. *In*: KITYK, V. I. (ed.) *Geology and geochemistry of Salt-bearing Formations of Ukraine*. Naukova Dumka, Kiev, 14–22 [in Russian].

KORENEVSKIY, S. M. 1982. Upper Jurassic salt bearing deposits of Ukraine and Moldaviya. *Litologiya i poleznyye iskopaemyye*, **4**, 107–116 [in Russian].

KORENEVSKIY, S. M. & DONCHENKO, K. B. 1963. Geology and conditions of formation of potash deposits of Soviet Forecarpathians. *Trudy Vsesoyuznoho geologicheskoho instituta N.S.*, **99**, 3–153 [in Russian].

KORENEVSKIY, S. M., BOBROV, V. P., SUPRONIUK, K. S. & KHRUSHCHOV, D. P. 1968. *Saliferous Formations of North-western Donbass and Dnieper–Donets Depression and their Potash-bearing*. Nedra, Moscow [in Russian].

KORENEVSKIY, S. M. & ZAHAROVA, V. M. & SHAMAHOV, V. A. 1977. *Miocene Saliferous Formations of the Carpathian Foredeep*. Nedra, Leningrad [in Russian].

KORENEVSKIY, S. M., VAKATCHUK, G. I., ZAHAROVA, V. M., KHRUSHCHOV, D. P. & SHAMAHOV, V. A. 1980. Correlation of sections, potash salts content and lithofacies complexes of Kramatorska suite of Dnieper–Donets depression. *Geologicheskiy zhurnal*, **40**(5), 48–56 [in Russian].

KORIN', S. S. 1992. Tectonic conditions of formation of potash deposit structure in the Boryslav-Pokutsky nappe. *Otechestviennaya geologiya*, **12**, 20–26 [in Russian].

KORIN', S. S. & MOSORA, T. M. 1990. The structure of the Vorotyshcha saliferous formations of the Forecarpathian (by example of Boryslav facies). *In*: PETRICHENKO, O. I. (ed.) *Geology and Geochemistry of Salt-bearing Deposits of Oil-and-gas Content Province*. Naukova Dumka, Kiev, 86–92 [in Russian].

KOROBTSOVA, M. S. 1955. Mineralogy of potash deposits of the Eastern Forecarpathian. *Voprosy mineralogii osadochnyh obrazovaniy*, **2**, 3–137 [in Russian].

KOVALEVICH, V. M. 1977. Rock salt of 'salting-out' origin in the Miocene saliferous deposits of the Eastern Forecarpathians. *In*: KITYK, V. I. (ed.) *Geology and Geochemistry of Salt-bearing Formations of Ukraine*. Naukova Dumka, Kiev, 48–53 [in Russian].

KOVALEVICH, V. M. 1978. *Physico-chemical Conditions of Salt Formation in the Stebnik Potassic-salt Deposit*. Naukova Dumka, Kiev [in Russian].

KOVALEVICH, V. M. 1979. The possibile utility of bromine content in halite as a prospecting criterion on potash salts in Forecarpathian. *In*: KITYK, V. I. (ed.) *The Questions of Geology and Geochemistry of Halogenous Deposits*. Naukova Dumka, Kiev, 35–44 [in Russian].

KOVALEVICH, V. M. 1982. Genesis of langbeinite in potassium salts of the Carpathian Foredeep based on study of inclusions of mineral-forming solutions. *In*: KITYK, V. I. (ed.) *Geology and Geochemistry of Non-metallic Mineral Resources*. Naukova Dumka, Kiev, 32–41 [in Russian].

KOVALEVICH, V. M. 1985. Physico-chemical conditions of salts accumulation of Lower Permian saliferous formations of Dnieper–Donets depression. *In*: KITYK, V. I. (ed.) *Evaporites of Ukraine*. Naukova Dumka, Kiev, 33–44 [in Russian].

KOVALEVICH, V. M. 1988. Phanerozoic evolution of ocean water composition. *Geochemistry International*, **25**, 20–27.

KOVALEVICH, V. M. 1990. *Salt Deposition and Chemical Evolution of the Ocean in Phanerozoic*. Naukova Dumka, Kiev [in Russian].

KOVALEVICH, V. M. & PETRICHENKO, O. I. 1997. Chemical composition of brines in Miocene evaporite basins of the Carpathian region. *Slovak Geological Magazine*, **3**, 173–180.

KOVALEVICH, V. M., PERYT, T. M. & PETRYCHENKO, O. I. 1998. Secular variation in seawater chemistry during the Phanerozoic as indicated by brine inclusions in halite. *Journal of Geology*, **106**, 695–712.

KOVALEVYCH, V. M. & PETRYCHENKO, O. Yo. 1994. Geochemistry of evaporite sedimentation in a Permian marine basin. *Earth Sciences and Resources Institute Occasional Publications*, **11B**, 41–47.

KOVALEVYCH, V. M. & VITYK, M. O. 1995. Correlation of sulfur and oxygen isotopes in evaporites with chemical composition of brines of Phanerozoic evaporite basins. *Dopovidi NAN Ukrainy*, **3**, 83–85 [in Ukrainian].

KOVALEVYCH, V. M., PERYT, T. M., CARMONA, V., SYDOR, D. V., VOVNIUK, S. V. & HALAS, S. 2002. Evolution of Permian seawater: evidence from fluid inclusions in halite. *Neues Jahrbuch für Mineralogie Abhandlungen*, **178**(1), 27–62.

KRUGLOV, S. S. & TSYPKO, A. K. (eds). 1988. *Tectonics of Ukraine*. Nedra, Moscow [in Russian].

KUDRYAVTSEV, U. E. 1971. Some new data about the geological structure of Stebnik potash salt deposit. *In*: KHODKOV, A. E. (ed.) *Materials on Hydrogeology and Geological Role of Subsurface Waters*. Izdatelstvo Leningradskoho universiteta, Leningrad, 71–79 [in Russian].

KULCHETSKAYA, A. A. 1987. *Genesis of gypsum and anhydrite from sedimentary rocks of Ukraine (on the basis of the data of fluid inclusions studies)*. Ph.D. thesis, Institut Geohkimii i Fiziki Mineralov, Kiev [in Russian].

KULCHYTSKIY, A. YA. 1986. *Geology and conditions of origin salt-bearing molasse deposits of Forecarpathian and Transcarpathian Neogene Foredeeps*. L'vivskyy universytet, L'viv [in Russian].

KUZNIETSOVA, S. V. 1970. *Mineralogy and some questions of genesis of mercury and lead-zinc ore-displays of north-western part of Donbass*. Ph.D. thesis, Institut Geohkimii i Fiziki Mineralov, Kiev [in Russian].

KUZNIETSOVA, S. V. 1971. About ore mineralization of north-western Donbass. *Mineralogicheskiy sbornik L'vovskogo universiteta* **25**, 2, 111–123 [in Russian].

KUZNIETSOVA, S. V., SOFRONOV, I. L., SKARZHYNSKIY, V. I. & ENTELIS, I. D. 1968. Main features of geological structure and endogenous ore mineralization of northwestern margin of Donbass. *Geologicheskiy zhurnal*, **24**(4), 32–44 [in Russian].

KUŹNIAR, Cz. 1939. Złoże solne w Kałuszu. *Prace Państwowego Instytutu Geologicznego*, **111**(3) [in Polish].

LAZARENKO, E. K., GABINET, M. P. & SLIVKO, O. P. 1962. *Mineralogy of Sedimentary Formations of the Forecarpathian*. Vydavnytstvo L'vivskoho universytetu, L'viv [in Ukrainian].

LAZ'KO, E. M. & REZVOY, D. P. 1962. About the tectonic nature of Carpathian 'cliff zone' range. *Visnyk L'vivskoho universytetu*, **1**, 60–65 [in Russian].

LOBANOVA, V. V. 1956. Questions of petrography of potash deposits of the eastern Forecarpathian. *Trudy Vsesoyuznoho instituta galurgii*, **32**, 164–214 [in Russian].

LOWENSTEIN, T. K., HARDIE, L. A., TIMOFEEFF, M. N. & DEMICCO, R. V. 2003. Secular variation in seawater chemistry and the origin calcium chloride basinal brines. *Geology*, **31**, 851–860.

LOWENSTEIN, T. K., TIMOFEEFF, M. N., BRENNAN, S. T., HARDIE, L. A. & DEMICCO, R. V. 2001. Oscillations in Phanerozoic seawater chemistry: evidence from fluid inclusions. *Science*, **294**, 1086–1088.

MALIKOVA, I. N. 1967. *The Regularities of Distribution of Rubidium, Thallium and Bromine in Potash Salt Deposits*. Nauka, Novosibirsk [in Russian].

MARKOVSKIY, V. I., KOTYK, V. A. & BEREZHYNSKAYA, L. F. 1979. Silurian evaporites of Volyno–Podoliya. *In*: KITYK, V. I. (ed.) *The Questions of Geology and Geochemistry of Saliferous Deposits*. Naukova Dumka, Kiev, 118–128 [in Russian].

MATKOVSKYY, O. I., BILONIZHKA, P. M. *ET AL*. 2003. *Minerals of the Ukrainian Carpathians. Borates, Arsenates, Phosphates, Molybdates, Sulphates, Carbonates, Organic Minerals and Mineraloids*. Vydavnychyy tsentr LNU, L'viv [in Ukrainian].

MERLICH, B. V. & SPITKOVSKAYA, S. M. 1965. The peculiarities of Neogene magmatism of deep faults of the Transcarpathian. *Geologicheskiy sbornik L'vovskogo geologicheskogo obshchestva*, **9**, 55–68 [in Russian].

MURATOV, M. V. 1960. *Short Studies of Geological Structure of the Crimea Peninsula*. Gosgeoltekhizdat, Moscow [in Russian].

NEVESSKAYA, L. A., GONCHAROVA, I. A. *ET AL*. 1984. The regional stratigraphic scale of the Neogene of East Paratethys. *Sovetskaya geologiya*, **9**, 37–49.

NOVIK, E. O., PERMIAKOV, V. V. & KOVALENKO, E. E. 1960. *The History of Geological Research of Donets Coal Basin (1700–1917)*. Izdatelstvo AN USSR, Kiev [in Russian].

OLIYOVYCH, O., YAREMCHUK, Ya. & HRYNIV, S. 2004. Clays of saliferous deposits and weathering zone of

Kalush–Holyn potash salt deposit (Miocene, Forecarpathians). *Mineralogichnyy zbirnyk*, **54**(2), 214–223 [in Ukrainian].

PANOW, G. & PŁOTNIKOW, A. 1996. Badeńskie ewaporaty ukraińskiego Przedkarpacia: litofacje i miąższość. *Przegląd Geologiczny*, **44**, 1024–1028 [in Polish].

PAVLENKO, V. V., KROPACHEVA, S. K. & POLTORAKOV, G. I. 1974. Regularities of native sulphur occurrence in Kerch peninsula. *Sovetskaya geologiya*, **8**, 149–152 [in Russian].

PERYT, T. M. 1996. Sedimentology of Badenian (middle Miocene) gypsum in eastern Galicia, Podolia and Bukovina (West Ukraine). *Sedimentology*, **43**, 571–588.

PERYT, T. M. 2001. Gypsum facies transitions in basin–marginal evaporites: middle Miocene (Badenian) of West Ukraine. *Sedimentology*, **48**, 1103–1119.

PERYT, T. M. & KOVALEVYCH, V. M. 1997. Association of redeposited salt breccias and potash evaporites in the Lower Miocene of Stebnyk (Carpathian Foredeep, West Ukraine). *Journal of Sedimentary Research*, **67**, 913–922.

PERYT, T. M., POBEREZHSKYY, A. V., JASIONOWSKI, M., PERYT, D., PETRYCHENKO, O. Y., LYZOON, S. O. & TURCHYNOV, I. I. 2004a. Correlation of Badenian sulphatic deposits in the Dnister river region. *Geologiya i geokhimiya goriuchyh kopalyn*, **1**, 56–69 [in Ukrainian].

PERYT, T. M., POBEREZHSKYY, A. V., PERYT, D., JASIONOWSKI, M. & DURAKIEWICZ, T. 2004b. Post-evaporitic restricted deposition in the Middle Miocene Chokrakian/Karaganian of East Crimea (Ukraine). *Sedimentary Geology*, **170**, 21–36.

PETRICHENKO, O. I. 1973. *Methods of Study of Inclusions in Minerals of Saline Deposits*. Naukova Dumka, Kyiv [in Ukrainian; translated in *Fluid Inclusions Research Proc. COFFI*, **12**, 214–274, 1979].

PETRICHENKO, O. I. 1988. *Physico-chemical Conditions of Sedimentation in the Evaporite Palaeobasins*. Naukova Dumka, Kiev [in Russian].

PETRICHENKO, O. I. 1989. *Epigenesis of Evaporites*. Naukova Dumka, Kiev [in Russian].

PETRICHENKO, O. I. & SHAIDETSKAYA, V. S. 1977. Mineralogical-geochemical peculiarities of rock salt of Kaplintsevsky salt dome (Dnieper-Donets depression). *In*: KITYK, V. I. (ed.) *Geology and Geochemistry of Salt-bearing Formations of Ukraine*. Naukova Dumka, Kiev, 84–94 [in Russian].

PETRICHENKO, O. I. & SHAIDETSKA, V. S. 1998. Chlorek wapnia w solankach górnodewońskich basenów ewaporatowych ryftogenu prypećko-dnieprowsko-donieckiego w swietle badań inkluzji w halicie. *Przegląd Geologiczny*, **46**, 689–699 [in Polish].

PETRICHENKO, O. I., SLIVKO, E. P. & SHAIDETSKAYA, V. S. 1974a. About conditions of formation Devonian rock salt of Dnieper-Donets depression. *In*: LAZARENKO, E. K. (ed.) *The Perspectives of Prospecting Minerals in Dnieper–Donets Depression*. Naukova Dumka, Kiev, 110–123 [in Russian].

PETRICHENKO, O. I., KOVALEVICH, V. M. & TCHALYY, V. N. 1974b. Geochemical environment of salt-formation in the Tortonian evaporite basin of north-western Forecarpathians. *Geologiya i geokhimiya goriuchih iskopaemyh*, **41**, 110–123 [in Russian].

PETRICHENKO, O. I., PANOW, G. M., PERYT, T. M., SREBRODOLSKIY, B. I., POBEREŻSKIY, A. W. & KOWALEWICZ, W. M. 1994. Zarys geologii miocenśkich formacji ewaporatowych ukraińskiej częsci zapadliska przedkarpackiego. *Przegląd Geologiczny*, **42**, 734–737 [in Polish].

PETRICHENKO, O. I., PERYT, T. M. & POBEREGSKI, A. V. 1997. Peculiarities of gypsum sedimentation in the Middle Miocene Badenian evaporite basin of Carpathian Foredeep. *Slovak Geological Magazine*, **3**(2), 91–104.

PETRYCHENKO, O. Y. & PERYT, T. M. 2004. Geochemical conditions of deposition in the Upper Devonian Prypiac' and Dnipro–Donets evaporite basins (Belarus and Ukraine). *Journal of Geology*, **112**, 577–592.

PETRYCHENKO, O. Y., SHAIDETSKA, V. S. & KOVALEVICH, V. M. 1976a. New geochemical criterion for potash salt prospecting. *Dopovidi AN URSR, Ser. B*, **6**, 512–515 [in Ukrainian].

PETRYCHENKO, O. Y., SLYVKO, O. P. & SHAIDETSKA, V. S. 1976b. Mineralogical–geochemical illustration of distinctions of rock salt of different ages. *Dopovidi AN USSR, Ser. B*, **9**, 778–781 [in Ukrainian].

POBEREZHSKYY, A. V. 2000. Physico-chemical conditions of gypsum sedimentation in the Badenian evaporite basin of Carpathian Foredeep. *Geologiya i geokhimiya goriuchyh kopalyn*, **4**, 38–55 [in Ukrainian].

POBEREZHSKYY, A. V. & KOVALEVYCH, V. M. 2001. Chemical composition of sea water in Cenozoic (by data of fluid inclusions research in sedimentary halite). *Mineralogichnyy zbirnyk*, **51**(2), 90–109 [in Ukrainian].

POBEREZHSKYY, A. V., PERYT, T. & PETRYCHENKO, O. Y. 2002. Facies and physico-chemical analysis of gypsum sedimentation conditions in the Badenian basin of Carpathian region. *Mineralogichnyy zbirnyk*, **52**(2), 106–110 [in Ukrainian].

POLIKARPOV, A. I. 1974. About some mineralogical peculiarities and origin of terrigenous and halogenous–terrigenous rocks of Eastern Holyn of Kalush–Holyn potash salts deposits. *Trudy Vsesoyuznoho instituta galurgii*, **68**, 65–72 [in Russian].

POVSTEN, E. F. & POVSTEN, G. M. 1980. Carnallite as indicator of faults in saliferous deposits of the Forecarpathian. *In*: KITYK, V. I. (ed.) *Lithology and Geochemistry of Salt-bearing Series*. Naukova Dumka, Kiev, 65–69 [in Russian].

RAUP, O. B. 1970. Brine mixing: an additional mechanism for formation for basin evaporites. *Bulletin American Association Petroleum Geologists*, **54**, 2246–2259.

ROMANOV, L. F. 1976. *Mesozoic Speckled Deposits of Country between Dniestr and Prut*. Shtiintsa, Kishinev [in Russian].

ROMANOV, L. F. & SLAVIN, V. I. 1970. Tectonic position and origin of the Jurassic Dobrogea Foredeep. *Vestnik Moskovskogo universiteta*, **5**, 77–87 [in Russian].

SANDLER, N. M. & VORONA, G. P. 1955. Short lithological description of the Upper Jurassic deposits of Western regions of the USSR. *Naukovi zapysky*

L'vivskoho naukovo-pryrodnychoho muzeyu AN URSR, IV, 55–58 [in Ukrainian].

SHAIDETSKAYA, V. S. 1990. The conditions of chemogenic sedimentation in Late Devonian salts-forming basins of Dnieper-Donets depression. In: PETRICHENKO, O. I. (ed.) *Geology and geochemistry of Salt-bearing Deposits of Oil-and-gas Content Province.* Naukova Dumka, Kiev, 101–106 [in Russian].

SHAIDETSKAYA, V. S. 1997. Geochemistry of Neogene evaporites of the Transcarpathian Trough in Ukraine. *Slovak Geological Magazine*, **3**, 193–201.

SHATSKIY, N. S. 1931. To a question about origin of Romny gypsum and rocks of Isachkovsky Hill in Ukraine. *Bulleten Mosovskogo obshchestva ispytateley prirody, otdelenie geologiyi*, **9**(39), 3–4, 156–163 [in Russian].

SHNIUKOV, E. F., NAUMENKO, P. I. & LEBEDEV, U. S. 1971. *The Mud Volcanism and Ore Formation.* Naukova Dumka, Kiev [in Russian].

SKVORTSOVA, K. V. 1964. About the tectonic structure of Kalush–Holyn potash salts deposit. *Trudy Vsesoyuznoho instituta galurgii*, **45**, 23–32 [in Russian].

SLAVIN, V. I. & KHAIN, V. E. 1965. The role of tectonic ruptures in construction and evolution of the Eastern Carpathians. In: *Materials of VI Congress of Carpathian-Balkan Geological Association: Contributions of Soviet geologists.* Naukova Dumka, Kiev, 255–276 [in Russian].

SLIUSAR, B. S. 1971. *Jurassic Deposits of North-western Part of Near-Black Sea Region.* Shtiintsa, Kishinev [in Russian].

SLIVKO, E. P. 1979. Questions about the possibility of correlation of potash salt deposits using their authigenic minerals. In: KITYK, V. I. (ed.) *The Questions of Geology and Geochemistry of Halogenous Deposits.* Naukova Dumka, Kiev, 61–76 [in Russian].

SLIVKO, E. P. & PETRICHENKO, O. I. 1967. *Accessory Lithium, Rubidium and Caesium in Salt Deposits of Ukraine.* Naukova Dumka, Kiev [in Russian].

SOBOLEVSKIY, YU. V. 1970. Gornostaivske showing of the native sulfur in Kerch peninsula. *Dopovidi AN URSR, Ser. B*, **9**, 789–792 [in Russian].

SOZANSKIY, V. I. 1973. *Geology and Origin of Salt Deposits.* Naukova Dumka, Kiev [in Russian].

ULMISHEK, G. F., BOGINO, V. A., KELLER, M. B. & POZNYAKEVICH, Z. L. 1994. Structure, stratigraphy and petroleum geology of the Pripyat and Dniper–Donets basins, Byelarus and Ukraine. In: LANDON, S. M. (ed.) *Interior Rift Basins.* American Association Petroleum Geologists Memoirs, **59**, 125–156.

VALYASHKO, M. G. 1962. *Geochemical Regularities of Origin of Potassium Salt Deposits.* Izdatelstvo Moskovskoho universiteta, Moscow [in Russian].

VUL, M. Y., DENEGA, B. I., KRUPSKYY, Y. Z., NIMETS, M. V., SVYRYDENKO, V. G. & FEDYSHYN, V. O. (eds). 1998. *Atlas of Oil and Gas Fields of Ukraine in Six Volumes. Vol. IV: Western Oil-and-Gas-bearing Regions.* Vydavnytstvo Tsentr Europy, L'viv [in Ukrainian].

VYALOV, O. S. 1980. Stratigraphic scheme of the Neogene deposits of the west regions of USSR. *Paleontologicheskiy sbornik*, **17**, 93–96 [in Russian].

VYSOTSKIY, E. A., GARETSKIY, R. G. & KISLYK, V. Z. 1988. *Potash-bearing Basins of the World.* Nauka i tekhnika, Minsk [in Russian].

WOJTOWICZ, A., HRYNIV, S. P., PERYT, T. M., BUBNIAK, A., BUBNIAK, I. & BILONIZHKA, P. M. 2003. K/Ar dating of the Miocene potash salts of the Carpathian Foredeep (West Ukraine) application to dating of tectonic events. *Geologica Carpatica*, **54**, 243–249.

WYSZYŃSKI, O. W. 1939. Przedgórze okolic Kosowa. *Przemysł Naftowy. (Lwów).* **14**, 7–13 [in Polish].

YARZHEMSKAYA, E. A. 1954. Composition of salt rock clays. *Trudy Vsesoyuznoho instituta galurgii*, **29**, 260–314 [in Russian].

ZHARKOV, M. A. 1984. *Paleozoic Salt Bearing Formations of the World.* Springer, Berlin.

ZATSIKHA, B. V., PETRICHENKO, O. I., DOLISHNIY, B. V. & LAS'KOV, V. A. 1973. Genetic peculiarities of mineral formation in Slaviansk mercuric deposits. *Mineralogicheskiy sbornik L'vovskogo universiteta*, **27**(4), 326–332 [in Russian].

Depth indicators in Permian Basin evaporites

S. D. HOVORKA[1], R. M. HOLT[2] & D. W. POWERS[3]

[1]*Bureau of Economic Geology, Jackson School of Geosciences, The University of Texas at Austin, Austin, TX, USA (e-mail: susan.hovorka@beg.utexas.edu)*

[2]*Department of Geology and Geological Engineering University of Mississippi, University, MS, USA*

[3]*Consulting Geologist, 140 Hemley Road, Anthony, TX 79821, USA*

Abstract: The Permian Basin of West Texas and New Mexico contains one of the world's best-preserved and most extensively studied evaporite basin-to-platform sequences. From analysis of fabrics and small-scale cycle patterns, reconstruction of the position of these elements in the basin-filling sequence and comparison to laboratory-grown and modern evaporite fabrics, we created a table of fabrics that serve as water-depth indicators. Evaporites formed in deeper water (more than a few to hundreds of metres) in both halite- and gypsum-precipitating settings in the Permian Basin are characterized by cumulate fabrics. Cumulates are fine crystals or rafts of fine crystals formed at the air – brine interface that fall though the water body and accumulate on the basin floor with fine lamination, draping relationships, dark colours and minimal early diagenesis. Intervals of coarser crystals precipitated on the basin floor are interpreted as evidence for episodic transport of saturated surface water to the basin floor during perturbation of stratified conditions. Shallow water evaporites in the Permian Basin are dominated by bottom-growth fabrics such as halite chevrons and near-vertically oriented gypsum crystals. Bands of fluid and other inclusions record high frequency changes in depositional rate. Truncated crystals document flooding by undersaturated fresh or marine water under shallow conditions where mixing was adequate to cause undersaturated low-density water to contact the basin floor. Formation of base-of-cycle insoluble residues is a strong indicator of shallow water during the flooding event that initiated each sedimentary cycle. In the Permian Basin, exposure is documented by formation of synsedimentary evaporite karst pits and pipes, truncation, dissolution and recrystallization of earlier fabrics, and precipitation of cements. Red siliciclastic mudstones are associated with the late stages of depositional cycles when the surface was subaerially exposed. Repetition of alternately exposed and saline water table conditions created an array of distinctive fabrics including chaotic mudstone–salt mixtures, karst fills, replacement and recrystallization fabrics, and cracks and salt polygons.

The Permian evaporites of the Permian Basin of Texas and New Mexico are among the best-known evaporite sequences in the world. Contributing to this extensive knowledge are numerous studies of the sequence because of its oil, gas and potash resources and extensive and accessible outcrops. During evaporite deposition, the Permian Basin underwent variable amounts of subsidence that created deep basin, shallow basin and shallow shelf settings in which accumulated a spectrum of evaporites. Post-depositional burial and structural deformation was modest, so Permian sedimentological fabrics are generally well preserved and can be confidently reconstructed in original relationships. However some complexity is introduced by dissolution of marginal, uppermost and locally lower halite units and folding in the Delaware Basin.

The authors have had the opportunity to participate in studies of the evaporite units as potential hosts for nuclear waste repositories and solution-mined cavern storage. This allowed us to examine these evaporites in wireline logs, cores, shafts and drifts. Based on this data we review the sedimentological history of the Permian Basin placing the evaporites in depositional context with regard to the evolution as deep basins, shallow basins and shelf environments, and we review the depositional cycles characteristic of these depositional settings. Then we document the evaporite sedimentary fabrics typical of these settings with a focus on how these fabrics can be used to interpret formative paleowater depth.

Geologic setting of evaporites in the Permian Basin

The greater Permian Basin consists of several basins (Fig. 1) that formed during late Palaeozoic Ouachita

From: SCHREIBER, B. C., LUGLI, S. & BĄBEL, M. (eds) *Evaporites Through Space and Time*.
Geological Society, London, Special Publications, 285, 335–364.
DOI: 10.1144/SP285.19   0305-8719/07/$15.00 © The Geological Society of London 2007.

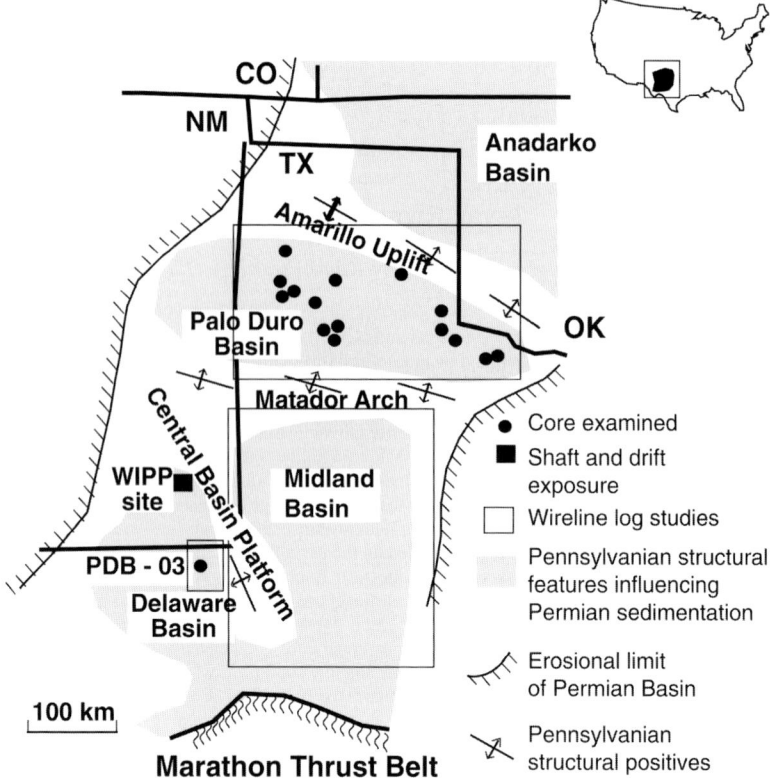

Fig. 1. Major basins and interbasin areas of reduced subsidence of the Permian Basin, focusing on elements described in the text.

deformation and were filled with Pennsylvanian and Permian siliciclastics, carbonates and evaporites (King 1942; McKee et al. 1967). The Permian Basin was on the southwest margin of the continental mass of Pangea, and palaeogeographic reconstruction suggests that marine brine entered the basin complex from the southwest (Adams 1944). Following the accumulation of carbonates and siliciclastics, each of the individual basins (the Palo Duro, Midland and Delaware Basins) accumulated evaporite sediments. The basins are separated by structurally positive regions (the Matador Arch and Central Basin Platform) that began to subside during the Permian and accumulated substantial thicknesses of sediments, including evaporites. Because of diverse evaporite depositional environments, the Permian Basin presents an opportunity to inventory the characteristics of a spectrum of evaporite fabrics formed in different water depths.

Basin infilling and initiation of evaporite sedimentation advanced fastest in the shallowest and most distal basins of the related group of basins (Fig. 2). The northeasternmost basin of the set described here, the Palo Duro Basin, first accumulated the shallow-water evaporites of the Wichita Group during the Leonardian. Later a thick Leonardian–Guadalupian evaporite sequence accumulated (San Andres Formation is a prominent example). The subsidence rate of the Palo Duro Basin diminished during the late Guadalupian and during the Ochoan it accumulated only minor thicknesses of evaporites during maximum highstands (Handford 1981; Presley 1981).

Facies geometries show that the Midland Basin remained a topographic depression through most of the Guadalupian. The Midland Basin and Central Basin Platform first accumulated shallow water evaporites during the late Guadalupian in the Seven Rivers Formation (Fig. 2). During the Ochoan, a thick sequence of evaporites including the Salado Formation was deposited (Hills 1972). Cycle geometries show that, by the latest Guadalupian and during the Ochoan, the former Midland Basin and structural positive of the Central Basin Platform subsided as a unit and acted as a broad shelf east of the Delaware Basin (Hovorka 2000a).

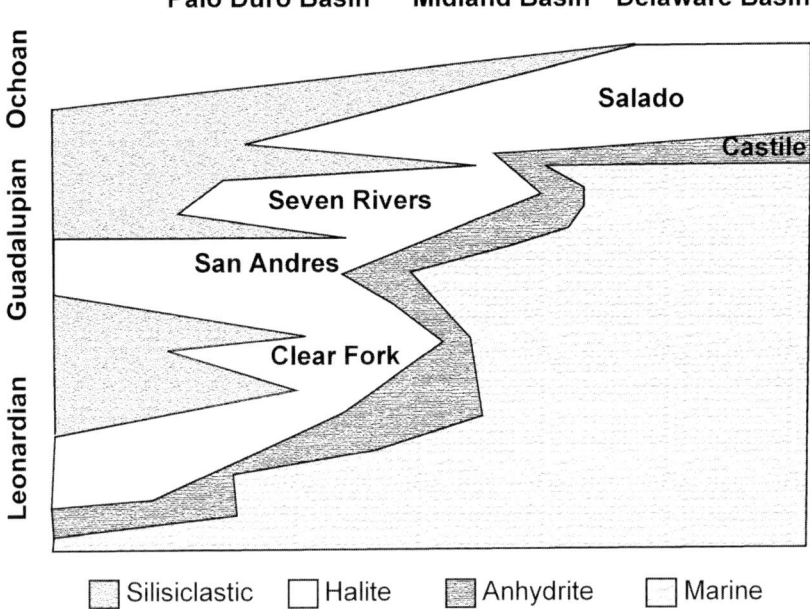

Fig. 2. Major sedimentary units of the greater Permian Basin, emphasizing units described in the text.

The deepest and southwesternmost basin of the set, the Delaware Basin, did not begin to accumulate evaporites until the Ochoan, when an initially deep basin accumulated a thick sequence of the Castile Formation. The Castile is overlain by a thick section of Salado evaporites, the approximate correlatives of the Midland Basin/Central Basin Platform unit of the same name and by the mixed sequence of carbonates, clastics, and evaporites in the Rustler Formation (Holt & Powers 1987). The Salado Formation in the Delaware Basin consists mainly of halite with lesser amounts of anhydrite, polyhalite, soluble potassium-bearing minerals and minor amounts of clay. Thin beds of anhydrite and polyhalite (0.1 to c. 5 m thick) alternate with thicker beds of halite throughout the Salado section and were historically considered a critical part of Salado cycles (Schaller & Henderson 1932; Jones 1954, 1972; Lowenstein 1982, 1988). Thickness relationships (Hovorka 2000a) show that the Delaware Basin subsided more strongly than the adjunct Central Basin Platform during most of the Salado deposition and therefore retained the geometry of a shallow basin within a broad shelf, although it was not adjacent to an open marine area.

Methods and sources of data

The primary sources of data are 11 cores drilled by the US Department of Energy as part of the High Level Waste Isolation Program from Palo Duro Basin, six cores collected from the eastern margin of the Palo Duro basin by the US Army Corps of Engineers (Hovorka & Granger 1988), one core (Gulf Research PDB-03) from the deepest part of the Delaware Basin (Hovorka 1990, 2000b), a series of wireline log cross sections though the Palo Duro, Midland Basin, Delaware basins (Fracasso & Hovorka, 1986; Hovorka 1990, 2000a), and a detailed study of unique exposures of shafts and drifts of the Waste Isolation Pilot Plant (WIPP) (Holt & Powers 1990a, b); (Fig. 1). We also examined core and extensive photo documentation of the Gulf Research PDB-04 core (Garber et al. 1989) and reconnaissance observations made of several cores though Central Basin Platform evaporites.

To compare Permian Basin evaporites to modern environments we use literature examples, supplemented with our reconnaissance field observations of Bristol Dry Lake and Death Valley. We also conducted simple experiments precipitating evaporites in 10 gallon tanks under various depth and stratification conditions using brines created by dissolving mixtures of halite and Epsom salt ($MgSO_4 \cdot 7H_2O$). The purpose of these experiments was to create controlled environments in which different sedimentary and depositional textures formed.

Determining water depths in basins

Before we inventory of the spectrum of evaporite fabrics formed in different water depths in the

Permian Basin and classify the characteristics that can be used to interpret depths, we first need to define the context for evaporite precipitation. We inventory four types of evidence, independent of the fabrics, that is used to establish the water depth of these evaporite units: (1) geometric relationships; (2) rate of sediment accumulation; (3) distribution of siliciclastics; and (4) evaporite geochemistry.

The evaporite sequences of the Permian Basin formed in response to sea-level variations and sediment aggradation superimposed on a pattern of regional subsidence (Meissner 1972). Each part of the facies tract has a somewhat different response to relative sea level change, with the largest contrast between shelf and basin settings.

Geometry of evaporite facies sequences and evaporite cycles

Relationships among facies provide evidence of topographic relief during deposition. Where structurally defined basin centres were topographically lower than shallow water shelves and basin margins, the sediments that accumulated in these settings can be differentiated because of contrasting lithology and thickness evident on cross sections. As infilling proceeded, levelling depositional topography to form extensive shallow water shelves, stratigraphic units can be traced with little change across former platforms and basins. Post-depositional evaporite dissolution has been considered elsewhere (for example Bachman 1984; Hovorka & Granger 1988) and estimated original geometry is substituted in facies conceptualizations.

Shelf settings of the Palo Duro and Midland basins. Geometric relationships show that the Palo Duro Basin was infilled to form a low-relief shelf area at the beginning of evaporite sedimentation in the Leonardian and that the Midland Basin similarly was infilled to sea level before evaporites began to accumulate during the Guadalupian. All of the evaporites in these basins are therefore interpreted to have formed in extensive and shallow shelf water bodies.

Strong cyclical variations between rock types occur in Palo Duro Basin shelf environments (Handford 1981; Fracasso & Hovorka 1986; Hovorka 1994). These cycles are conceptualized with the most marine-dominated facies (carbonate) deposited first, then evolve to facies deposited from more highly evaporated brines (anhydrite after gypsum, then halite). Cycles culminate in facies reflecting non-marine environments with infrequent marine influence (e.g. mudstone–halite mixtures and siliciclastic redbeds). Because cycles can be traced from marine to evaporite environments (Fracasso & Hovorka 1986; Hovorka 1994), it is clear that the evaporite shelf remained somewhat open to marine environments and water level was tied to marine conditions. Bars and shoals within carbonate-dominated environments (Chuber & Pusey 1972; Elliott & Warren 1989; Kerans & Fitchen 1995) combined with the width of the shelf to retain evaporated seawater sufficiently to precipitate sequences of gypsum and halite beds. Geochemical signatures based on by the bromine content in halite and $^{87}Sr/^{86}Sr$ in anhydrite support a dominantly marine origin for the brine that precipitated Palo Duro shelf evaporites (Hovorka 1994).

During optimum marine conditions on the shelf, thick carbonate facies including burrowed subtidal packstones and fossiliferous grainstones were deposited across the Palo Duro Basin as the initial sediment at the base of thick evaporite cycles (Fig. 3a). Typical cycles from the Palo Duro evaporite shelf have a thin dolomitic hypersaline carbonate at the base overlain by a thick section of gypsum and are capped by a thin interval of halite (Fig. 3b). As sediment aggradation filled the Midland Basin, Palo Duro environments become even more updip (near shore) and restricted. These updip cycles have a thin or absent carbonate, overlain by a thin gypsum, overlain by a thick sequence of halite containing minor mudstone beds (Fig. 3d). In even more restricted positions, a few gypsum partings mark the surface of maximum flooding, overlain by halite, in turn overlain by numerous layers of mudstone and siltstone. These patterns repeat with a large number of variations such as higher frequency sub-cycles (for example cyclic stacks of low halite, mudstone–halite mixture and halitic mudstones) or truncated cycles (for example dolomite–gypsum immediately overlain by dolomite–gypsum, as shown in Fig. 3c). Cycle patterns in the Midland Basin appear to be variations on the cycles seen on the Palo Duro Basin, but textures of these cycles are not well known because of limited geographic distribution of core.

Topographic relief in the Delaware Basin – Castile Formation. In contrast to the Palo Duro Basin, geometric relationships in the Delaware Basin show that prior to initial evaporite deposition as much as 300 m to 550 m of relief existed between the basin and the shelf (King 1942; Garber et al. 1989; Kirkland 2003). The Capitan reef complex rimmed the extensive shelf environments on all sides of the Delaware Basin while the basin accumulated distinctive slope and basinal deposits of the Delaware Mountain Group (Harms 1974; Harms & Williamson 1988; Gardner 1992). The initial evaporite within the Delaware Basin, the

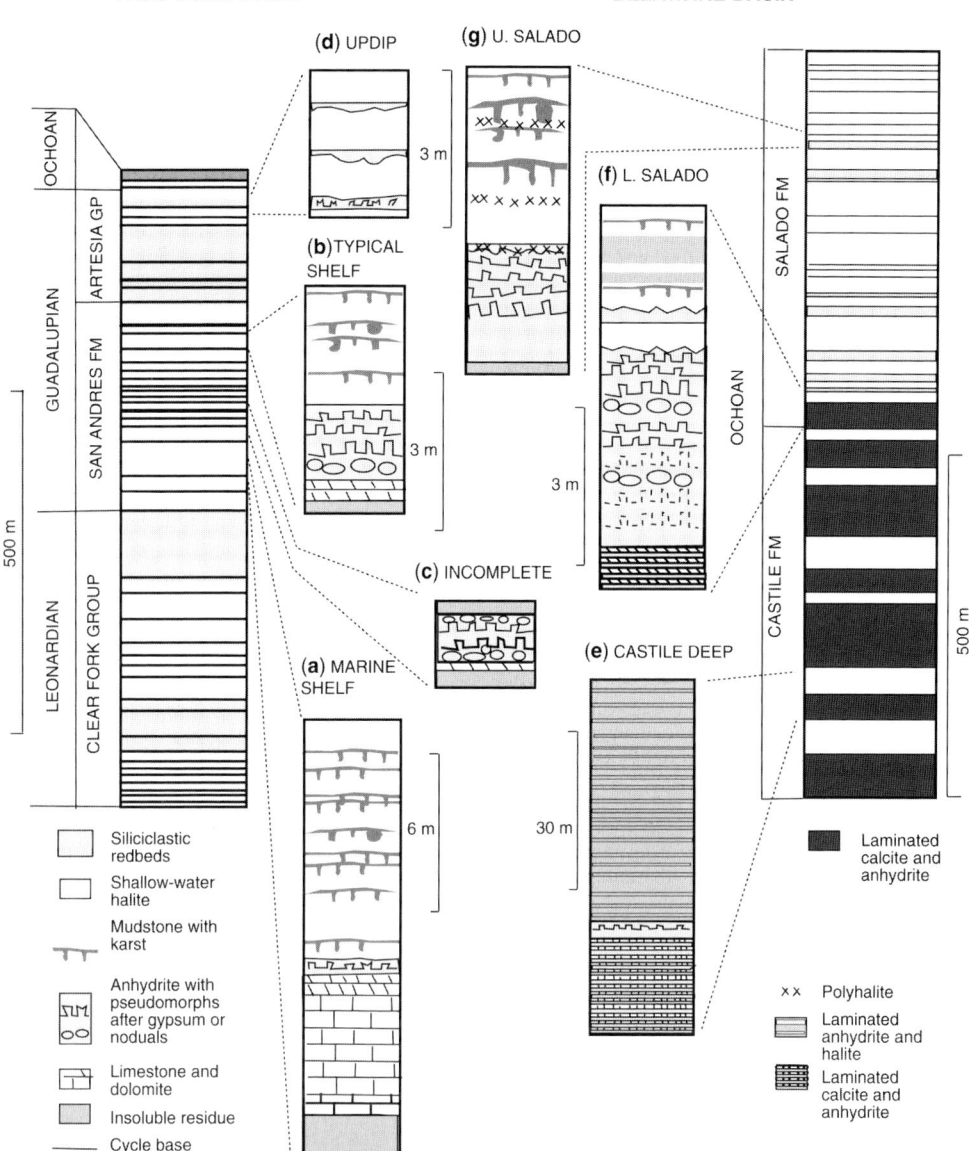

Fig. 3. Examples of Permian evaporite cycle types: (a) Palo Duro marine shelf, evaporite shelf cycle; (b) typical Palo Duro cycle; (c) incomplete cycle; (d) Palo Duro updip shelf cycle; (e) Delaware Basin deep water Castile cycle; (f) Delaware Basin Shallow Basin Lower Salado cycle; (g) Delaware Basin Shallow Basin middle to upper Salado cycle.

Ochoan Castile Formation, does not correlate with any similar sediments on the surrounding shelf, supporting the idea that topographic relief existed between the basin margins and basin floor during Castile Formation. This geometric relationship suggests to most workers that the evaporites of the Castile Formation were deposited on the floor of a deep basin.

Kendall & Harwood (1989) challenged this idea, suggesting the Castile might be a shallow-water deposit (within a deep basin) formed at times when marine influx was greatly restricted and water level was well below the basin rim, analogous to the scenarios once proposed by Hsü (1972) for the Mediterranean. This deep basin-shallow water setting (Kendall 1992) is similar to Death Valley where evaporite deposition is restricted to a basin floor below sea level. The geometrical relationship therefore leaves a question about water depth during Castile deposition.

Delaware Basin cycle patterns contrast with the shelf patterns (Fig. 3e–g). The Castile is composed of thick beds composed of calcite–anhydrite laminae alternating with anhydritic halite (Fig. 3e). The calcite–anhydrite beds have been informally numbered from base to top A-I, to A-IV (Snider 1966). A-I, II and III and corresponding thinner halite units H-I and H-II can be correlated across the Delaware Basin with relatively uniform thickness across significant distances. Upper units (H-III, A-IV and halite within anhydrite IV) have increased lateral variability and the intervening halite units are absent toward the west. Post-depositional dissolution and deformation as well as depositional facies variation play a role in the thickness of halite units.

We consider the evolution of these units from a depositional perspective. Geometries do not suggest progradation; units are interpreted as isochronous deposits with lateral and vertical facies geometries explained by evolution of a stratified water body. The lithologies typical of the Castile Formation are unlike those described elsewhere in the Permian Basin. Regionally correlative, less than meter-thick beds of thin calcite – organic laminae (the 'singlets' of Kirkland 2003), accumulated at times when evaporites were either not precipitated or were dissolved in the water column before or shortly after deposition. These are interpreted as deposits accumulated during maximum marine influx, when inflow was sufficient to sustain near marine salinity the surface water. Kirkland (2003) has argued that the observed mineralogies, setting and geochemistry of the Castile require a mass balance between inflow of marine water and reflux of dense brine out of an isolated Delaware Basin and that these flux variations could be climatically controlled.

The main anhydrite units of the Castile Formation consist of successions of couplets of a few millimetres or less of calcite overlain by 5–15 mm of anhydrite. Variations in thickness of the calcite–anhydrite couplets were analyzed by Anderson et al. (1972), and the patterns interpreted as annual depositional units that record orbital signatures. A climatic interpretation for evolution of such couplets is presented by Kirkland (2003). Anderson et al. (1972) correlated a small subset of couplets near the base of the Castile between two cores taken more than 100 km apart, and this has been taken as a significant indicator that the water column was little disturbed or stratified, and probably deep. Cores from less than 1 km apart, however, were not correlated. Sparse but significant units within these calcite/anhydrite laminated sequences are thicker (several centimetres) massive anhydrite beds with minor calcite enclosed. These beds are tentatively interpreted as evaporite sediments that slumped or were eroded and were redeposited down-slope as gravity deposits.

At the top of laminated calcite–anhydrite units, tens of beds with thicker anhydrite occur (thick couplets of Kirkland, 2003). These units commonly have nodular fabrics and locally contain anhydrite pseudomorphs after gypsum. Thick couplets occur under most halite units, but also occur between laminated calcite–anhydrite units. Halite units are composed of triplets of sparse and scattered crystals of calcite, a few millimetres of anhydrite, and millimetres to centimetres of halite. Halite units typically have been deformed or recrystallized.

These lithologies occur repetitively though the Castile Formation and are only recognized within the Delaware Basin, suggesting that the sedimentology of these units is related to the geometric setting in which they occur.

Reduction of topographic relief in the Delaware Basin–Salado Formation

The lowest evaporite unit above the Capitan Reef margin of the Delaware Basin is the Fletcher Anhydrite. Lang (1942) assigned the Fletcher to the Salado Formation, thereby defining the Castile as restricted to the Delaware Basin. With this geometry the Salado Formation is inferred to have formed when Castile deposition had filled the Delaware Basin and nearly eliminated topography, forming a regional flat shelf of minimal relief across the entire Permian Basin. The Salado Formation within the Delaware Basin is thicker than the lithologically similar Salado Formation on the shelf, showing greater subsidence continued in the basin, and that the Delaware Basin during Salado time can be described as a shallow basin within an extensive shelf.

Cycles in the Salado Formation in the Delaware Basin are composed of thin (0.1 to c. 5 m thick) anhydrite and polyhalite marker beds (Schaller and Henderson 1932; Jones 1954, 1972), overlain by thick sections of halite and minor amounts of mudstone and claystone mixed with halite forming stacked halite-and halite-mudstone cycles (Lowenstein 1982, 1988; Lowenstein & Hardie 1985). These metre-scale cycles can be correlated across the Delaware basin and onto adjacent shelves. Lithofacies stacking and continuity suggest relatively isochronous deposition of cycles over many tens of kilometres of essentially flat surface in response to relative changes in sea level, rather than progradation of a sedimentary sequence across a shelf ramp from marine to continental environments. Magnesite, polyhalite, sylvite and glauberite, absent in the Palo Duro Basin and sparse in the Midland Basin, occur abundantly in some cycle sets of the Salado Formation (for example the commercially mined McNutt Potash Zone).

Sediment accumulation rates

Sediment accumulation rates are sensitive to rate of sediment production and accommodation therefore can provide some indication of water depth. Sediment production is typically high in shallow water settings; however if subsidence does not keep pace with sedimentation, accommodation of continued sediment accumulation in shelf and shallow basin environments is limited by marine or local water level. As evaporite sediment accumulates rapidly to maximum water level, syndepositional dissolution and reworking become important. Our understanding of the absolute value of accumulation rates in the Permian Basin evaporites is limited by poor data for estimating ages. Two lines of evidence, however, suggest the Castile Formation accumulated much more rapidly than the other units. Anderson *et al.* (1972) interpreted the fine calcite–anhydrite laminae in much of the Castile as annual deposits (varves). The accumulation of couplet after couplet suggests a steady sediment build-up. Another line of evidence is the slope of the change stable isotopes of strontium. $^{87}Sr/^{86}Sr$ through the Castile and Salado decreases only about 0.00002 (data from Brookins 1988, plotted in Hovorka 1990; Kirkland *et al.* 2000). As a comparison, in the 900 m section of Leonardian to Guadalupian, marine evaporites of the Palo Duro Basin show a decrease in minimum (sea water) $^{87}Sr/^{86}Sr$ of 0.00025 (Hovorka 1990). These are similar thicknesses of evaporites; assuming a regular change in $^{87}Sr/^{86}Sr$ (Burke *et al.* 1982) and marine sources for the evaporites the sediment accumulation rate in the Castile was can be inferred to be about 10 times faster than in the evaporites of the Palo Duro Basin.

Sedimentation rates, *per se*, are not an indicator of water depth, as short-term sedimentation rates in shallow evaporite pans can be high. The inferences about continuing deposition of the couplets and general long-term high accumulation rates are helpful in establishing that much of the Castile was not dominated by the interruptions in deposition that are characteristic of Salado and Palo Duro shallow-water cycles.

Siliciclastics as depth indicators

Siliciclastic distribution as indirect evidence of water depth in Permian Basin evaporites is proposed here. Examination of facies stacking in the Palo Duro Basin shows that siliciclastic deposition on the shelf is related to exposure. In each cycle, during times when the shelf was flooded and carbonate or gypsum was deposited, mineralogy in cores and gamma-ray log signatures in other wells show that siliciclastics were suppressed. As exposure became more frequent in the upper part of cycles, siliciclastics begin to be deposited, forming mudstone–halite mixtures, siltstones and sandstones. A strong aeolian transport signature is seen in all siliciclastics in the Permian Basin evaporite section, even those that were deposited in non-aeolian settings (Fischer & Sarnthein 1988; Basham 1997). This aeolian signature includes excellent sorting, mineralogical immaturity, bimodal textures, frosted grains in sand fraction and aeolian sedimentary structures in preserved aeolian facies, including cross beds and infiltration of clays in aeolian flats (Mazzullo *et al.* 1985; Nance 1988). The dominant grain sizes are very fine sand to coarse silt, a fine tail represents transport of dust. While thickness of the unit may reflect an interaction between supply and accommodation, suppression of siliciclastics suggests that flooding of the shelf has limited the dominantly aeolian transport mechanism across the shelf.

Siliciclastic accumulation in the pre-evaporite Delaware Basin is also tied to the shelf conditions. Intermittently through the Guadalupian, siliciclastics were transported across the shelf as a result of limited accommodation, sediment by-pass and exposure; they accumulated on the slope and basin floor (Fischer & Sarnthein 1988; Gardner 1992). When the shelves were flooded, no effective transport mechanism for siliciclastics across the shelf to the basin was available, and siliciclastic deposition in the basin was suppressed (Fig. 4).

However, during deposition of the Castile, no siliciclastics accumulated in the basin. This suggests that, at this time, the shelf was mostly flooded, limiting aeolian transport of siliciclastics. Flooded conditions on the shelf require that the basin remained full to the shelf edge, and therefore the basin remained deep. The possibility that exposure episodes too brief to allow progradation of siliciclastics across the shelf may have occurred and been recorded in the basin cannot be eliminated.

As the basin shallowed, exposure of the shelves was again recorded by siliciclastic deposition in the Delaware Basin. The lowest occurrence of concentrated siliciclastics in the Ochoan is a minor mudstone in lower Salado anhydrite. Mudstone beds are a minor component in Salado halite of both the shelf and basin, but they are distributed with stratigraphic continuity and have significance as an indicator of water depth and exposure (Lowenstein & Hardie 1985; Holt & Powers

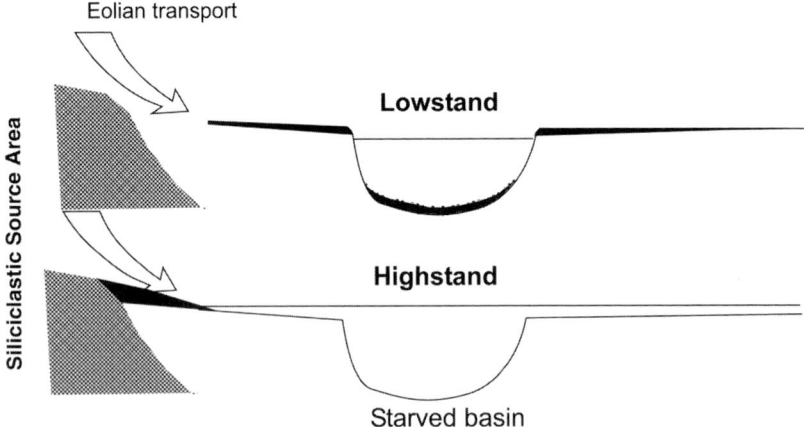

Fig. 4. Flooded shelves suppress transport of siliciclastics into the basin.

1990a, b). Powers et al. (1988) recognized that the natural gamma-ray signatures of slightly argillaceous units could be correlated over large areas. The first siliciclastic sandstone of the Ochoan evaporite sequence, the Vaca Trieste Sandstone Member of Adams (1944), is high in the Salado Formation and is a significant regional marker across both the shelf and basin, showing that the region was simultaneously exposed and therefore had little topographic relief by this time.

Geochemical evidence of depth

Geochemical data examined do not contain the indicators that would be expected if the basin had been evaporated and drawn down well below the shelf edge, as demonstrated below. Low values of $Sr/^{86}Sr$ throughout the Castile are consistent with minimal influx of waters from continental sources. If water levels in the Delaware Basin had episodically been drawn down significantly below the shelf edge by evaporation or reflux during Castile deposition, we would expect a groundwater contribution of connate brine stored within the shelf sediments into the basin. This groundwater would be evaporite-saturated because of the prevalence of evaporites in the shelf, but $^{87}Sr/^{86}Sr$ would be higher than Ochoan values because pre-Ochoan seawater (Burke et al. 1982) was more radiogenic and because diagenetic interaction between brine and siliciclastics pervasive in other Permian Basin groundwater settings would have occurred. Evidence for radiogenic groundwater was not found in the Castile (data from Brookins 1988, plotted in Hovorka 1990; Kirkland et al. 2000). However in the upper Salado Formation, associated with siliciclastics, the $^{87}Sr/^{86}Sr$ of anhydrite is variable and radiogenic.

An analysis of isotopic and trace element geochemistry of beds containing thick couplets, nodular fabrics and anhydrite pseudomorphs after gypsum within Anhydrite I of the Castile Formation (Leslie et al. 1997) did not identify conclusive proof of shallower water depth during these intervals, but showed that water input remained geochemically nearly constant.

Fabric criteria for depth determination

Comparing the Castile to the Palo Duro and Midland Basin shallow-water evaporites shows a distinctly different set of fabrics. Comparison of these fabrics to the inferred processes by which they formed, to laboratory analogues, and to analogues in other basins increases confidence that fabrics have depth significance. Other variables within the Permian Basin such as the source of water (marine or meteoric) and source of solute (marine or recycled evaporite) have been discussed elsewhere (for example Hovorka 1990, 1994). Although these factors certainly have an effect on the evolution of the evaporites, they do not have a primary control on the evaporite fabrics and sedimentary structures that seem to be somewhat independent of geochemistry and even mineralogy.

Deep water

Because evaporite brines are dense and have a high viscosity compared with marine water, depth is a relative term. For marine derived brine, chloride-precipitating brines are 10 times more viscous than sea water, gypsum-precipitating brines are 11% denser than sea water and halite brines are 22% denser (Sonnenfeld 1984, p. 102). Wave energy and vertical circulation are further damped by formation of density stratification with the result that bottom water and basin floor sediment are isolated from the surface at fairly shallow depths. Commonly, water masses that are too deep to be vigorously mixed become density stratified, as is seen in lake and ocean water masses. We can easily observe the strong effect of density stratification in an experimental tank (Fig. 5). Two solutions, one saturated with halite and Epsom salts and one saturated only with halite, poured into an aquarium will produce a strongly layered water mass. The halite saturated-brine floats on the halite-plus-Epsom salt brine. In natural evaporites, more highly evolved brines that have precipitated gypsum or halite accumulate in the lower part of the water mass. Less evolved brines derived from evaporation of seawater or dissolution of evaporites are generally less dense and float on top of more evolved brines. Under these conditions, the lower water body(s) is/are isolated from evaporation (Fig. 5a). Only after significant volumes of minerals have precipitated does the density of the upper water mass approach the density of the lower mass so that they can mix, as was observed in the Dead Sea (Neev & Emory 1967; Garber et al. 1987). An alternative mixing mechanism is that the upper water mass thins sufficiently to allow wave energy to mix the water masses. This effect is observed in the tank model (Fig. 5b) by introducing a thin upper water mass and using a fan to increase turbulence that causes the water masses to mix. The depth at which mixing will occur is variable depending on the density and viscosity of the stratified brines and energy and type of turbulence introduced.

The lower water mass of density-stratified brines is isolated from evaporation. Rapid precipitation of

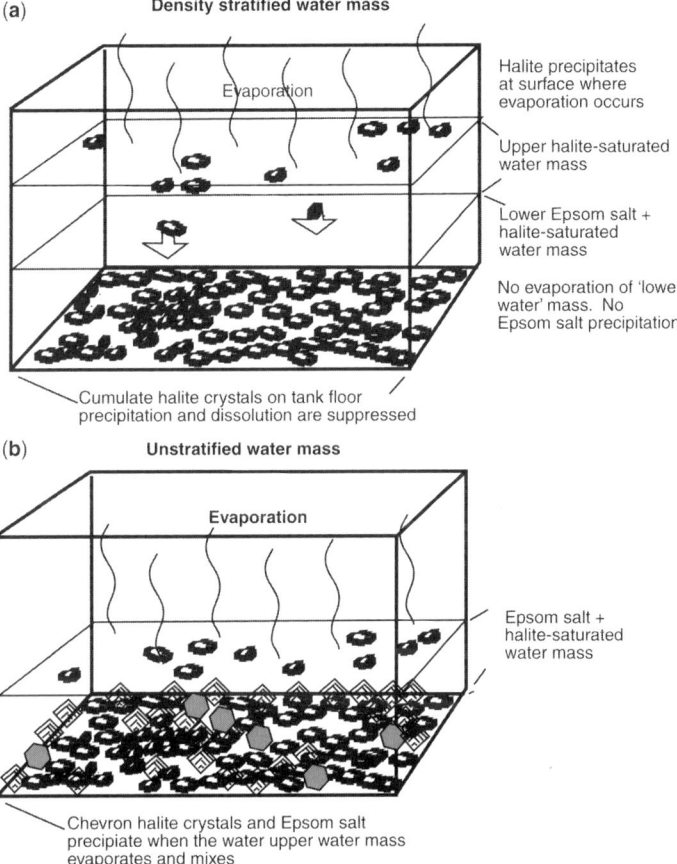

Fig. 5. Diagram of precipitation patterns observed in an experimental tank: (**a**) density stratification isolates the lower water mass; (**b**) only when the upper water mass has evaporated and mixed to form one shallow water body does precipitation occur on the floor of the tank.

evaporite minerals is not possible from these near stagnant brines. Accumulation of evaporites on the floor of a density-stratified water mass reflects allochthonous influences, such as transportation of minerals or brine from other parts of the water mass where concentration has occurred. We will examine these allochthonous process in the following sections.

Formation of cumulates

In essentially every evaporite environment, significant quantities of evaporite minerals precipitate at the point of evaporation – the air-brine interface. Halite precipitates at the surface as small floating 'hopper crystals', first described by Dellwig (1955). If undisturbed, these floating crystals are cemented together to form rafts (Shearman 1978). Crystal growth proceeds until gravitational force or turbulence exceeds the capillary strength of the surface water and the crystals or floating rafts founder, where they aggregate at the bottom and are known as cumulates (Fig. 6a & b). In density-stratified water environments, where bottom growth is suppressed, cumulates are preserved. Similar precipitation of fine gypsum at the surface or within the water mass of gypsum-saturated water has been observed in the Dead Sea (Neev & Emory 1967; Garber et al. 1987). The passive bottom accumulation of fine gypsum drapes on top of halite is described as 'snow-on-the-mountain fabric' and documents cumulate style deposition (Wardlaw & Schwerdtner 1966). Laboratory experiments show that cumulates form in water of any depth, however they are the dominant form of deposition in density-stratified water and have the best chance of preservation in that setting.

Castile anhydrite and halite textures have been recrystallized and most evidence of the size, shape and composition of the original precipitates is overprinted. However, some clues are preserved that allow the original texture to be inferred. Pervasive fine lamination defined by sharp lower contacts of couplets where calcite and organics overly fine blocky anhydrite (Fig. 7a) provides evidence that the original sediment was fine-grained. Within the Castile halite units pervasive fine anhydrite lamina suggest that halite also has also been recrystallized from a very fine-grained precursor. In most intervals, strong deformation including isoclinal folding and boudinage has modified halite textures; however halite beds in structurally isolated positions preserved fine lamination defined by dark-coloured fluid and organic inclusions within fine-grained halite mosaic (Fig. 7b). These layers with preserved fabrics provide a key for interpreting the original fabric of recrystallized beds through the similarity of the style of lamination even when fine crystals have been recrystallized.

Down-slope deposition

One indicator of deposition in a deep basin is an array of gravity-transported sediments. In evaporites, these are most commonly evaporites deposited on the shelf or basin slope that have slid or been transported downward. In parts of the Zechstein, a diverse suite of gravity fabrics are found (for example Peryt & Kovalovich 1997). Down-slope deposits include slump deposits that are folded, breccia and transported grains. A similar suite of resedimented fabrics are reported from the Messinian of Italy (Schreiber et al. 1976; Schreiber 1978; Manzi et al. 2005; Roveri & Manzi 2006).

In the Castile Formation, gravity deposits are found in thin beds in the PDB-03 core from the deepest part of the Delaware Basin. These beds appear to have been composed of clastic gypsum in a finer matrix of carbonate and gypsum and have been replaced by anhydrite and slightly compacted. Down-slope deposits in Castile anhydrite have also been mapped in the northwestern basin margin (Robinson & Powers 1987). These are associated with the Huapache Monocline that deformed some of the lower Castile and may be coeval with the deformation. Care must be taken to separate syndepositional breccias from widespread breccias formed by post-depositional dissolution of salts.

Shallow water

Three lines of evidence are recognized as diagnostic of shallow water deposition: (1) fabrics indicating dominant precipitation as layers of bottom grown crystals; (2) fabrics indicating physical energy in the environment; and (3) horizontal dissolution surfaces formed by flooding of the environment with undersaturated brine.

Bottom-growth fabrics

Shallow water evaporites in the Permian Basin are dominated by diagnostic bottom-growth fabrics. These fabrics have been widely recognized by previous evaporite sedimentologists and are: (1) textures indicating growth as layers of crystals on the brine pool floor; and (2) bands of fluid or other inclusions. Large volumes of crystals oriented with a preference for crystallographically favourable orientation shows that minerals precipitated relatively rapidly from the overlying water column (Arthurton 1973; Bąbel & Becker 2006). Bands of inclusions show high frequency variation in rate of growth that reflect daily or weather-related

Fig. 6. Cumulate halite: (a) cumulate halite from modern salt pan, Baja California; blue epoxy impregnated thin section in plane light of a sample collected by Robert Handford; (b) thin section in partly crossed polars of unusually well-preserved rafts entombed in anhydrite from the San Andres Formation, Palo Duro Basin (sample GF2512.3).

variations (Roedder 1982). Both of these processes suggest that the bottom fluids from which minerals are precipitated are in good communication with surface fluids. This is most likely to occur beneath an unstratified or a mixed water column, which is most easily maintained in shallow water. Note that we do not consider bottom growth uniquely characteristic of shallow water, but rather a typical fabric. Bottom precipitation occurs beneath deep stratified brine under conditions where supersaturation is produced, for example where saline springs discharge or supersaturated brines are transported into the deep basin.

Bottom-growth gypsum forms a wide array of fabrics, from rosettes of needles a few millimetres tall to giant palmate crystals (Hardie & Eugster 1971; Schreiber et al. 1976; Schreiber 1978). Variable textures (from primary gypsum) common though the Permian Basin include acicular (grass-like; Fig. 8a) or wide crystals (Fig. 8b), twinned crystals, well-defined bedding and crystals that renucleated repeatedly and grew though several laminae, large homogeneous crystals and crystals with well-defined growth bands. In this paper we will not attempt to further subclassify and interpret the sedimentological significance of the variations in gypsum morphology within bottom-growth fabrics. In the Permian Basin, sediments composed of variable mixtures of gypsum and carbonate sand and mud commonly occur as inclusions within crystals, beds draping bottom-grown crystals, and rippled gypsum sand (now

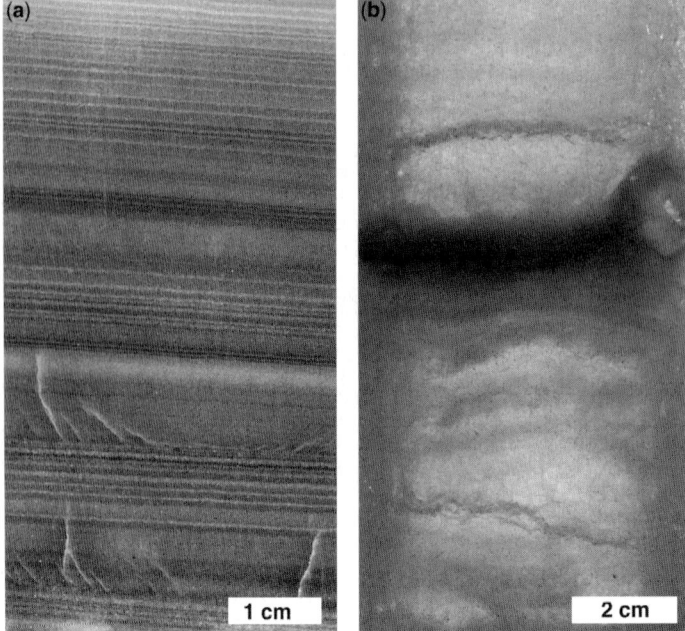

Fig. 7. Fine lamination in Castile suggests origin as cumulates: (a) lamination in calcite and anhydrite, Castile Formation unit I, Delaware Basin (sample G 4434.5); (b) rare well-preserved lamination in Castile halite, Delaware Basin, back lit core (sample G2954).

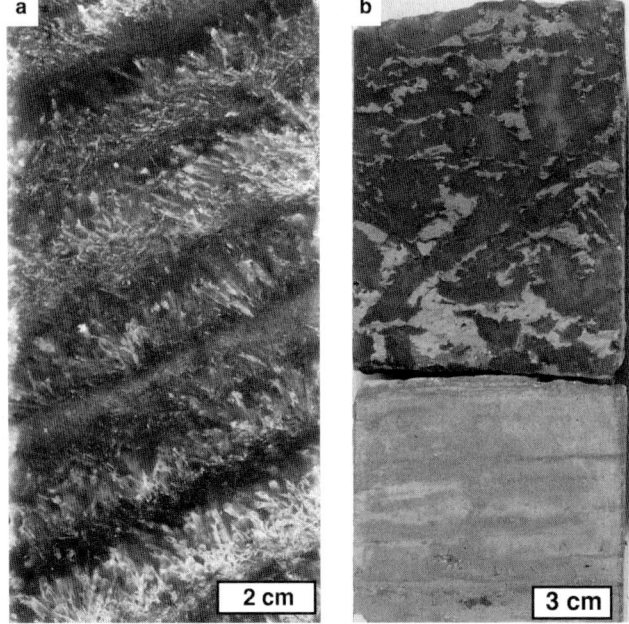

Fig. 8. Bottom growth fabrics in gypsum: (a) titled bedding of bottom-grown gypsum, presently anhydrite and dolomite pseudomorphs, from the lower Salado Formation of the Delaware Basin, sample G3232; (b) core slab of anhydrite pseudomorphs after bottom-grown gypsum at the base of a cycle, San Andres Formation of the Palo Duro Basin (sample O1323.9).

anhydrite). Gypsum has been pervasively replaced but locally original crystallographic fabric is partially preserved by preferred orientation of anhydrite (Fig. 9c). In the lower parts of thick anhydrite beds of the Palo Duro Basin, gypsum was most commonly replaced by anhydrite with loss of fine shape leading to a nodular appearance (Fig. 8b). In the upper part of thick anhydrite beds and in thin beds, pseudomorphs preserve fine gypsum textures, and partial replacement by halite is common (Fig. 9a & c). These variations are interpreted as evidence of increasing salinity of pore waters during early diagenesis (Hovorka 1992). In the Salado Formation, polyhalite has replaced some gypsum. Stratigraphic control of polyhalite replacement is interpreted as an effect of groundwater diagenesis (Holt & Powers 1990a; Holt 1993).

The equivalent bottom growth fabric in halite is formed by more rapid growth of halite crystals oriented with cube corners near vertical (Arthurton 1973; Shearman 1978). Fluid inclusions are trapped during periods of rapid precipitation, forming bands of inclusions at cube corners that are described as chevrons (Fig. 10a & b). Numerous researchers have extracted these fluids as aliquots of ancient brines (for example, Knauth & Beeunas 1985; Bein et al. 1991; Horita et al. 1996). In experimentally grown halite, investigated in our laboratory, halite grew skeletally, with the ribs of the crystals formed during rapid growth (Fig. 10c). Later clear cement filled in the inter-skeletal areas. Skeletal crystals have not been reported in natural bottom-growth environments where it is most commonly interpreted that inclusion rich zones precipitated rapidly and clear zones form more slowly, in response to diurnal cycles of variable heating, cooling and evaporation. Evidence relating depth to fabric at the time of precipitation in modern environments is imperfect. In gypsum-precipitating lakes of southern Australia, predominantly bottom-growth gypsum is found at depth of a few metres (Warren 1982). Presumably water level is closely tied to sea level and precipitation occurs at the depths observed. Bottom-growth halite is abundant in modern halite pans and precipitates beneath pools of centimetre to metre depths in most settings.

Bottom growth requires that saturated brines produced by evaporation at the air – brine interface flow to the bottom of the water mass to precipitate crystals on the bottom, a process favoured by shallow water. If dense brine is retained in the lower part of the water column, this flow will be prevented until the surface water density exceeds the bottom water density. Several mechanisms are suggested for overcoming density stratification. Down-slope flow of highly saturated saline brine from a shelf into the basin could produce large volumes of saturated brine dense enough to flow beneath a stratified water mass. Cooling or brine mixing with saturation reached through a common ion effect can also be a mechanism for attaining saturation and precipitating minerals on the floor of a deep basin. Brines discharged from springs can introduce saturated brines beneath a density stratified water mass, where they precipitate coarse crystals (Neev & Emory 1967). Lastly, density stratification can break up and brines mix as a product of brine evolution and density homogenization. Therefore bottom growth alone is a shallow-water indicator but should not be considered conclusive evidence of shallow water.

Wave reworked fabrics

Wave and current reworking of previously deposited evaporites results in deposition of clastic evaporites. These fabrics can be interpreted in terms of energy of the environment in the same way as other types of transported grains. Ripples and cross beds are characteristic of shallow water above wave base (Hardie & Eugster 1971), although down-slope deposition by density processes is possible and care must be taken to separate wave transported deposits from deep current deposits. Clastic fabrics are common in shallow-water gypsum deposits of the Permian Basin, where they most commonly can be recognized as laminated and cross-laminated sand-size detrital gypsum sediment deposited between large gypsum crystals (Fig. 9a). Fine sand- or silt-size detrital gypsum is common, forming laminated beds that drape bottom-grown fabrics. Depositional fabrics in these beds are most easily recognized where carbonate is present to outline depositional textures, because gypsum has been completely replaced by anhydrite, halite or polyhalite. Algal textures, cross-bedding and ridge forms have been documented in Salado sulphates (Holt & Powers, 1990a; Holt 1993). Equivalent transported halite in the form of ooids has been observed in modern environments (Weiler et al. 1974), Finely crystalline cumulate halite drapes bottom-growth halite formed during deposition of subaqeous Salado halite.

Bottom dissolution fabrics

A third strong indictor of shallow water is synsedimentary dissolution of previously deposited evaporite during flooding. Dissolution during flooding can be one of the least equivocal water depth indicators, because saturated water is normally denser and more viscous than undersaturated water, and it is

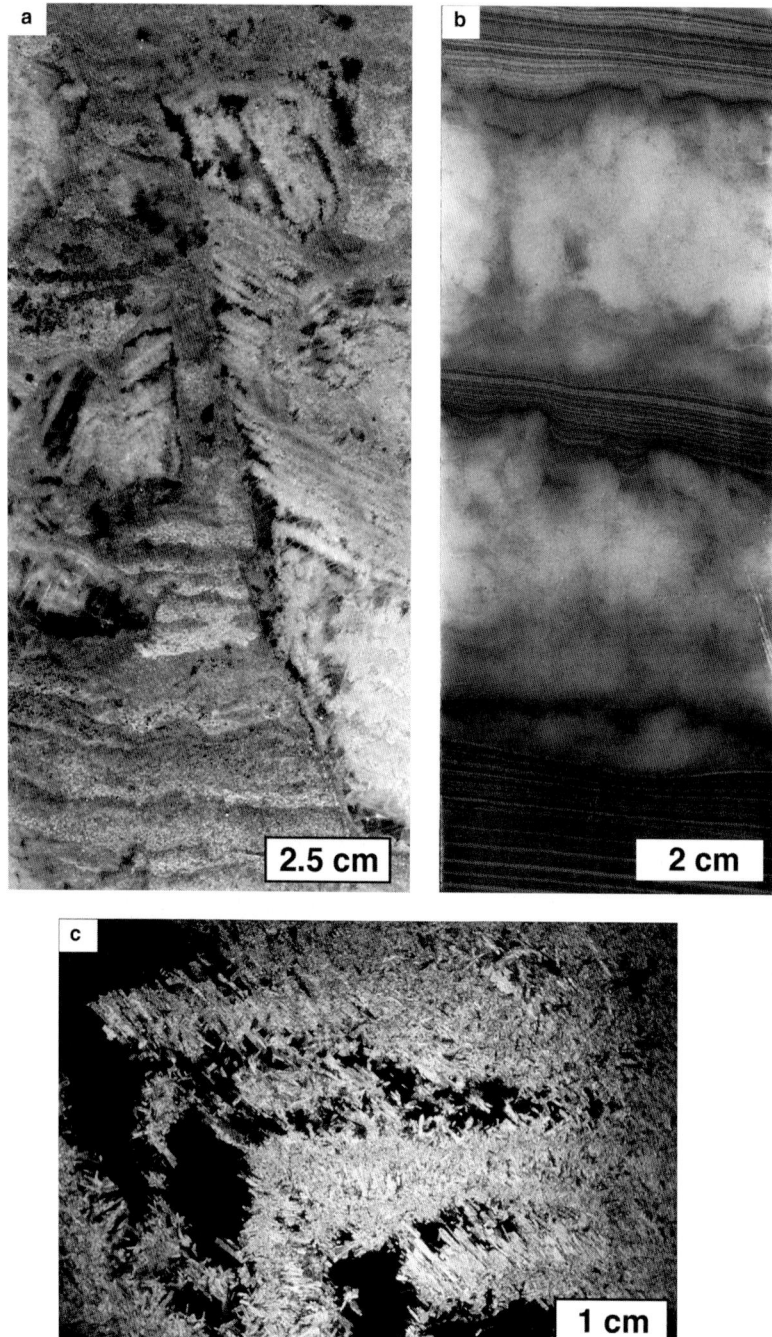

Fig. 9. Bottom growth fabrics in gypsum: (a) core slab of halite and anhydrite pseudomorphs after bottom-grown gypsum with ripple laminated detrital gypsum the middle of a cycle, San Andres Formation of the Palo Duro Basin (sample GF2009); (b) anhydrite pseudomorphs after bottom-grown gypsum at the top of calcite–anhydrite laminates, Castile Anhydrite I of the Delaware Basin, sample G4590; (c) photomicrograph of anhydrite (bright) and halite (dark) that have pseudomorphically replaced gypsum (seen in crossed polars), San Andres Formation of the Palo Duro Basin (sample O1540).

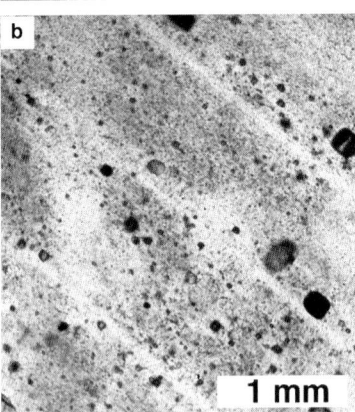
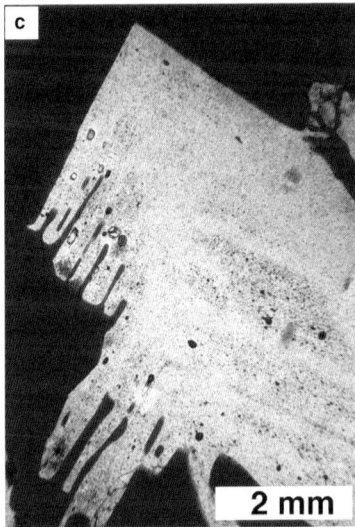

Fig. 10. (*Continued*)

difficult to mix them. Undersaturated water derived from marine or fresh water sources floats on top of gypsum- or halite-saturated water and cannot corrode the evaporite on the bottom unless the cumulative depth is shallow enough to mix the entire water column. Truncation of bottom-grown crystals during flooding of shallow evaporite environments by fresh or marine undersaturated water only occurs when the added brine mixes sufficiently well with any remaining saturated water to transport undersaturated brine to the basin floor. Bottom dissolution fabrics are recognized where the tops of crystals are truncated, forming horizontal bedding that cross-cuts crystal growth patterns (Schreiber 1978; Shearman 1978). In halite, the water depth indicated by this mixing is a metre or less because of the strong density contrast between halite-saturated brine and fresh brine is difficult to disturb by wave energy in deeper environments. In modern environments such as Baja California saline playas (Shearman 1970) and Bristol Dry Lake, California, truncation as a result of mixing undersaturated and saturated brine occurs in water a few centimetres deep. Observations of fabrics in artificial salt pans shows that low-density undersaturated water floats on halite-saturated brine of depths of only a few metres (Talbot *et al.* 1996). Halite on the floors of these deeper pools is therefore protected from dissolution.

Bottom-dissolution surfaces are repeated at intervals of 1–10 cm in shelf-deposited Permian halite of the Palo Duro Basin (Fig. 11a). Above most truncation surfaces, a layer of anhydrite that has replaced gypsum (Fig. 11b) shows that the water that flooded the brine pan brought additional sulphate; this in combination with the evaporite geochemistry (Hovorka 1987) shows that the flooding was by undersaturated marine-derived water. Halites with bottom-dissolution fabrics immediately underlie and are overprinted by exposure fabrics such as synsedimentary karst pits and chaotically bedded mudstone–halite mixtures.

Not all flooding events cause dissolution. In the Salado Formation some 10–20 cm beds of subaqueously deposited halite have very fine, opaque halite preserved at the top of the bed, beneath anhydrite or polyhalite. High bromine values of

Fig. 10. Bottom growth fabrics in halite: (**a**) thin section in partly crossed polars of chevron zonation defined by fluid inclusions in halite with anhydrite rimming grains, San Andres Formation of the Palo Duro Basin (sample O1377.0); (**b**) Thin section detail of zones of fluid inclusion in halite, partly crossed polars, San Andres Formation of the Palo Duro Basin (sample O1440.0); (**c**) skeletal halite crystal with fluid inclusion zones grown in a shallow-water tank, blue epoxy impregnated thin section in plane light.

80–120 ppm in these salts indicates that brines were retained in the basin and evolved without large amounts of dissolution. Some very flat surfaces developed on bottom-grown halite are attributed to abrasion by halite sand.

Truncation fabrics have been recognized in modern gypsum environments where tops of crystals have been truncated to form a pavement (Warren 1982). Truncation surfaces are less common in Permian Basin gypsum beds than they are in halite beds, possibly because gypsum is less soluble, but they have been recognized locally.

At the top of major cycles of the Palo Duro Basin where halite is overlain by carbonates, dissolution of the upper metre or two of evaporite by unstratified marine water left an accumulation of the insoluble materials – mudstone and gypsum or anhydrite (Hovorka 1987). This top-of-cycle dissolution created accommodation, increased circulation and helped to establish the open marine setting. Insoluble residue is recognized by compositional matching of the insoluble materials found in adjacent preserved halite and by distinctive upside-down-accreted fabrics formed as dissolution attacked successively older layers of previously deposited halite (Hovorka 1987). Insoluble residue colours are typically grey in contrast to the red colours typical of other siliciclastics.

Thin beds of insoluble residue (underclay) are found beneath the thin anhydrite beds of the Salado Formation in the Delaware Basin and on the shelf. In the Salado Formation of the Delaware Basin these layers contain clay and magnesite and also may contain a detrital component.

Exposure

Exposure is uniquely significant in the formation of evaporites as compared with other types of sediments because (1) many evaporites form in shallow pans with poor connection to the marine environments, therefore small perturbations in water circulation cause water level fall to fall below the evaporite surface, and (2) high solubility and reactivity of evaporites may record evidence of brief exposure episodes. Exposure is the most sensitive depth indicator in evaporites. Three components in the exposed environments are discussed: below the water table, above the water table in the vadose zone and at the surface.

Below water table

When the surface of the evaporite is exposed to the atmosphere, evaporite saturated brines remain within the sediment, forming a water table at depth of a few centimetres to a few metres below the surface. Brines below the water table remain highly active in modifying sediment. They precipitate evaporite cement and displacive crystals and drive replacement of one evaporite mineral by another.

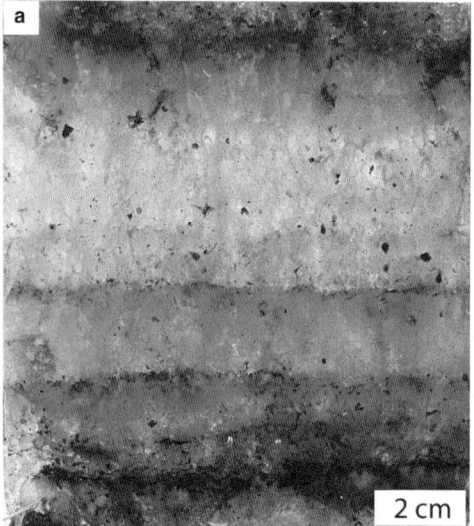
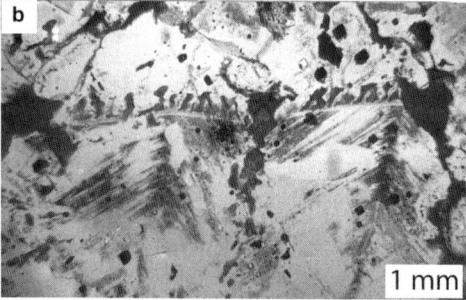

Fig. 11. Fabrics truncated by dissolution by undersaturated water: (**a**) back-lit core slab showing white chevron halite truncated by five syndepositional flooding events that formed corrosion surfaces marked by anhydrite or dark clay inclusions, San Andres Formation of the Palo Duro Basin (sample Z2859.8); (**b**) photomicrograph in partly crossed polars of chevron halite, truncated by a marine brine corrosion surface. Gypsum lamination with small-bottom grown crystals has been replaced by anhydrite (dark) San Andres Formation of the Palo Duro Basin (sample F2522).

Passive halite cement forms coarse, clear halite overgrowths and pore-filling cement. Porosity and permeability are reduced below the water table; more deeply buried evaporites have low porosity (Casas & Lowenstein 1989). Dissolution pits, pipes and macropores are filled with halite cement. Within the Salado Formation of the Delaware Basin, potash minerals (e.g. langbeinite) form as passive cements (Holt & Powers 1990a). Evaporite minerals also grow displacively, forming skeletal and euhedral forms with variable amounts of included host sediment. Anhydrite pseudomorphs of pseudohexagonal gypsum crystals with carbonate inclusions (zoned, sand crystals) document formation of highly reactive brines in the water table setting (Fig. 14b). Preferential growth of halite from the cube corners produces distinctive 'pagoda' halite that typically has a somewhat strongly skeletal form (Smith 1971; Gornitz & Schreiber 1981; Southgate 1982).

Large amounts of halite commonly grew displacively and incorporatively within mudstones, carbonate or gypsum sediments forming compressed fabrics within the host sediment (Smith 1971). Diagenetic replacement of one mineral by another occurs in the below-water table setting, for example replacement of gypsum by anhydrite, polyhalite, or halite is linked to cycle position (Hovorka 1992). The sequence of replacement is highly variable depending on original mineralogy, grain size, reactivity, brine composition and permeability.

If meteoric or marine water is introduced into the vadose environment during flooding, the upper part of the water table can become undersaturated, causing halite to dissolve along horizontal zones. These features have been observed in modern environments of Lake McLeod (Logan 1987). This type of top water-table dissolution creates a horizontal fabric in disrupted and extensively recrystallized halite and mudstone–halite mixtures. In other settings, dissolution is suppressed in the phreatic zone by dissolution of evaporites from the surface before the water reaches the water table.

Alternating precipitation and dissolution creates fabrics diagnostic of water-table evaporites. Multiple cycles of saturation then dissolution create highly disrupted fabrics, described as in the earlier stages of development of bedded and efflorescent salt as 'podular' halite (Holt & Powers 1990a, b) or in the later stages as chaotically bedded mudstone–halite (chaotic mud-salt of Handford 1982). Original bedded fabrics are destroyed and depositional bedding is eliminated; some layering related to the intensity of water-table alteration effects may be created. In gypsum, evidence of residence below water table environments is disruption of original bedding and bottom-growth fabrics. As alternation becomes more intense, the fabric becomes nodular and similar in appearance to classic sabkha fabrics. Classic sabkha fabrics, however, are formed by gypsum and anhydrite precipitation near the water table without a bedded gypsum precursor (Kinsman 1969).

Geochemical signatures in coarse pore-filling halite cement and fluid inclusions trapped with them retain evidence of precipitation in near-surface environments from brines similar to those that precipitated shallow water halite (Bein et al. 1991). The salinity of the shallow groundwater continues to evolve as a result of processes that occur in the vadose zone and at the surface, and brine alters interbedded fine siliciclastics, modifying the clay mineralogy (Bodine 1978, 1985) and releasing evidence of this alteration in the form of radiogenic $^{87}Sr/^{86}Sr$.

Vadose fabrics formed above the water table

Diagnostic fabrics are formed above the water table in the zone where evaporites and associated sediments are also subjected to alternating interaction with undersaturated water and hypersaline brine. Undersaturated water is contributed by rain, dew or marine flooding. Hypersaline brine is produced as shallow groundwater is drawn upward by capillary processes in response to evaporation at the surface. Vertical features formed by downward flow of undersaturated water are particularly diagnostic of this environment (Powers & Hassinger 1985). Stresses in rigid exposed salt (usually due to heating and cooling) cause polygonal cracking and heaving; these cracks then can form conduits to channel flow (Hunt et al. 1966; Tucker 1981; Lugli et al. 1999).

Syndepositional vertical features that cross-cut bedding are classified by geometry and timing of formation. Pits are features that are wider than they are deep (Figs 12a & c). Pits originate as voids formed by evaporite dissolution during water table drop and are formed below an exposure surface (Figs 12c & 13). Pits are partly filled by insoluble residue left when evaporite dissolves and by siliciclastics accumulated during exposure. Coarse halite cement precipitated when water-table rise re-established evaporite-precipitating environments filled the rest of the void (Fig. 12a). Deeper and larger pits are more commonly observed in the Salado Formation of the Delaware Basin than the Palo Duro Basin. We speculate that the large 'salt horses' of relatively pure halite encountered in Salado ore zones (Adams 1969) may be large halite-filled synsedimentary dissolution pits. Large pits mapped in mines in the Devonian Prairie evaporite (Baar 1974) may also be synsedimentary. Maximum pit size may be partly an artefact of exposure, as large-diameter shaft and drift exposures reveal larger features. However, it may also be a function of greater volatility of the water table in the more isolated Delaware Basin as compared with shelf settings where water level was more closely tied to sea level.

Pipes are narrower than they are deep and cut across several beds, with maximum observed more than 2 m (Fig. 12b). Pipes may be formed as voids during rapid water table fall or may originate at fractures where recrystallization was favoured.

Post-depositional karstic solution and collapse features are abundant in modern evaporite outcrops and can be very large. These features reflect later hydrologic conditions and should not be confused with syndepositional features. Post-depositional

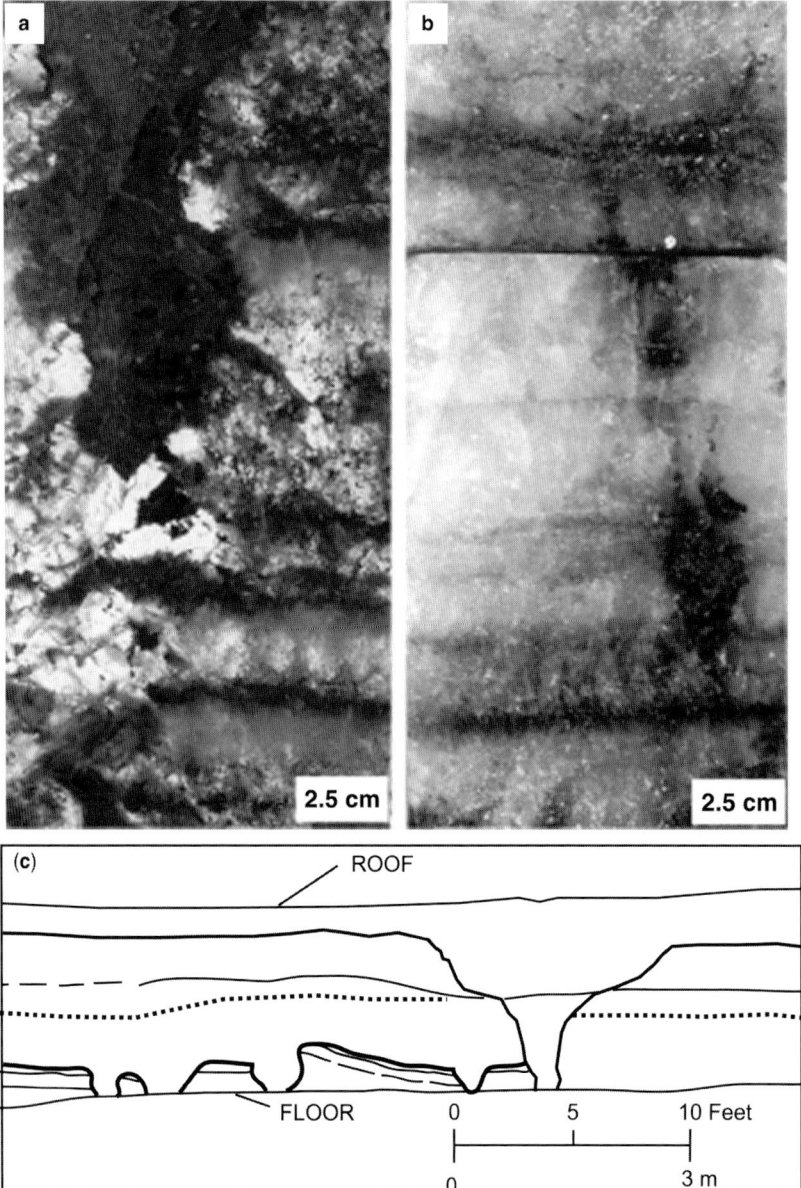

Fig. 12. Synsedimentary karst features: (**a**) small karst pit cross-cutting chevron halite with dissolution surfaces is filled with halite cement, mudstone and displacive halite documenting groundwater fluctuation, slabbed back-lit core sample San Andres Formation of the Palo Duro Basin (sample Z2039); (**b**) a narrow pipe filled with dark halite and mudstone cuts white laminated halite with corrosion surfaces; the narrow sinuous tube crosses the slabbed face of the core, San Andres Formation of the Palo Duro Basin (sample F2557); (**c**) drawing of sedimentary structures showing the abundance of vadose textures in the Salado Formation of the Delaware Basin exposed in the WIPP shaft. Modified from Holt and Powers (1990*a*). Drawing of the WIPP drifts modified from US Department of Energy (1983), sheet 61.

evaporite dissolution pits can be differentiated from syndepositional pits because post-depositional pits are filled with sediment of collapsed blocks of younger sedimentary units, typically lack halite cements, can involve younger units in the collapse (e.g. Snyder & Gard 1982), are related to post-depositional hydrologic and structural tends and can have cave-like geometries.

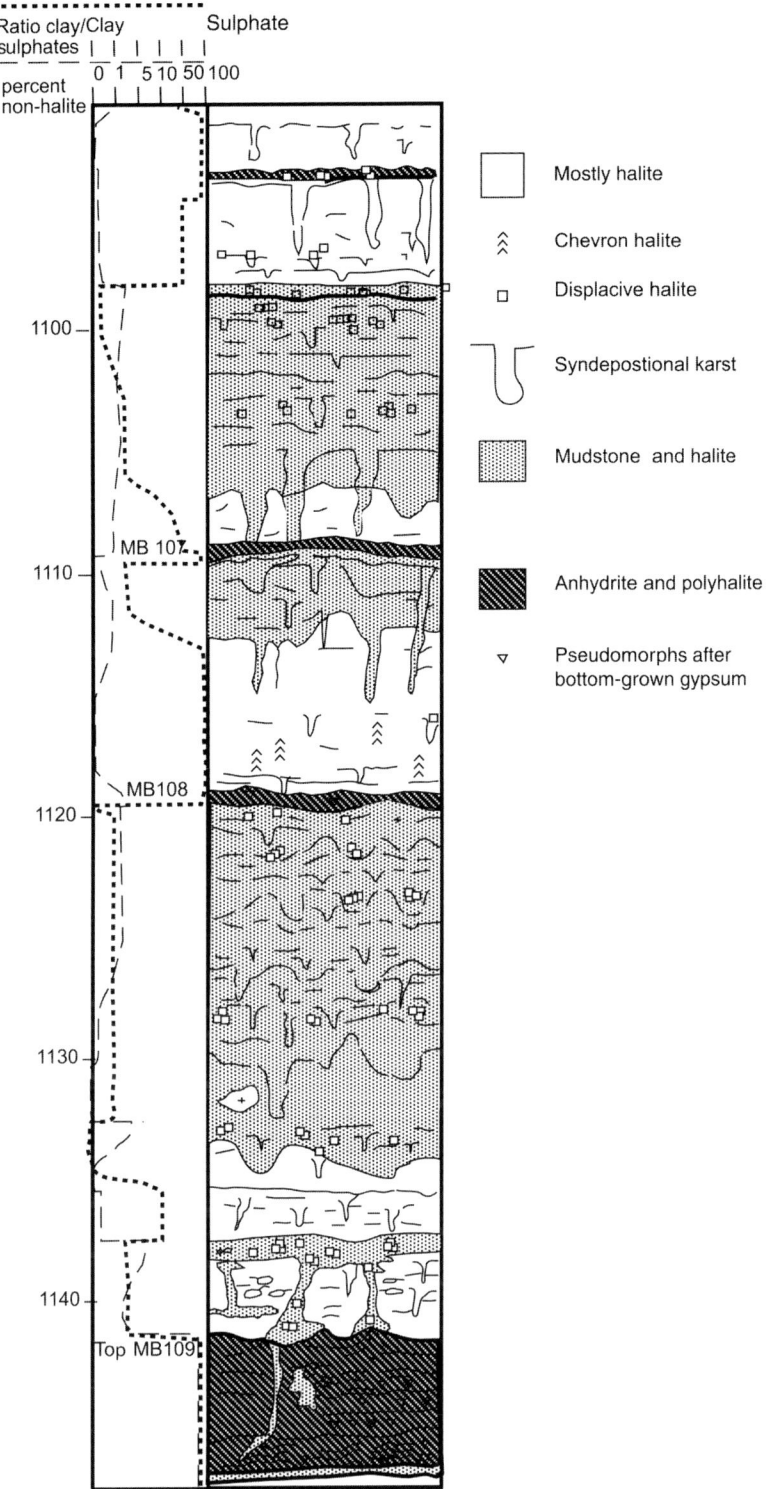

Fig. 13. Drawing of the WIPP drifts modified from Holt and Powers (1990a).

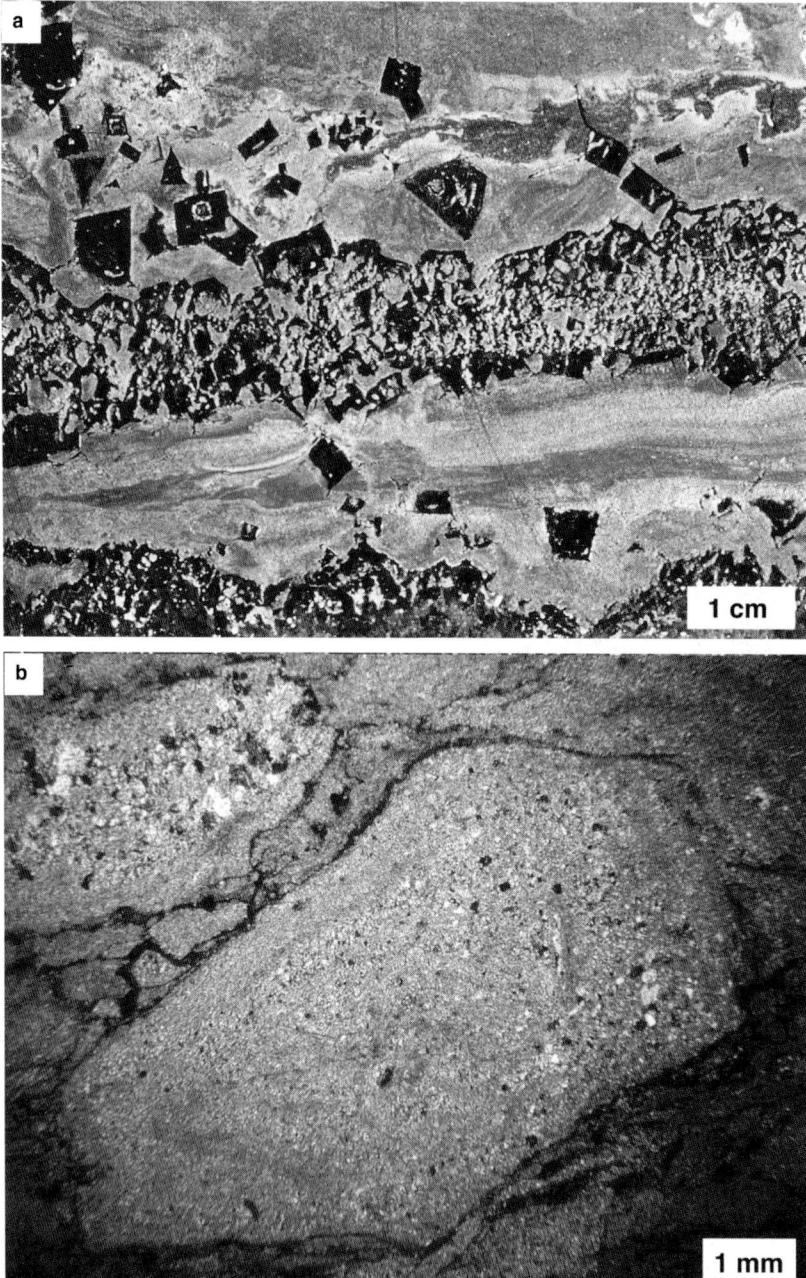

Fig. 14. Water table features: (a) slabbed core showing multiple episodes of deposition of siliciclastics by undersaturated water, followed by precipitation of partly incorporative–partly displacive halite from groundwater; deposition of each siliciclastic layer is accompanied by partial dissolution of previously deposited halite, creating casts of former halite crystals and brecciation, San Andres Formation, Palo Duro Basin (sample O1494);
(b) photomicrograph in crossed polars of sediment-inclusive gypsum sand crystals pseudomorphed by anhydrite formed near the top of the water table, lower Salado Formation of the Delaware Basin (sample G2762).

Deposition at the surface

Evaporite sediments at the exposed surface accumulate siliciclastic red mudstone (Hanford 1981; Presley 1987) and an efflorescence of fine evaporite crystals (Goodall et al. 2000). They also crack and heave, creating exposed ridges (Christiansen 1963), and are exposed to hygroscopic alteration.

Red siliciclastic mudstones commonly accumulate at the surface during exposure. The exposed dry flat allows effective transport of silt and clay as dust. Pauses in evaporite sedimentation also favour accumulation of dust. In shelf settings, siliciclastics typically show fine silt ripple lamination with clay drapes, indicating that they have been reworked by water, although association with evidence of exposure indicates that flooding was a transient condition (Fig. 14a). Red mudstones also contain abundant displacive halite. Casts of distinctive slightly skeletal halite cubes filled with silt of the next ripple set suggest episodic flooding, dissolution of previously deposited halite layers, followed by desiccation and reprecipitation of halite. Essentially identical fabrics can be created by introducing fresh water and mud into a tank containing a layer of previously deposited halite. First the top of the halite is dissolved and mud sags and slumps over the irregular surface. As dissolution and evaporation bring the pore fluids to saturation, the dissolved halite begins to precipitate as displacive crystals within the mud. The mud dries, shrinks and cracks and fibrous halite grows into mudcracks. The final stage is growth of finely crystalline efflorescent halite as saline fluids are drawn to the surface by capillary processes.

One of the most common types of deposition in modern evaporite environments is precipitation of evaporite efflorescent crusts (Hunt et al. 1966). Fluids saturated with halite and other bittern salts are drawn to the surface by capillary forces and precipitate as fine-grained crusts on the sediment surface. They coat previously formed evaporite sediments and are especially visible on non-evaporite clastic layers. The effect is especially pronounced (although not very important in the rock record) on features such as trees and posts protruding from evaporite flats. Efflorescences are generally ephemeral, because their fine crystal size, high porosity, and deposition at the surface makes them vulnerable to dissolution during deposition of the next layer of sediment. However, they may play a significant but not widely recognized role in brine chemistry, because the efflorescences may be chemically and isotopically fractionated from the bulk sediment composition, and can make a substantive contribution to brine composition when they are dissolved.

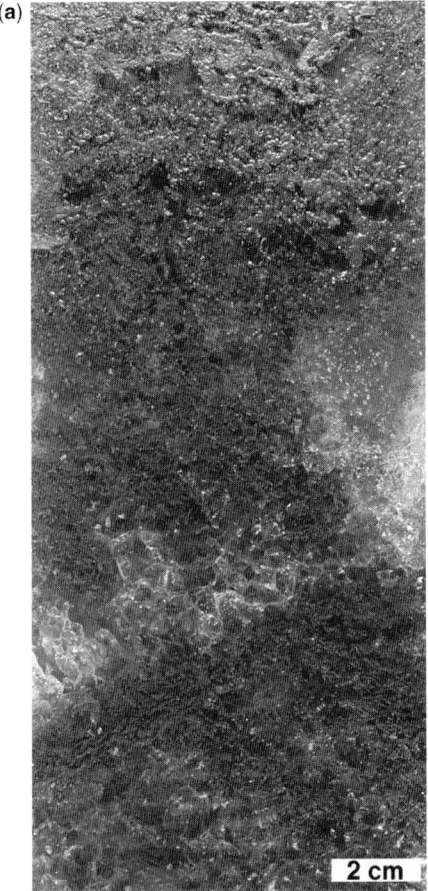

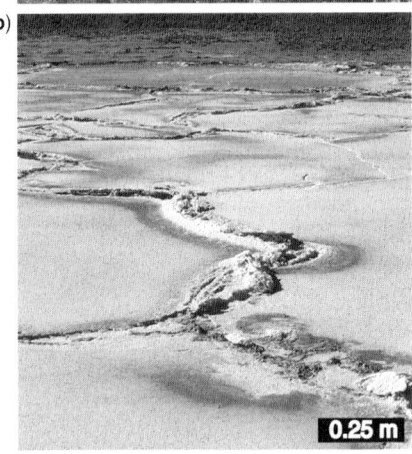

Fig. 15. Deformation and salt polygons: (**a**) tilted bedding in mudstone halite mixture shows deformation in vadose halite environments; slabbed core from the upper Salado Formation of the Delaware Basin (sample G1276); (**b**) early stages of formation of salt polygons with heaved edges and efflorescent salt growth, Bristol Dry Lake California.

Material exposed at the surface of a salt pan is subject to hygroscopic alteration that reduces crystal size and increases cementation. Halite is hygroscopic and moisture is attracted to the halite in the evening when the humidity is high, evaporation decreases, and the sediment cools. This process is accentuated if dew is precipitated. The moisture dissolves a small portion of halite that is then reprecipitated as very fine crystals during the following day when evaporation resumes. The newly created fine crystals offer more surface area for hygroscopically attracted water, providing a feedback mechanism to decrease the crystal size of large volumes of halite over time. This process dominates surface halite fabrics in the older surfaces of Devil's Golf Course in Death Valley, California, where nearly all surficial halite is finely crystalline and well-cemented, and is a common feature in Salado halite (Holt & Powers 1990a, b).

Discussion

Distribution of depth-related fabrics within cycles of the Permian Basin may best be described by putting all the described fabrics together in tabular form (Fig. 16), establishing a classification of typical gypsum and halite evaporite fabrics by whether they are deep and stratified, shallow unstratified or exposed and examining the sedimentary textures formed during flooding and evaporation. To examine how these depth indicators are developed in context, we examine the diagnostic fabrics in the Gulf PDB-03 core through complete Castile through Salado facies in the thickest (southeastern) part of the Permian Basin, Loving County Texas (Fig. 17).

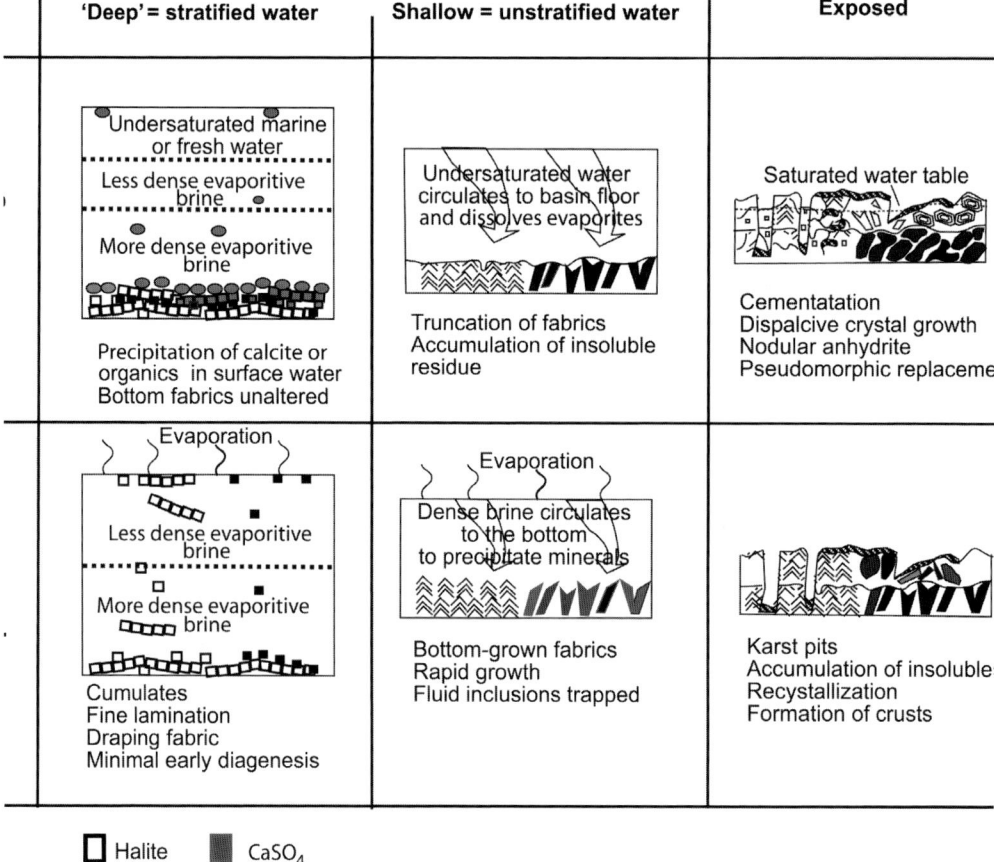

Fig. 16. Depth key to evaporite fabrics.

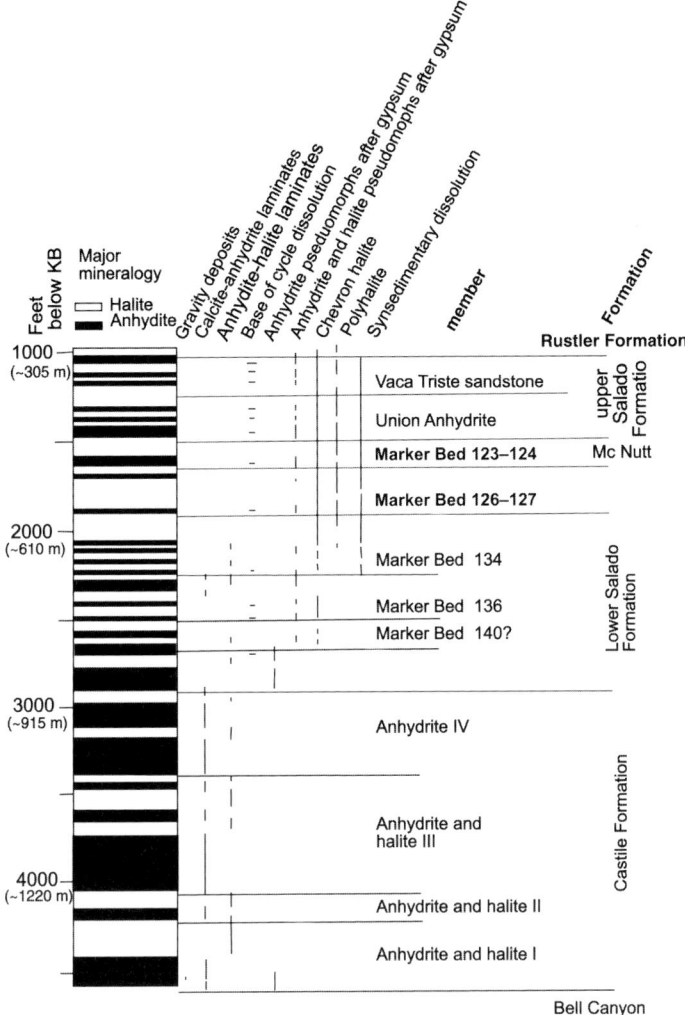

Fig. 17. Distribution of fabrics in Gulf Research PDB-03 core. Core location shown in Figure 1.

Delaware Mountain Group

We describe the Delaware Mountain Group to provide a context for interpreting Castile deposition. The Delaware Mountain Group was deposited in the deep sediment-starved Delaware Basin and is composed of very fine sandstone and siltstone alternating with calcareous, organic-rich siltstone and limestone. The very fine sandstone and siltstone were deposited during first-order lowstands when aeolian-sourced siliciclastics prograded across the shelf and into the basin (Harms & Williamson 1988; Gardner 1992). Calcareous, organic-rich siltstone and limestone was deposited during highstands, when the shelf was flooded and siliciclastic transport across the shelf was suppressed. High organic production and continued precipitation of carbonate shows that the marine water was supplied to the Delaware Basin; preservation of organics, absence of burrowing, and sediment transport characteristics suggest that the deeper part of the water column was hypersaline (or at least anoxic). The uppermost part of the upper organic-rich unit, the Lamar limestone of the Bell Canyon Formation is composed of finely laminated micritic limestone, minor silt and organics. Comparison of these fabrics with the overlying Castile Formation suggests that the upper Lamar deposition was almost identical to Castile deposition but without preservation of cumulate gypsum (Cys 1978). We interpret this change as the time at which the deep water mass became and was maintained as gypsum-saturated brine.

Castile Formation

Depositional cycles in the Castile Formation are related to systematic evolution of the stratified water body and not to progradation of environments. Genetically, Castile cycles begin with a Lamar-like unit of finely laminated organic-rich limestone. They are considered the most marine parts of the cycle, because at this time, any gypsum precipitated at the surface was dissolved in the water mass before it was deposited on the basin floor, suggesting that they correspond to sea level highstand, maximum marine influx and minimum salinity in surface water. High organic production and maintenance of calcite precipitation suggest that marine water recharged the Delaware Basin lamination shows a high frequency, possibly seasonal, fluctuation (Kirkland 2003). More than one genetic cycle is recognized per numbered unit.

The dominant unit of Castile cycles is composed of calcite–anhydrite cumulate couplets (Fig. 7a). These laminates are interpreted to represent a high frequency (possibly annual) alternation. First freshening of the surface layer with marine water produced calcite and organics. This was followed by evaporation of surface waters to gypsum saturatation and precipitation of gypsum at the surface as fine crystals that foundered and accumulated on the basin floor as cumulates. In Figure 16, cumulate textures are interpreted as formed in stratified water with surface water composition alternating between under-saturated and gypsum-saturated. The bottom water mass was fairly stable and maintained saturation with calcite, gypsum and halite irrespective of the mineral precipitated at the surface, so that fine crystals of these minerals were preserved as they fell though the water mass. Laminae in the range of 1 cm or less, with a ratio of <1 part calcite to 9 anhydrite, are reasonable products for episodic introduction of 1 m or less marine brine followed by evaporation. Down-slope resedimented beds occur at irregular intervals within the calcite–anhydrite cumulates sequence.

The upper metre or two of calcite–anhydrite cumulates below the onset of calcite–anhydrite–halite triplets are characteristically marked by thickening of the anhydrite part of the bed, and formation of a more irregular thickness and nodular texture in the anhydrite and lighter colour in the calcite lamina are interpreted as absence of organics. Local preservation of anhydrite pseudomorphs of bottom-grown gypsum within the thicker anhydrite beds (Fig. 9b) document that the change in depositional pattern corresponds to transport of increased volumes of gypsum-saturated brine to the basin floor. Calcite and anhydrite lamina drape the pseudomorphs, showing that cumulate-style deposition alternated with bottom growth, and that the bottom was isolated from undersaturated surface water during calcite precipitation episodes. Reduction in organic preservation is harmonious with introduction of more oxidized brine.

The cause of this change to bottom precipitation requires discussion. Kendall (1992) has suggested that the entire basin may have episodically shallowed to the point at which transport of brine from the surface was rapid (shallow water in a deep basin). The absence of an increase in siliciclastics in this bottom-growth interval and in overlying halite, however, suggests that the shelves were not exposed for an extended period at this time, because aeolian transport of siliciclastics remained suppressed. One alternative hydrologic mechanism for bottom-growth is evolution of the entire water mass to near-halite saturation resulting in elimination of density stratification. This is compatible with the position in the cycle beneath anhydrite–halite couplets, because deposition and preservation of these cumulates require that both surface and subsurface brines become saturated with calcite, gypsum and halite. Strong bottom transport of dense supersaturated brine from the shelves to the basin is another way to imagine introduction of a large volume of gypsum-saturated brine to the basin floor.

The upper part of some Castile cycles is composed of an interval of calcite–anhydrite–halite triplets similar to the calcite anhydrite couplets but with a thinner calcite component and a thicker halite layer. Deformation has extensively altered the fabric in many of these beds. A few minimally altered intervals and clasts preserve dark colours and millimetre layering in halite and release volatile organics when dissolved (Fig. 7b). Most other intervals have centimetre-size clear halite crystals that presumably formed by recrystallization of strained halite during deformation. The geometry of the better-preserved anhydrite lamina suggests accumulation as fine crystals deposited as cumulates, although fine original crystals have been recrystallized to coarser blocky anhydrite. Siliciclastics are suppressed throughout the interval.

Bromine concentration in halite is less than 65 ppm, suggesting that the brine originated by recycling of previously deposited halite by dissolution in marine water (Holser 1966; Hovorka 1990). Low Br concentrations are found even in minimally recrystallized halite, suggesting that low Br is not only a result of recrystallization (Adams 1969; Hovorka 1990). $^{87}Sr/^{86}Sr$ of anhydrite remains at low Ochoan values (Hovorka 1990;

Kirkland *et al.* 2000), and trace and stable isotopic composition are typical of marine derived evaporites (Leslie *et al.* 1997). Groundwater input would be expected if water level in the basin fell significantly below the shelf edge; the absence of radiogenic groundwater input is interpreted as evidence of basin-full conditions. Groundwater or surface water input in the basin would be more radiogenic than observed, because likely sources of groundwater or surface water would be Guadalupian connate brines stored in the shelf, brines that had equilibrated with Guadalupian or Leonardian carbonates or sulphates while moving through the shelf or basin floor, or brines that had been in contact with siliciclastics in the shelf or basin floor. However, minor episodes of water level fall below the shelf edge might have contributed only minor volumes of solute and not be identified by existing sample frequency. In Figures 16 and 17, the textures illustrated from Castile examples are interpreted to have formed in a stratified body water with surface water that was marine-sourced but alternating between states that were gypsum-saturated and halite-saturated.

The depositional environment of the Castile Formation is conceptualized as a setting where water levels in the Delaware Basin were hovering near the shelf break. The record of the geometry of the inlet at the southwest end of the basin has been lost by erosion, however we can infer from the Castile record that restriction was sufficient to retain a body of dense evaporite brine but open enough to allow continued or frequent recharge by marine brine and discharge of highly evolved brine throughout Castile deposition (e.g. see the model of Kirkland 2003).

The Castile Formation is usually interpreted as having no shelf equivalent. However, if the basin was full and the shelves flooded to suppress siliciclastic input, some shelf deposition should have occurred. The unit to examine for evidence of shelf conditions during Castile deposition is the lower part of the Fletcher anhydrite, usually considered basal Salado equivalent on the shelf. This unit was cored on the shelf by the Gulf Research PDB-04 core (Garber *et al.* 1989). The textures in the Fletcher Anhydrite in this core have been extensively altered; however close examination of core shows features such as small bottom-grown gypsum pseudomorphs, gypsum–carbonate insoluble residue, breccia, and displacively grown gypsum sand crystals. These textures including the extensive alteration and fabric destruction are diagnostic of shallow-water to exposed fabrics, and compatible with a history of halite–gypsum precipitation followed by exposure, dissolution and water table precipitation. We suggest that the lower part of the Fletcher Anhydrite is the shelf equivalent of the Castile Formation, and formed under conditions of limited accommodation and alternating evaporite precipitation and dissolution. Dissolution on the shelves could have been by marine brine that dissolved previously deposited halite and contributed brine to the Delaware Basin, producing a marine Ochoan signal in the $^{87}Sr/^{86}Sr$ in sulphate and a marine recycled Br signal to halite. A minor contribution by meteoric dissolution is possible; however it was not dominant (Leslie *et al.* 1997).

Lower Salado Formation

An interval in the PDB-03 core tentatively correlated with the base of the Salado Formation shown on the cross sections of Snider (1966) is marked by a gradual change in the dominant textures through a thick (43 m) anhydrite unit. Textures at the base of the sequence are typical regular, calcite–anhydrite laminates that grade upward into a well bedded sequence of calcite or dolomite laminae alternating with layers of bottom-growth gypsum. Upward though the sequence lamination becomes weak or absent, and crystals of 10 cm or more tall are dominant. Some of the interval appears to be massive, but patches of aligned anhydrite crystals suggest that this interval was deposited as large palmate crystals of nearly pure gypsum. Lower Salado beds are traditionally correlated with the Fletcher anhydrite of the shelf, showing that by Salado time the topography of the Delaware Basin had been nearly eliminated by deposition of the Castile Formation in the basin. The correlation of this interval across the basin is not well agreed upon, and the same textures are not represented everywhere. This suggests facies complexity developed as the Delaware Basin aggraded to sea level. However, everywhere in the basin, laminated calcite–anhydrite couplets transition to well-formed bottom grown gypsum textures. This evolution documents shallowing of the basin to depths at which stratification of gypsum-saturated water was not maintained. Bottom growth textures are visible in many gypsum beds throughout the Salado Formation, although massive and weakly laminated textures are also represented. In Figures 16 and 17, the lower Salado anhydrite textures illustrated are those typical of unstratified gypsum-saturated water.

Halite textures in the lower Salado in the PDB-03 core are a mixture of anhydrite–halite laminates and shallow-water halite, suggesting that, although the water depth was shallower, the basin became intermittently stratified when the basin brine salinity and density increased to halite saturation. Correlation within the lower Salado shows that stratigraphy of the interval is more complex than the lower Castile with halite beds pinching over distances of a few miles (Hovorka 1990). The lower Salado anhydrite is not finely folded like the Castile units, but the PDB-03 core shows intervals of steeply dipping core dip alternating with flat lying core. These complexities are tentatively interpreted as evidence that, for the first time in basin evolution, undersaturated water dissolved halite on the basin floor. This is part of pattern documenting decreasing water depth. Textures observed in the lower Salado halite illustrated in Figures 16 and 17 include under-saturated unstratified water dissolving evaporite, and halite-saturated unstratified and stratified water. Cycle patterns are complex, suggesting the basin may have become somewhat isolated from marine sea level, allowing water levels in the basin to become more variable.

Middle and Upper Salado Formation

Most of the Salado Formation is characterized by textures indicating upward-shallowing shallow water unstratified to exposed environments in both anhydrite and halite (Fig. 16). Underclay insoluble residues are recognized at the base of thicker anhydrite beds documenting the absence of stratification during flooding. Bottom-grown gypsum is common in thicker beds, and has been extensively replaced by halite and polyhalite, showing the strong diagenetic overprint typical of saline groundwater environments. Bedded mud-poor halite in the lower part of the cycles reflects shallow water subaqueous deposition and includes bottom-growth chevron structures defined by fluid inclusions. Sedimentary structures higher in the cycles indicate increasing duration of exposure. Siliciclastic mudstone is significant within podular and dilated halite, and halitic mudstone beds appear in the upper part of halite cycles at the same time as the sedimentary evidence of shallowing. Mudstone increase in the same stratigraphic interval as shallowing shows that exposure of shelf areas and reduction of accommodation and exposure in the shallow basin are linked. Synsedimentary dissolution pits also appear at the same interval. Exposures in drifts and shafts of the WIPP site show that these karst pits are several metres deep, deeper than seen in the Palo Duro Basin, suggesting that groundwater drawdown had a larger magnitude in this isolated shallow basin than in the more open Palo Duro Basin shelf setting. Partial dissolution of halite from halite pseudomorphs after gypsum shows the prevalence of a thick unsaturated zone during exposure. The middle and upper Salado Formation also preserves more exposure fabrics than are recognized in the Palo Duro Basin, which may reflect episodes of rapid sediment accumulation with minimal recrystallization and also may be an effect of an isolated basin with variable water levels.

Conclusions

Sedimentary structures in Permian Basin halite and anhydrite provide readily interpretable evidence of water depth at the time of deposition. Deeper-water fabrics show evidence of density stratification. The absolute depth at which density stratification occurs is dependant on density and viscosity contrasts in the water mass. Modern examples suggest that anything over a few metres deep is likely to have characteristic fabrics of 'deep water' evaporites. Deep water fabrics in both gypsum and anhydrite are dominated by deposition of fine-grained cumulate crystals precipitated at the air–brine interface. Cumulates are recognized by fine size, rafts, and 'snow-on-the-mountain' textures. Because cumulates can form in any water depth, the key indicators of deep water is suppression of shallow-water fabrics. Other indicators of down-slope deposition are resedimented fabrics, very similar to those in non-evaporitic sediments.

The transition to shallow water is marked by greatly increased volumes of bottom-growth fabrics. The transition to dominant-bottom growth takes place at different depths in gypsum than in halite, because maintenance of density stratification can require greater water depth in gypsum-saturated brine. Dissolution surfaces provide another clear indicator of shallowing. Horizontal dissolution of previously deposited halite or gypsum is favoured in relatively shallow water when a newly introduced water mass is able to mix with the previous water mass and dissolve bottom sediments.

The most unequivocal depth indicator in halite and gypsum is an array of distinctive features formed during exposure and water table drop. Abundant and diagnostic exposure indicators are dissolution features including vertical and horizontal synsedimentary karstic dissolution, deformation forming polygons and cracks, and aggressive replacement of one fabric by another with or without mineralogic change in fluctuating saline groundwater and vadose settings.

We thank Kathy Benison for a review of the draft manuscript of this paper and Charlotte Schreiber for her persistence in reviewing, editing and seeing the preparation to completion. The senior author appreciates partial support for manuscript preparation by a fellowship from the John K. and Katherine G. Jackson School of Geosciences.

References

ADAMS, J. E. 1944. Upper Permian Ochoa Series of Delaware Basin, West Texas and southeastern New Mexico. *American Association of Petroleum Geologists Bulletin*, **28**, 1596–1625.

ADAMS, S. S. 1969. Bromine in the Salado Formation, Carlsbad Potash District, New Mexico. *New Mexico Bureau of Mines and Mineral Resources Bulletin*, **93**.

ANDERSON, R. Y., DEAN, W. E., KIRKLAND, D. W. & SNIDER, H. I. 1972. Permian Castile varved evaporite sequence, West Texas and New Mexico. *Geological Society of America Bulletin*, **83**, 59–86.

ARTHURTON, R. S. 1973. Experimentally produced halite compared with Triassic layered halite-rock from Cheshire, England. *Sedimentology*, **20**, 145–160.

BAAR, C. A. 1974. Geological problems in Saskatchewan potash mining due to particular conditions during deposition of potash beds. *In*: *Fourth Symposium on Salt*. Northern Ohio Geological Society, Cleveland, OH, **1**, 101–118.

BĄBEL, M. & BECKER, A. 2006. Cyclonic brine flow pattern recorded by oriented gypsum crystals in the Badenian evaporite basin of the northern Carpathian foredeep. *Journal of Sedimentary Research*. **76**,

BACHMAN, G. O. 1984. Regional geology of Ochoan evaporites, northern part of the Delaware Basin. *New Mexico Bureau of Mines and Mineral Resources Circular*, **184**.

BASHAM, W. L. 1997. Provenance of the Delaware Mountain Sandstone Group of the Delaware Basin, Texas and New Mexico. *American Association of Petroleum Geologists Bulletin*, **81**, 865.

BEIN, A., HOVORKA, S. D., FISHER, R. S. & ROEDDER, E. 1991. Modification of seawater derived brines in evaporite brine pools and early diagenesis: evidence from fluid inclusions in Permian bedded halite (Palo Duro Basin, Texas). *Journal of Sedimentary Petrology*, **61**, 1–14.

BODINE, M. W. JR. 1978. Clay–mineral assemblages from drill core of Ochoan evaporites, Eddy County, New Mexico. *In*: AUSTIN, G. S. (compiler), *Geology and Mineral Deposits of Ochoan Rocks in Delaware Basin and Adjacent Areas*. New Mexico Bureau of Mines and Mineral Resources, Circular, **159**, 21–30.

BODINE, M. W. JR. 1985. Trioctahedral clay mineral assemblage in Paleozoic marine evaporite rocks. *In*: SCHREIBER, B. C. & HARNER, H. L. (eds) *Sixth International Symposium on Salt*. The Salt Institute, Alexandria, VA, **1**, 267–284.

BROOKINS, D. G. 1988. Seawater $^{87}Sr/^{86}Sr$ for the late Permian Delaware Basin evaporites (New Mexico, U.S.A.). *Chemical Geology*, **69**, 209–214.

BURKE, W. H., DENNISON, R. E., HETHERINGTON, E. A., KOEPNICK, R. B., NELSON, H. F. & OTTO, J. B. 1982. Variation of seawater $^{87}Sr/^{86}Sr$ throughout Phanerozoic time. *Geology*, **10**, 516–519.

CASAS, E. & LOWENSTEIN, T. K. 1989. Diagenesis of saline pan halite: comparison of petrographic features of modern, Quaternary and Permian halites. *Journal of Sedimentary Petrology*, **59**, 724–739.

CHRISTIANSEN, F. W. 1963. Polygonal fracture and fold systems in the salt crust, Great Salt Lake Desert, Utah. *Science*, **136**, 607–609.

CHUBER, S. & PUSEY, W. C. 1972. Cyclic San Andres facies and their relationship to diagenesis, porosity, and permeability in the Reeves Field, Yoakum County, Texas. *In*: ELAM, J. G. & CHUBER, S. (eds) *Cyclic Sedimentation in the Permian Basin*, 2nd edn. West Texas Geological Society, **72-60**, 135–150.

CYS, J. M., 1978. Transitional nature and significance of the Castile-Bell Canyon contact. *In*: AUSTIN, G. S. compiler, *Geology and Mineral Deposits of Ochoan Rocks in Delaware Basin and Adjacent Areas*. New Mexico Bureau of Mines and Mineral Resources, Circular **159**, 53–56.

DELLWIG, L. F. 1955. Origin of the Salina Salt of Michigan. *Journal of Sedimentary Petrology*, **25**, 83–110.

ELLIOTT, L. A. & WARREN, J. K. 1989. Stratigraphy and depositional environment of Lower San Andres Formation in Subsurface and equivalent outcrops: Chaves, Lincoln, and Roosevelt Counties, New Mexico. *American Association of Petroleum Geologists Bulletin*, **73**, 1307–1325.

FISCHER, A. G. & SARNTHEIN, M. 1988. Airborn silts and dune-derived sands in the Permian of the Delaware Basin. *Journal of Sedimentary Petrology*, **58**, 637–643.

FRACASSO, M. A. & HOVORKA, S. D. 1986. *Cyclicity in the Middle Permian San Andres Formation, Palo Duro Basin, Texas Panhandle*. Texas Bureau of Economic Geology, Report of Investigations, **156**.

GARBER, R. A., GROVER, G. A. & HARRIS, P. M. 1989. Geology of the Capitan shelf margin – subsurface data from the northern Delaware Basin. *In*: HARRIS, P. M. & GROVER, G. A. (eds) *Subsurface and Outcrop Examination of the Northern Capitan Shelf Margin, Northern Delaware Basin*. SEPM Core Workshop, **13**, 3–269.

GARBER, R. A., LEVY, Y. & FRIEDMAN, G. M. 1987. The sedimentology of the Dead Sea. *Carbonates and Evaporites*, **2**, 43–58.

GARDNER, M. 1992. Sequence stratigraphy of eolian-derived turbidites: deep water sedimentation patterns along an arid carbonate platform and their impact on hydrocarbon recovery in the Delaware Mountain Group reservoirs, West Texas. *In*: MRUK, D. & CURRAN, B. (eds) *Permian Basin Exploration and Production Strategies: Application of Sequence Stratigraphic and Reservoir Characterization Concepts*. West Texas Geological Society publication, **92-91**, 7–11.

GOODALL, T. M., NORTH, C. P. & GLENNIE, K. W. 2000. Surface and subsurface sedimentary structures produced by salt crusts. *Sedimentology*, **47**, 99–118.

GORNITZ, V. M. & SCHREIBER, B. C. 1981. Displacive halite hoppers from the Dead Sea: Some implications for ancient evaporite deposits. *Journal of Sedimentary Petrology*, **51**, 787–794.

HANDFORD, C. R. 1981. Coastal sabkha and salt pan deposition of the lower Clear Fork Formation (Permian), Texas. *Journal of Sedimentary Petrology*, **51**, 761–778.

HANDFORD, C. R. 1982. Sedimentology and evaporite genesis in a Holocene continental-sabkha playa basin, Bristol Dry Lake, California. *Sedimentology*, **29**, 239–253.

HARDIE, L. A. & EUGSTER, H. P. 1971. The depositional environment of marine evaporites: a case for shallow, clastic accumulation. *Sedimentology*, **16**, 187–220.

HARMS, J. C. 1974. Brushy Canyon Formation, Texas: a deep-water density current deposit. *Geological Society of America Bulletin*, **85**, 1763–1784.

HARMS, J. C. & WILLIAMSON, C. R. 1988. Deep-water density current deposits of Delaware Mountain Group (Permian), Delaware Basin, Texas and New Mexico. *American Associations of Geologists Bulletin*, **72**, 299–317.

HILLS, J. M. 1972. Late Paleozoic sedimentation in West Texas Permian Basin. *American Association of Petroleum Geologists Bulletin*, **56**, 2303–2322.

HOLSER, W. T. 1966. Diagenetic polyhalite in Recent salt from Baja, California. *American Mineralogist*, **51**, 99–109.

HOLT, R. M. 1993. *Sedimentary Textures, Structures, and Lithofacies in the Salado Formation: A Guide for Recognition, Classification, and Interpretation*. DOE/WIPP 93-056, US Department of Energy, Carlsbad, NM.

HOLT, R. M. & POWERS, D. W. 1987. The Permian Rustler Formation at the WIPP site, southeastern New Mexico. *In*: POWERS, D. W. & JAMES, W. C. (eds) *Geology of the Western Delaware Basin, West Texas and Southeastern New Mexico*. El Paso Geological Society Guidebook, **18**, 140–156.

HOLT, R. M. & POWERS, D. W. 1990a. *Geologic Mapping of the Air Intake Shaft at the Waste Isolation Pilot Plant*. DOE-WIPP, **90–051**.

HOLT, R. M. & POWERS, D. W. 1990b. Halite sequences within the late Permian Salado Formation in the vicinity of the Waste Isolation Pilot Plant. *In*: POWERS, D., REMPE, N., HOLT, R. & BEAUHEIM, R. L. (eds) *Geological and hydrological studies of the evaporites in the northern Delaware Basin for the Waste Isolation Pilot Plant (WIPP), New Mexico*. Geological Society of America 1990 Annual Meeting Field Trip Guidebook, **14**, 45–78.

HORITA, J., WEINBERG, A., DAS, N. & HOLLAND, H. D. 1996. Brine inclusions in halite and the origin of the Middle Devonian Prairie evaporites of Western Canada. *Journal of Sedimentary Research, A: Sedimentary Petrology and Processes*, **66**(5), 956–964.

HOVORKA, S. D. 1987. Depositional environments of marine-dominated bedded halite, Permian San Andres Formation, Texas. *Sedimentology*, **34**, 1029–1054.

HOVORKA, S. D. 1990. *Sedimentary processes controlling halite deposition, Permian Basin, Texas*. PhD Dissertation, The University of Texas at Austin, 393 p.

HOVORKA, S. D. 1992. Halite pseudomorphs after gypsum in bedded anhydrite – clue to gypsum – anhydrite relationships. *Journal of Sedimentary Petrology*, **62**, 1098–1111.

HOVORKA, S. D. 1994. *Water-level controls on halite sedimentation: Permian cyclic evaporites of the Palo Duro Basin*. The University of Texas at Austin, Bureau of Economic Geology Report of Investigations, **214**.

HOVORKA, S. D. 2000a. *Characterization of bedded salt for storage caverns – a case study from the Midland Basin, Texas*. The University of Texas at Austin, Bureau of Economic Geology Geological Circular, **00–1**.

HOVORKA, S. D. 2000b. Deep-water to shallow-water transition in evaporites in the Delaware Basin, Texas. *In*: LINDSAY, R. F., TRENTHAM, R. C., WARD, R. F. & SMITH, A. H. (ed.) *Classic Permian Geology of West Texas and Southeastern New Mexico: 75 Years of Permian Basin Oil and Gas Exploration and Ddevelopment*: West Texas Geological Society Publication, **00-108**, 273–299.

HOVORKA, S. D. & GRANGER, A. 1988. Subsurface to surface correlation of Permian evaporites – San Andres–Blaine–Flowerpot relationships, Texas Panhandle. *In*: MORGAN, W. A. & BABCOCK, J. A. (eds) *Permian Rocks of the Midcontinent*. Society of Economic Paleontologists and Mineralogists, Special Paper, **1**, 137–159.

HSÜ, K. J. 1972. Origin of saline giants – a critical review after discovery of the Mediterranean evaporite. *Earth-Science Reviews*, **8**, 371–396.

HUNT, C. B., ROBINSON, T. W., BOWLES, W. A. & WASHBURN, A. L. 1966, *Hydrologic Basin Death Valley, California*. U.S. Geological Survey Professional Paper, **494B**.

JONES, C. L. 1954. The occurrence and distribution of potassium minerals in southeastern New Mexico. *New Mexico Geological Society, Fifth Field Conference*, 107–112.

JONES, C. L. 1972. *Geology and hydrology of the Carlsbad potash area, Eddy and Lea Counties, New Mexico*. US Geological Survey Open-File Report, **4339-1**.

KENDALL, A. C. 1992. Evaporites. *In*: WALKER, R. G. & JAMES, N. P. (ed.) *Facies Models*. Geological Association of Canada, St Johns, Newfoundland, 375–409.

KENDALL, A. C. & HARWOOD, G. M. 1989. Shallow-water gypsum in the Castile Formation – significance and implications. *In*: HARRIS, P. M. & GROVER, G. A. (eds) *Subsurface and Outcrop Examination of the Capitan Shelf Margin, Northern Delaware Basin*. SEPM Core Workshop, **13**, 451–457.

KERANS, C. & FITCHEN, W. 1995. *Sequence Hierarchy and Facies Architecture of a Carbonate Ramp System: San Andres Formation of Algerita Escarpment and Western Guadalupe Mountains, West Texas and New Mexico*. The University of Texas at Austin, Bureau of Economic Geology Report of Investigations, **235**.

KING, B. 1942. Permian of West Texas and southeastern New Mexico. *American Association of Petroleum Geologists Bulletin*, **26**, 535–763.

KINSMAN, D. J. 1969. Modes of formation, sedimentary association, and diagenetic features of shallow-water and supratidal evaporites. *American Association of Petroleum Geologists Bulletin*, **53**, 830–840.

KIRKLAND, D. W. 2003. An explanation for the varves of the Castile evaporites (Upper Permian), Texas and New Mexico, USA. *Sedimentology*, **50**, 899–920.

KIRKLAND, D. W., DENISON, E. E. & DEAN, W. E. 2000. Parent brine of the Castile evaporites (Upper Permian), Texas and New Mexico. *Journal of Sedimentary Research*, **70**, 749–761.

KNAUTH, L. & BEEUNAS, M. A. 1985. Isotope geochemistry of fluid inclusions in Permian halite with implications for isotopic history of ocean waters and the origin of saline formation waters. *Geochimica et Cosmochimica Acta*, **50**, 419–433.

LANG, W. E. 1942. Basal beds of Salado formation in Fletcher potash core test, near Carlsbad, New Mexico. *American Association of Petroleum Geologists Bulletin*, **26**, 63–79.

LESLIE, A. B., HARWOOD, G. M. & KENDALL, A. C. 1997. Geochemical variations within a laminated evaporite deposit: evidence for brine composition during formation of the Permian Castile Formation. *Sedimentary Geology*, **110**, 223–235.

LOGAN, B. W. 1987. *The MacLeod Evaporite Basin, Western Australia*. American Association of Petroleum Geologists Memoir, **44**, Tulsa, OK.

LOWENSTEIN, T. K. 1982. Primary features in a potash evaporite deposit, the Permian Salado Formation of West Texas and New Mexico. *In*: HANDFORD, C. R., LOUCKS, R. G. & DAVIES, G. R. (eds) *Depositional and Diagenetic Spectra of Evaporites – A Core Workshop*. SEPM Core Workshop, **3**, 276–304.

LOWENSTEIN, T. K. 1988. Origin of depositional cycles in a Permian 'saline giant': The Salado (McNutt zone) evaporites of New Mexico and Texas. *Geological Society of America Bulletin*, **100**, 592–608.

LOWENSTEIN, T. K. & HARDIE, L. A. 1985. Criteria for recognition of salt-pan evaporites. *Sedimentology*, **32**, 627–644.

LUGLI, S., SCHREIBER, B. C. & TRIBERTI, B. 1999. Giant polygons in the Messinian salt of the Realmonte Mine (Agrigento, Sicily): implication for modeling the 'Salinity Crisis' in the Mediterranean. *Journal Sedimentary Research*, **69**, 764–771.

MADSEN, B. M. & RAUP, O. B. 1988. Characteristics of the boundary between the Castile and Salado Formations near the western edge of the Delaware Basin, southeastern. *New Mexico Geology*, **10**, 1–9.

MANZI, V., LUGLI, S., RICCI LUCCHI, F. & ROVERI, M. 2005. Deep-water clastic evaporites deposition in the Messinian Adriatic foredeep (northern Apennines, Italy): did the Mediterranean ever dry out? *Sedimentology*, **52**, 875–902.

MAZZULLO, S., MAZZULLO, J. & HARRIS, P. M., 1985. Eolian origin of quartzose sheet sand in Permian shelf facies, Guadalupe Mountains. *In*: CUNNINGHAM, B. & HEDROCK, C. (eds) *Permian Carbonate/Clastic Sedimentology, Guadalupe Mountains An Analog for Shelf-Basin Reservoirs*, SEPM Publication, **85–24**.

MCKEE, E. D., ORIEL, S. S. *ET AL*. 1967. *Paleotectonic investigations of the Permian System of the United States*. United States Geologic Survey Professional Paper, **515**.

MEISSNER, F. F. 1972, Cyclic sedimentation in middle Permian strata of the Permian Basin, West Texas and New Mexico. *In*: ELUM, J. G. & CHUBER, S. (eds) *Cyclic Sedimentation in the Permian Basin*, 2nd edn, West Texas Geological Society, 203–232.

NANCE, H. S. 1988. Interfingering of evaporites and red beds: an example from the Queen/Grayburg formation, Texas. *Sedimentary Geology*, **56**, 357–381.

NEEV, D. & EMORY, K. O. 1967. *The Dead Sea, Depositional Processes and Environments of Evaporites*. Geological Survey of Israel Bulletin, **41**.

POWERS, D. W. & HASSINGER, B. W. 1985. Synsedimentary dissolution pits in halite of the Permian Salado Formation, Southeastern New Mexico. *Journal of Sedimentary Petrology*, **55**, 769–773.

POWERS, D. W., MARTIN, M. & HOLT, R. M. 1988. *Siliciclastic-rich units of the Permian Salado Formation, Southeastern New Mexico*. Abstracts with programs, Geological Society of America, **20**, A174.

PERYT, T. M. & KOVALEVICH, V. M. 1997. Association of redeposited salt breccias and potash evaporites in the lower Miocene of Stebnyk (Carpathian Foredeep, West Ukraine. *Journal of Sedimentary Research*, **67**, 913–922.

PRESLEY, M. W. 1981. *Middle and Upper Permian Salt-Bearing Strata of the Texas Panhandle: Lithologic and Facies Cross Sections*. The University of Texas at Austin, Bureau of Economic Geology Cross Section Set.

PRESLEY, M. W. 1987. Evolution of the Permian Basin in Texas Panhandle. *American Association of Petroleum Geologists Bulletin*, **71** 167–190.

ROBINSON, J. Q. & POWERS, D. W. 1987, A clastic deposit within the lower Castile Formation, western Delaware Basin, New Mexico. *In*: POWERS, D. W. & JAMES, W. C. (eds) *Geology of the Western Delaware Basin, West Texas and Southeastern New Mexico*. El Paso Geological Society Guidebook, **18**, 69–79.

ROEDDER, E. 1982. Possible Permian diurnal periodicity in NaCl precipitation, Palo Duro Basin, Texas. *In*: GUSTAVSON, T. C., BASSETT, R. L. *ET AL*. *Geology and Geohydrology of the Palo Duro Basin, Texas Panhandle, a report on the progress of nuclear waste isolation feasibility studies*. The University of Texas at Austin, Bureau of Economic Geology Geologic Circular, **82–7**, 101–104.

ROVERI, M. & MANZI, V. 2006. The Messinian Salinity Crisis: looking for a new paradigm? *Palaeogeography, Palaeoclimatology, Palaeoecology*, (in press).

SCHALLER, W. T. & HENDERSON, E. P., 1932, *Mineralogy of Drill Core from the Potash Field of New Mexico and Texas*. US Geological Survey Bulletin, **833**.

SCHREIBER, B. C. 1978. Environments of subaqueous gypsum deposition. *In*: DEAN, W. E. & SCHREIBER, B. C. (eds) *Marine Evaporites*. SEPM Short Course Notes, **4**, 43–73.

SCHREIBER, B. C., FRIEDMAN, G. M., DECIMA, A. & SCHREIBER, E. 1976. The depositional environments of the Upper Miocene (Messinian) evaporite deposits of the Sicilian Basin. *Sedimentology*, **23**, 729–760.

SHEARMAN, D. J. 1970. Recent halite rock, Baja California, Mexico. *Transcription, Institute of Mining & Metallurgy*, **79**, 155–162.

SHEARMAN, D. J. 1978. Halite in sabkha environments. *In*: DEAN, W. E. & SCHREIBER, B. C. (eds) *Marine Evaporites*. SEPM Short Course Notes, **4**, Tulsa, OK, 30–42.

SMITH, D. B. 1971. Possible displacive halite in the Permian Upper Evaporite Group of Northeast Yorkshire. *Sedimentology*, **17**, 221–232.

SNIDER, H. I. 1966. *Stratigraphy and associated tectonics of the upper Permian Castile–Salado–Rustler evaporite complex, Delaware Basin, West Texas*. The University of New Mexico, PhD dissertation.

SNYDER, R. P. & GARD, L. M. JR. 1982. *Evaluation of breccia pipes in southeastern New Mexico and their relation to the Waste Isolation Pilot Plant (WIPP) site, with section on drill-stem tests*. Open-file Report **82-968**. US Geological Survey, Denver, CO.

SONNENFELD, P. 1984. *Brines and Evaporites*. Academic Press, Orlando, FL.

SOUTHGATE, N. 1982. Cambrian skeletal halite crystals and experimental analogues. *Sedimentology*, **29**, 391–407.

TALBOT, C. J., STANLEY, W., SOUB, R. & AL-SADOUN, N. 1996. Epitaxial salt reefs and mushrooms in the Southern Dead Sea. *Sedimentology*, **43**, 1025–1047.

TUCKER, R. M. 1981. Giant polygons in the Triassic salt of Cheshire, England: a thermal contraction model for their origin. *Journal of Sedimentary Petrology*, **51**, 779–786.

US Department of Energy, 1983. *Geologic Mapping of Exploratory Drift: Geotechnical Field Data Report 7. In*: Results of Site validation experiments, **II**, TME 3177. US Department of Energy, Carlsbad, NM.

WARDLAW, N. C. & SCHWERDTNER, W. M. 1966. Halite–anhydrite seasonal layers in the Middle Devonian Prairie Evaporite Formation, Saskatchewan, Canada. *Geological Society of America Bulletin*, **77**, 331–342.

WARREN, J. K. 1982. The hydrological setting, occurrence and significance of gypsum in Late Quaternary salt lakes in South Australia. *Sedimentology*, **29**, 609–630.

WEILER, Y., SASS, E. & ZAK, I. 1974. Halite oolites and ripples in the Dead Sea, Israel. *Sedimentology*, **21**, 623–632.

Index

Page numbers in *italics* refer to Figures, while those in **bold** denote Tables.

accumulation rates 2–3, 341–2
acicular gypsum 201, *203*
Africa *see* North Africa; South Africa; West African margin
age data 24, 28, 31–2
alabaster facies 133–4, 227
Algeria, Berkine/Ghadames Basin 87–105
allochthonous gypsum 117–18, 200, 207, 209
analogues *see* modern analogues
Angola margin *26*
anhydrite
    Bilche–Volytsya zone, Ukraine 322
    breccia 317
    Carpathian Foredeep Basin 108, *109*
    Dhiban Formation 59, *60, 61*, 62
    Great Kavir Basin 76, 78
    isotopic composition **257**
    nodules 175, *176*
    Permian Basin *337*, 340
    Poland 277
    pseudomorphs *346*, 347
    rehydration 203–5
    Ukraine 267, 269, 270, 317, 320
    Zbudza Formation 254, **257**
anhydrite–carbonate couplets *8*
anhydrite–carbonate–halite triplets *8*
anhydrite–halite laminates 359–60
anoxia 236–7
anoxic monimolimnion 236–7
antiform structures, calcite *293, 300, 302*
Anzano Molasse 194–5, *196*
Apennines *see* Southern Apennines
Aptian salt basin 24–8, 31
aragonite 147
Aral Sea 236
Argilles Vertes Formation 20, 24
Artemivs'k rock salt deposit, Ukraine 314
authigenic mineralization 313
autochthonous gypsum 117–18, 200, 207–8, 209
autoclastic breccias, calcite *291*

Badenian evaporite (meromict) basins
    Carpathian Foredeep *4*, 219–46
        brine transport 238–40
        evaporites distribution 220, *221*
        halite crystallization 221, 240
        hydrographical model 230–1
        mixolimnion *231*, 235–6
        modern analogues **232–4**, 236
        monimolimnion 236–7
        subbasins 221
    East Slovakian Basin 247–64
    Ukraine 268, 319
Badenian gypsum facies *4*, 107–42, *116*
    clastic gypsum (allochthonous) 117–18

coarse-crystalline selenite 118–24
    glass-like selenite 124–7
    lithosomes 117–18
    microbialite 117–18
    stratigraphy *119*, 240–1
    see also salina-type evaporite basin
banded halite facies 171, 173
Barremian sediments 31
base-level oscillations *210*, 211
basement morphology, South Atlantic 25–8
basins
    Badenian evaporite basin 219–46
    Berkine/Ghadames Basin 87–105
    Carpathian Foredeep Basin 107–42, 219–46, 265–73, 318–23
    East Slovakian Basin 247–64
    Great Kavir Basin, Iran 69–85
    Kirkuk Basin, Iraq 53–68
    Permian Basin, USA 335–64
    pre-salt sag basins 15–35
    salina-type 107–42
bassanite 322
Berkine/Ghadames Basin 87–105
    evaporite cycles 93–6
    lithostratigraphy 90, *91*
    palaeogeography/evolution 98–9, *100–1*
    seismic stratigraphy 91–3, *94–5*
    sequence stratigraphy 96–8
Bilche–Volytsya zone, Ukraine 320–3
bloedite 150, 152
bottom dissolution fabrics 347–50
bottom-growth deposits
    calcite–anhydrite 358
    fabrics 344–7, *348–9*
    gypsum 200, 225, 237
    selenite 112–15
Brazilian continental margin *17*
    pre-salt sag basin deposits 19–20
    seismic reflection profiles *21–3*
    topography 28–31
breccias
    anhydrite 317
    autoclastic *291*
    calcite *291*, 302
    carbonate 155–68, 198–9
    collapse 302
    from dissolution 159, *160, 166*
    halite facies 172, *175*
    microbreccias 161
    Monte Castello evaporites 209
brine flows
    downslope transport 238–9
    halite zone 228–30
    Halych, Ukraine 227–8
    meromict basins 239–40

brine flows (*Continued*)
   mixolimnion *231*
   orientation 219–20
   swirl pattern 228
brine inclusions *see* fluid inclusions
brine sheets, majanna-type shoals *111*, 115–17
brines
   density stratification 343–4, 360
   Great Kavir Basin 81–3
   transport concepts 238–40
   Ukraine 270, 329
bromine
   fluid inclusions 280, 282
   in halite 78–9, **81**, 358–9
   rock salt *255*, 257, 268, 317, 329
Burdigalian stage
   basin configuration 55–64
   Kirkuk Basin 53–68
   marine transgression 66

$CaCl_2$ hydrothermal brines 80, 81–3
Calabria, Italy, Messinian halite facies 169–78
calcite
   herringbone structure *289*, *294*, *296*–301, 303–5
   mosaics, Neoarchaean 296–301
   nodules *288*, *292*, *296*
   pseudomorphs 287
   see also carbonates
calcite–anhydrite cumulate couplets *346*, 358, 359
calcite–anhydrite–halite triplets 358
calcium sulphate *see* anhydrite; gypsum; selenite
Campbellrand Subgroup, South Africa *286*
Campos basin, Brazil 20, 28–31
cap rocks, residual halite facies 174, *175*, 176
carbon isotopes, carbonates 183, 188
carbonate–anhydrite cycles 96
carbonates
   'B marker', Berkine/Ghadames Basin 93–6
   Bilche–Volytsya zone, Ukraine 320
   breccias 155–68, 198–9
   conglomerates 155–68
   former evaporite features 285–308
   Great Kavir Basin 78
   Kirkuk Basin 53–68
   oxygen and carbon isotopes 182–3, 188
   see also calcite; dolomite; limestones
carnallite 76–7, 267, 268, 323, 324
Carpathian Foredeep Basin
   Badenian gypsum facies *4*, 107–42
   Badenian meromict basin 219–46
   evaporites distribution *116*, 220, *221*
   selenite facies *4*, 118–37
   Ukraine 265–73, 318–23
Castile Formation, USA 358–9
cauliflower (cavoli) structures *292*, 301
Central Ebro Basin, Spain 143–54
chambersite 324–5
channel structures 133–5
channel-mouth lobe deposits 37–52
Chela unconformity 22, 28
chemocline 230, 231
   see also pycnocline
chevron structures, halite 78, 253, 254, 257, 275, 278–80

chicken-wire structures *295*
chlorides
   Bilche–Volytsya zone, Ukraine 322–3
   brines, Ukraine 270
   Great Kavir Basin 74–8
   Zbudza Formation 248, 249–54
clastic evaporites 169, 347
   gypsum *136*, 205, *206*
   halite 249–54
clastic lobe deposits 37–52
   depositional environment 46–50
   ellipsoidal mounds 39–41, 43, *46*
   ribbon-shaped bodies 41–3, *46*, *48*
clay laminae 322
clear halite facies 171–2, *173–4*
coarse-crystalline gypsum 322, 328
coarse-crystalline selenite 114, 118–24
collapse breccias 302
columnar structures 302–3
conceptual models, salina-type
   basin 107–11
constructional clastic depositional body 43–6
continental extension, South Atlantic 20–3, 25–7
continental red bed facies 70–3
Coriolis effect 238
Crimea 319–28, *326*
Crotone basin, Calabria 169–78
   halite facies 171–4
   residual facies 174–6
crustal thinning, South Atlantic margins 20–3, 25–7
crystallization *see* individual minerals
cumulate deposition 344, *345*, *346*, 358–60
cyclicity
   Berkine/Ghadames Basin 93–6
   Castile Formation, USA 358–9
   Messinian evaporites 181
   Permian Basin *339*
   selenites 212
   Zbudza Formation 254–6, 261
   Zechstein evaporites 277

d'ansite 76, 77, 78
Daunia tectonic unit 196
debris flows 20
dedolomitization 165
deep-brine pans
   coarse-crystalline selenite 118–24
   depositional model *123*
deep burial alteration 6
deep water
   facies 7, *8*, 31, 93
   fabric criteria 342–4, *356*
   see also monimolimnion
deformation
   calcite 302, 304–5
   halite 173, 177
Delaware Basin, USA 336
Delaware Mountain Group, USA 357–8
density stratification in brines 343–4, 360
deposition
   deep water 342–4, *356*
   models 80–1, *82*, *123*, *260*
   rates 2–3
   shallow water 344–50

styles 5–6
see also redeposition
depth indicators 335–64, *356*
    accumulation rates 341–2
    fabric criteria 342–56
    methods of determination 337–41
detrital pseudocarniole 161–3, *166*
Devonian evaporites, Ukraine 312–15, 328
Dhiban Formation, Iraq 59–64
diachronous basin development 18–19, 24, 31
diagenetic features, Neoarchaean carbonates 287, *288*, 301–4, *305*
diapirs
    Crotone basin 170, *171*, 174, 177
    Dnipro–Donets depression 311–12, 313, **314**
    Iran 69, *71*
    Transcarpathian trough 326
    see also salt domes
diatomite–carbonate–gypsum sequence 211
directional structures *224*, 225–8
dissolution
    bottom fabrics 347–50
    carbonate breccias 159, *160*, *166*
    dolomite 165
    gypsum/anhydrite 164–5
    halite 177
    pipes 351, *352*
    pits 351–2, *352–3*, *356*
    residual pseudocarniole 159–61, *166*
dissolution surfaces
    bottom growth 349–50
    coarse-crystalline selenite 118–24, *119*, *122*
    microbial mats 122–3, *124*
distal sector evaporites 211
Djeno Formation 19, 20
Dnipro–Donets depression, Ukraine 310–15
    Devonian evaporites 312–15
    Permian evaporites 313–15
dolomite *60*, *62*, 165
    see also carbonates
dolomitization 287, *290*, 301
downslope deposition 344
downslope transport 238–9
drawdown
    Aptian salt basin 27–8
    Badenian evaporite basin 108, *109*
    Late Messinian 211
    see also water-level fluctuations

East European platform *316*, *319*
East Slovakian Basin
    geology 247–8
    salt facies deposition model *260*
    Zbudza Formation 247–64
economic deposits *see* industrial deposits
El Arish–Afiq Canyon 40, 42, 43, 44, 48–9, *48*, *49*
ellipsoidal mounds, clastic lobe deposits 39–41, *43*
emersion events, shallow-brine pans 130–3
encrusting pseudocarniole 163, *164*, *166*
enterolithic structures 295–9
Eocene
    continental red bed facies 70–3
    marine regression, Iran 70
ephemeral (seasonally drying) lakes 131–3

epsomite 147–8, 152, 268
Erva Formation 20
Euphrates Formation, Iraq 55–9
euxinic monimolimnion 237
evaporation rates 3
Evaporiti di Monte Castello Formation 191–218
    depositional setting 209–12, *210*
    diatomic and euxinic facies 198
    evaporitic limestones *197*, 198–9
    geological setting 194–5
    gypsum lithofacies 200–5, 211
        stratigraphic relations 207–9
        strontium geochemistry 205–7
    pre-evaporitic lithofacies *198*
    regional tectonic control 212, *213*
    stratigraphy 196–200
    tectonic setting 213–14
experimental evaporation 143–54
exposure depth indicators 350–6
    above water table 351–2, *356*
    below water table 350–2
    surface deposition 355–6
extensional faulting 26
Ezanga evaporites 20–5

fabric criteria depth indicators 342–56
    above and below water table 350–6
    deep water 342–4, *356*
    distribution *357*
    shallow water 344–50, *356*
facies
    Kirkuk Basin 56–9
    'pseudocarniole' 158–63, *166*
    tectonically active/passive basins 6–9, *8*
    see also gypsum facies; residual facies; selenite facies
Faeto Flysch 194, *196*
Famennian evaporites, Ukraine 313
faunal assemblages, Kirkuk Basin **57**
filamentous laminates 287–90
fine crystalline halites 253, *254*
fine-grained gypsum microbialite 118
flamboyant calcite 298, 301, *303*
floral assemblages, Kirkuk Basin **57**
flows
    debris flows 20
    halite 173, 177
    mud flow 261–2
    see also brine flows
fluid inclusions
    halite 79–80, 277–9
    sylvite 79–80, 275–84
    Ukraine evaporites 309, 315
fold-and-thrust belt 194
folded flysch 318
foraminifera **57**, 58–9, 65
Forecarpathian region, Ukraine 317, 323–5
Foredobrogean trough, Ukraine 317–18
Frasnian evaporites, Ukraine 312–13

Gabon–Angola continental margin *16*
geochemistry
    depth indicator 342
    Great Kavir Basin 78–9

geochemistry (*Continued*)
  isotopes *187*, 188, 205–7
  modelling, natural brines 143–4
  PHRQPITZ code program 144, 147
  Zbudza Formation 256–7, 262
geology
  Badenian evaporite basin 220–5
  Carpathian Foredeep Basin 266–8
  Crotone basin 169–70
  East Slovakian Basin 247–8
  Great Kavir Basin 69–74
  Monte Castello evaporites 194–5
  Romagna Apennines 180–1
graben structures 70, 248, 310–11
grass-like selenite facies 124–7
  long-distance correlation 127, *128*
  sedimentary features *126*
gravity deposits 20, 344
Great Kavir Basin, Iran 69–85
  brine origin and evolution 81–3
  carbonate unit 78
  chloride unit 74–8
  depositional model 80–1, *82*
  geochemistry 78–9
  geology and stratigraphy 69–74
  siliclastics 73, 78
  sulphate beds 78
Great Salt Lake, Utah 236
gypsarenites 205, *206*
gypsiltites 205
gypsrudites 205, *206*
gypsum
  Bilche–Volytsya zone, Ukraine 322
  bottom-growth 345–7, *348*
  cement 176–7
  clastic *136*, 205, *206*
  crusts 112–15
  Dhiban Formation 59
  Ebro Basin brines 147–8, 150, 152
  enterolithic folds *295*–9
  Great Kavir Basin 76, 78
  Kerch peninsula 327–8
  monimolimnion 237
  Monte Castello 199–200, 201, *203*
  nodules 175, *176*, 203–5, *296*
  replacement 305
  sulphur isotopes 183
  turbidites *10*
  Ukraine 270, 317, 320
gypsum facies
  Carpathian Foredeep Basin *4*, 107–42, *109*, 220
  Monte Castello evaporites 200–5, 207–9
  see also Badenian gypsum facies
gypsum-anhydrite deposits 320–3

halite
  Badenian basin 221, 239–40
  Bilche–Volytsya zone, Ukraine 322
  bottom-growth 347, *349*
  brine flows 228–30
  bromine content 78–9, **81**
  Carpathian Foredeep Basin 108, *109*
  clastic 249–54
  cumulate deposition *345*

deformation and flow 173, 177
Dhiban Formation 59, *60*, 62
diapirs 170, *171*, 174, 177
dissolution 177
facies, Crotone basin 171–4
fluid inclusions 79–80
Great Kavir Basin 72, 73, *75*, 76–8
La Playa brines 147–8, 152
majanna flats 115
non-deposition 61
Permian Basin *337*
Poland 277
primary 78–9, 268, 322
redeposition 249–54, 258–61
sedimentary structures 277–9, **278**, *280*
Ukraine 267
see also rock salt
halite arenites 253, 254
halite rudites 253, 254
Halych, Ukraine, palaeocurrent analysis 223–8
herringbone calcite *289*, *294*, 296–301, 303–5
hexahedrite 268
high-amplitude bodies 39–41, *43*–8
hinge zone, South Atlantic margin *19*, 20, *21*
holomictic pans 112–13
horst-and-graben structures 70, 248
  see also grabens
horsts 311
hydrographical model, meromictic basin 219, 230–1
hydrothermal fluids
  'pseudocarniole' origin 156, 163, 164–5, *166*
  Ukraine salt domes 309, 313

industrial deposits
  potash 265, 323–4
  rock salt 249, 261
ionic strength of brines 147, 148, *150*
Iran, Great Kavir Basin 69–85
Iraq, Kirkuk Basin 53–68
Irpinia–Daunia Mountains, Italy 191–218, *195*
isochronous deposition 127–30
isopach analysis 55–6, 59–64, 65
isotopes
  carbon 183, 188
  fractionation 272
  geochemistry *187*, 188
  oxygen 183, 188
  stratigraphy 179–90
  strontium 181, 182–3, 207, 272
  sulphur 183, 265–73
Italy
  Crotone basin, Calabria 169–78
  Monte Castello evaporites 191–218
  Vena del Gesso evaporites 179–90

Jeribe Formation, Iraq 65, *66*–7
Jurassic evaporites
  Berkine/Ghadames Basin 87–105
  Ukraine 316–18, 328–9

kainite rocks 266, 267, 268, 323–4
kainite–langbeinite rocks 266, 323–4
Kalush–Holyn potash deposit, Ukraine 265–72, 319, 320, 323–4

karst features *199*
    infill 161–3, 165
    post-depositional 351–2
    synsedimentary 351, *352–3*
Kenya, Lake Magadi evaporites 128
Kerch peninsula (Crimea) *326*, 327–8
kieserite 266, 267, 324
Kirkuk Basin, Iraq 53–68
    Burdigalian configuration 55–64
    Langhian configuration 64–5

La Playa/La Salina saline systems, Spain 143–53, *144*
    brines 147–8
    chemical data **145**, **146**
    mineral precipitation sequence 148
    saturation indexes 147, 149–50, *151*
laminar deposits 6–7
laminated gypsum
    Badenian basin 220–1, 237, 240
    channel structures 134–5
    Monte Castello evaporites 201–3, *204*
laminites, calcite 287–90
langbeinite 76, 77, 78, 266, 267, 324
langbeinite–kainite rocks 266
Langhian stage 53–68
Levant continental margin 37–52
limestones 59, *61*, 64, *197*, 198–9
    see also carbonates
lithology and environment 9–11
lithospheric mantle thinning 29
lithostratigraphy 90, *91*, 223–5
Loeme evaporites 20–5
lowstand deposits 270–2
Lukunga Sandstone Formation 19–20

magnesium sulphates 80, 81–3, 265–73
majanna-type shoals *111*, 135–7
    depositional environment 115–17
    gypsum microbialites 133–5
Malmani Subgroup, South Africa *286*
Mansuriya oilfield, Iraq *63*
mantle thinning 29, *30*
marine deposits
    Kirkuk Basin 58, 68
    Miocene evaporites 325, 328
    Permian Basin *337*
    Zechstein salts 281–3
marine recharges 188
marine regression 70
marine transgressions 66, 99–101
Marnes Noires Formation 20, 23–4
mass mineral precipitation 153
Mediterranean region
    Levant margin 37–52
    Lower Evaporites 192, 193
    Messinian sedimentary cycles 192
    Messinian Salinity Crisis 191–2
    regional stages *327*
    Upper Evaporites 192
Melheh salt pit, Iran 72, *73*, *75*, **79**, *80*
meromictic basins
    Badenian evaporite basin 219–46
    brine accumulation 239–40
    classification 111–12
    hydrographical model 230–1
    mixolimnion 231–6
    monimolimnion 236–7
Messinian
    clastic lobe deposits 37–52
    Crotone basin, Calabria 169–78
    evaporites pinch-out *38*, *39*, *40*
    Monte Castello evaporites 191–218
    reworked evaporites *10*
    tectonic activity 192, 193
    Vena del Gesso evaporites 179–90
Messinian Salinity Crisis (MSC) 37, 48–9, 169, 191–2, 212
micro-breccias 161
microbial mats 130, 133
    calcite 287–90, *300*
    gypsum dissolution surfaces 122–3, 124
microbialites 304–5
    architecture 135–7
    Badenian 118
    shallow-brine pans *125*, 133–5
Middle Miocene *see* Badenian
Midland Basin, USA 336
mineral precipitation sequence 148, 150
mineralization
    authigenic, Ukraine 313
    hydrothermal 309, 313
    rock salt residues, Ukraine **314**
mineralogy
    La Playa/La Salada brines *152*
    Ukraine evaporites 268–9
Miocene
    geochronology and biostratigraphy *249*
    Kirkuk Basin 53–68
    palaeogeography *108*
    stratigraphy 53–4, *320–1*
    see also Badenian...
Miocene evaporites
    active tectonic setting *10*
    Carpathian Foredeep 107–42, 219–46, 265–73, 318–23
    clastic deposits 37–52
    East Slovakian Basin 247–64
    Great Kavir Basin 69–85
    Ukraine 265–73, 318–28, 329
    Zbudza Formation 247–64
mirabilite 148, 150, 152
mixolimnion (mixed layer)
    Badenian evaporite basin *231*, 235–6
    brine flows *231*
    meromictic basin model 230–1
    modern analogues 231–5
    stratification-mixing pattern *230*, 231–6
models
    depositional 80–1, *82*, *123*, *260*
    meromictic selenite basin 219–46
    Pitzer's model 143
    salina-type evaporite basin 107–11
modern analogues
    Badenian evaporite basin **232–4**, 236
    mixolimnion 231–5
    shallow-brine pans 131–3
Moldova *119*, *121*, 223

monimolimnion
  Badenian evaporite basin 236–7
  halite 240
  meromictic basin model 230–1
  selenite deposition 237
monogenetic breccias 159, *160*, 163
monomictic pans 112–14
Monte Castello evaporites 191–218
MSC *see* Messinian Salinity Crisis
mud flow 261–2
mudstones 93, 355

Na–K–Mg–Cl–SO$_4$ brines 281, 282–3
needle-like gypsum 201, *203*
Neoarchaean carbonates 285–308
  calcite mosaics 296–301
  diagenesis 301–4, 305
  evolution 305
  granular facies 292–6
  laminites 287–90
  sedimentary structures *295–9*
Neogene 325–6
  see also Miocene
New Mexico, Permian evaporites 335–64
Nile Delta 49
nodular 'fenestrate' fabric 303–4
nodules
  anhydrite 175, *176*
  calcite *288, 292, 296*
  gypsum 175, *176*, 203–5, *296*
non-deposition, halite 61
non-selenite deposition 130–3
North Africa, Berkine/Ghadames Basin 87–105
nucleation cones 301–2

oligotrophic pans 133
onlap, Messinian 50
ophiolitic mélange zones 70, *71*
organic matter 182, *186*
orientation
  brine flows 219–20
  selenite deposits *224*, 225–8
oxygen isotopes 183, 188

palaeocurrent analysis
  Badenian basin *231*
  Halych, Ukraine 223–8
  Zolota Lipa, Ukraine *229*
palaeogeography
  Badenian evaporite basin 220–2
  Berkine/Ghadames Basin 98–9, *100–1*
  East Slovakian Basin 248
  Kerch peninsula *326*
  Mediterranean, Messinian *192*
palaeokarst features *199*
Palo Duro Basin, USA 336
Paraná basin, Brazil 28–31
Paratethys 247–8, *249*, *327*
Permian Basin, USA 335–64
  depositional history 335–7
  evaporite cycle types *339*
  fabric criteria 342–56
    deep water deposition 342–4, *356*
    exposure 350–6

    shallow water deposition 344–50, *356*
  sediment accumulation rates 341–2
  water depth determination 337–41
Permian evaporites
  depth indicators 335–64
  Dnipro–Donets depression 313–15, 328
  primary sylvite 275–84
PHRQPITZ geochemical code program 144, 147
pipes, dissolution 351, *352*
pits, dissolution 351–2, *352–3*, *356*
Pitzer's model 143
playa–lake systems 131–3
  experimental evaporation 143–54
  mass mineral precipitation 153
Poland
  Badenian evaporite basin *221*, *223*
  Badenian gypsum deposits *116*, *119*, *121*, *126*
  primary sylvite generation 275–84
polygenetic breccias 159, 161, *162*
polyhalite
  Great Kavir Basin 76, 77, 78
  Ukraine 266, 267, 268, 270, 324
polyhalite–anhydrite bed 269–70, **271**
polymictic pans 112–14
Porto Seguro Formation 20
post-depositional karst features 351–2
post-rift deposition 15, 17–18, 23, 25
potash salts
  Carpathian Foredeep 265–73, 309
  depositional model 80–1
  Forecarpathian region 323–5
  Frasnian, Ukraine 312
  Great Kavir Basin 72
  Miocene, Ukraine *319*, 320
  Permian, Ukraine 314–15
  Poland 277
  precipitation 272
  Ukraine 318
pre-salt sag basins 15–35
  Barremian to Aptian sediments 31
  basement morphology and structure 25–31
  Brazilian margin *17*, 18, 19–20
  capping sequence 20–5
  depositional packages 19–20
  depositional space problem 18, 26–8, 31
  tectonic accommodation 18
  West African margin *16*, 18, 19–20
precipitation sequence 148, 150
primary halite 78–9, 268, 322
primary sylvinite 76
primary sylvite 275–84
proximal sector evaporites 211
'pseudocarniole' 155–68
  chronology 165–6
  facies and sub-facies 158–63, *166*
  genesis 163–5, *166*, 167
  hydrothermal fluids 156, 164–5, *166*
pseudomorphs
  anhydrite *346*, 347
  selenite *292*, 301
pycnocline 3–5, 228, 231

recrystallization, halite 173, 177
red algae **57**, 58–9

Red Formation, Iran 70–3
red siliciclastic mudstones 355
redeposition 347
 gypsum 205, *206*, 207, 209
 halite 249–54, 258–61
regional tectonic control 212, *213*
relative humidity (RH) 3, 24
replacement, gypsum 305
residual facies, Crotone basin 174–6
residual pseudocarniole 159–61, *166*
reworked evaporites 5–6, 7, *10*
 see also redeposition
RH see relative humidity
rifting
 deformed evaporites 9, *10*
 evaporite deposition 80–1
 Great Kavir Basin 69–85
 pre-salt sag basins 18–19, 31
 Saharan evaporite basin 101–3
 syn/post-rift deposition 15, 17–18, 23, 25, 32
rock salt
 Bilche–Volytsya zone, Ukraine 320–3
 geochemistry, Zbudza Formation 256
 mining 318
 Transcarpathian trough 326
 Ukraine 312–13, 314, 317
 see also halite
Romagna Apennines 179–90
rose diagrams, selenite orientation *226–7, 228*

sabkha-type evaporites 96
Saharan evaporite basin 99–103
Saharan Platform, North Africa 87–105
Salado Formation, USA 359–60
salina-type evaporite basin 107–42
 deep-brine pans 120–4
 definition 108–9
 lithosomes 117–18
 main features 109–11
 majanna-type shoals *111*, 115–17
 saline pans classification 111–15
 shallow-brine pans 124–37
 stratification-mixing cycles 112–14
 water-level fluctuations *110*, 120–2, *123*
 see also Badenian gypsum facies
saline clays 319
saline pans
 hydrographical classification 111–15
 selenite deposition 112–15
 stratification-mixing cycles 112–14
salinity 3–5
 see also pycnocline
salt domes see diapirs
Santos Basin, Brazil 20, 28–31
 depocentre thickening 24
 seismic reflection profiles *21–3*
saturation indexes, La Playa/La Salada brines 147, 149–50, *151*
saturation shelf concept 238
SCC see sedimentary chaotic complex
sea-level position 49–50
seafloor spreading, South Atlantic 18, 20–3, 25
seawater evolution 309–10
sediment accumulation rates 2–3, 341–2

sedimentary chaotic complex (SCC) 169
sedimentary structures
 calcite 285–308
 cauliflower structures *292*, 301
 enterolithic structures *255–9*
 halite 277–9, **278**, *280*
 herringbone structure *289*, *294*, 296–301, 303–5
 selenite *124*
 Zbudza Formation *252*
sedimentology, Zbudza Formation 247–64
seismic control 7, 260–1
seismic data
 Berkine/Ghadames Basin 88–90
 clastic lobe deposits 38–42
 interpretation 42–6
 ribbon-shaped bodies 41–2, 43, *46, 48*
seismic geomorphology techniques 37–52
seismic stratigraphy 91–3, *94–5*
selenite
 crystal aggregates *294*
 pseudomorphs *292*, 301
selenite deposition
 cycles 212
 meromictic basin 219–46
 mixolimnion 237
 orientation *224*, 225–8
 below pycnocline 114, 120–2, *123*, 127, *129*
 rose diagrams *226–7, 228*
 saline pans 112–15
selenite facies *4*, 118–37
 architecture 135–7
 coarse-crystalline 114, 118–24
 dissolution surfaces 118–24, *122*
 environmental interpretation 121
 grass-like 124–7
 marker beds *129*
 stratigraphic relations *119*
selenitic gypsum 200–1, *202*
sequence stratigraphy 96–8
Serikagni Formation, Iraq 55–9
shallow-brine flat-bottomed pans
 channel structures 133–5
 depositional model *123*
 grass-like selenite facies 124–7
 isochronous deposition 127–30
 marker beds *129*
 microbialite deposition 133–5
 modern analogs 131–3
 non-selenite deposition 130–3
 tectonic control 137
shallow salina-type evaporite basin 107–42
 saline pans 111–15
 selenite facies 118–37
shallow water
 evaporites 7, *8*
 fabric criteria 344–50
shelf settings, Permian Basin 338
shoals see majanna-type shoals
Sialivakou Formation 19
siliciclastic deposits
 constructional clastic body 44–5
 depth indicators 341–2
 Great Kavir Basin 73, 78
 Levant margin 37, *39–40*

siliciclastic deposits (*Continued*)
  Permian Basin *337*
  red mudstones 355
  Zbudza Formation 254–6, 261
slope slides, clastic halite 254–6, 261
Slovakia *see* East Slovakian Basin
sodium sulphate salts 153
soft-sediment deformation processes 304–5
South Africa, vanished evaporites 285–308
South Atlantic pre-salt sag basins 15–35
Southeastern Brazilian highlands 28–31
Southern Apennines
  fold-and-thrust belt 194
  Irpinia–Daunia sector 194, *195*
  Monte Castello evaporites 191–218
space problem, evaporite deposition 18, 26–8, 31
Spain, natural playa–lake systems 143–54
Stebnyk potash deposit, Ukraine 265–72, *268*, 270, **271**, 323–4
stratification-mixing cycles 219
  deep-brine pans 120–2
  mixolimnion *230*, 231–6
  saline pans 112–14
stratified brines 3–5, 230–1
stratigraphy
  Badenian gypsum facies 240–1
  Berkine/Ghadames Basin 90–3, *94–5*, 96–8
  Great Kavir Basin 69–74
  isotopes 179–90
  Miocene 53–4, *249*, *320–1*
  Monte Castello evaporites 196–200, 207–9
  Neoarchaean carbonates *286*
  Saharan evaporite basin *102*
  seismic 91–3, *94–5*
  Zechstein *279*
stromatoclasts *288*, 294, *300*
stromatolites *8*, 290–1, 294
strontium
  geochemistry 205–7
  isotope ratios 181, 182–3, 207
structure
  Iraq 54–5
  Western Alps *156*
subaerial exposure 48, 355–6
subaqueous fans 261
submarine channel-mouth lobes 49–50
subsidence, salina-type basin 137
subterranean karst infill 161–3, 165
sulphates
  Bilche–Volytsya zone, Ukraine 320–2
  brines, Ukraine 270
  Great Kavir Basin 78
  lowstand deposits 270–2
  magnesium 80, 81–3, 265–73
  sodium 153
  sulphur isotopic composition 265–73
  Ukraine 312, 317
  Zbudza Formation 254–6
  see also anhydrite; gypsum; selenite
sulphur isotopic composition
  materials sampling 269
  methods 269–70
  polyhalite–anhydrite bed **271**
  sulphates, Ukraine 265–73
  Vena del Gesso evaporites 182, 183
Susa Valley, 'pseudocarniole' 155–68
swirl flow pattern 228
sylvinite 76–7
sylvite
  fluid inclusions 79–80, 277–83
  Great Kavir Basin 76–8
  homogenization temperatures 281
  primary 275–84
  Ukraine 266, 267, 268, 324
syn-rift deposition 15, 17–18, 25, 32
syn-rift faulting 23
synsedimentary karst features 351, *352–3*

tachyhydrite 83
TAG-I *see* Triassic Argilo-Greseux Inferieur
tectonic activity
  deposition styles 5–6
  evaporite deposition 80–1
  facies diversity 7–9
  Kirkuk Basin 61, 66–8
  Messinian 192, 193
  passive basins 5–7, *8*, *10*
tectonic control
  halite redeposition 259–61
  Monte Castello evaporites 212, *213*
  pre-salt sag basin accommodation 18
  salina-type basin deposition 137
tectonic pseudocarniole 163, *165*
thenardite 147–8, 150, 152
thermal plume, South Atlantic 29, *30*
thermal subsidence 32
thick-bedded selenite facies 114, 118–24
thin-bedded selenite facies 124–7
thinned continental crust 25–7
tholeiitic basalts 29, *30*
timing, pre-salt sag basins 15–35
topography
  Permian Basin 338–41
  Walvis Ridge 24–5, 28–31
Toppo Capuana Formation 194, *196*
Torrente Fiumarella unit 194, *196*
Tortonian, reworked evaporites *10*
Transcarpathian trough, Ukraine 325–6
transgressive lag deposits 22
travertine 163, *164*, *166*
Trias Carbonaté 99–101
Triassic Argilo-Greseux Inferieur (TAG-I) 88, 99
Triassic, 'pseudocarniole' 156, *157*
Triassic–Jurassic evaporites, Berkine/Ghadames Basin 87–105
Tristan da Cunha plume 29, *30*
Tunisia, Berkine/Ghadames Basin 87–105
turbidites, gypsum *10*

Ukraine, palaeocurrent analysis *229*
Ukraine evaporites 309–34
  Badenian basin 219–46, *221–2*
  Badenian gypsum deposits *116*, *119*, *121*, *126*, *129*
  Bilche–Volytsya zone 320–3
  Carpathian Foredeep 265–73, 318–20
  Devonian 312–15
  Dnipro–Donets depression 310–15

Forecarpathian region 317
Foredobrogean trough 317–18
Jurassic 316–18
Kerch peninsula (Crimea) 327–8
Miocene 265–73, 318–28
origin 315, 325, 328
palaeocurrent analysis 225–8
Permian 313–15
sulphur isotopic composition 265–73
United States, Permian basin 335–64

vadose fabrics 351–2
vanished evaporites 285–308
Vena del Gesso evaporites 179–90
    facies description **181**
    isotope geochemistry *187*, 188
    isotope stratigraphy 179, 181–8
    lithology and isotope data **184–5**
    organic matter 182, *186*
volcaniclastic layers 195, *196*
Vorotyshcha potash suite, Ukraine 266, 268, 320
vuggy carbonate rocks 155–68
    see also 'pseudocarniole'

Walvis ridge 24–5, 28–31
water depth determination
    accumulation rates 341–2
    fabric criteria 342–56
    methods 337–41

water-level fluctuations
    playa lakes 131–3
    salina-type basins 109–11, 120–2, *123*
    shallow-brine pans 130
    tectonic control 137
    see also drawdown
water-table evaporites 350–2, *354*
water transport, detrital pseudocarniole 161–3
wave-reworked fabrics 347
weld rocks 174, 175–6
West African margin, pre-salt deposits 19–20, *21*
West Texas, Permian evaporites 335–64
Western Alps, 'pseudocarniole' 155–68
white halite facies 171, *172*, 173
Wieliczka salt mine *259*, 261
wireline log data 90, *94*

Zbudza Formation 247–64
    borehole lithology profiles *251*
    cyclicity 254–6, 261
    genesis 258–61
    geochemistry 256–7, *262*
    lithology and facies 249–54, 261
    locations *250*
    sedimentary structures *252*
Zechstein evaporites 275–84
    halite inclusions 277–9
    marine origin 281–3
    stratigraphy *279*